T0202347

THE OXFORD HANDBOOK OF THE HISTORY OF MODERN COSMOLOGY

Frontispiece: The whole-sky map of the cosmic microwave background radiation (CMB) as observed by the Planck satellite of the European Space Agency (ESA). The image shows large-scale structures in the Universe only 380,000 years after the Big Bang. It encodes a huge amount of information about the cosmological parameters which describe our Universe. (Courtesy of ESA and the Planck Collaboration)

The Oxford Handbook of the History of Modern Cosmology

Edited by

Helge Kragh

Malcolm S. Longair

OXFORD
UNIVERSITY PRESS

OXFORD
UNIVERSITY PRESS

Great Clarendon Street, Oxford, OX2 6DP,
United Kingdom

Oxford University Press is a department of the University of Oxford.
It furthers the University's objective of excellence in research, scholarship,
and education by publishing worldwide. Oxford is a registered trade mark of
Oxford University Press in the UK and in certain other countries

Published in the United States of America by Oxford University Press
198 Madison Avenue, New York, NY 10016, United States of America

British Library Cataloguing in Publication Data

Data available

Library of Congress Cataloging in Publication Data

Data available

ISBN 978–0–19–881766–6 (Hbk.)
ISBN 978–0–19–889654–8 (Pbk.)

DOI: 10.1093/oxfordhb/9780198817666.013

Printed and bound by
CPI Group (UK) Ltd, Croydon, CR0 4YY

MIX
Paper | Supporting
responsible forestry
FSC® C013604

Oxford Handbooks offer authoritative and up-to-date surveys of original research in a particular subject area. Specially commissioned essays from leading figures in the discipline give critical examinations of the progress and direction of debates, as well as a foundation for future research. Oxford Handbooks provide scholars and graduate students with compelling new perspectives upon a wide range of subjects in the humanities, social sciences, and sciences.

Also published by
OXFORD UNIVERSITY PRESS

The Oxford Handbook of Soft Condensed Matter
Edited by Eugene M. Terentjevand David A. Weitz

The Oxford Handbook of Nanoscience and Technology
Volumes 1, 2, and 3
Edited by A.V. Narlikar and Y.Y. Fu

The Oxford Handbook of the History of Physics
Edited by Jed Z. Buchwald and Robert Fox

We live in a rapidly expanding universe originating some 14 billion years ago. How did astronomers and physicists arrive at this stunning picture of the universe as a whole? The Oxford Handbook of the History of Modern Cosmology offers a comprehensive and authoritative answer in 13 chapters written by experts in the field. It describes in detail discoveries such as the expansion of the universe, the Big Bang theory, and the cosmic background radiation. But history is more than a string of successes. The book also pays attention to the many erroneous ideas of the past, from Einstein's static universe to Fred Hoyle's belief in a steady-state universe with no beginning in time. Moreover, it includes sections on some of the modern and still controversial theories, such as the idea of many universes known as the multiverse hypothesis.

Helge Kragh gained doctoral degrees in physics (1981) and in philosophy (2007). He was a high school teacher in physics and chemistry (1970–87), and then later a professor of history of science at Cornell University, USA; University of Oslo, Norway; and University of Aarhus, Denmark (1987–2015). After retiring, he is now emeritus professor at the Niels Bohr Institute, University of Copenhagen, Denmark.

Malcolm Longair is Director of Development and Outreach, Jacksonian Professor Emeritus of Natural Philosophy at the Cavendish Laboratory, University of Cambridge, UK. He also holds the position of Editor-in-Chief of the Biographical Memoirs of Fellows of the Royal Society (2016–2020). Longair has previously held positions as a university lecturer in the Department of Physics, University of Cambridge (1970–80); Astronomer Royal for Scotland, Regius Professor of Astronomy, University of Edinburgh, Director of the Royal Observatory, Edinburgh (1980–90); Jacksonian Professor of Natural Philosophy, University of Cambridge (1991–2008); and Head of Cavendish Laboratory (1997–2005).

Preface

In 1967 and 1975 the two pioneering Russian astrophysicists Yakov Zeldovich and Igor Novikov published two monographs under the common title *Relyativistskaya Astrofizica* (Relativistic Astrophysics) which were subsequently translated into English. The second of the volumes, titled *The Structure and Evolution of the Universe*, was a comprehensive account of modern cosmology with an emphasis on the victorious big-bang model. In their introduction Zeldovich and Novikov stated that 'Our philosophy is that the history of the Universe is infinitely more interesting than the history of the study of the Universe.' This is a defensible view but not the philosophy underlying the present handbook on the history of modern cosmology. This volume aims at describing in detail how our present understanding of the universe has emerged through a long and complex series of investigations with roots back in the nineteenth century and even earlier.

There are indeed two ways in which one can speak of the history of the universe. We know today and have known since about 1930 that the universe has its own physical history, in the sense that it evolves over time and can be ascribed a definite age. The details of this evolution, ranging from the big bang to the present and even into the future, are the business of astronomers, astrophysicists and cosmologists. But evolution is not the same as history in the ordinary sense of the term. Darwin's masterwork of 1859 carried the title 'The Evolution of Species' and not 'The History of Species'. History of cosmology is not primarily concerned with stars, galaxies, curved space, and dark energy, but with the scientists who explored the universe and the collective results of whom have provided the picture of the universe accepted at any given time. The 'history' of the universe is without a doubt highly interesting but not necessarily more interesting than the history of the study of the universe. Moreover, the two kinds of history are evidently closely connected, their relationship being synergetic rather than contradictory. It is only through the historical approach that we recognize that our present picture of the universe is not inevitable but the result of a long, bumpy and contingent development which might conceivably have resulted in a very different picture.

It is worthwhile contemplating the terminology and changed meanings of some of the key concepts of cosmology. First, to the ancient Greeks the term cosmos ($\kappa o \sigma \mu o \varsigma$) carried connotations such as order, regular behaviour, and beauty—it is no coincidence that the term appears also in cosmetics and cosmetology. The Greeks boldly claimed that the universe as a whole is a cosmos rather than a chaos and for this reason it must be possible to understand it rationally. Cosmology must be possible. It was a wildly, almost reckless claim but more than two thousand years later it turned out to be more than just a beautiful dream. While cosmology was for a long time the playground of speculative natural philosophy, today it is no less scientific than other branches of the physical sciences. And yet it is and will always remain a different science, principally because the domain of cosmology is the universe at large, a unique and epistemically extraordinary concept.

The universe or cosmos is everything that has, has had, or will have physical existence, whether matter, energy, space, or time. There is nothing outside the universe and literally everything of a physical nature is inside it. Cosmology in the traditional sense refers to the study of the structure of the universe, what in the seventeenth century was often known as cosmography, a term which stresses the mapping of the universe and which could also refer to what we would consider as geography today. Although the term is rarely used today, much of modern cosmology is in the older cosmographical tradition except that the universe under study is no longer assumed to be static. Edwin Hubble, aptly described by his biographer Gale Christianson as a 'mariner of the nebulae', was as much a cosmographer as a cosmologist. The dynamic conception of the universe is largely a product of the twentieth century, and yet scientists of the past often dealt with cosmogony, which literally means the study of how the universe came to be what it is. The earliest cosmological views we know of, those of the Mesopotamian and Egyptian cultures, were cosmogonies rather than cosmographies. They were mythical tales of how the world and the gods came into existence, not science.

While the name cosmology does not in itself include a temporal dimension, cosmogony does. However, from a modern perspective older terms such as cosmography and cosmogony can be misleading as they often referred to the description and formation of the planetary system and not to the universe as a whole. Henri Poincaré's important work of 1913 entitled *Hypothèses Cosmogoniques* (Cosmogonic Hypotheses) was primarily concerned with theories of the formation of the planetary system and had almost nothing to say about cosmology in the modern meaning of the term.

These and other aspects of a semantic nature are not unimportant when it comes to a historical understanding of how the science of the universe has progressed. They illustrate the simple point known from all branches of culture that words come and go. Moreover, although the same word may be used during a very long period of time, in many cases the meaning of the word changes and sometimes drastically so. Just as the 'electron' of 1900 was very different from what physicists called an electron a few decades later, so the name 'galaxy' changed its meaning during roughly the same period. For practical reasons we retain the name universe, but we should not believe that when astronomers wrote about the universe in the early twentieth century they referred to the same concept that the name encompasses today.

There is another point with regard to the name cosmology that needs to be commented upon. While in the present context the name refers to the science of the universe, in other and equally legitimate contexts it refers to a world view or ideology common to a particular society or culture. Historians and anthropologists speak of the romantic ideology or the communist ideology; and they investigate the cosmology of Australian aboriginals or that of the Hopi Indians. In a broad historical perspective the two widely different meanings of cosmology were originally connected, but this is no longer the case. We generally have no problem distinguishing scientific cosmology from what may be loosely called philosophical cosmology. This volume is about the former and not about the latter kind of cosmology. Nonetheless, even today one cannot completely separate the philosophical-religious aspects of cosmology from the scientific aspects. As illustrated in some of the chapters, changes in scientific cosmology have often resulted in discussions of a much wider nature related to philosophical, political and religious issues.

When did the universe become modern? The question is largely equivalent to asking when the study of the universe as a whole became a science based on mathematical models combined with astronomical observations and deductions from physical theory. The seeds for a new and immensely fruitful chapter in the history of cosmology were planted in the second half of the nineteenth century when the galaxies moved to the forefront of astronomical research and the art of spectroscopy made it possible to study the heavenly bodies from the perspective of physics and chemistry. During the same period the second law of thermodynamics suggested that a cosmic arrow of time might be ascribed to the universe. And yet all this was only prolegomena to the later development. It was Einstein's cosmological field equations of 1917 which provided a solid theoretical framework for understanding the universe at large and which today are often hailed as the true beginning of modern cosmology at least as far as theory is concerned. Based on the theory of general relativity and Hubble's measurements of extragalactic nebulae the traditional static picture of the universe was eventually replaced by the then mind-boggling idea of a universe in expansion.

The further development from the Friedman–Lemaître expanding universe to the present standard model was far from smooth as it followed a route with many false trails and ideas later recognized to be wrong. These false trails and wrong ideas are as much part of the history of modern cosmology as are the developments which today enter textbooks as the milestones that led to our present picture of the universe. From a proper historical perspective the important developments were those which were considered important at that time and not those which current cosmologists recognize as the preferred cosmological model. Consequently a substantial part of the chapters in this volume deals with theories and observational claims which are not only outdated but wrong.

Until recently the historiography of modern cosmology and in particular the development in the post-World War II period was an underdeveloped field of history of science. The low priority was evident not only if compared to the history of other branches of the modern physical sciences but even more so in comparison to the hundreds or thousands of scholarly works devoted to ancient cosmology or the Copernican revolution. Still by 1980 and possibly even today the situation was that our knowledge of how the heliocentric system of the world came into existence was more complete and fine-grained than our knowledge of the emergence of big-bang theories in modern cosmology. And yet it would be difficult to argue that the twentieth-century picture of the evolution of the universe is less revolutionary than the pictures constructed by Ptolemy, Copernicus, and Kepler.

There were a few works which covered in a comprehensive of scholarly manner the post-war development, such as John North's *The Measure of the Universe* and Jacques Merleau-Ponty's *Cosmologie du XXeme Siècle*. But unfortunately both of these works were completed just before the magic year of 1965 when the cosmic microwave background (CMB) radiation was discovered. During the last two or three decades the situation has improved markedly with the publication of many books and articles describing the developments from about 1965 to the present. Although today there is no shortage of literature on the history of modern cosmology, by far most of it is either of a specialized nature or belonging to the genre of popular history of science.

The present volume aims at presenting a comprehensive overview of the development of cosmology from the late nineteenth century to the early twenty-first century and to include not only the necessary scientific details but also the broader contexts. Although comprehensive, of course it does not give a complete account of the history. One of the issues which are missing is a sociologically oriented analysis of how and when cosmology became professionalized and accepted as a proper scientific discipline with its own journals, associations, and reward system. After all, modern cosmology is more than just a subfield of astronomy and physics. Unfortunately the social aspects of modern cosmology are a subject which has received almost no serious attention from historians and sociologists of science.

The volume describes and explains the historical background to what we know about the universe today but does so in a non-linear, more historically authentic way that pays attention also to the many developments that turned out to be unfruitful or just wrong. The book is organized in thirteen roughly chronologically ordered chapters, with some focusing on theory and others more on observations, experiments and technological advances. A few of the chapters are of a more general nature, relating to larger contexts such as politics, economy, philosophy and world views. While each chapter can be read separately it also connects to the other chapters, and the book thus aims at describing the history coherently and as a whole. The content and style of the chapters differ, of course, but they all form part of the same grand story.

The authors come from diverse backgrounds, some professional astrophysicists and cosmologists who have lived through the exciting and dramatic developments of the last 50 years. Others come from a background in the history and philosophy of science where the perspective is different from and complementary to that of the practitioners at the coalface. As editors, we have made no to effort to smooth out the difference in approach and perception. We find the synergies and tensions between the approaches stimulating, revealing, and thought-provoking.

The contents

The first chapter in this 'Handbook of the History of Modern Cosmology', written by Helge Kragh, surveys cosmological ideas in the pre-Einsteinian era from approximately 1860 to 1910. During this period cosmology was not recognized as a distinct field of scientific study and yet there was a lively discussion of subjects which were de facto cosmological. Some of these discussions were stimulated by Olbers' paradox and others by the controversial application of the law of entropy increase to the universe at large. Importantly, astrophysics emerged as a new and powerful transdisciplinary science which substantially changed classical astronomy and eventually also cosmology. During the same period the idea of non-Euclidean geometry aroused great mathematical interest and was even embraced by a few astronomers and physicists. But throughout the half-century cosmological theory remained characteristically speculative and qualitative, largely without a stable foundation in the form of relevant observations. Most professional astronomers either adopted an agnostic attitude or simply ignored the universe at large as a possible domain of science.

Chapter 2 by Robert W. Smith covers the same period as Chapter 1 but from the perspective of observational astronomy and cosmology and extending the period up to about 1940. With the establishment of state-of-the-art observatories in the US in the early twentieth century astronomers began to address the nebulae, one of the major results being V.M. Slipher's discovery of redshifts in the spectra of spiral nebulae. The so-called Great Debate in 1920 was primarily concerned with the relationship of spiral nebulae to the Milky Way Galaxy, but the result of the debate was inconclusive. It was only with E. Hubble's discovery of Cepheids in the Andromeda Nebula that a distance indicator was established and the question resolved. With the new distance scale and more data for nebular redshifts astronomers looked for a redshift–distance relation that would eventually become the main evidence for the expanding universe. To a large extent through the work of Hubble, extragalactic astronomy became a reality in the 1930s and immediately applied to test cosmological models.

As detailed in Chapter 3 by Matteo Realdi, during the 1920s cosmological models based on general relativity theory meant essentially the matter-filled Einstein model or the empty universe of Willem de Sitter. Only in 1930 was it recognized that none of the static models were adequate and that a better choice was the expanding universe which had earlier been proposed by Alexander Friedman and Georges Lemaître. With the new revelation physicists and astronomers such as Otto Heckmann, Howard P. Robertson, and Richard C. Tolman systematized the various models satisfying the requirements of general relativity. Some of the models led to a finite age of the universe smaller than the age of the Earth and much smaller than the age of stars and galaxies. The notorious timescale difficulty continued to haunt relativistic cosmology until the mid-1950s. Nonetheless, by the late 1930s the expanding universe was generally accepted and disseminated to a broader public in the form of popular books and articles. On the other hand, Lemaître's suggestion of a kind of big-bang model was unfavourably received.

It is tempting but unhistorical to identify cosmological models with solutions to Einstein's field equations. As discussed in Chapter 4 by Helge Kragh, since the 1920s a series of alternative cosmological theories have been proposed, some of them radically alternative and others mere modifications of relativistic models. Apart from discussing the general issue of what 'alternative' means, the chapter focuses on a small number of more or less heterodox theories. Some of them, such as E.A. Milne's kinematic cosmology, once played an important role but now belongs to history. The same is the case with the steady-state theory which, because of its historical importance, is considered separately (Chapter 5). Other ideas, such as oscillating models and the introduction of time-varying constants of nature, are still discussed if mainly outside mainstream cosmology. The chapter also describes plasma models and tired-light hypotheses, and it briefly relates to more recent sceptical views regarding cosmology's status as a proper science. Such views, found in the recent literature, have a long historical heritage but are increasingly marginalized by the remarkable developments related in this Handbook.

Still in 1960 it was far from evident that we live in a big-bang universe. Chapter 5, also by Helge Kragh, deals with the extensive and on occasion dramatic cosmological controversy that raged throughout the period 1948–1965 between two radically different conceptions of the universe. In the late 1940s George Gamow and collaborators revived the idea of a finite-age, exploding universe which they developed into a quantitative model of the early hot universe. At the same time Fred Hoyle, Thomas Gold, and Hermann Bondi proposed the steady-state model resting on assumptions quite different from those of relativistic cosmology. The chapter follows the controversy up to the discovery of the CMB radiation which essentially sounded the death knell of the steady-state theory. And yet, according to a minority of cosmologists the death sentence was premature. They continued to fight the big-bang orthodoxy and suggest modified steady-state models in accordance with observations. But their fight was unsuccessful and, some may say, misguided. The case of the cosmological controversy is not only of historical importance, it also provides a nice illustration of the classical problem of theory choice in science.

Chapter 6 by Malcolm Longair is devoted to observational and astrophysical cosmology in the period from 1940 to 1980 and thus continues the earlier story covered in Chapter 2. The timescale problem which had plagued most relativistic models was solved with Walter Baade's revision of Hubble's constant in about 1953, but later in the century a new dispute emerged regarding the value of the constant. Other disputes of mainly an observational nature concerned the value of the deceleration parameter and the possibility of a non-zero cosmological constant. Perhaps the leading observational cosmologist of the period, Allan Sandage, believed that the correct cosmological model could be nailed down with the 200-inch telescope on Mount Palomar. Other issues dealt with in the chapter, apart from the determination of cosmological parameters, are the formation of large-scale structures in the universe and the role played by radio astronomy in the resolution of the cosmological controversy. The chapter also discusses the work that led to a reliable thermal history of the universe based on observations and begins the search for the density fluctuations in the CMB radiation which are further dealt with in other chapters.

The subject of Chapter 7, also by Malcolm Longair, is relativistic astrophysics and the role it has played as a tool in the cosmological arena. Although astrophysics in the relativistic domain began in the 1930s with work on supernovae and neutron stars, it was only three decades later that the new research area attracted massive attention and became of crucial importance in cosmology and in the testing of theories of gravity. The discovery of extragalactic radio sources and quasars entered the controversy over the steady-state theory and it was followed in 1967 by the discovery of pulsars. The new and strange objects provided valuable input for cosmological and gravitational theories. The same was the case with the surprising discovery in 1997 that the high-redshift gamma bursts are at cosmological distances and thus must have enormous energies. The chapter also recounts the story of celestial X-rays and gamma rays which dates back to rocket flights in the early 1960s, and it discusses the attempts to detect gravitational waves which culminated in the recent successful LIGO experiments rewarded with the 2017 Nobel Prize in physics. Through the whole of this period, general relativity changed from a specialised mathematical discipline to an everyday tool of the astrophysicist.

The discovery and further study of the CMB has been of unrivalled importance to our understanding of the early universe and to cosmology generally. Chapter 8 by Bruce

Partridge is devoted to the epic story of the CMB with a focus on the years from the mid-1960s to the present day. Although predicted by Ralph Alpher and Robert Herman as early as 1948, the CMB became a reality only with its serendipitous discovery by Arno Penzias and Robert Wilson in 1964. Initially it was questioned whether or not the CMB was cosmic, but measurements at other wavelengths soon silenced the scepticism. The chapter deals with the advances in instrument technologies that turned the CMB to a phenomenon of quantitative studies and ultimately discovered the tiny anisotropies predicted by theory. The detailed study of the radiation's power spectrum opened up a Pandora's box full of new insights and new problems. The CMB became an foundational tool for calculations of nucleosynthesis and led to the necessity of dark matter and dark energy as major constituents of our universe, not to mention hypotheses of an early inflationary phase. This study reveals the enormous effort needed from many physicists, engineers and cosmologists, as well as funding agencies, to determine unambiguously the central role which the CMB occupies in modern cosmology.

The era of big science space programmes coincided with the Cold War. As suggested by Silvia De Bianchi in Chapter 9, even cosmology became involved in the political, military, and technological competition between the USA and the Soviet Union. Rather than focusing on the ideological scene, she tells the story, based on archival sources, of how the two superpowers developed research programmes of both scientific and military significance to test theories of cosmology and gravitation. Radio astronomy and radar imaging techniques were in part developed for the purpose of planetary astronomy and in part for military purposes. Advanced laser technology, of obvious military importance, provided a tool for testing general relativity and determining whether or not the constant of gravitation varied slowly in time. Although the general picture during the 1960s and 1970s was characterized by competition, there were also collaborative space science programmes. Generally the chapter argues that the historiography of cosmology and gravitation in the period needs to take into account Cold War space science and its associated technologies.

Chapter 10 by Malcolm Longair continues the story from Chapter 6, now bringing the development in observational and astrophysical cosmology up to the present. Since about 1980 our empirical knowledge of the universe has progressed tremendously and what was once a dream, precision cosmology, has become a reality. The remarkable development has largely been technology-driven, the result of increased computer power, new generations of telescopes for all wavebands, semiconductor-based detectors (CCDs), and much more. The discipline has also benefitted from the influx of experimental and theoretical physicists into the cosmological arena. The chapter describes how the accuracy and reliability of the cosmological parameters have increased drastically and established the ΛCDM consensus model. The age of the universe has been determined by different techniques resulting in the same age of about 13.8 billion years. A major advance in this period involved the supernovae projects which in the late 1990s led to the conclusion that the universe is in a state of acceleration. Most of its gravitating mass is non-baryonic and most of its total energy density is due to vacuum energy. The chapter also surveys several other crucial areas of modern cosmology, including light element nucleosynthesis and the formation of galaxies and large-scale structures.

Chapter 11, written by Malcolm Longair and Chris Smeenk, deals with concepts that are key components of modern cosmology but the nature of which are not well understood. While dark non-baryonic matter is generally accepted on observational grounds, there is no sure knowledge of what this kind of matter consists of. The history of dark energy is part of the history of Einstein's cosmological constant, or perhaps it is the other way around. It may well be that the dark energy is the cosmological constant, but several alternatives are still discussed. With the proposal of the inflationary scenario for the very early universe in the early 1980s, a major revision of the very early phases of the big bang model saw the light of day. Studies of the CMB later resulted in a more sophisticated picture of inflation and its role in structure formation. Although many cosmologists assert that inflation has become part of the present consensus model, critics argue that it is too early to claim with confidence that the universe started in an explosive inflationary phase.

Some of the themes in Chapter 11 are further discussed in Chapter 12 by Milan M. Ćirković if from a different, theoretical and to some extent speculative perspective. The history of the universe supposedly started with the big bang at $t = 0$ but theories of quantum gravity make it possible to extend the history arbitrarily into the past, so-called pre-big-bang theories. And as to the far future, physicists have produced scenarios of the evolution of the universe and its constituent objects zillions of years from now, what is sometimes known as physical eschatology. The notorious anthropic principle entered cosmology in 1974 as a selection effect for observations and an argument for other possible universes. The controversial idea of a multiverse has several roots, including string theory, the many-worlds interpretation of quantum mechanics, and self-reproducing inflationary models. Thus, cosmology in the twenty-first century is not only characterized by precise measurements and an observationally confirmed standard model but also by an increased engagement with other fields of fundamental physics, resulting in fields such as string cosmology and its bedfellows. Once again the development has stimulated discussions of a philosophical nature.

Although cosmology has long ago become a genuine physical science, it is not and cannot be totally divorced from philosophical concerns. The final Chapter 13 by Chris Smeenk takes up some of the pertinent philosophical issues relating to modern cosmology. Theology has a place too, as in the classical and still living discussion of the ultimate origin of the universe. During the controversy over the steady-state theory in the 1950s (Chapter 5) the very nature and aims of cosmology were hotly debated and not all of the questions raised then have received a final answer. Cosmology rests on extrapolations. It is empirically underdetermined in a strong sense as observational knowledge is restricted to just a small part of the universe. More recently the question of the initial state has been reconsidered from the perspective of theories of quantum gravity. Anthropic reasoning and the epistemic status of multiverse hypotheses have attracted as much philosophical as scientific interest. In short, for philosophers, aspects of modern cosmology question traditional modes of scientific inquiry related to fundamental issues such as explanation, prediction, and falsification. And there is more.

A number of the authors of the papers which include images reproduced in this handbook are deceased or uncontactable. We are most grateful to the publishers who have been helpful in giving permission for the use of the figures from journals, books, and other media for which they hold the copyright. Every effort has been made to track down the copyright owners for all the pictures, but a very few of them have proved to be beyond what we consider to be reasonable effort.

Helge Kragh and Malcolm S. Longair

Contents

List of Contributors

Milan M. Ćirković. Research Professor, Astronomical Observatory of Belgrade, Serbia.
mcirkovic@aob.rs

Silvia De Bianchi. Ramón y Cajal Fellow, Department of Philosophy and Centre for the History of Science, Autonomous University, Barcelona, Spain.
silvia.debianchi@uab.cat

Helge Kragh. Emeritus Professor, history of science, Niels Bohr Institute, University of Copenhagen, Denmark.
helge.kragh@nbi.ku.dk

Malcolm S. Longair. Emeritus Professor, natural philosophy, Cavendish Laboratory, Cambridge, UK.
msl1000@cam.ac.uk

R. Bruce Partridge. Emeritus Professor, astronomy, Haverford College, PA, USA.
bpartrid@haverford.edu

Matteo Realdi. Postdoc, history of astronomy, Institute for History and Social Aspects of Science, Vrije Universiteit Amsterdam, The Netherlands.
m.realdi@vu.nl

Chris Smeenk. Associate Professor, history and philosophy of physics, Department of Philosophy, University of Western Ontario, London, Canada.
csmeenk2@uwo.ca

Robert W. Smith. Professor, history of science, Department of History and Classics, University of Alberta, Edmonton, Canada.
rwsmith@ualberta.ca

1

Cosmological theories before and without Einstein

Helge Kragh

During the ages there seems to be an oscillating between two different types of cosmologies: during some periods it was believed that the structure of the universe can be understood through religious-philosophical-theoretical speculations, during others it is claimed that an empirical-observational approach is preferable.

Alfvén (1983), p. 323

1.1   Introduction

It is a common misunderstanding, not least among physicists and astronomers, that cosmology only became a science, as distinct from philosophy, in the twentieth century. Many modern scientists will be inclined to point to Einstein's pioneering theory of 1917 as a precondition for a truly scientific cosmology, perhaps the very beginning of the field. Others will identify the beginning with the discovery of the cosmic microwave background in 1965 and others again with the COBE satellite experiments starting in 1989. But attempts to understand the entire universe in rational and scientific terms go much farther back in time. One can reasonably speak of scientific (or at least proto-scientific) cosmology in the cases of Aristotle, Hipparchus, Ptolemy, and other thinkers of Greek antiquity. The later achievements of Copernicus, Kepler, and Newton were more than just contributions to theoretical planetary astronomy. They were also concerned with the structure of the universe as a whole and therefore qualify as cosmology.[1]

Although with roots that go back to the earliest civilizations, cosmology in essentially the way we understand the term today only emerged in the enlightenment era. Perhaps characteristically, it is in this period we find the first books carrying the word "cosmology" in their titles. The first time may have been in 1731 when the influential German philosopher Christian Wolff published his *Cosmologia Generalis*, a book that was primarily in the older philosophical tradition.

Kragh, H., 'Cosmological theories before and without Einstein' in *The Oxford Handbook of the History of Modern Cosmology*, edited by Kragh, H. and Longair, M. S. © Oxford University Press 2019.
DOI: 10.1093/oxfordhb/9780198817666.013.1

Twenty-four years later another and more famous philosopher, Immanuel Kant from Königsberg in Germany (the present Kaliningrad), published anonymously a small book entitled *Allgemeine Naturgeschichte und Theorie des Himmels* (General Natural History and Theory of the Heaven; Fig. 1.1). Although Kant referred frequently to God and presented his theory as theistic, in reality his references to the Creator were largely rhetorical except for the original creation of matter. Contrary to the much-admired Newton he found no place for divine miracles in the universe. "A world-constitution, which without a miracle does not maintain itself does not have the steadiness which is the hallmark of God's choice," he piously wrote.[2] Perhaps the most important innovation in Kant's cosmology was that he provided the universe with an evolutionary perspective. He started with a primeval, divinely created chaos of particles at rest, distributed throughout an infinite void. The initial chaos was unstable, he said (but did not prove), and out of the chaos condensations were formed by gravitational attraction. He claimed that the result must necessarily be a regular and orderly cosmos consisting of planetary systems and huge nebulae like the Milky Way. Ostensibly on the basis of Newtonian mechanics he

Fig. 1.1 *Title page of Kant's treatise on the constitution and evolution of the world.*

pictured the present world as an "island universe," the islands being nebulae floating in the infinite sea of void space. The term, which became popular since the late nineteenth century, cannot be found in Kant's work. It may have been coined by the American astronomer and evangelist Ormsby Mitchel who used it in 1846 in a journal he edited, *The Sidereal Messenger*.

Another novel aspect of Kant's grandiose cosmological scenario was its dynamic and evolutionary nature. Of course, God had originally created the universe, but not in its present form for it had slowly developed from the primeval chaos solely governed by the laws of nature. The evolutionary perspective implied that the universe has a history and then an age, a concept that only appeared in modern cosmology in the 1930s. Indeed, not only did it have a beginning, it also might have an end, for Kant speculated that the universe would eventually return to its chaotic state and that this would possibly happen an infinity of times. The idea of an eternally recurrent universe, sometimes known as a "phoenix universe," can be traced back to the earliest mythological world views and was particularly common in Indian culture.[3] Although Newton considered the possibility, he rejected it as theologically dangerous. With Kant and later thinkers the fascinating scenario of a phoenix universe experienced a revival.

Interesting as Kant's combined cosmology and cosmogony is from the point of view of history of ideas it did not live up to our standards of science and not even to the standards of his own time. It was still philosophical cosmology although now dressed in scientific language. To speak of the beginning of modern cosmology we need to jump ahead another century, to a period when more was known observationally of the stellar universe and when the laws of thermodynamics had entered alongside Newtonian gravitation as universally valid laws of nature. This chapter focuses on the conceptual and theoretical problems in cosmology ca. 1860–1910 rather than the astronomical observations relevant to the structure of the universe as a whole.

1.2   Olbers' so-called paradox

Heinrich Wilhelm Olbers was trained in medicine and during most of his career he earned his living as a practising physician in Bremen, Germany (Fig. 1.2). However, he considered his true vocation to be astronomy in which area he was recognized as one of the most eminent researchers of the early eighteenth century. His reputation was based in particular on his pioneering observations of comets, asteroids, and meteoric showers. In 1823 Olbers submitted a theoretical paper to the *Astronomisches Jahrbuch* (Astronomical Yearbook) which contained his most well-known contribution to astronomy but which only appeared in print three years later. Olbers introduced his paper with lengthy quotations from Kant's almost forgotten book of 1755 and in this way he contributed to a revival of interest in the cosmological vision of the Königsberg philosopher. Indeed, like Kant he firmly believed that the universe was infinite and throughout populated with stars distributed approximately uniformly. A devout Christian, Olbers argued that only an infinite universe agreed with God's omnipotence. But then he was faced with

Fig. 1.2 *The German astronomer H. W. Olbers (1758–1840). Wikimedia Commons.*

an old puzzle concerning the darkness of the night sky which first had been noticed two centuries earlier.

The puzzle became eventually known as "Olbers' paradox" although Olbers did not consider it to be paradoxical at all. The eponymous label may first have been used by Hermann Bondi, who in his textbook *Cosmology* from 1952 dealt with the subject in some detail and of course with the benefit of hindsight.[4] According to Bondi, the riddle of cosmic darkness could be explained as a result of the redshift of distant stars and galaxies caused by the expansion of the universe.

The so-called paradox arises from the assumption that the infinite or just hugely large universe is uniformly filled with stars that shine in the same way. The light received from a star varies with the inverse square of the distance and so is negligible for distant stars; but the number of stars at a certain distance increases with the square of the distance with the result that the integrated starlight received on Earth will make the sky at night as bright as or brighter than on a sunny day. This is the brief version of Olbers' famous paradox. More formally, consider the Earth to be surrounded by spherical shells of radius r and thickness dr; assume that the distribution of stars, each of the same luminosity L, is uniform and unchanging in time. If ρ denotes the average stellar density the number of stars in a shell is $4\pi\rho r^2\,dr$ and the total apparent luminosity at the Earth is

$$L_{\text{total}} = 4\pi\rho L \int_0^R \mathrm{d}r. \tag{1.1}$$

For R approaching infinity the expression becomes infinitely large.

The history of Olbers' paradox has been traced back to Johannes Kepler who in a work of 1610, *Dissertatio cum Nuncio Sidereo* (Conversation with the Starry Messenger), used it as an argument for the finite and surprisingly small universe that he favoured. Edmond Halley, on the other hand, was convinced that the stellar universe was infinite and in a paper of 1720 he claimed to have defused the paradox by means of an argument of a somewhat obscure nature; it rested on the claim that the intervals between the stars in an infinite system decreased linearly with distance while the intensity of light from them decreased with the double square of the distance.

Olbers was aware of Halley's much earlier paper in *Philosophical Transactions* but found its explanation to be obscure and unsatisfactory. As an alternative he assumed that the intensity of light suffered a slight absorption during its journey through interstellar space. Referring to the tails of comets, the zodiacal light, and other astronomical phenomena, he concluded that "Without doubt the universe is not absolutely transparent." According to Olbers, "Only the slightest degree of non-transparency suffices to refute the conclusion—so contrary to our experience—that if the fixed stars stretch away to unlimited distance, the entire sky must blaze with light."[5] To Olbers and many of his contemporaries the proposed explanation was yet another proof of God's infinite wisdom. In the words of Olbers: "The Almighty with benevolent wisdom has created a universe of great yet not quite perfect transparency and has thereby restricted the range of vision to a limited part of infinite space."

From a priority point of view Olbers' paradox should perhaps be called Chéseaux's paradox, a reference to the Swiss astronomer Jean-Philippe Loys de Chéseaux who in an essay of 1744 analyzed more lucidly and in greater detail than Olbers the paradox of the night sky. Following Halley in conceiving the stars to be situated in concentric spherical shells, he stated that the quantity of light emitted from each shell was proportional to the sum of the squares of the apparent diameters of the stars in the shell. He did not consider the possibility that stars may fall on the same line drawn from the Earth and thus shield for some of the incoming light. Chéseaux concluded that the paradox would arise not only if the number of stars was infinite, but also if it was exceedingly large but still finite, namely larger than a value corresponding to 76×10^{13} concentric shells.

Rather than opting for a finite, relatively small universe he suggested that the intensity of starlight decreased at a greater rate than given by the squared-distance law and that the physical cause was absorption in an interstellar ethereal fluid. Chéseaux argued that the problem would disappear if the intensity of light were to diminish by 3 percent when passing through a layer of thickness equal to the diameter of the solar system. For the nearest stars he estimated a distance 120,000 times greater, implying that we observe only the foreground stars and not the more distant ones. Olbers essentially duplicated the work of the Swiss astronomer of which he was apparently unaware. Olbers and Chéseaux had in common that neither of them questioned the assumption of more or less uniformly distributed stars. Another assumption they implicitly agreed upon was that the stars are not subject to systematic motion in time.

Priority aside, the Chéseaux–Olbers hypothesis of a light-absorbing medium was widely accepted in the nineteenth century. It was not without problems, though. In 1848 the famous British astronomer John Herschel pointed out that radiant energy from the stars, when absorbed in the hypothetical interstellar medium, would heat up the medium until it reached a state of thermal equilibrium and the medium itself became radiant. Herschel realized that the infinite universe could be reconciled with the dark night sky even without assuming light absorption, namely if the stars were arranged in a suitable, non-uniform way. Nothing is easier, he wrote, "than to imagine modes of systematic arrangement of the stars in space...which shall strike away the only foundation on which [the problem] can be made to rest, while yet fully vindicating the absolute infinity of space."[6] The idea of a specially arranged hierarchic stellar universe was another solution to Olbers' paradox, but one for which there was as little observational evidence as there was for the absorption hypothesis.

And there were more solutions, including one that did not assume interstellar absorption, a hierarchic universe, or a finite number of stars. In 1858 the German astronomer Johann Mädler, at the time director of the Dorpat Observatory in present Estonia, came up with a novel solution in his book *Der Fixsternhimmel* (The Heaven of the Fixed Stars). He argued that Olbers' paradox would not arise if we only receive light from stars within a certain horizon. This might happen not as a result of absorption but as a result of the universe having a finite age:

> The world is created, and hence is not eternal. Thus no motion in the universe can have lasted for infinite time; in particular, this applies to a beam of light. In the finite amount of time it could travel before it reached our eye, a light beam could pass though only a finite space no matter how large the speed of light. If we knew the moment of creation, we would be able to calculate its boundary.[7]

Mädler's proposal failed to attract attention among his colleagues in astronomy except that it was restated by the Leiden astronomer Frederik Kaiser in a popular book of 1860.

More importantly, in 1901 the great physicist William Thomson (who by then was Baron Kelvin of Largs) independently examined how to "test an old and celebrated hypothesis that if we could see far enough into space the whole sky would be seen occupied with discs of stars...and that the reason why the whole of the night-sky and day-sky is not as bright as the sun's disc, is that light suffers absorption in travelling through space."[8] Thomson considered a stellar Milky Way universe uniformly filled with stars up to the enormous distance of 3.3×10^{14} light years and took into consideration that each star has a finite lifetime. Based on what he claimed was the "irrefragable dynamics" underlying his own and Hermann von Helmholtz's contraction theory of solar energy emission, Thomson assured that a star could shine for at most 100 million years.[9] This implied that the time it would take for light to travel from one of the stars farthest away would be about 3.3 million times the lifetime of a star. "Hence," he wrote, "if all the stars through our vast sphere commenced shining at the same time... at no one instant would light be reaching the earth from more than an excessively small proportion of the stars." This, a much more sophisticated version of Mädler's argument, was essentially Thomson's solution to Olbers' paradox. But Thomson did not refer to either Olbers or Chéseaux and was undoubtedly unaware of Mädler's suggestion.

Olbers' paradox was not taken very seriously during the second half of the nineteenth century, in part because it was generally assumed that interstellar space was filled with a rarefied medium that absorbed part of the starlight. The medium was sometimes identified with the physicists' world ether, the existence of which was generally assumed. Only during the early part of the new century did observational evidence indicate that space was much more transparent than previously assumed and the problem therefore real. In 1917 Harlow Shapley formulated the paradox as support of a finite stellar universe: "Either the extent of the star-populated space is finite or 'the heavens would be a blazing glory of light' ... Then, since the heavens are not a blazing glory, and since space absorption is of little moment throughout the distance concerned in our galactic system, it follows that the defined stellar system is finite."[10] Models of the Milky Way proposed at the time by J.C. Kapteyn, H. von Seeliger, and K. Schwarzschild all assumed that the absorption of starlight by interstellar matter was negligibly small.

Only in the late 1920s did astronomers begin to reconsider the reality of absorption, which was proved observationally by the Swiss–American astronomer Robert Trumpler at the Lick Observatory. As Trumpler demonstrated in 1930, the interstellar absorption effect amounted to an average change in apparent magnitude of 0.67 per kiloparsec.[11] However, in his meticulous investigation of open star clusters Trumpler did not refer specifically to Olbers' paradox and he did not relate his work to the cosmological problem. Incidentally, a value roughly similar to the one obtained by Trumpler followed from observational estimates reported much earlier by Friedrich Struve, the first director of the Pulkovo Observatory, in a paper of 1847. Convinced of the absorption of starlight by interstellar matter Struve praised Olbers' earlier solution to the problem of the dark night sky.

To summarize, there were in the period basically five proposals of how to avoid Olbers' paradox. Apart from interstellar absorption of light one could assume a hierarchic structure of the stellar universe or that there existed only a finite number of luminous bodies in space; or one could assume the stellar universe to be of finite age. The last possibility, of which more below, was to assume that space itself was finite (but unbounded) and filled uniformly with stars.

1.3   The rise of astrophysics

Although one can meet the term "physical astronomy" in the scientific literature of the enlightenment era, astrophysics is a child of the nineteenth century. In 1866 Friedrich Zöllner was appointed professor of astrophysics at the University of Leipzig, the first academic position of its kind. Thirty years later astrophysics was flourishing and about to gain status as a scientific discipline, such as indicated by *The Astrophysical Journal* founded in 1895 by George E. Hale and James E. Keeler. The new interdisciplinary branch of science significantly changed the course not only of astronomy but also of physics generally. Moreover, astrophysics turned out to have very important consequences for cosmology, although these were only fully recognized much later. The emergence of the field was closely related to the introduction of spectroscopy, which from its very beginning was applied to the study of the stars.

For a couple of decades astrophysics and astrospectroscopy were nearly synonymous terms. And then we should not forget that spectroscopy was as much related to chemistry as to physics. Astrochemistry was another product of the spectroscopic revolution and no less important than astrophysics. In 1887 the prominent British amateur astronomer J. Norman Lockyer—also known as the founder of the journal *Nature*—published a voluminous monograph significantly titled *The Chemistry of the Sun*.

Advancements in the design of spectroscopes and other optical instruments proved for the first time that analysis of light received from the stars could give valuable information about the stars' physical and chemical composition. That such information was possible at all, was a novelty. The very notion of a "star" changed from a geometrical to a physical concept, a change which implied a profound transformation of the astronomical sciences.

Still to Friedrich Bessel and his contemporaries the business of astronomy was restricted to precise measurements of the positions and orbits of celestial bodies whether planets, comets or stars. In a letter of 1832 to the great naturalist Alexander von Humboldt, Bessel wrote "Everything else that one may learn about the [heavenly] objects, for example their appearance and the constitution of their surfaces, is not unworthy of attention, but is not the proper concern of astronomy."[12] Only very few scientists imagined that experimental physics and chemistry would be of astronomical importance, such as Hans Christian Ørsted,the discoverer of electromagnetism, prophesied in a lecture of 1807. "Some day chemistry will have just as much influence on astronomy as mechanics so far," he said.[13] Thanks to advances in astrophysics, what Bessel called the "proper concern of astronomy" became much broader, incorporating important parts of the physical and chemical sciences.

Disregarding still earlier developments, astrospectroscopy began in 1814 with the Bavarian instrument maker Joseph Fraunhofer's systematic investigation of the solar spectrum. Fraunhofer found in the spectrum a large number of dark lines which he carefully measured but without being able to explain them. What was the cause of the mysterious lines? It took more than three decades until the Fraunhofer lines attracted wide attention and it was gradually realized that the bright emission lines of chemical elements coincided with the dark lines. The real breakthrough in spectrum analysis occurred in 1859 as a result of a unique collaboration between the physicist Gustav Robert Kirchhoff and the chemist Robert Wilhelm Bunsen, both at the University of Heidelberg.[14] In experiments using Bunsen's new gas burner the two scientists demonstrated that the emission spectra could be used to identify small amounts of known chemical elements and also to predict new ones. Within a year the chemical power of the spectroscope was dramatically demonstrated with the discovery of two new metallic elements, cesium and rubidium. Kirchhoff immediately pointed out that the new technique had important astronomical consequences. In a paper of 1860 he demonstrated that sodium was a constituent of the solar atmosphere and soon thereafter he also identified the presence of iron, magnesium, copper, zinc, nickel, and barium.[15]

Not only did the work of Bunsen and Kirchhoff establish stellar astrophysics and astrochemistry as promising fields of research, it also led Kirchhoff to a theoretical study of the thermodynamics of radiant heat and what he called blackbody radiation. This fundamental work became the starting point of a process that in 1900 culminated with Max Planck's celebrated quantum hypothesis. By the 1920s quantum theory and the new atomic physics based on it had become a *sine qua non* for progress in astrophysics. Quantitative astrophysics based on quantum theory took its start in the early 1920s with important contributions of Meghnad Saha, Ralph Fowler, E. Arthur Milne, and others. Saha's pioneering work on the temperature and pressure of stellar atmospheres, building on a combination of quantum theory, statistical mechanics and chemical equilibrium theory, was particularly important. As Eddington generously wrote, "Saha's theory has dominated all recent progress in the observation and interpretation of stellar spectra."[16]

But even earlier the new quantum and atomic physics had on occasions been applied to problems of astronomy and astrophysics.[17] Thus, in Niels Bohr's epoch-making theory of atomic structure of 1913 he famously explained some mysterious spectral lines from the star ζ Puppis which astronomers attributed to a new hydrogen series but according to Bohr were due to ionized helium in the form He^+. It is less known that Bohr also predicted what came to be known as "Rydberg atoms" and on this basis explained why astronomers had observed many more Balmer lines than physicists had found in the laboratory. Bohr's prediction of large and highly excited atoms in areas of space at very low density was verified in 1965 when such atoms were detected by means of radio-astronomical methods in interstellar gas clouds.[18]

Fig. 1.3 *Early spectroscope used by W. Huggins for astronomical studies.*
Source: *H. Schellen,* Spectrum Analysis *(New York: Apleton, 1872), p. 322.*

The versatile spectroscope was used for a variety of purposes, some chemical, others physical, and others again astronomical (Fig. 1.3). In 1842 the Austrian physicist and mathematician Johann Christian Doppler read a paper to the Royal Bohemian Society of Sciences in which he announced the principle named after him. As indicated by the title of his paper, "Über das farbige Licht der Doppelsterne" (On the colored light of double stars), he originally thought of the principle in an astronomical context.[19] However, he wrongly thought that the color of a star would undergo a perceptible change because of its motion relative to the Earth. He even suggested that for large stellar velocities the light would be shifted to such an extent that the star would become invisible, a line of reasoning that he used to explain the color of double stars. What became known as the Doppler effect was first verified for sound waves three years later, and in 1848 the French physicist Armand Hippolyte Louis Fizeau, who was at the time unaware of Doppler's treatise, interpreted it as a shift in wavelength for spectral lines. The principle thus originated in astronomy and after been proved in acoustics it returned to astronomy. History apart, according to the Doppler or Doppler–Fizeau principle, if a light source moves relative to the observer with radial velocity v there will be a change in wavelength given by

$$z \equiv \frac{\Delta\lambda}{\lambda} = \left(\frac{\lambda' - \lambda}{\lambda}\right) = \frac{v}{c}, \tag{1.2}$$

where c is the velocity of light and λ' the measured wavelength, redshifted relative to the emitted wavelength λ. As Einstein proved in his seminal paper of 1905 in which he introduced the special theory of relativity, if the recession velocity v is very large the formula becomes

$$z = \sqrt{\frac{(1 + v/c)}{(1 - v/c)}} - 1, \tag{1.3}$$

In the limit $v \ll c$ Einstein's expression reduces to Doppler's.

The first successful attempt to observe a stellar redshift was made by the wealthy English gentleman astronomer William Huggins in 1868, in stiff competition with another pioneer of astrospectroscopy, the Italian Jesuit astronomer Angelo Secchi. Comparing the H_β line in the spectrum of the bright star Sirius with that produced in the laboratory by a discharge tube filled with hydrogen, Huggins found a shift in wavelength of about one angstrom. The displacement was at the limit of observability and Huggins had to convince himself that it was real and not due to some artefact caused by the instrument. On the assumption that the shift was a Doppler effect he cautiously concluded that it implied a recession velocity of Sirius of around $+40\,\mathrm{km\,s^{-1}}$. To many of his contemporaries the reported value was inconceivably high and contradicting common sense, not to mention the consensus view of a static stellar universe. Nonetheless, Huggins' measurements were accepted as the first proof of stellar Doppler shifts. Ironically, it later turned out that the result was wrong both in amount and sign (the currently accepted value is about $-8\,\mathrm{km\,s^{-1}}$).

The validity of the optical Doppler effect was for a period controversial and firmly demonstrated only in the 1880s by examinations of the rotation of the Sun. It took another twenty years until the Doppler effect for light was detected in the laboratory. Attempts to do so were made by the Russian astronomer Aristarch Belopolski at the end of the century but without success. The first unequivocal laboratory demonstration of the effect was made by the German physicist and future Nobel Prize laureate Johannes Stark, who in 1905 proved it for so-called canal rays or positive rays (which are rays of positive ions). By that time measurements of Doppler shifts had become an important tool in stellar astronomy but their significance for cosmology was not yet recognized. Nor did Vesto Melvin Slipher's discovery of the first nebular Doppler shift in 1912 immediately lead to such recognition, but two decades later the insight of the Austrian physicist had become an indispensable tool for the new cosmology theoretically based on the general theory of relativity.

Astrospectroscopy also became crucially important in relation to one of the most discussed issues of nineteenth-century astronomy, the so-called nebular hypothesis with roots in Kant's cosmogony of 1755 and Pierre–Simon Laplace's 1796 theory of the formation of the planetary system. According to the nebular hypothesis as understood by William Herschel and others in the early part of the nineteenth century, some of the nebulae were composed of hot gaseous clouds that would undergo different stages of condensation and eventually end up as stars. The hypothesis was widely associated with the fashionable but also controversial view of nature being in a state of continual evolution.[20] But if all nebulae could be resolved into discrete stars nothing would be left of the nebular hypothesis, and by the mid-nineteenth century this seemed to be the result of observations performed with the most powerful telescopes of the period. The apparently refuted nebular hypothesis was dramatically revived by astrospectroscopic observations made by Huggins in the 1860s.

Huggins knew from Kirchhoff's work that emission line spectra were produced only by gaseous bodies, whereas hot solids yielded a continuous spectrum. Observations of planetary nebulae in 1864 convinced him that they were composed of glowing gas, and the following year he turned to the Orion Nebulae which he considered a crucial test of the nebular hypothesis. His spectroscope revealed no continuous spectrum but only three bright emission lines. Huggins later recalled:

> The riddle of the nebulae was resolved. The answer, which had come to us in light itself, read: Not an aggregation of stars, but a luminous gas.... There remained no room for doubt that the nebulae, which our telescopes reveal to us, are the early stages of long processions of cosmical events, which correspond broadly to those required by the nebular hypothesis in one or other of its forms.[21]

By the end of the nineteenth century, the nebular hypothesis was widely accepted if still being criticized by a minority of astronomers.

As mentioned, chemistry was no less part of the spectroscopic revolution than were physics and astronomy. During the late Victorian era it was generally assumed that terrestrial elements were all over in the universe but also that there possibly existed chemical elements in the stars that are not found on the Earth. Moreover, some researchers such as William Crookes and Norman Lockyer in England believed that stellar spectra provided evidence for the complexity of the chemical atom. Perhaps, they speculated, our present elements are the products of evolutionary processes going on in the hot stars. "Let us start at the moment when the first elements came into existence," Crookes said in a lecture of 1886. He continued:

> Before that time matter, as we know it, was not. It is equally impossible to conceive of matter without energy, as of energy without matter; from one point of view the two are convertible terms. Before the birth of atoms all those forms of energy which become evident when matter acts upon matter could not have existed—they were locked up in the *protyle* as latent potentialities only. Coincident with the creation of atoms all those attributes and properties which form the means of discriminating one chemical element from another start into existence fully endowed with energy.[22]

Does Crookes' speculation count as an anticipation of the big-bang scenario?

Although most spectral lines from the stars could be identified with lines known from laboratory experiments, some unidentified lines might indicate the presence of elements, or states of elements, particular to the stars and not known on Earth. On the basis of stellar, solar and cometary spectra, scientists claimed the existence of several spurious elements, among them "coronium", "nebulium," and "asterium." The hypothesis of the first element was based on a line in the coronal spectrum first observed by Charles Young in 1869. Coronium was sometimes believed to be an element lighter than hydrogen and in a few cases placed in the periodic system. The mystery of the spectral line was only solved in 1939, when Walter Grotrian and Bengt Edlén independently concluded that the line was due to highly ionized iron (Fe^{13+}). The existence of nebulium was first proposed in 1864 and it took until 1927 before Ira Bowen showed that the pair of nebular lines could be explained as "forbidden transitions" in doubly ionized oxygen (O^{2+}).[23]

Most of the claims of celestial elements were wrong, but not all were. In 1868 Lockyer studied spectral lines from a solar prominence noticing a yellow line with a wavelength of 5876 angstroms. The new line, which he named D_3 (because it was close to sodium's D doublet), did not correspond to any line from a known element (Fig. 1.4). Contrary to what is usually stated, Lockyer did not originally claim that he had detected a new element that might exist in the Sun only and nor did he suggest the name "helium" for it.[24] That only came some years later. For more than two decades helium remained a ghost element that the majority of chemists refused to accept, but in 1895 its status changed abruptly when the chemist William Ramsay identified the gas in cleveite, a uranium mineral. Ramsay, and not Lockyer, discovered the element.

Helium was originally believed to be exceedingly rare, and since it was as chemically inert as the more abundant argon it was considered little more than a curiosity. In the early part of the twentieth century nobody could foresee helium's central role in the universe, just as nobody could foresee the equally central role that the Doppler redshifts would come to play in cosmology. The cosmological significance of helium was dimly recognized in the 1930s, but became of crucial importance only in the 1960s when the

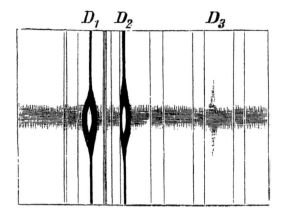

Fig. 1.4 *Emission spectrum of D lines from sunspot, showing the weak D₃ or helium line. The two stronger lines to the left are the D lines from sodium.*
Source: H. Kayser, Lehrbuch der Spektralanalyse *(Berlin: Springer, 1883), p. 188.*

study of helium isotopes in the universe became an important part of the development that led to and consolidated the new hot big-bang theory.[25]

Speculations concerning the composite nature of atoms and their evolution from a common primeval substance were *en vogue* at the turn of the nineteenth century and during the first decade of the new century. As one of many examples consider the American geologist and astronomer Thomas C. Chamberlin, professor at the University of Chicago and a leading critic of the nebular world view. In a paper of 1899 Chamberlin pointed out that gravity was not the only force of astronomical significance and that the fate of the stars, and of the entire universe, might depend on the still unknown structure of the atom. "It is not improbable that they [atoms] are complex organizations and the seats of enormous energies," he said, elaborating:

> Certainly, no competent chemist would affirm either that the atoms are really elementary or that there may not be locked up in them energies of the first order of magnitude.... Nor would he probably be prepared to affirm or deny that the extraordinary conditions which reside at the center of the sun may not set free a portion of this energy.... Why should not atoms, atomecules [sic], and whatever lies below, one after another have their energies squeezed out of them; and the outer regions be heated and lighted for an unknowable period at their expense?[26]

As a later example consider the British physicist John Nicholson, a rival to Bohr in the art of atom-building and an early advocate of the nuclear atom. Foreshadowing much later developments in the physics of elementary particles, Nicholson suggested that, in the future, fundamental physics would have to rely on astrophysics. "Astronomy in the wider interpretation of its scope which is now general, owes much to Physics" he pointed out. "A point appears to have been reached in its [astronomy's] development at which it becomes capable of repaying some of this debt, and of placing Physical Science in its turn under an obligation."[27] Nicholson believed that astrophysics might become "an arbiter of the destinies of ultimate physical theories" because the astrophysicist studied nature in a more primordial form than the laboratory physicist examining terrestrial matter and the laws governing them.

Fin-de-siècle speculations of a more or less cosmological nature spanned a wide range. The British physicist Arthur Schuster suggested in 1898 that there might exist in the universe a hitherto unknown form of matter with the remarkable property that it would be repelled gravitationally by ordinary matter. He coined the word "anti-matter" for the hypothetical substance. "Worlds may have formed of this stuff, with elements and compounds possessing identical properties with our own, undistinguishable in fact from them until they are brought into each other's vicinity." Schuster further speculated that atoms and anti-atoms might enter into chemical combinations, with the short-range attractive forces dominating over the long-range gravitational repulsion. "Large tracts of space might thus be filled unknown to us with a substance in which gravity is practically non-existent, until by some accidental cause…unstable equilibrium is established, the matter collecting on one side, the anti-matter on the other until two worlds are formed separating from each other, never to unite again."[28] The hypothetical stuff made up of matter and anti-matter, he went on, "we cannot call a substance because it possesses none of the attributes which characterize matter ready to be called into life by the creative spark? Was this the beginning of the world?" Schuster was well aware that his cosmological scenario was nothing but an innocent speculation. About seventy years later a different sort of antimatter, the real antimatter based on quantum mechanics, turned out to be of great interest to cosmologists (see Section 4.6).

1.4   Thermodynamics and the universe

The term "thermodynamics" only entered the scientific language through a manual of steam engines published by the Scottish engineer and physicist William Rankine in 1859. Rankine adapted it from the phrase "thermo-dynamic engine" coined by William Thomson four years earlier. The new general theory of heat that emerged in the mid nineteenth century was claimed to be universally valid, and for this reason it was bound to have important cosmological implications. The two laws on which the theory was founded were supposed to work not only for steam engines and chemical processes but also for the solar system and beyond, indeed for the universe as a whole.

From the law of energy conservation it followed that although the energy of the universe might change in form, overall it would remain constant. The second law with its claim of a common direction for all natural processes had consequences for both extremes of the cosmic timescale. If extrapolated to the far future, it indicated that the world would come to an end; and if extrapolated to the far past, it might lead to the conclusion that the world had not always existed but had a beginning in time. Neither of these predictions could be tested (nor can they today), but this only made them more popular as subjects of discussion far outside the small communities of professional scientists.[29]

In an important paper of 1850, Rudolf Clausius introduced the second law of thermodynamics as a natural and inevitable tendency for heat to equalize temperature, and four years later he reformulated his theory by basing it on a function that in 1865 reappeared under the new name "entropy." For two states A and B at absolute temperature T Clausius defined the difference in entropy as

$$S_A - S_B = \int_A^B \frac{\delta q}{T},$$
(1.4)

where δq denotes an infinitesimal change in heat. Armed with his entropy concept, Clausius now stated the second law as "the entropy of the world tends towards a maximum." He similarly expressed the first law globally, as "the energy of the world is constant." The cosmological connection was cultivated more fully in William Thomson's alternative route to the second law. In a paper of 1852 Thomson derived as a consequence of the law that there would come a time when the Earth became "unfit for habitation of man as at present constituted." While Thomson suggested a terrestrial "heat death" in 1852, Helmholtz was the first to extend the idea to the entire universe, such as he did in a famous lecture given in Königsberg in 1854. According to Helmholtz's gloomy prediction, the universe would necessarily approach a state of equilibrium and, when this state had been reached, it would be condemned to eternal rest.

At the Liverpool meeting of the British Association for the Advancement of Science in 1854 Thomson invited his audience to trace backwards in time the actions of the laws of physics. He speculated that the source of mechanical energy in the universe might be sought in "some finite epoch [with] a state of matter derivable from no antecedent by natural laws." However, such an origin of matter and motion, mechanically unexplainable and different from any known process, contradicted his sense of both causality and uniformitarianism. "Although we can conceive of such a state of matter," he wrote, "yet we have no indications whatever of natural instances of it, and in the present state of science we may look for mechanical antecedents to every natural state of matter which we either know or can conceive at any past epoch however remote."[30] Here we have the second law used not to predict the far future but to speculate about a singular and primordial state in the distant past.

From about 1860 the heat death scenario was expounded by several leading physicists and significantly entered, either explicitly or implicitly, the scientific literature. Clausius, who coined the term "heat death" (Wärmetod), formulated it in terms of the continual increase of entropy. Imagine, he said, a time when the entropy of the universe had attained its maximum value. Then "no further change could evermore take place, and the universe would be in a state of unchanging death."[31] Here we have the idea of "physical eschatology," a speculation that would return in a different disguise in late twentieth-century cosmology (see Chapter 12). Clausius further emphasized that the second law, whether formulated in terms of entropy or dissipation, contradicted any idea of a cyclic universe. Not only for Clausius, but even more so for Thomson and his circle of Christian scientists, was it an appealing feature of the second law that it countered what they considered the materialistic and un-Christian notion of a world evolving cyclically.

Expectedly, the claim of a future heat death did not go uncontested. Many scientists and even more non-scientists felt it unbearable that life in the universe should one day cease to exist, never to return. They came up with a variety of suggestions to avoid the scenario, either by devising counter-entropic processes or by questioning the premises upon which the heat death prediction rested. Some of the critics flatly denied the validity of the second law; not a few failed to understand it.

Among scientists a common objection was to question that the second law was valid for the universe as a whole. After all, how can such a huge extrapolation possibly be justified? Reputed physicists such as Pierre Duhem in France and Ernst Mach in Austria did not believe in the heat death prediction. As Duhem argued in an important book of 1905, *The Aim and Structure of Physical Theory*, it did not follow from the entropy law that there is any lower or upper limit of the entropy of the universe. It only follows that the entropy increases endlessly. The entropy could increase asymptotically for an infinite number of years, following for example a logistic curve as given by

$$S(t) = \frac{S^*}{1 + e^{-kt}}. \tag{1.5}$$

The aim of Duhem, an orthodox Catholic, was not to argue against a universe of finite age but to warn that physical theory does not justify such long-term extrapolations. "It is absurd," he wrote, "to question this theory [thermodynamics] for information concerning events which might have happened in the extremely remote past, and absurd to demand of it predictions of events a very long way off." More generally Duhem argued that physics and cosmology were radically different sciences and that the former could not answer questions belonging to the latter:

> The methods by which the physicist develops his theories are without force when it comes to proving that a certain proposition of cosmology is true or false; the propositions of cosmology, on the one hand, and the theorems of theoretic physics, on the other hand, are judgments... [which] can neither agree with nor contradict one another.... [And yet] it is very plain that a cosmological system cannot be reasonably constituted without any knowledge of physics.[32]

Much of the cosmological literature in the second half of the nineteenth century was of a philosophical and speculative nature. It was often associated with the rise of materialism and positivism as alternatives to the traditional way of conceiving nature, culture, and society. In this kind of literature astronomical observations played no substantial role compared to speculations of what the world ought to look like to be in harmony with the author's philosophical and political desires. Although it was not cosmology in our present scientific sense, at the time it was an important and widely read genre of literature.

To mention but one example, a German author by the name Hermann Sonnenschmidt published in 1880 a book titled *Kosmologie* which was largely based on the popular Kant–Laplace nebular world view.[33] In Sonnenschmidt's universe, time had always existed and would remain for ever, and space was infinite, continuously filled with matter or ether. Although he accepted that the entropy law led to the decay of individual parts of the universe, such as stars and nebulae, he denied the cosmic heat death and argued for a cyclic universe. Sonnenschmidt explicitly denied that the world had come into existence by a divine act, a view he considered to be plainly unscientific and nothing but superstition. In German-speaking Europe literature of the same genre as Sonnenschmidt's was no less popular than the more scientifically informed books and articles written by the astronomers. The common denominator in much of the literature was the claim of an eternally cyclic universe.

According to Ludwig Boltzmann's statistical theory of thermodynamics, dating from the 1870s, there was a non-zero probability that the entropy of a closed system would decrease rather than increase. And yet he believed that all attempts to save the universe from the heat death were futile because the probability was ridiculously small. Indeed, for a gas volume of 1 cm^3 he calculated that it would take 10^{109} years until the molecules returned to their initial state! However, in 1895 he came up with a new, speculative answer to the old problem of why we have not already suffered the heat death. His answer involved a remarkable scenario of anti-entropic pockets in a universe which as a whole is in thermal equilibrium:

> If we assume the universe great enough we can...also make the probability great that, though the whole universe is in thermal equilibrium, our world is in its present state. But can we imagine, on the other side, how small a part of the whole universe this world is? Assuming the universe great enough, the probability that such a small part of it as our present world be in its present state, is no longer small. If this assumption were correct, our world would return more and more to thermal equilibrium; but because the whole universe is so great, it might be probable that at some future time some other world might deviate as far from thermal equilibrium as our world does at present.[34]

In this passage Boltzmann assumed implicitly that the universe as a whole had existed eternally or for an exceedingly long time. He did not consider the possibility of a beginning. In general he suggested a many-worlds picture, although the worlds were apparently taken to be just different parts of the universe and not, as in some later many-worlds or multiverse theories, causally separated regions. In the second volume of his influential textbook on gas theory, published in 1898, Boltzmann included a section in which he repeated and amplified his ideas of many worlds, entropy fluctuations and time reversal.[35] He was of course fully aware that these cosmological considerations were highly speculative, but found them to be consistent and also useful because they opened up new perspectives on the range of fundamental physics.

Less discussed than the heat death, but of no less cosmological importance, the second law of thermodynamics was also taken to imply that the universe had a finite age. What has been called the "entropic creation argument" is simple, if not necessarily convincing. The first premise is that according to the fundamental second law the entropy of the world increases continually towards an equilibrium state. The second premise is the simple observation that the present world is far from this state. Order, structure, and life do exist, proving that we do not live in a high-entropy world. But, and this is the conclusion, had the world existed in an eternity of past time then, given the first premise, we should not live in the world we experience—we should not live at all. The discussion that raged from about 1870 to 1910 concerning thermodynamics and cosmology involved a mixture of science, philosophy and religious belief, categories that in this case cannot be easily separated.

For some commentators in the Victorian era the argument implied that the world was created supernaturally in accordance with the Bible. After all, if the universe had an absolute beginning, what else could have caused it except a creative and transcendent being, in other words God? It is no coincidence that the entropic argument attracted much positive attention among theologians and Christian scientists; nor was it a coincidence that it was seriously criticized by atheists scientists and philosophers. Maxwell and Thomson were among those who associated the second law with divine creation, and yet they and most other scientists refrained from speculating about the origin of everything. The general attitude, as expressed by the British astronomer Edward Walter Maunder, was that the scientist "cannot go back to the absolute beginning of things, or forward to the absolute ends of things.... Not merely because these beginnings of things were of necessity outside his experience, but also because beginnings, as such, must lie outside the law by which he reasons."[36] While several prominent physicists participated in the thermodynamic controversy it is remarkable that the astronomers kept a low profile. Very few professional astronomers showed an interest in cosmological questions. Even fewer referred to the implications of thermodynamics, a branch of science which somehow failed to attract astronomical interest. Charles Augustus Young, a respected Princeton astronomer and author of astronomical texts, was one of the few exceptions. In a widely read textbook of 1893, *General Astronomy*, he included a section on the cosmological consequences of the second law. He had this to say:

> If we carry our imagination backwards we reach at last a "beginning of things," which has no intelligible antecedent: if forwards, an end of things in stagnation. That by some process or other this end of things will result in "new heavens and a new earth" we can hardly doubt, but science has as yet no word of explanation.[37]

If entropy was a problematic concept to use in arguing for a finite-age universe, after about 1900 there was the possibility of replacing it with another cosmic clock, the newly discovered radioactivity. Although radioactivity was studied as a terrestrial phenomenon, it was generally agreed—incorrectly it turned out—that the material composition of the Earth was largely representative of the cosmos. How could there still be radioactive elements if the world had existed in an eternity? Even uranium and other elements with a very long half-life would have disappeared. In 1911 the Austrian physicist Arthur Haas addressed the question, his answer being that the universe must have had a beginning in time. To avoid the conclusion one might object that uranium and thorium were perhaps decay products of some heavier, hypothetical elements, but Haas showed that the objection would only lead to new difficulties. His suggestion of linking together radioactivity and cosmology attracted but little attention at the time but it played some role in later cosmological thought. Notably, when Georges Lemaître proposed the first cosmological big bang model ever in 1931, he based it in part on the radioactive creation argument.[38]

A few *fin de siècle* scientists contemplated other ways in which radioactivity might be relevant to cosmological questions. Radioactivity not only produced new atoms but also liberated a large amount of energy apparently stored in the material structure of the atom. Moreover, at the time it was generally assumed that radioactivity was a common property of all matter, only appearing in different degrees of intensity. The British chemist Frederick Soddy was a pioneer of radioactive research and the discoverer, together with Ernest Rutherford, of the fundamental law of radioactive decay. In the first textbook ever on radioactivity he speculated that the strange irreversible phenomenon indicated an end of evolution, namely "when all the available energy shall have run its course to exhaustion." And when he looked backwards in time the phenomenon might indicate a beginning:

> Correspondingly, a sudden beginning of the universe—the time when present laws began to operate—is also fixed. It is necessary to suppose that the universe, as a thing in being, had its origin in some initial creative act, in which a certain amount of energy was conferred upon it sufficient to keep it in being for some period of years. It is possible, it is true, to avoid the end indefinitely, since the rate of change will diminish as the end is approached, and theoretically the end will require an infinite time to be attained. But the difficulty with the beginning cannot be so avoided.[39]

We have here, possibly for the first time, the idea of a known physical mechanism responsible for the origin of matter in the universe. However, it was a speculation only and no more, and no more an anticipation of the big-bang scenario than Crookes' earlier speculation. In fact, as Soddy (who was an atheist) revealed shortly later in his book, he was in favor of an eternally cyclic universe in which destructive decay processes were balanced by constructive creation processes. He suggested that this kind of cosmology was in better accord with the sciences of thermodynamics and radioactivity. "The universe," he concluded, is "limited with reference neither to the future nor the past, and demanding neither an initial creative act to start it nor a final state of exhaustion as its necessary termination." To many contemporary scientists this was a more appealing world view.

1.5   Gravitational collapse

At least on a relatively short timescale the near universe is stable. The sky we observe today is approximately the same that people observed in ancient Greece and the Babylonians before them. The question of the stability of the stellar universe became an issue with Newton's universal law of gravitation which supposedly secured the stability and order of the universe. In a famous correspondence with the theologian Richard Bentley from the early 1690s Newton argued in qualitative terms that although a finite collection of stars would coalesce gravitationally, this would not be the fate of an infinite number of uniformly distributed stars. However, he did not give any proof and instead appealed to

God's wisdom in placing the stars so delicately in space that a tiny perturbation would not cause the system to collapse.[40]

For two centuries Newton's cosmological argument was accepted without being rigorously examined. This only happened in 1895 when the German astronomer Hugo von Seeliger, professor and director of the Munich Observatory, reconsidered the argument.[41] Seeliger proved that an infinite Euclidean space with a roughly uniform mass distribution could not be brought into agreement with Newton's law of gravitation. He showed that calculation of the gravitational force exerted on a body by integration over all the masses in the infinite universe does not lead to a unique result, as the integral diverges. Hence the conclusion, "Newton's law, applied to the immeasurably extended universe, leads to insuperable difficulties and irresolvable contradictions if one regards the matter distributed through the universe as infinitely great."[42] Seeliger's concern was not really to save Newton's infinite stellar universe, for he rejected the notion of an actual infinity and tended to believe that the universe was finite. The following year Seeliger framed the gravitation paradox differently, by showing that the Newtonian universe allowed motions that start with finite speed and accelerate to infinitely great speeds in a finite time. As he pointed out, such motions are no less inadmissible than a collapsing universe. In a popular presentation of 1898 Seeliger summarized that whatever the mass distribution in the universe, there must occur infinitely great accelerations in it. His remedy for the conceptual sickness was to suggest that Newton's law should be modified at very large distances (Fig. 1.5). A body of mass m moving in the gravitational field of a central mass M will, according to Newton, experience a gravitational pull given by

$$F(r) = G\frac{mM}{r^2}.$$ (1.6)

Seeliger now suggested that for very large distances, the body would move as if there were a repulsive force in addition to the attractive gravitational force. The suggestion amounted to introducing an attenuation factor of the form $\exp(-\Lambda r)$, which leads to the new law of force

$$F(r) = G\frac{mM}{r^2}\,\mathrm{e}^{-\Lambda r}.$$ (1.7)

Seeliger contemplated that the constant Λ might be important also in planetary astronomy and that its value might be inferred by assuming that it was responsible for the anomalous motion of Mercury's perihelion. He found that the anomaly could be accounted for if $\Lambda = 3.8 \times 10^{-7}$ but also that this value would lead to unobserved disturbances in the motions of the other planets. For this reason he did not seriously pursue the idea. Nor did he pursue the radical idea that the law of gravity might conceivably vary in space. Yet it is interesting that he mentioned it, if only as a remote possibility. "Though highly probable," he wrote, "it is far from self-evident that the forces of attraction follow the same law in all places throughout the universe."[43]

In a book of 1896 the German physicist Carl Neumann repeated the conclusion of Seeliger, claiming that he had come to the same result of a modification of Newton's

Die einfachste Formel, welche eine Absorption berücksichtigt, erhält man, wenn man dem Newton'schen Gesetze $\dfrac{k^2 m m'}{r^2}$ den Factor $e^{-\lambda r}$ hinzufügt, wo e die Basis des natürlichen Logarithmensystems ist. λ wird nicht eine Constante zu sein brauchen, aber sie soll als solche gelten. Die Anziehungskraft A wird also ausgedrückt durch

$$A = k^2 m m' \cdot \frac{e^{-\lambda r}}{r^2} \qquad (2)$$

Es ist ersichtlich, dass man λ stets so klein wählen kann, dass innerhalb unseres Planetensystems mit beliebiger Annäherung das Newton'sche Gesetz hervorgeht. Andrer-

Fig. 1.5 *Seeliger's passage in* Astronomische Nachrichten *from 1895 in which he suggested his modified law of gravitation including a repulsive λ-constant.*

law many years earlier. In the literature of the history of cosmology one can sometimes read of the "Neumann–Seeliger paradox." However, as shown by John Norton this is a misunderstanding.[44] The paradox belongs to Seeliger alone.

Seeliger's constant can be considered a classical version of the cosmological constant that Einstein famously introduced in his cosmological field equations of 1917. In fact, at the very beginning of his paper of that year Einstein dealt with the same problem that Seeliger had identified, the incompatibility of Newtonian gravitation and an infinite stellar universe with a constant mass density. At the time Einstein was probably unaware of Seeliger's old paper in *Astronomische Nachrichten*, but he later acknowledged the German astronomer's modification of Newton's law.[45] The modified force law was essentially *ad hoc* and also arbitrary, since many other modifications might resolve the gravitation paradox in a similar way. The idea of modifying Newton's inverse-square law was not, by itself, very original, as many such modifications were proposed in the nineteenth century. The exponential correction factor can be found in 1825 in Laplace's *Mécanique Céleste* which can hardly have avoided Seeliger's attention. However, what was original in Seeliger's approach was that he used it in a cosmological context and not, as in most other proposals, just to solve problems of planetary astronomy such as Mercury's anomalous revolution around the Sun.

William Thomson, apparently unaware of Seeliger's work, arrived at basically the same results in two papers published 1901–1902. In the first paper he proved that in an infinite universe with a non-zero density of matter, "a majority of the bodies in the universe would each experience infinitely great gravitational force."[46] Considering a homogeneous model universe of radius 3×10^{16} km, he concluded that the number of stars, assumed to be of solar mass on average, must be neither too great nor too small. He found it "highly probable" that there were fewer than two billion stars within the sphere, and more than one hundred million. Thomson also calculated the time it took for the stellar system to collapse to a zero radius. This collapse time turned out to be independent of the initial size of the universe and to depend only on its mass density ρ_0:

$$t_{\text{collapse}} = \frac{1}{4}\sqrt{\frac{3\pi}{2} G\rho_0}. \qquad (1.8)$$

For a universe containing one billion stars, the collapse time would be 17 million years, a figure of the same order as that which Thomson had found for the age of the Earth and the lifetime of the Sun. Contrary to Seeliger, Thomson did not present his investigation clearly as a problem for the infinite Newtonian universe, and he did not propose a way out of the problem.

There were other ways to escape gravitational collapse than modifying Newton's law of gravitation. One could leave the law intact and change some of the cosmological assumptions of the Newtonian universe, such as the homogeneous distribution of matter. This is what the astronomy author Richard Proctor did in his widely read *Other Worlds than Ours* first published 1870, although he developed his model in the context of Olbers' optical paradox and not the gravitational paradox. Proctor conceived of a hierarchic model universe in which the higher star systems were separated by increasingly larger distances from the lower ones. In that case, the contributions of light from stars lying in successive shells would not be equal, but successively less, and the total amount of light received from the infinite number of stars could be quite small.[47]

In 1908, Carl Charlier, professor of astronomy at Lund University, Sweden, developed in mathematical detail a hierarchic model in a paper titled "Wie eine unendliche Welt aufgebaut kann" (How an infinite world can be constructed).[48] This was not the first time he expressed his interest in the grander aspects of cosmology. Unusually for professional astronomers at the time, Charlier transcended the divide between scientific astronomy and philosophical reflections on the more speculative aspects of cosmology. In 1896 he argued that Olbers' and Seeliger's paradoxes indicated that the universe was finite, a solution he preferred to Seeliger's modification of Newton's law of gravity. However, he also pointed out that the paradoxes rested on the assumption of uniformly distributed stars, an assumption that might be questioned. If the distribution of stars followed a law such that "the density of the stars decreases faster, as we move out in space," the paradoxes did not need to arise.[49] As mentioned, this idea, as far as it is related to Olbers' paradox, goes back to John Herschel in 1848 and was revived by Proctor in 1870. As to the temporal extension of the universe Charlier thought that a beginning and end were ruled out by the law of mass conservation as well as common sense. "A finite time is a contradiction," he wrote, "an infinite time may be difficult to conceive, but it is not contradictory."

Let the Milky Way S_1 be composed of N_1 stars and let N_2 Milky Way galaxies form a second-order galaxy S_2; N_3 galaxies of type S_2 form a third-order system S_3, and so on. The system S_1 has a radius R_1. Charlier showed that if the mean density of a structure of size R decreased as

$$\rho \sim R^{-\alpha} \text{ with } \alpha > 2,$$ (1.9)

then the inequality

$$\frac{R_{i+1}}{R_i} \geq \sqrt{N_{i+1}}$$ (1.10)

was satisfied. As a consequence Seeliger's gravitation paradox would disappear and there would be no infinite velocities. In Charlier's hierarchical model the mass of the universe was infinite but its average density zero. Charlier further derived an inequality for the limiting angular diameter of a spiral nebula, namely

$$\theta < N^{-1/6}. \tag{1.11}$$

From this he found values for the angular diameter and apparent magnitude of the nearest spiral that agreed qualitatively with the observed values of Andromeda.

Charlier's ideas received support from Franz Selety, a Viennese physicist, who in 1922 developed them as a Newtonian alternative to the relativistic cosmology that Einstein had introduced five years earlier.[50] Selety criticized Einstein's theory of a closed universe and argued that if matter became diluted with distance in a suitable way (faster than $1/r^2$), Newtonian theory allowed an infinite universe completely filled with matter. The local matter density in Selety's "hierarchical molecular world" was everywhere finite and yet the average density would tend to zero. A brief debate followed between Einstein and Selety in the pages of *Annalen der Physik*, but after his first reply in 1922 Einstein chose not to reply to Selety's papers.

The non-uniform universe proposed by Charlier continued to attract interest among cosmologists who did not accept the cosmological uniformity principle. One of them, the Polish-American physicist Ludwik Silberstein, dismissed in a book of 1929 the closed Einstein model and its apparent agreement with Edwin Hubble's count of nebulae. Although Silberstein stayed on the ground of relativistic cosmology he much preferred a hierarchic world model of the type suggested by "the great Swedish astronomer."[51]

One year after Charlier's 1908 paper had appeared, another Swedish scientist, the Nobel laureate Svante Arrhenius, examined the gravitation paradox. As mentioned, Arrhenius was an ardent advocate of a homogeneous and infinite universe, and consequently he felt it necessary to criticize Charlier's hierarchic model. One might have thought that he would be sympathetic to Seeliger's solution, but this was not the case. He could see no problem with the Newtonian universe and therefore found Seeliger's work to be irrelevant. (As Seeliger was quick to point out, Arrhenius had partly misunderstood his work.) "There really is no weighty reason why the world would not be sown uniformly with stars," Arrhenius concluded.[52] Arrhenius' denial that there was a problem at all hardly convinced anyone except himself. When it came to the notion of a finite-age universe Arrhenius shared the view of Charlier, namely that it contradicted the fundamental conservation laws of physics and for this reason alone was ruled out. As Arrhenius phrased it in a textbook of 1903, a universe with a beginning in time "is hard to bring into agreement with the indestructability of energy and matter."[53] Indeed it is, and it was not the last time that the difficulty was pointed out.

1.6 Non-Euclidean astronomy

Whereas curved space as a mathematical concept dates from the early nineteenth century, it took more than a century before it permeated to the physical and astronomical sciences. The eminent mathematician and polymath Karl Friedrich Gauss arrived as early as 1816 to the conclusion that the ordinary, flat or Euclidean geometry was not true by necessity. As he wrote in a letter to Olbers, "Maybe in another life we shall attain insights into the essence of space which are now beyond our reach. Until then we should class geometry not with arithmetic, which stands purely a priori, but, say, with mechanics."[54] Gauss realized that if the question could be settled at all it would require astronomical measurements over very large, stellar distances. Although Gauss seems to have been aware that information about the curvature of space could in principle be obtained from data of stellar parallaxes, he did not pursue this kind of reasoning. In a letter to the German-Danish astronomer Heinrich C. Schumacher of 12 July 1831 he communicated the formula for the circumference of a circle of radius r in the new geometry, stating it to be

$$l = \frac{\pi}{2} k \left[\exp\left(\frac{r}{k}\right) - \exp\left(-\frac{r}{k}\right) \right].$$

(1.12)

The quantity k is a constant to be determined observationally and which, in the Euclidean case, is infinite. Should space not be Euclidean, Gauss said in his letter to Schumacher, from an observational point of view we can only say that the curvature measure k must be incredibly large.

The true founders of non-Euclidean geometry were the Hungarian mathematician János Bolyai and the Russian Nikolai Ivanovich Lobachevsky, both of whom published their independent discoveries that a geometry different from and as valid as Euclid's is possible. Whereas Bolyai did not refer to the skies, the astronomy-trained Lobachevsky did. Although primarily a mathematician, as a young man he had studied astronomy and in the 1820s he served as director of the Kasan University Observatory. Already in his 1829 paper in the *Kasan Messenger* Lobachevsky suggested that one consequence of his "imaginary" (or hyperbolic) geometry might be tested by astronomical means.[55] As he pointed out, the angle sum of a triangle would always be less than 180° and the more so the bigger the triangle becomes. From astronomical data he concluded that the angle sum of the triangle spanning the Sun, the Earth, and Sirius deviated from the Euclidean value of 180° by at most 0.000372 arcsec, evidently much less than the observational error. The modern value of Sirius' parallax is 0.37 arcsec, which is less than a third of the value adopted by Lobachevsky.

The Russian mathematician realized that while it could in principle be proved that astronomical space is non-Euclidean, it could never be proved to be Euclidean, and for this reason he tended to see his comparison as inconclusive. In a later work, titled *Pangeometry* and translated into French in 1856, he argued that, assuming space to be hyperbolic, there must be a minimum parallax for all stars irrespective of their distances from the Earth. This is contrary to flat space, where the parallax tends toward zero as the distance increases toward infinity. Consider a triangle spanned by a star and the two positions of the Earth half a year apart in its orbit around the Sun. Let the angle at the star be denoted α and the two angles at the positions of the Earth be β and γ. Lobachevsky showed that the parallax angle p can be expressed as

$$p = \pi - (\beta + \gamma) = \alpha - K\beta,$$

(1.13)

where the curvature is $K = 0$ in flat space. The ideas of non-Euclidean geometry pioneered by Gauss, Bolyai, and Lobachevsky circulated but slowly in the mathematical community. Only about 1870 did they truly enter the world of mathematics if not yet the world of the astronomical and physical sciences.[56] The new ideas then resulted in a revolution in geometry. Of particular importance was the work of the Göttingen mathematician and physicist Bernhard Riemann, who in a famous address of 1854 (published only 1867) put the concept of curvature as an intrinsic property of space on a firmer basis. Importantly, his address led to the now standard distinction between the three geometries of constant curvature. The curvature K, which has the dimension of an inverse area, relates to the radius of curvature R by

$$R^2 = \frac{k}{K} \text{ with } k = 0, \pm 1. \tag{1.14}$$

The three possibilities correspond to Euclidean space (curvature constant $k = 0$), spherical space ($k = +1$), and hyperbolic space ($k = -1$).

Riemann referred only to astronomy in passing, pointing out that on the assumption of a constant-curvature space it follows "from astronomical measurements that it [the curvature] cannot be different from zero."[57] Although he accepted the idea of just one physical space, like Gauss he did not accept that the geometry of this space could be known a priori or with absolute certainty. Our experience of the physical world, he said, could be consistent with any of the three possible geometries. Moreover, he left open the possibility that on a microphysical scale the curvature of space might vary in such a way that the averaged curvature over measurable distances becomes inappreciably close to zero.

Although Riemann's emphasis on the possibility of an unbounded yet finite space implicitly addressed an old cosmological conundrum, it failed to attract interest among astronomers. It took nearly two decades before a scientist made astronomical use of Riemann's insight. When it happened it made no impact at all on the astronomical community. The Leipzig astronomer Johann C. Friedrich Zöllner was a pioneer astrophysicist known in particular for his contributions to astrophotometry. After 1877 he focused increasingly on what he called "transcendental physics," the study of spiritualist phenomena based on the postulate of a fourth space dimension. As one might expect, this line of work created so much public attention that it damaged his scientific reputation.[58] A few years before his conversion to spiritualism, Zöllner published a controversial book with the title *Natur der Cometen* (The Nature of Comets; Fig. 1.6) which included a chapter of considerable interest to the history of cosmological thought. By arguing that cosmic space might be positively curved in accordance with Riemann's idea, Zöllner offered an original solution to Olbers' paradox.

In his systematic discussion of the finite versus the infinite in the universe Zöllner assumed, for the sake of discussion, that there is only a finite amount of matter in the world. He then argued that in an infinite Euclidean space any finite amount of matter would eventually evaporate and dissolve to zero density. Given the actual existence of matter he concluded that either is space finite or the universe has only existed

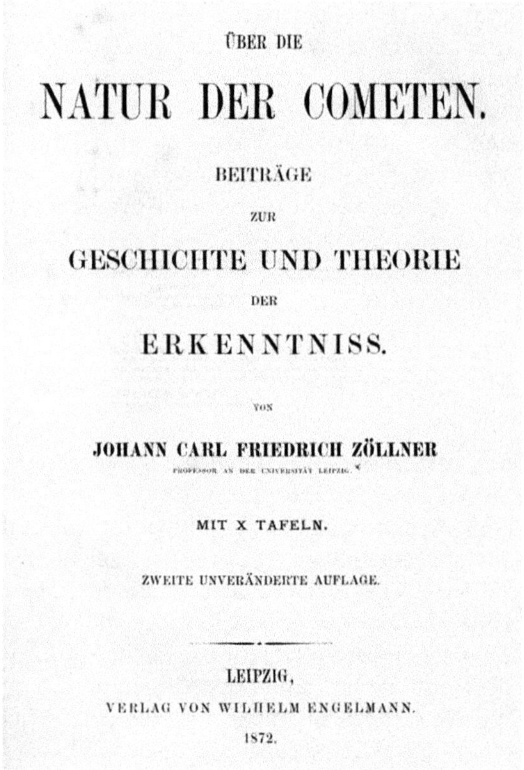

ÜBER DIE

NATUR DER COMETEN.

BEITRÄGE

ZUR

GESCHICHTE UND THEORIE

DER

ERKENNTNISS.

VON

JOHANN CARL FRIEDRICH ZÖLLNER

PROFESSOR AN DER UNIVERSITÄT LEIPZIG.

MIT X TAFELN.

ZWEITE UNVERÄNDERTE AUFLAGE.

LEIPZIG,

VERLAG VON WILHELM ENGELMANN.

1872.

Fig. 1.6 *Title page of Zöllner's 1872 book on the nature of comets, including his proposal of a geometrically closed universe.*

in a limited period of time. Unwilling to accept the latter hypothesis he suggested that Riemann's geometry might provide the key that would unravel the secrets of the universe. "It seems to me," he wrote, "that any contradictions will disappear...if we ascribe to the constant curvature of space not the value zero but a positive value. The assumption of a positive value of the spatial curvature measure involves us in no way in contradictions with the phenomena of the experienced world if only its value is taken to be sufficiently small."[59] In this way he could explain Olbers' paradox without having to assume interstellar absorption of starlight or taking recourse to a limitation of either cosmic time or space. Zöllner's innovative cosmological speculations attracted some interest in German philosophical circles, but were ignored by almost all physicists and astronomers. They were essentially forgotten.

During the last quarter of the nineteenth century, non-Euclidean geometry became a hot topic in mathematical and philosophical circles, and it was discussed in hundreds of books and scientific papers. On the other hand, the number of astronomers who expressed interest in the topic can be counted on the fingers of one hand. Moreover, the interest rarely went beyond uncommitted comments and it appeared more often in popular than scientific works. The Irish astronomer Robert Stawell Ball, who from 1892 to 1913 served as professor of astronomy and geometry at Cambridge University, mentioned the possibility of a positively curved space at some occasions but without endorsing it as a reality. This cautious and ambiguous attitude was shared by his distinguished American colleague Simon Newcomb, who discussed the hypothesis in the widely read *Popular Astronomy* first published in 1878. Like Lobachevsky many years earlier, Newcomb pointed out that the hypothesis was testable although this might perhaps be more in principle than in practice. As he wrote, "Unfortunately, we cannot triangulate from star to star; our limits are the two extremes of the earth's orbit. All we can say is that, within those narrow limits, the measures of stellar parallax give no indication that the sum of the angles of a triangle in stellar space differs from two right angles"[60]

A few physicists and mathematicians wondered if the puzzle of the anomalous motion of Mercury's perihelion might be solved on the assumption that the geometry of space is non-Euclidean.[61] Thus, in 1885 the German mathematician Wilhelm Killing derived the orbit of Mercury moving in spherical space by means of a modified law of gravitation that he found suitable for that kind of space. His investigation was of mathematical interest only and the same was the case with a similar work by Carl Neumann. The later approach of the Austrian physicist Josef Lense was different in so far that he assumed Newton's law to be universally true and applied it to spherical space. In a paper of 1917 he derived in this way an expression for the perihelion advance but unfortunately his expression could not be translated into a numerical answer. In any case, by that time the motion of Mercury's perihelion was no longer anomalous. The problem had been solved on the basis of Einstein's general theory of relativity.

In the period before general relativity only two astronomers, both of them German, examined more systematically the possibility of a Newtonian stellar universe embedded in non-Euclidean space. One of them was 27-year-old Karl Schwarzschild, who in a lecture of 1900 discussed how to determine the geometry of space from astronomical observations. He summarized his results as follows:

> One may, without coming into contradictions with experience, conceive the world to be contained in a hyperbolic (pseudo-spherical) space with a radius of curvature greater than 4,000,000 earth radii, or in a finite elliptic space with a radius of curvature greater than 100,000,000 earth radii, where, in the last case, one assumes an absorption of light circumnavigating the world corresponding to 40 magnitudes.[62]

Schwarzschild saw no way to go further than this rather indefinite conclusion and decide observationally whether space really has a negative or positive curvature, or whether it really is finite or infinite. Nonetheless, from a philosophical and emotional point of view he definitely preferred a closed universe. His reason was that "then a time will come when space will have been investigated like the surface of the Earth, where macroscopic investigations are complete and only the microscopic ones need continue."

It is noteworthy that some later cosmologists, including Eddington and Lemaître, expressed their preference for closed universe models in similar philosophical language.

Eight years later Paul Harzer, professor of astronomy at the University of Kiel, developed a more detailed model of a closed stellar universe.[63] Harzer's universe was enclosed in a finite cosmic space with a volume about 17 times that of the stellar system. This system contained the same number of stars but was compressed to a size approximately one half of what it had in other models based on flat space. The size of the entire universe was given by the time it took for a ray of light to circumnavigate it, which Harzer estimated to be 8,700 years. The Schwarzschild–Harzer suggestion of a closed space filled with stars had the conceptual advantage that it did away with the infinite empty space, but it made almost no impact on mainstream astronomy. The cosmological problem that moved to the forefront of astronomy in the 1910s was concerned with the size of our Galaxy and the question of whether the spiral nebulae were external objects or belonged to our own Galaxy. This was a problem in which the geometry of space was considered to be irrelevant. In short, the main reason why astronomers were reluctant to consider the consequences of space being non-Euclidean was that they had no need for the hypothesis.

Among the very few scientists who referred to curved space as more than just a mathematical curiosity before Einstein was also the Austrian meteorologist Wilhelm Trabert, a professor in Vienna. In a textbook of 1911 on what at the time was known as "cosmical physics" he included a chapter on cosmology in which he discussed Olbers' paradox, the anomalous motion of Mercury's perihelion, and other issues. In this context he contemplated that the universe might be finite because space is finite. It is conceivable, he wrote, that astronomical observations would one day demonstrate that "light, gravitation and electricity do not propagate in straight lines, but in circles."[64]

Schwarzschild did not consider his investigation of 1900 to be very important, but it was well known and its main conclusions re-appeared in the 1911 edition of the recognized and widely read Newcomb–Engelsmanns *Populäre Astronomie*. Schwarzschild's early work was also of some significance in the earliest phase of relativistic cosmology. When Willem de Sitter developed his model universe in 1917 based on Einstein's field equations, he referred to the limit of stellar parallaxes which Schwarzschild had found earlier on the assumption of cosmic space being either elliptic or hyperbolic. "The limit found by Schwarzschild still corresponds to our present knowledge," de Sitter wrote.[65] Most likely, it was through de Sitter that Einstein first became acquainted with the astronomical works of Seeliger and Schwarzschild.

1.7   Finite or infinite?

Conceptual and speculative ideas about the universe as a whole did not occupy a major role in nineteenth-century astronomy and cosmology. In so far that such ideas were discussed the question of the size of the universe permeated much of the discussion. Briefly put, is the universe finite or infinite in extension? The "universe" might in this context refer to either space or, what was more common, to the material stellar system embedded in it. As we have seen, the question was at the basis of the discussions concerning Olbers' paradox, the gravitation paradox, and also of the cosmic consequences derived from the second law of thermodynamics. It was not, however, a question that kept most astronomers awake at night; and they were even less concerned with the possibility of an origin of the universe.

At the end of the nineteenth century it was generally recognized that if an observationally grounded cosmology was ever to be established, a first step would be to understand the size and structure of the Milky Way. Space was usually thought to be infinite, but there was no consensus at all regarding the distribution of stars and nebulae in the universe. Many astronomers hesitated to admit it as a problem that could be solved scientifically by means of the usual combination of theory and observations. In an article of 1869 Proctor gave voice to the dilemma: "The only question for us is between an infinity of occupied space and an infinity of vacant space surrounding a finite [material] universe. Either idea is equally incomprehensible; but the former is merely beyond, the latter seems contrary to reason."[66]

On this issue Thomson agreed, and with basically the same argument as Proctor. He was convinced that matter was distributed endlessly through space, which was one reason why he did not accept the cosmic version of the heat death as real. "Finitude is incomprehensible, the infinite in the universe is comprehensible," he claimed in a popular lecture of 1884. The illustration of the claim made popular sense but was far from a scientific argument: "What would you think of a universe in which you could travel one, ten or thousand miles, or even to California, and then find it come to an end? Can you suppose an end of matter or an end of space?"[67] While only the few scientists acquainted with Riemannian geometry could conceive a finite space, to many astronomers an end of matter was perfectly intelligible. This kind of subjective argument—based on what individual scientists could comprehend or not comprehend—colored much of the debate not only in the pre-Einstein era but also in later periods.

In a book of 1890 the Irish astronomer and historian of astronomy Agnes Mary Clerke made clear that the nebulae were parts of the Milky Way rather than separate galactic systems. What, then, about the universe beyond the Milky Way? Clerke raised the question only to dismiss it as unworthy of scientific consideration: "With the infinite possibilities beyond, science has no concern." In a later book, the massive *Problems of Astrophysics*, she likewise refrained from going beyond what she poetically called "the equatorial girdle of a sphere containing stars and nebulae."

As she wrote:

> The whole material creation is, to our apprehension, enclosed within this sphere. We know nothing of what may lie beyond. Thought may wander into the void, but observation cannot follow. And where its faithful escort halts, positive science comes to a standstill. Fully recognizing the illimitable possibilities of omnipotence, we have no choice but to confine our researches within the bounds of the visible world.[68]

There were many voices like Clerke's, Newcomb's being one of them. In an essay on unsolved problems in astronomy Newcomb asked if the universe was populated with stars all over, or if they were largely contained in the system of the Milky Way, itself floating in infinite empty space. The question, he wrote, "must always remain unanswered by us mortals . . . Far outside of what we call the universe might still exist other universes which we can never see." For all practical purposes, the Milky Way "seems to form the base on which the universe is built and to bind all the stars into a system." In an age of positivism, the general attitude was that theories and hypotheses were put forward to explain facts, and hence, when there were no facts to be explained, no theory was required. "As there are no observed facts as to what exists beyond the farthest stars, the mind of the astronomer is a complete blank on the subject," Newcomb wrote. "Popular imagination can fill up the blank as it pleases."[69]

To mention yet another example of the agnostic tradition prevailing in astronomy, consider the presidential address that the physicist and astronomer George H. Darwin, a son of Charles Darwin, gave to the British Association in 1905. Does it not seem as futile to imagine that man "can discover the origin and tendency of the universe as to expect a housefly to instruct us as to the theory of the motions of the planets?" Although Darwin admitted that great progress had occurred in the sciences, he concluded that "the advance towards an explanation of the universe remains miserably slight." His prediction of future knowledge was as pessimistic as it was wrong: "We may indeed be amazed at all that man has been able to find out, but the immeasurable magnitude of the undiscovered will throughout all time remain to humble his pride. Our children's children will still be gazing and marveling at the starry heavens, but the riddle will never be read."[70]

The question of the spatial and material finitude of the universe was hotly debated in the last decades of the nineteenth century, if more among philosophers, theologians and social critics than among professional astronomers.[71] The Austrian amateur astronomer Rudolf Falb, founder of the popular-astronomy magazine *Sirius*, argued that the very concept of space had only meaning in connection with matter. From this premise followed the possibility of limited space, as the question of what is beyond the end of space was deemed illegitimate; for according to Falb there was no way to define "beyond." His view of a necessary connection between the concepts of space and matter was a curious return to Aristotelian natural philosophy. It allowed Falb to conceive of space as limited without questioning the implicit assumption of Euclidean geometry.

While the consensus view was that temporal finitude followed spatial finitude, this was not the opinion of Falb. On the contrary, he was convinced that the universe must be as eternal as matter. A universe consisting of a limited number of stars and nebulae would eventually contract gravitationally to a giant sun, but this sun would immediately evaporate and form an expanding gas, a nebulous body of the same kind as the one that had originally given rise to the celestial bodies. Falb contended that the process would go on endlessly: "The life of the world is to be conceived as a recurrence of expansion and contraction, like the breaths of a monstrous colossus. In this way the eternity of processes becomes understandable, that is, the infinite duration of the universe. . . . The end of the world is at the same time the beginning of the world."[72]

If only in a speculative and classical sense Falb's scenario included the idea of a finite expanding universe with expansion and contraction in eternal change. A somewhat similar idea was later proposed by Ludwig Zehnder, a Swiss-German professor of physics in Freiburg and a student of Wilhelm Röntgen. In a book of 1897 Zehnder assumed that all celestial bodies were confined within a gigantic sphere of ether. Like several other scientists at the time, he conceived the ether as a kind of rarefied gas consisting of tiny atoms endowed with mass and other physical properties. A sidereal universe including a limited number of stars—a picture favored by Olbers' paradox— would be subject to gravitational collapse. However, according to Zehnder this would not be the end of the world, for the cooling of the central stellar body would lead to an increase of the kinetic energy of the ether atoms and to an increased electrical repulsion between atoms of matter. The result would be an expansion of the material universe, eventually to be followed by a new contraction when the gravitational attraction came to dominate the electrical repulsion. In this way Zehnder believed he had constructed on a scientific basis an eternal and cyclic universe.[73]

The speculations of Falb and Zehnder referred to what can possibly be regarded as anticipations of the expanding universe, although in this case the expanding phase was just part of a longer cyclic evolution. But although one can find in their works the phrase "expanding universe" (Ausdehnung der Welt) it was of course in a sense quite different to the one that later entered the cosmological vocabulary. In this context it may be relevant to mention that the German word for expansion (Ausdehnung) may also have the different meaning of expanse. The great mathematician David Hilbert gave in 1925 a famous address on the infinite in which he stressed that although the concept made mathematical sense it had no counterpart in real nature. In this connection he spoke of *die Ausdehnung der Welt*, but it was not an allusion to what we call the expanding universe. Hilbert undoubtedly referred to the expanse and not to the expansion of the world.[74]

Whereas the notion of infinite space was largely uncontroversial it was commonly claimed that the corresponding notion of an infinity of material objects, whether atoms of stars, was absurd. The claims were mostly based on philosophical and logical arguments and in many cases they were on the rhetorical level only. Since the time of Aristotle the majority of mathematicians and philosophers had taken for granted that so-called realized or actual infinities, as distinct from potential infinities, cannot exist. There typically was a strong element of theology in these claims. For example, according to Augustin Cauchy, a brilliant mathematician and a passionate Catholic, infinity and eternity were divine attributes not to be found in nature. Many thinkers of a Christian orientation continued to hold this view. To claim that the physical universe is infinite would obscure a crucial difference between God and nature and, they suggested, open up for pantheism and like heresies.

On the other hand, it was a dogma among materialists, atheists, and positivists that the universe is infinite in extension, which was usually taken to imply spatial as well as material infinitude. When it came to infinitism, the problematic issue was usually the number of things in the universe, not the more abstract concept of spatial extension. It was relatively easy to conceive space as infinite, such as Thomson did, but much harder to conceive it as filled with matter. For in this case one was faced not only with astronomical problems, such as Olbers', but also with the philosophical problems arising with of an actual infinity of numbers of objects or processes. According to the German mathematician Georg Cantor's theory of transfinite numbers, dating from the 1880s, the actual infinite existed in the same sense that finite numbers exist, namely, as a logically consistent and operational concept. But one thing is mathematical consistency; another question is whether or not an actual infinite belongs to the real world as examined by physicists and astronomers. Cantor's theory of numbers attracted the attention of philosophers but was thought to be too abstract to apply to the real cosmos.

In the mid-1920s another brilliant German mathematician, David Hilbert, discussed the possible relevance of Cantor's actual infinities to the real world. Although Hilbert admired Cantor's theory, he argued that the infinite was merely an idea and that it had no role to play whatever in the empirical sciences. It was in this context that he introduced the imaginary "Hilbert's hotel" with infinitely many rooms. Even though all rooms are occupied in the hotel there will always be free rooms for new guests—and even infinitely many of them! Only much later did Hilbert's bizarre hotel enter cosmological thinking, first in George Gamow's popular book *The Creation of the Universe* from 1952 and subsequently, if only peripherally, in relation to the steady-state theory of the universe.[75]

In the late nineteenth century ideas of space-only infinitism depended on one's conception of space. If space was merely thought of as a container of matter and physical phenomena, as was typically the picture of the Milky Way universe, problems did not necessarily arise. The situation was different if space was thought to be filled up with ether, and especially if the ether was conceived to be endowed with energy or otherwise being substantial in nature. As mentioned, during the second part of the nineteenth century belief in the infinite universe was correlated with atheism, and finitism with theism. And yet the correlation was not generally valid. In fact, for centuries there had been an opposite tradition of linking God's omnipotence to a divinely created, infinite universe. Descartes, Newton and Leibniz conceived the infinite universe as the only one worthy for the almighty God, and a similar view was held by Kant and Olbers. A Christian scientist as he was, Thomson nonetheless concluded that a spatially finite universe was impossible. However, at the time Thomson was an exception. Astronomers in the *fin de siècle* period realized that the question of the size of the universe could not be decided observationally and for this reason they rarely discussed it.

However, cosmology in the first decade of the twentieth century was far from restricted to astronomers' analyses of the Milky Way. In the eyes of the public, Arrhenius may have been the period's best known "cosmologist," a reputation due to a large extent to his highly successful *Worlds in the Making* of 1908. While many astronomers conceived the material universe to be roughly identical to the Milky Way, Arrhenius based his cosmology on what was later called the cosmological principle, namely, that the universe is uniformly populated throughout with stars and nebulae (Section 4.3). And not only that, for Arrhenius' guiding principle was "the conviction that the Universe in its essence has always been what it is now. Matter, energy, and life have only varied as to shape and position in space."[76] In other words, he built his theory on what in the later steady-state theory of the universe would be called the perfect cosmological principle (see Section 5.5). Like several other writers at the time, Arrhenius advocated an infinite, eternally cyclic universe. He argued that such a world view did not necessarily disagree with the second law of thermodynamics since radiation pressure provided a mechanism to keep the universe going for ever. After Arrhenius' hypothesis had been shown to be untenable by Poincaré in 1911 and by Schwarzschild two years later, little more was heard of it in scientific circles. But although most astronomers ignored his ideas, they were widely discussed by laymen and amateur astronomers.

The Norwegian physicist Kristian Birkeland, an authority in research of the aurora borealis, shared with Arrhenius his ambition of establishing cosmology on the basis of physical mechanisms. In articles published between 1911 and 1916 he extrapolated his early ideas of the origin of the aurora into a grand cosmological sketch of a universe dominated by cathode rays and other electrified particles (Fig. 1.7). According to Birkeland:

> One of the most peculiar features of this cosmogony is that space beyond the heavenly bodies is assumed to be filled with flying atoms and corpuscles of all kinds in such density that the aggregate mass of the heavenly bodies... would only be a very small fraction of the aggregate mass of the flying atoms there. And we imagine that an average equilibrium exists in infinite space, between disintegration of the heavenly bodies on the one hand, and gathering and condensations of flying corpuscles on the other.[77]

For a brief period of time Birkeland's theory attracted international attention, if mostly outside the astronomical community. For example, *New York Times* covered in detail

Fig. 1.7 *Kristian Birkeland (1867–1917) to the left, demonstrating discharge phenomena with his model of the Earth placed in a vacuum chamber and exposed to cathode rays. The person to the right is his assistant Karl Devik.*

Source: K. Birkeland, The Norwegian Aurora Polaris Expedition, vol. 1 *(Christiania: Aschehough, 1913), p. 667.*

his "amazing picture of the future development of the universe." The important point in Birkeland's cosmology was the hypothesis that the stars and nebulae were potent generators of electrons and ions, with the consequence that interstellar and intergalactic space was filled with a tenuous ionized gas. In effect, he introduced the notion of what was later called space plasma. Many years after Birkeland's death his ideas were reconsidered by the Swedish physicist Hannes Alfvén, a Nobel Prize laureate of 1970, who turned them into a theory of plasma cosmology.[78]

Einstein's theory of 1917 is rightly considered a watershed in modern cosmology, but it took a long time before the majority of astronomers recognized that a revolution had occurred. In the spring of 1924 Charlier gave a series of lectures on cosmology at the University of California which were published the following year under the title "On the Structure of the Universe." A supporter of the island-universe theory Charlier had no problem in dealing with "the unknown world at the other side of the Milky way" which he presented in terms of his hierarchic theory of an infinite system of nebulae. His lectures were in the style of traditional cosmology built on stellar statistics, with no mention of nebular redshifts and no indication that the general theory of relativity had changed the foundation of cosmology. Although Charlier ignored the theories of Einstein and de Sitter, he was undoubtedly aware of them. His only, indirect reference to the closed Einstein universe was in relation to the heat death: "You know that there are also speculative men in our time who put the question whether space itself is finite or not, whether space is Euclidean or curved (an elliptic or hyperbolic space)." But Charlier had no like for these speculative men whom he compared to the philosopher Kant—and he did not mean it as praise. Their speculations, he said, "must be discussed on the support of facts and at present they must be considered as lying outside such a discussion."[79]

Still in the early 1930s Charlier's classical universe was alive and taken seriously by a few astronomers. The distinguished astronomer Heber Curtis, director of the University of Michigan observatories and a specialist in the study of spiral nebulae, preferred the hierarchical system over the new expanding models of the universe based on the field equations of general relativity. "The remarkably close parallelism between the structure of the suggested Charlier universe and the observed arrangement of the accessible exterior universe, is too exact and detailed to be cast lightly aside," he wrote in the authoritative *Handbuch der Astrophysik* (Handbook of Astrophysics): "For this and other reasons, therefore, the writer prefers, solely as a matter of personal choice and belief, to adhere, until more evidence may be gathered, to a cosmogony based on the scheme of Charlier, rather than to accept one or another of those which seem as yet somewhat nebulously based upon a four-dimensional frame of reference."[80] And this was not the last time that a leading astronomer found inspiration in Charlier's classical picture of the universe, if as a complement rather than an alternative to the expanding universe based on the theory of general relativity. As late as 1970 the French-American astronomer Gérard de Vaucouleurs suggested that "a hierarchical structure *à la* Charlier must be included in any realistic cosmological model."[81]

....

NOTES

1. There are only few modern works which describe the historical development of cosmology in its entirety. See Kragh (2007) and North (2008). This chapter draws on material in the first source. Merleau-Ponty (1983), an important work on nineteenth-century cosmology and astronomical philosophy, only exists in its French original.
2. See Kant (1981), p. 152, an annotated translation by Stanley Jaki.
3. See Jaki (1974) for a history of the idea of the cyclic or oscillating universe. For a review of the idea in modern cosmology, see Kragh (2011), pp. 193–215.
4. Bondi (1952), p. 21. The history and substance of Olbers' paradox is detailed in Jaki (1969) and Harrison (1987).
5. Olbers (1826), p. 119.
6. Quoted in Jaki (1969), p. 147.
7. See Tipler (1988).
8. Thomson (1901); Harrison (1986), pp. 155–65, 227–8.
9. According to the Helmholtz–Thomson theory first proposed by Helmholtz in 1854, the source of stellar energy was gravitational. In modern astrophysics the KH or Kelvin–Helmholtz timescale refers to the time for a stellar body (mass M and radius R) to radiate away all of its potential energy. The KH time is given by GM^2/RL, where L is the average luminosity. See Kragh (2016a) for details of the classical contraction theory and its cosmological implications.
10. Quoted in Berendzen, Hart, and Seeley (1984), p. 183. For the history of interstellar absorption, see also Seeley and Berendzen (1972a, 1972b).
11. Trumpler (1930).
12. Quoted in Kragh (2007), p. 89.
13. Ørsted (1998), p. 253.
14. For details on the history of astrospectroscopy, see Hearnshaw (2014), and for spectroscopy generally, Hentschel (2002).
15. Kirchhoff (1861).
16. Eddington (1926a), p. 347.
17. For early quantum astrophysics and references to the literature, see Tassoul and Tassoul (2004).
18. See Dalgarno (1983).
19. For Doppler's life and work, see Schuster (2005).
20. The nebular hypothesis was as much a world view as a scientific theory. For aspects of the hypothesis and its reception, see Brush (1987) and Schaffer (1989).
21. Huggins and Huggins (1909), p. 107. See Becker (2011) for a critical analysis of Huggins' work and retrospective accounts.
22. Crookes (1886), p. 568.

23. The riddles of the spectral lines attributed to coronium and nebulium are described in Hufbauer (1991), pp. 112–115 and Hirsh (1979).
24. Kragh (2009b) is a detailed account of the history of helium up to and including its discovery in 1895. The role of helium in the birth of astrophysics is described in Nath (2013).
25. Tayler (1995). See also Chapter 8 in the present volume.
26. Chamberlin (1899), p. 12. The term "atomecule" was invented by Chamberlin, perhaps in the meaning of a hypothetical subatomic entity. It was not used by other scientists either then or later.
27. Nicholson (1913), pp. 103–105.
28. Schuster (1898).
29. This section is largely taken from Kragh (2007). A more detailed account of the discussion concerning the cosmological consequences of the second law of thermodynamics can be found in Kragh (2008a) which includes references to the primary sources.
30. Thomson (1882–1911), vol. 1, p. 514.
31. Clausius (1868), p. 405.
32. Duhem (1974), p. 288 and p. 299. See also Kragh (2008c).
33. Sonnenschmidt (1880). Kragh (2008a), pp. 120–1.
34. Boltzmann (1895). See also Ćirković (2003a).
35. Boltzmann (1898), pp. 256–9.
36. Maunder (1908).
37. Young (1893), p. 524.
38. "The idea of this hypothesis arose when it was noticed that natural radioactivity is a physical process which disappears gradually... If it were not for a few elements of average lifetimes comparable to T_H [Hubble time], natural radioactivity would be completely extinct now." Lemaître (1949), p. 452. On Haas, see Kragh (2008a), pp. 70–2. See also Chapter 4.
39. Soddy (1904), p. 188. In 1921 Soddy was awarded the Nobel Prize in chemistry for his work on radioactivity and isotopes.
40. See Harrison (1986).
41. The cosmological problems associated with Newton's law of gravitation are described in Jaki (1979) and in mathematical details in Norton (1999).
42. Seeliger (1895), p. 132.
43. Seeliger (1898a), p. 547.
44. Norton (1999).
45. Einstein (1956), p. 65, in the 17th edition of a book originally published by Einstein in 1917, wrote of a "fundamental difficulty... which, to the best of my knowledge, was first discussed in detail by the astronomer Seeliger."
46. Thomson (1901), and see also Norton (1999).
47. Proctor (1896), p. 285.
48. Charlier (1908) and also Charlier (1922). For his cosmological views, see Norton (1999), Jaki (1969), pp. 198–204, and Holmberg (1999), pp. 73–78.
49. Charlier (1896), p. 486.

50. On Selety's work, see Jung (2005).

51. Silberstein (1929), pp. 171–173.

52. Arrhenius (1909), p. 226.

53. Arrhenius (1903), p. 232.

54. Quoted in Kragh (2012a), which gives details about curved-space astronomy before Einstein.

55. Daniels (1975). It is a popular and persistent myth that Gauss made high-precision geodetic measurements of a triangle extending between three mountain peaks in Germany in order to test whether physical space is flat or not. See Breitenberger (1984) for the correct story.

56. The rise of non-Euclidean geometry during the nineteenth century is illustrated by Duncan Sommerville's bibliography of 1911 which includes about 4,000 titles published between 1830 and 1909. By far most of the publications date from after 1870 and only a few dozens of them refer to physics and astronomy. See Sommerville (1911).

57. Riemann (1873), p. 36.

58. See Kragh (2012c) for Zöllner as astronomer, cosmologist, and spiritualist. See also Jaki (1969), pp. 158–64.

59. Zöllner (1872), p. 308

60. Newcomb (1898), p. 7.

61. Roseveare (1982) is a detailed account of the many attempts to solve the Mercury anomaly in the period before Einstein. See pp. 162–4 for the use of curved space.

62. Schwarzschild (1900), p. 345. An English translation appears in *Classical and Quantum Gravity* **15** (1998): 2539–2544. See also Schemmel (2005).

63. Harzer (1908).

64. Trabert (1911), p. 259.

65. De Sitter (1917b), pp. 231–234. See Chapter 3 for the cosmological models of Einstein and de Sitter.

66. Proctor (1869b), p. 112.

67. Thomson (1891), p. 322. As the historian of astronomy John North remarked: "It is easy to speak of the infinite, as every theologian knows, but it is difficult to speak of it meaningfully." North (1990), p. 23.

68. Clerke (1890), p. 368 and Clerke (1903), p. 538.

69. Newcomb (1906), pp. 5–6. The latter quotation is from a review Newcomb wrote in *The Observatory* **30** (1907), p. 362.

70. Darwin (1905), p. 32.

71. The issue is discussed and provided with examples in Kragh (2008a).

72. Falb (1875), p. 202.

73. Zehnder (1897), who in 1914 published an updated version of his cosmological theory in a book with the characteristic title *Der Ewige Kreislauf des Weltalls* (The Eternal Cycle of the Universe).

74. Hilbert was well aware of Einstein's curved-space cosmology to which he referred and he might conceivably have read also Friedman's paper of 1922 in which the

expanding universe was introduced. But there is no evidence that Hilbert knew of Friedman's work or had it in his mind.

75. Hilbert introduced his remarkable hotel in unpublished lectures that he gave in Göttingen in 1924–1925. For the history of Hilbert's hotel, see Kragh (2014c). The general problem of infinities in cosmology is dealt with in North (1990), pp. 371–383.

76. Arrhenius (1908), p. xiv. For Arrhenius' theories of radiation pressure, the aurora borealis, and the universe, see Kragh (2013a).

77. Book of 1913, quoted in Kragh (2013a), p. 8.

78. Peratt (1995). See Section 4.6 on plasma cosmology.

79. Charlier (1925b, 1925c) and Charlier (1925a), p. 182. He probably referred to the theories of Einstein and de Sitter but may also have had in mind Schwarzschild's article of 1900.

80. Curtis (1933), p. 908. Contrary to some other critics of the expanding universe, Curtis was well acquainted with the theories of Lemaître, Eddington, Tolman and other pioneers of relativistic cosmology.

81. De Vaucouleurs (1970), p. 1211. See also Section 4.9

2

Observations and the universe

Robert W. Smith

2.1 Introduction

In the course of the middle decades of the eighteenth century, a small number of thinkers—including Immanuel Kant, Johann Heinrich Lambert, and Thomas Wright—considered the existence of what we would term galaxies beyond our galaxy, the Milky Way, as elements of speculative cosmologies in which religious concerns and ideas on extraterrestrial life were central. Kant had even suggested that some of the nebulae visible in the sky might be such distant galaxies.[1]

The nature of the nebulae and broad cosmological questions were of little interest for professional astronomers. As Friedrich Wilhelm Bessel, arguably the leading professional astronomer, would assert in 1832, "What astronomy must do has always been clear—it must lay down the rules for determining the motions of the heavenly bodies as they appear to us from the earth. Everything else that can be learned about the heavenly bodies, e.g., their appearance and the composition of their surfaces, is certainly not unworthy of attention; but it is not properly of astronomical interest."[2] Observational astronomy, then, as performed in professional observatories, was focused on the precise determination of the positions of astronomical objects. The telescopes and ancillary instruments in these observatories were designed for accuracy, not great light grasp. The directors of observatories, therefore, emphasized transit instruments.

William Herschel was an entirely different sort of astronomer from a typical professional. His approach to problems was more that of a natural philosopher with particular concerns in classification than a "mainstream" positional astronomer. And it was Herschel who, in the late eighteenth century, made the study of the nebulae a serious subject for research, and it was the nebulae that, it would turn out, were to be the crucial objects for observational cosmology, both in the nineteenth century and in the first decades of the twentieth. Herschel excelled too as a telescope maker, and the story of the study of the nebulae was also, we will see, very much shaped by the development of powerful telescopes.

Smith, R. W., 'Observations and the universe' in *The Oxford Handbook of the History of Modern Cosmology*, edited by Kragh, H. and Longair, M. S. © Oxford University Press 2019.
DOI: 10.1093/oxfordhb/9780198817666.013.2

2.2   The Herschels and nebulae

Herschel's first career was music, but he developed a passion for astronomy that had become an obsession by the late 1770s. His life was transformed in 1781 when, in the course of a systematic hunt for double stars, he observed what he believed to be a comet. When other astronomers calculated the comet's orbit, Herschel's object proved to be the first major planet to be discovered in recorded history, Uranus. Uranus opened the door for Herschel to patronage from King George III of England, support that enabled him to abandon his musical career in Bath to become a full-time astronomer.

Herschel founded his astronomical career on his skills as a telescope-builder. Through his efforts, he turned the reflecting telescope from little more than a scientific toy into a serious tool of astronomical research. Although James Short had made telescopes with sizable primary mirrors, Short's instruments were far surpassed by Herschel's reflectors regarding size and effectiveness. With the aid of large sums from George III, William even built in 1789 a monster reflector with a 40-foot focal length and a primary mirror 48-inches in diameter. This telescope, however, was a major disappointment. Regarding light grasp and ease of operation, Herschel's best telescope was his 20-foot reflector which had a primary mirror with a diameter of 18-inches, and which he completed in 1783.

Contrary to widespread belief, Herschel spent much of his time observing objects within our solar system. But he, along with his observing companion and long-time collaborator, his sister Caroline, devoted more attention to hunting for and examining nebulae and star-clusters than any other astronomers had ever done. Through their systematic sweeps of the skies, William and Caroline increased the total of known nebulae from the 100 or so previously identified to around 2,500.[3]

In 1811, Herschel explained that "A knowledge of the construction of the heavens has always been the ultimate object of my observations..."[4] By the "construction of the heavens" he meant the arrangement of our star system (the Milky Way), as well as the arrangement and development of other star systems and nebulae. Herschel, then, was not content with discovering nebulae and star clusters and charting their positions. He also sought to classify them in the manner of a natural historian of the heavens, to employ a term he coined to describe himself, and to understand their nature.[5]

For Herschel, there were two crucial questions about the nebulae he sought to answer at the telescope. First, are large telescopes able to resolve nebulae into collections of stars? And, second, do the nebulae change? If the nebulae displayed changes over (astronomically) short time periods, this would argue against them being distant systems of stars. If he could answer these two questions, he could decide if the nebulae were truly clouds of luminous material, or if they were distant star systems whose milky appearance was the result of the light of the stars they contained merging together because the system was so remote that the individual stars could not be resolved.[6] Herschel's questions and issues would echo down the nineteenth century.

Herschel's views on, and interpretations of, the nebulae shifted over time.[7] At one stage he regarded all the nebulae as stars systems. In 1811, toward the end of his career, however, he recalled that although he had earlier

> surmised nebulae to be no other than clusters of stars disguised by their very great distance...a longer experience and better acquaintance with the nature of nebulae, will not allow a general admission of such a principle, although undoubtedly a cluster of stars may assume a nebulous appearance when it is too remote for us to discern the stars of which it is composed.[8]

Now he believed that many nebulae are single stars or comets in the process of formation.

For a period Herschel viewed the Andromeda Nebula (Fig. 2.1) as a detached Milky Way. Indeed, he had calculated that its light takes about two million years to reach us. As he had reasoned in 1802:

Fig. 2.1 *Isaac Robert's photograph of the Andromeda Nebula, taken in 1888. At the time this photograph was generally regarded as showing a solar system in formation. Observational cosmology for the three decades after this photograph was taken can be understood in part as the transformation of the Andromeda Nebula into an external galaxy.*

A telescope with a power of penetrating into space, like my 40-feet one, has also, as it may be called, a power of penetrating into time past. To explain this, we must consider that, from the known velocity of light, it may be proved, that when we look at Sirius, the rays which enter the eye cannot have been less than 6 years and $4\frac{1}{2}$ months coming from that star to the observer. Hence it follows, that when we see an object of the calculated distance at which one of these very remote nebulae may still be perceived, the rays of light which convey its image to the eye, must have been more than nineteen hundred and ten thousand, that is, almost two millions of years on their way; and that, consequently, so many years ago, this object must already have had an existence in the sidereal heavens, in order to send out those rays by which we now perceive it.[9]

But, by 1811, he was unsure about the Andromeda Nebula's nature, and in that year placed it in a list "of objects of an ambiguous construction."

The extent to which Herschel changed his position on the existence of star systems beyond the Milky Way was debated during the nineteenth century. Nevertheless, when in 1813 the poet Thomas Campbell met William and his son John on holiday in Brighton on the south coast of England, William told Campbell that

I have looked further into space than ever human beings did before me. I have observed stars of which the light, it can be proved, must have taken millions of years to reach the earth.[10]

But Herschel was also widely reckoned to have demonstrated the existence of true nebulae.

2.3   John Herschel and nebulae

Pierre-Simon Laplace's nebular hypothesis—which he had first advanced at the end of his *Exposition du système du monde* of 1796 as an extended note which he elaborated on for his *Mécanique Céleste*—on the formation of the solar system relied on the existence of nebulous matter. Laplace maintained that the planets and their satellites had been born out of the shrinking outer, rotating nebulous atmosphere of the sun. Some astronomers, as we shall see later, like Laplace, looked to Herschel's observations as proof of such material.[11]

William's fellow astronomers had admired his novel researches, but they were not inclined to go where he had led. William's son John, however, followed his father's example in observing and mapping the nebulae. For his observations, John exploited the 20-foot reflector, which had been William's most successful telescope, and which he and his father had refurbished. In 1820, when he addressed the recently founded Astronomical Society of London (what would later become the Royal Astronomical Society), John Herschel claimed that one

of the first great steps towards an accurate knowledge of the construction of the heavens, is an acquaintance with the individual objects they present: in other words, the formation of a complete catalogue of stars and other bodies, upon a scale infinitely more extensive than any that has yet been undertaken; and that shall comprehend the most minute objects visible in good astronomical telescopes.[12]

As for the nebulae, John maintained in 1826 that the

nature of nebulae, it is obvious, can never become more known to us than at present; except in two ways—either by the direct observation of changes in the form or physical condition of some one or more among them, of from the comparison of a great number, so as to establish a kind of scale or gradation from the most ambiguous, to objects of whose nature there can be no doubt.[13]

John extended William and Caroline's sweeps to the southern hemisphere by transporting the 20-foot reflector to the Cape of Good Hope for a four-year stay. He also, much more so than his father, was interested in extremely detailed examinations and drawings of individual nebulae.[14]

John, however, was an altogether more cautious theorist than his father, and the different positions he adopted on the nebulae at different times are not easy to discern. What is clear, however, is that the four years he spent observing at the Cape of Good Hope from 1834 to 1838 were critical ones for his views on the nebulae, clusters of stars and our galactic system.

He was fascinated by the two Magellanic Clouds, and he scrutinized them with the refurbished 20-foot reflector, the first big telescope to be taken to the southern hemisphere. In presenting his results, Herschel argued that

the two Magellanic clouds, Nubecula Major and Minor, are very extraordinary objects. The greater is a congeries of stars, clusters of irregular form, globular clusters and nebulae, of various magnitudes and degrees of condensation, among which is interspersed a large portion of irresolvable nebulae, which may be, and probably is star dust, but which the powers of the twenty foot telescope show only as a general illumination of the field of view, forming a bright ground on which the other objects are scattered....

Further,

The Nubecula Major, like the Minor, consists partly of large tracts and ill-defined patches of irresolvable nebula, and of nebulosity in every stage of resolution, up to perfectly resolved stars like the Milky Way, as also of regular and irregular nebulae properly so called, of globular clusters in every stage of resolvability, and of clustering groups sufficiently insulated and condensed to come under the designation of "clusters of stars"...[15]

Herschel claimed that in the Large Magellanic Cloud he could see "every imaginable size of star and of nebulosity, all gathered together in the large Cloud and so at essentially the same distance from the observer..." This meant that "it was evident that no limits could be set to the variety to be found in celestial objects."[16] In this remarkable report, then,

John, as Hoskin has emphasized, broke with his father's position that celestial objects form distinct "species" and that within each species, the variations among the objects are limited.

2.4   The nebular hypothesis, star systems, and spirals

When John Herschel returned to England from the Cape of Good Hope, his career as an astronomical observer came to an end. By this time the question of the nature of the nebulae had also become a topic of wide interest. John Pringle Nichol, the Regius Professor of Astronomy at the University of Glasgow, was also an ardent political economist and no less a figure than John Stuart Mill was one of his nominators for the post of professor of political economy at the Collège de France. Nichol also wrote on astronomy for broad audiences. His best-known book was probably *Architecture of the Heavens and Phenomena of the Solar System,* first published in 1837. It was as a result of his writings that the term "nebular hypothesis" came into widespread use, and through his *Architecture of the Heavens* the link between Laplace's Hypothesis and William Herschel's observations was strengthened. In this book, Nichol interpreted the different sorts of nebulae as different stages that nebulae reached on the way to contracting into groups of stars. As I have argued, following Schaffer,

> Nichol and his allies fastened on the nebular hypothesis as an object of both a natural and moral science. They emphasized the stellar progress of the nebular hypothesis which they then exploited as a general model of universal progress, one that could be pressed into service to argue for political reform. Nichol, as the author of *Vestiges [of the Natural History of Creation]* was to do later, also wove together the nebular hypothesis, geological theorizing and speculations on the transformation of species.[17]

The now infamous *Vestiges of the Natural History of Creation* was published in 1844, and its anonymous author (who in fact was Robert Chambers, a Scottish publisher, writer, and political radical) provoked an enormous controversy. The book's large number of readers even included Queen Victoria. Chambers explained the development of the Earth, the emergence of life upon it, and, in time, the appearance of human beings. Chambers contended, however, that the "nebulous matter of space, previously to the formation of stellar and planetary bodies, must have been a universal Fire Mist, an idea which we can scarcely comprehend, though the reasons for arriving at it seem irresistible."[18]

As Chambers' volume reached bookshops, the third Earl of Rosse, William Parsons, was constructing a remarkable telescope at Parsonstown in central Ireland. Observations made with Rosse's 72-inch reflector would take the debate on nebulae in a new direction. William Herschel's largest instrument was his 40-foot reflector, at the heart of which was a 48-inch primary mirror. Until the 1860s, telescope mirrors were generally made of speculum metal, an alloy of copper and tin. With the casting, grinding and polishing of his 48-inch speculum metal mirrors, Herschel had confronted severe problems. He found that such a size of mirror, for example, could not be worked by hand by one individual, and so he had devised a scheme to enable several men to work the mirror simultaneously. The telescope was not a success. William instructed his son John to never repeat his mistake by attempting to construct a telescope as large as the 48-inch.

But Lord Rosse's ambitions went further. He bent his energies, and directed very considerable sums of money, toward fashioning a 72-inch reflector. He had completed a 36-inch reflector at Parsonstown in 1839, and work began on the giant 72-inch shortly after that, and first light for what would in time be known as the Leviathan of Parsonstown took place in 1845.

Even before the Leviathan began operations, Thomas Romney Robinson, director of the Armagh Observatory in Ireland and with whom Rosse worked closely and who would be one of the main observers to use the 72-inch, argued that the Leviathan would be capable of resolving the supposed clouds of nebulous matter into collections of stars. For the politically conservative Robinson, a blow against the nebular hypothesis would be welcomed.[19] The early observations by Rosse, Robinson, and others with the Leviathan gave evidence for Robinson's hopes. Robinson even told a cheering crowd at a meeting in Dublin at the Royal Irish Academy that all nebula observations had to be reformed, and the nebular hypothesis rejected.[20] Robinson reported that he and Sir James South had employed the Leviathan to examine forty nebulae from John Herschel's catalogue of nebulae. In his opinion,

> no REAL nebulae seemed to exist among so many of these objects chosen without any bias: all appeared to be clusters of stars...If it prove to be the case that all the brighter nebulae yield to this telescope, it appears unphilosophical not to make universal Sir J. Herschel's proposition, that a "nebula, at least in the generality of cases, is nothing more than a cluster of discrete stars."[21]

Observations of the resolution of nebulae, including the Orion Nebula, were reported from the Harvard College Observatory, and so supported the findings of the Parsonstown observers. Armed with a new 15-inch refractor, George P. Bond, the Observatory's director, told the Harvard President: "You will rejoice with me that the great Nebula in Orion has yielded to the powers of our incomparable telescope!"[22] Nichol, who we saw was one of the leading advocates of the nebular hypothesis and who had also observed with the "Leviathan of Parsonstown," was persuaded. In 1848, he decided that the

> supposed distribution of a self-luminous fluid, in separate patches, through the Heavens, has beyond all doubt been proved fallacious by the most remarkable of telescopic achievements—the resolution of the Great Nebula in Orion into a superb cluster of stars.[23]

Rosse was more cautious and avoided sweeping claims. Astronomers understood that it was very challenging to portray nebulae accurately, and to decide if nebulae could be resolved or if changes had occurred, given the delicate form of the nebulae and apparent changes in them due to shifting conditions in the earth's atmosphere. The performance of an observer's telescope could also vary. When Rosse addressed a scientific meeting in 1845, he explained how he had executed his sketch of the spiral-shaped M51 to secure reliable results:

> He first laid down, by an accurate scale, the great features of the nebula as seen in his smallest telescope, which, being mounted equatorially, enabled him to take accurate measurements; he then filled in the other parts, which could not be distinguished in that telescope, by the aid of the great telescope; but as the equatorial mounting of this latter was not yet complete, he could not lay these smaller portions down with rigorous accuracy; yet as he had repeatedly gone over them, and verified them with much care, though by estimation, he did not think the drawing would be found to need much future correction.[24]

Rosse also made a striking but puzzling discovery early in the Leviathan's observing career. He noticed a spiral pattern in one of the target nebulae, M51 (or number 51 in Charles Messier's famous catalogue, the first version of which he had published in 1771, as a tool to spare observers who were hunting for comets mistakenly identifying a nebula or star cluster for a comet). In time, Parsonstown observers found 79 spiral nebulae.[25] But writing on the nebulae in 1850, Rosse cautioned there was "no fair ground even for plausible conjecture" on the laws that produced "these wonderful systems," and "as observations have accumulated the subject has become, to my mind at least, more mysterious and more inapproachable."[26] In the 1840s, and for decades later, astronomers were baffled by the meaning of these spiral patterns. Nor, until the 1880s, were photographic techniques sensitive enough for photography to be a worthwhile tool in studies of nebulae.

2.5   Astrophysics changes the debate

There was still no consensus on the nature of the nebulae by the early 1860s. Astronomers had to weigh the claimed resolutions of some nebulae against other observations of the variability of some nebulae over extremely short (by astronomical standards) periods. If a nebula were truly a remote star system containing a vast body of stars, the idea that it could substantially change form or brightness, or perhaps both, in a relatively brief period seemed impossible.

One variable nebula was named for John Russell Hind. Hind, Superintendent of the Nautical Almanac Office and director of George Bishop's private observatory at South Villa in London, explained in a letter to *The Times* newspaper in 1862 that a nebula he had discovered some years earlier was no longer visible. Other observers confirmed the nebula's disappearance, and this result and similar ones were well known.[27]

The nature of the nebulae, however, was not a pressing matter for the great majority of professional astronomers. For them, astronomy still dealt with the issues that Bessel laid out in the quotation we saw earlier. Not surprisingly, the next significant development in the debate on the nature of the nebulae came from one of the many "gentlemen of science" who would continue to play a major role in scientific research, at least in Britain, until the end of the nineteenth century, and not a professional astronomer. This person was William Huggins, a one-time silk mercer and linen draper with no university training who had retired from business in his early 30s to pursue his scientific interests.

In 1862, Huggins began to collaborate with his near neighbor in the London suburb of Tulse Hill, William Allen Miller. Miller, a Chair of Chemistry and a dean at King's College, London, was an accomplished spectroscopist and author of a textbook on *Elements of Chemistry* in which he discussed spectroscopy, including the dark lines in the solar spectrum. Miller had addressed the British Association for the Advancement of Science in 1861 and spoken on the advances in chemistry in the previous year. He highlighted the discoveries of the physicist Gustav Kirchhoff and chemist Robert Bunsen at the University of Heidelberg of the elements cesium and rubidium through spectroscopic techniques. Kirchhoff and Bunsen persuaded their colleagues that particular chemical elements and compounds are evidenced by particular groups of spectral lines. The pair also offered a convincing interpretation of the production of the lines. Here was a startling and far-reaching development. As another gentleman of science/astronomer well known to Huggins, Warren de la Rue, put it, the

> physicist and the chemist have brought before us a means of analysis that…if we were to go to the sun, and to bring away some portions of it and analyze them in our laboratories, we could not examine them more accurately than we can by this new mode of spectrum analysis.[28]

Huggins, who was eager to explore the possibilities raised by Kirchhoff and Bunsen, joined forces with Miller to undertake spectroscopic observations of a range of stars. These researches drew wide-spread attention and made Huggins's scientific reputation in this new field of scientific research that would come to be known as astrophysics.

Huggins possessed an 8-inch refractor housed in a substantial observatory attached to his house in Upper Tulse Hill in London. The observatory now began to take on something of the form of a physical laboratory. A spectroscope was added to his refractor, and batteries and extra apparatus were installed so that Huggins and Miller could produce comparison spectral lines, as well as observe spectral lines from celestial objects. When Miller, who was extremely busy with his teaching duties as well as committee work for the Royal Society and government bodies, decided within a couple of years to abandon his collaboration with Huggins, Huggins pressed on alone.

In 1864, Huggins expanded his interests to the nebulae.[29] The first nebula to which he directed his telescope and its attached spectroscope was a planetary nebula, the Cat's Eye Nebula in the constellation of Draco. For Huggins, the planetary nebulae were the oddest of all the nebulae, given their blue/green light, the lack of central condensation, and their round or oval shapes. He would later recall, not entirely accurately, his initial expectations and actual findings. In a passage which is now perhaps one of the most famous in the entire history of astronomy, Huggins asked the reader:

to picture to himself to some extent the feeling of excited suspense, mingled with a degree of awe, with which, after a few moments of hesitation, I put my eye to the spectroscope. Was I not about to look into a secret place of creation?

I looked into the spectroscope. No spectrum such as I expected! A single bright line only! At first, I suspected some displacement of the [spectroscope's] prism, and that I was looking at a reflection of the illuminated slit from one of its faces. This thought was scarcely more than momentary; then the true interpretation flashed upon me. The light of the nebula was monochromatic, and so, unlike any other light I had as yet subjected to prismatic examination, could not be extended out to form a complete spectrum…A little closer looking showed two other bright lines on the side towards the blue, all the three lines being separated by intervals relatively dark.

The riddle of the nebulae was solved. The answer, which had come to us in the light itself, read: Not an aggregation of stars, but a luminous gas. Stars after the order of our own sun, and of the brighter stars, would give a different spectrum; the light of the nebula had clearly been emitted by a luminous gas.[30]

Huggins explained in a paper on his first observations of the Cat's Eye Nebula that he wanted to establish if there was what could be termed a similarity of plan between the stars and the "distinct and remarkable class of bodies known as nebulae." He and Miller had already argued for the similarity of the composition of the Sun and stars from the similarity of their spectra, and so they had further contended that the Sun and stars display a similarity of plan. Moreover, in Huggins's view:

The importance of bringing analysis by the prism to bear upon the nebulae is seen to be greater by the consideration that increase of optical power alone would probably fail to give the desired information; for, as the important researches of Lord Rosse have shown, at the same time that the number of clusters may be increased by the resolution of supposed nebulae, other nebulous objects are revealed, and fantastic wisps and diffuse patches of light are seen, which it would be assumption to regard as due in all cases to the united glare of suns still more remote.[31]

Huggins's position, then, was the nature of the nebulae could not be determined by big telescopes alone. As astronomers increased the number of resolved nebulae, so more unresolved ones would be discovered. And for Huggins, the spectroscope had demonstrated that the planetary nebulae are not composed of stars, but are constituted by vapor or luminous gas. Although he would later drop this position, in 1864 Huggins was of the opinion that "We have in these objects to do no longer with a special modification only of our own type of suns, but find ourselves in the presence of objects possessing a distinct and peculiar plan of structure."[32]

Huggins examined further planetary nebulae as well as star clusters, and then directed his telescope and spectroscope to some nebulae claimed to be composed of stars. The key example was the Orion Nebula, which various observers, including as we have seen Lord Rosse and the Parsonstown observers, had decided could be resolved into a multitude of stars. For Huggins, here was "a crucial test of the correctness of the usually received opinion that the resolution of a nebula into bright stellar points is a certain and trustworthy indication that the nebula consists of discrete stars after the order of those that are bright to us."[33]

But wherever Huggins looked within the nebula with his spectroscope, he could find no indication of a continuous spectrum of the sort that was to be expected if the Nebula was stellar in nature. Instead, he observed only the three emission lines, which indicated the Nebula was composed of a luminous gas or vapor. The conclusion for Huggins was obvious:

> the detection in a nebula of minute closely associated points of light, which has hitherto been considered as a certain indication of a stellar constitution, can no longer be accepted as a trustworthy proof that the object consists of true stars. These luminous points, in some nebulae at least, must be regarded as themselves gaseous bodies, denser portions, probably, of the great nebulous mass, since they exhibit a constitution which is identical with the fainter and outlying parts which have not been resolved. These nebulae are shown by the prism to be enormous gaseous systems; and the conjecture appears probable that their apparent permanence of general form is maintained by the continual motions of these denser portions which the telescope reveals as lucid points.[34]

For the great majority of astronomers, these spectroscopic investigations, would, in time, be taken to mean that the nebulae are not clusters of stars or distant star systems, as Robinson had both claimed and hoped in the 1840s.

2.6   No external galaxies

Observational studies of nebulae during the first half of the nineteenth century were largely the domain of astronomers in England and Ireland with the leading figures, as we have seen, William and John Herschel, Lord Rosse and William Huggins. And it was in Britain and Ireland that there was the strongest interest in building the biggest reflecting telescopes, a tradition sparked by William Herschel's success in building formidable reflectors. But what would in time prove to be a major innovation for reflectors was introduced in 1856 and 1857 by Léon Foucault and Karl August von Steinheil when, independently of each other, they succeeded in depositing very thin layers of silver onto astronomical mirrors. These silver-on-glass mirrors reflected more of the incident light than the equivalent speculum metal mirrors, and as only very thin layers of silver were needed the cost was not exorbitant. Silver-on-glass mirrors were also lighter, which, compared to speculum metal mirrors, lessened the difficulties of fashioning mechanical systems to maintain their surface figures as they were directed to different parts of the sky. Glass disks were less challenging to grind and figure than speculum metal ones. The silver-on-glass mirrors had the disadvantages that they could be damaged easily and tarnished relatively rapidly in damp air, but given "all the trouble of repolishing metal mirrors," replacing silver-on-glass mirrors was a "comparatively simple and inexpensive process."[35]

Despite this advance, when the Royal Society Committee responsible for planning and overseeing the construction of a large telescope for the Melbourne Observatory to be paid for by the government of Victoria in Australia, the main scientific charge of which was to be the observation of nebulae, the committee opted for the conservative choice of speculum metal as the material for the 48-inch primary mirror. The Melbourne telescope, which went into service in 1869, was, however, the last large telescope to have a speculum metal mirror. For a variety of reasons, the Melbourne Telescope was not an effective instrument in its early years, and contributed little to the debate on the nature of the nebulae.[36]

The first large silver-on-glass reflector had a 47-inch primary mirror, and was completed by Adolfe Martin, Wilhelm Eichens, and Paul Gautier in 1877, but in the opinion of the leading historian of the telescope, Henry King, it was "comparatively useless."[37] A more impressive big silver-on-glass reflector was completed by Andrew Ainslie Common in 1879, at the heart of which was a 36-inch mirror fashioned in the workshop of George Calver. Another significant silver-on-glass reflector completed in this period was the 20-inch reflector purchased by Isaac Roberts from the famous telescope maker Howard Grubb. It went into operation in 1885 at Roberts's private observatory. Both Common and Isaac Roberts took photographs of nebulae with their telescopes that would be important for the debate on the nature of the nebulae.

Roberts was a Welsh-born engineer and businessman who had retired to southern England and there built an observatory, at the heart of which was the 20-inch silver-on-glass reflector, with which he pursued a program of research that won him a reputation as a skilled astronomical photographer. In 1888, Roberts photographed the Andromeda Nebula. When he exhibited one of his images of the Nebula at a meeting of the Royal Astronomical Society that year, it made a deep impression. For Roberts as well as others, it was

> no exaggeration to say that the nebulae were now for the first time seen in intelligible form, and no verbal description could add to the information which the eye obtained at a glance. Those who accepted the nebular hypothesis would be tempted to appeal to the constitution of these nebulae for confirmation, if not demonstration, of the hypothesis. Here one apparently might see a new solar system in process of condensation from the nebula. The central sun was now in the midst of nebulous matter, which in time would be either absorbed or separated into rings more or less symmetrical with the nucleus, and presenting a general resemblance to the rings of Saturn. The two [accompanying] nebulae, Herschel 44 and Herschel 55, seemed as though they were already undergoing their transformations into planets.[38]

William Huggins, lecturing in 1891, explained that the photograph displayed "cosmical evolution on a gigantic scale" and that it indicated a development of the sort to be expected "following broadly upon the lines of Laplace's theory."[39] In the view of many astronomers, with his photograph of the Andromeda Nebula, Roberts had offered compelling evidence for the nebular hypothesis, and so against the Andromeda Nebula as an external galaxy.

For astronomers interested in the debate on the nebulae another telling argument involved the distribution of the nebulae in different regions of the sky. In 1867, Cleveland

Abbe, who went on to become a distinguished meteorologist in the United States but who early in his career focused on astronomy, wrote "On the Distribution of Nebulae in Space."[40] In so doing he offered an analysis of the nebulae and star clusters in John Herschel's General Catalogue, published in 1864. Abbe, as others had noted before, found the nebulae strongly avoided the Milky Way. Two years after Abbe, the British astronomer and popular writer Richard Proctor conducted a similar study. Proctor concluded that the nebulae overwhelmingly cluster around the galactic poles and shun our stellar system. There is, as he put it, a zone of dispersion and the position of the zone matches that of the Milky Way.[41] Any theory of the nebulae, Proctor stressed, had to take this fact into account. And for him, no "theory, which looks upon our galaxy as simply a member of the nebular system can possibly be reconciled with the view that the position of the zone of nebular dispersion is otherwise than accidental..."[42] Further, "I cannot see how we can look upon the coincidence I have spoken of as not accidental, without being led to the conclusion that the nebular and stellar systems are part of a single scheme," Proctor argued. "If the nebulae were external star-galaxies, the coincidence might have happened, but it would be accidental." But John Herschel had observed the intermingling of stars and nebulae in the Magellanic Clouds, and so the stellar and nebular systems form "one great scheme." At the same time, he reckoned that there "may be individual nebulae which are true external universes—it is possible, for example, that the spiral nebulae may be of this class, as also perhaps the Andromeda Nebula, and a few others,—but I consider that for one nebula which is really external there are hundreds which are associated with the stellar system."[43] The distribution of the nebulae and the zone of dispersion—it would later generally be termed the zone of avoidance—was often used to argue against external galaxies in the late nineteenth and early twentieth centuries.

The truly nebulous nature of the nebulae was further underscored for astronomers by other important observations made in the 1880s. The first centered on a remarkable nova that flared in the Andromeda Nebula.[44] First reported publicly in August 1885, the nova was for a short period approximately one-tenth as brilliant as the entire nebula. If the Andromeda Nebula was an enormous body of stars that appeared cloudy because of its great distance, how could a single star suddenly reach a brightness of 1/10th of such a huge assemblage? The well-known author of astronomical works in the late nineteenth century and first decade of the twentieth, Agnes Clerke, made the key points with respect to the cosmological implications of the 1885 nova:

> ...if the Andromeda Nebula were a universe apart of the same real extent as the Galaxy, it should be situated, in order to reduce it to its present apparent dimensions, at a minimum distance of twenty-five galactic diameters. And a galactic diameter being estimated by the same authority at thirteen thousand light-years, it follows that, on the supposition in question, light would require 325,000 years to reach us from the nebula. The star then which suddenly shone out in the midst of it in August 1885 should have been at 564 times the distance inferred from its effective brightness. In real light it should have been equivalent to 318,000 stars like Regulus, or to nearly *fifty million* such suns as our own! But even this extravagant result inadequately represents the real improbability

of the hypothesis it depends upon; since the Andromeda Nebula, if an external galaxy, would almost certainly be at a far greater remoteness from a sister-galaxy than would be represented by twenty-five of its own diameter.[45]

For astronomers at the end of the nineteenth century, then, it was clear that the observational evidence told very strongly against the nebulae as external galaxies. Although the nebular hypothesis as it was originally advanced by Laplace was no longer credible, that some form of nebular hypothesis did explain at least the great majority of the nebulae was generally agreed. Agnes Clerke spoke for just about all astronomers when in 1890 she passed judgement on the island universe theory: "No competent thinker, with the whole of the available evidence before him, can now, it is safe to say, maintain any single nebula to be a star system of co-ordinate rank with the Milky Way."[46]

2.7   Observational astronomy remade

In the previous section, we discussed Common and Roberts. Here it is worth noting that neither Common nor Roberts was a professional astronomer: Common was a sanitary engineer and Roberts a businessman and engineer. Both of them employed reflectors. In contrast, throughout the nineteenth century, professional astronomers almost invariably worked with refractors, which they regarded as better suited than reflectors for the sort of positional astronomy that still occupied most professionals even at the end of the century. The focus of most professionals was the sort of astronomy that emphasized determining ever more accurate positions and motions for celestial objects, including the objects in the solar system, for which the positions of the stars provided in effect a reference grid.[47]

Positional astronomy and astrophysics were generally separate in the first decades after 1860, so, as Lankford had pointed out, it is incorrect to argue that the pre-existing astronomy, was "transformed by the emergence of astrophysics."[48] But by the end of the century, astrophysics had grown in importance in the eyes of an increasing number of professional astronomers, and was pursued in a growing number of observatories, including some observatories established specifically for the study of astrophysics. For astrophysicists, the main goal was to understand the evolution of stars. This meant that, as it was generally agreed that the nebulae are either proto-solar systems or small star clusters in formation, they drew increasing attention from astrophysicists.

Here it is important to note that by the end of the nineteenth century, it was also almost universally agreed that it was certain that the planetary nebulae and the diffuse nebulae are within our own stellar system. The key pieces of evidence in reaching this conclusion were their locations in space, the stars within them, as well as their motions.

We have noted that one feature of the divide between astrophysics and positional astronomy was that large reflectors became the telescope of choice for astrophysicists. In the years around the turn of the century, builders of reflectors also made major innovations. A significant instrument in terms of pointing the way to future telescopes, was the compact 24-inch silver-on-glass reflector with an eight foot focus put into operation in 1897 at the newly established Yerkes Observatory by the optician, George

Ritchey.[49] For George Ellery Hale, the Yerkes director, Ritchey's 24-inch underlined the enormous potential of reflectors.

The Yerkes Observatory was also one marker of a major shift in terms of observational astronomy that saw the rise to prominence of American observatories and a relative decline of those in Europe. Indeed, by the late 1910s, the most powerful telescope in the world, a 100-inch reflector, was sitting at an altitude of 6,000 feet at the Mount Wilson Solar Observatory of the Carnegie Institution of Washington. In 1902, the steel magnate Andrew Carnegie founded the Carnegie Institution of Washington. Although Carnegie's advisors had raised the possibility of founding a university, the bulk of the Institution's monies were directed to a series of research institutes. The Mount Wilson Solar Observatory was one of them. In short order, the Observatory, established in 1904, soon became the leading astrophysical observatory in the world, with, by the standards of other big observatories, plentiful funds for telescopes and equipment, as well as staff members.[50]

Carnegie was one of the private patrons who transformed American observational astronomy in the late nineteenth century and early twentieth century. The chief developments were the establishment of Yerkes (named for the traction magnate Charles Yerkes) and Mount Wilson, as well as the Lick Observatory (named for the real estate magnate James Lick) which had begun operations in 1888. Another major American observatory founded in 1894 by a patron was the Lowell Observatory in Flagstaff, Arizona. Lowell, however, was unusual in that its patron, Percival Lowell, was also an astronomer and user of its telescopes, as we shall see later. All of these institutions were to play significant roles in the debates on the nature of the nebulae that were to be fundamental to the shifting views on cosmology in the first half of the twentieth century. The professional astronomers based at these observatories would dominate in these debates.

2.8 Spiral nebulae

James E. Keeler of the Lick Observatory was the first of this new group of professionals to secure path-breaking results on the nebulae. A skilled spectroscopist, Keeler was appointed the director of the Lick Observatory in 1898. He also took charge of the Observatory's Crossley Reflector.[51] With the aid of the Crossley, Keeler embarked on a program of photographing the nebulae that had been listed as spirals by Lord Rosse and the Parsonstown observers.

When in 1895 John Louis Emil Dreyer (a one-time assistant at Parsonstown) published his now famous New General Catalogue,[52] he recorded over 9,000 nebulae. Of this number, the great majority had been discovered by observers who had put their eyes to telescope eyepieces, including thousands by the Herschels but also 1,400 that had been identified by Stéphane Javelle by using a 30-inch visual refractor at the Nice Observatory.[53]

It is against this background, then, that we need to place Keeler's photographic investigations. As he photographed different regions of the sky with the Crossley, far more nebulae revealed themselves than he had anticipated. He calculated that, based on the numbers on his photographic plates of limited areas of the key, the total number for the entire sky had to be counted in the hundreds of thousands. He was also able to "invariably" confirm the spiral nature of the nebulae so catalogued by Lord Rosse and his colleagues. Further, Keeler soon decided that so many of the other nebulae he examined displayed a spiral structure that "I have now come to regard a small, compact nebula, which is not a spiral, as the object of greater interest."[54] Keeler thereby elevated enormously the importance of spirals.

Keeler did not regard himself as engaged on cosmological investigations. Rather, he, like almost everyone else, reckoned the spiral nebulae he photographed represented the first stage in the development of stars. This belief was in line with the theoretical researches of the American geologist Thomas Chamberlain and the astronomer Forest Ray Moulton with their account of how solar systems come into being, what became known as the Chamberlain–Moulton hypothesis. While it was not widely supported, particularly outside of the United States, it served to emphasize both the failings in Laplace's nebular hypothesis and the growing importance of spiral nebulae. Chamberlin and Moulton claimed that Keeler's photographs displayed the sorts of spiral nebulae to be anticipated from their hypothesis, although in 1905 Moulton warned that "those spirals which have been photographed are immensely larger than the one from which our system may have developed, and as a rule have relatively less massive centers."[55]

Percival Lowell also fashioned an alternative to what by 1900 was the widely discredited nebular hypothesis of Laplace. One of the last of the "gentlemen of science" to play a major role in astronomical investigations, he had used part of his private wealth to establish the Lowell Observatory in Flagstaff, Arizona in 1894. He had wide-ranging interests in the development of the solar system, although they centered on the possible existence of life on the planet Mars, evidence for which Lowell found in the system of canals that he, and some other observers, believed criss-crossed the planet's surface.[56]

The Lowell Observatory's main telescope was a fine 24-inch refractor built by the well-known American maker Alvan Clark. In 1901, Lowell purchased a state-of-the-art spectrograph for use with the telescope, and five years later Lowell tasked one of his observatory's staff members, Vesto Melvin Slipher, with examining the spectra of the "white nebulae," what were generally termed spiral nebulae by other astronomers. Lowell wondered if the spectra might yield information on the solar system's history (Fig. 2.2).

The spectra of spiral nebulae had been investigated earlier by a few other astronomers. Julius Scheiner, of the Potsdam Observatory, reported in 1899 absorption lines in the spectrum of the Andromeda Nebula. These lines even suggested to him that as "the previous suspicion that the spiral nebulae are star clusters is now raised to a certainty, the thought suggests itself of comparing these systems with our stellar system, with especial reference to its great similarity to the Andromeda Nebula."[57] Scheiner's findings, however, seem to have carried little weight with other astronomers. In 1905, Hermann C. Vogel, Scheiner's director at Potsdam and himself a distinguished spectroscopist, reckoned all that was known about the spectrum of the Andromeda Nebula was "that it does not show the gaseous spectrum that is generally observed in nebulae."[58]

Edward Arthur Fath at Lick and Max Wolf at Heidelberg also examined the spectra of spirals. For his Ph.D. (which he defended in 1909), Fath directed a spectrograph of his

Fig. 2.2 *V.M. Slipher with his spectrograph attached to the 24-inch Clark Refractor at the Lowell Observatory, about 1912. (Courtesy of the Lowell Observatory)*

own design to seven spirals and three globular clusters.[59] He decided the Andromeda Nebula has a solar type spectrum, and that five of the other spirals also evidenced absorption lines of the solar type. In one spiral, however, NGC 1068, Fath found both absorption and emission lines, and in another he detected two absorption lines as well as a bright band. None of the spirals, though, exhibited a "truly continuous spectrum." Fath was clearly inclined to interpret the spirals as collections of stars.[60] At Heidelberg, Wolf also detected absorption lines in spirals, as well as indications of emission lines in some of them.[61] Fath cautioned, however, that the only direct measurement of the distance to a spiral—in 1907 by the Swedish astronomer Karl Bohlin of the Andromeda Nebula— had led to an answer of around 19 light years. He therefore wrote that it would be of "great interest" if another measurement were attempted of the nebula's trigonometric parallax.[62]

When in 1906 Lowell asked Slipher to turn to the spectra of the spirals, Slipher concentrated on the Andromeda Nebula as the brightest spiral and the one with the largest apparent diameter. By late 1912, and after trials with various configurations of his spectrograph, Slipher had in hand four spectrograms of the nebula. Not only did he record absorption lines, but he also saw that the lines were shifted toward the blue end of the spectrum from the positions they would have held if the nebula had been at rest relative to the earth. Interpreting these shifts as Doppler shifts, Slipher calculated that the Andromeda Nebula is racing toward the Earth at about 300 km s^{-1}. The prevailing theories of stellar evolution suggested he should have found a much lower velocity. Here, then, was a highly surprising result. As Slipher believed the Andromeda Nebula to be a solar system in formation, the Nebula's very high speed suggested to him that the 1885 nova might have been the consequence of the Nebula smashing into a 'dark star,' thereby provoking a sudden and massive outpouring of light.[63]

Lowell's controversial claims on the existence of a system of canals criss-crossing the surface of Mars cast something of a shadow over the credibility of the researches issuing from his observatory. William Wallace Campbell, a top-rank spectroscopist and director of the Lick Observatory, like some other astronomers, had publicly criticized Lowell and he also well understood the challenges posed by the observation of the spectra of the sprials.[64] He was skeptical of Slipher's unexpected results.[65] Other astronomers too were doubtful.

But Slipher plugged away. By 1914, he had results for 15 spirals. By this time, Slipher had decided that his first result, for the Andromeda Nebula, was anomalous. Most of the spirals exhibited redshifts, and so were receding from the Earth, if the spectral shifts were interpreted as Doppler shifts.[66] By 1917, Slipher had pushed the total of spirals measured to 25, and the predominance of redshifts over blueshifts was even clearer.[67]

Shortly after Slipher announced his first results, some astronomers had concluded that the spirals were moving so swiftly that they could not be part of the galactic system. Rather, they had to be extra-galactic. Given their angular sizes, they could not be stars in formation or sparse star-clusters associated with our Galaxy, so had to be distant star systems, perhaps far-off island universes.[68]

2.9   Internal motions

The leading champion of external galaxies in the second half of the 1910s was the Lick astronomer Heber Doust Curtis. He had photographed many spirals at Lick, and in what he took to be edge-on spirals he noticed dark lanes running along their length. These photographs suggested to him that the zone of avoidance might be explained by obscuring matter positioned in a ring around our Galaxy that blocks low-lying spirals from view. If so, the clustering of the spirals around the galactic poles was merely an optical effect and carried no force as an argument against the island universe theory.[69]

As part of his program of photographing spirals, Curtis also attempted to determine the motions of the spirals. He compared two photographs of a spiral taken some years apart, and measured the differences, if any, in the positions of the spiral's constituents. Curtis found no evidence for the rotation of a spiral or internal motions within a spiral, such as, for example, motion along the arms. He did, however, calculate the average annual proper motion for 66 large spirals to be 0.033". Then, by assuming the spirals move randomly, using the mean radial velocity for the spirals that had been obtained by Slipher, and the method of statistical parallax, Curtis computed an average distance to these spirals of 10,000 light years. But Curtis had no confidence in his result due to, among other reasons, what he judged to be the sub-standard quality of some of the plates he had used for comparison purposes. The best he could say for his results was that they provided a minimum distance to the spirals.[70] Curtis's own experience of the challenges of making such measurements—how to compare the positions of a knot of light in a spiral in photographs taken at two different times, for example—plus his acceptance of the island universe theory, also made him doubtful of the positive results claimed by a few other astronomers for internal and proper motions of spirals.[71]

By far the most influential and significant of the other studies of the motions of spirals were by the Dutch born astronomer Adriaan van Maanen. Van Maanen had completed a Ph.D. at the University of Utrecht in 1911, and shortly thereafter secured a post at Mount Wilson, where he established a reputation as a skilled and careful observer, noted chiefly for determining stellar parallaxes and proper motions. He also measured more spirals than other observers, and the main venue for his results was the *Astrophysical Journal*, a prestigious publication. Van Maanen's results therefore had to be taken seriously, and they proved to be the chief barrier for several years around 1920 to the general acceptance of the island universe theory.

Van Maanen began to measure internal motions in 1915 when he analyzed the giant spiral M101. When he presented his findings, he followed Curtis and suggested the nebula is about 10,000 light years away (though Curtis, as we have seen, took 10,000 light years as the minimum distance).[72] In 1915, such a figure did not bluntly contradict the island universe theory. In fact, for a few years the British applied mathematician and astronomer James Jeans, whose results were regarded highly at Mount Wilson, claimed van Maanen's results supported both his own researches on the nature of, and motions to be expected in, spiral nebulae, as well as the island universe theory.[73]

Jeans' claims underscore that the status of the island universe theory depended on theories of the Galaxy, as any external galaxies obviously had to be beyond the boundaries of our own galactic system. Determinations of the size of our Galaxy, then, had important cosmological implications. Here too it is important to note that in the 1910s and early 1920s, there were two main versions of the island universe theory in play. The first stated that there are external galaxies beyond the Milky Way. In the second, and stronger, version, not only are there external galaxies, but they are of a similar size to our own stellar system, in other words, this was a comparable galaxy theory.[74]

When van Maanen published his papers on M101, there was no consensus on the dimensions of our own Milky Way galaxy, though it was generally agreed that the Sun is quite close to the center. A small group of so-called statistical astronomers had developed sophisticated mathematical techniques to analyze the counts of stars as seen in different directions in order to map the stellar system. Hugo von Seeliger, one of the statistical astronomers, had advanced in 1898 an ellipsoidal, sun-centered model of the stellar system that had a diameter of about 3,000 parsecs and extended for about 1,300 parsecs at right angles to the galactic plane.[75] In his 1914 *Stellar Movements and the Structure of the Universe*, Arthur Stanley Eddington had argued, for example, that it is believed that the great mass of the stars...are arranged in the form of a lens- or bun-shaped system. This is to say, the system is considerably flattened towards one plane...In this aggregation the Sun occupies a fairly central position...The thickness of the system, though enormous compared with ordinary units, is not immeasurably great. No definite distance can be specified, because it is unlikely that there is a sharp boundary; there is only a gradual thinning out of stars. To give a general idea of the scale of the system, it may be stated that in directions towards the galactic poles the density continues practically uniform up to a distance of about 100 parsecs; after that the falling off becomes noticeable, so that at 300 parsecs it is only a fraction (perhaps a fifth) of the density near the Sun. The extension in the galactic plane is at least three times greater.[76]

When, therefore, van Maanen used a distance of 10,000 light years for M101 in 1915, the implication was that it was close to the boundary of, if not beyond, our own stellar system. Van Maanen had thereby provided some measure of support for the spirals as external galaxies provided the galactic system was of the sort of size suggested by von Seeliger and Eddington. In the next few years, van Maanen measured several more spirals, with consistent results.[77]

The context of van Maanen's observations shifted in the late 1910s when his Mount Wilson colleague Harlow Shapley advanced a radical new theory of the galactic system, the "Big Galaxy." After completing a Ph.D. at Princeton, Shapley joined the Mount Wilson staff in 1914. There he began a study of globular clusters with the aid of the Mount Wilson 60-inch reflector. The 60-inch had been completed in 1908 and was for a decade the most powerful telescope in the world. With its aid, Shapley constructed a set of distance indicators that enabled him to calculate the distances to the clusters. By the end of 1917, Shapley had fashioned a radical new theory, the Big Galaxy, in which our stellar system has a diameter of around 300,000 light years and the Sun is tens of thousands of light years from the centre.[78]

The key distance indicators for Shapley were Cepheid variables. Henrietta Swan Leavitt, at the Harvard College Observatory, had demonstrated in 1912 that the Cepheids exhibit a relationship between their absolute brightness and the period of their light variations (Fig. 2.3). Leavitt's result would prove to be one of the most important finds in early twentieth century observational astronomy.[79]

Leavitt had examined numerous photographic plates of the Small Magellanic Cloud secured at the Harvard southern station in Peru. In so doing, she identified 25 Cepheids (these would now be regarded as Classical Cepheids). As all these Cepheids are probably at effectively at the same distance, she attributed their differences in apparent brightness to their variations in their absolute brightness. There was, Leavitt emphasized, "a remarkable relation between the brightnesses of these variables and the length of their periods ..."[80]

She pointed out that it would be possible to calculate the distances to those groups of stars that contained Cepheids if the relationship was calibrated. Leavitt, however, did not perform such a calibration herself.

The first astronomer to do so was Ejnar Hertzsprung in 1913, and he used the calibrated relationship to determine a distance of 30,000 light years to the Small Magellanic

Fig. 2.3 *Henrietta Swan Leavitt of the Harvard College Observatory. Leavitt made the key discovery in 1912 of the period–luminosity relationship.*

Cloud.[81] The leading American astrophysicist Henry Norris Russell explained his own calculation of the distance to the Small Magellanic Cloud in a letter to Hertzsprung. Russell told Hertzsprung that his own answer of 80,000 light years was "an enormous distance, but it is not intrinsically incredible." If the galaxy has a diameter of some 5,000 to 10,000 light years, these distance estimates clearly meant the Small Magellanic Cloud was an external galaxy.[82]

Given Shapley's "Big Galaxy," however, a distance to the Small Magellanic Cloud of 30,000 or even 80,000 light years, put it within or close to the limits of the galactic system. Certainly if the dimensions Shapley was advancing for the Big Galaxy were anything like correct, then the sizes of the internal motions and so the distances van Maanen was obtaining for the spiral nebulae, told very strongly against them as external galaxies, and especially against them as comparable galaxies.

By the late 1910s, then, the significant observations that were shaping the debate on the nature of the spiral nebulae were being made largely from relatively well-funded and well-staffed observatories in the American west. Astronomers in the United States were not affected by World War I to anything like the extent of astronomers in Europe. American observatories had also not been burdened by the Carte du Ciel project. Many European observatories participated in the project, the goals of which were to photograph the whole of the sky, and to produce both a catalogue and chart of the observed stars. The Carte, however, proved to be of very limited astronomical value.[83]

It is against the rapid observational, cognitive, and institutional developments that we need to see the various events leading to what has become known as the "Great Debate" on the "Scale of the Universe" of 1920 between Heber Curtis of the Lick Observatory and Harlow Shapley of the Mount Wilson Observatory. That Curtis and Shapley would discuss this topic at that time was far from inevitable.[84] Nor was the Debate itself on the evening of 26 April 1920 in Washington DC of particular historical significance. Curtis spoke at a somewhat technical level, but Shapley had an eye on the members of the audience involved in the selection of the next director of the Harvard College Observatory, a position that was then open and to which he aspired, and he pitched his talk at a very popular level.[85]

Curtis and Shapley did, however, engage with each other's evidence and arguments in the published (and much longer and quite different) versions of their talks. The published papers are especially notable as we see two very well-informed astronomers on opposite sides of the debate on island universes giving their in-depth assessments of the available observational evidence and explained their arguments in detail. Curtis was in effect the champion of the "Lick School," as the Lick Observatory was the center of support for island universes and whose astronomers often took a somewhat skeptical attitude to some of the latest astrophysical theories. Curtis was doubtful about Shapley's 'Big Galaxy,' and the distance indicators Shapley had employed to establish it. He was especially critical of Shapley's calibration of the period–luminosity relationship, based as it was on the proper motions of eleven Cepheids. Nor did he set much store on Jacobus Cornelius Kapteyn's rival theory of the galaxy, what would become known as the Kapteyn universe. Kapteyn was the most prominent of the statistical astronomers who had developed an ellipsoidal model. He would calculate in 1922 that the Galaxy has a diameter of around 60,000 light years, but Curtis was dubious of Kapteyn's results as he was unconvinced of the reliability of Kapteyn's data. Instead, Curtis suggested the Galaxy is lens-shaped and has a diameter of around 30,000 light years. In his Great Debate paper, Curtis also exploited observations of novae in some spiral nebulae to estimate their distances. His answers using this method put the spirals far beyond the Galaxy (we will discuss novae in spirals more later).

Shapley, however, found it difficult to conceive of the spirals as even close to comparable in size to his Big Galaxy. Nor did he regard them as star systems. Rather, he suggested they might be clouds of nebulous matter driven-off from the Galaxy in some manner. For Shapley, the observations by his friend at Mount Wilson, van Maanen, of internal motions in the spirals—which argued for distances to the measured spirals of around 5,000 to 10,000 light years—was persuasive evidence against the spirals as external star systems. As recently as 1917, Shapley had preferred the island universe theory, and admitted to H.N. Russell (his former Ph.D. advisor) his private doubts about the reality of van Maanen's motions. But in 1919, and then in his "Great Debate" paper, he argued that van Maanen's measures, which he thought were in line with those of other astronomers, when combined with his Big Galaxy, made the island universes untenable. The sizes of the internal motions, if the spirals were at the distances required of comparable galaxies, would have to be greater than the velocity of light. As Shapley told Russell in 1920, "I am as much baffled by spiral nebulae as you, except I see no reason for thinking them stellar or universes. What monstrous assumptions that requires before you get done with it."[86]

By the early 1920s, then, a body of observations—particularly Slipher's radial veloc-ities, observations of novae in, and the spectra of, spirals—had influenced the views of astronomers so that most likely favored the island universe theory. But some, like Shapley, were far from persuaded, and van Maanen's observations could not be dismissed easily. With the benefit of hindsight, we can see that what was needed to settle matters was a means to determine distances to at least some spirals that astronomers could agree was accurate.

2.10   Hubble and the Cepheids

The means, in fact, was already to hand, in the form of the Cepheid period–luminosity relationship. Indeed, when the Mount Wilson astronomer Edwin Hubble started to detect Cepheids in the Andromeda Nebula in 1923, the debate over the distances of the spirals ended quickly on the side of the spirals as galaxies. Hubble was born in Evanston, Illinois in 1889, and he is now usually regarded as the most important observational cosmologist of the first half of the twentieth century. He began graduate studies at the University of Chicago in 1914, and employed the Yerkes Observatory's great 40-inch refractor for his Ph.D. on "Photographic Investigations of Giant Nebulae." In his dissertation, he showed himself to favor the spirals as external galaxies.[87] After a two-year spell in the U.S. Army (he had enlisted following the U.S. declaration of war in 1917), Hubble joined the Mount Wilson staff in 1919. Here he had access to the 60-inch reflector, as well as the just completed 100-inch Hooker telescope, easily the most powerful in the world.

By 1923, Hubble was engaged on an observing programe to identify novae in the Andromeda Nebula. S Andromedae was the most famous such example, and, as we saw earlier, had been regarded as strong evidence against the Nebula as an island universe. A decade later, Z Centauri had flared in NGC 5253.[88] But in July 1917, George Ritchey at Mount Wilson, while photographing spirals for van Maanen's internal motions research, had stumbled on a nova in NGC 6946. In the same month Curtis discovered three novae in two different spiral nebulae.[89]

Several astronomers now began to search old photographic plates for novae in spirals, and more were spotted. Employing the new evidence that had come to light, Curtis calculated the distances to the spirals using the novae in two different ways. In the first computational scheme, his answer was that the average distance to the spirals was around 20,000,000 light years. Curtis's second scheme also pointed clearly to the spirals as external galaxies.[90]

Hubble's plan in 1923 was to hunt for novae in the Andromeda Nebula with the aid of the 100-inch reflector. The more novae that were observed, the more reliable would be the distance estimates to the nebula based on their brightnesses. As part of his program, Hubble centered his photographic plates on the outer regions of the Nebula to minimize the telescope aberrations in those regions.

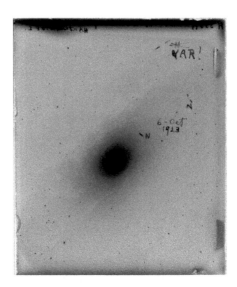

Fig. 2.4 *1923 plate of the Andromeda Nebula taken by Edwin Hubble. In the top right is the object he marked initially as a nova, an identification he later changed to "VAR" for variable star, in particular a Cepheid variable. (Courtesy of the Carnegie Observatories)*

In October 1923, Hubble discovered what he took to be a nova in the nebula's outer regions (Fig. 2.4). He marked it as such on his photographic plate. But as Hubble examined other plates that exhibited the nova and constructed a light curve for its varying brightness, he decided it was a Cepheid variable, not a nova. Variables had already been found in the spiral M33, in 1920 by John Charles Duncan using Mount Wilson plates, and by Max Wolf at Heidelberg in 1922. Duncan had found three variables, and Wolf one. Neither Duncan nor Wolf, however, had established if any of these four variables displayed a regular periodicity.[91] Hubble measured the period and apparent brightness of the Cepheid. Then, by employing the period–luminosity relationship, he arrived at a distance of around 900,000 light years, putting the Andromeda Nebula well outside the limits of even Shapley's model of the Galaxy.[92]

The possibility that the Andromeda Nebula was much nearer than the Cepheid, and that the nebula and Cepheid appeared close together on the sky due to a chance alignment, became even more remote when Hubble found other Cepheids that supported his findings on the first. Hubble soon detected Cepheids in the remote stellar system NGC6822 and the spiral M33, and his measured distances argued very strongly for them as external galaxies too. With the Cepheid results, Hubble brought the debate over island universes to a swift end.[93]

Van Maanen's distances to the spirals were at odds with Hubble's Cepheids, as well as what for astronomers was an imposing body of other evidence. Indeed, it was by the mid-1920s, so imposing a body of evidence that it is better to regard Hubble as having clinched the case for island universes, rather than discovering them in any meaningful sense. But how to explain the contradiction between van Maanen's distances to the spirals and the distances to the spirals as island universes? For astronomers, the cause of the contradiction, despite van Maanen's reputation as a skilled measurer of photographic plates and a position at the world's leading astrophysical observatory, was an error or errors in the measurements, perhaps an unknown effect in the photographic emulsion of the plates. There was now a difficult situation for the Mount Wilson managers as there was a personal conflict between Hubble and van Maanen. The public end to the dispute between the two arrived in a pair of brief papers in the *Astrophysical Journal* in 1935 when van Maanen conceded his results had to be treated with caution.[94]

By 1935, Hubble was seen by many other astronomers as the leading authority on extragalactic astronomy. In an era when the most powerful telescopes were seldom available to astronomers who worked outside of the observatories at which those telescopes were based, observational astronomy was far from a meritocracy. Hubble, given his extensive access to the 100-inch Hooker telescope (Fig. 2.5), and by serving as a staff member of an institution devoted solely to research, had big advantages over other astronomers. But it was one thing to have those advantages, another to make effective use of them. And Hubble did so throughout the 1920s and first half of the 1930s. Hubble followed-up his initial exploration of M31 with a more detailed study of the Andromeda Nebula, as well as in-depth investigations of NGC 6822, what he termed as a remote stellar system, the spiral M33 and a general survey of what he termed extragalactic

Fig. 2.5 *The 100-inch Hooker telescope atop Mount Wilson. (Courtesy of The Observatories of the Carnegie Institution for Science Collection at the Huntington Library, San Marino, California)*

nebulae.[95] To avoid confusion, we will not use Hubble's "extragalactic nebulae," but will refer only to galaxies.

2.11   Redshift–distance relationships

Hubble had decided in 1927 to address what had become one of the leading questions for extragalactic astronomy: Was there a relationship between the distance of a galaxy and the size of its redshift? For those astronomers who regarded the redshifts as Doppler shifts, the question became, Was there a relationship between velocity and distance?

The possibility of a redshift–distance relation had first been raised by Willem de Sitter. De Sitter had discovered in 1917 a solution to Einstein's field equations of general relativity that became known as Solution B, to distinguish it from Einstein's solution, Solution A. As discussed elsewhere in this volume, de Sitter's solution had very important cosmological consequences. The key point for us is that de Sitter's researches encouraged some astronomers to search for observational evidence of a redshift–distance relation. These sorts of analyses were in line with what has been called the "Correlation Era" or the "Great Correlation Era," in that astronomers sought to establish relationships between the different properties of stars and in this case what proved to be galaxies. The chief success of this enterprise was what would come to be called the Hertzsprung–Russell diagram.[96]

By the mid-1920s, despite the efforts of several astronomers—including Knut Lundmark (Fig. 2.6), Carl A. Wirtz, and Ludwik Silberstein—astronomers were not persuaded of the reality of a redshift–distance relation.[97] General relativity theory, as interpreted by de Sitter, suggested one should exist, but the observational studies had not provided evidence astronomers regarded as convincing. The chief issue was the distances to those galaxies for which Slipher had measured the redshifts, given what other astronomers judged to be the unconvincing nature of the different distance scales used before Hubble discovered Cepheids in the Andromeda Nebula.

Following discussions with other astronomers, Hubble embarked on a program in which he would determine the distances to the targeted galaxies, while his Mount Wilson colleague Milton Humason measured their redshifts. By this date, we should note, Slipher's own program of measuring the redshifts of distant galaxies had come to an end.[98] Hubble's aim was to fashion a convincing distance-scale, moving from galaxies in which he could detect various sorts of stars, through to galaxies in which no individual stars were visible, but for which he could establish other distance indicators, say the integrated brightness of an entire galaxy.

The spectral shifts of 46 galaxies had been measured by 1929, the great majority by Slipher. Hubble wrote a short paper in that year to announce his early results in which he employed his own distance estimates and Slipher's redshifts.[99] In what he called his primary proof, Hubble employed the 24 galaxies (20 of the redshifts for these 24 had been determined by Slipher) that were in groups (Fig. 2.7), a point that enabled him to use average apparent magnitudes (for his secondary proof, Hubble used the 22 galaxies whose redshifts had also been measured, but which were not part of a group). As Hubble

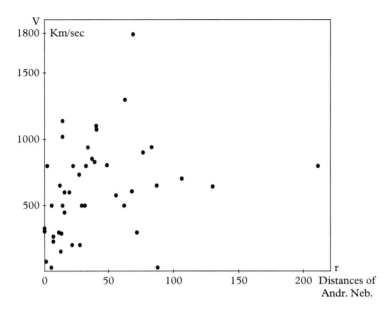

Fig. 2.6 *1924 plot by Knut Lundmark of velocity against distance, from his paper on "The Determination of the Curvature of Space-Time in the de Sitter world" (1924). Lundmark was one of a number of astronomers who during the 1920s researched the issue of a redshift–distance relation.*

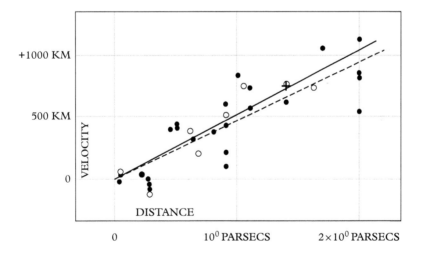

Fig. 2.7 *Hubble's 1929 plot of velocity against distance. For Hubble, "velocity" was to be understood as "apparent velocity." The black dots represent 24 individual galaxies (the redshifts for all but four of these were determined by Slipher).*

explained in his 1929 paper, calculations of the solar motion with respect to the galaxies had led astronomers to employ "a K term of several hundred kilometres which appears to be variable. Explanations of this paradox have been sought in a correlation between apparent radial velocities and distances, but so far the results have not been convincing. The present paper is a re-examination of the question, based on only those nebular distances which are believed to be fairly reliable."[100]

Hubble contended that his plot of redshift against distance was best represented by a linear relation, at least to the first approximation. Instead of putting the relation into the form that would soon become usual ($V = H \times r$ with H a constant and following what would later be termed the Hubble law),[101] Hubble wrote it in terms of the solar motion equations where the K term now depended on distance in a linear fashion. That is, the equations describing the solar motion took the form: $V = X \cos\alpha \cos\delta + Y \sin\alpha \cos\delta + Z \sin\delta + kr$. For those astronomers who interpreted the redshifts as Doppler shifts, the result indicated a simple, linear relation between distance and velocity, $V = kr$. For Hubble, the "outstanding feature...is the possibility that the velocity–distance relation may represent the de Sitter effect, and hence that numerical data may be introduced into discussions of the general curvature of space."[102]

We have noted that several other astronomers had pursued a redshift–distance relation some years before Hubble. In Hubble's view, however, his 1929 paper provided "first presentation of the data where the scatter due to uncertainties in distance was small enough as compared to the range in distances, to establish the relation," and in future years other astronomers would generally agree with this judgement.[103] In 1930, de Sitter also published a plot of redshift versus distance (Fig. 2.8). His result too was a linear relation.[104]

Astronomers soon accepted that there was a linear relation, and, as described elsewhere in this volume, it swiftly became a crucial relation for mathematical cosmologists. These researchers had also, by the early 1930s, confronted the failings of the static solutions of the field equations of general relativity. It was in this climate that Hubble's observational results and the mathematical investigations of Georges Lemaître became decisive for the widespread, but certainly not universal, acceptance of the idea of an expanding universe.

Lemaître's theorizing and Hubble's handling of the redshift/distance observations convinced many that the static solutions had failed because the universe is expanding. The redshift–distance relation was, then, a result of that expansion.[105] Hubble, however, was one astronomer who balked at identifying the redshifts as Doppler shifts, and when he referred to "velocity," he used it in the sense of "apparent velocity."

Hubble gave a thorough account in 1931 of the methods he used to obtain distances when he extended his distance determinations to more distant galaxies, and in collaboration with Humason, who had secured 40 additional redshifts to those galaxies with the aid of a very efficient new spectrograph, presented a revised plot of redshift versus distance (Fig. 2.9). Hubble calculated the most remote galaxy to be at a distance of 32×10^6 pc (a big jump compared to the most far-off galaxy in the 1929 paper, 2×10^6 pc). Hubble and Humason were also at pains in the 1931 paper to argue for a separation of interpretation and observation. The redshifts measured by Humason had

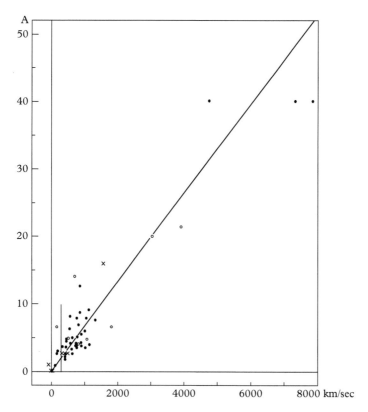

Fig. 2.8 *Plot of distances against velocities, as drawn by Willem de Sitter in 1930.*

not been proved to be velocity shifts, they argued. "The present contribution," Hubble and Humason emphasized, "concerns the correlation of empirical data of observation. The writers are constrained to describe the 'apparent velocity-displacements' without venturing on the interpretation and its cosmologic significance. Further observations are desirable and will be carried on, although it seems probable that the general features of the relation are already sketched in outline nearly to the limit of existing equipment."[106]

Many astronomers were soon agreed that the redshift-distance relation disclosed that the universe is expanding. But the sizes of the redshifts Humason was finding—interpreting the spectral shift of one galaxy he measured in 1934 as a Doppler shift indicated a speed of 39,000 km s^{-1}, a strikingly high value—made it difficult for some others to agree.[107] Further, the expanding universe interpretation led, on the basis of a straightforward interpretation of the expansion, to what was called the time-scale problem. This problem is discussed more fully elsewhere in the current volume, but the issue was that the age of the universe appeared to be shorter than the ages of the oldest stars, as well as shorter than James Jeans' estimate for the age of the Galaxy of 10^{12}–10^{13}

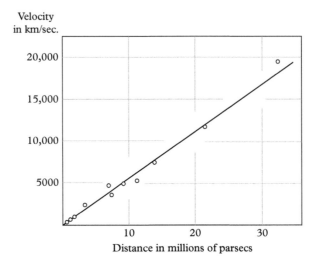

Distance in millions of parsecs

Fig. 2.9 *Hubble and Humason's 1931 plot of velocity against distance. Again the term "velocity" was used in the sense of "apparent velocity". The black dots close to the origin of the plot show the data points from Hubble's 1929 plot.*

years. Hubble would prove to be the leading skeptic about the Doppler interpretation of the redshifts.

2.12 Models

During the 1920s, astronomers had searched for a possible redshift-distance relation. In the 1930s, researchers investigated the observational consequences to be expected from different cosmological models.[108] Perhaps the most important of these studies were those by Richard C. Tolman, a mathematical physicist and physical chemist at Caltech. Tolman had published on the cosmological consequences of general relativity starting in the late 1920s, including the 1934 book *Relativity, Thermodynamics and Cosmology*. He also collaborated with Hubble in the mid-1930s, and in their collaboration we see the growth of the links between theory and observations of the distant universe, which although very limited by later standards, was a significant development, and an indication of the "division of labor" between observers and theorists that had emerged.

Some years earlier, in 1926, Hubble had examined some 400 'extragalactic nebulae.' He concluded that, with the exceptions of the Galaxy and the Virgo Cluster, this group of galaxies indicated that there is a uniform density of galaxies in all directions.[109] In 1935, Hubble and Tolman published a paper in which they established relationships between the dimensions and luminosities of galaxies, and between counts of galaxies, to different limiting magnitudes for the cases of the redshifts as velocity shifts and the redshifts as non-Doppler shifts. They decided that the observational data, uncertain as it was, could not decide between the two cases. Instead, they argued that:

...it might be possible to explain the results on the basis of either a static homogeneous model with some unknown cause for the red-shift or an expanding homogeneous model with the introduction of effects from spatial curvature which seem unexpectedly large but may not be impossible. This statement must be made very tentatively, however, and may need serious modification in the light of further counts, of new auxiliary data for treating the counts, or of better analysis of the present data.[110]

The historian of science Norriss Hetherington,[111] as well as Allan Sandage, who worked as an observing assistant with Hubble for a year at the end of Hubble's life and who was himself a very distinguished cosmologist, have scrutinized Hubble researches on the number counts of galaxies.[112] For Sandage, Hubble's 1936 analysis of the number counts (and these were to a fainter magnitude than in the 1935 paper with Tolman),

is a most fascinating study of how an interpretation, without caution concerning possible systematic errors, led to a conclusion that the systematic redshift effect is probably not due to a true Friedman–Lemaître expansion, but rather to an unknown, then as now, unidentified principle of nature.[113]

Two years later, in 1938, Hubble argued that "the results, at the moment, seem to favour the concept of a stationary universe, but they do not definitely rule out the possibilities of an expanding universe."[114]

By the time of Hubble's death in 1953, astronomers had reached an overwhelming consensus that the redshifts are Doppler shifts, and that the universe is expanding. By the mid-1950s too, revisions to the distance-scale Hubble had established meant that the timescale problem had effectively disappeared. The two main changes were due to, first, Sandage, who showed that the brightest stars Hubble had identified in distant galaxies were in fact HII regions (that is, clouds of ionized hydrogen), and, second, Walter Baade, who revised the Cepheid period-luminosity relationship.[115]

2.13   Clusters of galaxies

J.H. Lambert had speculated in the eighteenth century on the degree to which the universe exhibits a hierarchical arrangement.[116] Early in the twentieth century, Carl V.L. Charlier had further explored the idea of a hierarchical universe, but his researches did not attract much interest.[117]

The related issue that was of crucial importance to cosmologists by the end of the 1920s was the extent to which the universe, when viewed on a large scale, is homogeneous and isotropic. As with much else in terms of observational extragalactic astronomy in the 1920s and 1930s, a leading investigator on this subject was Hubble. We noted when discussing the nature of the redshifts, that Hubble had analyzed the space density of galaxies. The results he published in 1936 were based on number counts in sample areas to different limiting magnitudes with the 60-inch and 100-inch Mount Wilson reflecting telescopes, as well as the counts performed by Nicholas U. Mayall at Lick Observatory with the 36-inch Crossley reflector. Hubble's chief finding was that "although nebulae are found singly, in groups and even occasionally in great clusters, when very large regions of space were compared the tendency to cluster averaged out and one region was very much like any other."[118]

In contrast to Hubble's "deep but narrow" approach, at Harvard, Harlow Shapley, with no access to the sorts of very powerful reflectors employed by Hubble, scrutinized bigger areas of the sky than Hubble to study local irregularities in the distribution of galaxies. By 1926, Shapley and his collaborator Adelaide Ames had examined the Virgo Cluster of galaxies, and had concluded that much "remains unknown concerning the distribution over the sky of the [extragalactic] nebulae of the spiral family, notwithstanding the rather confident assertions and generalizations that are frequently made. The uncertainty is founded, on the one hand, on the incompleteness of the available surveys, and, on the other, on the prevalence of large and irregular clouds of nebulae of the spiral or 'extragalactic' type."[119]

In 1932, Shapley and Ames published what would become a very well-known catalogue of the magnitudes and positions of the brighter galaxies, complete, they believed, to magnitude 12.9 over the whole sky. They hoped that the catalogue would provide "a suitable basis for a study of the uniformity of distribution of the galaxies, the clustering of the nearer systems, the relation of the apparent distribution to the obscuring clouds in low galactic latitude, and similar problems."[120] For Shapley, the catalogue indicated that galaxies tend to be found in clusters and groups. In fact, by 1936, when Hubble wrote his book *The Realm of the Nebulae*, it had become widely agreed that clusters are an important feature of galaxy distribution. Hubble's own view, that there is a general field of galaxies that is in places disturbed by clusters or groups, had no support by the early 1950s. Hubble's interpretation was viewed by than as a product of selection effects, the result of a limited sampling of galaxies due to his employment of powerful, but narrow-field, telescopes.[121]

2.14   Conclusion

The roughly 150 years between the late eighteenth century and the middle decades of the twentieth century saw enormous changes in observational cosmology. Not only have we seen that observational cosmology was transformed in terms of its content and what astronomers believed was evident from observations, but also in terms of the methods and techniques astronomers employed, as well as who made the observations, how they were supported, and from where they made their observations. In the period we have examined, the study of nebulae was eventually changed in the 1910s, 1920s, and 1930s into extragalactic astronomy. In some respects, for much of our period of interest, astronomers had been engaged on 'cosmology by accident' rather than design, as the nebulae as evidence of some version of the nebular hypothesis, were remade into external galaxies, in large part through observations made by professional astronomers with powerful telescopes and ancillary equipment housed in the American west. These efforts had reached their high points with the establishment of external galaxies and the expanding universe in a relatively short space of time, both of which surely count as among the major discoveries of twentieth century science. Although the linkages between theory and observation that had been established by the late 1930s were decidedly limited by later standards, with a growing number of practitioners with access to first-rate instruments at good observing sites, and strong institutional commitments to the pursuit of cosmology (although the term itself was not much used until later), observational cosmology had come of age.

. . . .

NOTES

1. Crowe (1986), pp. 41–59.
2. Quoted in Hufbauer (1991), p. 43.
3. On the William/Caroline relationship and Caroline's indispensable role in William's astronomy, see Fara (2004), Chapter 8. For a biography of Caroline, see Hoskin (2013).
4. William Herschel (1811), p. 269.
5. Hoskin (2012).
6. Hoskin (1982), p. 128.
7. Michael Hoskin (2011) provides an excellent overview of the Herschels, including William's changing ideas on the nebulae.
8. William Herschel (1811), p. 270.
9. William Herschel (1802), p. 498.
10. Holmes (2008), p. 210.
11. On the early development of the nebular hypothesis, see, among others, Jaki (1977), Brush (1996), and Schweber (1990).
12. John Herschel (1822), p. 4.
13. John Herschel (1826), p. 487.
14. On John Herschel's astronomy and cosmology, see, among others, Hoskin (1987), Nasim (2014), and Case (2015).
15. John Herschel (1847), p. 146.
16. Hoskin (1987), p. 19.
17. R. W. Smith (2014b), p. 116. See also Schaffer (1989).
18. Anonymous, but see Chambers (1844), p. 30, and Oglivie (1975). On *Vestiges* in general, indispensable is Secord (2000).
19. Schaffer (1998).
20. Hoskin (1990), p. 337.
21. T.R. Robinson (1845), p. 130.
22. G.P. Bond to President Everett, 22 September 1847, Bond Papers, Harvard University Archives.
23. Nichol (1848), p. vi.
24. The Earl of Rosse (1846), p. 4.
25. Dewhirst and Hoskin (1991).

26. The Earl of Rosse (1850), p. 503.
27. J.R. Hind, "Changes among the stars," *The Times*, 4 February 1862, p. 8b and Hoskin (1967), p. 80. See also Barbara Becker (2011) pp. 66–9.
28. De la Rue (1861), p. 130.
29. For a fuller account of Huggins' investigations of stars and nebulae in the 1860s, see R.W. Smith (2014b).
30. William Huggins (1897), p. 916. Huggins' account should be read in combination with that of Becker (2011), pp. 72–5. See also R.W. Smith (2014b), pp. 113–15.
31. William Huggins (1864), p. 437. See also R.W. Smith (2014b).
32. Huggins (1864), p. 442.
33. Huggins (1865), p. 40.
34. Huggins (1865), p. 41.
35. King (1979), p. 262.
36. Gillespie (2011).
37. King (1979), p. 274.
38. Report of meeting of the Royal Astronomical Society (1888), p. 51.
39. Huggins (1891), p. 523.
40. Abbe (1867).
41. Proctor (1869a), p. 342.
42. Proctor (1869a), p. 341.
43. Proctor (1869a), p. 342.
44. The 'nova' would later be reclassified as a supernova, but this class had not been established at the time.
45. Clerke (1890), p. 369.
46. Clerke (1890), p. 368.
47. R.W. Smith (2003).
48. Lankford (1997), p. 36.
49. King (1979), p. 327.
50. Van Helden (1984), pp. 138–9.
51. This 36-inch reflector had been donated by the English carpet manufacturer Edward Crossley. Crossley had purchased it from A. A. Common who, as we noted earlier, completed it in 1879, and had used it in his photographic studies of nebulae. On Keeler, see Osterbrock (1984).
52. Dreyer (1895).
53. R.W. Smith (2008), p. 99.
54. Keeler (1900), col. 1.
55. Moulton (1905), p. 169. The Chamberlin–Moulton hypothesis is examined in-depth by Brush (1978).
56. There is a very large literature on Lowell and Mars. See, for example, Hoyt (1976) and Lane (2011).
57. Scheiner (1899), p. 150. The original German version of this paper is "Ueber das Spectrum des Andromedanebels," *Astronomische Nachrichten*, **148** (1898–99), cols. 325–8.

58. H. Vogel, revision of [Simon] Newcomb–[Rudolf] Engelmann's Populäre Astronomie (Leipzig, 1905), p. 577. Scheiner's observations and the response by Vogel are examined in Duerbeck (2002).
59. Donald E. Osterbrock discusses Fath's researches in some detail in Osterbrock (1991).
60. Fath (1909).
61. Wolf (1912).
62. Bohlin (1907). The prominent Swedish astronomer Karl Lundmark later argued that Bohlin's spurious result was the product of mistakes in measuring the nebula's hour angle; see Lundmark (1927).
63. Slipher (1913). On Slipher's radial velocity studies, see R.W. Smith (2013).
64. Campbell (1896).
65. R.W. Smith (1982), pp. 17–22.
66. V.M. Slipher (1914), "Spectrographic Observations of Nebulae." Manuscript in Lowell Observatory Archives.
67. Slipher (1917).
68. Ejnar Hertzsprung to V.M. Slipher, 14 March 1914, Lowell Observatory Archives, and Campbell (1917).
69. Heber Doust Curtis, "Report 1 July 1913–15 May 1914," manuscript in Lick Archives.
70. Curtis (1915).
71. R.W. Smith (1982), pp. 30–2.
72. Van Van Maanen (1916a, 1916b).
73. J.H. Jeans (1917, 1919).
74. R.W. Smith (1982), pp. 15 and pp. 86–9. For an overview of the studies of the galaxy from 1900 to 1952, see R.W. Smith (2006).
75. Seeliger (1898b). On statistical astronomy, see Paul (1993).
76. Eddington (1914), p. 31.
77. For a listing of the papers where van Maanen published his results, see R.W. Smith (1982), p. 209, as well as Hart (1973).
78. R.W. Smith (1982), pp. 59–77.
79. Pickering (1912). As Pickering wrote at the start of this Circular, "The following statement regarding the periods of 25 variable stars in the Small Magellanic Cloud has been prepared by Miss Leavitt."
80. Pickering (1912).
81. Hertzsprung (1913).
82. Russell to Hertzsprung, 8 July 1913, Aarhus, and R.W. Smith (1982), pp. 71–3.
83. Lankford (1997), pp. 394–400. But see also DeVorkin (1998).
84. R.W. Smith (1982), pp. 77–87. This section also includes a more detailed discussion than is presented here of the astronomical arguments advanced by Curtis and Shapley in their talks and later published papers. The published papers of Curtis and Shapley are both titled "The Scale of the Universe," and appear in the *Bulletin of the National Research Council*, **2** (1921), pp. 194–217 and pp. 171–93.

85. R.W. Smith (1982), pp. 78–9 and Hoskin (1976a).

86. H. Shapley to H.N. Russell, 30 September 1920, Harvard University Archives. On Shapley's shifting positions, see R.W. Smith (1982), pp. 65–66.

87. Hubble (1920).

88. E. Pickering, "A new star in Centaurus," *Harvard College Observatory Circular*, No. 4, 20 December 1895.

89. R.W. Smith (1982), pp. 42–5.

90. Curtis (1917) and Hoskin (1976b).

91. R.W. Smith (1982), p. 122.

92. Hubble's results were first announced publicly in the *New York Times*: "Finds spiral nebulae are stellar systems; Dr. Hubbell [sic] confirms view that they are 'island universes' similar to our own," *New York Times*, November 23, 1924, p. 6. Just over a month later, H.N. Russell read a paper on the results on Hubble's behalf to a joint meeting of the American Astronomical Society and the American Association for the Advancement of Science. See Berendzen and Hoskin (1971).

93. R.W. Smith (1982), pp. 114–21.

94. Hubble (1935) and van Maanen (1935). Van Maanen's measures have led to a large literature. The works on them include Berendzen and Hart (1973), Hart (1973), and Hetherington (1990).

95. Hubble (1925, 1926a, 1926b, 1929b).

96. The "Great Correlation Era" is discussed in Robert W. Smith, "Astronomy in the Time of Pannekoek and Pannekoek as an Astronomer of his Times," forthcoming.

97. For a short overview of these other researches, see R.W. Smith (2009), pp. 10–15. See also Way (2013), pp. 106–11.

98. The issue of whether or not Slipher could have fruitfully continued his researches on redshifts after the early 1920s given the capabilities of the Flagstaff instruments is taken up in various places in Way and Hunter (2013).

99. In recent years, a number of contemporary astronomers have been critical of the fact that Hubble did not reference Slipher in this paper. See, for example, Way (2013). Way's chapter is also an important guide to a range of works that seek to re-evaluate Hubble's various contributions, particularly on the classification of galaxies and the expanding universe.

100. Hubble (1929a), p. 168.

101. For an account of Hubble's law, see R.W. Smith (2015).

102. Hubble (1929a), p. 173.

103. E. Hubble to W. de Sitter, 21 August 1930, Huntington Library; see also Hubble (1929a), p. 173.

104. De Sitter (1930).

105. On the discovery of the expanding universe, see Kragh and Smith (2003). See also Way (2013).

106. Hubble and Humason (1931), p. 80. See also R.W. Smith (2009), pp. 14–19. Edward Harrison argued that later researches usually failed to separate carefully the redshift-distance relation published in 1929 by Hubble, and the velocity-distance law that was derived from theoretical grounds. Although "both laws are indiscrim-

inately referred to as Hubble's law...no general proof exists demonstrating the two laws are equivalent, and perhaps for that reason many astronomers since the time of Hubble have viewed with reservation the velocity interpretation of extragalactic redshifts. The failure to distinguish between the linear redshift-distance law (an empirical approximation of limited validity) and the velocity-distance law (a theoretical derivation of unlimited validity) leads to confusion and obscuration of the fundamental concepts of modern cosmology." See Harrison (1993), p. 28.

107. Humason (1934).
108. Kohler (1933) and McCrea (1935). See also Kragh (1996), pp. 269–87.
109. Hubble (1926a).
110. Hubble and Tolman (1935), p. 335.
111. Hetherington (1982).
112. Sandage (1998), p. 5.
113. Sandage (1998), p. 5.
114. Hubble (1938), p. 123. See also Hubble (1937).
115. Sandage (1958), pp. 520–6, and Osterbrock (2001), pp. 162–176, detail Baade's researches. See also R.W. Smith (2006), pp. 334–6, and Baade and Gaposchkin (1963).
116. J.H. Lambert (1761). Stanley L. Jaki translated this book as *Cosmological Letters on the Arrangement of the Universe* (New York: Science History Publications, 1976). See also Hoskin (1982), pp. 117–23.
117. See, among others, Charlier (1922, 1925a). On Charlier, see Holmberg (1999), pp. 90–131.
118. R.W. Smith (2009), pp. 22–5. For Hubble's publications, see also Hubble (1936b). Also, Hubble (1934a, 1936a, 1937).
119. Shapley and Ames (1926).
120. Shapley and Ames (1932), p. 43.
121. In 1956, for example, Humason, Mayall, and Sandage, all of whom had worked closely with Hubble, argued that "it is becoming increasingly evident that the nebular distribution is characterized by a predominant tendency to cluster..." Humason et al. (1956), p. 135.

3

Relativistic models and the expanding universe

Matteo Realdi

> *My subject disperses the galaxies, but it unites the earth. May no "cosmical repulsion"*
> *intervene to sunder us!*
>
> Eddington (1933), p. vii.

3.1 Introduction

This chapter is devoted to the history of the early phase of relativistic cosmology, from the formulation of the first theoretical models based on general relativity in 1917, through the recognition of the expanding universe in 1930 to the early systematization of relativistic cosmology in the late thirties.[1]

The first relativistic world models were proposed when the commonly accepted view was that the universe was static. Section 3.2 analyzes the formulation in 1917 of the first model that could consistently describe the universe in its entirety. This cosmological model was proposed by Albert Einstein. To comply with the principle of the relativity of inertia and to make possible the stasis of the universe through his model, Einstein introduced λ into his equations, a new constant that later became known as the cosmological constant. Einstein can rightly be said to have inaugurated the transformation of cosmology into a modern discipline. Since then, general relativity has dramatically changed the scientific understanding of the origin, structure and evolution of the universe.

In the same year, 1917, Willem de Sitter proposed an alternative relativistic model, which describes a static universe devoid of matter. De Sitter's cosmological solution and Einstein's reaction to it are described in Section 3.3. The path opened by Einstein and de Sitter in relativistic cosmology was followed by a number of astronomers, physicists and mathematicians. As described in Section 3.4, most of them initially focused on de Sitter's cosmological solution and its astronomical consequences. They proposed static and non-static versions of this universe, and they formulated various

Realdi, M., 'Relativistic models and the expanding universe' in *The Oxford Handbook of the History of Modern Cosmology*, edited by Kragh, H. and Longair, M. S. © Oxford University Press 2019.
DOI: 10.1093/oxfordhb/9780198817666.013.3

redshift–distance relations predicted by de Sitter's cosmological solution. A general tendency of particles to recede in the universe of de Sitter, first derived in 1923 by Arthur Eddington, seemed to explain the large radial velocities measured in spiral nebulae. In those years, astronomical observations revealed the extragalactic nature of spiral nebulae, which were widely recognized as galaxies. The incorporation of Edwin Hubble's observations into relativistic cosmological models culminated in 1930 with the recognition of the expanding universe, which was a breakthrough in the scientific investigation of the universe as a whole.

As from 1930, the dynamical solutions formulated in 1922 by Alexander Friedman and independently in 1927 by Georges Lemaître became widely known, and the expanding universe was largely accepted. The recognition in 1930 of the expanding universe is described in Section 3.5. Section 3.6 presents a number of speculative proposals related to the origin of the universe, with a focus on Lemaître's primeval-atom hypothesis. The early systematization of the formal structure of relativistic cosmology is the subject of Section 3.7, which summarizes the variety of evolutionary models formulated until the late thirties on the basis of the so-called Friedman–Lemaître equations and the Robertson–Walker metric.

The work carried out during the early phase of relativistic cosmology was mainly oriented towards a consistent geometric and mathematical description of the universe as whole, on the basis of cross-fertilization between general relativity and the first astronomical observations of cosmological relevance. Scientific disciplines such as thermodynamics and subatomic physics were included in the cosmological framework in the following decades, when astronomical observations at different wavelengths also contributed to turning cosmology into a physical discipline. Relevant topics entered the modern cosmological scene during the early phase of relativistic cosmology, such as the cosmological principle, the definition of cosmic time, the interpretation of extragalactic redshift, the existence of singularities and horizons in the universe and, of course, the expanding universe. The protagonists of the first two decades of relativistic cosmology may be regarded in this sense as the first modern cosmologists, who contributed to shaping the modern scientific understanding of the universe through the application of general relativity to cosmology.

A number of theoretical aspects of general relativity will be referred to in the following sections. A brief description of these aspects is presented later.[2] The general theory of relativity is the metrical and physical theory of gravitation formulated by Einstein. One of the innovative aspects of Einstein's theory, if not the most innovative, lies in the interpretation of gravity not as a force, as in Newtonian theory, but as pure geometry. Gravitation is the manifestation of the curvature of space-time. The metrical properties of space-time are described by the quantities $g_{\mu\nu}$, which are the components of the metric tensor g (the indices μ and ν are the coordinates adopted for the geometrical representation of space-time). These components represent the gravitational field of a region of space-time, and they are determined by the distribution of matter and energy in that region. In general relativity, space-time is modeled as a four-dimensional (pseudo-Riemannian) manifold. Locally, the gravitational field can be approximated to the Minkowski space-time of special relativity through the choice of an inertial system in which the metric tensor g has the values of flat space-time. The space-time interval between two events, or the line element, is written as $ds^2 = g_{\mu\nu}\, dx^\mu dx^\nu$, in which the summation convention is used. The relativistic field equations, also known as the Einstein equations, show how curvature and matter-energy influence each other. These equations are a set of covariant tensor equations that relate the geometry of a region of space-time to the matter and energy distribution, and they are written as:

$$G_{\mu\nu} - \tfrac{1}{2} g_{\mu\nu}\, G = -\kappa\, T_{\mu\nu}. \tag{3.1}$$

In this equation, $G_{\mu\nu}$ is the so-called Ricci tensor (representing the curvature of space-time), G is the Ricci scalar, $T_{\mu\nu}$ is the energy-momentum tensor, and κ is a constant equal to $8\pi G/c^2$, in which c is the velocity of light and G the gravitational constant. The matter and energy distribution is described by the energy-momentum tensor. For a perfect fluid with four-velocity U_μ, pressure p and energy density ρ, this tensor is:

$$T_{\mu\nu} = -pg_{\mu\nu} + (p + \rho c^2)\, U_\mu U_\nu. \tag{3.2}$$

In special relativity the path of a light ray is given by $ds = 0$, and the path of material particles is given by the stationary values of the integral of ds along the path itself. The paths of material particles are straight lines when no external forces apply, and forces such as gravitation and electromagnetism cause the deviation of particle tracks from a straight line. In general relativity the notion of straight lines is replaced by that of geodesics. A geodesic is the world line of a freely falling particle. If $ds^2 > 0$ the interval is time-like, and the time measured by a clock following that world line is called *proper time*. Geodesics are no longer straight lines because of the presence of the gravitational field contained in $g_{\mu\nu}$. Conversely, the metric $g_{\mu\nu}$ is determined by the presence of matter in a certain region of space-time. The Einstein equations therefore make it possible to describe the relationships between the geometry of space-time and the distribution of matter.

3.2   Einstein's cosmological considerations in general relativity

The quest for a covariant formulation of the laws of physics and the issue of an "epistemological defect" of special relativity, that is, the privileged role of inertial frames in special relativity, motivated Einstein in developing his general theory of relativity.[3] On 25 November 1915, Einstein presented to the Prussian Academy of Sciences the final form of the covariant field equations of general relativity, the first systematic exposition of which was published in May 1916. Already in late 1915 Einstein had explained through his theory of gravity the advance of 43" per century of the perihelion of the planet Mercury, an anomaly unexplained in Newtonian theory. In 1916, the physicist and astronomer Karl Schwarzschild found the first non-trivial exact solution of Einstein equations, which describes a spherically symmetric gravitational field in vacuum. Schwarzschild subsequently published the exact solution for the gravitational field of the interior of a non-rotating sphere of fluid with uniform energy density. One of the predictions of general relativity concerns the bending of light in the gravitational field of a massive body. This prediction was confirmed by the observations of the total solar eclipse of 29 May 1919. Eddington and the astronomer Frank Dyson directed two expeditions of British astronomers to test the prediction of the bending of star light by the gravitational field of the Sun.[4] The measurement of star positions near the Sun in the photographic plates obtained during the eclipse roughly confirmed the prediction of general relativity. The publication of the report of the eclipse expedition in fact made Einstein a celebrity. Headlines in the *London Times* of 7 November 1919 read "Revolution in science. New theory of the universe: Newtonian ideas overthrown."[5]

The foundational principles upon which Einstein formulated general relativity are the principle of relativity, the principle of equivalence and the principle of the relativity of inertia. In 1918, Einstein summarized these principles in the article "On the foundations of the general theory of relativity." The principle of relativity assumes the invariant formulation of the laws of nature. These laws, as described in Einstein's 1918 article, are "merely statements about temporal-spatial coincidences; therefore, they find their only natural expression in generally covariant equations." The principle of equivalence assumes that "[i]nertia and gravity are phenomena identical in nature. From this and from the special theory of relativity it follows necessarily that the symmetric 'fundamental tensor' ($g_{\mu\nu}$) determines the metric properties of space, the inertial behavior of bodies in this space, as well as the gravitational effects." The third basic assumption concerns the relativity of inertia, for which Einstein coined in 1918 the term "Mach's principle:" "The *G*-field is completely determined by the masses of the bodies. Since mass and energy are—according to the results of the special theory of relativity—the same, and since energy is formally described by the symmetric energy tensor ($T_{\mu\nu}$), it follows that the *G*-field is caused and determined by the energy tensor of matter."[6]

In elaborating the concept of inertia and formulating general relativity, Einstein was inspired by the examination of the foundations of classical mechanics proposed by the physicist and philosopher Ernst Mach. In Mach's critical analysis, mechanics should be reformulated on the basis of the relative motion of material bodies with respect to each other. In his 1916 obituary of Mach, Einstein acknowledged him as a precursor of general relativity, for Mach had "clearly recognized the weak points of classical mechanics, and thus came close to demand a general theory of relativity."[7] After the publication of the final version of the field equations, investigation of the universe as a whole and the origin of inertia was for Einstein a natural, even necessary, continuation of his work on general relativity. In particular, the principle of the relativity of inertia played a crucial role in Einstein's lines of thoughts, in view of the need to satisfy the requirement that inertia should be entirely determined by matter at the global scale: "In a consistent theory of relativity there can be no inertia *relatively to 'space,'* but only an inertia of masses *relatively to one another*."[8]

Einstein presented his reflections on the universe as a whole in the article "Cosmological considerations in the general theory of relativity," which was submitted to the Prussian Academy of Sciences on 8 February 1917. In this article, Einstein proposed a model that describes the geometrical and physical properties of the universe in its entirety. The solution of the relativistic field equations that had been found by Einstein represents a universe spatially closed, temporally infinite and static. His cosmological model satisfied the requirement of the relativity of inertia inspired by Mach's ideas. To Einstein, this solution therefore proved that general relativity was a consistent theory free of contradictions; however, this achievement was obtained—Einstein admitted— after traveling through a "rough and winding road."[9]

The relativity of inertia and the cosmological consequences of general relativity were debated particularly in an extensive correspondence between Einstein and de Sitter. It was in the course of this debate that each of the two scientists propounded his own cosmological solution of the field equations, which are in fact the first two relativistic models of the universe. Actually, Einstein and de Sitter first faced these issues during a visit by Einstein to the Netherlands in the autumn of 1916. The debate initially developed around Einstein's attempts of assigning boundary conditions at infinity that complied with the principle of the relativity of inertia. This assumption required that the inertia of a test particle at very large distances from all other masses should fall to zero, and it corresponded to the appropriate values of the $g_{\mu\nu}$ at infinity being invariant for all transformations of coordinates. In this way, the whole of the $g_{\mu\nu}$ could be consistently considered to be of material origin. Einstein postulated for the metric field at infinity a set of degenerate values that was the same for all systems of coordinates, its components being either 0 or ∞. On the other hand, the metric field of stellar regions at large but finite distances could be well approximated by using the Minkowski metric values of flat space-time. Einstein thus postulated the existence of some distant masses to explain the variation of the gravitational potential from the degenerate values to the Minkowski's. Beyond such distant masses there would be just the field of the $g_{\mu\nu}$ degenerating at infinity to the values proposed by Einstein. De Sitter objected that these unknown masses formed a kind of material envelope of the physical world that could never be observed. This unobservable envelope could actually be viewed as the absolute space of Newton's theory under another name. His objections persuaded Einstein to abandon the idea.

Einstein was not successful in assigning suitable boundary conditions at infinity. He nevertheless found an ingenious solution by eliminating spatial infinity. Einstein assumed the spatial closure of the universe, and he considered in his model a density of matter ρ that was uniformly and homogeneously distributed through space. This hypothetical and extremely small (but non-zero) density of matter represented the mean density of stars in the universe. In his 1917 cosmological article, Einstein explained that the introduction of a constant density of matter was also a viable solution to the cosmological problems of Newtonian theory. By an appropriate modification of the Poisson equation, which is the equation that makes it possible to determine the gravitational potential from the mass distribution, the introduction of a constant density of matter solved the problem of the gravitational instability of an infinite Newtonian universe. A uniform density of matter ensured that the gravitational potential remained constant everywhere, and that there was no center in the universe with respect to the gravitational field. As described in Chapter 1 of this volume, a solution to the problem of the gravitational instability of the Newtonian universe had been proposed by Hugo von Seeliger at the end of the nineteenth century. Seeliger had introduced an exponential factor in Newton's law of force that was meant to counterbalance the attraction of gravitation at large scales. At the time of writing his 1917 cosmological considerations, Einstein was unaware of Seeliger's proposal, with which he became acquainted later, and he considered in fact a similar modification for the Newtonian universe.

A further assumption made by Einstein in constructing his relativistic model was the stasis of the universe: "The most important fact that we draw from experience as to the distribution of matter," Einstein pointed out, "is that the relative velocities of the stars are very small as compared with the velocity of light." It was therefore possible to consider a reference frame in which the stars were on average at rest. The introduction of the mean density of stars also avoided the difficulty of taking into account local non-homogeneous distributions of matter like stars and planets. The procedure emulated the approach of the geodesists, "who, by means of an ellipsoid, approximate to the shape of the earth's surface, which on a small scale is extremely complicated."[10] A spatially finite, yet unbounded, universe filled with a uniform distribution of matter complied with the requirement that inertia be totally reduced to interaction between material bodies, and it avoided the problem of finding values of the $g_{\mu\nu}$ at infinity that were invariant for all transformations. In a letter to his friend Michele Besso, Einstein explained his solution as follows:

> I do not seriously consider believing that the universe is statistically and mechanically at equilibrium, even though I argue as I do. The stars would all have to conglomerate, of course (if the available volume was finite). But closer reflection reveals that statistics can be legitimately applied to the problems of importance to me. It could also be done without statistical considerations, by the way. It is certain that infinitely large differences in potential would have to yield stellar velocities of very significant magnitude that probably would have ceased long ago. Small differences in potential, in conjunction with infinite extension of the world, require emptiness in the universe at infinity (constancy of the $g_{\mu\nu}$'s at infinity for a suitable choice of coordinates), in contradiction to a sensible interpretation of relativity. Only the closure of the universe frees us from this dilemma; this also suggests itself *in that the curvature has the same sign throughout because, according to experience, the energy density does not become negative.*[11]

The assumption of a quasi-static and uniform distribution of matter resulted in a simple form of the energy-momentum tensor, on the assumption that the pressure contribution can be neglected. The uniformity of the distribution of matter ρ required the curvature of space to be constant. Einstein assumed that the spatial continuum was a three-dimensional spherical space, the curvature radius R of which was independent of the time coordinate. The volume V of Einstein's spherical universe is finite, and its value is determined by the curvature radius through $V = 2\pi R^3$. The universe that Einstein modeled is spatially uniform and isotropic at large scales, that is, it is identical everywhere and looks the same in every spatial direction. Its properties do not change with time because of the stasis assumed by Einstein on the basis of observed small stellar velocities. Einstein's model is often referred to as the "cylinder universe." When two spatial dimensions are suppressed, such a model can be represented by the surface of a cylinder, in which the coordinate of the vertical axis is the time coordinate (Fig. 3.1).

Concerning the line element of his world model, Einstein noted an intriguing consequence of the assumption of matter being uniformly and homogeneously distributed through space: "The odd thing," he wrote to Paul Ehrenfest, "is that now a quasi-absolute time and a preferred coordinate system do reappear in the end, while fully complying with all the requirements of relativity."[12] Einstein's assumption resulted in a simplified model of the universe, in which space-time is split into a three-dimensional uniform and isotropic space and a one-dimensional time. It is worth noting that in the subsequent phases of modern cosmology, Einstein's approach of making basic assumptions in the construction of a cosmological model was emulated and further developed by other scientists. Assumptions of certain symmetries have facilitated the theoretical understanding of the universe as a whole. Also, guiding principles have been used in constructing theoretical models. A remarkable example in this regard is the cosmological principle. Homogeneity and isotropy of space were explicitly advocated by the American mathematician and physicist Howard Percy Robertson as intrinsic physical properties of the universe. This was the assumption that guided Robertson in deriving in 1929 all static and non-static line elements of relativistic cosmology that

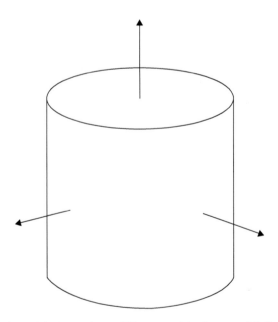

Fig. 3.1 *Einstein's static model of the universe. When two of the three spatial dimensions are suppressed, Einstein's spherical world corresponds to the surface of a cylinder. The time coordinate is along the vertical axis.*

satisfy this condition. The claim of the homogeneity and isotropy of space at large scales is known as the cosmological principle, and this principle has been crucial in shaping the conceptual and methodological framework of relativistic cosmology. Actually, the term was introduced in 1935 by the British astrophysicist Arthur Milne as one of the fundamental principles of kinematic cosmology, which is Milne's alternative theory to relativistic cosmology.[13]

Turning to Einstein's 1917 cosmological considerations, the relativistic field equations were not satisfied by the metric components and the energy-momentum tensor of the spherical model. Einstein thus had to modify the field equations. He introduced in the left-hand side of the field equations an additional term containing λ, a new constant which became later known as the cosmological constant:

$$G_{\mu\nu} - \tfrac{1}{2} g_{\mu\nu} G - \lambda g_{\mu\nu} = -\kappa T_{\mu\nu}. \tag{3.3}$$

The new term did not disturb the covariant character of the field equations and was compatible with the laws of conservation of energy and momentum. It did not alter the relativistic predictions for the solar system either, provided that λ was sufficiently small. The density of matter ρ and the spatial curvature radius R were related to λ by the relation: $\lambda = \kappa\rho/2 = 1/R^2$. In a way, these relations display in formulae the road followed by Einstein in his cosmological investigations in general relativity. The introduction of λ into the field equations, the value of which had to be different from zero, was necessary to guarantee a non-zero density of matter ρ, which determined the curvature of space. The cosmological constant can be interpreted as a repulsive force, which is able to counterbalance the effects of gravitation on large scales and maintain the stasis of the universe. Eddington demonstrated in 1930 that Einstein's spherical world is unstable. Any slight disturbance in the density of matter, whether a decrease or an increase, would cause the Einstein universe to depart from the equilibrium and to expand or contract, respectively. Einstein dismissed the cosmological constant in 1931, in the light of the observation of systematic extragalactic recession as the empirical evidence which had led in 1930 to the acceptance of expanding world models.[14]

In the final part of his 1917 cosmological article, Einstein remarked that the cosmological term had the purpose of making possible a quasi-static distribution of matter, and that the modification of the field equations was not justified by the actual knowledge of gravitation. He did not touch upon estimates of R and ρ. On the one hand, the modification of the field equations represented to Einstein a complication of his theory. He later commented that the introduction of the cosmological term was "gravely detrimental to the formal beauty of the theory," but it was the only viable solution he had found in 1917 to formulate a consistent view of the universe as a whole based on general relativity.[15] On the other hand, Einstein estimated in 1917 a value of the world radius of about $R = 10^7$ light-years $\approx 6 \times 10^{11}$ astronomical units or AU. He obtained this result by considering the value of the density of matter $\rho = 10^{-22}$ g cm^{-3}, as calculated by astronomers from star counts. The estimate of the world radius was reported by Einstein in letters written to Besso, de Sitter, Ehrenfest, and Erwin Freundlich.[16] This value was for Einstein uncomfortably large compared to the farthest distances of stars obtained from astronomical observations, which were approximately 10^4 light-years. "One day," Einstein commented to de Sitter, "our actual knowledge of the composition of the fixed-star sky, the apparent motions of fixed stars, and the position of spectral lines as a function of distance, will probably have come far enough for us to be able to decide empirically the question of whether or not λ vanishes. Conviction is a good mainspring, but a bad judge!"[17] In any case, Einstein felt he had reached the end of the road. He explained to de Sitter:

> From the standpoint of astronomy, of course, I have erected but a lofty castle in the air. For me, though, it was a burning question whether the relativity concept can be followed through to the finish or whether it leads to contradictions. I am satisfied now that I was able to think the idea through to completion without encountering contradictions.[18]

Einstein's satisfaction was however short-lived, in the light of the cosmological solution proposed shortly after by de Sitter. De Sitter formulated his solution upon assumptions substantially different from those of Einstein, and his solution corresponded to a universe devoid of matter.

3.3   The empty universe of de Sitter

De Sitter considered Einstein's theory of gravitation "an enormous progress over the physics of yesterday" towards the understanding of the laws of nature.[19] He appreciated the full explanation that general relativity provided of the universal nature of gravitation, as well as the tight connections the theory provided between scientific domains until then considered almost independent of each other. De Sitter especially appreciated the explanation given by general relativity of the anomalous precession of Mercury's perihelion, because this explanation had been obtained without the introduction of any new constant or hypothesis.

The approach of de Sitter was strongly empirical and critical of extrapolations. The fact that de Sitter formulated a cosmological model may therefore appear in a way contradictory, because of the extrapolations involved in modeling the universe in its entirety. Moreover, the solution of the modified field equations proposed by de Sitter in 1917 corresponds to an empty universe, something that seems physically untenable. Actually, de Sitter approached the study of the universe as a whole and the issue of spatial infinity from a different perspective than Einstein. As he would later explain in a work entitled "The astronomical aspect of the theory of relativity:"

> Infinity is not a physical but a mathematical concept, introduced to make our equations more symmetrical and elegant. From the physical point of view everything that is outside our neighborhood is pure extrapolation, and we are entirely free to make this extrapolation as we please to suit our philosophical or aesthetical predilections—or prejudices. It is true that some of these prejudices are so deeply rooted that we can hardly avoid believing them to be above any possible suspicion of doubt, but this belief is not founded on any physical basis.[20]

To de Sitter, the postulate of Machian inspiration that inertia should be fully determined by matter had no physical foundation, but was essentially metaphysical. The cosmological model constructed by Einstein on the basis of this postulate was objectionable for several reasons. A first issue of criticism was Einstein's assumption of a quasi-static universe, which Einstein had extrapolated only from a "snapshot" of the world that should not be considered definitive.[21] Also, the hypothesis of a uniformly distributed density of matter, which de Sitter called "world matter," was objectionable because it referred to unobserved regions of the universe. As in the case of his criticism of the hypothesis of distant masses that Einstein had initially made and later abandoned, world matter for de Sitter took the place of Newtonian absolute space. Furthermore, in Einstein's universe the time coordinate had a different role than the spatial coordinates. This almost privileged position was objectionable as well, since it elevated the time coordinate to something like universal time.

A few weeks after the publication of Einstein's article "Cosmological considerations in the general theory of relativity," de Sitter found that a coherent solution of the modified field equations could be obtained by assuming a null contribution of world matter:

$$G_{\mu\nu} - \lambda g_{\mu\nu} = 0. \tag{3.4}$$

Elaborating a suggestion by Ehrenfest, de Sitter extended Einstein's hypothesis of finite three-dimensional space to four-dimensional space-time. That is, de Sitter considered the entire space-time as a four-dimensional spherical manifold. This space-time could also be represented as a four-dimensional hyperboloid.[22] The boundary conditions at infinity represented to de Sitter a mathematical problem, which he addressed without the physical import attached to it by Einstein. In the solution found by de Sitter, all $g_{\mu\nu}$ vanished at infinity, thus they were invariant for all transformations. The vanishing of $g_{\mu\nu}$ at infinity was equivalent to the finiteness of space, the curvature of which was positive. To obtain a coherent solution of the modified field equations de Sitter retained the cosmological constant, although he considered this new constant physically undesirable. The relation between λ and the curvature radius R that resulted in de Sitter's solution was $\lambda = 3/R^2$. De Sitter called the requirement of null $g_{\mu\nu}$ at infinity the "mathematical postulate of relativity of inertia." As he wrote to Einstein about his own cosmological solution: "I do not know if it can be said that 'inertia is explained' in this way. I do not concern myself with explanations. If a single test particle existed in the world, that is, there were *no* sun and stars, etc., it would have inertia."[23]

De Sitter further elaborated his solution and worked out its line element to be comparable to that of Einstein's spherical universe. Through new coordinates introduced by de Sitter, the metric of the two cosmological solutions is the same except for the temporal component. De Sitter labeled Einstein's model "A" and his own model "B":

$$ds_A^2 = -dr^2 - R^2 \sin^2(r/R)\left(d\psi^2 + \sin^2\psi\, d\theta^2\right) + c^2 dt^2, \tag{3.5}$$

$$ds_B^2 = -dr^2 - R^2 \sin^2(r/R)\left(d\psi^2 + \sin^2\psi\, d\theta^2\right) + \cos^2(r/R)\, c^2 dt^2. \tag{3.6}$$

De Sitter also added a model "C", which corresponded to the cosmological solution of the field equations without the cosmological term, that is, the flat Minkowski space-time of special relativity. The line element of system B in the static coordinates introduced by de Sitter became known as the static frame of the de Sitter universe. De Sitter chose to use elliptical geometry to describe his solution. This choice did indeed avoid the antipodal points of spherical space. The volume of the elliptical universe (πR^3) is half the volume of the spherical universe. Einstein admitted that the possibility of considering elliptical geometry had escaped him in constructing his model, and he conceded that this geometry appeared more likely to him.

In comparing the properties of solution A to solution B, de Sitter especially analyzed their astronomical consequences. To be sure, the lack of matter in solution B did not prevent his studying theoretical predictions of possible cosmological relevance, through the working hypothesis of investigating what happened when "we can take an empty universe and put in the galactic systems later."[24] De Sitter described cosmological solutions A and B in articles that appeared in the *Proceedings of the Royal Netherlands Academy of Arts and Sciences* in 1917. He also described the two models in the last article of a series of three detailed papers devoted to Einstein's theory of gravitation and its astronomical consequences, which were published in the *Monthly Notices of the Royal Astronomical Society* between 1916 and 1917. This trio of reports introduced general relativity to the English-speaking world, as direct communication between the British and German communities was not possible because of the First World War. Eddington, at the time Secretary of the Royal Astronomical Society, arranged for an exposition by de Sitter of Einstein's theory of gravity: "Hitherto," Eddington wrote to de Sitter in June 1916, "I had only heard vague rumours of Einstein's new work. I do not think anyone in England knows the details of his paper."[25]

To Einstein, de Sitter's universe could not correspond to any physical possibility: "In my opinion, it would be unsatisfactory if a world without matter were possible. Rather, the $g^{\mu\nu}$-field should *be fully determined by matter and not be able to exist without the matter*."[26] The debate between Einstein and de Sitter continued with Einstein's attempts to rule out de Sitter's counterexample of his own interpretation of the relativity of inertia.[27] This debate was a goldmine of argumentations, which inspired further investigations of the empty de Sitter universe (Fig. 3.2). Different representations of this universe were proposed in the following years, and different interpretations of it were discussed as well. In retrospect, such a variety of interpretations reflects the complexity and the related difficulties that scientists faced in disentangling inherent properties of the de Sitter space-time from the properties due to particular coordinate representations. Matters of discussion such as the existence of singularities and horizons in de Sitter's universe are summarized later, whereas the variety of interpretations of spectral shifts in this universe is described in Section 3.4.

Einstein raised several objections to de Sitter's cosmological solution, and he finally focused on a particular behavior of the metric of solution B that revealed the existence of a singularity.[28] In the static form of the line element of de Sitter's universe, the time component g_{44} varies according to position, and it vanishes at $r = \frac{\pi}{2}R$. Einstein argued that such a surface was a singularity the existence of which was independent of the

Fig. 3.2 *A number of the protagonists of the early phase of relativistic cosmology (Leiden, 26 September 1923). Standing from left to right are Albert Einstein, Paul Ehrenfest, and Willem de Sitter. In the first row are Arthur Eddington (left) and Hendrik Lorentz (right). Courtesy of the Erfgoed Leiden en Omstreken (Groepsfoto van hoogleraren in de werkkamer van Professor W.A. de Sitter).*

choice of coordinates, and that it revealed the presence of matter in de Sitter's space-time. He found support from similar results obtained by Hermann Weyl. Einstein and Weyl agreed that in de Sitter's universe the "equator" at $r = \frac{\pi}{2}R$ corresponded to a "mass-horizon," that is, a surface where matter tended to aggregate. In 1918 Einstein published a critical comment on the universe of de Sitter, in which he argued that no space-time continuum could exist without matter, and that the de Sitter solution was neither free of singularities nor free of matter: "it seems that no choice of coordinates can remove this discontinuity... [W]e have to assume that the De Sitter solution has a genuine singularity on the surface $r = \frac{\pi}{2}R$ in the finite domain... The De Sitter system does not look at all like a world free of matter, but rather like a world whose matter is concentrated entirely on the surface $r = \frac{\pi}{2}R$."[29]

The interpretation advocated by de Sitter was different. De Sitter argued that the equator was placed at a finite distance in space, but that this surface was physically inaccessible. Time needed by a light ray or a material particle to travel to that surface was infinite, therefore the equator could not affect physical experiments: "All these results sound very strange and paradoxical. They are, of course, all due to the fact that g_{44} becomes zero for $r = \frac{1}{2}\pi R$. We can say that on the polar line the four-dimensional time-space is reduced to the three-dimensional space: *there is no time*, and consequently no motion."[30] De Sitter's argumentation contains the first hint of the existence of horizons in cosmological models, that is, surfaces that divide observable regions of the universe from regions that are not observable. Eddington too took the singularity and the horizon in solution B into account. These issues were mentioned in the analysis of the curvature of space-time presented in Eddington's influential book *The Mathematical Theory of Relativity*, first published in 1923. About the universe of de Sitter, Eddington wrote: "A singularity of ds^2 does not necessarily indicate material particles, for we can introduce or remove such singularities by making transformations of coordinates. It is impossible to know whether to blame the world-structure or the inappropriateness of the coordinate-system...I believe then that the mass-horizon is merely an illusion of the observer at the origin, and that it continually recedes as we move towards it."[31]

The question of the existence of the singularity in solution B was addressed and solved by the German mathematician Felix Klein. In letters to Einstein, Klein demonstrated that the singularity in de Sitter's solution was an artefact due to the choice of coordinates. It could "simply be transformed away" by an appropriate coordinate change in the hyperboloid representation of de Sitter's space-time, which showed that the hyperboloid was regular.[32] Conversely, the static coordinates used by de Sitter cover only a portion of the hyperboloid, and the time slices intersect each other at the equator, that is the surface interpreted by Einstein and Weyl as a mass-horizon (see Fig. 3.3). In the light of Klein's

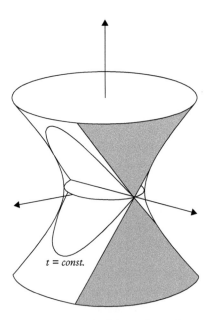

Fig. 3.3 *Hyperboloid representation of de Sitter's universe in static coordinates. When only one spatial coordinate and the time coordinate are used, such a universe corresponds to the surface of the hyperboloid. Static coordinates cover only part of the hyperboloid, in which the world lines at constant time intersect each other at the equator.*

explication, Einstein accepted that the singularity was only apparent and that de Sitter's solution was free of matter. Einstein nevertheless maintained that this cosmological solution was not physically admissible because it was non-static, in the sense that "in this world, time t cannot be defined in such a way that the three-dimensional slices $t = $ const. [constant] do not intersect one another and so that these slices are equal to one another (metrically)."[33] However, it was clear that the field equations with the cosmological term admitted a coherent solution that corresponded to an empty universe. It can be said that the acceptance by Einstein of de Sitter's solution as a counterexample of the relativity of inertia marked the beginning of Einstein's gradual abandonment of Mach's principle. In the early fifties, Einstein would write: "In my view one should no longer speak of Mach's principle at all. It dates back to the time in which one thought that the 'ponderable bodies' are the only physically real entities and that all elements of the theory which are not completely determined by them should be avoided."[34]

Regarding the de Sitter universe, the singularity of the metric in static coordinates was still the source of different interpretations proposed during the twenties. De Sitter related the particular behavior of the time component of the metric of solution B to spectral shifts. His work in this direction inspired a number of studies on redshift-distance relations predicted in the de Sitter universe that were related to radial velocities measured in spiral nebulae.

3.4 Motion without matter: redshift in the universe of de Sitter

As from 1917, the study of the universe in its entirety could be addressed through the geometrical and physical theory of general relativity. In analyzing the properties of the universe of Einstein (solution A) and those of his own cosmological solution (solution B), de Sitter inaugurated the confrontation between theoretical predictions of relativistic world models and observational data. The path opened by Einstein and de Sitter was initially followed only by a small group of scientists, comprising astronomers, theoretical physicists and mathematicians. The extrapolations to the universe as a whole proposed by Einstein and de Sitter appeared too extreme to many, especially because observations of cosmological relevance were scarce at the time. As the American mathematician Archibald Henderson wrote in 1925, "[m]uch additional data will be required and many further researches made before it will be possible categorically to decide between the infinite, limitless, Euclidean universe of Newton and the finite, unbounded, non-Euclidean universe of Einstein or of de Sitter."[35] The identification of the relationships between the formal structure of general relativity and the physical phenomena that it describes was no easy task, and the cosmological consequences of the theory were no exception in this regard. In the early twenties there was no disciplinary framework that might be called "relativistic cosmology," and the scientists who initially investigated the universe through the lens of general relativity approached the cosmological problem with different trainings and methods.

This section describes reactions to de Sitter's cosmological solution that appeared before the general acceptance of the expanding universe in 1930. The empty universe of de Sitter, considered as a limiting case of a universe with an infinitely low density of matter, seemed preferable to the truly static model of Einstein because it seemed to offer explanations of the large redshifts measured in spiral nebulae: de Sitter's cosmological solution did indeed predict motions of particles in the universe. Spiral nebulae were widely accepted as extragalactic systems, that is galaxies, from the mid-twenties on, after Hubble's determination of the distance of the Andromeda nebula (now the Andromeda Galaxy) which demonstrated that this nebula was located far beyond the known limits of our galaxy, the Milky Way.[36]

Relativistic predictions of spectral shifts and astronomical observations of stars and nebulae were first connected in a cosmological framework by de Sitter himself. In comparing, in 1917, the astronomical consequences of his solution and that of Einstein, de Sitter considered different methods of estimating the curvature radius.[37] In particular, the curvature radius of de Sitter's universe (R_B) could be estimated through measurements of redshifts in stars and spirals.

In the metric of solution B in static coordinates, that is, the line element proposed by de Sitter, the time component g_{44} diminishes with increasing distance, and clocks would appear to run more slowly with increasing r. "The lines in the spectra of very distant stars or nebulae," de Sitter argued in 1917, "must therefore be systematically displaced towards the red, giving rise to a spurious positive radial velocity."[38] De Sitter discussed with the Dutch astronomer Jacobus Kapteyn the reliability of observational evidence of systematic stellar recession. Kapteyn commented: "These are hard nuts you are giving me to crack. I will answer to the best of my knowledge, without pretending this 'best' is good enough."[39] Following Kapteyn's suggestion, de Sitter referred to the spectroscopic analysis of B-class stars carried out by the American astronomer William Campbell, which revealed a systematic displacement towards the red. Computing this spectral shift in terms of velocity, the result was a systematic positive velocity of B stars of about 4.5 km s^{-1}. One of the theoretical predictions of Einstein's general relativity theory concerns the gravitational redshift: spectral lines originating in a strong gravitational field would appear displaced to the red to an observer placed in a weaker field. De Sitter calculated that the gravitational redshift produced by B stars would roughly correspond to 1.4 km s^{-1}. He then considered that the systematic redshift due to the variation of g_{44} in solution B was superposed on the spectral shift produced by the gravitational field of stars themselves. For B stars, this systematic redshift thus corresponded to a velocity of about 3 km s^{-1}. Assuming an average distance of B stars of $r = 3 \times 10^7$ AU, the curvature radius of solution B could be obtained through g_{44} and resulted in $R_B = \frac{2}{3} \times 10^{10}$ AU.

In the static frame of de Sitter's universe, the world lines of freely falling particles are not geodesics. Actually, the only curve that is a geodesic is the curve that passes through the origin of coordinates. De Sitter calculated the general form of the equations of motion of freely falling particles, and he noted that for objects moving relatively close to the origin one should "expect radial and transversal velocities of the same order, but for objects at very large distances we should expect a greater number of large or very large radial velocities."[40] He also noted that these velocities could equally likely be positive or negative. Superposed on these velocities were the spurious positive velocities due to time dilatation. Astronomical observations had revealed that a number of nebulae had large velocities compared to the usual stellar velocities. De Sitter took those nebulae into account the velocities of which had been determined by more than one observer. These radial velocities were usually determined through the classic Doppler formula that relates observed spectral shifts z to velocities v. A blueshift indicates an approaching motion (negative velocity), while a redshift indicates a receding motion (positive velocity): $z \equiv (\lambda_0 - \lambda_e)/\lambda_e = v/c$. In this formula λ_0 and λ_e denote the wavelengths measured by the observer and the wavelength emitted by the observed object, respectively; the velocity v is small compared to the speed of light c. The Andromeda Nebula showed a large negative velocity, while the nebulae NGC 1068 and NGC 4594 had very large positive velocities. It is reasonable to assume that, at the time, de Sitter was unaware of further spectroscopic measurements of spiral nebulae determined by Vesto Slipher because of communication difficulties during the First World War. De Sitter computed the mean velocity of the three nebulae and assumed an average distance of them of about $r = 2 \times 10^{10}$ AU. Through g_{44} a value of R_B was obtained of 3×10^{11} AU.

Regarding these results, de Sitter remarked that "[a]ll this of course is very vague and hypothetical. Observation only gives us certainty about the existence of our own galactic system, and probability about some hundreds more. All beyond this is extrapolation."[41] Nevertheless, if future observations confirmed systematic positive radial velocities, solution B would turn out to be a better approximation of the actual universe than solution A. Radial velocities of twenty-five spiral nebulae were made available in 1920, and positive velocities were preponderant. To de Sitter, however, the choice between solutions A and B was "purely a matter of taste. There is no physical criterion as yet available to decide between them…The decision between these two systems must, I fear, for a long time be left to personal predilection."[42] De Sitter, who was the director of the Leiden Observatory from 1919 until his death in 1934, did not deal with cosmological issues for nearly ten years. He would resume his work on solution B in the late twenties and play a pivotal role in the recognition of the expanding universe.

Redshifts in the de Sitter universe were analyzed by Eddington in 1923.[43] Eddington had previously dealt with the cosmological solutions of Einstein and de Sitter in several publications. In a report on the theory of relativity published in 1918, Eddington argued that the existence of antipodal points in spherical space made Einstein's model a "rather picturesque theory of anti-suns and anti-stars. It suggests that only a certain proportion of the visible stars are material bodies; the remainder are ghosts of stars, haunting the places where stars used to be in a far-off past."[44] The cosmological constant introduced by Einstein was also objectionable because it seemed to be a very artificial adjustment; however, in subsequent years the cosmological constant gained great importance in Eddington's search for a fundamental theory that could unify quantum theory, relativity and cosmology through the constants of nature. Eddington was not particularly attracted by de Sitter's solution either. As he reported in his book *Space, Time and Gravitation*, first published in 1920, "[i]t might seem that this kind of fantastic world building [de Sitter's solution] can have little to do with practical problems."[45]

In his 1923 book *The Mathematical Theory of Relativity*, Eddington presented the world systems of Einstein and de Sitter "as two limiting cases, the circumstances of the actual world being intermediate between them. De Sitter's empty world is obviously intended only as a limiting case; and the presence of stars and nebulae must modify it, if only slightly, in the direction of Einstein's solution."[46] Actually, the possibility of distinguishing the cosmological solution that better approximated to the actual universe was strongly influenced by the preponderance of positive velocities of spirals, which favored de Sitter's solution. Eddington worked out the line element of system B, in which he used a different coordinate – here denoted as r_e – rather than de Sitter's r. They are related by $r_e = R\sin(r/R)$. From the analysis of the motion of freely falling particles, Eddington found that a particle initially at rest would be repeled from the origin with an acceleration increasing with the distance: $dr_e^2/ds^2 = \frac{1}{3}\lambda r_e$.

The result of Eddington's analysis was that the de Sitter metric predicted a general tendency of particles to scatter as soon as matter was inserted into this system. Eddington pointed out that solution B actually offered a double explanation of the receding motion observed in spiral nebulae. Besides the general scattering that resulted in Doppler motions, there was also the systematic redshift due to time dilatation, which corresponded to a spurious velocity of recession and would be erroneously interpreted as an actual motion of recession. This resulted in an approximately quadratic dependence between redshift and distance. The interpretation given by Eddington in connecting these two effects, which may have appeared to his contemporaries no less puzzling than the predicted redshifts themselves, was in terms of "*an anticipation* of the motion of recession which will have been attained before we receive the light."[47] That is, Eddington proposed that nebulae did not have motion when they emitted light, but they acquired motion during the time light was traveling to the observer. Regarding astronomical observations, Eddington reported in his book a table prepared by Slipher containing radial velocity measurements of forty-one spiral nebulae. These measurements showed a great preponderance of positive velocities, but the corresponding receding motion could not be considered systematic because a number of spirals had approaching motions (negative velocities). Therefore, to quote Eddington, "the cosmogonical difficulty is perhaps not entirely removed by de Sitter's theory."[48]

Different representations of the de Sitter universe were proposed during the twenties. A number of scientists formulated non-static versions of its line element, in which the spatial sections depended on time. Accordingly, they proposed different theoretical redshift-distance relations, and they connected these relations to radial velocities measured in spiral nebulae. The comment by the Hungarian physicist and mathematician Kornel Lanczos exemplifies the varied aspects of de Sitter's universe: "It is interesting to observe how one and the same geometry can lead to totally different interpretations, depending on the choice of the coordinate system, and how the individual coordinates are interpreted."[49] In publications of the time, the term "de Sitter effect" was often used to denote the spurious positive velocity predicted in the static frame of de Sitter's universe, although in a number of cases this term was used with reference to the predicted Doppler motion too.[50]

The variety of representations of the de Sitter universe, as well as of the redshift-distance relations, can be regarded in retrospect as a kind of laboratory for shaping the theoretical basis of relativistic cosmology in which the researchers worked under dim light.[51] They dealt with novel issues compared to the usual questions associated with Newtonian cosmology, issues the complexity of which made it difficult to find clear interpretative frames. Some of these issues are the interpretation of redshift through relativistic world models, the differentiation between static and non-static cosmological frames of general relativity, and the identification of invariant properties of models against the features due to particular coordinate representations.[52]

The de Sitter universe in itself is not a fully-fledged cosmological model. Weyl recognized in this regard the necessity of assuming a time-like geodesic flow—a congruence of time-like geodesic curves—to represent the motion of stars and nebulae and to unambiguously derive a redshift relation. His work was rooted in his search for global causality, and it resulted in the postulate formulated in 1923 that became known as Weyl's principle. Such a postulate later became a basic assumption for specifying cosmological models.[53]

Weyl shifted his support from Einstein's cosmology to de Sitter's: "The principle that matter produces the metric field cannot be maintained in the sense that far from any matter, or if matter has been annihilated, there will be no guiding field, that is the field will be undetermined...The status of rest of the metric field is the homogeneous metric, as it reigns on de Sitter's hyperboloid."[54] In the hyperboloid representation of de Sitter's universe he considered the geodesic lines cut out by planes through a specified surface line (generatrix) of the asymptotic cone to the hyperboloid. These geodesics are oriented towards the future temporal direction and cover half the hyperboloid. They do not overlap with each other, and they diverge towards the infinite future and converge in the infinite past. This system of geodesics is causally connected through the common origin in the infinite past. To Weyl, this system formed the "'undisturbed state' of stars, if anything in the theoretical line regarding the displacement to red is to be formulated."[55] Each point of space-time can be uniquely associated to one geodesic (except at the origin), and the observers which are assigned to each geodesic are called the *fundamental observers*. *Cosmic time* is defined as the proper time measured by any fundamental observer, given the possibility of synchronizing the standard clocks of the fundamental observers at the time when the geodetic world lines meet at one point in the common past. In 1923 Weyl derived a relation of general validity between the frequency v_s and the proper time s of the light source, and the frequency v_0 and the proper time σ of the observer: $v_0/v_s = d\sigma/ds$. In terms of the redshift z, the general relation proposed by Weyl was: $z = d\sigma/ds - 1$.

Concerning the interpretation of the redshift, in Weyl's view the gravitational shift and the Doppler shift were "inseparably connected."[56] He formulated a relation between redshift and distance in the form: $z = \operatorname{tg} r/R$. In this relation, r is the proper distance between the light source (for example, a star or a galaxy) and the observer, and R is the curvature radius of de Sitter's universe. Weyl used this redshift relation and the observational data of spiral nebulae to calculate R. He estimated the curvature radius to be $R_B = 10^9$ light-years (6×10^{13} AU), a value that he reported in the appendix of the fifth edition of his book *Raum, Zeit, Materie*. Weyl confirmed in 1930 that his analysis of de Sitter's cosmology corresponded to the analysis of the non-static frame of the de Sitter universe carried out by Robertson in 1928. In turn, Robertson had in 1928 independently found the same result obtained by Lemaître three years before.

The first contribution that Lemaître devoted to relativistic cosmology appeared in 1925 and was an analysis of de Sitter's universe. To Lemaître, the metric of this universe in static coordinates was objectionable because the symmetry between space and time was broken in it, and the natural homogeneity of space-time was not satisfied. Such a metric had a singularity at the "horizon" that was not simply a coordinate artefact. It attributed special properties to the point at $r = \frac{\pi}{2} R$, because world lines were not geodesics except for the one that passed through this "center." "It is clear that such an introduction of an apparent center in a universe which, by definition, has none is objectionable for a study of the properties of this universe."[57] Lemaître introduced new coordinates that removed this inhomogeneity, and he formulated a non-static version of the line element in which space and time were separated without spoiling the homogeneity of space-time. The spatial sections in this form have null curvature and they depend on time through an exponential factor. By choosing geodesics converging towards the future and the light source at the origin of coordinates, Lemaître obtained for the Doppler effect the relation $z = -\sin(r/R)$. The same relation with the opposite sign could be obtained by reversing the temporal direction. The new coordinates demonstrated the non-static character of de Sitter's universe, which offered a possible interpretation of the motion of spiral nebulae. However, the infiniteness of space due to the null spatial curvature was to Lemaître "a very unsatisfactory feature of any conception of the whole universe," which for him was a reason to discard de Sitter's world.[58]

Independently of Lemaître, Robertson in 1928 also formulated the line element of the de Sitter universe in the exponential form. Contrary to Lemaître, the null curvature of spatial sections was regarded by Robertson as an advantage of simplicity of this version of de Sitter's solution. This form of the line element eliminated the paradoxes of the mass-horizon and the arrest of time at the horizon. As described in Section 3.3, these aspects had been previously discussed by Einstein, de Sitter, Weyl and Klein in the case of the static frame of de Sitter's universe. Robertson pointed out that in his version of the line element "space-time is isotropic in time in the sense that at any time the line element has the same form as at any other, provided the spacial [sic] scale is appropriately chosen."[59] In this sense, the universe of de Sitter was stationary. However, the explicit dependence on time of the spatial line element made natural processes non-reversible. Although spatial sections were unlimited, this was not the case for the observable region of the universe. The existence of a spherical region centered on the observer and determined by the value of R_B did in fact delimit the region in which events could be observed. This kind of cosmological horizon was later called an *event horizon*. The line element of de Sitter's universe in the exponential form proposed by Robertson was:

$$ds^2 = -e^{2\kappa ct}\left(dr^2 + r^2\,d\theta^2 + r^2\sin^2\theta\,d\phi^2\right) + c^2\,dt^2. \tag{3.7}$$

In this formula $\kappa = \sqrt{\lambda/3}$. Robertson took into account the contribution of several sources of redshift in elaborating a relation between velocity and distance. First Robertson assumed that the contribution of the gravitational shift could be neglected. He then analyzed the velocity–distance effect in the case of the observer and the light source both being stationary, which he regarded as a "distance effect." Finally Robertson obtained a formula for the general case of the source having a non-vanishing proper (coordinate) velocity. Through the additional assumption that there was no systematic correlation of coordinate velocity with distance from the origin, the Doppler effect would indicate a residual positive radial velocity of distant objects. However, the Doppler effect could not be used to distinguish between the proper motion of the source and redshift due to the distance effect. Robertson concluded that an approximately linear correlation between the radial velocity v and the distance l could nevertheless be expected for small distances, according to the formula $v \simeq c(l/R)$ in which $R = R_B$. By using the data furnished by Hubble in 1926 for distances, and Slipher's measurements for radial velocities published in Eddington's 1923 book, Robertson estimated the curvature radius to be $R_B = 2 \times 10^{27}$ cm (1.3×10^{14} AU).

After the recognition of the expanding universe, it became clear that the non-static frame analyzed by Lemaître and Robertson corresponded in fact to de Sitter's exponentially expanding model, which was later used in the framework of steady state cosmology and has been introduced in inflationary cosmology as well. The diversity of redshift-distance relations they obtained is due to the different choice of geodesics in de Sitter's space-time (see Fig. 3.4).

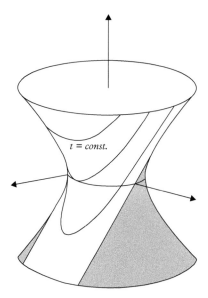

$t = const.$

Fig. 3.4 *The exponential form of de Sitter's universe, which was proposed by Georges Lemaître in 1925 and independently by Howard Robertson in 1928. Spatial sections at constant time have flat geometry (null curvature).*

A different representation of the de Sitter universe was proposed by Lanczos in 1922. It corresponds to a non-static universe in which the spatial sections are closed (the spatial curvature is positive) and proportional to the factor $\cosh^2 t$. As Lanczos emphasized, the mass-horizon in solution B was an artefact due to the static coordinates. The coordinates adopted by Lanczos cover the whole hyperboloid, which in this representation is fully regular (see Fig. 3.5). According to Lanczos, the static frame of de Sitter's space-time was not stationary. Lanczos defined a cosmological solution as stationary "if the coefficients of its metric are independent of time in a coordinate system in which the masses are at rest on average…A necessary and sufficient condition for this is that the time lines of our coordinate systems are geodesics. Therefore the static solution given by de Sitter is not an example of a stationary world."[60]

In an article published in 1923, Lanczos acknowledged Weyl's analysis devoted to the invariant redshift relation. He added "that my investigation has not become superfluous even in its special part, because our assumptions on the distribution of the material world lines are not congruent."[61] Lanczos did indeed consider different geodesics and derived a redshift–distance relation with a time-dependent additional term, according to the formula: $z + 1 = \cos a/R - \sin a/R \sinh \tau_0$. In this relation, a is the geodesic distance between the observer and the light source, R is the curvature radius of the de Sitter universe and τ_0 the time of light reception. In Lanczos' analysis, any redshift had to be uniquely interpreted as due to the Doppler effect. However, he did not relate his result to astronomical observations.

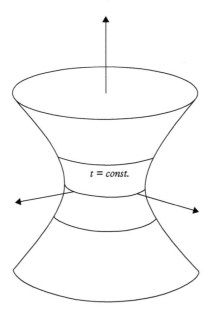

t = const.

Fig. 3.5 *The universe of de Sitter in the version proposed by Kornel Lanczos in 1922. Spatial sections at constant time are closed (positive curvature). The coordinates cover the whole hyperboloid.*

In 1924, the German astronomer Carl Wirtz related de Sitter's cosmology to observations of spiral nebulae. Distance measurements of these objects were very scarce; therefore Wirtz considered their apparent diameter and found a linear relation between the velocities and the logarithm of the apparent diameters. This relation made it possible to conclude that radial velocities increased with increasing distances, as predicted by de Sitter's cosmology.[62] On the other hand, the general recession expected in de Sitter's universe was strongly criticized by the Polish-American physicist Ludwik Silberstein.

Because of the skeptical approach that he maintained to many aspects of general relativity, Silberstein can be viewed as one of the main critics of Einstein's theory of gravitation during its early phase.[63] As from 1924, Silberstein produced several articles devoted to the determination of the curvature radius of de Sitter's space-time. This cosmological solution represented to him an attractive piece of reasoning, more so than Einstein's universe. Silberstein analyzed the spectral shift effects in the static frame of solution B, in which he assumed that the gravitational shift due to time dilatation was "insolubly amalgamated with the velocity or usual Doppler effect."[64] The complete effect merging the two effects was appropriate for investigating freely falling particles that could not be at rest as their world lines were not geodesics. To Silberstein, it was Weyl's merit to have introduced a theoretical relation of general validity between spectral shift and proper time. However, Silberstein objected that "Dr. Weyl introduced the perfectly gratuitous assumption, more or less disguised as a necessary feature of de Sitter's world, according to which the world lines of all the stars belong to a 'pencil of geodesics diverging towards the future'," an assumption that to Silberstein was a "sublime guess…entirely undesirable."[65]

Silberstein also criticized Eddington's prediction of universal scattering, which he depicted as a "fallacy based upon a hasty analysis."[66] The negative velocities measured in a number of nebulae, among them the Andromeda nebula, were to Silberstein the tell-tale proof contradicting general recession. This was in agreement with the globular clusters showing both approaching and receding motions; they were equally distributed through space, and the estimate of their distances could be considered more reliable than those of spirals. Therefore Silberstein investigated the astronomical possibilities offered by de Sitter's solution in relation to globular cluster observations.

Silberstein criticized Weyl's assumption of the choice of time-like geodesics, but he nevertheless used Weyl's general redshift relation ($z = d\sigma/ds - 1$) and derived from the static line element of de Sitter's universe a relation between redshift and distance for freely falling particles. Weyl later criticized this method, which he found "quite abstruse."[67] Silberstein obtained for what he called the complete Doppler effect a relation that admitted both positive velocities (receding motions) and negative velocities (approaching motions). For very distant objects, the relation had the form $z \simeq \pm r/R$. By using data of seven globular clusters, Silberstein calculated $R_B = 6 \times 10^{12}$ AU. In turn, this value of the curvature radius made it possible to estimate the distance of the spiral that had the largest velocity known at the time (NGC 584), which resulted in $r = 3.6 \times 10^{10}$ AU. He noted, "Huge as this may seem it will be remembered that Shapley's latest estimate of the semi-diameter of our galaxy is only four times smaller."[68] Following a suggestion by Henry Russell, Silberstein later found a linear relation between velocity and distance of ten objects, namely eight clusters and the two Magellanic Clouds, by plotting the modulus of the redshift: $r = |z|R$. Silberstein, however, discarded data belonging to three other globular clusters, the velocities of which seemed to him "suspiciously small."[69]

Silberstein's work on globular clusters raised several objections among theoreticians like Eddington and Weyl, and astronomers like the Swedes Knut Lundmark and Gustaf Strömberg. In particular, Lundmark objected that Silberstein "has not given, and will probably not be able to give, any justification for the use of the velocities of the globular clusters."[70] If observations clearly showed that globular clusters were closer than spiral nebulae, then the systematic redshift due to time dilatation would not affect these objects very much. No definite correlation between velocity and distance of globular clusters resulted from the analysis by Lundmark. On the contrary, Lundmark used other astronomical objects as distance indicators in order to estimate the curvature radius of de Sitter's universe, such as Cepheid variable stars, novae and eclipsing variable stars. Regarding spiral nebulae, Lundmark remarked that "[a] rather definite correlation is shown between apparent dimensions and radial velocity, in the sense that the smaller and presumably more distant spirals have the higher space-velocity."[71] The curvature of space was the subject of a number of studies proposed by Strömberg. Reliable estimates of spiral distances were lacking, so that it was reasonable to estimate the distance by assuming that spiral nebulae had the same absolute brightness. However, no definite correlation resulted between their velocity and distance. As Strömberg commented in 1925, "de Sitter's effect can be regarded as disproved by the clusters if their distances are of the same order as those of the nebulae. Silberstein's effect seems possible, but cannot be established by the data."[72]

In 1929 the American cosmologist and physical chemist Richard Chace Tolman proposed a comprehensive analysis of the motion of particles and light rays in de Sitter's universe. He considered the metric in static coordinates in the form investigated by Eddington, and he found some of the results previously published by Silberstein. Tolman took into account the shape of the particle path and the particle velocity and acceleration as well. He investigated the Doppler shift both in the case of the observer at the origin, and in the case of the source at the origin. Tolman noted that the general recession suggested by Eddington, which resulted from Eddington's assumption of particles initially at rest, referred to the specific case of the source placed at the point that corresponded to the perihelion of its curved path. That is, the Doppler velocity at perihelion was positive even though the source had no radial velocity at the time of emission. The general prediction of Tolman's analysis was that one should expect in solution B both receding motions and approaching motions, although receding motions would be preponderant.

The various interpretations associated with the universe of de Sitter and the redshift predictions resulted in a "curious complex of uncertainties."[73] They were superseded in 1930 by the interpretation of extragalactic redshifts as due to the expansion of the universe.[74] To quote Eddington, once the expansion of the universe was generally accepted, the de Sitter effect was just "regarded as an imitation recession."[75]

3.5   Friedman–Lemaître models and the expanding universe

The discovery of the expanding universe is one of the major breakthroughs in twentieth-century science, which marks the turning point from conceiving the cosmos as static and unchanging to viewing the universe as a whole as having a *history*: the universe expands, and its evolution can be investigated through physical laws.

The recognition by the scientific community at large of the expanding universe took place in early 1930, when the observational evidence of galaxy recession was properly incorporated into relativistic dynamical models.[76] The common framework in which theoretical and observational work was carried out until 1930 was that of a static or stationary universe. The interpretation in terms of a stationary frame is evident in the case of Robertson, who in 1929 formulated static and non-static line elements of relativistic cosmology that satisfied the assumption of homogeneity and isotropy of space. Robertson was aware of the work Friedman had published in 1922. He considered Friedman's dynamical models of the universe to be non-stationary, and he found the assumptions on which Friedman had based his analysis unsatisfactory.[77] As described below, both Tolman and de Sitter analysed the static form of de Sitter's universe, now in the light of the velocity-distance relation among galaxies established by Hubble in 1929.

On 17 January 1929, Hubble communicated the results of his observations of spiral nebulae in the article "A relation between distance and radial velocity among extra-galactic nebulae." The article was published in March 1929 in the *Proceedings of the National Academy of Sciences of the United States of America*. As related in Chapter 2 of this volume, Hubble showed that the distances and velocities of forty-six galaxies satisfied an approximately linear relation: $v = zc = Kd$. Hubble's analysis established that the recession of galaxies was systematic, apart from a few counterexamples. The linear relation empirically found by Hubble later became known as Hubble's law, and the parameter of proportionality K between velocity and distance, later denoted by the symbol H, is known as Hubble's constant. In his 1929 article, Hubble proposed the value $K = 500 \text{ km s}^{-1} \text{ Mpc}^{-1}$. He also briefly mentioned that this result could be related to the cosmology of de Sitter, "and hence that numerical data may be introduced into discussions of the general curvature of space."[78]

The line element of de Sitter's universe was the subject of a detailed analysis carried out by Tolman early in 1929. Tolman considered several hypotheses that could reconcile Hubble's velocity–distance relation with the motion of freely falling particles in the static form of de Sitter's universe. As had already been found by Silberstein in 1924, the particle equations of motion in the static form of solution B "are completely reversible, and there seems to be no reason why the *a priori* probability for approaching and receding nebulae should not be the same."[79] To Tolman, the most natural assumption to explain the observed systematic recession of galaxies was the "continuous entry," namely the hypothesis that nebulae continually entered and left the range of observation.

Theoretical shift–distance relations investigated by Tolman made it possible to consider Doppler receding motions as more common than approaching motions. However, a systematic recession and a linear relation could be obtained only through the imposition of several restrictions on the parameters which determined the motion of the nebulae. For this reason, Tolman concluded that "the de Sitter line element for the universe does not appear to afford a simple and unmistakably evident explanation of our present knowledge of the distribution, distances, and Doppler effects for the extra-galactic nebulae." In a further analysis of the line element of the Einstein and de Sitter solutions, Tolman remarked on the difficulty of recognizing evolutionary processes if the metric is assumed to be static. "The investigation of non-static line elements would be very interesting," Tolman suggested.[80]

De Sitter and Eddington discussed spiral nebulae at the meeting of the British Association for the Advancement of Science that took place in South Africa in July 1929. De Sitter reported his renewed interest in cosmological problems in letters to the American astronomer Frank Schlesinger. The issue faced by de Sitter is basically the same as that of Tolman, that is, how to reconcile the motion predicted by the line element in static coordinates with Hubble's systematic recession.

In de Sitter's analysis of solution B the paths of freely falling particles observed from the origin of coordinates are hyperbolas. The approximate formula for the velocity-distance relation obtained by de Sitter was: $v/c = \pm r/R + \frac{1}{2}(r/R)^2$. The first term, that is, the linear term, represented Doppler velocities and could be either positive or negative, corresponding to motions on the receding and the approaching branch of the hyperbola, respectively. The quadratic term corresponded to spurious velocities. The problem arose from the measured velocities, which were all positive. De Sitter asked: "Why are all the spirals found on the receding branches and none on the approaching? The evident solution (evident once you think of it) is: some of them (say one half) have originally been on the approaching branches, but have long since past [sic] their nearest point, and are now receding."[81] Eddington suggested in this regard that all spirals could form at small distances from the perihelion. To de Sitter, a further intriguing issue was the following:

> Now there comes a most unexpected and curious coincidence. If we suppose the whole world to be filled with spiral nebulas... the density becomes $2 \cdot 10^{-28}$ in C.G.S. units, which on first sight might be thought not to differ much from emptiness. But if we now take Einstein's own solution A of his field-equations, i.e. the solution for a full world, then there is a relation between the radius of the universe and the density, and taking for the density $2 \cdot 10^{-28}$, the radius is found to be $2 \cdot 1/4 \cdot 10^9$ light years, i.e. practically the same as for the empty world! In the solution A, however, there is no radial velocity... We are thus confronted with the mathematical problem: what becomes of the empty world B if you fill it with matter. I have not yet been able to solve this problem completely but I have reasons to expect that the solution will be intermediate between the solutions A and B.[82]

It was in the spring of 1930 that the models of an expanding universe formulated by Friedman and Lemaître became widely known within the scientific community. During the Meeting of the Royal Astronomical Society that took place in January 1930, de Sitter and Eddington commented on the possibility of considering solutions intermediary between those of Einstein and de Sitter. "The question now arises," de Sitter noted about spiral nebulae, "how can we account for the linear connection between the velocities and the distances?" Eddington then remarked about relativistic world models that "[o]ne puzzling question is why there should be only two solutions. I suppose the trouble is that people look for static solutions. Solution A is such a static solution. Solution B is, on the contrary, non-static and expanding, but as there isn't any matter in it that does not matter."[83]

When the report of the meeting was published in *The Observatory* in February 1930, Lemaître, who had been a student of Eddington at the Cambridge Astronomical Observatory in 1923, called Eddington's attention to the analysis of dynamical solutions of the field equations he had already carried out in 1927. In that year, Lemaître had investigated the properties of the expanding universe in the article "Un univers homogène de masse constante et de rayon croissant, rendant compte de la vitesse radiale des nébuleuses extra-galactiques" ("A homogeneous universe of constant mass and increasing radius, accounting for the radial velocity of extra-galactic nebulae"), which had been published in the *Annales de la Société Scientifique de Bruxelles* (Fig. 3.6). Actually, at the time Lemaître had sent a copy of his article to Eddington, who apparently had not taken notice of it. In 1930 Lemaître sent Eddington copies of his 1927 article together with a letter in which he said that he had investigated the intermediary solution de Sitter and Eddington were searching for: "I consider a universe of curvature constant in space but increasing with time. And I emphasize the existence of a solution in which the motion of the nebulae is always a receding one from time minus infinity to plus infinity."[84]

As Lemaître explained in his 1927 article, by using just two coordinates, one spatial and the other temporal, the universe of de Sitter could be represented as the surface of a sphere in which meridians corresponded to spatial lines and parallels to time lines. The equator of such a sphere corresponded to the geodesic, while the pole of the sphere represented the singularity due to the time coordinate. Space-time homogeneity was achieved by exchanging the two coordinates. Time lines now corresponded to meridians, and spatial lines to parallels: "Therefore the radius of space varies with time."[85] On the one hand, the universe of de Sitter offered a natural explanation of the observed receding velocities of spiral nebulae; on the other hand, Einstein's universe had a finite radius that could be properly related to the total mass of the universe. "In order to find a solution combining the advantages of those of Einstein and de Sitter, we are led to consider an Einstein universe where the radius of space or of the universe is allowed to vary in an arbitrary way."[86]

Lemaître analyzed the properties of a dynamical model of the universe, the material content of which he approximated by a rarefied gas uniformly and homogeneously distributed through space. Accordingly, the corresponding hypothetical average density of matter δ depended on time. While the matter contribution to pressure could be considered negligible, the radiation pressure p could not be ignored. The total energy density was thus $\rho = \delta + 3p$. The field equations reduced to two equations which described the variation with time of the world radius R including the cosmological constant λ as well as the matter and radiation contribution:

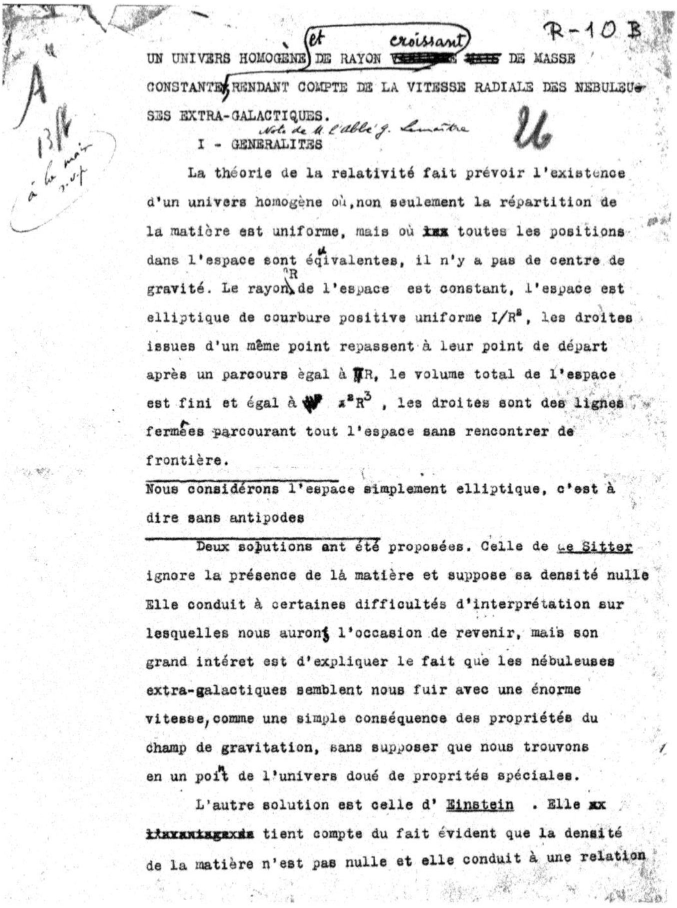

Fig. 3.6 *The first page of the draft of Lemaître's article "Un univers homogène de masse constante et de rayon croissant, rendant compte de la vitesse radiale des nébuleuses extra-galactiques" (1927). Handwritten revisions show that Lemaître initially used the expression "rayon variable," which he later substituted by "rayon croissant." Courtesy of the Archives Georges Lemaître, Université catholique de Louvain, Louvain-la-Neuve, Belgique.*

$$3\frac{\dot{R}^2}{R^2} + \frac{3}{R^2} = \lambda + \kappa\rho,$$ (3.8)

$$\frac{\dot{R}^2}{R^2} + 2\frac{\ddot{R}}{R} + \frac{1}{R^2} = \lambda - \kappa p.$$ (3.9)

In the above relations $\dot{R}$ and $\ddot{R}$ refer to the first and second time derivatives: $\dot{R} \equiv dR/dt$ and $\ddot{R} \equiv d\dot{R}/dt$ where the speed of light $c = 1$. Friedman had already formulated these equations in 1922, without however considering the contribution of the radiation pressure. The two equations have become known in modern cosmology as the Friedman–Lemaître equations. By assuming the adiabatic expansion of the universe, the second equation could be obtained from the condition of energy conservation. Given the constant mass of the universe, and by introducing the new constants α and β related to the density of matter and radiation pressure, Lemaître obtained from the first equation:

$$t = \int \frac{\mathrm{d}R}{\sqrt{\dfrac{\lambda R^2}{3} - 1 + \dfrac{\alpha}{3R} + \dfrac{\beta}{R^2}}}. \qquad (3.10)$$

The case $\alpha = \beta = 0$ corresponded to de Sitter's solution, while setting $\beta = 0$ plus the condition of the constancy of the world radius led to Einstein's model. The cosmological constant was related to R_0 through $\lambda = 1/R_0^2$. R_0 was the asymptotic value that the curvature radius R had at $t \to -\infty$, from which R increased with time. The resulting world radius was $R^3 = R_E^2 R_0$, where R_E was the radius of Einstein's universe which depended on the density of matter and increased with time.

In a section of his 1927 article devoted to the "Doppler effect due to the variation of the radius of the universe," Lemaître explained the cosmological nature of spectral shifts. Redshifts observed in spiral nebulae were not due to a relative motion between the observer and the observed object, nor to the slowing down of atomic vibrations. The large recession velocities of the extragalactic nebulae, Lemaître pointed out, were a cosmic effect due to the expansion of the universe: $\delta t_2/R_2 - \delta t_1/R_1 = 0$. In this formula, R_1 and R_2 denote the radius of the universe at the time of emission t_1 and that at the time of reception t_2, respectively, and δt_1 and δt_2 are the period of the emitted light and received light, respectively. The relation with the velocity v that resulted was $v/c = R_2/R_1 - 1$.

In 1927 Lemaître not only proposed the expanding model of the universe and the cosmological interpretation of redshift. He also found a theoretical linear relation between velocity and distance, and by using astronomical data he calculated the value of the constant of proportionality which was later, in 1929, empirically established by Hubble and which is known as Hubble's constant. For light sources which were near enough, Lemaître obtained indeed a theoretical relation between velocity and distance that was approximately linear: $v/c = (\dot{R}/R)\,r$. Lemaître was aware of the previous attempts to find a relation between velocity and distance of extragalactic nebulae proposed by Eddington, Weyl, Lanczos, Silberstein, and Lundmark. Investigating the properties of de Sitter's universe in 1925, Lemaître had found a relation that was approximately linear for small distances. In 1927, with the help of the radial velocities of spirals analyzed by Strömberg in 1925 and the apparent magnitudes of these objects measured by Hubble in 1926, Lemaître calculated for the proportionality factor $\dot{R}/R$ the value of 625 km s^{-1} Mpc^{-1}. However, Lemaître pointed out that the errors in determining the distances of spiral nebulae did not permit the accurate establishment of the linearity of such a relation. He obtained for R_0 a value of 2.7×10^8 parsec (5.6×10^{13} AU), and for R a value of 6×10^9 parsec (1.2×10^{15} AU). In the final part of his article, Lemaître suggested that the cause of the velocity of expansion $\dot{R}/R$ might be attributed to the pressure of radiation.

Eddington now realized the importance of Lemaître's ingenious solution (Fig. 3.7). On 19 March 1930, Eddington sent de Sitter a copy of Lemaître's article, in which he pencilled "this seems a complete answer to the problem we were discussing." On a postcard to de Sitter attached to Lemaître's paper, Eddington added: "by the way it was the report of your remarks and mine at the R.A.S. [Royal Astronomical Society] which caused Lemaître to write to me about it... A research student McVittie and I had been worrying about the problem and made considerable progress; so it was a blow to us to find it done much more completely by Lemaître (a blow softened, as far as I am concerned, by the fact that Lemaître was a student of mine)."[87] At the time, Eddington was indeed investigating with George McVittie the stability of Einstein's spherical universe. Eddington acknowledged Lemaître's analysis as a "remarkably complete solution... [A]lthough not expressly stated, it is at once apparent from his formulae that the Einstein world is unstable, an important fact which, I think, has not hitherto been appreciated in cosmogonical discussions."[88] De Sitter also appreciated that Lemaître had discovered a simple and elegant solution, which appeared to him entirely satisfactory. An English translation of Lemaître's paper was published in March 1931 in the *Monthly Notices of the Royal Astronomical Society*. It was Lemaître himself who prepared the translation of the French text. In fact, he decided not to insert in the English version the analysis of the linear velocity-distance relation he had proposed in 1927. In the light of the observational results that Hubble had published in 1929, Lemaître wrote in 1931 to the astronomer William Marshall Smart: "I did not find advisable to reprint the provisional discussion of radial velocities which is clearly of no actual interest."[89]

In letters to Eddington and de Sitter, Lemaître also informed them about the dynamical cosmological models investigated by Friedman, which as a result became widely known within the scientific community. Friedman had in 1922 investigated the curvature of space, with the purpose of showing that Einstein's and de Sitter's solutions were "special cases of more general assumptions, and secondly to demonstrate the possibility of a world in which the curvature radius of space is independent of the three spatial coordinates but does depend on time." In his article, which was published in the *Zeitschrift für Physik*, Friedman considered dynamical solutions of the universe with positive spatial curvature. He systematically analyzed the cosmological solutions of the relativistic field equations that satisfied the condition of homogeneity and isotropy of space, that is, the cosmological principle. Friedman assumed that the pressure contribution could be ignored and that stellar velocities were small compared to the speed of light. The curvature radius was a function of the time coordinate. Furthermore, by an appropriate choice of coordinates "space can be made orthogonal to time. We cannot offer any philosophical or physical justification for these assumptions; they simplify the calculations." The general form of the metric was written as:

$$ds^2 = R^2 \left(dx_1^2 + \sin^2 x_1 \, dx_2^2 + \sin^2 x_1 \sin^2 x_2 \, dx_3^2 \right) + M^2 \, dx_4^2. \qquad (3.11)$$

M was a function of all four coordinates. The case of constant R and $M = 1$ corresponded to Einstein's universe, whereas the case of constant R and $M = \cos x_1$ corresponded to de Sitter's solution. These two cases were the only solutions that corresponded to stationary universes. Regarding the dynamical solutions that Friedman denoted as non-stationary worlds, R depended on time, and M was only a function of the time coordinate. Without loss of generality, M could be taken to be equal to 1. With these assumptions, the field equations reduced to two equations which related R, $\dot{R}$, and $\ddot{R}$ to the cosmological constant and the density of matter, the so-called Friedman–Lemaître equations mentioned above. The integration of the first equation made it possible to obtain different dynamical solutions, either expanding or cyclical, depending on the value of the cosmological constant. Friedman noted that "[o]ur knowledge is insufficient for a numerical comparison to decide which world is ours. It is possible that the causality problem and the centrifugal force problem will illuminate these questions. Finally, we may remark that the cosmological constant in our formulae is undetermined...Perhaps electrodynamics will help to define it."[90]

In 1924, one year before his death, Friedman also considered hyperbolic models in which the spatial curvature was negative. He investigated the issue of the finiteness of the world, and with an eye to this question he analyzed solutions of Einstein's field equations that led to a universe with a constant negative curvature of space. Friedman found that stationary world models were only possible with either a zero or a negative density of matter. Conversely, non-stationary solutions were admissible with constant negative spatial curvature and positive matter density. However, Friedman pointed out the difficulty in dealing with the finiteness of space. The field equations alone, he wrote, "are not enough to allow a decision as to whether or not the world is finite...To definitively decide about its finiteness one needs additional conditions."[91]

While Lemaître's 1927 article was ignored until 1930, the analysis of dynamical models that Friedman proposed in 1922 was mentioned by a few scientists. Robertson referred to it in 1929, in his study of cosmological solutions that satisfied the cosmological principle. In 1928, A. Freedericksz and A. Schechter investigated aberration and parallax formulae in Friedman's models. Moreover, Einstein had written already in 1922 a brief note in which he objected that Friedman's result regarding a non-stationary world "seems suspect" to him: "From the field equations it follows necessarily that the divergence of the matter tensor T_{ik} $[T_{\mu\nu}]$ vanishes. This...implies that the world-radius R is constant in time."[92] Thanks to Aleksander Krutkoff, Einstein could shortly afterwards read a letter in which Friedman set out his calculations. In a subsequent note, Einstein acknowledged that Friedman's results were both correct and clarifying. However, Einstein still thought that Friedman's results were not physically admissible. In fact, it was Einstein himself who informed Lemaître about Friedman's analysis, with which Lemaître was not acquainted at the time of writing his cosmological paper on the expanding universe. The occasion was a meeting between Lemaître and Einstein during the fifth Solvay Conference, which took place in Brussels in 1927. Lemaître could also briefly present to Einstein his own analysis of the expanding universe, which, however, appeared to Einstein "tout à fait abominable" from the physical point of view.[93]

In 1930, it was clear to Eddington and de Sitter that non-static and non-empty world models such as those of Friedman and Lemaître should be investigated so as to find close approximations to the actual universe. In May 1930, de Sitter wrote to Tolman, "The

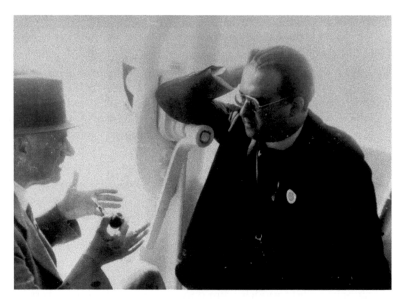

Fig. 3.7 *Arthur Eddington and Georges Lemaître in Stockholm (1938). Courtesy of the Archives Georges Lemaître, Université catholique de Louvain, Louvain-la-Neuve, Belgique.*

important point is that we must look for a dynamical solution of the field equations as the true interpretation of nature."[94]

3.6   The origin of the universe and Lemaître's primeval-atom hypothesis

During the first decades of twentieth century, concepts such as the beginning and the end of the universe were matters of debate mainly in the framework of thermodynamics and geophysics (see Chap. 1). Among the astronomical sciences, work on celestial mechanics had led to the formulation of different hypotheses on the formation and evolution of cosmic structures such as the Milky Way and the nebulae. As Henri Poincaré commented in his *Leçons sur les Hypothèses Cosmogoniques*, the question of the origin of the world could not be answered currently, and "we can only conclude with a question mark."[95] In any case, the cosmogonic hypothesis concerned the formation and evolution of observable heavenly bodies in past epochs, not a common beginning of space, time and matter. Debates about the origin and age of the universe as a whole were essentially absent within the astronomical community.

Relativistic cosmology opened new horizons in dealing with the origin and evolution of the universe. It is important to emphasize in this context that relativistic models that describe cosmic evolution do not necessarily encapsulate the concept of a beginning or an end of space-time. For instance, the static cosmological solutions proposed in

1917 by Einstein and by de Sitter, which are limiting cases of dynamical Friedman–Lemaître universes, are temporally infinite. The 1927 dynamical universe of Lemaître and the 1948 steady state model do not involve a temporal beginning either. A physical theory describing the evolution of the universe from an initial state was proposed by George Gamow in 1948. This theory, widely known as the hot Big Bang model, combines relativistic cosmology and nuclear physics, and it has been largely accepted from the late sixties on. Conversely, few ideas and speculative proposals on the origin of the universe appeared in the framework of early relativistic cosmology. The first contributions were made by Friedman and Lemaître.[96]

The possibility that the universe has a beginning in time was first mentioned in relativistic cosmology by Friedman in 1922. The model that he called "monotone world of the first class" describes a homogeneously expanding universe. Friedman noted for this model: "Since R cannot be negative there must be, as one decreases the time, a time when R vanishes." He added in a footnote: "The time since the creation of the world is the time interval between which $R = 0$ and $R = R_0$; this time might be infinite." A certain age could also be ascribed to the universe in a specific case of Friedman's "monotone world of the second kind," namely the case of the universe eternally expanding from a non-zero curvature radius x_0. In this case, the age of the universe is the time elapsed from the moment when $R = x_0$ to the moment when $R = R_0$.[97]

Friedman's article is the first publication on relativistic world models in which a beginning of the universe is mentioned. Actually, he used for such an event the word "creation," a word that has perhaps certain metaphysical or religious connotations, although in his article there is no trace of any inspiration of this kind. Friedman did not elaborate on the origin of the world as a physical possibility for the actual universe. He rather presented the model with a singularity at $R = 0$ as one of the different mathematical solutions that resulted from his analysis of non-static cases. In the final part of his article he suggested that the "world period" of his cyclic cosmological model could be estimated at about ten billions years. Friedman also mentioned the age of the universe in his semi-popular book *The World as Space and Time*, which was first published in Russian in 1923. In this book he offered an estimate of the time that had elapsed since the creation of the world to its present state. Friedman suggested that the age of the universe was of the order of "tens of billions of ordinary years," but he did not specify the calculation leading up to this result and reported the value "for the sake of curiosity." Friedman inserted this estimate in a summary of a number of non-static cosmological cases that he had found the previous year. He depicted these possibilities as basically equivalent to each other, because the lack of reliable observational data made it impossible to discriminate between these different scenarios. Nevertheless, a certain preference for the oscillatory model emerges from Friedman's words:

> A non-static Universe represents a variety of cases. For example, it is possible that the radius of curvature constantly increases from a certain initial value; it is also possible that the radius changes periodically. In the latter case the Universe compresses into a point (into nothingness), then increases its radius to a certain value, and then again compresses into a point. Here one may recall the teaching of Indian philosophy about "periods

of life." It also provides an opportunity to speak about the world "created from nothingness." But all these scenarios must be considered as curiosities which cannot be presently supported by solid astronomical experimental data.[98]

Friedman's suggestion apart, very few scientists referred to the origin of space-time in the framework of relativistic cosmology before the recognition in 1930 of the expanding universe. Einstein assumed that the universe was static and temporally infinite, and he commented in 1929 that "[t]hrough the general theory of relativity...the view that the [space-time] continuum is infinite in its time-like extent but finite in its space-like extent has gained in probability."[99]

Eddington reflected upon the beginning of the universe in 1930, now in the light of the dynamical solution formulated in 1927 by Lemaître. When he studied the instability of Einstein's model, Eddington analyzed the different cases that resulted from the comparison of the mass of the universe M to M_E, which was the total mass of Einstein's universe. Eddington considered the limiting case $M = M_E$ as the most attractive. He actually endorsed the model proposed in 1927 by Lemaître, in which the curvature radius increases indefinitely from the non-zero radius that the universe had in the infinite past. Eddington commented that "[t]here is at least a philosophical satisfaction in regarding the world as beginning to evolve infinitely slowly from a primitive uniform distribution in unstable equilibrium."[100] The expanding model of the universe that was originally proposed in 1927 by Lemaître and supported by Eddington has become known in modern cosmology as the Eddington–Lemaître model. Eddington rejected the cases of $M > M_E$ and $M < M_E$, which both involved a beginning of space-time. He strongly disliked the association of the state of minimum entropy of the world to that of a sudden and rather peculiar beginning of the universe. Thus, in a popular address of early 1931, "philosophically the notion of a beginning of the present order of Nature is repugnant to me," a view that Eddington held until his death in 1944.[101]

Lemaître's view was the opposite of Eddington's. In a brief letter published in *Nature* on 9 May 1931, Lemaître proposed in an imaginative qualitative way that the world began with a single original quantum, a suggestion that he would later denote as the primeval-atom hypothesis. Eddington's criticism of a natural beginning of the world can be viewed as a stimulus for Lemaître to make public his own idea, rather than the very cause of Lemaître's elaboration of the primeval-atom hypothesis. Lemaître had already mentioned in a footnote in his 1927 cosmological paper the possibility of a null value of the curvature radius, which at the time he related to the temporal extent of R_0. He later became acquainted with Friedman's 1922 concept of the creation of the world, as well as with the finite-age universes starting from a singularity that de Sitter had included in an analysis of various types of expanding world models published in 1930. Shortly before the 1931 letter to *Nature*, Lemaître had further worked out his own dynamical model, and he had introduced the notion of stagnation as the mechanism at the basis of the expansion of the universe. The stagnation phase was due to the variation of pressure; in particular, a diminution of the pressure would cause the universe to expand from the original state of (unstable) equilibrium of an Einstein-type universe. According to

Lemaître, such a rupture of equilibrium would have happened at an epoch of at most 10^{11} years ago.

The picture of the primordial state of the universe that Lemaître proposed in his letter to *Nature* was that of a single quantum. The very initial state of the universe at $t = 0$ was not a singularity of infinite density of energy, but a material entity physically inaccessible. Lemaître considered that such a primordial entity was equivalent to a unique "atom," the atomic weight of which was the total mass of the universe. To Lemaître, a kind of super-radioactive process caused the primordial atom to explode and disintegrate into smaller atoms. The association with quantum theory and the light-to-matter crystallization allowed him to consider that photons would transform into matter, and that this matter, concentrated in a single atom, was the primordial state of the universe. In his proposal, Lemaître related quantum theory to thermodynamic arguments, in the sense that the "[e]nergy of constant total amount is distributed in discrete quanta" and that the "number of discrete quanta is ever increasing. If we go back in the course of time we must find fewer and fewer quanta, until we find all the energy of the universe packed up in a few or even in a unique quantum." In addition, the application of Heisenberg's uncertainty principle to the single quantum made it possible to avoid the philosophical problem of a deterministic interpretation of the beginning (and evolution) of the universe. Lemaître explained it as follows:

> Clearly the initial quantum could not conceal in itself the whole course of evolution; but, according to the principle of indeterminacy, that is not necessary. Our world is now understood to be a world where something really happens; the whole story of the world need not have been written down in the first quantum like a song on the disc of a phonograph. The whole matter of the world must have been present at the beginning, but the story it has to tell may be written step by step.[102]

Later in 1931, Lemaître further worked out the primeval-atom hypothesis in a relativistic model that corresponded to a finite-age universe. In this cosmological model, which is known as the Lemaître model, the evolution of the universe consists of three phases. The first phase is that of the explosion of the primordial atom, the radius of which was assumed by Lemaître to be about 1 AU. According to Lemaître, cosmic rays were the remnants of what he denoted as a fireworks beginning, which was followed by a rapid expansion. The second phase is characterized by a long and slowing stagnation, in which galaxies formed through condensation processes. During the stagnating phase, gravitational attraction was counterbalanced by the repulsive action of the cosmological constant. The third phase is that of accelerated expansion up to the present era.

Lemaître's primeval-atom hypothesis was an innovative attempt aimed at formulating the idea of the origin of the universe in a scientifically satisfactory way. However, he did not manage to formulate a convincing theory, and the early reactions were negative. On the one hand, Lemaître's suggestion that cosmic rays were the remnants of the initial atomic explosion was in general objected to by his contemporaries. On the other hand, scientists like Eddington, Tolman, de Sitter and Robertson were not inclined to accept the notion of a beginning of the universe. As a matter of fact, in subsequent years Lemaître did not further elaborate his hypothesis, nor did he try to relate it to nuclear physics and Gamow's Big Bang scenario. To be sure, Lemaître persistently defended the interpretation of the primeval atom in purely scientific terms, and he clearly distinguished the scientific concept of a natural beginning of the universe from a supernatural creation in religious terms. As a priest-scientist, Lemaître always held the view that two independent, separate roads should be followed simultaneously to reach the same truth, the scientific road and the religious road. As Lemaître clarified in 1958 at the Solvay Conference, the primeval-atom hypothesis "remains entirely outside any metaphysical or religious question. It leaves the materialist free to deny any transcendental Being...For the believer, it removes any attempt to familiarity with God...It is consonant with the wording of Isaias speaking of the 'Hidden God' hidden even in the beginning of creature [sic]."[103]

3.7   The many universes of relativistic cosmology

From 1930 onwards, the choice between Einstein's and de Sitter's models was replaced by the investigation of a variety of models intermediate between these two cosmological solutions, now considered as limiting cases of Friedman–Lemaître world models. The reorientation of the analysis of dynamical cosmological solutions of the field equations resulted in systematizing the approach to the cosmological problem and identifying the basic properties of relativistic world models.[104] Several articles were published on the expanding universe in the early thirties. Detailed reviews were published by Robertson, de Sitter, Tolman, and the German astronomer Otto Heckmann. Besides these scientific publications, the idea of the expanding universe was widely circulated in popular accounts written by a number of the protagonists of the early phase of relativistic cosmology. This was the case, for instance, of the popular books and articles published by Eddington, de Sitter, and James Jeans. In May 1932, Jeans offered explanations of the latest developments in modern cosmology and the expanding universe in the columns of *The Times*.

One of the issues confronted before the recognition of the expanding universe was the definition of the spatial and temporal coordinates to be used in the geometrical representation of cosmological solutions, an issue that had been tackled in particular in the analysis of de Sitter's empty universe. Robertson had in 1929 obtained a general form of the line element describing all static and non-static, isotropic and homogeneous models of the universe, without however justifying the introduction of a cosmic time. In his influential review of relativistic cosmology published in 1933, Robertson identified the cosmological principle and Weyl's principle as the basic assumptions concerning the natural state of motion of matter, the definition of cosmic time and the formulation of dynamical solutions of the relativistic field equations. Robertson adopted from Weyl both the existence of a congruence of time-like geodesics and the existence of spatial sections which are orthogonal to this congruence. To quote Robertson, "the universe is a coherent whole rather than the fortuitous superposition of two or more incoherent parts...The

possibility of thus introducing in a natural and significant way this *cosmic time t* we consider as guaranteed by Weyl's postulate, which is in turn a permissible extrapolation from
the astronomical observations."[105] Through the additional fundamental assumption of
spatial isotropy and homogeneity, Robertson obtained the form of the metric for any
isotropically expanding *substratum*, the particles of which represent galaxies. In 1935, the
mathematician Arthur Walker also formulated this line element, now in the framework
of Milne's kinematic cosmology. As a result, the space-time metric that describes an
expanding universe and that complies with the cosmological principle has become known
as the Robertson–Walker metric. This line element can be written as:

$$ds^2 = -R^2(t) \left[\frac{dr^2}{1 - kr^2} + r^2 \, d\theta^2 + r^2 \sin^2\theta \, d\phi^2 \right] + c^2 \, dt^2. \qquad (3.12)$$

The line element of the Friedman–Lemaître cosmological models described in
Section 3.5 is in fact the Robertson–Walker metric. Conversely, the Friedman–Lemaître
models can be obtained as a solution of the field equations when the Robertson–Walker
metric is inserted into these equations. The first letters of the names of the four scientists
who investigated these geometric and physical properties of the universe are used to
denote the world models based on the Friedman–Lemaître equations and the Robertson–
Walker metric: they are called FLRW cosmological models.

In the formula for the Robertson–Walker metric, the coordinate t is the proper time.
Polar coordinates (r, θ, ϕ) refer to a comoving reference frame, in the sense that they
are constant for each particle of the perfect fluid which is used as an approximation
for the content of the universe and which is at rest with respect to this reference
frame. The function $R(t)$, also denoted as $a(t)$, is the *cosmic scale factor* or *expansion
parameter*, which absorbs the physics of the expansion and varies according to the
expansion. The scale factor R is related to the cosmological redshift z through the relation
$R = (1 + z)^{-1}$. This relation therefore makes it possible to directly obtain information
about the scale that the universe had at the time of a certain light emission. The expansion
rate of the universe is represented by the parameter $H(t) = \dot{R}(t)/R(t)$, that is, Hubble's
constant. In particular, H_0 is the value of Hubble's constant at the present time t_0.
The parameter k describes the constant curvature of spatial sections. This parameter
can be negative $(k = -1)$, zero $(k = 0)$, or positive $(k = +1)$, and it corresponds to an
open universe (three-dimensional hyperbolic space), flat universe (Euclidean space), or
closed universe (three-dimensional spherical space), respectively. The geometry of space
depends on its energy–matter content. The spatial sections of the universe can be closed,
flat, or open, provided that the *density parameter* $\Omega(t) = \rho/\rho_c$ is greater than, equal to, or
less than 1, respectively. The so-called critical density of matter ρ_c is the density of a flat
universe: $\rho_c = 3H^2/8\pi G$.

Distinct types of expanding world models were investigated during this period,
according to the different values that could be assigned to the curvature parameter k and
the cosmological constant λ. Further options resulted when conditions for the density ρ
and the pressure p were introduced. Some of these solutions are illustrated in Fig. 3.8:
the scale factor $R(t)$ contracts to a minimum and then expands monotonically towards

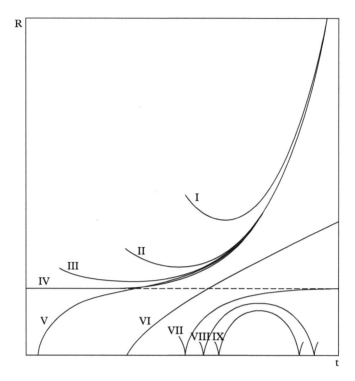

Fig. 3.8 *Different relativistic models of the universe, the curvature radius of which (vertical axis) varies with cosmic time (horizontal axis).*

a de Sitter-type universe (for instance case I in Fig. 3.8); it expands asymptotically from the Einstein universe to the de Sitter universe (Eddington–Lemaître model, case IV in Fig. 3.8): it starts from a singularity and then expands passing through a point of inflexion (Lemaître model with a stagnation phase, case V in Fig. 3.8).

Further cases correspond to oscillatory models, which can have both expanding and contracting phases (see for instance case VIII in Fig. 3.8). Following Friedman's analysis of 1922, Einstein proposed in 1931 a cyclic model, in which the value of the cosmological constant is zero.[106] In light of the evidence for the expansion of the universe constituted by Hubble's observations, Einstein in fact rejected the cosmological constant that he had introduced in 1917. Einstein would later explain his feelings in a letter to Lemaître as follows:

> Since I have introduced this [cosmological] term, I had always a bad conscience. But at that time I could see no other possibility to deal with the fact of the existence of a finite density of matter. I found it very ugly indeed that the field law of gravitation should be composed of two logically independent terms which are connected by addition. About the justification of such feelings concerning logical simplicity it is difficult to argue. I cannot help to feel it strongly and I am unable to believe that such an ugly thing should be realized in nature.[107]

The oscillating model advocated by Einstein has become known as the Friedman–Einstein universe, the thermodynamic properties of which were investigated by Tolman in 1931. Tolman was a pioneer in investigating the connections between general relativity and thermodynamics, and in elucidating the cosmological consequences. In his book *Relativity, Thermodynamics and Cosmology*, first published in 1934, Tolman analyzed issues such as the thermodynamic equilibrium in the static universe of Einstein, the conditions for reversible and irreversible changes in non-static models, and models filled either with incoherent matter exerting no pressure or black body radiation as examples of reversible behavior.

Contrary to Einstein, Lemaître always supported the existence of a non-zero cosmological constant. By changing its value it was possible indeed to adjust the duration of the stagnation phase and consequently the age of the universe. The cosmological constant could be used to explain the formation of galactic structures, an issue that was matter of debate between Lemaître, McVittie and the English astronomer William McCrea. Eddington also considered that λ was an essential ingredient of cosmological models, but for motivations different from those of Lemaître. To him, the cosmological constant was not only responsible for the recession of galaxies. No, the constant was one of the fundamental entities in nature, together with the fine structure constant, the number of particles expected in an expanding universe, and the ratio between electrostatic and gravitational forces. Eddington firmly pursued a program aimed at formulating a fundamental theory that could unify processes happening on different scales, such as those on the scale of protons and electrons and those on cosmological scales. However, his approach appeared abstruse and obscure to many, and it was essentially abandoned after Eddington's death in 1944.[108]

A cosmological model of great importance is the so-called Einstein–de Sitter model of the universe, which describes a monotonic expansion of the universe from a singularity without the stagnation phase (case VI in Fig. 3.8). Einstein and de Sitter, the very first protagonists of relativistic cosmology, collaborated in 1932 in formulating a simple cosmological solution of the relativistic field equations. In the Einstein–de Sitter model the value of the cosmological constant is zero, the spatial curvature is zero ($k = 0$) and the pressure is also zero. The density of matter has the value of the critical density of matter mentioned above. The Einstein–de Sitter model is thus often referred to as the critical model, since this case separates the ever-expanding open models with hyperbolic geometry from the closed models with spherical geometry that will eventually contract and collapse into a singularity. By assuming for Hubble's constant the value of $H_0 = 500$ km s^{-1} Mpc^{-1}, Einstein and de Sitter obtained in 1932 the value of $\rho = 4 \times 10^{-28}$ g cm^{-3}, a value that they considered to be in agreement with the current estimate of the density of matter empirically calculated by Hubble. For its simplicity, the Einstein–de Sitter universe was considered a benchmark in the subsequent developments of relativistic cosmology. This cosmological model also involves a beginning of the universe. However, Einstein and de Sitter did not touch upon this feature in the brief article that they published in 1932, nor did they explicitly mention the temporal evolution of the scale factor. Very likely, these omissions were due to the so-called timescale problem.

One of the main cosmological problems that followed the recognition of the expanding world models of Friedman and Lemaître was the age of the universe, also known as the timescale difficulty.[109] As mentioned in Section 3.6, the early responses to Lemaître's finite-age universe were not enthusiastic at all. In the following years the notion of the beginning of the universe gradually entered the cosmological debates, and by the end of the thirties the age of the universe was considered one of the relevant parameters in the FLRW cosmology. The inverse of H_0, Hubble's constant at the present time, is the so-called Hubble time, and it gives an approximate value of the age of the universe, here denoted by t_0. When Hubble's constant was taken to be $H_0 = 500$ km s^{-1} Mpc^{-1}, which was the value generally accepted in the thirties, the resulting Hubble time was $1/H_0 = 2 \times 10^9$ years. In the case of the Einstein–de Sitter model, the time elapsed since the beginning of the expansion of the universe is shorter than the Hubble time. According to the relation $t_0 = 2/3H_0$, the value of t_0 is about 1.3 billion years.

An issue of concern among the modern cosmologists was the discrepancy between the age of the universe calculated through relativistic cosmological models and the age of the constituents of the universe themselves, because according to the models the universe was younger than the elements that compose it. The problem arose by comparing the estimated value of t_0 in the Einstein–de Sitter model with the age of the earth or that of the stars and galaxies, and it also remained unsolved in most cosmological models different from the Einstein–de Sitter solution. The analysis of radioactive elements such as thorium and uranium had furnished in the twenties estimates of the age of the earth ranging from 1.5×10^9 years up to four billion years. The discrepancy with the age of the universe was even worse in the case of the age of the stars obtained from the analysis of stellar dynamics. According to Jeans, who at the time was one of the most authoritative experts in this area, a timescale of about 10^{13} years should be expected for the formation and evolution of less massive stars. This long stellar timescale contrasted dramatically with the much shorter Hubble time. Early in the thirties the temporally infinite Eddington–Lemaître model gained support among cosmologists, because it avoided the timescale paradox. A kind of bouncing model was supported by de Sitter to escape the issue of an absolute physical beginning of the universe and the corresponding short cosmic timescale. In 1933, de Sitter considered a model in which the universe contracted during an infinite time, passed through a minimum value of its curvature radius—something that, according to de Sitter, happened around five billion years ago—and then started to expand. Thus the beginning of the universe was for de Sitter different from the beginning of the expansion. Tolman also advocated a picture in which the concept of the beginning of the universe was avoided: "Indeed it is difficult to escape the feeling that the time span for the phenomena of the universe might be most appropriately taken as extending from minus infinity in the past to plus infinity in the future."[110]

A better understanding of stellar evolution and galactic dynamics would make it possible in later years to properly tackle and downsize the timescale difficulty. In the late fifties the value of Hubble's constant was dramatically reduced, in the light of the recalibration of the Cepheid distance scale produced in 1952 by Walter Baade. In 1958, Hubble's constant was estimated to be about 75 km s^{-1} Mpc^{-1}, a value that offered a proper solution to the timescale dilemma (see Chapter 6).

By the end of the thirties, the main questions of relativistic cosmology had been identified. They concerned the determination of parameters defining the FLRW models such as Hubble's constant, the curvature of space, the density of matter in the universe, the age of the universe and the cosmological constant. The observational programm of cosmological relevance that were implemented in the subsequent years were aimed at tackling these issues. During the early phase of modern cosmology described in this chapter, the application of general relativity to the universe as a whole had made it possible to inaugurate an ambitious field of study that resulted in a new picture of an expanding universe. To quote Tolman,

> The non-static models which we shall now study are, to be sure, mathematically more complicated than our previous static ones; nevertheless, the history of human endeavours to understand the universe would certainly indicate no *a priori* right to demand mathematical simplicity of nature...It is appropriate to approach the problems of cosmology with feelings of respect for their importance, of awe for their vastness, and of exultation for the temerity of the human mind in attempting to solve them. They must be treated, however, by the detailed, critical, and dispassionate methods of the scientist.[111]

Acknowledgments

I am grateful to Helge Kragh and Malcolm Longair for inviting me to contribute to this volume. I would like to thank Erhard Scholz and Chris Smeenk for their useful comments. I am sincerely grateful to Tjeerd B. Jongeling for comments and for helping with the revision of the text. I would like to thank Liliane Moens for the collaboration in consulting archival material at the Georges Lemaître Archives (Louvain-la-Neuve), and for permission to quote from archival sources and reproduce images. I would also like to thank Kasper van Ommen for helping in the research at the Leiden University Library (Leiden), and for permission to quote from archival sources. I am grateful to Monica Rocco and Luca Realdi for helping with the preparation of the figures. This research has received funding from the European Union's Horizon 2020 research and innovation program, under the Marie Skłodowska-Curie grant agreement no. 656139.

. . . .

NOTES

1. On the history of the early phase of relativistic cosmology, see Ellis (1989, 1990), Kerszberg (1989), Kragh (2015a), Kragh and Smith (2003), North (1965), Nussbaumer and Bieri (2009), and Smith (1982).
2. See Hartle (2003) for an introduction on the general theory of relativity. See Coles and Lucchin (2002) and Harrison (2000) for the case of modern scientific cosmology.

3. Einstein (1916a), p. 148. On the history of Einstein's formulation of general relativity, see Janssen (2014). On the genesis of general relativity, see Renn and Schemmel (2007).

4. Eddington led the British eclipse expedition to Principe Island in Africa, accompanied by Edwin Cottingham, while Andrew Crommelin and Charles Davidson went to observe the solar eclipse in Sobral, Brazil. On the history of the 1919 eclipse expedition, see Almassi (2009) and Kennefick (2012). On the history of testing Einstein's theory of relativity, see Crelinsten (2006).

5. The *London Times*, 7 November 1919.

6. Einstein (1918b), pp. 33–4.

7. Einstein (1916b), p. 144. On Mach's role in the genesis of general relativity, see Barbour (1990) and Renn (2007). On Mach's principle, see Barbour and Pfister (1995).

8. Einstein (1917a), p. 424.

9. Einstein (1917a), p. 423. On Einstein's formulation of his cosmological model, see O'Raifeartaigh et al. (2017), Realdi and Peruzzi (2009), and Smeenk (2014).

10. Einstein (1917a), p. 428.

11. Letter from Einstein to Besso, after 9 March 1917. In Schulmann et al. (1998b), pp. 296–7.

12. Letter from Einstein to Ehrenfest, 14 February 1917. In Schulmann et al. (1998b), p. 285.

13. The cosmological research carried out by Robertson is described in Sections 3.4, 3.5, and 3.7 of this chapter. Milne's kinematic cosmology is described in Chapter 4.

14. On the history of the cosmological constant, see Earman (2001).

15. Einstein (1919), p. 83.

16. See O'Raifeartaigh et al. (2017).

17. Letter from Einstein to de Sitter, 14 April 1917. In Schulmann et al. (1998b), p. 316.

18. Letter from Einstein to de Sitter, before 12 March 1917. In Schulmann et al. (1998b), p. 301.

19. De Sitter (1916), p. 183

20. De Sitter (1933), p. 154.

21. Letter from de Sitter to Einstein, 1 April 1917. In Schulmann et al. (1998b), p. 313.

22. On the different representations of de Sitter's universe, see Moschella (2006) and Schrödinger (1957).

23. Letter from de Sitter to Einstein, 20 March 1917. In Schulmann et al. (1998b), p. 303.

24. De Sitter (1933), p. 158.

25. Letter from Eddington to de Sitter, 11 June 1916. Leiden University Libraries— Leiden Observatory Archives, Directorate W. de Sitter, AFA-FC-WdS-14.

26. Letter from Einstein to de Sitter, 24 March 1917. In Schulmann et al. (1998b), p. 309.

27. On the debate between Einstein and de Sitter and its aftermath, see the Editorial Notes "The Einstein–de Sitter–Weyl–Klein debate" in Schulmann (1998a), pp. 351–7 and Janssen (2014).

28. A singularity is as a point (or region) of space-time where some metric components have infinite or null values, so that the metric tensor is undefined or not suitably differentiable. There are singularities that are artifacts of the chosen coordinate system. Such *coordinate singularities* can be removed by making appropriate coordinate transformations. Conversely, an *intrinsic singularity* at a point (or region) reflects an ill-behavior of the space-time curvature that does not depend on the coordinate system. An example of intrinsic singularity is the point at $r = 0$ in Schwarzschild geometry (the center of a black hole). The definition of singularities would be clarified in the second half of twentieth century. On the history of singularities in general relativity, see Earman (1999) and Earman and Eisenstaedt (1999).
29. Einstein (1918a), pp. 37–8.
30. De Sitter (1917a), p. 17.
31. Eddington (1923), pp. 165–6. The questions associated with cosmological horizons turned out to be of certain relevance in the theoretical understanding of cosmological models. A clarification of the subject would appear in the late fifties through the work of Wolfgang Rindler. On cosmological horizons, see Harrison (2000).
32. Letter from Klein to Einstein, 16 June 1918. In Schulmann et al. (1998b), p. 593.
33. Letter from Einstein to Klein, 20 June 1918. In Schulmann et al. (1998b), p. 594.
34. Letter from Einstein to Pirani, 2 February 1954. Quoted in Janssen (2014), p. 207.
35. Henderson (1925), p. 223.
36. The rise of observational cosmology is described in Chapter 2.
37. On the measure of the curvature radius of the universe, see Peruzzi and Realdi (2011).
38. De Sitter (1917a), p. 26.
39. Letter from Kapteyn to de Sitter, 28 June 1917. Quoted in Van der Kruit and van Berkel (2000), p. 96.
40. De Sitter (1917a), p. 27.
41. De Sitter (1917b), p. 237.
42. De Sitter (1920), p. 868.
43. On the role played by Eddington in the early phase of relativistic cosmology, see Stanley (2013).
44. Eddington (1918), p. 87.
45. Eddington (1920a), p. 159.
46. Eddington (1923), p. 160.
47. Eddington (1923), p. 164.
48. Eddington (1923), p. 162.
49. Lanczos (1922). English translation in Nussbaumer and Bieri (2009), p. 78.
50. On the history of the de Sitter effect and the early debates on relativistic cosmology, see Ellis (1990), North (1965), Rowe (2016a), and Rowe (2016b).
51. This image is borrowed from Eddington, who wrote about the dawn of the expanding universe: "We have been going round a workshop in the basement of the building of science. The light is dim, and we stumble sometimes. About us is confusion and mess which there has not been time to sweep away. The

workers and their machines are enveloped in murkiness. But I think that some-thing is being shaped here—perhaps something rather big." Eddington (1933), pp. 125–6.

52. See Eisenstaedt (1989).
53. On Weyl's principle and Weyl's contributions to cosmology, see Bergia and Mazzoni (1999), Ehlers (1988), Goenner (2001), and Rugh and Zinkernagel (2011).
54. Weyl (1923a), p. 296. English translation in Nussbaumer and Bieri (2009), p. 81.
55. Weyl (1924), p. 348.
56. Weyl (1923b), p. 230. English translation p. 1662.
57. Lemaître (1925a), p. 188.
58. Lemaître (1925b).
59. Robertson (1928), p. 847.
60. Lanczos (1924). English translation p. 363.
61. Lanczos (1923), pp. 188–9. English translation in Goenner (2001), p. 122.
62. On Wirtz's astronomical observations, see Duerbeck and Seitter (2005).
63. On Silberstein's research in general relativity and cosmology, see Flin and Duerbeck (2006).
64. Silberstein (1924c), p. 350.
65. Silberstein (1924a), p. 909.
66. Silberstein (1924c), p. 350.
67. Weyl (1924), p. 349.
68. Silberstein (1924a), pp. 916–7.
69. Silberstein (1924b), p. 602.
70. Lundmark (1924), p. 750.
71. Lundmark (1925), p. 867.
72. Strömberg (1925), p. 361.
73. "Curieux complexe d'équivoques." Merleau-Ponty (1965), p. 61.
74. On the different kinds of redshift in astronomy and cosmology, see Ellis (1989) and Harrison (2000).
75. Eddington (1933), p. 2.
76. On the history of the discovery of the expanding universe, see in particular Kragh and Smith (2003). The cosmological research carried out by Lemaître is detailed in Holder and Mitton (2012) and Lambert (2015). On Friedman's contributions to cosmology, see Belenkiy (2013).
77. See Robertson (1929), pp. 827–8.
78. Hubble (1929a), p. 173. On Hubble's contributions to cosmology, see Chapter 2, Christianson (1995), and Hetherington (1996).
79. Tolman (1929a), p. 267.
80. Tolman (1929b), p. 304.
81. Letter from de Sitter to Schlesinger, 8 November 1929. Leiden University Libraries—Leiden Observatory Archives, Directorate W. de Sitter, AFA-FC-WdS-52.

82. Letter from de Sitter to Schlesinger, 28 December 1929. Leiden University Libraries—Leiden Observatory Archives, Directorate W. de Sitter, AFA-FC-WdS-52.
83. Royal Astronomical Society (1930), pp. 38-9.
84. Letter from Lemaître to Eddington, before 19 March 1930. Archives Georges Lemaître, Université catholique de Louvain, Louvain-la-Neuve, Folder D17.
85. "Alors le rayon de l'espace varie avec le temps." Lemaître (1927), p. 51.
86. Lemaître (1927), p. 51. English translation p. 484.
87. Letter from Eddington to de Sitter, 19 March 1930. Leiden University Libraries—Leiden Observatory Archives, Directorate W. de Sitter, AFA-FC-WdS-14.
88. Eddington (1930), p. 668.
89. Letter from Lemaître to Smart, 9 March 1931. Quoted in Livio (2011).
90. Friedman (1922), pp. 385–6. English translation p. 58.
91. Friedman (1924), pp. 331–2. English translation p. 65.
92. Einstein (1922), p. 326. English translation p. 66.
93. Letter from Lemaître to Eddington, before 19 March 1930. Archives Georges Lemaître, Université catholique de Louvain, Louvain-la-Neuve, Folder D17.
94. Letter from de Sitter to Tolman, 7 May 1930. Leiden University Libraries—Leiden Observatory Archives, Directorate W. de Sitter, AFA-FC-WdS-150.
95. "Nous ne pouvons donc terminer que par un point d'interrogation." Poincaré (1911), p. XXV.
96. On Lemaître's formulation of the primeval-atom hypothesis, see Kragh and Lambert (2007).
97. Friedman (1922), pp. 384–5. English translation pp. 56–7.
98. Quoted in Belenkiy (2013), pp. 80–1.
99. Quoted in Kragh (2008b), p. 10.
100. Eddington (1930), p. 672.
101. Eddington (1931), p. 453.
102. Lemaître (1931a).
103. Lemaître (1958), p. 7. On Lemaître's approach to the relationship between science and religion, see Kragh (2004) and Lambert (2007).
104. On the history of the research in relativistic cosmology carried out during the thirties, see Ellis (1989), Kragh (1996), Longair (2006), and North (1965). See also Gale (2017) for a review of the methodological debates in cosmology during the thirties and forties.
105. Robertson (1933), p. 65.
106. On Einstein's model of 1931, see O'Raifeartaigh and McCann (2014).
107. Letter from Einstein to Lemaître, 26 September 1947. Quoted in Earman (2001), p. 197.
108. On Eddington's fundamental theory, see Kilmister (1994) and Kragh (2017a).
109. On the timescale problem, see Brush (2001) and Nussbaumer and Bieri (2009).
110. Tolman (1934b), p. 486.
111. Tolman (1934b), p. 361 and p. 488.

4

Alternative cosmological theories

Helge Kragh

Those who seek a revolution in cosmology must bear in mind that the days are gone when it was easy to think of viable alternatives.

Peebles (1993), p. 197.

4.1 Introduction

This chapter covers a select subset of so-called alternative theories which in the period from about 1930 to 1970 were of some importance in the development of modern cosmology. They belong to the history of the field as much as do the more successful theories that in a more direct way paved the way to the modern view of the universe. The meaning of a scientific theory being alternative is briefly and in a general manner discussed in the section below. If a theory is alternative, what is it alternative to? The first and oldest of the alternatives, Milne's theory of kinematic relativity, was highly ambitious and radically heterodox. It exerted considerable impact on cosmological thinking in the 1930s and 1940s but then disappeared, as if it had never been there. In part inspired by Milne's theory, and also by the equally heterodox ideas of Eddington, in 1937 Dirac suggested that the strength of gravity had been greater in the cosmic past. Cosmologies based on Dirac's hypothesis or modifications of it are surveyed in Section 4.4. They also turn up in Section 4.5 on the extension of general relativity known as scalar-tensor theories.

Plasma cosmology as developed by Alfvén in particular is a somewhat later and quite different cosmological alternative as its picture of the universe is primarily based on concepts of plasma physics rather than gravitation. This kind of theory still exists but only as a heterodox view not taken seriously by mainstream cosmologists. Neither Milne nor Alfvén denied the expansion of the universe, a concept which was however resisted by many scientists in the early days. Section 4.7 describes Zwicky's non-recession hypothesis of galactic redshifts as well as some related hypotheses. Ideas of a static universe are still being proposed, if in forms quite different from those of the past.[1]

Kragh, H., 'Alternative cosmological theories' in *The Oxford Handbook of the History of Modern Cosmology*, edited by Kragh, H. and Longair, M. S. © Oxford University Press 2019.
DOI: 10.1093/oxfordhb/9780198817666.013.4

The expansion of the universe may in the future be succeeded by a contraction, which is the general assumption of the class of cyclic or oscillating models. Such models, first proposed in a relativistic context as early as 1922, have played a remarkable role but always been considered heterodox or of little plausibility. Although declared dead several times they continue to attract attention, in part because they combine elements of big-bang theory with a universe existing eternally. The final section discusses the broader question of cosmology's status as a science through various episodes in which critics have objected that cosmology generally, or the current standard theory in particular, does not live up to the epistemic and ethical standards of science.

4.2 On orthodoxy and heterodoxy

The alternatives considered in this chapter are limited to non-mainstream scientific theories and do not include broader theories or ideas which may more properly be characterized as counter-movements or anti-science.[2] On the contrary, advocates of alternative cosmological ideas are keen to stress that these fully comply with the established norms and methods of science. Their ideas are alternative science, but not alternatives *to* science.

The concept of an alternative scientific theory or idea is not well defined except that it must be different from and meant as a rival to some generally accepted theory. A theory, view or idea can never be alternative *per se* but only if compared to what is typically called a standard or consensus theory. In so far that there exists such a standard theory—and this is not always the case—it is accepted by the large majority of the relevant scientific community. The standard theory is an important part of mainstream science or what the philosopher of science Thomas Kuhn called normal science; it is orthodox and rarely questioned. The orthodoxy may have the consequence that the theory obtains an almost paradigmatic status. In periods of crisis, on the other hand, alternatives flourish, again according to Kuhn.[3]

It is worth keeping in mind that concepts such as standard, orthodox, mainstream, and consensus are purely sociological, and that a standard theory is not necessarily more correct or truer than an alternative and more heterodox theory. Moreover, what is accepted as a standard theory naturally changes over time and for this reason labels such as "standard," "orthodox," and "alternative" may be reversed as science progresses. In 1600, the Copernican view of the universe was a minority view, an alternative to the still dominant view of the geocentric universe. A century later the status of the two positions had reversed. More to the point in the present context, Lemaître's audacious idea of an explosive beginning of the universe remained for more than a decade a fringe hypothesis; it was considered a heterodox alternative until it finally reappeared as the core of the new standard cosmological theory in the later 1960s. Within the context of the hot big-bang theory models with a positive cosmological constant were still considered non-standard in the 1980s, but later on they became part of the standard ΛCDM theory.[4]

To get an impression of what the orthodox or standard theory is in some area of science, one has to look at textbooks. These deal by their very nature with mainstream science although in some cases they may briefly mention alternatives. Jim Peebles' influential *Principles of Physical Cosmology* of 1993 was solidly founded on the hot big-bang relativistic theory but also included sections on alternative theories, in this case Milne's kinematic relativity, steady-state theory, plasma cosmology, and hierarchical or fractal cosmologies. A much earlier textbook, Paul Couderc's *The Expansion of the Universe* from 1952 disregarded ideas of a static universe but discussed in an appendix what was called "three heterodox theories" (namely Milne's theory, the steady-state theory, and Jordan's cosmology).

When a standard theory is confronted with an alternative it may or may not result in a controversy, depending on the circumstances. Only if the alternative enjoys a certain degree of support in the scientific community and is taken seriously by at least some of its members will a controversy or dispute arise. A classic example is the epic controversy between the steady-state theory and relativistic evolution cosmologies in the 1950s described in Chapter 5. But in many cases alternative theories are not perceived as threats to the consensus view and are therefore more or less ignored. They are not tested or critically discussed, but just rejected as implausible or uninteresting. A dissenting view may either be absorbed into the standard theory or allowed to live its own life apart from it, without any kind of controversy emerging.

Whether in cosmology or some other branch of science, scientists are supposed to be open-minded but they also have to defend their authority and the high standards on which their discipline rests. According to Michael Polyani, a distinguished physical chemist and philosopher of science, an element of dogmatism is justified and desirable. Without referring to cosmology in particular, Polanyi wrote:

> Journals are bombarded with contributions offering fundamental discoveries in physics, chemistry, biology and medicine, most of which are nonsensical. Science cannot survive unless it can keep out such contributions and safeguard the basic soundness of its publications. This may lead to the neglect or even suppression of valuable contributions, but I think the risk is unavoidable.[5]

Alternative cosmological views are many and confusingly diverse not only today but have also been so in the past. They cover a broad spectrum, both as regards of how general and fundamental they are and as regards of how seriously they are taken and intended to be. While some theories are put forward by scientists who think their theories are candidates for the real structure of the universe, others are suggestions of a much less committed nature. They may be toy models or just loose ideas thrown up to see if they work or not. And then there is a large number of alternative cosmological theories which belong to the fringe of science and typically (but not always) are suggested by amateurs outside the scientific community.[6]

Theories or models of this kind have proliferated in recent time but the genre is far from new. To mention but one example, amateur cosmological speculations were popular in the late nineteenth century in connection with the debate concerning the heat death of the universe (Sections 1.4 and 1.7). It may be difficult to distinguish clearly between speculations of this sort and sober scientific contributions to cosmology, but in practice the site of publication serves as an indication and a means of demarcation. Whether the author is a scientist or an amateur, contributions to *Journal of Scientific Exploration, Apeiron, Journal of Cosmology*, and *Meta Research Bulletin* are deemed to be outside scientific cosmology as tacitly defined by its practitioners. They are considered to be more pseudoscience than science and in this chapter they are disregarded.

The sample of alternative cosmological ideas considered below have all been discussed in the scientific literature and played some role in the development of twentieth-century cosmology up to about 1980. While some of them deny the expansion of the universe and its origin in a big bang, others are alternatives with respect to the role played by Einstein's theory of general relativity. By far the most important alternative from a historical perspective, the steady-state theory of the universe, is discussed in Chapter 5 and therefore not part of the present chapter. At least as seen from a modern point of view, an alternative cosmological theory means a theory in conflict with what is currently the standard hot big-bang model and which typically includes inflation, dark matter, and dark energy. The cases referred to below are from an earlier period when the latter subjects had not yet entered cosmology. We do not include post-1980 alternatives such as multiverse models and theories based on quantum gravity as these will be dealt with in Chapter 12.

4.3   Milne's kinematic cosmology

Today, the cosmological system developed by the British astrophysicist Edward Arthur Milne in the period 1932–1950 is largely forgotten. His system was an ambitious, original and highly unorthodox attempt to reconstruct the very foundation of physics and to build from general principles a new science of the universe in the most comprehensive sense. Although it soon fell into oblivion, in the 1930s and 1940s Milne was considered a leading reformer of cosmological thought and his theory of "world physics" attracted a great deal of scientific attention as a radical alternative to the cosmology based on general relativity. For example, it was much more discussed than Lemaître's theory of the exploding universe and in England in particular, it was received with positive interest. It has been estimated that from 1932 to 1940 there appeared about 70 papers related to Milne's theory, which means that it had a very strong position in the period.[7] The papers were published not only in physics and astronomy journals, but also in journals devoted to mathematics and philosophy.

When Milne began focusing on cosmology in the early 1930s he already had a name as an outstanding astrophysicist and specialist in stellar structure and thermodynamics. In 1927, at the age of 31, he was appointed professor of mathematics at the University of Oxford. Dissatisfied with the state of relativistic cosmology, in 1933 he published his alternative in a book-length article in *Zeitschrift für Astrophysik*. Two years later he presented a systematic exposition of his cosmology in the impressive monograph *Relativity, Gravitation and World-Structure*, and in 1948 he extended the theory in another monograph with the title *Kinematic Relativity*. Milne believed that his theory was more than just an alternative scientific cosmology and that it had important implications also for natural philosophy and even theological thought. These implications he spelled out in a book published posthumously in 1952, *Modern Cosmology and the Christian Idea of God*, in which he argued that the notion of God unavoidably belonged to the scientific study of the universe. In the second edition of *Principia*, Newton famously declared that "to treat of God from phenomena is certainly part of natural philosophy [physics]." Milne very much agreed.

Milne's world system built on two fundamental principles or postulates. The one was the principle of the constancy of the velocity of light, which it shared with the theory of special relativity. Milne's other postulate was that the world would appear the same to all observers who thus would see the same events and agree upon the laws of nature. In his 1933 paper he introduced what he first called the "extended principle of relativity" as follows:

> Not only the laws of nature, but also the events occurring in nature, the world itself, must appear the same to all observers, wherever they be, provided that their space-frames and time scales are similarly oriented with respect to the events which are the subject of observation.[8]

The cosmological principle (as Milne called it in his later writings) corresponded to the commonly held assumption of large-scale homogeneity and isotropy, as formulated by Einstein and H.P. Robertson. But Milne believed that his version differed from the one appearing in relativistic cosmology. And indeed it did, for in standard cosmology it was an extrapolation from observations, a hypothesis which could be abandoned if future observations spoke against it. In Milne's theory the cosmological principle had an *a priori* character as it was considered a precondition for having knowledge of the universe. It was only with Milne's theory that the cosmological principle became a basic postulate of cosmology and was widely discussed. The origin of the Robertson–Walker (or FLRW) metric, the most general form of metric for a space-time satisfying the cosmological principle, was indebted to Milne, in Robertson's case critically and in the case of Arthur G. Walker more directly and sympathetically.[9] Walker worked with Milne, whose approach to cosmology fascinated him. Characteristically, Walker's 1936 paper carried the title "On Milne's Theory of World Structure."

Milne admired and mastered Einstein's theory of general relativity, but he did not accept it as a theory of the real world. The cosmological constant Λ was controversial in the 1930s, abandoned by Einstein (i.e., $\Lambda = 0$) but maintained by Eddington, de Sitter and Lemaître. In Milne's view the Λ-constant was a mistake, a quantity that represented the inadequacy of relativistic cosmology. What is more important, Milne denied the central message of general relativity that space is a deformable entity subject to the action of mass and energy and he similarly denied the existence of a cosmic time common to all observers. For him the question of space curvature did not enter as he considered space to be nothing but an abstract system of reference. It thus could have no structure, curved or not, and nor could space itself expand or contract. In agreement with the conventionalist philosophy of Henri Poincaré, the French mathematician, he wrote:

No observation can distinguish whether space is Euclidian or non-Euclidian, simply because there is no such thing as physical space, but only the space, of our arbitrary choice, which we create and use for purposes of description. . . . It follows immediately from Poincaré's view as to the arbitrariness of geometry that no knowledge whatever concerning the universe can be obtained by making assumptions about its geometry.[10]

In his 1933 paper Milne contrasted his own theory to "the current theory which involves the notion of curved, finite, 'expanding' space . . . [and which] attributes the expansion phenomenon to an effect of gravitation."[11] All facts of observation, he claimed, could be accounted for by adopting flat, infinite and static Euclidean space as the seat of physical events. But this kind of space was chosen for reasons of convenience and not because he attached any reality or ontological priority to it. What were real were the galaxies and other celestial objects, and according to Milne the number of objects in the universe was infinite. Milne characterized expansion as "inevitable" and as a natural phenomenon taking place in the absence of gravitation and proceeding in spite of gravitation. To say that the universe expands simply meant that the distance between any two galaxies increases.

In Milne's theory, galaxies were represented by randomly moving particles, much like the molecules of a gas. Neglecting collisions and gravitational forces he showed by simple kinematic arguments that such a system of concentrated particles would naturally evolve in such a way that the fastest moving particles created a densely populated spherical front near a distance ct from the point of origin. At this distance the system was bounded by an impenetrable barrier or singularity of infinite particle density (Fig. 4.1). Remarkably, without further assumptions apart from the cosmological uniformity principle Milne found an outward velocity v which related to the distance r between any two galaxies as

$$r = v(t - t_0). \tag{4.1}$$

Milne commented: "If we reckon t, our present epoch, at which according to this evaluation the nebulae were close together, we may take $t = 0$ and then $v = r/t$, where the 'present' value of t is 2×10^9 years."[12] In other words, Milne derived Hubble's distance–velocity relation from kinematic considerations and without appealing to general relativity or any other law of gravitation. The Hubble constant was simply the inverse of the time elapsed since the original expansion, $t = 1/H$. As Milne pointed out, his explanation was much simpler than the one based on relativistic cosmology. Moreover, it had the advantage that it predicted an expansion, whereas the cosmology based on the Friedman–Lemaître theory was unable to differentiate between expansion and contraction without further assumptions. Had it not been for the observations of Slipher, Hubble, and Humason, the equations of relativistic cosmology would not, strictly speaking, lead to an expanding universe.

In his book of 1935 Milne deduced that the value of Newton's gravitational constant depended on the epoch according to

$$4G\rho t^2 \approx 1. \tag{4.2}$$

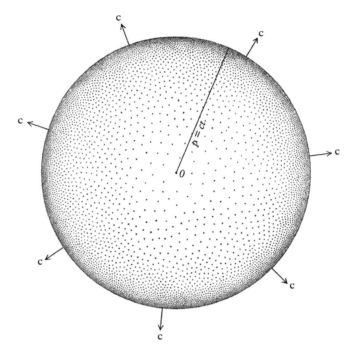

Fig. 4.1 *Milne's expanding universe. In Milne's kinematic model an observer placed in* O *would see the nebulae as bounded at distance* ct *at an epoch* t *by a barrier of infinite density, which he would take to represent the creation of the universe. Each dot represents a nebula in outward motion from the observer. The boundary is receding from the observer at the speed of light.*

Source: Milne (1935), Plate I.

Since the matter density ρ varied inversely as t^3, it meant that G increased linearly with the epoch. The result had the advantage that shortly after $t = 0$, when the particle-galaxies were closely packed, there would be no gravitation to brake the rapid expansion; with increasing epoch G would grow, but now the galaxies would be so far apart that gravitation could be neglected. Milne argued that the $G(t)$ relation did not imply that local gravitation, as in the solar system, increased in strength. In fact, he did not consider the relation to be experimentally testable.

Milnes's model corresponded to a uniform expansion of $v = c$ for all times, implying that the age of the universe was the same as the Hubble time, about 2 billion years. As a consequence it would seem that his model of the universe faced the same timescale problem as most relativistic models, but in fact this problem was avoided in Milne's theory, or so he claimed. The reason why Milne could declare it a non-problem lay in his profound analysis of the concept of time which led him to distinguish between two scales of time, so-called kinematic time (t) and dynamic time (τ). The two measures were logarithmically related according to

$$\tau = t_0 \log \left(\frac{t}{t_0} \right) + t_0. \tag{4.3}$$

The constant t_0 is the present epoch, making the two times equal at present. The relation can also be written as

$$\frac{d\tau}{t_0} = \frac{dt}{t}. \tag{4.4}$$

Based on this idea and his general, Poincaré-inspired conventionalist philosophy of physics Milne argued that the two descriptions were merely two different ways of picturing the same universe. It was meaningless to ask if the universe was *really* expanding or not. On the t-scale the universe would expand from a point-origin, hence involving a kind of big bang, but on the τ-scale the system would be static and stretch back infinitely in time.

Milne's idea of two timescales was further elaborated in collaboration with Gerald Whitrow and William McCrea and generally played an important role in cosmological thinking. For example, it influenced Dirac's view of cosmology (see below). Some later cosmologists have used Milne's idea of the two timescales to get rid of the big bang or to argue that the concept of the beginning of the universe does not follow from its age being numerically finite.[13] These later ideas have not won acceptance or proved scientifically fertile.

While ordinary cosmological theories are empirically testable, at least in principle, this was not the case with Milne's system of world physics which rested on concepts of a non-observational nature. His ideal of theoretical physics was geometry where propositions follow logically from definitions, and he often compared the results of his kinematic cosmology with geometric theorems. As he stated in a paper of 1937, within his system the laws of nature were "no more arbitrary than geometrical theorems."[14] All the same, Milne was not unconcerned with observations and especially not so if they supported his deductions. For example, he argued that his theory was superior to conventional astrophysics and cosmology in areas such as cosmic rays, the spiral structure of galaxies, and the distribution of galaxies in space. But what really mattered to him were the theory's logical structure and its rational explanation of the cosmos as a whole.

Of course, Milne had no idea of the cosmic microwave background (CMB) discovered much later and his cosmological theory was unprepared for a blackbody-distributed background radiation. Still it is interesting to note that in 1945, when suggesting his own version of big-bang cosmology, he speculated that at $t = 0$ photons would have an almost infinite frequency. Although the spectrum of the photons would now be in the low-frequency range, some of the photons of very high frequency might have survived as fossils of the primordial radiation. Milne suggested that they were part of the presently observed cosmic rays.[15]

The kinematic theory of cosmology aroused considerable interest and even more criticism. The worst thing that can happen to an alternative cosmological theory is that the scientific community ignores it. This was far from the fate of Milne's theory.

Early textbooks in cosmology, such as George McVittie's *Cosmological Theory* from 1937 and Otto Heckmann's *Theorien der Kosmologie* (Theories of Cosmology) from 1942 contained detailed if critical chapters on Milne's kinematic cosmology. The same was the case with Hermann Bondi's influential *Cosmology* of 1952 which praised Milne's theory from a methodological point of view but nonetheless concluded that "grave difficulties stand in the way of interpreting it as a valid description of the universe."[16] Bondi's deductive version of steady-state cosmology was to some extent indebted to Milne's theory and in a general way shared some of its methodological characteristics (see Chapter 5). Mainstream cosmologists in the United States such as Richard Tolman and Howard Robertson had no confidence at all in Milne's grand project. Tolman relegated Milne's theory to a single footnote in his 1934 textbook (*Relativity, Thermodynamics and Cosmology*) and Robertson concluded that the theory, although interesting and challenging, was too weird to represent the real universe. Einstein may have agreed. He reportedly "regarded Milne's brilliant mathematical mind as lacking in critical judgment."[17]

Contrary to most of his American colleagues Hubble found the kinematic theory to be valuable and even had some sympathy for it. He had met Milne on several occasions in Oxford and elsewhere. As mentioned in Section 4.7, during the 1930s Hubble vacillated between a static and an expanding universe, both of which possibilities he saw represented in Milne's theory. In his 1934 Halley Lecture, Hubble paid tribute to "Professor Milne's fascinating kinematical theory of the expanding universe" and in private correspondence he assured Milne that his theory was widely discussed in Pasadena. For his part, Milne believed that Hubble's recent observations provided empirical support for his theory: "Hubble's observations disclosed a density-distribution of nebulae increasing outwards if recession is adopted, and a homogeneous distribution is recession is denied. This is just what is predicted on the present treatment."[18]

Milne was aware of the new steady-state theory of the universe presented in 1948. Given that this theory was in sharp opposition to relativistic cosmology, just as his own alternative was, one might expect that the Hoyle–Bondi–Gold theory appealed to him. But it did not. On the contrary, he rejected in strong language the foundation of the steady-state theory and in particular its central postulate of continuous creation of matter. His dislike of the theory was in part methodological but also had roots in his Christian conviction.

The development of Milne's alternative cosmology is unique in so far that it was widely discussed and most influential for a decade or so, but then interest sharply declined. By the mid-1950s it was nearly forgotten and since then no serious attempt has been made to revive it. Still of interest to historians and philosophers, the theory has long ago vanished from the scientific scene. What is currently known as the Milne model is a special and physically unrealistic case of the Robertson–Walker model corresponding to a negative space curvature and a completely empty universe.

4.4   Varying constants of nature

In Einstein's general theory of relativity Newton's gravitational constant G, or the corresponding Einstein constant $\kappa = 8\pi G/c^2$, is just that—a constant. However, in 1937 the famous quantum physicist Paul Dirac suggested that G varies slowly in cosmic time, and the following year he developed the unorthodox idea into a cosmological model. Dirac's hypothesis was not quite the first suggestion of time-varying constants of nature, for in 1933 the Russian physicist Matvei Bronstein had suggested that the cosmological constant varied in time; and two years later, J.A. Chalmers and B. Chalmers came up with a similar hypothesis with regard to Planck's constant.[19] But it was only with Dirac that the possibility of varying natural constants became widely discussed. Although Dirac's cosmological model based on decreasing gravity attracted little initial attention, much later it inspired a whole class of varying-G cosmologies.

In a brief note to *Nature* of February 1937 Dirac discussed the significance of two very large dimensionless combinations of constants of nature. With T_0 denoting the Hubble time, or approximately the age of the universe, Dirac called attention to the combinations (in cgs units):

$$\frac{T_0}{e^2/mc^3} \simeq 2 \times 10^{39} \quad \text{and} \quad \frac{e^2}{GmM} \simeq 7 \times 10^{38}. \tag{4.5}$$

Here m and M refer to the mass of the electron and the proton, respectively, and e is the elementary charge. The denominator e^2/mc^3 is the time it takes light to pass the classical electron radius given by $r = e^2/mc^2$. The second of Dirac's numbers he took over from Eddington according to whom it was related to the number of electrons and protons in the closed universe, or what Eddington called the "cosmical number" N^*. Eddington, who since 1929 had been occupied with a new theory based on the constants of nature, derived the relationship

$$\frac{e^2}{GmM} = \frac{2}{\pi}\sqrt{N^*}. \tag{4.6}$$

A firm believer in the cosmological constant Λ, Eddington also expressed this quantity in terms of other constants of nature.[20] In 1931 he came up with the result

$$\Lambda = \left(\frac{2GM}{\pi}\right)^2 \left(\frac{mc}{e^2}\right)^4 \simeq 10^{-54} \, \text{cm}^{-2}. \tag{4.7}$$

While this was believed to be of the right order, unfortunately (or fortunately for Eddington) there were no astronomical determinations of Λ with which the theoretical value could be compared. Eddington further used the value of Λ to calculate from first principles the Hubble recession constant, for which he obtained $H_0 = 528 \, \text{km s}^{-1} \text{Mpc}^{-1}$. While Eddington's so-called fundamental theory was highly unorthodox, he did not develop it into an alternative cosmological theory. As far as cosmology was concerned his views were orthodox and based on the equations of general relativity including the cosmological term.

Dirac's interpretation of the large numbers was quite different from Eddington's. According to what Dirac called the "fundamental principle" and since 1972 the "large numbers hypothesis" (LNH), "We may take it as a general principle that all large numbers of the order $10^{39}, 10^{78}, \ldots$ turning up in general physical theory are, apart from simple numerical coefficients, just equal to $t, t^2 \ldots$ where t is the present epoch expressed in atomic units."[21] From this principle it followed that

$$G \sim \frac{1}{t} \quad \text{or} \quad \frac{1}{G}\frac{dG}{dt} \sim -\frac{1}{t}. \tag{4.8}$$

Moreover, using the same kind of numerological argument Dirac derived from the LNH that the number of protons in the universe would increase with its age according to

$$N \sim t^2. \tag{4.9}$$

In a follow-up paper of 1938 Dirac developed his numerological arguments into a quantitative cosmological model with testable predictions.[22] But he now decided to disregard the spontaneous and accelerating matter creation that he had originally derived from the LNH.[23] He thus took the average density of matter ρ to vary proportionally to R^{-3}, where R is the cosmic scale factor. Dirac's cosmological model showed distinct traces of Milne's theory and incorporated the latter's cosmological principle. From the LNH and matter conservation Dirac found that the galaxies receded according to the law

$$R(t) \sim t^{1/3}. \tag{4.10}$$

It followed that

$$\rho \sim t^{-1}. \tag{4.11}$$

He further argued that the only geometry compatible with the LNH was flat space and also that the LNH ruled out a non-zero cosmological constant. In this sense his model was somewhat similar to the Einstein–de Sitter model of 1932 with a recession law given by $R(t) \sim t^{2/3}$. But whereas the Einstein–de Sitter model was a special case of the cosmological field equations, Dirac's was not.

Dirac's cosmological theory yielded two consequences which were empirically testable in principle, but at the time not in practice, and one consequence which was contradicted by measurements. First, the gravitational constant decreased as

$$\frac{1}{G}\frac{dG}{dt} = -3H_0 \sim 10^{-11} \text{ year}^{-1}. \tag{4.12}$$

Second, Dirac's recession law led to a deceleration parameter $q_0 = 2$ for which there was no observational support. Third, it predicted an age of the universe given by

$$t_0 = \frac{1}{3}T_{\text{H}}. \tag{4.13}$$

As Dirac pointed out with an understatement, this gave the "rather small" age of 7×10^8 years or less than one third of the age of the Earth as determined by reliable radioactive methods. This was a serious problem indeed, but most cosmological models based on general relativity shared the same problem if not quite as dramatically. For example, the age of the Einstein–de Sitter universe was 1.4×10^9 years. Moreover, Dirac thought that the problem might disappear if the rate of radioactive decay varied with the epoch and was greater in the past than it is now.

With only five citations in the period 1938–1947 Dirac's cosmological theory was far from a success in the scientific community (Fig. 4.2). The only scientist who supported it wholeheartedly was Pascual Jordan, who like Dirac was a quantum theorist and an outsider to cosmological research. The theory was occasionally criticized on methodological grounds, as when the astrophysicist and philosopher Herbert Dingle accused it of being a rationalistic fantasy with no foundation in either experiment or observation. According to Dingle, Dirac's theory was a regrettable example of a "pseudo-science of invertebrate cosmythology."[24]

Only in 1948 did the nuclear physicist Edward Teller arouse some interest in the theory by arguing that it contradicted established knowledge of the surface of the Earth in the geological past.[25] From astrophysical theory Teller took the solar luminosity to vary as

$$L \sim G^7 M^5, \tag{4.14}$$

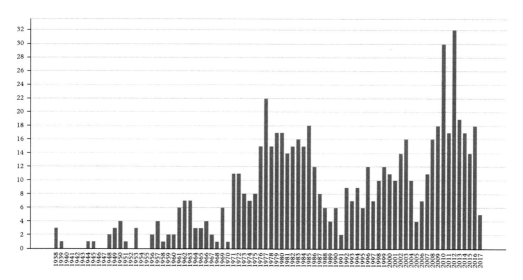

Fig. 4.2 *Citations to Dirac (1938). Notice the very few early citations, reflecting the lack of popularity of the G(t) hypothesis ca. 1938–1960. Based on data from Web of Science.*

where M is the mass of the Sun. From this and a few other assumptions, including that M = constant, he derived that on Dirac's hypothesis the temperature of the Earth's surface in the past would be related to its present temperature T by

$$T = T_0 \left(\frac{t_0}{t}\right)^{9/4}. \tag{4.15}$$

Here t_0 denotes the present epoch or approximately the Hubble time. For the Cambrian era this yielded a temperature about 110°C, which disagreed with paleontological evidence of life on Earth at that time. Consequently Teller concluded that Dirac's hypothesis was most likely wrong. With the extension of the cosmological timescale since the mid-1950s Teller's objection seemed less conclusive and especially so if the effect of cloud formation in the atmosphere was also taken into account. On the other hand, as shown by the Princeton astronomers Philip Pochoda and Martin Schwarzschild in a paper of 1964, Dirac's $G(t)$ hypothesis conflicted with the well-established standard model of the Sun and for this reason was hardly acceptable.[26] Although aware of the problems, Dirac maintained his faith in a decreasing gravitational constant as implied by the LNH.

"I am the only one who has been ready to take Dirac's world model seriously …and to consider its more precise formulation."[27] Thus wrote the German physicist Pascual Jordan, one of the founders of quantum mechanics, in a book of 1952. Jordan had been fascinated by Dirac's theory ever since the late 1930s and after World War II he developed it into his own cosmological theory inspired by the LNH and also by Lemaître's explosion hypothesis of the origin of the universe. At the same time he became increasingly interested in the connections between cosmology and geophysics, a subject that dominated much of Jordan's later scientific work.[28] According to Jordan there once had been a primeval explosion (a kind of big bang) and since then the universe had expanded linearly with time, implying that the age of the universe equals the Hubble time. Contrary to Dirac, he argued that cosmic space must have a small positive curvature and consequently be finite. On the other hand, he followed Dirac and most other cosmologists in assuming a zero cosmological constant. As Jordan presented his cosmology in the 1950s it was not only governed by a decreasing G but also by mass and average density varying as

$$M \sim t^2 \quad \text{and} \quad \rho \sim t^{-1}. \tag{4.16}$$

Jordan formulated his theory in the form of field equations that differed from those of general relativity and which he thought of as an extension of Einstein's equations (see Section 4.5). Although Jordan's extended gravitation theory only attracted limited attention in the English-speaking world, it was taken up and further developed by several German physicists, among them Kurt Just, Jürgen Ehlers, Wolfgang Kundt, and Engelbert Schücking.[29]

When Dirac returned to cosmology he retained the LNH and its consequence of G decreasing in time, but now he also returned to his original hypothesis of continual creation of matter. He vaguely described the creation process as "a new physical process, a kind of radioactivity, which is quite different from all the observed radioactivity."[30] By the 1970s an alternative cosmological theory would have to explain the existence of a Planck-distributed microwave background. Since in Dirac's theory photons were created continuously according to $N_\gamma \sim t^2$ (or in another version $N_\gamma \sim t^3$) he was unable to account in a natural manner for the blackbody spectrum. Dirac fell back on the not very convincing hypothesis that the microwave radiation was due to a recent decoupling of photons from an intergalactic medium. As Gary Steigman, an astrophysicist at Yale University, concluded in a critical investigation, "The application of the LNH to physics and cosmology is fraught with ambiguity."[31] Nonetheless, Dirac stuck to his theory.

Although Dirac's revived $G(t)$ theory was not very successful it inspired other physicists to develop more sophisticated theories along the same line. For example, Vittorio Canuto and collaborators at the NASA Institute for Space Science favored a theory consistent with the LNH and decreasing gravity. They claimed in a series of papers that their theory explained the CMB, but critics disagreed. Yet another version of $G(t)$ cosmology was suggested by Fred Hoyle and Jayant Narlikar who in the early 1970s discussed the astrophysical and geophysical consequences of their "conformal gravity" theory in much the same way as Jordan did with his theory.[32]

Whether in one version or another, Dirac-like cosmologies rested on the crucial assumption of G decreasing at a relative rate of 10^{-10} or 10^{-11} per year. During the 1970s and 1980s much experimental and observational work was done on testing the $G(t)$ hypothesis.[33] It was generally realized that geophysical and paleontological tests were unable to deliver decisive proof for or against the hypothesis. The accuracy of laboratory tests was limited to the level $10^{-7} < dG/G < 10^{-8}$, far outside the range of the theoretical predictions. Astronomical methods based on measurements of the distances to the Moon and the nearby planets over long periods of time were more promising. Using the technique of radar-echo time delays, in 1971 Irwin Shapiro and collaborators concluded that G varied less than 4 parts in 10^{10} years. However, four years later Thomas Van Flandern created a minor sensation by announcing from a study of the orbit of the Moon a positive result, namely

$$\frac{1}{G}\frac{dG}{dt} = (-8 \pm 5) \times 10^{-11} \text{ year}^{-1}. \tag{4.17}$$

Van Flandern interpreted the result as evidence for Dirac's hypothesis, but it soon turned out to be discordant with more precise data from the Viking landers on Mars.[34] To cut a long story short, data from the Lunar Laser Ranging Project showed convincingly that G is in fact constant, in agreement with Einstein's theory of general relativity. By 2007 the best result was

$$\frac{1}{G}\frac{dG}{dt} = (2 \pm 7) \times 10^{-13} \text{ year}^{-1}. \tag{4.18}$$

Despite the refutation of Dirac's hypothesis the general idea of varying gravity is still alive and continues to be explored by a minority of physicists and cosmologists.

If the gravitational constant is allowed to vary in time, why not consider the variation of other constants of nature? Speculations of this kind and attempts to incorporate them in cosmological models go back to the 1930s, the same decade in which Dirac proposed his $G(t)$ hypothesis. Scientists did not seriously expect the dimensionless fine structure constant

$$\alpha \equiv \frac{2\pi e^2}{hc} \qquad (4.19)$$

to change in cosmic time, and thus it caused much excitement when a team led by the Australian astrophysicist John Webb reported results indicating a smaller value of α in the past (Fig. 4.3).[35] Webb and his collaborators concluded in 2001 that for quasars with redshifts in the range $0.5 < z < 3.5$ the fine structure constant varied as $\Delta\alpha/\alpha = (0.72 \pm 0.18) \times 10^{-5}$. Theorists quickly produced cosmological models accommodating a varying α, but enthusiasm cooled when it was realized that the discovery claim was premature. More precise measurements made in 2004 strongly indicated that the α-constant is in fact constant. At present the most stringent constraint on the time variation of α is[36]

$$\frac{1}{\alpha}\frac{d\alpha}{dt} = (1.6 \pm 2.3) \times 10^{-17} \text{ year}^{-1}. \qquad (4.20)$$

The case of a varying speed of light is different from the cases of $G(t)$ and $\alpha(t)$, except that it can also be traced back to the 1930s. In 1987 the Russian astrophysicist V. Troitskii suggested a heterodox static model of the universe based on the idea of a decreasing speed of light. His suggestion attracted very little attention. The class of cosmological models presently known as VSL (varying speed of light) was introduced in the 1990s principally as an alternative to the popular inflation theory of the early universe. The first to suggest such a model was the Canadian physicist John Moffat in 1993 but it was only six years later that Andreas Albrecht and João Magueijo put VSL cosmology on the map of cosmological theories.[37] According to the early version of VSL theory there was no inflationary phase in the early universe which expanded in agreement with the classical Friedman–Lemaître equations. But the local speed of light was assumed to decrease drastically at a time close to the Planck epoch 10^{-43}s, from an exceedingly large value of perhaps 10^{38} km s^{-1} to its current value of just 3×10^5 km s^{-1} (other VSL models assume a very different decrease in the speed of light). It further followed from their theory that energy conservation was violated in the very early universe, so that matter can be created and destroyed without energy compensation. What matters is that Albrecht, Magueijo, and their followers were in this way able to reproduce most of the predictions of the inflation theory.

Modern VSL cosmology is definitely a heterodox theory but one which has attracted much attention and been developed into a variety of theories by many physicists. It has also been severely criticized from an observational and a methodological point of view. After a brief period of popularity, interest in the theory declined somewhat. Until the

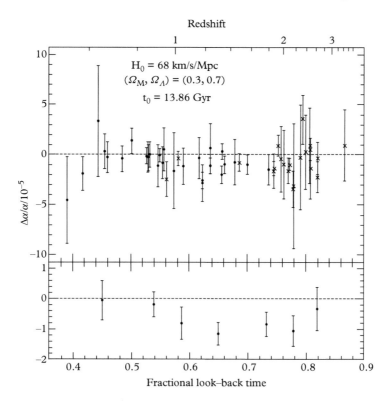

Fig. 4.3 *Data from 2001 showing the relative change $\Delta\alpha/\alpha$ of the fine structure constant as inferred from the light received from distant quasars. The upper graph shows all raw data with error bars, while in the lower one the same data are collected in groups. They are shown as a function of redshift and look-back time, that is, the time it takes before light reaches us. The data indicates a negative shift in the redshift range $1 < z < 3$. Reproduced with permission from M. T. Murphy et al., "Possible evidence for a variable fine-structure constant from QSO absorption lines,"* Monthly Notices of the Royal Astronomical Society *327 (2001): 1208–22.*

end of 2016 the Albrecht–Magueijo paper of 1999 has received 250 citations (Web of Science). Although there are still a fair number of physicists working within the framework of VSL theory, the theory has not succeeded in seriously challenging the inflation paradigm of the very early universe. However, work on or inspired by VSL theory continues to this day. Thus, in 2016 Magueijo and Niayesh Afshordi published a "critical model" in which photons and other massless matter particles travel faster than gravity.[38] Characterizing the standard hot big-bang model as "an unfinished work of art" they derived several precise predictions from the model, including that no gravitational waves are produced during the big bang. The two authors stressed the predictivity and vulnerability of their model: "Improved observations will soon vindicate or disprove this model."

4.5 Brans–Dicke–Jordan theory

The basis of all standard cosmological models is Einstein's general theory of relativity, where gravitation is fully described by the geometry of space-time as given by the metrical tensor. Although Einstein's theory works very well indeed, it is not unreasonable to assume that it can be extended and generalized to work even better. According to the class of scalar-tensor (ST) theories developed in the 1950s and 1960s a scalar field ϕ must be added to the tensor equations to account for the measurable value of the gravitational constant G. Whereas the scalar field is generated by matter in the universe, it does not exert any direct influence on matter and only indirectly influences the geometry of space-time. The value of the new ϕ field depends on the point in space-time, $\phi = \phi(x, y, z, t)$. The same is the case with the effective value of $G = 1/\phi$, for which reason there is a connection to theories of the Dirac type where G varies in time. ST theories are not specifically or only cosmological theories but alternative theories of gravity in general. Although they lead to alternative models for the universe, these are not particular heterodox and contrary to some other alternative models they are not considered outside mainstream cosmology. We limit our historical review to the period ca. 1950–1970.

The ST formalism became well known only after the two Princeton physicists Carl Brans and Robert Dicke published an important paper on it in 1961. Within a few years the paper became recognized as a most interesting alternative to standard general relativity. Until the end of 2017 it has received a total of 3,320 citations in the scientific literature, of which 344 were in the years 2016 and 2017 (Web of Science). However, the mathematical framework of the ST formalism was established two decades before the Brans–Dicke paper. As mentioned in Section 4.4, it was part of Jordan's project of extending general relativity.[39] The kind of theory is often known under the names of Brans and Dicke alone, but Jordan–Brans–Dicke or Brans–Dicke–Jordan is also used.

While Jordan had originally assumed that the mass of the universe increased according to $M \sim t^2$, in 1959 he concluded that the hypothesis of matter creation was untenable and instead proposed a new mass-conserving version of his ST theory. For nearly a decade Jordan maintained that gravity decreased in Dirac's form $G \sim 1/t$, but when confronted with objections based on the CMB he once again revised his theory. As pointed out by two German astrophysicists, Heinz Dehnen and Helmut Hönl, the problem with Jordan's cosmological theory was that it resulted in an energy density of the background radiation increasing in time and with a spectrum that deviated from the blackbody curve.[40] Consequently, in 1968 Jordan modified the ST field equations in such a way that G varied more slowly, namely as

$$G \sim t^{-n} \quad \text{with} \quad n < 1. \tag{4.21}$$

The move did not imply that Jordan now abandoned Dirac's hypothesis, but only that he separated it from the gravitational field equations and instead considered it to be an independent hypothesis. Jordan was pleased to note that his new formulation was equivalent to the one of Brans and Dicke and, he thought, the only plausible generalization of Einstein's gravitation theory.

To construct a new theory of gravitation "which is more satisfactory from the standpoint of Mach's principle than general relativity" Brans and Dicke replaced the gravitational constant with a long-range scalar field ϕ generated by the celestial bodies.[41] The reference to Ernst Mach's principle was significant, for Dicke was a strong believer in the principle of the Austrian philosopher-physicist which served as the philosophical beacon guiding much of his work in gravitation and cosmology. Dicke argued that it followed from Mach's principle that the mass M of the visible universe was related to its space curvature R by

$$\frac{GM}{Rc^2} \simeq 1. \tag{4.22}$$

Thus, as the universe expands and M and R change, G changes accordingly. Jordan did not share Dicke's fascination of Mach's principle and flatly denied that it was of any value in theoretical cosmology. On the other hand, he accepted the relationship $GM/Rc^2 \simeq 1$ but interpreted it as indicating a universe with zero total energy, such as he had suggested as early as 1939. The same relationship had been derived by Dennis Sciama in 1953 and later, more or less independently, by other scientists.[42]

Contrary to the early version of Jordan's theory, Brans and Dicke constructed their ST theory in such a way that it accommodated energy conservation in agreement with standard general relativity. According to the two Princeton physicists the locally measured value of G did not only depend on ϕ but also on a dimensionless parameter (ω) sometimes known as the Brans–Dicke coupling constant. The ω parameter was not given by theory, but could only be determined by observation. For this reason and also because the Brans–Dicke theory admits more solutions than the Einstein theory, it is less stringent than standard general relativity.

Assuming that "the universe expands from a highly condensed state," Brans and Dicke examined various cosmological models modified by the hypothesis of a slowly decreasing gravitational constant. In the lower limit $\omega = 0$ the equations of the Brans–Dicke theory described a model universe in which distance and gravity varied as

$$R \sim t^{1/3} \quad \text{and} \quad G \sim t^{-1/2}. \tag{4.23}$$

On the other hand, in the limit $\omega \to \infty$ the equations passed asymptotically over into those of general relativity. To secure agreement with astronomical observations and the anomalous precession of Mercury's perihelion in particular, Brans and Dicke settled on a value $\omega \approx 6$ or possibly $\omega > 6$. Unfortunately the value of ω consistent with observations turned out to be much larger, but this was not yet known in the early 1960s. With $\omega = 6$ and H_0 the present value of the Hubble parameter Brans and Dicke found gravity's variation to be given by

$$\frac{1}{G}\frac{dG}{dt} = -\frac{H_0}{7} = 1.2 \times 10^{-11} \text{ year}^{-1}. \tag{4.24}$$

This was about twenty times less than in Dirac's theory, where it was 2.4×10^{-10} per year. For the age of the universe expressed in terms of the Hubble time they obtained $T_0 = 14T_H/22$ or close to the value of $2T_H/3$ of the Einstein–de Sitter cosmological model. For $\omega > 6$ their model for a flat-space universe was practically the same as the Einstein–de Sitter model based on ordinary general relativity.

Dicke was greatly interested in the terrestrial consequences of the ST theory which he examined in a variety of ways. One of them was the problem of the Earth's surface temperature in the geological past mentioned in the previous section. Dicke concluded that the problem of the hot young Earth was much less serious in his and Brans' theory than it was in Dirac's. He also concluded that the Brans–Dicke theory was the only theory able to explain the disagreement between the age of the universe and various evolutionary ages of stars and galaxies.[43] Whereas Jordan did not take part in the transformation that changed cosmology in the 1960s—the discovery of the CMB came as an unwelcome surprise to him—Dicke was crucially involved and one of the fathers of the new hot big-bang theory. It was to a large extent the Brans–Dicke theory which guided his thinking about the early universe and in 1963 led Dicke to consider an initial hot phase filled with an intense blackbody radiation.

The significance of the Brans–Dicke theory is evident from an extensive review paper that Dicke and Peebles submitted in early March 1965, shortly before they became aware of the CMB detection by Arno Penzias and Robert Wilson the previous year. Discussing the possibility that the present universe might have a radiation temperature well below 10 K the two authors referred to the decreasing gravitational constant of the Brans–Dicke theory. "The universe," they wrote, "would have expanded through the early phase very much faster than is implied by general relativity [and] thus reduce the time available for helium production, thus reducing the lower limit on the present radiation temperature."[44] A passage to the same effect appeared in the seminal paper in the July 1965 issue of *Astrophysical Journal* in which Dicke and his co-authors analyzed and interpreted the CMB. A few years later Dicke returned to the problem of helium production in ST cosmology, but his result was discouraging.

The Brans–Dicke ST theory was not necessary for the revival of hot big-bang cosmology in the mid-1960s but it nonetheless happened to play an important role. In Dicke's long-term research program the core was the ST theory of gravitation and the universe one of several testing grounds.

Other testing grounds were geophysical and others again astronomical in the more traditional sense. The problem of the assumed oblate shape of the Sun and its possible effect on Mercury's perihelion advance had first been noticed by Simon Newcomb in 1865. A century later it was reconsidered by Dicke, who realized that it might be used to discriminate between general relativity and the Brans–Dicke ST alternative. Together with his Princeton colleague Mark Goldenberg he designed a new telescope by means of which the solar oblateness could be measured with greater precision than previously. The measurements published in 1967 were interpreted as a flattening of the Sun's disk of such a magnitude that it would contribute significantly to Mercury's perihelion shift. The result seemingly supported the Brans–Dicke theory with $\omega \simeq 6$, whereas it disagreed with general relativity. However, the Dicke–Goldenberger paper was received with skepticism and not generally accepted.[45]

Subsequent measurements and theoretical analyses made by Henry Hill and others showed that the solar oblateness is too small for exerting an effect on Mercury's orbit around the Sun. Within a decade the consensus view was that the best data were fully consistent with general relativity. To be sure, the data could be reproduced on the basis of the Brans–Dicke theory, but then it would require $\omega > 500$ in which range the two theories were indistinguishable. Understandably, the large value of ω required to satisfy the experimental constraints made the Brans–Dicke theory unattractive since it amounted to a fine-tuning of the ω parameter.

And yet theories of the Jordan–Brans–Dicke type continued to attract interest and other theories of a similar kind, extensions of general relativity, were developed in the 1970s. These included a theory of "conformal gravity" proposed by Fred Hoyle and Jayant Narlikar and a "scale-covariant" theory proposed by Vittorio Canuto and others. ST theories and related theories are still being investigated, but they are no longer considered serious rivals to Einstein's theory of general relativity. As Brans commented in a review from 2014:

> In fact, the results of increasingly accurate tests in the solar system and beyond have continually increased the lower limit on ω to such an extent that interest in the BD [Brans–Dicke] variation from standard Einstein theory has been greatly diminished. Einstein's original formulation seemed to be as well suited to solar system observations as alternative theories. Consequently by the 1970s interest in ST had greatly declined.[46]

Despite the decline in interest, ST theories of gravity are still being investigated by a minority of physicists and cosmologists who study the use of scalar fields in areas as diverse as superstring theory and inflationary scenarios of the early universe.[47]

4.6   Plasma models, antimatter, and the metagalaxy

Dirac's prediction in 1931 of what he called the anti-electron, which was soon to be renamed the positron, had its basis in pure quantum physics, namely his relativistic wave equation of the electron. Two years later, when receiving the Nobel Prize in Stockholm, Dirac suggested that hypothetical antiprotons might combine with positrons and form anti-atoms. At the time the atomic nucleus was still considered to consist of protons and electrons, so he needed only two antiparticles. Perhaps, Dirac speculated, there might even be stars made up entirely of antiparticles, the number of anti-stars being equal to the number of ordinary stars. The concept of antimatter (if not the name, see Section 1.3) was thus introduced by Dirac, but it was obviously highly speculative and for more than two decades it attracted very little interest. When Dirac in 1937 took up cosmology he did not mention antimatter.

The hypothetical antiproton became a real particle in 1955 when physicists at Berkeley's Bevatron accelerator succeeded in producing and detecting antiprotons.

Shortly later they were also found in cosmic rays, and proton–antiproton annihilations according to

$$\bar{p} + p \rightarrow 2h\nu \qquad (4.25)$$

were first detected. Two years later the existence of antineutrons was established. The discoveries encouraged physicists and now also astrophysicists to take the question of cosmic antimatter seriously. How is it that our world seems to consist of matter only? Is it possible that antimatter exists in abundance elsewhere in the universe as Dirac had originally speculated? Yet another reason why antimatter became fashionable in the 1950s was the contemporary formulation of the fundamental invariance principle known as the *CPT* theorem. The theorem combines particle-antiparticle symmetry or charge conjugation (*C*) with parity invariance (*P*) and time reversal (*T*). Given the elevated status of the *CPT* theorem there were strong theoretical reasons to believe in the existence of antimatter as symmetric to matter.

Antimatter entered cosmology in the year of 1956, when two papers on the subject appeared. As pointed out by Maurice Goldhaber, an Austrian-born physicist at Brookhaven National Laboratory, all existing theories of physical cosmology made the "unsatisfactory tacit assumption" that matter creation is not followed by the creation of antimatter.[48] In what he thought of as a modification of Lemaître's primeval-atom model, Goldhaber speculated that perhaps our cosmos has a counterpart in the form of an "anticosmos" which was separated off from the cosmos at the beginning of time. At a conference on astrophysics held in May 1957 in the Vatican gardens, Edwin Salpeter asked Lemaître if the creation of antimatter was part of his picture of the early universe. Lemaître's response was evasive except that he stated that his theory was "not reconcilable with a large region of anti-mass."[49]

In a somewhat less speculative manner than Goldhaber, Geoffrey Burbidge and Fred Hoyle suggested that half of the particles created continually in the steady-state theory were antiparticles. However, the calculated rate of proton-antiproton annihilation processes, or the resulting flux of high-energy photons, disagreed with measurements of the cosmic gamma rays intensity made a few years later.[50] Burbidge and Hoyle quickly shelved the idea of a symmetric steady-state universe.

Much work on the cosmological role of antimatter was done in the 1960s and 1970s. A main problem was the separation mechanism which seemed necessary to keep matter and antimatter apart instead of annihilating. The French physicist Roland Omnès examined this and related problems within the framework of the new hot big-bang theory, suggesting a separation mechanism operating in the very early and very dense universe.[51] Although Omnès' cosmological matter–antimatter model differed from most models based on the standard theory, it shared with them the assumption of a hot big bang expanding in agreement with general relativity theory. In this respect it was orthodox or at least not very unorthodox. But other antimatter models developed in the same period questioned the big-bang assumption and were in part designed as alternatives to the standard theory. The most important of the alternative theories was developed by the

Swedish physicist Hannes Alfvén who soon emerged as a leading critic of what he called the myth of the big bang.

Alfvén had started his career with studies of cosmic rays, magnetic storms and the theory of the aurora. He considered himself the heir of the long deceased Norwegian physicist Kristian Birkeland, a specialist in the aurora borealis and a pioneer of plasma physics *avant la lettre*.[52] Much like Birkeland he thought that electromagnetic fields were as important as or even more important than gravitation in the understanding of cosmic phenomena. In the 1940s Alfvén developed an important if controversial theory of the formation of the solar system and in 1970 he crowned his scientific career with the reception of the Nobel Prize in physics for his work on magneto-hydrodynamics, the study of plasmas in magnetic fields. He ended his Nobel Lecture with the words, "in the beginning was the plasma." By the time he received the prize he had become seriously involved in cosmology, a branch of science he thought was on a false track and had deteriorated into metaphysics. Alfvén's turn to cosmology was indebted to his compatriot and colleague at Stockholm's Royal Institute of Technology, the theoretical physicist Oskar Klein.

An important figure in the early development of quantum mechanics and a co-founder of five-dimensional (Kaluza–Klein) relativity theory, Klein had in the 1950s become interested in problems of cosmology and cosmogony. Critical to both relativistic evolution theories and the rival steady-state theory, Klein emphasized that cosmology, like any other science, should be based on known laws of physics. At a conference on nuclear astrophysics held in Liège in 1953 he suggested a cosmological theory which was in part motivated by Eddington's numerical relations between the constants of nature and which he knew about from Bondi's 1952 textbook in cosmology. But while the relations were usually regarded as expressing some hidden connection between microphysics and macrophysics, "I thought that it would be worthwhile to look for a natural origin of them, which might tell us something about the main forces at work in the development of the original contracting cloud to the present expanding state of the metagalaxy."[53]

Klein assumed as a starting point for his cosmological scenario an extremely dilute cloud of very cold hydrogen. As a result of the opposite forces of gravitational contraction and radiation pressure the observed "supergalactic system" would be formed and its present expansion explained as an effect of the radiation pressure. In Klein's scenario the initial state of the supergalactic system of density $\rho_0 = 6 \times 10^{-27}$ g cm^{-3} had expanded by a factor of 2.7 and the outermost galaxies now had a recession velocity of 0.4 times the velocity of light. Although Klein was at the time thinking about the role of antimatter, it was only later on that he assumed that the initial cloud was a matter–antimatter mixture. It was the problem of separating the two components that brought him into contact with Alfvén and his expertise in plasma physics. The result was a paper of 1962, the first version of Klein–Alfvén cosmology or what became known as plasma cosmology or sometimes symmetric cosmology.[54] In a series of later publications this kind of cosmological theory was mostly developed by Alfvén and some of his Swedish collaborators.

The universe that Klein and Alfvén had in mind was the observable universe and not the unbounded universe of relativistic cosmology. The two Swedes preferred to speak of the *metagalaxy* (corresponding to the supergalactic system), a term which has a meaning different from what is usually called the universe. The term was introduced by Harlow Shapley in a paper of 1934, but it was only some twenty years later that it became widely used and then typically associated with unorthodox cosmological views. Whether known to Klein and Alfvén or not, it was commonly used in the Soviet Union where the proper domain of cosmology was considered to be the assemblage of observable galaxies and clusters of galaxies rather than the universe as a whole. Indeed, the concept of the universe in the wider sense was held to be metaphysical and bourgeois. There is a surprising analogy between the views and rhetoric of the few Western scientists in favor of plasma cosmology and the views held by Soviet scientists until the late 1960s. For example, the distinguished astrophysicist Victor Ambartsumian (who was a devout Marxist) denied any extrapolation from the metagalaxy to the universe as a whole and he also denied that the metagalaxy was a homogeneous cosmic system.[55] Alfvén and his small group of plasma cosmologists agreed.

The characteristic feature of Klein–Alfvén cosmology as it existed in the late 1960s was the assumption of a universe with equal amounts of matter and antimatter separated by cosmic electromagnetic fields. Ordinary matter was called "koinomatter" (koinos = common or well known) and for a plasma consisting of both koinomatter and antimatter Alfvén coined the term "ambiplasma." Alfvén suggested that the metagalaxy consisted originally of a very dilute and cold mixture of protons and antiprotons but that the two components were separated into distinct regions by means of a plasma mechanism. Due to gravitational condensation, annihilation processes would increase and produce a radiation pressure that eventually would halt the collapse and turn it into the observed expansion (Fig. 4.4). In the words of Alfvén:

> The annihilation generates radiation, particularly in the form of gamma rays and radio waves. The pressure imparted by this "radiation explosion" stops the contraction and changes it to expansion, which is occurring at the present time, as is manifested in the red shift of the galaxies.... Data from nuclear physics tell us how great a density ambiplasma must have to produce an annihilation forceful enough to reverse the process of contraction to expansion. It appears that the density at the turning point must have been of the order of 1,000 particles per cubic meter.[56]

The plasma universe thus included a kind of cosmic collapse and a subsequent explosion, but the explosion did not come from a singularity or a primeval atom but from a region of vast size, estimated to be at least 10^9 light years. Moreover, the explosion was a process occurring over a long period of time; and it did not cover the entire universe but only what became the metagalaxy. The question of where the protons and antiprotons in the original cloud came from was outside the framework of Alfvén's theory, which postulated an eternal universe.

While the standard models of relativistic cosmology assumed the cosmological uniformity principle and hence that the universe is homogeneous on a very large scale, Alfvén emphatically disagreed. Relying in part on an influential paper of 1970 by the French astronomer Gérard de Vaucouleurs, advocates of the plasma cosmology denied

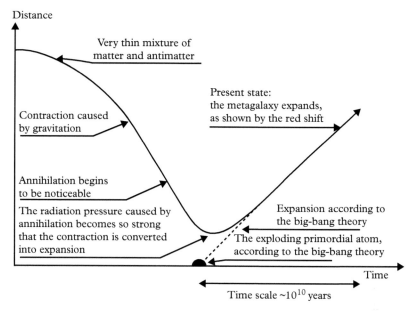

Fig. 4.4 *Alfvén's plasma universe.*
Source: Alfvén (1966), p. 78.

the meaningfulness of one of the parameters of standard cosmology, the average density of matter in the universe as a whole (see also Section 4.9). Yet another way in which the plasma universe differed from the one of standard cosmology was that it largely disregarded general relativity, the very foundation of conventional cosmological theory. Klein and Alfvén were not anti-relativists but they argued that for all practical purposes Einstein's theory could be ignored. Alfvén and his followers consequently described the evolution of the metagalaxy in terms of three-dimensional Euclidean space, seeing no reason to make use of curved space-time and other difficult concepts based on general relativity. They assumed space as well as time to be infinite in extension. The metagalaxy was however a finite and bounded system perhaps floating around in an infinite sea of emptiness.

In his early papers on the metagalaxy Klein had considered the possibility that there might be other structures in the universe of a similar enormous size. That is, he had speculated of a kind of multiverse. Inspired by Klein and de Vaucouleurs' recent data Alfvén introduced in the late 1970s a revised version of plasma cosmology according to which there had not been just a single metagalactic explosion but a whole series of them. The picture he outlined was now a hierarchical universe in which the explosion of the primeval cloud of ambiplasma resulted in its fragmentation into many smaller clouds or metagalaxies. First secondary clouds would be ejected and these would also explode, giving rise to tertiary clouds and so on.[57] In all cases the explosive power was provided by matter-antimatter annihilation. Ironically, Alfvén called his new model the "fireworks universe," possibly unaware that Lemaître had used the same metaphor for his original big-bang theory in the early 1930s. Of course, Alfvén's fireworks model was entirely different from Lemaître's model.

In whatever version of plasma cosmology the theory's explanatory and predictive power was limited. Alfvén and his group claimed that the theory agreed better with observations than the standard big-bang theory, but the large majority of astrophysicists and cosmologists was unimpressed. There was no evidence at all of large amounts of antimatter in the observed universe. Moreover, could the plasma cosmologists offer an alternative explanation of the CMB and the nucleosynthesis of the light elements? If so, how did their explanations compare to the ones derived from big-bang cosmology? In his publications from the late 1960s Alfvén disregarded the CMB and also largely the formation of helium and other light isotopes. Klein briefly addressed the CMB problem in a paper of 1971, arguing that the existence of the "so-called fireball radiation" was compatible with the model he and Alfvén preferred. But it was only in the 1990s that Eric Lerner came up with an alternative explanation of the CMB and the small variations in isotropy found in the COBE satellite experiment.[58] Using concepts from space plasma physics and tired-light cosmologies Lerner hypothesized a mechanism where magnetized plasma filaments in the intergalactic medium produced the observational results. His explanation was ignored or rejected as artificial and inadequate by mainstream cosmologists.

According to one of the mainstream cosmologists, Jim Peebles, "there is no way that the results [of plasma cosmology] can be consistent with the isotropy of the cosmic microwave background and X-ray backgrounds."[59] Peebles further pointed out that the scenario of plasma cosmology was "highly unlikely" because it presupposed that we are near the center of the metagalaxy and that the Milky Way system thus have a special status in the universe. One more reason not to take plasma cosmology seriously was that it denied the existence of dark matter. There is no doubt that Peebles' view was shared by most cosmologists, who preferred to ignore the plasma alternative. At least in the eyes of some mainstream cosmologists, plasma cosmology was not only heterodox but also tended towards pseudo-science. Characteristically, most of the contributions to this theory were published in the journals of the IEEE (Institute of Electrical and Electronic Engineers) and not in the established research journals devoted to astrophysics and cosmology. Anthony Peratt, a plasma physicist at the Los Alamos National Laboratory, was active in organizing the opposition and promoting the plasma approach as an alternative to mainstream cosmology. He and a few others succeeded in raising some interest in the public media, but their efforts made no impact on mainstream relativistic cosmology.[60]

In Alfvén's vociferous critique of the big-bang theory scientific arguments appeared along with arguments of a sociological, philosophical and ideological nature, much as they did in the critique launched by other big-bang opponents. While plasma cosmology differed completely from the late versions of steady-state theory, the two alternatives had much in common on the rhetorical level. The statements of Alfvén in the 1980s and those of Hoyle and Geoffrey Burbidge a decade later were remarkably similar. In a paper titled "Cosmology: Myth or Science?" Alfvén firmly placed the big-bang theory in the category of myths, suggesting that it presupposed divine creation. He further complained that with the advent of relativistic cosmology the study of the universe had become monopolized by big-bang believers and specialists in general relativity:

> No one else is allowed to have any views about cosmology. Textbooks on "modern cosmology" start with general relativity and often do not even mention the existence of heretical views. Still more serious ...the increasing number of observations which prove the Big-Bang hypothesis to be wrong are swept under the rug.[61]

Alfvén was not only a controversialist with a reputation as an eccentric (so was Hoyle), he was also an outsider in the fields of astrophysics and cosmology. This was undoubtedly one of the reasons why plasma cosmology remained a peripheral theory.

4.7   Redshifts without expansion

The expanding universe, as it emerged in the early 1930s, was observationally based (and only based) on the galactic redshifts and the Hubble law relating redshifts to distances. If the redshifts could be explained otherwise it might still be possible to retain the traditional concept of a static universe. Although not all alternative ideas of the origin of the redshifts deny cosmic expansion, most "tired light" hypotheses do. The name generally refers to the idea that photons slowly lose energy on their long journey through space and therefore arrive at the observer with an increased wavelength. According to this view the galactic redshifts are not cosmological in nature and not peculiar to the galaxies. The name "tired light" may have been coined by Howard P. Robertson who in a semi-popular address of 1932 on the expanding universe referred to the hypothesis that "the observed red shift would be due to the properties of 'tired' light rather than the nebulae themselves."[62] Robertson found explanations of this kind to be unsatisfactory and *ad hoc*. In this section we review some of the tired-light hypotheses with an emphasis on the early period. Since in most cases hypotheses of this kind did not involve specific cosmological models it will be more appropriate to call them tired-lights mechanisms rather than tired-light cosmologies.

Evidence based on Hubble's publications suggests that he never embraced a universe in expansion. He consistently adopted an agnostic and empirical attitude, stressing that the explanation of the redshift-distance relation could not yet be decided by means of observation. Although Hubble only reluctantly entered what he, in his *The Realm of the Nebulae* of 1936 called the "dreamy realms" of cosmological theory, on some occasions he suggested that a non-recession explanation was preferable to one based on expansion. "It is easier," he wrote in 1929, "to suppose that the light-waves are lengthened and the lines of the spectra are shifted to the red, as though the objects were receding, by some property of space or by forces acting on the light during its long journey to the Earth."[63] In his later work his was more ambiguous. In fact, he had no particular sympathy for tired-light explanations, which he found suspicious for methodological reasons.

The first suggestions of tired-light explanations appeared shortly after Hubble's announcement of the empirical law named after him. Aristarkh Belopolski, a respected Russian astronomer, suggested in a note of 1929 that a photon emitted with energy $h\nu$ would be received by an observer at distance r from the source as $h\nu/r$.[64] More importantly, Fritz Zwicky discussed slightly earlier several possible explanations of the spectral shifts, among which he focused on what he called "a gravitational analogue of the Compton effect." A photon of frequency ν has a gravitational mass $h\nu/c^2$ and therefore should be able to transfer momentum and energy to an atom; as a result of the recoil the photon's frequency would be diminished. Zwicky calculated that a photon travelling a distance r would be redshifted by the amount

$$\frac{\Delta\nu}{\nu} = \frac{1.4\,G\rho D}{c^2}\,r. \tag{4.26}$$

The quantity $D \gg r$ is a measure of the distance over which the gravitational "drag" operates, and ρ is the average density of matter in the universe. Assuming D to be of the order 3×10^{27} cm he arrived at frequency shifts in "qualitative accordance with all of the observational facts known so far."[65] At the time he did not present the theory as an alternative to the relativistic view of the expanding universe, which was still in the future. Zwicky solely proposed a rival interpretation of the redshifts and not, either in 1929 or in his later papers, a cosmological model; nor did he use the term "tired light." Although the literature on cosmology contains references to "Zwicky's model" there never was such a model.

Methodological and observational arguments entered side by side in the early discussion concerning the origin of the galactic redshifts. As to the first, Zwicky claimed in 1935 that it is "scientifically more economical *not* to link the redshift from nebulae with any *purely hypothetical* curvature and expansion of space."[66] However, as Zwicky argued for his theory in methodological terms, so Robertson criticized it by appealing to similar standards of evaluation. "I should prefer," he wrote, "to wield Occam's razor on all ad hoc explanations of the red shift and accept that one which follows so naturally from our present views of the nature of the physical world."[67]

As to observations, Zwicky was keen to point out that his theory had predictive power and led to empirically testable consequences. Thus, an initially parallel beam of light would, on this theory, gradually open itself because of small angle scattering. Moreover, it followed that the redshift should depend on the distribution of matter in space and that redshifts from within the Milky Way should therefore depend on the direction. For observational support Zwicky referred to discussions with the young German astronomer Paul ten Bruggencate, who at the time was working at the Mount Wilson Observatory and was acquainted with the results obtained by Hubble and Humason. In a study of the radial velocities of globular clusters on the assumption of the gravitational drag hypothesis Bruggencate concluded, that the number of stars required to bring the observed redshifts into agreement with the hypothesis was justified.[68]

What came to be known as the Tolman surface brightness test was originally conceived by Tolman in 1930. In an important paper of 1935 Hubble and Tolman derived expressions for the variation of surface brightness S with redshift $z = \Delta\lambda/\lambda$. They found different relations for simple expanding models and tired-light models, namely

$$S_{\text{EM}} \sim (1+z)^{-4} \quad \text{and} \quad S_{\text{TL}} \sim (1+z)^{-1}. \tag{4.27}$$

However, due to lack of reliable data for the surface brightness the test was ineffective at the time.[69] Roy Kennedy and his collaborator Walter Barkas at the University of Washington devised an interferometer experiment to test tired-light hypotheses of the kind proposed by Zwicky, concluding in 1936 that the hypotheses disagreed with the Hubble law.[70] However, although the Kennedy–Barkas experiment was probably well known, it had little impact on the cosmological debate.

Zwicky's hypothesis was the best known and most elaborate alternative to the expanding universe, but it was far from the only one. More than a dozen physicists, astronomers and amateur scientists proposed in the 1930s tired-light ideas having in common the assumption of nebular photons interacting with intergalactic matter to which they transferred part of their energy. If a photon lost its energy in proportion to its distance r from the source according to

$$\nu(r) = \nu_0 \exp\left(-\frac{H}{c}\right) r, \tag{4.28}$$

the linear Hubble law could be reproduced in the form

$$\frac{\Delta\nu}{\nu} \simeq \frac{H}{c} r. \tag{4.29}$$

Among those who argued for a static universe and redshifts produced by tired-light mechanisms were the physicist John Q. Stewart, the astronomer William D. MacMillan, and the chemist Walther Nernst.

A decade after the discovery of the expanding universe most physicists and many astronomers accepted that the galactic redshifts were due to recession, if not necessarily to the relativistic expansion of space. Tired-light theories were not highly regarded by mainstream astronomers and physicists, who tended to conceive them as speculative and based on arbitrary assumptions with no support in known physics. Few of the suggestions were specific enough to be testable and those which could be tested, such as Zwicky's theory or ideas based on a varying speed of light, turned out to disagree with experiments and observations. Astronomers had for decades been used to stellar Doppler shifts and could therefore regard the galactic redshifts as just an extension of previous practice. Alternative theories were typically mentioned as a possibility, not because they were attractive but because they offered an explanation to the serious timescale problem of the expanding universe.

There was also a social dimension involved in that many of the alternative proposals were made by amateur scientists or people with no recognition in the fields of cosmology and extragalactic astronomy. A few of them, including Zwicky and Stewart, were recognized scientists who published their theories in, for example, *Physical Review*. More commonly articles on alternative explanations appeared in less reputed journals, often in the letter section of *Nature*, in *Astronomische Nachrichten*, in *Popular Astronomy*, or in *Journal of the Franklin Institute*. Notably, none of the alternatives were published in the leading journals of astrophysics (such as *Astrophysical Journal*, *Zeitschrift für Astrophysik*, and *Monthly Notices*). The lack of scientific interest and visibility of tired-light publications is illustrated by the very few citations to them in the scientific literature.

By and large, and despite Hubble's reservations, by 1940 the static universe was no longer part of mainstream astronomy. On the other hand, it had not yet been replaced by the expanding universe in the sense of relativistic cosmology. There is some truth in Peebles' statement, referring to the early 1930s, that "even then people appear not to have found it [the Hubble phenomenon] too exciting because it is so hard to think of any physical effect other than expansion that would fit the observations."[71] But it depends on two provisos. First, that "people" is limited to professional scientists acquainted with extragalactic research and, second, that "expansion" is understood as also referring to recession. Moreover, hard as it is to think of non-expansion alternatives this did not discourage people from doing just that.

After World War II the reputation of tired-light hypotheses as support of a static universe was low. Describing those who denied the expansion of the universe as "conservative spirits," the French astronomer Paul Couderc wrote:

> The vanity and sterility of twenty years' opposition to recession is characteristic of a poor intellectual discipline. To hunt for an ad hoc interpretation, to search for a means of sidestepping a phenomenon which is strongly indicated by observation simply because it leads to "excessive" conclusions is surely contrary to scientific method worthy of the name.[72]

But only two years later the German-British astronomer Erwin Finlay-Freundlich revived interest in tired-light hypotheses. For stellar redshifts he suggested the formula

$$\frac{\Delta\lambda}{\lambda} = AT^4 r. \tag{4.30}$$

Here T denotes the temperature of the radiation field through which the light passes and r is the length of its path; A is a constant of value 2×10^{-29} cm^{-1} deg^{-4}. Although Finlay-Freundlich's proposal was aimed at stellar redshifts, he believed that the formula also applied to the cosmological redshift and that its physical mechanism might be a kind of photon-photon interaction.[73] The paper attracted considerable attention, positive as well as negative, and was followed by several other theories of tired-light mechanisms proposed by, among others, Jean-Pierre Vigier, Jean-Claude Pecker, Paul LaViolette, Grote Reber, and Sergey Stepanov.

The many tired-light hypotheses of the last half century or so have occasionally been published in mainstream science journals but more frequently in less reputed journals or, more recently, on internet sites of a dubious academic quality. They all have a high degree of *ad hoc*-ness and have been shown either to be in conflict with observations or rest on unverifiable assumptions. The last time that a tired-light paper was published in one of the premier journals of astrophysics and cosmology may have been in 1986.[74] While non-expansion hypotheses were considered a possibility in the 1930s, if generally viewed with some suspicion, and a remote possibility in the 1950s, this is no longer the case. According to the large majority of modern astrophysicists and cosmologists they belong to the fringe of science.

4.8   Cyclic models of the universe

As mentioned in Chapter 1, the idea of a cyclic or oscillating universe was popular in the last part of the nineteenth century. With the advent of evolutionary cosmology based on the field equations of general relativity theory, the idea returned in a new shape, now formulated as mathematical models that might or might not correspond to the real universe.[75] Milne was among the majority of cosmologists who did not believe that the universe evolved in cycles. His rejection of cyclic models was emphatic and emotional: "They are the fantastic weavings of the mathematical loom, orgies of mathematical license, divorced from experience. They are possible only in the sense that in a dream everything is possible."[76]

   An oscillating universe—also known as cyclic, pulsating or periodic—consists of several cycles and perhaps an infinite number of them. A single cycle from a big bang to a big crunch—sometimes known as the Friedman–Einstein model—is one of the solutions to the Friedman equations, but since there are no repeated rebirths of the universe in this model, it does not count as truly cyclic. Nor is this the case with models in which the universe contracts from past eternity to a singularity and subsequently expands indefinitely into the future. Such bouncing models, which occur in some models built on quantum gravity (Chapter 12), cannot be ascribed a period. Cyclic models considered in this section are alternative only in the sense that they are not solutions to the canonical field equations which do not, by themselves, justify such models.

   In a series of works from 1931 to 1934 Richard Tolman investigated in detail cyclic models from the points of view of general relativity and thermodynamics. Written in a parametric form he found for the time variation of the scale factor $R(t)$ that it represented a cycloid in the $R - t$ plane.[77] With α denoting the constant quantity $8\pi\rho R_{\mathrm{E}}^3$, where R_{E} is the radius of the static Einstein model, the expression was

$$R = \frac{\alpha}{3}(1 - \cos \psi), t = \frac{\alpha}{6}(1 - \sin \psi). \tag{4.31}$$

The radius will thus oscillate between $R = 0$ at $t = 0$ and $R = \alpha/3$ at $t = \pi\alpha/6$. Tolman concluded that a model universe such as the Friedman–Einstein universe with $\Lambda = 0$ could expand without increase in entropy and undergo a continual series of cycles. The entropy would increase from one cycle to another, but without ever approaching a limit. Moreover, the cycles would not be identical, for the maximum value of R

would increase with the number of cycles and so would the period. Strictly speaking, Tolman's mathematical analysis of a cyclic universe was valid only for a single cycle, as in the Friedman–Einstein model, but he argued that an extension to an endless series of successive expansions and contractions made sense from a physical point of view. More generally he found it "difficult to escape the feeling that the time span for the phenomena of the universe might be most appropriately taken as extending from minus infinity in the past to plus infinity in the future."[78]

The oscillating model was theoretically problematic because of the two singularities it contained per cycle, corresponding to a beginning and an end. From an observational point of view it led to an age of the present universe that was much too short and it required a high average density of matter to reverse the motion of the universe at R_{max}. These problems were shared by some other models, though, and Tolman did not consider them to be fatal. Although Tolman was clearly fascinated by the eternally cyclic universe, he was not, as some other scientists, emotionally or philosophically committed to it. His lack of strong commitment is illustrated by his later publications in cosmology. From 1934 to his death in 1948 he continued doing cosmological research, but without paying much attention to oscillating models of the kind he had analyzed in his earlier work.

Although cyclic universes were never part of mainstream cosmology, interest in the subject did not stop with the death of Tolman. Philosophers continued to suggest speculative cyclic cosmologies in the older tradition, but they had practically no impact on physicists and astronomers.[79]

During the 1950s the possibility of a cyclic universe was investigated in particular by Herman Zanstra in the Netherlands and William Bonnor in England. With their contributions two important innovations were introduced, one of which concerned the number of cosmic cycles and the other concerned the physical state near the bounce from contraction to expansion. Zanstra, a professor of astronomy at the University of Amsterdam, confirmed Tolman's result that with each new cycle the universe would grow bigger; but he also found from thermodynamical arguments that if the universe was closed and cyclic it could only have been preceded by a finite number of cycles (Fig. 4.5).[80] That is, it could not have existed prior to a certain time corresponding to the origin of the first cycle. With regard to the bounce Zanstra speculated that during the final phase of a contraction the universe would consist of a hot gas exerting a high negative pressure. From this maximally compressed but non-singular state of $R = R_{min}$ a new cycle would start.

The idea of a negative pressure built up near R_{min} was independently proposed by Bonnor, who favored a universe which bounced periodically and smoothly between non-singular states of high density. But contrary to Zanstra, he did not admit that thermodynamics prohibited an infinite number of cycles. For Bonnor it was imperative that the universe had an unlimited past and future, for only then could the question of an absolute beginning be avoided. He believed that the hypothetical origin of the universe as $t = 0$ constituted a miracle and was for this reason contrary to science. As Bonnor saw it, an eternally cyclic universe governed by the laws of known physics had some of the advantages of the steady-state universe without relying on its unacceptable features such as continual creation of matter.[81]

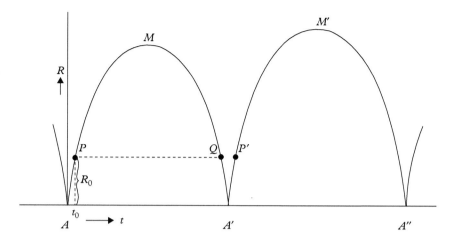

Fig. 4.5 *Zanstra's pulsating universe. The cycle starts in A, where* t $= 0$ *arbitrarily, and bounces at maximum density* (A′, A″). *The state of the present universe is represented by P, at radius r_0 and age t_0. Q marks the later contracting state. The second cycle* A′M′A″ *is greater than the first cycle* AMA′.

Source: Zanstra (1957), p. 292. With permission from The Royal Netherlands Academy of Arts and Sciences.

The hypothesis of an ideal fluid with negative pressure came to play an important role in later cosmology. In an address of 1933 Lemaître had pointed out that according to general relativity the vacuum corresponds to a physical state with pressure p and energy density ρ given by

$$p = -\rho c^2 = -\frac{\Lambda}{8\pi G} c^2. \tag{4.32}$$

Lemaître's proposal attracted no immediate attention and when the negative pressure eventually entered cosmology it was at first in different versions[82].

Among the cyclic models making use of negative pressure to avoid a singular state was a theory by the Polish physicist Jaroslav Pachner according to whom the extreme state of maximum contraction corresponded to a density of $\rho \geq 10^{16}$ g cm^{-3}. "Our universe is oscillating and no singularity with an infinite density of matter occurs during the periodical pulsations between two extreme values," he summarized.[83] The appeal of the cyclic universe is further exemplified by Robert Dicke, one of the fathers of modern big-bang cosmology. Although Dicke never published on the subject, it is known that he had a predilection for the eternally cyclic universe. The idea entered briefly in the seminal paper in *Astrophysical Journal* in which the CMB was first explained as a relic of the big bang. As Dicke (and formally his three co-authors) wrote, "a closed universe, oscillating for all time ... relieves us of the necessity of understanding the origin of matter at any finite time in the past."[84] However, most cosmologists considered models of the cyclic universe to be contrived and speculative. In what he called an "eschatological study" of the closed universe, Martin Rees of Cambridge University noted in 1969 that in each cycle the entropy per baryon would increase by a finite amount. He thus revived Zanstra's entropy problem. As Rees pointed out, the entropy increase per baryon "raises difficulties for the view that there has been an infinite number of oscillations in the past."[85] The result was confirmed by other cosmologists, including Peebles and Igor Novikov.

Despite all problems cyclic models continued to be explored by a minority of physicists and astronomers. For example, according to Mark Israelit and Nathan Rosen our present universe began in a "cosmic egg" with an energy density $\rho \sim 10^{94}$ g cm^{-3} and a correspondingly enormous negative pressure. The period of the cycles in the Israelit–Rosen model was of the order 10^{12} years.[86] However, oscillating models of this classical kind depended on the assumption of a spatially closed universe. With the recognition at the end of the century that the universe is accelerating and space most likely flat, realistic models of a cyclic universe seemed out of the question. And yet, a few years after the discovery of the accelerating universe Paul Steinhardt and Neil Turok introduced a radically new cyclic theory in part inspired by string physics. According to the two physicists:

> Space and time exist forever. The big bang is not the beginning of time. Rather, it is a bridge to a pre-existing contracting era. The universe undergoes an endless sequence of cycles in which it contracts in a big crunch and re-emerges in an expanding big bang, with trillions of years of evolution in between. The temperature and density of the universe do not become infinite at any point in the cycle.[87]

The Steinhardt–Turok model does not rely on a positive space curvature but takes space to be flat. Also contrary to the Tolman model and later models in this tradition, the entropy remains constant and the cycles are identical. As Steinhardt and Turok emphasized, their ambitious theory was falsifiable, contained no inflation phase, and offered a complete history of the universe stretching arbitrarily far back into the past. Since 2002 the theory has been developed in a series of papers, but its impact on mainstream cosmology remains limited. It has not succeeded in replacing the inflation paradigm any more than Magueijo's varying-speed-of-light theory has.

4.9   Is cosmology a science?

As mentioned in the chapters 1 and 5, for a long time it was a matter of debate whether or not the universe as a whole could be subject to scientific analysis. Would cosmology by its very nature remain a philosophical discourse, which was related to the sciences but still essentially different from them? According to Immanuel Kant's masterpiece, the influential *Critique of Pure Reason* from 1781, the notions of age and extent were meaningless when applied to the universe. Since he concluded that the concept of the world at large was contradictory, it followed that it cannot cover a physical reality but only be a concept of heuristic value. Kant and his followers effectively ruled out cosmology as a field of scientific study.

A century later doubts of a similar kind were raised in connection with the universal heat death, the prediction from the second law of thermodynamics that the universe irreversibly will tend towards a state of high-entropic eternal sleep (cf. Section 1.4). The Russian physicist Orest Chwolson was a respected figure in international physics and the author of much-used textbooks. He insisted that the universe, which he took to be infinite, was outside the domain of science:

> *Physics has nothing to do with the universe*; it is not an object of scientific research as it is not accessible to any observation. World-laws are laws that are valid in all parts of the world, i.e. the physicists' world. There may be universal laws, and perhaps world-laws are just special cases of the universal laws. When the physicist speaks of the "world" he means his limited [and observable] world. . . . To identify this world with the universe is a proof of either thoughtlessness or madness, and in any case lack of scientific understanding.[88]

Chwolson's attitude was not uncommon at the turn of the century. In his 1905 presidential address to the British Association for the Advancement of Science, the eminent astronomer and geophysicist George Howard Darwin (a son of Charles Darwin) denied that science would ever be able to explain the riddle of the universe. Darwin was a scientific optimist, but when it came to cosmology he was an agnostic.

Half a century after Darwin's pessimistic forecast cosmology had made great progress, both theoretically and observationally; and yet doubts remained whether the study of the universe could be counted as a genuine physical science. Under the impact of the new steady-state theory many physicists, astronomers and philosophers discussed cosmology's scientific status. Parts of this discussion are summarized in Chapter 5, and here we just call attention to the public discussion of 1954 between Hermann Bondi and Gerald Whitrow concerning cosmology's position between science and philosophy. Tellingly, the title of the discussion was "Is Physical Cosmology a Science?" As seen in hindsight, Bondi's more optimistic view of the future of physical cosmology was closer to the mark than the doubts expressed by Whitrow (Section 5.6).

Despite the unique nature of cosmology and the awe-inspiring concept of its domain, the universe in its totality, physical cosmology has grown into a proper science both in an epistemic respect and from a sociological point of view. With the successes in the late 1960s of the big-bang theory and the elimination of its steady-state rival, there emerged a consensus among cosmologists on the main problems to be solved and criteria to be used. The hot big-bang relativistic theory became the paradigm of cosmology, and alternative interpretations were marginalized. At the same time that cosmology became cognitively institutionalized, it achieved a social institutionalization that made the subject a full-time professional occupation rather than a part-time hobby for astronomers, physicists and mathematicians. Philosophers stopped playing any significant role. On the quantitative side, the number of publications on cosmology grew dramatically. Whereas some 100 papers were published worldwide on the subject in 1965, in 1980 the number had increased to about 400.[89] On the other hand, impressive as the growth is, compared to the total number of physics publications cosmology remained a very small field of research (Fig. 4.6).

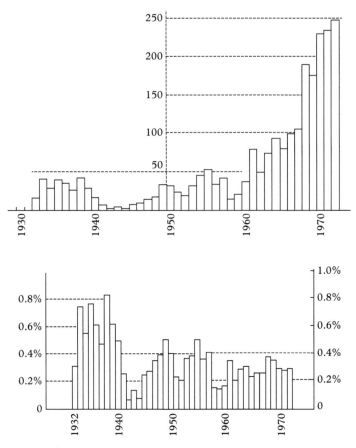

Fig. 4.6 *Annual number of publications in cosmology according to* Physics Abstracts. *The upper figure shows the absolute number of publications, the lower one the percentage of all publications abstracted in* Physics Abstracts. *Reproduced from M. Ryan and L. Shepley, "Cosmology," American Journal of Physics 44 (1976): 223–240, p. 223, with the permission of The American Association of Physics Teachers.*

The development was followed by a growing integration of cosmology into university departments, and courses and textbooks in cosmology became common. For the first time, students were taught standard cosmology and brought up in a research tradition with a shared heritage. Cosmology was (and still is) regarded as solidly founded on Einstein's theory of general relativity and some kind of big-bang scenario, elements that were largely taken for granted and conceived as defining features of cosmological theory. As far as textbooks are concerned, from the late 1960s onwards they multiplied and, with a few exceptions, shared the same foundation. Significantly, Jim Peebles' *Physical Cosmology*, dating from 1971 and based on a graduate course Peebles gave in Princeton two years earlier, was the first textbook with this title. Other books from the same period include Dennis Sciama's *Modern Cosmology* (1971), Steven Weinberg's *Gravitation and Cosmology* (1972), and *Gravitation* (1973) written by Charles Misner, Kip Thorne, and John Wheeler.

As to status and recognition it is worth pointing out that the first Nobel Prize ever honoring a cosmological discovery was awarded in 1978 to Arno Penzias and Robert Wilson for their discovery of the CMB. Among the earlier generation of cosmologists only Hubble and Lemaître were nominated for a Nobel Prize, Hubble in 1953 and Lemaître in 1954 and 1955 (in the latter case for the chemistry prize). None of the nominations were taken seriously by the Nobel Committee. Hoyle too was unsuccessfully nominated, but for his contributions to stellar nucleosynthesis and not for his work in cosmology; similarly, Zwicky's single nomination of 1935 was unrelated to his contributions to astronomy. The Penzias-Wilson prize has been followed by two more cosmology prizes, in 2006 to John Mather and George Smoot (COBE experiment and CMB anisotropies) and in 2011 to Saul Perlmutter, Brian Schmidt, and Adam Riess (accelerating universe).[90]

Whereas astronomy and cosmology are not part of the Nobel system, and the prizes are therefore physics prizes, there also exist a few scientific prizes specifically for cosmological research. One of them is the Gruber Cosmology Prize awarded since 2000, the first to Jim Peebles and Allan Sandage. The two scientists were also awarded the prestigious Crafoord Prize in astronomy, Sandage in 1991 and Peebles in 2005. In 1997 Hoyle and Edwin Salpeter received the Craaford astronomy prize for their work on stellar nuclear processes.

Not everybody is or has been happy with the situation in modern cosmology and the almost complete predominance of the hot big-bang consensus model. Very few scientists reject cosmology as a worthwhile science *per se*, but ever since the hot big bang became the standard model several researchers have vehemently criticized the model and its near-monopoly in cosmological thought. Some are in favor of alternative models, while the critique of others is unrelated to such models and focus exclusively on observations, predictions and methodology. Fred Hoyle, Jayant Narlikar, Geoffrey Burbidge and a few other advocates of a modified steady-state theory continued for a couple of decades to attack the big-bang consensus model, accusing it of conformism and disregard of falsifying observations (see Section 5.9). As we have seen, Alfvén and proponents of the plasma universe made somewhat similar charges, although in their case they went farther, in so far that they denied that cosmology was a science of the universe as a whole.

Some of the early criticism came from observational astronomers of high reputation with no sympathy for the steady-state theory or the plasma universe. An example is the French-American astronomer Gérard de Vaucouleurs, who was first of all an indefatigable observer and cataloguer of galaxies. In the 1970s he got engaged in a controversy with Allan Sandage over the value of Hubble's constant, with Vaucouleurs advocating a small constant ($H_0 \sim 100 \text{ km s}^{-1} \text{ Mpc}^{-1}$, $T \sim 9.5 \times 10^9$ yr) and Sandage a large one ($H_0 \sim 50 \text{ km s}^{-1} \text{ Mpc}^{-1}$, $T \sim 19 \times 10^9$ yr). Vaucouleurs also took an interest in the interpretation of the cosmological data, which he believed supported a universe of a hierarchical or fractal structure and with a non-zero cosmological constant. Both elements were heterodox.

In a paper of 1970 Vaucouleurs complained that many of the basic problems of cosmology had been answered by "aesthetic prejudices or considerations of mathematical simplicity" rather than a critical study of the empirical evidence. Finding "disturbing parallelisms between modern cosmology and medieval scholasticism," he wrote: "If nature refuses to cooperate, or for a time remains silent, there is a serious danger that in the constant repetition of what is in truth merely a set of *a priori* assumptions ...will in time become accepted dogma that the unwary may uncritically accept as established fact or as an unescapable logical requirement."[91]

The advent of precision cosmology did not silence this kind of criticism, which in a variety of different shapes has continued until the present. A more recent example is the American astronomer Halton Arp, a long-time staff member at the Palomar Observatory. From the late 1960s and until his death in 2013 Arp criticized big-bang cosmologies on an observational basis and suggested a non-cosmological origin of the redshifts of quasars and so-called peculiar galaxies.[92] As mentioned in Section 4.2, the general inclination of many mainstream cosmologists is to ignore rather than criticize theories which are offered as radical alternatives to the standard big-bang theory. After all, it is hard to believe that the present theory is wrong in some fundamental sense. Commenting on the balance between tolerance and dogmatism, Peebles wrote: "The darker side to the tradition in cosmology are the overreactions of those who consider themselves the guardians of the true and canonical faith... and the tendency for dissent to slip into pathological science." Admitting that there had been "documented examples of frivolous criticism of dissenting arguments," Peebles still believed that "the establishment view has [not] been overly reactionary on average."[93]

In the year 2004 critics of the standard big-bang model united as signers of an "open letter" addressed to the scientific community and published in the journal *New Scientist*. The letter summarized the case against the big bang as follows:

> The big bang today relies on a growing number of hypothetical entities, things that we have never observed—inflation, dark matter, and dark energy are the most prominent examples. Without them, there would be a fatal contradiction between the observations made by astronomers and the predictions of the big bang theory. In no other field of physics would this continual recourse to new hypothetical objects be accepted as a way of bridging the gap between theory and observation.[94]

The 33 signing scientists complained that alternative theories of the universe were deliberately silenced and that big-bang cosmology had acquired its role as an undisputed standard theory more for sociological reasons than for its scientific merits: "In cosmology today doubt and dissent are not tolerated, and young scientists learn to remain silent if they have something negative to say about the standard big bang model....As a result, the dominance of the big bang within the field has become self-sustaining, irrespective of the scientific validity of the theory." The following year a group of scientists, most of them signers of the open letter, convened at a conference in Portugal devoted to what they called the crisis in cosmology. In 2008 it was followed by another conference on the same subject.

While some of the critical views are concerned with the basis of the standard cosmological theory, such as the "hypothesis" of a big bang, others relate to the more recent and more speculative extensions of the theory. They have much in common with the critique of other branches of fundamental physics and of many-dimensional string theory in particular. Peter Woit's *Not Even Wrong* and Lee Smolin's *The Trouble with Physics*, both published in 2006, were popular books which gave rise to much debate and media coverage. The target of the two physicists was primarily string theory, but their critique was also aimed at modern ideas of the very early universe which, they argued, did not live up to established standards of physics. Could these ideas be falsified? Had they resulted in empirical progress or would they ever lead to such progress?

The theoretical cosmologist George Ellis has in several papers questioned the scientific nature of inflation theories and multiverse ideas, pointing out that issues of a philosophical nature unavoidably arise in these parts of cosmology. In a paper with Tony Rothman of 1987 dealing with inflation theories, Ellis suggested that "cosmology is approaching the frontier where science is no longer based on experimental evidence and makes no testable predictions. Once this border is crossed, we have left the world of physics and have entered the realm of metaphysics."[95] Once again, the continuity from past controversies in cosmology is striking.

According to some critics of modern cosmology, the field is dominated by a "snowball effect" which ensures that talented young researchers, eager to get a tenured position, uncritically accept the standard model and work within its confines. But belief in the standard model is just that – a belief. "We might well wonder whether cosmology, our knowledge of the Universe as a whole, is a science like other fields of physics or a predominant ideology," says Martín López-Corredoira, a Spanish astrophysicist.[96] Exceptionally, in the year 2000 objections of this kind appeared in the recognized journal *General Relativity and Gravitation*, where the British astronomer Michael Disney broadly criticized current cosmological theory and methodology. "Much of cosmology is unhealthy self-referencing," Disney claimed, fairly or unfairly, adding that "cosmological fashions and reputations are made more by acclamation than by genuine scientific debate." Not unlike earlier big-bang critics he referred to cosmology's "unspoken parallel with religion." The parallel caused Disney to suggest that "the word 'cosmologist' should be expunged from the scientific dictionary and returned to the priesthood where it properly belongs."[97] Disney did not refer to Herbert Dingle's much earlier critique of "cosmythology" (Section 4.4), but his message was essentially the same.

Interesting as the case against cosmology is from the perspectives of history, philosophy, sociology, and psychology of science, it is of no direct scientific relevance. The earlier critical voices were heard and in a few cases responded to by physicists and astronomers engaged in cosmological research, but their impact was very limited. Today the impact seems to be even more limited.

....

NOTES

1. See Wetterich (2013) which proposes a singularity-free big-bang cosmological model in which the current universe is not expanding. According to this model the redshifts are produced by a hypothetical decrease of the size of atoms. The model offers a new picture of the universe without claiming that it represents reality.
2. On anti-science, pseudoscience, and alternatives to science, see Nowotny and Rose (1979).
3. Kuhn (1962).
4. For alternatives to standard big-bang theory as seen from the perspective of the early 1980s, see Ellis (1984a).
5. Polanyi (1967), p. 539.
6. See the website http://rationalwiki.org/wiki/Alternative_cosmology for a list of about 40 alternative cosmological models. Even scientists of high distinction sometimes propose ideas of a heterodox or even pseudoscientific nature. Some of Hoyle's ideas belonged to this category. The British physicist Brian Josephson, a Nobel laureate of 1973 for the prediction of the quantum effect named after him, turned to quantum mysticism and paranormal science.
7. According to Urani and Gale (1994). On Milne's cosmology and its historical context, see also North (1990), pp. 149–85 and Kragh (2004), pp. 200–29.
8. Milne (1933), p. 4.
9. The historical connection between Milne's cosmology and the Robertson–Walker metric is discussed in Urani and Gale (1994).
10. Milne (1935) pp. 128–9.
11. Milne (1933), p. 2.
12. Milne (1935), p. 80.
13. See, for example, Lévy-Leblond (1990) according to whom "linear time [Milne's dynamic time] makes it obvious that never did the Big Bang begin."
14. Milne (1937), p. 999.
15. Untitled paper in *Nature*, **155**, 135–6, (1945).
16. Bondi (1952), p. 136. Bondi argued that the timescale problem was most severe in Milne's theory because radioactivity follows t-time and not τ-time. But in the revised 1960 edition of his book he changed his mind, concluding that Milne's theory completely resolved the timescale problem (p. 138).
17. Douglas (1956), based on a conversation with Einstein of January 1954.
18. Hubble (1934a), p. 17; Milne (1938), p. 344.
19. Early ideas of the constants of nature and their possible variation in time are described in Barrow (2002) and Kragh (2011), pp. 175–89.
20. For Eddington's theory and references to his work, see Kragh (2017a).
21. Dirac (1937).
22. Dirac (1938). For details on Dirac's cosmological theory and his view on cosmology, see Kragh (1990), pp. 223–46.

23. Dirac did not publish on cosmology between 1938 and 1972, but during the 1960s he discussed his theory privately with Gamow. For the interesting correspondence between the two physicists, see Kragh (1991).
24. Dingle (1937). Dingle's criticism was directed not only at Dirac but also at Milne and Eddington.
25. Teller (1948). See Kragh (2016b) for a historical account of Teller's argument and later attempts to test the $G(t)$ hypothesis by means of paleontological and geological evidence.
26. Pochoda and Schwarzschild (1964); Gamow (1967).
27. Jordan (1952), p. 137.
28. Jordan (1971). For details on and references to Jordan's theories, see Kragh (2015b, 2016b).
29. Heckmann and Schücking (1959).
30. Dirac (1974), p. 440.
31. Steigman (1978) who also pointed out that Dirac's theory was unable to explain the large abundance of helium in the universe.
32. Hoyle and Narlikar (1972). See also Wesson (1978) for a careful discussion of the theories of Dirac, Jordan, Hoyle, Canuto, and others.
33. See Gillies (1997) for a detailed report on gravitational studies made in the period.
34. Van Flandern (1975); Kragh (2016b), pp. 134–41.
35. Murphy et al. (2001).
36. Rosenband et al. (2008).
37. Albrecht and Magueijo (1999). For a historical perspective on VSL cosmology, see Kragh (2006).
38. Afshordi and Magueijo (2016).
39. As expressed by Brans (2014), "If a list of names were to be used for people who independently proposed ST modifications of standard Einstein theory, the resulting compound title would be extravagantly unwieldly." For the origins and early development of ST theories, see Goenner (2012). This section relies on material in Kragh (2016b).
40. Dehnen and Hönl (1968).
41. Brans and Dicke (1961), p. 928. Brans was a former Ph.D. student of Dicke and had recently completed his doctoral thesis.
42. The idea of a zero-energy universe satisfying $GM/Rc^2 = 1$ was first proposed by Arthur Haas in 1936. For its curious history, see Kragh (2016b), pp. 29–30.
43. Dicke (1961). See Peebles (2017) for a detailed account of Dicke's work written by one of his former collaborators.
44. Dicke and Peebles (1971b), p. 451. On the CMB discovery, see Chapter 8.
45. See Pinch (1985) which offers an interesting sociological and philosophical perspective on the episode.
46. Brans (2014).
47. See Faraoni (2004) for modern theories of scalar-tensor cosmology.
48. Goldhaber (1956). See also the critical comments in Alpher and Herman (1958).

49. O'Connell (1958), p. 448. Other participants in the conference included A. Sandage, F. Hoyle, W. Baade, W. Fowler, J. Oort, M. Schwarzschild, and B. Strömgren.
50. Burbidge and Hoyle (1957). The cosmic gamma rays of energy larger than 100 MeV corresponding to two proton masses were measured in 1961 by means of a detector installed in the artificial satellite Explorer XI.
51. See Stecker (1974) for a brief and general account of Omnès' work.
52. For Birkeland as a cosmologist, see Section 1.7. According to Peratt (1995), p. 3, plasma astrophysics and cosmology "may be traced back to the seminal research first published by Kristian Birkeland in 1896." The name "plasma" for a highly ionized gas was coined by the physical chemist Irving Langmuir in 1928.
53. Klein (1971), p. 343; Klein (1954); Bondi (1952) pp. 59–62, 157–60. Klein's model of 1954 attracted little attention.
54. Alfvén and Klein (1962). See also Kragh (1996), pp. 223, 382–4.
55. Ambartsumian's view is documented in Graham (1972), pp. 165–71.
56. Alfvén (1966), p. 89.
57. Alfvén (1983); Lerner (1995), pp. 220–5.
58. Lerner (1995); Lerner (1991), pp. 30–2.
59. Peebles (1993), pp. 207–9.
60. On the occasion of an international workshop on plasma cosmology held in San Diego in 1989 and sponsored by the IEEE, *New York Times* ran an article announcing that the new theory posed a serious threat to the big-bang paradigm. See also Lerner (1991), pp. 272–5. For the current state of plasma cosmology, see https://www.plasma-universe.com/ (accessed April 2017).
61. Alfvén (1984).
62. Robertson (1932), p. 226. The name "tired light" is sometimes ascribed to Tolman, but always without a proper reference.
63. Hubble (1929a). For contrasting views concerning Hubble's attitude to the expanding universe, compare Hetherington (1982) and Kragh and Smith (2003).
64. Belopolski (1929). See also http://luchemet.narod.ru/hubble.html (accessed October 2016).
65. Zwicky (1929). For the tired-light hypotheses of Zwicky and other scientists, see Kragh (2017d).
66. Zwicky (1935), p. 803.
67. Robertson (1932), p. 226.
68. Bruggencate (1930).
69. On the Tolman surface brightness test and its history, see Peebles (1971b) and Chapter 6 in the present handbook.
70. Kennedy and Barkas (1936). Kennedy is known for the Kennedy–Thorndike experiment of 1932 confirming some of the predictions of the special theory of relativity See Kennedy and Thorndike (1932).
71. Peebles (1971b), p. 175.
72. Couderc (1952), p. 97.
73. Finlay-Freundlich (1954). As a young man Erwin Freundlich collaborated with Einstein in matters of astronomy and tests of the general theory of relativity. He left

Germany in 1933 and six years later he came to St. Andrews University, Scotland. See Hentschel (1997).

74. LaViolette (1986).
75. See Kragh (2009c) for a general history of the cyclic universe in modern cosmology.
76. Milne (1935), p. 337. Although Milne did not refer to Tolman, he most likely had him in mind.
77. Tolman (1934b), p. 413. See also Heller and Szydlowski (1983) for later cosmological models inspired by Tolman's work. Cyclic models of the universe were independently and a little earlier investigated by the Japanese physicist Tokio Takeuchi. His paper "On the cyclic universe" in the *Proceedings of the Physico-Mathematical Society of Japan* **13**, 166–77 (1931) attracted very little attention.
78. Tolman (1934b), p. 486.
79. For example, Reiser (1954), who defended a "pantheistic cyclic-creative cosmology."
80. Zanstra (1957). The work, published in the proceedings of the Dutch Academy of Sciences, was not (and still is not) widely known.
81. Bonnor (1954, 1957).
82. Lemaître (1934), lecture given to the National Academy of Sciences on 20 November 1933.
83. Pachner (1965).
84. Dicke et al. (1965), p. 415.
85. Rees (1969), p. 197.
86. Israelit and Rosen (1989). Nathan Rosen was a former collaborator of Einstein and co-author of the famous EPR (Einstein–Podolsky–Rosen) paper of 1935.
87. Steinhardt (2004), p. 466. See also the popular exposition Steinhardt and Turok (2007) and Kragh (2011), pp. 202–8.
88. Chwolson (1910), p. 52.
89. See Kaiser (2006) and Marx and Bornmann (2010).
90. On astronomy and cosmology in relation to the Nobel Prize system, see Kragh (2017c).
91. De Vaucouleurs (1970), p. 1205. Note the similarity in rhetoric to the earlier critique of the steady-state theory as exposed by H. Dingle, G. McVittie, and others (Chapter 5).
92. Arp (1989). At the time Arp had been committed to non-cosmological redshifts for more than twenty years.
93. Peebles (1993), p. 197.
94. *New Scientist* **182** (22 May 2004), p. 23. Among the signers were H. Arp, H. Bondi, T. Gold, E. Lerner, J. Narlikar, A. Peratt, and J.-C. Pecker.
95. Rothman and Ellis (1987), p. 22.
96. López-Corredoira (2014), p. 86.
97. Disney (2000), pp. 1131–2. See also Disney (2007). As pointed out in Ćirković (2002), Disney's critique is to a large extent a repetition of earlier sceptical views concerning the scientific status of cosmology.

5

Steady-state theory and the cosmological controversy

Helge Kragh

> *In one corner you have burly, pun-making Russian-American physicist George Gamow. He says the universe did have a beginning and that beginning was a very big bang.... In the other corner you have piano-playing, novel-writing, baggy-tweeded English astronomer Fred Hoyle. His side says that there was no instant creation. The universe is in a steady state.*
>
> Martin Mann (1962), p. 29.

5.1 Introduction

For about half a century the standard conception of the universe has been the hot big-bang model as described by general relativity and elementary particle physics. While there have been earlier standard models, it was only with the general acceptance of the hot big bang in the late 1960s that a stable framework for physical cosmology was established. The route toward this framework was anything but smooth and it illustrates the features of contingency and irregularity that often characterize scientific progress. If modern cosmology was born in the interwar period, it went through a turbulent period of adolescence in the first two decades after World War II. In this period of slow maturation the revised big-bang theory, which had by then become essentially a theory of the early universe based on nuclear physics, was challenged by the entirely different theory of the steady-state universe.

The historical trajectory of the big-bang model is peculiar, as the model was proposed three times, and largely independently, over a period of more than thirty years. Lemaître's semi-qualitative primeval-atom hypothesis of 1931 played no significant role for Gamow when he developed his theory of an explosive beginning of the universe in the late 1940s. Likewise, when Dicke, Peebles, and others developed their version of the hot big-bang theory in 1965 and the following years, they did not build on Gamow's earlier work and not at all on Lemaître's. Given that the scenario of a hot big bang existed

Kragh, H., 'Steady-state theory and the cosmological controversy' in *The Oxford Handbook of the History of Modern Cosmology*, edited by Kragh, H. and Longair, M. S. © Oxford University Press 2019.
DOI: 10.1093/oxfordhb/9780198817666.013.5

in a developed form in the early 1950s, and that it predicted the existence of the cold microwave background found in 1964 and understood the following year, it is remarkable that the earlier theory was simply forgotten and had to be reinvented in the mid-1960s.

Directly and indirectly, the steady-state alternative proposed by Hoyle, Bondi and Gold in 1948 greatly influenced the cosmological scene for more than fifteen years. Although eventually proved wrong by observations, from a historical perspective it was no less important than what today is recognized to be the roots of the standard big-bang picture of the universe. What was at stake during the cosmological controversy— basically the confrontation between the steady-state theory and relativistic evolution cosmologies—was not merely an astronomical world picture but also the very standards on which scientific cosmology should be judged. Foundational issues were part and parcel of the controversy. According to Bondi, "cosmology is the most fundamental of the physical sciences [and] ... the long-sought link between philosophy and the physical world."[1] And yet the steady-state theorists shared with their opponents the belief that in the end only observations, and not philosophical preferences, can tell us what kind of world we live in. And they all agreed that the most pressing problem facing cosmological research was the scarcity of relevant and reliable observational data.

5.2   From primeval atom to nuclear archaeology

Some of the ideas which led to the first versions of the big-bang theory can be found in non-cosmological contexts even before the discovery of the expanding universe. One of the avenues was nuclear astrophysics and the problem of element formation in the stars. Concerning the very light elements, could one understand why there was so much hydrogen in the universe and—as it was thought in the 1920s—so little helium? In 1928 a Japanese physicist by the name Seitaro Suzuki studied theoretically the problem of the hydrogen-helium ratio by means of the methods of physical chemistry that had proved useful for stellar atmospheres. He found that an appreciable thermal dissociation of helium into hydrogen, He $\rightarrow$ 4p, would require unrealistic physical conditions in the interior of the stars, including temperatures larger than 10^9 K. This was 10^2 to 10^3 times as high as Eddington's value for the central temperature of stars. The result was disappointing as it indicated an almost complete combination of hydrogen nuclei to helium nuclei. Nonetheless, Suzuki suggested that the observed helium-hydrogen ratio might be explained "if the cosmos had, at the creation, the temperature higher than 10^9 degrees."[2]

Seen in retrospect the comment is remarkable by referring to the creation of the universe as a problem related to nuclear reactions. However, Suzuki's paper was meant as a contribution to astrophysics and not to cosmology. In any case, published in an obscure Japanese journal of physics it made no impact on Western physics and astronomy. It was only with the introduction of quantum mechanics to the study of nuclear fusion processes, as pioneered in a paper of 1929 by Robert d'Escourt Atkinson and Fritz Houtermans that nuclear astrophysics slowly took off. The two physicists, one a Briton and the other a German, still worked with the proton–electron model of the atomic nucleus. The same was the case with a greatly expanded version of the theory which Atkinson presented in 1931 and in which he assumed that "in its initial state any star, or indeed the entire universe, was composed solely of hydrogen."[3] With the discoveries of the neutron and the deuteron in 1932 new possibilities were opened for understanding stellar nuclear reactions and the energy production in the stars. The latter problem was essentially solved with Hans Bethe's celebrated theories of 1938–1939 hailed as a breakthrough not only by physicists but also by astronomers. However, the associated problem of the synthesis of heavier elements was not part of Bethe's theory which explicitly disregarded the building up of elements heavier than helium. For this reason the theory was of no direct relevance to cosmology.

Quantum theory entered cosmological models at a surprisingly early date, for the first time in a paper of 1925 written by the Hungarian theorist Kornel Lanczos. In a study of an Einstein-like world model, Lanczos was led to introduce a certain "world period" T that would depend on the radius of the universe. For this period, and with m denoting the mass of the electron and h Planck's constant, he derived the relation

$$T = \frac{4\pi^2 m R^2}{h^2} \simeq 10^{41} \text{ years.} \qquad (5.1)$$

Lanczos' interpretation of the world period was unclear and neither he nor others developed the idea. And yet the paper deserves mention because of Lanczos' attempt to relate the quantum phenomena of the atomic world to the structure of the universe at large. "The solution to the quantum secrets are hidden in the spatial and temporal closedness of the world," he suggested.[4]

In the early 1930s the atomic nucleus entered cosmological theory in a way entirely different from the one associated with stellar nuclear reactions. What can reasonably be called the first version of the big-bang universe was proposed by Georges Lemaître in 1931, explicitly conceiving the "primeval atom" or "unique quantum" as a huge and highly radioactive atomic nucleus comprising the total mass of the universe (Figs. 5.1 and 5.2). Although Lemaître used the term "atom" in reality he was thinking of a nucleus or, as he did in some of his later writings, an "isotope of the neutron." In 1931 the neutron had not yet been discovered and Lemaître chose to express himself vaguely and metaphorically when describing in words the initial state of the universe. His primordial atom or nucleus was changeless and completely undifferentiated, devoid of physical qualities and without a measure of time. It was an atom in the sense of the ancient Greek philosophers. Although the primeval atom was inaccessible to science, Lemaître stressed that it was physically real.[5]

Lemaître's picture of the primeval atom was something like a huge and extremely dense collection of protons, electrons, and alpha particles. Such an unusual state of nuclear matter had in the late 1920s been considered in a stellar context (white dwarfs) by Wilhelm Anderson, Edmund Stoner, Edward Milne, and a few others and would later lead to the idea of neutron stars. However, although the primeval atom was in some

Fig. 5.1 *Lemaître in 1934. Reproduced with permission from Archives Georges Lemaître, Université Catholique de Louvain, Louvain-la-Neuve, Belgium.*

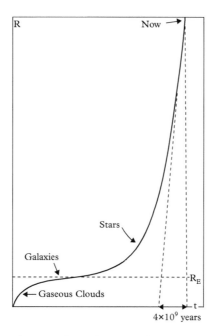

Fig. 5.2 *Lemaître's big-bang universe as shown in a paper from 1958.*

Source: G. Lemaître, "Instability in the expanding universe and its astronomical implications," Pontificiae Academiae Scientiarum, Scripta Varia no. 16 (1958): 475–486, on p. 476. Reproduced with permission from the Pontifical Academy of Sciences, Vatican City.

respects similar to a supermassive star, there is no evidence that the Belgian cosmologist was inspired by the tradition in astrophysics.[6]

The 1931 model of the primeval atom was a bold attempt to describe the origin of the universe realistically, in terms of physics and not only in terms of mathematics. As the result of the disintegration or explosion of the primeval atom stars and galaxies would be formed. Lemaître argued that the presently observed cosmic rays were fossils from the early disintegration of "atomic stars" formed from the original explosion approximately ten billion years ago. The charged particles making up the rays were "the hieroglyphs of our astronomical library," as he brilliantly expressed it.[7] However, his attempts through the 1930s to justify the primeval atom theory in terms of the cosmic rays were unsuccessful.

Although the violent disintegration of the primeval atom at $t = 0$ was a big bang, Lemaître was convinced that the mathematical appearance of $R = 0$ in the field equations did not correspond to physical reality. In an important paper of 1933, only translated into English many years later, he insisted that "matter has to find ... a way of avoiding the vanishing of its volume." The solution, he thought, was to appeal to the hypothesis of the primeval atom:

> When the distances between the atomic nuclei and the electrons become of the order of 10^{-12} cm, the non-Maxwellian forces which prevent the mutual interpenetration of elementary particles must become predominant and are without doubt capable of stopping the contraction. The universe would then be comparable to a colossal atomic nucleus. ... We thus conclude that only the subatomic nuclear forces seem capable of stopping the contraction of the universe, when the radius of the universe is reduced to the dimensions of the solar system.[8]

Lemaître's primeval atom hypothesis attracted early attention in newspapers and popular-science magazines but failed to win scientific approval. To the extent that the theory was known, physicists and astronomers tended to dismiss it or they just ignored it. The Canadian astronomer John Plaskett characterized it as "the wildest speculation of all" and "an example of speculation run mad without a shred of evidence to support it."[9] Among the few who found the hypothesis appealing was the Mount Wilson astronomer Paul Merrill, who called attention to Lemaître's unusual explanation of our chemical elements as descendants of much heavier elements. "Perhaps," Merrill said in a paper of 1933,

> ... we are already too late for some of the original heavier elements, but just in time for uranium, thorium and radium which will, in turn, soon be exhausted. Future chemists may speculate about them just as we speculate about elements heavier than uranium. ... Carried to its logical limit the theory postulates an original universe in the form of one immense super-radioactive cosmic atom. It is a daring speculation, but a beautiful and suggestive one.[10]

The primeval-atom theory did not appeal to the nuclear physicists who were not impressed by the crude imagery and vague allusions to quantum and nuclear theory.

Lemaître was curiously uninterested in developing his picture of the primeval atom in accordance with the progress in nuclear physics, a branch of science which did not appeal to him and which he considered inferior to general relativity. The version of the theory as he discussed it at the 1958 Solvay conference was almost identical to his 1931 version. He showed no interest at all in Gamow's revived big-bang theory based on advanced nuclear physics. And it was Gamow's version, rather than Lemaître's, that served as the blueprint for the later so successful theory of the hot big bang. And yet, while Gamow's theory was never nominated for a Nobel Prize, Lemaître's was. Curiously, the nomination for the 1956 prize, specifically referring to his primeval-atom hypothesis, was not in physics but in chemistry![11]

The Russian physicist George Gamow was a pioneer of nuclear theory and the author of the first textbook ever on the subject, *Constitution of Atomic Nuclei and Radioactivity* published in 1931.[12] Two years later he wrote his first paper, together with Lev Landau, on the constitution of stars and after having migrated to the United States in 1934 he focused increasingly on nuclear astrophysics (Fig. 5.3). Like a few other physicists at the time Gamow emphasized that neutron capture was probably the basic mechanism in the formation of heavy elements in the interior of stars. In a paper of 1938 he suggested a scheme involving neutrons and deuterons by means of which he hoped to build up all the elements from an original stellar state of pure hydrogen. Although the scheme did not work, a decade later he adopted a somewhat similar approach, but now in a cosmological and not a stellar context. Element formation by means of thermonuclear processes remained a central theme in Gamow's approach to cosmology as it first emerged in a vague form in the late 1930s. For example, in a popular book on the Sun he concluded that the formation of the very heavy elements could only have occurred in the "early superdense and superhot stages of the universe."[13]

Gamow's speculation was indebted to a most important work by the 26-year-old German physicist Carl Friedrich von Weizsäcker who had not only proposed a theory of stellar energy production but also, contrary to Bethe, emphasized the formation of the heavier elements. Weizsäcker argued that these elements must have been formed cosmologically, in an early and very hot pre-stellar state of the universe. He suggested tentatively that in the earliest state of the universe the density was about that of an atomic nucleus, $\rho = 10^{15}$ g cm^{-3}, and the temperature about $T = 10^{11}$ K. Such conditions might arise from a primeval aggregation of hydrogen collapsing under the influence of gravity, resulting in an explosive act with element formation taking place. "Our imagination," Weizsäcker wrote, "has the freedom to conceive not only the Milky Way system but also the whole cosmos known to us combined in it." He saw his speculation justified because it promised "a concrete cause" for the expansion of the universe:

> The energy released in nuclear reactions is about 1% of the rest energy of matter and imparts to the nuclei on the average a velocity of the order of magnitude a tenth of the velocity of light. At approximately this speed the fragments of the [primordial] star should fly apart. If we ask where today speeds of this order of magnitude may be observed, we find them only in the recessional motion of the spiral nebulae. Therefore, we ought at least to reckon with the possibility that this motion has its cause in a primeval catastrophe of the sort considered above.[14]

Fig. 5.3 *Georges Gamow (1904–1968). Photograph from the late 1930s, at a time when he had not yet begun his work in cosmology. The dedication is in Danish. Courtesy of the Niels Bohr Archive, Copenhagen.*

Weizsäcker's cosmic scenario combined for the first time nuclear synthesis with a universe expanding from a primordial state. It came close to the later big-bang scenario except that it did not make use of the relativistic models developed by Friedman, Lemaître and other cosmologists. On the other hand, it pioneered the approach known as "nuclear archaeology," meaning the attempt to reconstruct the history of the universe by means of hypothetical cosmic or stellar nuclear processes, and to test these by examining the resulting pattern of element abundances. As Weizsäcker expressed it, to get information about the primordial universe it was necessary to "draw from the frequency of distribution of the elements conclusions about an earlier state of the universe in which this distribution might have originated." Although the historical style in modern cosmology was only fully developed by Gamow, its roots can be found a little earlier.

The style eventually became commonly accepted and considered almost self-evident, but originally it was controversial.[15]

Fortunately, the Norwegian geochemist Victor Goldschmidt had recently, after many years of painstaking work, compiled extensive tables of element distributions based on solar, stellar, and meteoritic data.[16] His diagrams showed not only the variation of element abundance with the atomic number but also with the mass and neutron numbers (Fig. 5.4). Moreover, he suggested that the variations might be explained in terms of astrophysics and cosmology, possibly by Weizsäcker's still unpublished theory of which he was aware. Goldschmidt's abundance tables played a most important role in later theories of physical cosmology, not least in the theories developed by Gamow and his collaborators after World War II.

To return to Gamow, in April 1942 he arranged a conference in Washington D.C. on problems of stellar evolution and cosmology with participation of, among others, Atkinson, Subrahmanyan Chandrasekhar, Harlow Shapley, Svein Rosseland, and Edward Teller. According to the conference report it was concluded that to explain the abundance of the heavier elements "one must necessarily assume that two or three billion years ago the density of matter in space exceeded ten million times that of water and the temperature was as great as several billion degrees." Moreover, "It seems…plausible that the elements originated in a process of explosive character, which took place at the 'beginning of time' and resulted in the present expansion of the universe."[17] The

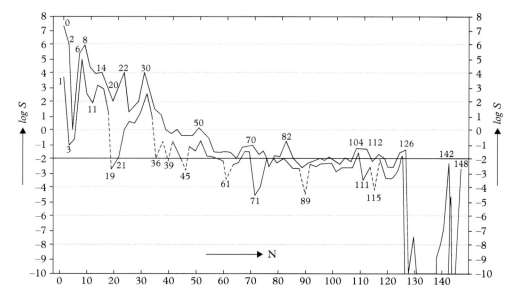

Fig. 5.4 *Goldschmidt's 1938 summary data for the cosmic abundances of the elements as a function of their neutron number.*

Source: V. Goldschmidt, "Geochemische Verteilungsgesetze der Elemente, IX," Skrifter, Norske Videnskaps-Akademiet, Mat.-Nat. Klasse, no. 4 (1938). With permission from The Norwegian Academy of Sciences and Letters.

big-bang picture was gaining momentum and three years later Gamow was ready to take the final step, such as he reported in a letter to Niels Bohr. I am, he wrote, trying to study "the problem of the origin of elements at the early stages of the expanding universe [which] means bringing together the relativistic formulae for expansion and the rates of thermonuclear and fission reactions."[18]

What Gamow had reported to Bohr he fleshed out in a paper in *Physical Review* appearing a year later. In this first version of his big-bang theory he calculated from the Friedman equations that, as a result of the rapid expansion, a mean density of the universe of the order 10^6 g m^{-3} could only be maintained for a very short time. During this brief time interval he assumed that the elements were formed out of a primordial soup of comparatively cold neutrons. The 1946 version was thus a cold big bang and it did not take into account the component of electromagnetic radiation. In a letter to Einstein from the same time he wrote that the original creation of matter in the universe took place at a mean density of 10^7 g m^{-3} and a temperature of 10^{10} K.[19]

5.3   The hot big bang

1948 was a great year in the history of modern cosmology. Not only did the new steady-state theory enter the scene, it was also the year in which the first model of the hot big-bang universe took shape.[20] Gamow was at the time supervising the doctoral thesis of Ralph Alpher on the subject of element formation in the primeval universe. The two physicists were soon joined by Robert Herman, a physicist who had mostly worked in molecular spectroscopy but also had a solid knowledge of relativistic cosmology.

In the brief but famous "$\alpha\beta\gamma$" paper published in 1948, Gamow and Alpher concluded that cosmological element formation must have taken place within the first half hour of expansion.[21] They described the very early universe as a hot, highly compressed neutron gas which at some time, set to be $t = 0$, started decaying into protons and electrons. Some of the protons would recombine with electrons and form neutrons, but with continued expansion and cooling this process would become rare and soon stop, whereas neutrons would continue decaying at a constant rate. The protons generated from radioactive decay would then combine with neutrons to form deuterons, and higher nuclei would be built up by further neutron capture followed by beta decay. Disregarding neutrinos the beginning of matter creation was thus:

$$n \rightarrow p^+ + e^- \quad \text{and} \quad p^+ + n \rightarrow d^+. \tag{5.2}$$

By working out this scenario semi-quantitatively, but without making detailed calculations of the nuclear processes, Gamow and Alpher found a reasonably close fit with Goldschmidt's abundance curve. In an account of Alpher's Ph.D. thesis the weekly *Science News Letter* reported in plain words the substance of the new theory of the early universe:

At the very beginning of everything, the universe had infinite density concentrated in a single zero point. Then just 300 seconds—five minutes—after the start of everything, there was a rapid expansion and cooling of the primordial matter. The neutrons—those are the particles that trigger the atomic bomb—started decaying into protons and building up the heavier chemical elements. ... This act of creation of the chemical elements took the surprisingly short time of an hour. (The Bible story said something about six days for the act of creation.)[22]

Whereas Gamow had originally assumed the early universe to be matter-dominated, in 1948 he and Alpher more or less independently reached the conclusion that in this early stage the universe was in fact dominated by radiation. This follows from the high temperature of about 10^9 K and the Stefan–Boltzmann law according to which the radiation density varies as

$$\rho_r = aT^4. \tag{5.3}$$

According to the equations of relativistic cosmology the radiation density decreases faster than the matter density ρ_m, namely as

$$\rho_r \propto t^{-2} \quad \text{and} \quad \rho_m \propto t^{-3/2}. \tag{5.4}$$

It follows that there must have been an era, the crossover or decoupling time, when the two densities were equal. Rather than considering either a matter-dominated or a radiation-dominated universe, in the autumn of 1948 Alpher and Herman included both components in the relevant Friedman equation (Fig. 5.5). With R_0 denoting the present curvature of space they wrote it as

$$\frac{dR}{dt} = \sqrt{\frac{8\pi G}{3}(\rho_r + \rho_m)R^2 - \left(\frac{\text{const}}{R_0}\right)^2}. \tag{5.5}$$

In their letter to *Nature* appearing in the issue of 13 November, they remarked that "the temperature in the universe at the present time is found to be about 5°."[23]

Only in a subsequent paper did Alpher and Herman clarify the somewhat cryptic remark. They now pointed out that the product of ρ_r and $\rho_m^{-4/3}$ would remain constant during the expansion, meaning that

$$\rho_{(r,0)}\,\rho_{(m,0)}^{-4/3} = \rho_r\,\rho_m^{-4/3}. \tag{5.6}$$

The quantities to the left denote initial densities and those to the right are values of a later period, for example the present one. For the present mean density of matter Alpher and Herman adopted Hubble's value of 10^{-30} g cm^{-3}, and they estimated the values at the time of element formation to be $\rho_m \simeq 10^{-6}$ g cm^{-3} and $\rho_r \simeq 1$ g cm^{-3}. As a rough value of the present radiation density, they thus got 10^{-32} g cm^{-3}, corresponding to

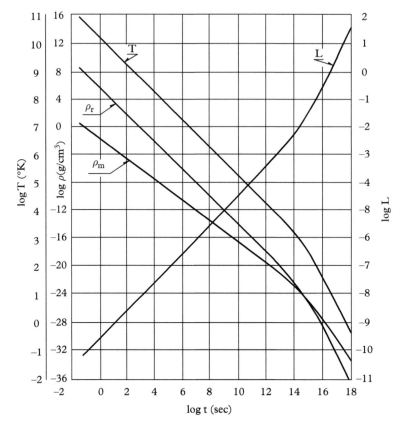

Fig. 5.5 *The "divine creation curve" showing the variation of proper distance* L(t), *the* ρ(t) *curves for matter and radiation, and the temperature curve* T(t). *Reprinted with permission from* Physical Review *75 (1949), p. 1093, copyright by the American Physical Society.*

a temperature $T \simeq 5$ K. They commented: "This mean temperature for the universe is to be interpreted as the background temperature which would result from the universal expansion alone. However, the thermal energy resulting from the nuclear energy production in stars would increase this value."[24] Although Alpher and Herman did not mention that the remnant heat radiation consisted mainly of microwaves, this follows immediately from Wien's displacement law going back to 1894. According to this law, a special case of Planck's radiation law, the wavelength for maximum intensity is given by

$$\lambda_{max} T = 2.9 \times 10^{-3} \text{ m K.} \tag{5.7}$$

The peak wavelength for $T = 5$ K becomes 0.06 cm.

The mentioned papers of Alpher and Herman undeniably contained a prediction of the cosmic microwave background (CMB) radiation, perhaps the most important phenomenon in twentieth-century cosmology. Strangely, at least as seen in retrospect, the prediction failed to attract the interest of physicists and astronomers, and that although Alpher, Herman, and Gamow mentioned it in their publications seven times between 1948 and 1956. Moreover, Alpher and Herman referred to it in many talks and colloquia. Gamow, who first mentioned the CMB in a paper of 1950, did not initially support the efforts of his two colleagues and in general understood the prediction differently.[25] And yet he clearly conceived it as a residual radiation left over from a hot and superdense state of the universe. One of the reasons for the neglect of the prediction was undoubtedly that it was widely regarded a theoretical claim without the possibility of experimental confirmation. In fact, Herman made several attempts to get observational astronomers interested in their prediction and, if possible, to detect the radiation, but their efforts bore no fruit. It is reasonable to assume that apart from scientific factors also sociological factors played a role in their failure (see Section 5.4).

In a certain but rather artificial sense the CMB may be said to have been detected as early as 1941 when the Canadian astrophysicist Andrew McKellar studied a new interstellar absorption line which he ascribed to the cyanogen molecule CN. McKellar interpreted the line as due to a thermal excitation of CN from its ground state to a rotational state and determined the excitation temperature to be $T \simeq 2.3$ K. As it was realized only 25 years later, the source of the excitation was the CMB.[26] Of course, McKellar did not discover the background radiation in 1941 any more than Vesto M. Slipher discovered the expanding universe in about 1912. In the period during which Alpher and Herman unsuccessfully promoted their prediction no one thought of linking the spectroscopic observation to questions of cosmology. The CMB remained undiscovered until 1965.

On the occasion of Einstein's 70th birthday a special issue of *Reviews of Modern Physics* was published in July 1949. Among the invited contributions were review articles by Gamow and Lemaître, both dealing with problems of relativistic cosmology. While the title of Gamow's paper was "On Relativistic Cosmogony," Lemaître's paper was titled "Cosmological Applications of Relativity." Despite the similarity of the titles the two papers were very different and with none of the authors referring to the other. They strikingly illustrate the different interests and scientific styles of the two founding fathers of big-bang cosmology.

Like most other cosmologists Gamow was acutely aware of the timescale problem, which he considered a "serious disagreement between the conclusions of relativistic cosmology and the observed facts."[27] It was not the first time that he considered the problem, for at the 1942 Washington conference he had discussed the small Hubble time of 1.8 billion years in relation to the larger value of the age of the Earth. He suspected at the time that Hubble's data were wrong because they did not take into account the decreasing luminosity of the galaxies. In the 1949 paper Gamow chose another strategy, namely to reintroduce the cosmological constant Λ which he otherwise disregarded. He consequently suggested changing the Friedman equation to

$$\frac{\mathrm{d}R}{\mathrm{d}t} = \sqrt{\frac{8\pi G}{3}\rho R^2 + \frac{\Lambda}{3}R^2 - kc^2}. \tag{5.8}$$

Gamow's solution to the timescale problem was to assign to Λ a very small positive value, namely $\Lambda = 8.6 \times 10^{-34}$ s^{-2}. The result indicated that "the space of the universe has a negative curvature and is therefore infinite in extension, and also that the value of the cosmological constant ... is still sufficiently small to be of no importance in consideration of planetary motion." Contrary to Lemaître, who was committed to the cosmological constant for methodological and aesthetic reasons, for Gamow it was merely a saving device about which he had no strong feelings. After 1952, when Walter Baade and others re-evaluated Hubble's constant, he quietly dropped Λ and never returned to it. On the other hand, he remained faithful to a negatively curved space.

Although the universe of Gamow, Alpher, and Herman was a prototypical big-bang universe, it did not include a creative event and nor did the singularity at $t = 0$ play any real role. The three physicists simply started their calculations at a certain moment in a pre-existing mini-universe without caring about the difficult question of an absolute beginning. Theirs was a creation cosmology, but not in the *creatio ex nihilo* sense, only in the sense of explaining matter and radiation as the creation—or better evolution—from an earlier state. It merely meant "making something shapely out of shapelessness," as Gamow light-heartedly said in *The Creation of the Universe*, his popular and widely read book of 1952. Likewise, Gamow and his collaborators refrained from speculating about a cosmic singularity and held with Einstein, Lemaître, and others that the gravitational field equations would break down very near $t = 0$. Gamow did not like the term "big bang" which Fred Hoyle had introduced to characterize exploding-universe models and which in the popular mind was associated with Gamow's cosmology. In an interview of 1968 Gamow expressed his dislike of the term because it emphasized the original explosion rather than the evolutionary feature of cosmological models.[28] He considered it a cliché which wrongly likened the very early universe to the explosion of an atomic bomb.

On the other hand, Gamow could not entirely resist the temptation to speculate about the origin of the universe and what was possibly before it. In some of his popular works from the 1950s he pictured the history of the universe not as beginning at $t = 0$ but as an eternally existing, rebounding universe. That is, he imagined a hypothetical collapse—a "big squeeze" (or big crunch) as he called it—that had preceded the present expansion. As he noted, although such a previous state cannot be known physically, it is allowed mathematically since it corresponds to a solution of the Friedman equations. Gamow's speculation did not belong to the tradition of cyclic or oscillating universes (Section 4.8), for models of this kind would be inconsistent with Gamow's insistence on the present universe being open and ever expanding. He thought of a cosmic one-cycle process where the universe evolved from infinite rarefaction over a super-dense but non-singular state toward a new state of continual expansion. In this way the universe would be infinitely old and a causal explanation of the primordial state could be avoided. Gamow found such a model to be "much more satisfactory" than the finite-age explosion model of Lemaître. His picture was this:

Thus we conclude that our Universe has existed for an eternity of time, that until about five billion years ago it was collapsing uniformly from a state of infinite rarefaction; that five billion years ago it arrived at a state of maximum compression in which the density may have been as great as that of the particles packed in the nucleus of an atom ..., and that the Universe is now on the rebound, dispersing irreversibly toward a state of infinite rarefaction.[29]

Realizing that it was just a speculation, he cautiously added that "from the physical point of view we must forget entirely about the pre-collapse." Bouncing models of Gamow's type attracted no scientific interest at the time, but many years later they re-appeared in the context of cosmologies based on theories of quantum gravity (see Chapter 12).

5.4   The Gamow approach to cosmology

Nuclear processes in the early universe were at the heart of the big-bang theory developed by Gamow and his collaborators, but Gamow was also much concerned with the issue of galaxy formation. As early as 1939, in a paper written jointly with Teller, he had addressed the problem from the standpoint of the expanding universe and he continued to worry about it. In general terms, the problem of structure formation was to establish a balance between the gravitational forces keeping matter together and the thermal motion and the expansion tending to disperse matter and hence preventing the formation of galaxies. What was needed seemed to be a sufficiently low temperature combined with a sufficiently high density and low expansion rate.[30]

Within the context of the new big-bang theory Gamow concluded that in the early, radiation-filled universe there would be no possibility of condensations, the highly diluted gas being completely uniform. Only when matter began to dominate, after some thirty million years, would condensations take place in accordance with the mechanism proposed by James Jeans in 1928 (and in less detail in 1902). However, calculations showed that galaxies could not be formed by Jeans' gravitational mechanism within the life span of the expanding universe if condensations did not already exist. Inspired by a theory of the formation of the solar system proposed by Weizsäcker in 1943, Gamow developed in the early 1950s a theory in which initial turbulent motion caused contractions of matter in the cosmic gas. Yet, although Gamow had confidence in his theory it was soon shown to be untenable. According to the British physicist William Bonnor, who investigated the problem in 1956, the primordial turbulences postulated by Gamow would not persist during the initial phase of expansion. By the late 1950s it was realized that big-bang cosmology, as known at the time, did not provide a convincing solution to the problem of structure formation.

The $\alpha\beta\gamma$ approach was to build up the elements sequentially by neutron capture, but Gamow and Alpher recognized at an early stage that this was not enough, and that detailed thermonuclear reactions between the light elements would also have to be included. The relevant reactions were investigated by Enrico Fermi and Anthony Turkevich at the University of Chicago. Although their extensive calculations were never published they were communicated to Gamow and known by the small group of nuclear physicists interested in the problem. Fermi and Turkevich first considered the building up of isotopes of hydrogen and helium from an initial hot state of neutrons which, due to neutron decay, would soon turn into a state of neutrons mixed with protons. Among the reactions they studied were

$$n + d \rightarrow t + \gamma \quad \text{and} \quad p + t \rightarrow {}^4\text{He} + \gamma. \tag{5.9}$$

Here d is a deuteron and t a triton, the nucleus of the ^{3}H isotope. Fermi and Turkevich found that after half an hour the number ratio between hydrogen and helium was approximately H/He = 6.7, corresponding to 24% helium by weight (Fig. 5.6).

This was an encouraging result as it broadly agreed with the observed ratio, but the work failed to account for the existence of the heavier elements. The problem was the difficulty in synthesizing elements heavier than helium from the particles existing five minutes after the big bang. Nuclei with mass numbers of 5 and 8 were needed as bridges to build up the elements, and no such nuclei were known to exist.

The problem of the missing nuclei He-5 and Be-8 was first noticed in the early 1930s. For a while it was uncertain if the nuclei were stable or not, but by 1950 nuclear physicists agreed that they were in fact unstable and thus non-existing under normal conditions.[31]

The Austrian-American physicist Edwin Salpeter suggested in 1952 that at the high temperature in a red giant a small fraction of helium nuclei (or alpha particles) would form Be-8 nuclei and that some of these would absorb another alpha particle and form C-12. Salpeter's triple-alpha process was

$$\alpha + \alpha \rightarrow {}^8\text{Be} + \gamma \quad \text{followed by} \quad \alpha + {}^8\text{Be} \rightarrow {}^{12}\text{C} + \gamma. \tag{5.10}$$

However, not only was the yield of carbon very low, the mechanism was also restricted to the interior of very hot stars. It would not work under the conditions assumed to govern the early big-bang universe and was therefore not considered to be relevant for cosmology. The "mass gap problem" remained. It deeply concerned Fermi, Turkevich, Gamow, Alpher, and Herman, who investigated a number of ways to bridge the gap in a realistic manner. None of the proposals worked and by 1953 at the latest it was clear that the problem was a serious one indeed, and one that might even jeopardize the entire Gamow approach to big-bang cosmology.

As Gamow and his collaborators saw it, the situation was not quite as bad. The mass gap problem might weaken their theory as they originally conceived it, but it did not suggest that it was wrong in a fundamental sense. For one thing, although no satisfactory solution had been found, this did not imply that no such solution existed. Perhaps the difficulty could be resolved if more realistic and even more detailed reaction schemes were considered. Moreover, even if the mass gap could not be bridged and heavy elements thus not be produced in the big bang, it did not contradict the hot big-bang theory. After all, it was compatible with this theory that some or even most elements were built up in the stars. Gamow remained optimistic, retreating to the statement that

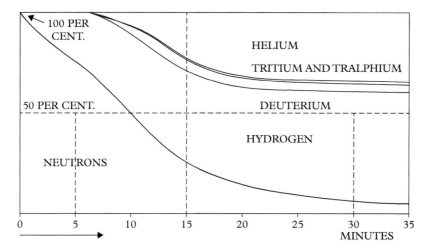

Fig. 5.6 *The result of the Fermi–Turkevich calculations concerning the chemical composition of the first half-hour of the universe. "Tralphium" was Gamow's name for helium-3.*

Source: G. Gamow, The Creation of the Universe *(New York: Viking Press, 1952), p. 69. Reproduced with permission of Igor Gamow.*

"I would agree that the lion's share of the heavy elements may well have been formed later in the hot interior of stars."[32]

As far as helium was concerned, the Fermi–Turkevich result was in broad agreement with what the Japanese astrophysicist Chushiro Hayashi, at the Nanikawa University, found in 1950. Hayashi suggested that at the very high initial temperature of 10^{12} K, corresponding to a time after $t = 0$ of the order of 10^{-4} s, also electrons and positrons would contribute importantly to the nuclear processes.[33] These light particles would be created from the electromagnetic radiation field and result in induced beta processes such as

$$n + e^+ \rightarrow p + \bar{\nu} \quad \text{and} \quad n + \nu \rightarrow p + e^-. \tag{5.11}$$

The symbol ν denotes a neutrino, and $\bar{\nu}$ an antineutrino. Hayashi concluded that the original matter of the universe could be neither purely neutronic nor much dominated by neutrons at the time element formation began; it would be composed of a mixture of neutrons and protons in the ratio 4:1. According to his calculations, helium would be built up in the early neutron-proton inferno in such a way that the present hydrogen–helium ratio would be close to 6:1.

Inspired by Hayashi's work, in 1953 Alpher, and Herman, now collaborating with James Follin, developed their theory into a much improved version. The three authors made innovative use of the most recent advances in nuclear and particle physics, for example by taking into account elementary particles such as muons and pions (then called mesons). They even considered the possible role of gravitons, the hypothetical quanta of gravitation which here appeared for perhaps the first time in a cosmological context. "We have carried our study," they wrote,

> ... back to a temperature of ~100 Mev (~1.2 × 10^{12} K), corresponding to an epoch of 10^{-4} s. For temperatures below this value one can treat reactions among elementary particles with some confidence. Furthermore, below ~100 Mev the energy stored in the gravitational field is a negligible part of the total energy ... Finally, at ~100 Mev one has a state of thermodynamic equilibrium among all the known constituent particles and radiation so that a knowledge of the previous history of the universe is not required. [34]

The Alpher–Follin–Herman scenario of the very early universe was a soup of photons, neutrinos, electrons, positrons, and muons, with only traces of neutrons and protons. After a few minutes' expansion, most electrons and positrons would have annihilated, leaving mainly photons and neutrinos. After 100 million years had elapsed, Alpher, Herman, and Follin found that the universe had cooled to 170 K and the mass density decreased to 10^{-26} g cm^{-3}. For the present hydrogen–helium ratio they derived a value between 10:1 and 7:1, or a mass content of helium between 29% and 36%. However, the observed amount of helium was not known with any certainty and for this reason the prediction did not count much as a confirmation of the theory. Only in the early 1960s could astrophysicists claim with some confidence that about 30% of the matter in the universe consisted of helium and it was only from that time that the hydrogen–helium ratio became established as an important cosmological test (see Section 5.8 and in Chapter 8).

By the early 1950s the big-bang theory of Gamow and his collaborators had developed into a minor and apparently thriving industry disseminated through the scientific and popular literature and also through many meetings and conferences. It was almost exclusively an American industry with Hayashi in Japan as the only exception. Alas, the appearance of a healthy state was deceiving, for interest soon faded and with the Alpher–Follin–Herman paper of 1953 the theory came to an almost complete halt. The three physicists sought to revive interest in the theory, and gave several presentations of new calculations of light-element formation at scientific meetings, but to no avail. In fact, a 1956 paper by Hayashi and his collaborator Minoru Nishida on nucleosynthesis in the hot early universe was the only contribution to big-bang theory for a decade. Given the fact that the big-bang theory was resurrected about 1965 in almost the same form as it had twelve years earlier, the low interest after 1953 is puzzling.

There is no easy answer to the disregard of Gamow's big-bang program, but an answer must presumably be sought in a combination of scientific and social factors.[35] One reason was the failure of bridging the mass gap, which made the theory look less appealing to many physicists. Some scientists considered it to be suspicious because they thought of it as a theory offering an account of the creation of the universe. But social and disciplinary reasons were probably more important. Although Gamow's research program had important astronomical implications it was primarily oriented towards nuclear physics and not towards traditional astrophysics and astronomy. Remarkably, during the 1950s no research papers on big-bang cosmology appeared in astronomical journals such as the American *Astrophysical Journal* and the *Monthly Notices* published by the Royal Astronomical Society.

The careers and reputations of the small core group of big-bang cosmologists also mattered. Gamow, with whom the theory was generally associated, largely left cosmological research after about 1952 and for the rest of his career he did not enjoy a high reputation among his peers. While the public listened to him, few scientists did. For a few years Alpher and Herman carried on the torch, but when they both left the academic world for research careers in private industry—General Electric and General Motors, respectively—they no longer had strong connections to the scientific community of physicists and astronomers. In general, it should be recalled, cosmology did not yet exist as a scientific discipline or sub-discipline. Nuclear physics was flourishing but to combine it with general relativity theory, such as Gamow, Alpher, and Herman did, was highly unusual and added to the difficulty of recruiting new people to research in big-bang cosmology. One might believe that the competition from the steady-state theory of the universe was another reason for the lack of interest in big-bang cosmology, but this was hardly the case. The steady-state theory did not enjoy general support, and in America it was not even taken seriously as a cosmological model.

5.5   The steady-state alternative

At about 1950, two decades after the discovery of the expanding universe, most astronomers agreed that the universe is in a state of evolution as described by the equations of relativistic cosmology. Although there was no consensus view, many found the Lemaître–Eddington model to be an appealing picture of the universe and its history. Astronomers and physicists were aware of models with a singular origin in time, such as the Einstein-de Sitter model, but models of this kind were rarely discussed as realistic candidates for the universe. At the same time as Gamow and Alpher reformed big-bang theory, there appeared a diametrically opposed cosmological theory which became known as the steady-state theory of the universe or, initially, the "new creation theory." According to this theory, there was neither a cosmic beginning nor a cosmic end. The large-scale features of the universe had always been, and would always be, the same. To make an eternal universe of this kind agree with observations it was necessary to assume that matter is continuously created throughout the universe. The new theory clashed with the more traditional finite-age models based on Einstein's equations, and the ensuing controversy deeply influenced the cosmological scene for more than fifteen years.[36]

The steady-state theory appeared in 1948 in two papers in *Monthly Notices of the Royal Astronomical Society* written by three young Cambridge physicists. Fred Hoyle was trained in theoretical physics and had by 1940 moved into astronomy. In relation to his war-time research on radar he came to know two young Austrian-born physicists, Hermann Bondi and Thomas Gold, and the three eventually discussed the state of relativistic cosmology. None of them had previously dealt with questions of cosmology, but they soon reached the conclusion that models based on the expanding solutions of Einstein's field equations were unsatisfactory and in need of fundamental revision. Among other reasons, they found it objectionable that the class of relativistic models was so wide that it could accommodate almost any observation. They also considered the age paradox a grave problem. For scientific as well as philosophical reasons, the three physicists wanted an unchanging yet dynamic universe in agreement with the observations of Hubble and other astronomers.

To make the two desiderata meet, they were forced to postulate that matter was continuously created through the universe. This essential part of the new theory originally came from Gold and it was also Gold who coined the pompous name "perfect cosmological principle" (PCP) as an extension of the ordinary cosmological principle. According to this cornerstone of classical steady-state cosmology, the universe is not only spatially but also temporally homogeneous. In other words, the large-scale appearance of the universe is the same at any location and at any time. The PCP implied an eternal universe and hence disposed of the timescale problem as well as problems related to the origin of the primeval universe or big bang. There also could be no big crunch or final heat death. The steady-state theorists considered the continual creation of matter to be a deduction from the PCP rather than a separate postulate.

Hoyle, Bondi, and Gold came to their theory independently of other scientists, not knowing that in a qualitative sense important parts of it had previously entered cosmological thought. As mentioned in Chapter 1, as early as 1908 Svante Arrhenius based his view of the universe on what was in essence the PCP, and in the 1920s he was followed by several other scientists. Thus, the Chicago astronomer William MacMillan not only assumed that stars and galaxies were distributed uniformly throughout infinite space, he also denied "that the universe as a whole has ever been or ever will be essentially different from what it is today."[37]

Similar ideas of an eternally regenerating universe, including new matter created by radiation or ether, were entertained by Robert Millikan in the United States, Oliver Lodge in England, and Walther Nernst in Germany. However, they all thought of the universe in terms of classical physics and disregarded or were unaware of relativistic cosmology. Their ideas were restricted to the qualitative level and peripheral to mainstream astronomy. Of more interest is perhaps that Einstein in a manuscript of early 1931 explored a model of what can reasonably be called the steady-state universe.[38] In Einstein's model the universe expanded exponentially and yet its matter density remained constant. He associated the expansion with a continual creation of matter the source of which was the cosmological constant. However, Einstein never published his model and when the steady-state theory appeared two decades later he dismissed it as a speculation (Section 5.6).

As mentioned, the new steady-state theory of the universe was introduced in the form of two papers, one written by Bondi and Gold and the other by Hoyle.[39] The Bondi–Gold paper was much more qualitative and philosophical than Hoyle's and also more critical to the use of general relativity in cosmology. From a methodological point of view the approach of Bondi and Gold was to some extent inspired by Milne's kinematic system (Section 4.3) which Bondi admired because of its deductive nature and emancipation from general relativity. Whereas Hoyle based his version as closely as possible on Einstein's field equations, his two friends deliberately avoided a field-theoretic formulation and based their version directly on the PCP. Nonetheless, different as the two papers were in formulation, style and outlook, they were about the same theory. Bondi and Gold proved by means of a simple argument based on the PCP that the space of the steady-state universe must be flat and expand exponentially as

$$R(t) = R_0 \exp(Ht) = R_0 \exp(t/T). \tag{5.12}$$

Whereas the Hubble time $T = 1/H$ is a measure of the age of the universe in models of relativistic cosmology, it follows from the PCP that it must be a constant timescale. Bondi and Gold further showed that if mass density is conserved, new matter in a volume V must be created at a rate given by

$$\frac{1}{V}\frac{dm}{dt} = 3\rho H \simeq 10^{-43} \text{ g s}^{-1} \text{ cm}^{-3}. \tag{5.13}$$

"This rate corresponds to one new atom of hydrogen per cubic metre per 3×10^5 years. Accordingly we cannot expect the process to be directly observable." The two authors further pointed out that there should be no systematic difference in age between galaxies nearby and far away. In any large volume of space there will be old and young galaxies, the ages being distributed according to a statistical law. The distribution law was derived by William McCrea, an early advocate of the steady-state theory who had previously worked with both Eddington and Milne. As McCrea showed in 1950, it follows from steady-state assumptions that the average age of galaxies in any large region of space is $T/3$; the fraction of galaxies which are older than a certain age t is given by $\exp(-3Ht)$.[40]

Hoyle's approach was based on field equations modified to accommodate matter creation, which he did by formally replacing the cosmological term $\Lambda g_{\mu\nu}$ in Einstein's equations of 1917 with a tensor $C_{\mu\nu}$ responsible for the gentle creation of matter. The equations could then be written as

$$R_{\mu\nu} - \tfrac{1}{2} g_{\mu\nu} R + C_{\mu\nu} = -T_{\mu\nu}. \tag{5.14}$$

Hoyle scornfully dismissed big-bang theories with their "creation of the universe in the remote part" as being "against the spirit of scientific enquiry." From his equations Hoyle derived a definite value for the constant matter density, namely

$$\rho_{SS} = \frac{3H^2}{8\pi G} \simeq 5 \times 10^{-28} \text{ g cm}^{-3}. \tag{5.15}$$

The expression happened to be exactly the same as the critical density appearing in the Einstein–de Sitter model. The numerical value was much larger than Hubble's value of 5×10^{-31} g cm^{-3} estimated from observations. But, like Einstein and de Sitter in 1932, Hoyle did not consider the discrepancy to be a problem worth worrying about. It merely suggested that "only about one part in a thousand of the intergalactic medium is at present in a condensed state." Altogether, the new cosmological theory of a universe in a steady state was remarkably precise and yielded a number of definite consequences (Section 5.8).

As regards the form in which the continuously produced matter appeared, the theory was much less precise. In principle it could be matter of any size and complexity, but the three physicists agreed, mostly for reasons of simplicity, that it must be electrically neutral and in some elementary form. Whether it was neutrons, or protons and electrons, or hydrogen atoms was less important at this stage of the theory. Contrary to earlier ideas of cosmic matter creation, Bondi and Gold stressed that the new particles were created *ex nihilo* and not out of energy or some other agency. For this reason the steady-state theory in its classical version of 1948 violated the fundamental principle of energy conservation. Bondi and Gold did not consider this a cardinal sin, but thought it was a small price to pay for a satisfactory cosmological theory. After all, energy non-conservation of the tiny rate predicted by the theory did not contradict experimental facts. Rhetorically they asked if energy conservation had ever been tested to an indefinite accuracy.

The steady-state theory received wide attention with the publication of two books, one authored by Bondi and the other by Hoyle. Bondi's concise textbook *Cosmology* from 1952—a second edition appeared in 1960—was one of the first of its kind and widely used for two decades. It covered not only the steady-state theory but also relativistic cosmology, Milne's kinematic theory, Newtonian cosmology, and the more unorthodox ideas (such as Dirac's) based on the combinations of natural constants. On the other hand, Bondi did not refer to the $\alpha\beta\gamma$ big-bang theory or even mentioned Gamow by name. Bondi admitted in the preface his own preference, namely that "in my opinion the steady-state theory agrees best with observation and has the simplest and most logical basis." He further stated that "I do not regard cosmology as a minor branch of general relativity or as a branch of philosophy and logic."[41] While the first reference was obviously addressed to his colleagues working in the Einstein tradition, the second was presumably a reference to Milne's system. Of an entirely different kind and addressed to a very different audience was Hoyle's best-selling *The Nature of the Universe* first published in 1950.

Hoyle's popular book was based on a series of BBC broadcasts given in early 1950 and printed verbatim in *The Listener*, a widely sold BBC magazine. It was in this context that Hoyle coined the name "big bang" as the belief that "the Universe started its life a finite time ago in a single huge explosion, and that the present expansion is a relic of the violence of this explosion ... the whole creation process taking the form of one big bang." This belief he found to be highly unsatisfactory and "very much queerer than continuous creation."[42] In fact, the origin of the cosmological term big bang dates a little earlier, for Hoyle first used it in another BBC talk on 28 March 1949. Although he may have thought of Gamow's recent theory, or perhaps of Lemaître's earlier theory, he did not mention any names. The catchy name for the explosion theory eventually stuck, but not immediately. It took more than two decades until big bang became a household word in the scientific community. Hoyle only used it again in 1965.

At first the steady-state theory failed to arouse much attention among mainstream cosmologists outside Great Britain. Cosmological models flourished and wasn't the new one just another short-lived fancy? Early commentators tended to reject the theory without examining it seriously. Continual matter creation surely was preposterous and the theory an artificial attempt to avoid the finite-age universe based on the authoritative equations of general relativity. Many agreed with the French astronomer Paul Couderc, who in 1952 dismissed the steady-state model as "risky and over-imaginative." Couderc undoubtedly spoke for the majority of astronomers when he warned against theoretical excesses in cosmology, and argued that observations rather than theory should be given priority:

> At the present time, new ideas—some mere readjustments, others quite fantastic in their novelty—are experiencing an extraordinary efflorescence.... Everything possible has been extracted from existing observations; until the appearance of new data we would be wise to give our imaginations a rest. Let the giant telescopes be set to work at the limits of the observable region, while the humbler instruments ... patiently investigate the nearer horizons, where there is still plenty of unexplored ground.[43]

Much of the work on steady-state theory that took place during the 1950s was motivated by the wish of modifying the theory in such a way that the objectionable concept of matter creation could be accounted for within the mathematical framework of general relativity. In 1951 McCrea presented a new version of Hoyle's steady-state theory with this aim in mind.[44] Rather than assuming the pressure term to be zero, as Hoyle did, McCrea suggested to write it as a stress term proportional to the mass density:

$$p = -\rho c^2. \tag{5.16}$$

In Hoyle's model the expansion of space was caused by an outward pressure produced by the created matter, whereas McCrea explained the expansion as a result of a hypothetical negative pressure. This pressure was of no direct physical significance because its gradient vanishes everywhere and thus would have no mechanical effects. Apparently unknown to McCrea, in a different context a negative vacuum pressure had been suggested by Lemaître as early as 1934 (see Section 4.8). McCrea's formal reinterpretation remained within the steady-state theory and led to the very same observational results as Hoyle's theory and the Bondi–Gold theory. But instead of making use of the creation process itself as a primary postulate, the introduction of a negative pressure in empty space shifted the focus of the theory away from the mysterious creation of matter. It was no longer seen as a genuine creation process *ex nihilo*, but rather as a kind of transformation. In the new interpretation, the creation process was a consequence of space being endowed with negative pressure.

McCrea's alternative version of the steady-state theory inspired other British physicists to take up similar investigations, including George McVittie, Felix Pirani, and William Davidson. They all wanted to formulate matter creation in conformity with the energy-conserving theory of general relativity. In a series of papers during the 1960s Hoyle developed his *C*-field theory in collaboration with Jayant Narlikar, making use of McCrea's work. According to the extended *C*-field theory, matter creation took place without violating the principle of energy conservation, as the energy of the created particles would be compensated by the negative energy of the C-field. Although the Hoyle–Narlikar theory led to interesting mathematics and promised a unification of particle physics and gravitation, most scientists found it barren as a physical theory describing the real universe.

In 1959 Bondi and the British astronomer Raymond Lyttleton, a former collaborator of Hoyle, suggested a cosmological theory based on the idea that McCrea's stress was a manifestation of the electromagnetic field owing to a slight inequality between the numerical charges of the proton and the electron.[45] Assume that a hydrogen atom carries a tiny charge ϵe, where $-e$ is the elementary charge of an electron; the proton's charge is then

$$q_\mathrm{p} = (1 + \epsilon)e. \tag{5.17}$$

Bondi and Lyttleton found that with $\epsilon = 2 \times 10^{-18}$ the Hubble law could be explained as a result of electrostatic repulsion and without recourse to gravitation theory. If matter was created out of nothing, so was electric charge, which meant that they had to modify Maxwell's charge-conserving field equations. The Bondi–Lyttleton theory not only led to an explanation of Hubble's law but also to values of the cosmic density and the creation rate of matter in broad agreement with those of the original steady-state theory:

$$\rho = (3H^2)/4\pi G \quad \text{and} \quad 3H\rho = 10^{-46} \text{ g s}^{-1}\text{cm}^{-3}. \tag{5.18}$$

More importantly the electric cosmology provided McCrea's negative pressure with a physical cause, namely, the charge excess. However, the ingenious theory rested on the hypothesis of a minimum charge excess of $\epsilon e = 2 \times 10^{-18}e$, which quickly was contradicted by experiments. Already in 1960 precision experiments proved that $\epsilon < 10^{-19}$ in disagreement with the electric universe hypothesis and later experiments confirmed that atoms and molecules are electrically neutral. Consequently the electric version of the steady-state universe was abandoned and no-one felt tempted to revive it. In spite of being an almost instant failure the theory is worth noting, as it may have been the first time ever that a theory of the universe was shot down by ordinary laboratory experiments.

5.6   Creation and controversy

During the period from the late 1940s to the early 1960s much of the cosmological scene was colored by the controversy between the steady-state theory of the universe and the rival class of relativistic evolution theories. The new theory of Hoyle and his colleagues provoked interesting discussions concerning the scientific status of cosmology in which methodological and other philosophical arguments were no less important than arguments based on observations. This was at a time when it was still a matter of some debate whether cosmology should be counted as a proper science in the first place.

The radio astronomer Antony Hewish, a Nobel Prize laureate for his discovery of pulsars, recalled about the state of cosmology around 1950: "The whole thing was so nebulous...cosmological theory was sort of there but people didn't bother about it because they didn't think you could really ever know the answer to these things."[46]

Given the scarcity of observations to distinguish between the two main conceptions of the universe it is not surprising that meta-scientific and other philosophical views played an important role in the debate. It was not uncommon to highlight subjective factors and to characterize the choice of cosmological model as due mainly to taste. In 1954 the Estonian-Irish astronomer Ernst Öpik described the choice as "a matter for aesthetic judgment," and the previous year the Swedish physicist Oskar Klein admitted that cosmology was a field where "personal taste will greatly influence the choice of basic hypotheses."[47] Neither Klein nor Öpik rated the steady-state theory highly by such standards, but others did. Dennis Sciama, a young British physicist and later an important figure in astrophysical theory, turned to the steady-state theory primarily because of its simplicity and what he considered its philosophical appeal. One element in the theory's appealing structure was, in his interpretation, the promise of a non-contingent and self-consistent cosmology. Sciama remained an important ally of the steady-state theory from about 1954 to 1966 when he finally decided that the theory, although appealing and aesthetically attractive, was wrong.[48]

The skeptical attitude with regard to cosmology's scientific status, such as aired by Klein and Öpik, was held by many scientists not engaged in astrophysical or cosmological research. Years before the emergence of the steady-state theory, Percy Bridgman expressed himself in terms that were still common in the 1950s. A Nobel Prize laureate and noted philosopher of physics, Bridgman stressed that in cosmology, subjective elements of an emotional and metaphysical kind must necessarily play an important role. He doubted if the study of the universe would ever become a decent science solidly based on empirical evidence:

> To the untutored critic it must appear a trifle rash to peer 10^{16} years into the past or even greater distances into the future on the basis of laws verified by not more than 300 years' observation. The only justification for such hair-raising extrapolations is to be found in the tacit assumption of some system of metaphysics; we are convinced that nature obeys mathematically exact rules, and that we have found some of them.[49]

Cosmology's lack of professional maturity is further illustrated by an illuminating discussion of 1954 between Bondi and Gerald Whitrow, a respected British astronomer and mathematician. The title of the discussion, as it appeared in the *British Journal for the Philosophy of Science*, was simply "Is Physical Cosmology a Science?" Whitrow was skeptical but not clearly dismissive to the steady-state theory which he believed led to consequences that were highly bizarre and perhaps even contradictory. However, the discussion did not concern the steady-state theory specifically but cosmology more generally. According to Whitrow cosmology *anno* 1954 was not truly scientific, for other reasons because cosmologists did not share common standards and methods. The study of the universe, he suggested, would remain a borderline subject between science and philosophy. Bondi viewed the situation more optimistically, arguing that the still immature science of physical cosmology was rapidly on its way to become a scientific discipline on par with other sciences such as physics and chemistry. The key issue entering the discussion was the role played by philosophical considerations. As Bondi saw it, observation and physical theory had now largely replaced philosophy with the result that cosmologists, despite their differences, agreed on the basic methodological criteria.

Bondi suggested that the hallmark of science was empirical falsifiability of theories and that on this criterion—"the cardinal test of any science"—cosmology was indeed a science. Although not mentioning Popper by name, Bondi was clearly referring to the philosophy of Karl Popper, whom he knew personally and whose philosophy of science he was acquainted with and much admired. Bondi's enthusiastic support of Popper's philosophy is evident from a glowing review essay he wrote in 1959 of Popper's *The Logic of Scientific Discovery* together with the mathematical physicist Clive Kilmister. Not only did Bondi and Kilmister praise the method of falsificationism, they also used it as an argument for the steady-state theory: "For here the correct argument has always been that the steady-state model was the one that could be disproved most easily by observation. Therefore, it should take precedence over other less disprovable ones until it has been disproved."[50] Although Popperian criteria of science played a considerable role in the cosmological controversy, they were rarely an issue of dispute. By and large, such criteria were accepted also by many cosmologists favoring an evolving universe governed by the laws of general relativity.

Steady-state cosmology was widely considered controversial, primarily because of its element of matter creation but also because it, and especially in the version of Bondi and Gold, rested deductively on a general principle, the PCP. The theory was attacked fiercely by Herbert Dingle, an astrophysicist who since 1949 had served as professor of history of science at University College London. In his 1953 presidential address to the Royal Astronomical Society, Dingle dismissed in strong words the new theory as a ridiculous piece of pseudoscience whose basis in the PCP was deeply unscientific. As regards the continual creation of matter, "it has no other basis than the fancy of a few mathematicians who think how nice it would be if the world was made that way." What Dingle presented to the audience of astronomers, Lemaître among them, was a caricature of the steady-state theory which in the Bondi–Gold version was in fact remarkably non-mathematical.

Dingle's attack was not very effective in scientific circles and was ignored by the steady-state proponents. On the other hand, it was welcomed by the Argentine physicist and philosopher Mario Bunge who in similar unrestrained language accused the theory of being closer to magic than to science. Again it was the spontaneous creation of matter in space which caused his anger. Also the famous physicist Wolfgang Pauli, a specialist in general relativity and a believer in immutable conservation laws, rejected the theory because it violated the principle of energy conservation. "I consider his [Hoyle's] continuous creation of matter from nothing to be pure madness," he wrote in a letter of 1951.[51] Einstein agreed, such as evidenced from a letter of 1952: "The cosmological speculations of Mr. Hoyle, which presume the creation of atoms from empty space, are in my view much too poorly grounded to be taken seriously."[52] Two years later Einstein confirmed in an interview that he disliked the steady-state theory and its postulate of matter creation.[53]

Although the majority of physicists, astronomers, and philosophers found continual creation of matter to be inadmissible, there were exceptions. Remarkably, Gamow was one of them. He rejected the steady-state, of course, but not for this reason. In an interview of 1968 he remarked, "It's perfectly logical. Matter is not conserved—so what?"

The poor reputation of the steady-state theory was not only caused by the unorthodox introduction of matter creation but also by the methodological features that Bondi and Gold stressed in particular. The theory was received with a mixture of hostility and indifference in the strong community of American astronomers. Hubble presumably knew about it from Hoyle, who stayed at Caltech in 1953 and had conversations with Hubble, but the esteemed astronomer never mentioned the new theory of the universe. Hubble's successor Allan Sandage found the steady-state theory to be contrived, hopelessly speculative, and disagreeing with observational astronomy. He and other Mount Wilson astronomers felt that the advocates of the steady-state theory consistently valued theoretical predictions higher than observations and that they tended to ignore sound observational work. Their methods, they complained, contradicted astronomical research practice and for this reason they could not be taken seriously.

The suspicion that Hoyle and his colleagues were playing a game with rules different from those of established science was shared by the leading cosmologist George McVittie. After he moved from England to the United States in 1952, his initially positive interest in the steady-state theory increasingly turned into hostility.[54] He felt that matter creation was a violation of the basic rules of scientific reasoning. A supporter of mainstream relativistic cosmology McVittie warned against rationalist excesses without a proper empirical foundation such as he saw them in the steady-state theory. From a long career in cosmology he knew that "the temptation to substitute logic for observation is peculiarly hard to resist in astronomy and especially so in cosmology," as he wrote in *General Relativity and Cosmology*, a textbook of 1965. According to McVittie, the proper method in cosmology was the same as in other physical sciences, namely empirical-inductive and not hypothetical-deductive. It was to "discover how much can be found out about the universe through measurements that yield non-null results rather than by consideration of logical possibilities which might conceivably be the case." The latter method was unfortunately the one followed by what he privately called "the steady-state boys."[55]

What was perhaps the most comprehensive and detailed critique of the hypothesis of continual creation of matter came from a philosopher and not from a physicist. In a book published in 1957 with the title *Space, Time and Creation* the American philosopher Milton Munitz analyzed carefully and critically the concept of matter creation *ex nihilo*, arguing that it was unscientific because it was unexplainable even in principle. The very concept of *ex nihilo* creation, he suggested, presupposed some kind of creative agent and therefore opened the door to supernaturalism—much the same objection that Hoyle had raised against big-bang cosmology. What worried Munitz was that continual matter creation defied explanation, which to his mind implied that it was scientifically meaningless. Neither Hoyle nor other supporters of the steady-state theory responded to Munitz's philosophical critique, perhaps because it focused on a concept, *ex nihilo* creation, which they no longer subscribed to. The only scientific response came from William Bonnor, himself a critic of steady-state cosmology, who disagreed with Munitz's claim that matter creation was beyond explanation.

From what has been said one might get the impression that steady-state theory was taken seriously only by a small group of British astronomers and physicists. But this would be to underrate the significance of the theory which in the period from about 1954 to 1960 was considered a serious rival to the relativistic theory of the evolving universe and was better known than the big-bang theories of Lemaître and Gamow. One piece of evidence is the Solvay conference of 1958 where the theme was the structure and evolution of the universe, the first time that astrophysics and cosmology featured in a Solvay conference.[56] If this prestigious meeting can be taken as an indication of the strength of the steady-state theory at the time, it must be concluded that it was in a strong position. Among the invited participants were all the leading steady-state theorists, namely Hoyle, Gold, Bondi, and McCrea. Other participants were Heckmann, Sandage, Baade, and Klein, none of whom had any sympathy for either the steady-state theory or the big-bang theory. Indeed, the big-bang idea was defended only by Lemaître and then in his own version quite different from Gamow's. Although Gamow had much wanted to participate, he was not invited.

Physical big-bang theory was also not represented in a symposium on rival theories of cosmology arranged by the BBC in 1959.[57] Apart from Whitrow, who served as chairman and moderator, the symposium involved three speakers of whom two belonged to the steady-state camp, namely Bondi and Lyttleton; the third speaker was Bonnor, who defended the relativistic evolution theory and his own idea of a cyclic universe in particular. In the publication based on the symposium one looks in vain for the name of Gamow.

While Gamow was nearly absent from the cosmological scene in Europe, he was highly visible in the United States. The only time that Hoyle and Gamow engaged in a kind of direct confrontation was in 1956, when both scientists argued their case in companion articles in a special issue of *Scientific American*. Other contributors to the issue included Robertson, Baade, Sandage, Ryle, and Dingle. While Hoyle explained the steady-state universe, Gamow described the relativistic theory of the expanding universe and also briefly referred to the view that the universe is in "a steady-state equilibrium which has always existed and will go on through eternity." Gamow disbelieved Hoyle's theory "because of the simple fact (apart from other reasons) that the galaxies in our neighbourhood all seem to be of the same age as our own Milky Way."[58] To repeat, what the steady-state theory challenged was not specifically or only big-bang conceptions of either Gamow's or Lemaître's type, but more generally the class of evolution theories based on the Friedman equations, and especially those with a singular beginning or what Hoyle and Bondi called source-point models. Gamow's theory belonged to this class, but so did many other models.

During the period from 1948 to 1966, when steady-state cosmology was effectively abandoned, about forty physicists and astronomers contributed to various aspects of the theory. Most of them supported the theory or sympathized with it, but less than ten were committed to the steady-state universe and worked on it over an extended period of time. The core group can be narrowed down to seven scientists, all Britons or living in the United Kingdom: T. Gold, F. Hoyle, H. Bondi, W. McCrea, D. Sciama, J. Narlikar, and C. Wickramasinghe. These are small numbers as seen from a modern perspective, but then it should be recalled that at the time cosmology was a small field of research. Although the steady-state theory was at no time the accepted view of the universe, not even in England, in the 1950s there were as many or more publications on this theory as there were on relativistic theories of the finite-age universe. There is no doubt that the steady-state theory was very important, and that it made up a considerable part of the cosmological landscape for more than a decade.

5.7   Some broader contexts of cosmology

The cosmological controversy did not only unfold in the pages of the scientific research journals but also, and even more so, in the form of popular books and magazine articles intended for a broad readership. Indeed, the controversy appears more clearly through the lenses of the period's popular writings. Gamow and Hoyle were seasoned and successful writers of popular science and they both used the genre to disseminate and promote their views of the evolution of the universe. Neither of them respected the strict border that is supposed to distinguish research writings from popular writings, and they also were not much concerned with the lack of academic respectability that may follow such a view. They were not, of course, the first scientists to use popular writings in discussing questions of cosmology, for the same had been done by Jeans, Eddington, and de Sitter before World War II and the tradition goes still longer back.

Gamow's *The Creation of the Universe* was a remarkable hybrid between popular and scientific, whereas Hoyle used his more general *The Nature of the Universe* to expose views that would be unfit for a scientific paper. Another of Hoyle's influential and popular books was *Frontiers of Astronomy* from 1955, a work which included chapters on the steady-state universe and continual creation of matter. Hoyle used his book to question big-bang theories and their reliance on arbitrary starting conditions, which he thought was "quite characteristic of the outlook of primitive peoples."[59] Hoyle and Gamow were the most prominent of the popular-writing cosmologists but far from the only ones. To mention but two more examples, in 1960 Bondi published *The Universe at Large* and four years later Bonnor came out with *The Mystery of the Expanding Universe*. Both books dealt with the ongoing controversy between relativistic models and the steady-state theory, and both authors stressed that a resolution of the controversy would depend on the results of observational tests in the near future.

The cosmological controversy was one of those rare episodes in the history of modern science in which philosophers interacted with researchers in physics and astronomy. While some of those involved in the controversy were professional philosophers others, such as Dingle, Whitrow, Bunge, and Richard Schlegel, may be best labeled as physicist-philosophers. We have already referred to Popper and his considerable if indirect influence on the development and also to Munitz's critique of the concept of continual creation of matter. Although the philosophers intervening in the controversy over the universe studied the most recent scientific research and dealt with questions that were high on the scientific agenda, they had relatively little impact on the work of the cosmologists. The relationship was asymmetric, in the sense that the philosophers eagerly analyzed the new ideas in cosmology whereas the cosmologists, with a few exceptions, paid no attention to the philosophers' contributions and in many cases were just unaware of them.

One of the earliest and most interesting philosophical interventions was due to the American Adolf Grünbaum, who in 1952 wrote a comprehensive and insightful review of the situation in modern cosmology including the steady-state theory, Milne's cosmology, and the $\alpha\beta\gamma$ big-bang theory.[60] Grünbaum objected to Dingle's attack on the steady-state theory and his objection was acknowledged by Bondi. The age of the universe was a key issue in the controversy and one of equal interest to scientists and philosophers. Is the universe finite or infinite? The issue was addressed by the British *Society for the Philosophy of Science* in the form of a prize competition resulting in 26 essays, most of them from philosophers but also some from astronomers and physicists. To the latter group belonged G. Whitrow and E. Öpik. Somewhat remarkably, the first prize was awarded to the American philosopher Michael Scriven according to whom the question of the age of the universe could not be answered on the basis of science.

A similarly skeptical view of the situation in cosmology and cosmogony was held by the Oxford philosopher Rom Harré and also by Stephen Toulmin, a prominent British-American philosopher. They both suggested that statements about the universe as a whole were scientifically illegitimate. The American Norwood Russell Hanson, another leading figure in the new empirically oriented philosophy of science, was no less skeptical about the conceptual foundation of modern cosmology. He considered it a subject eminently suitable for philosophical analysis and less so for scientific insight. Unfortunately Hanson completely misrepresented the central issue of the cosmological controversy: "Big Bangers and Continual Creators all agree that the evidence suggests our universe, in its very early youth, was considerably different in constitution and appearance from what it is now."[61]

Among the logical and conceptual reasons for doubting the reality of the steady-state universe were its infinite past and also the infinite number of created particles (see also Section 1.7). Some philosophers and scientists argued that actual infinities cannot possibly occur in nature and that this rules out the steady-state theory.[62] This kind of conceptual critique has continued to attract interest, if more among philosophers and mathematicians than among astronomers. Philosophically interesting as this category of arguments is, they were mostly discussed after the demise of the steady-state theory and played only a limited role during the cosmological controversy. And yet, the ghost of the infinity continues to be discussed in twenty-first century cosmology.[63]

Religious issues rarely appeared in the scientific literature on cosmology but they nonetheless played a considerable role in the controversy and especially in how it was perceived by the lay public. The only cosmologist who significantly and openly introduced God in his arguments was Milne, who seriously believed that he had found God in the universe and that divine rationality precluded certain cosmological models, including the steady-state model (Section 4.3). The title of his last book, the posthumously published *Modern Cosmology and the Christian Idea of God* (1952), speaks for itself. However, Milne's version of cosmo-theology was highly idiosyncratic and his own world model played very little role during the cosmological controversy in the 1950s and early 1960s. Nor did his adventure into apologetic cosmology make much of an impact in theological circles.

To put it briefly and crudely, in the controversy between the relativistic big-bang theory and the steady-state theory the former was often associated with theism while the latter was widely regarded as an atheistic alternative. The first part of the equation was spelled out in no uncertain terms by pope Pius XII in an address of 1951 to the Pontifical Academy of Sciences in which the big-bang picture of the universe was presented as scientific support of divine creation. The pope claimed that the size and age of the universe, as given by theory and astronomical data, were in full agreement with Christian faith:

> Indeed, it would seem that present-day science, with one sweep back across the centuries, has succeeded in bearing witness to the august instant of the primordial *Fiat Lux*, when, along with matter, there burst forth from nothing a sea of light and radiation, and the elements split and churned and formed into millions of galaxies . . . [Science] has confirmed the contingency of the universe, and also the well-founded deduction as to the epoch when the world came forth from the hands of the Creator. We say: therefore, there is a Creator. Therefore, God exists![64]

At the time the big-bang theory was far from generally accepted and cosmologists did not speak with the one voice that the pope suggested. Given the steady-state challenge and the general uncertainty in cosmology the propaganda carried little conviction, and even less so in regard of the fact that Gamow, the leading big-bang cosmologist, was a non-believer. As Lemaître was keen to point out, the Bible is not a textbook where answers to cosmological questions can be looked up.

In his book of 1950 Hoyle did not make any specific association between big-bang cosmology and theism, nor between steady-state theory and atheism, and yet he made it clear that he considered religious belief to be childish and anti-scientific. His remarks concerning religion and cosmology aroused antagonistic feelings in many people and caused several sharp replies. Hoyle and Bondi were both atheists but generally careful not to present their favored view of the universe as an argument against religion. Only later in life did Hoyle explicitly claim that big-bang cosmology was religious fundamentalism in scientific disguise. More or less the same claim was made by Bonnor in the 1960 BBC symposium mentioned above. On the other hand, the atheist Bonnor had no sympathy at all for Hoyle's steady-state theory.

As indicated, there is no one-to-one correspondence between views about cosmology and views about religion. The big-bang theory was originally founded by Lemaître, a Catholic priest, and it was developed into the modern version by Gamow, an atheist or agnostic. While Hoyle and Bondi were atheists, another leading steady-state theorist, McCrea, was a Christian. At the 1958 Solvay Congress Lemaître emphasized that his primeval-atom theory remained entirely outside religious questions and that it left the materialist free to deny any divine being. He even went as far as describing the primeval atom as the antithesis of a supernatural creation of the universe.

The point is further illustrated by Bernard Lovell, the British radio astronomer and director of the Jodrell Bank Observatory who was also a devoted Christian. As he explained in a series of BBC broadcasts published as *The Individual and the Universe*, the steady-state theory was not a threat against the belief in a divine creator. The so-called creation of the world, Lovell pointed out, does not imply that the world was brought into existence at some moment in the past. To his mind it was of no great significance whether creation occurred abruptly, as in the big-bang theory, or continually as in the steady-state theory. What Lovell had in mind was *creatio continua*, a theological concept established in the middle ages and according to which divine creation means that God preserves the world or causes it to exist.

At the time of the cosmological controversy theologians were quick to point out that an infinitely old universe is perfectly compatible with Christian belief. Whether of a finite age or infinitely old, the universe is and must be created. There must be a cause for why it exists and this cause does not need to imply a beginning in time. According to Eric Mascall, a physics-trained priest and philosopher, Hoyle's cosmology did not conflict with the Christian view but could, ironically, be taken as support of this view: "An uninstructed person, when told that Hoyle is a fervent advocate of the view of continuous creation, might not unreasonable suppose that Hoyle was a convinced theist who saw the hand of God in every being and event in the world's history."[65]

Given the close connection between religion and politics in the era of the Cold War it is not surprising that the rival cosmological views became part of the ideological battlefield in the 1950s between the liberal-capitalist world and the communist world. In the Soviet Union, and later also in Red China, cosmological models of the big-bang type came to be seen as politically suspect and contrary to the atheistic world view of Marxism-Leninism.[66] According to this world view the universe must not only be eternal but also infinite in space and content of matter. Theories including a big bang, and especially closed models such as Lemaître's, were seen as apologies for divine creation, much as the pope had suggested in his speech of 1951. One might believe that for these reasons the steady-state universe—eternal, spatially infinite, and associated with atheism—was welcomed by communist party ideologues. But this was not the case, for the PCP and the continuous creation of matter were unacceptable features from the point of view of orthodox Marxism-Leninism. The consequence was that for a decade, cosmological research as practiced in the Western countries practically ceased to exist in the Soviet Union. More about this sad chapter in the history of modern cosmology can be found in Chapter 9.

5.8   Predictions and cosmological tests

Despite the role played by aesthetic and philosophical considerations in the choice between cosmological models, all scientists involved in cosmological research agreed that questions concerning the structure and evolution of the universe should preferably be answered by appeal to observation and experiment. This was the case with cosmology in general and particularly so with regard to the two rival systems during the 1950s. The problem was the embarrassing paucity of observations that realistically could distinguish between the steady-state theory and the class of relativistic evolution models. The two theories agreed that the universe expands according to the Hubble law, but in most other respects they led to different predictions. Were the predictions of such a nature that they, if actually tested, would favor one theory over the other, or perhaps even disprove one of the theories?

Hoyle, Bondi, Gold, and their followers were no less committed to testing than their opponents in the big-bang camp. But as they often pointed out, the relationship between the two rival cosmologies was asymmetric in the sense that their own theory was more vulnerable to new observations, more falsifiable, than the evolution theories. Whereas the steady-state theory led to a number of sharp predictions, this was not the case with the theories based on general relativity if taken as a whole. For example, the steady-state theory led uniquely to the conclusion that the universe expands exponentially at a rate given by the deceleration parameter

$$q_0 \equiv - \left(\frac{\ddot{R}}{RH^2} \right)_0 = -1. \tag{5.19}$$

If observations proved without a doubt that the expansion followed another pattern, say that $q_0 > 1$, the theory was proved wrong. End of story (at least in principle). The relativistic evolution models, on the other hand, encompassed almost any kind of expansion, including $q_0 = -1$ (which was possible in the Lemaître model), and could therefore not be similarly falsified.

While most of the important tests were empirical, meaning based in observations or experiments, cosmological theories were also evaluated according to non-empirical criteria. Is the theory incoherent? Is it based on *ad hoc* hypotheses or is it otherwise methodologically flawed? Does it lead to consequences that are paradoxical or too fantastic to be taken seriously? Testing of this sort was a fairly important part of the cosmological controversy, but it was never assigned the same epistemic authority as reliable observations. Again, observational testing of a cosmological theory necessarily involves theoretical assumptions. As a contemporary philosopher reflecting on the situation in cosmology put it: "Evidence is relative to a theory, and if the theory is not formally acceptable, it cannot be factually confirmed.... The same data represent different evidence according to the theory for which they are employed."[67] But cosmologists did not need to be taught this lesson known as theory-ladenness in philosophical jargon, for they knew very well that theory and observation cannot be cleanly separated. The steady-state theorists in particular were keenly aware that observational facts cannot be taken for granted without critical scrutiny and that the transformation from a believed fact to a true fact requires some measure of theoretical interpretation.

The various empirical tests that were considered during the period of the cosmological controversy were basically the following:

- the timescale problem;
- the Stebbins–Whitford effect;
- nucleosynthesis, that is, formation of helium and heavier elements;
- redshift–magnitude relationship;
- redshift–diameter relationship;
- radio-astronomical source counts;
- gamma and X-ray backgrounds;
- quasars, redshifts, and distribution;
- the CMB.

Of these, we shall mainly deal with the first three. Other tests will be dealt with in Chapters 6, 7, and 8.

Another kind of test, which cannot be classified as empirical and which has been alluded to above, was the formation of galaxies. It was generally agreed that if one of the rival cosmologies could offer a reasonable account of galaxy formation, while the other cosmological theory could not, the former would *ceteris paribus* be preferable. As mentioned, Gamow's attempts to explain galaxy formation on the basis of big-bang theory were unsuccessful. Within the context of steady-state theory the problem was radically different, for here galaxies had always existed. The problem of how galaxies were initially formed simply did not enter. What had to be explained was the formation of new galaxies from sources of matter condensations and here the gravitational perturbations of existing galaxies on interstellar dust could be used as a mechanism. The first steady-state-based theory along this line was proposed by Sciama in 1955. For a while the theory seemed to provide an adequate framework for understanding how galaxies were formed and distributed in clusters.

Sciama considered his theory a triumph for steady-state cosmology but closer examination showed that the triumph was illusory. According to work done by the Czech-American astronomer Martin Harwit in 1961, the theory suffered from problems that could only be solved by introducing *ad hoc* forces other than gravity. By the early 1960s the general impression was that steady-state cosmology offered better explanations of the formation of galaxies than the theories based on relativistic cosmology, but also that the entire problem was too complex to act as an effective test in discriminating between the two rival conceptions of the universe.[68]

The small age of the universe compared with the age of its constituent bodies was a major motivation for the steady-state theory, where the problem just did not exist, and gave it an advantage over the relativistic evolution theories. Although there were various ways to avoid the age paradox in the latter class of theories, such as introducing the cosmological constant or assuming the universe to be inhomogeneous, the problem was recognized to be serious indeed. As Einstein wrote in a book of 1945:

The age of the universe . . . must certainly exceed that of the firm crust of the earth as found from the radioactive minerals. Since determination of age by these minerals is reliable in every respect, the cosmologic theory here presented would be disproved if it were found to contradict any such result. In this case I see no reasonable solution.[69]

The accepted value of the Hubble time in about 1950 was 1.8 billion years, corresponding to a recession constant of $H = 540$ km s^{-1} Mpc^{-1}. Hubble's value was considered to be authoritative, but in 1951 the German astronomer Albert Behr concluded that all intergalactic distances had to be doubled and that the Hubble time consequently was closer to 4 billion years than to 2 billion years. Behr's result did not attract much attention, and it was only when Walter Baade took up the question that things changed. During the 1952 meeting of the International Astronomical Union in Rome he announced that the Cepheids used as distance indicators had a higher luminosity than hitherto assumed.

Baade did not relate the new galactic distances to problems of cosmology, a branch of science for which he had no respect.[70] But others did make the connection, realizing that the cosmic time scale had to be doubled (just as Behr had suggested). Support for the new Hubble time followed quickly, and new measurements increased it further. From observations made with the Hale telescope, in 1956 Milton Humason, Nicholas Mayall, and Sandage found $T = 5.4$ billion years, and two years later Sandage concluded that the age of the universe, based on the Einstein–de Sitter model, was between 6.5 and 13 billion years (Fig. 5.7).

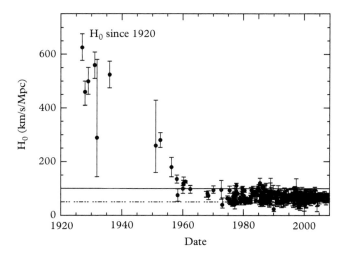

Fig. 5.7 *Estimations of the Hubble recession constant since the 1920s. The first data point is due to Lemaître (1927) and the second to Robertson (1928). However, Robertson did not actually calculate a value for the recession constant.*

Source: *https://www.cfa.harvard.edu/ dfabricant/huchra/hubble/*

Although the revised timescale was good news for theories of the big-bang class, it did not make the age paradox disappear. For one thing, the age of the Earth continued to increase until it stabilized at the value $(4.55 \pm 0.07) \times 10^9$ years obtained by the American geochemist Clair Patterson in 1956. For another thing, theories of stellar evolution indicated ages of the oldest stars of 15 billion years or more. And yet, although the age problem remained, the revised timescale had the psychological effect that it was no longer considered an important part of the cosmological controversy, and by the late 1950s it was rarely mentioned as an argument in favor of the steady-state theory. In fact, the increased value of the Hubble time was sometimes considered a problem for this theory because of the effect it had on the age distribution of galaxies.

The "Stebbins–Whitford effect" does not appear in modern textbooks in astrophysics and cosmology, but it played an important role in the 1950s when it, for a period, was regarded a strong argument against the steady-state theory. According to the core of this theory, the PCP, there could not be any systematic dependence of the ages of galaxies on their distance. But in 1948 the American astronomers Joel Stebbins and Albert Whitford concluded from photometric measurements that the light from elliptical galaxies exhibited a reddening in excess of that expected from the ordinary redshift due to recession. Moreover, Stebbins and Whitford argued that the more distant galaxies were redder than those closer by and that it possibly indicated that there were more red giant stars in the former galaxies. This could be explained as an age effect in the evolutionary view, but no such effect was possible within the framework of steady-state cosmology. To Gamow and other relativist cosmologists the Stebbins–Whitford effect amounted to nearly a disproof of the steady-state theory, but of course it rested upon the assumption that the effect was real. In a paper of 1954 Bondi, Gold and Sciama questioned the assumption, arguing that the effect was spurious. For a couple of years the situation was unsettled but after about 1957 most experts agreed that there was not, in fact, any color excess in the spectra, and therefore also that the threat against the steady-state theory had vanished. The result bolstered the confidence of the steady-state theorists but it did not change much in the overall balance between this theory and the rival evolution theory.

As far as nucleosynthesis was concerned, the situation in steady-state theory was quite different from the one in big-bang theory. Gamow's research program of nuclear archaeology had not and could not have a counterpart in the theory of Hoyle and his allies, for in this theory element formation had to go on in existing stellar objects. Although element formation played no central role in the steady-state theory, of course it was realized that the formation of elements had to be explained in terms of stellar nuclear processes and without the assumption of a hot primordial state of the universe. The problem was investigated by Hoyle in particular. In 1953, while working at Caltech, he took a fresh look at Salpeter's triple-alpha mechanism, realizing that the carbon yield would increase immensely if there existed in the carbon nucleus a resonance level of 7.68 MeV. The net result of Hoyle's triple-alpha process was

$$\alpha + \alpha + \alpha \rightarrow {}^{12}\text{C} + 2\gamma + 7.28\,\text{MeV}. \tag{5.20}$$

Experiments made by Ward Whaling and collaborators at the Kellog Laboratory resulted in a value of 7.68 ± 0.03 MeV—a stunning confirmation of Hoyle's prediction. It should be noted that Hoyle's work in nucleosynthesis was not motivated by cosmology and also not by a wish to explain that carbon-based life can exist.[71] But indirectly it became relevant to the cosmological controversy as it initiated an important collaboration with Fowler and the two British astrophysicists Margaret Burbidge and Geoffrey Burbidge.

The main result of the four physicists was a comprehensive and detailed theory of 1957 in which they explained the abundances of all elements on the basis of nuclear reactions in stars and supernovae.[72] The Burbidge-Burbidge-Fowler-Hoyle theory, or B^2FH theory for short, was astrophysical and not cosmological. The more than 100-page long paper in *Reviews of Modern Physics* did refer to Gamow's "primeval theory," but only to suggest that it was unnecessary. Although the B^2FH theory was not explicitly associated with steady-state cosmology, indirectly it gave support to the view held by Hoyle and other steady-state theorists. The success of the B^2FH theory seemingly implied that there was no need to look for an alternative theory of element formation based on big-bang assumptions. But the success was incomplete, for it was doubtful if the theory of Hoyle and his collaborators was able to provide a sufficient amount of helium and deuterium. Nonetheless, in the second edition of his *Cosmology*, Bondi evaluated the progress in stellar nuclear synthesis as support for the steady-state universe:

> In the course of constructing this theory [B^2FH], Hoyle was forced to postulate the exact value of a previously ill-determined nuclear energy level in ^{12}C which has since been verified experimentally. Since it has also been shown that any hot dense early stage of the universe could not have left us any nuclei heavier than helium, the origin of such nuclei is no longer a question of cosmology. It might however be said that the abundance of helium may conceivably be greater than would be accounted for by ordinary stellar transmutation and so might have to be explained on a cosmological basis, but the evidence as yet is far too slight to merit serious consideration now.[73]

The question of the light elements became an issue only after observations indicated a cosmic helium abundance of about 25-35%, in good agreement with the big-bang calculations of Alpher, Herman, and Follin. In a paper of 1961 the two American astrophysicists Donald Osterbrock and John Rogerson showed that the helium abundance in the Sun was essentially the same as that in planetary nebulae and the Orion nebula, namely 32%. They suggested that the amount of helium could be "the original abundance of helium from the time the universe formed, for the build-up of elements to helium can be understood without difficulty on the explosive formation picture."[74] Despite this reminder of Gamow's big-bang theory, to which they referred, it took a few more years until the helium problem was recognized to be cosmologically significant. This was made clear in a 1964 paper by Hoyle and Roger Tayler in which the two authors realized that sufficient amounts of helium needed calculations based on what was in effect big-bang assumptions. But they did not take their result as evidence that the universe in fact started in a big bang.

The classical observational tests focused on determination of the deceleration parameter q_0 which, for relativistic evolution models with $\Lambda = 0$ and no pressure term, relates to the space curvature by

$$\frac{kc^2}{R^2} = H_0^2 \left(2q_0 - 1\right).$$ (5.21)

If $q_0 > \frac{1}{2}, k = +1$ and space is closed (elliptic); if $q_0 = \frac{1}{2}, k = 0$ and space is flat (Euclidean); and if $q_0 < \frac{1}{2}, k = -1$ and space is open (hyperbolic). The method was cultivated by Sandage in particular, who once described cosmology as a search for two numbers, one being q_0 and the other the Hubble constant H_0.[75] In 1956 Sandage, Humason, and Mayall reported as a best value $q_0 = 2.5 \pm 1$ which, if reliable, ruled out the $q_0 = -1$ predicted by the steady-state model. However, other measurements gave rather different results. To Hoyle and his followers the muddled situation indicated that the method was inconclusive and unable to refute the theory they preferred. In spite of much work, it proved disappointingly difficult to get an unambiguous value for the deceleration parameter. Sandage, McVittie, and most other astronomers were confident that the accumulated redshift-magnitude observations spoke against the steady-state alternative, but the alleged refutation was not clear enough to convince those in favor of the theory.

As the redshift–magnitude test was unable to resolve the cosmological controversy, so was it the case with the redshift–diameter test (or angular diameter test) proposed by Hoyle in 1959. The idea of this method is that the angular diameter of distant objects of the same absolute size will depend on their distances, and hence on their redshifts, in a way that depends on the geometry of space. The optical version of the method was to observe the redshifts and angular diameters of very distant clusters of galaxies, and it was supplemented by a radio-astronomical version where the objects were radio sources. However, it soon turned out that the data were not precise enough and did not cover large enough distances to make the redshift-diameter method an effective test in discriminating between the rival world models.

It was primarily two other methods, of a very different nature, which settled the matter and proved that we do not live in a steady-state universe. The method of radio-astronomical source counts strongly suggested that the steady-state theory was wrong, and with the discovery and interpretation of the CMB in 1965 the suggestion was turned into something close to a fact.

5.9   Swan songs of the steady-state universe

Radio cosmology, the use of radio-astronomical methods for cosmological purposes, was pioneered by the British astronomer Martin Ryle, who in 1954 suggested that observations of discrete radio sources might function as a test of the steady-state model of the universe.[76] His method was to count the number N of radio sources with a flux density larger than a certain value S. If it is assumed that the sources are distributed uniformly in flat space, the two quantities will be related as

$$\log\ N(\geq S) = -1.5\log\ S + \text{const.} \tag{5.22}$$

Cosmological models with different geometries and expansion rates will lead to different predictions, meaning that a number count results in a $\log N$–$\log S$ distribution which can be compared with a straight line with slope -1.5. It followed from the steady-state model that the data points must be located beneath the line, whereas there was no corresponding prediction from the relativistic models.

Ryle and his team of radio astronomers at Cambridge University concluded in 1955 that data from nearly 2000 radio sources disagreed with the steady-state prediction. However, the results obtained by Bernard Mills and his group in Sydney, Australia, were quite different from those obtained in Cambridge and the slope much closer to the one expected from the steady-state theory. Still in the late 1950s radio-astronomical observations failed to distinguish unequivocally between the competing models of the universe. New and better data were needed. The point of no return was reached in 1961, when Ryle presented data that clearly contradicted the steady-state slope of -1.5. The measured slope of -1.8 ± 0.1 was soon confirmed by the Sydney group (slope -1.85 ± 0.1) and accepted by other specialists in radio astronomy. Although the consensus view did not amount to a proper refutation of the steady-state theory, it weakened it very considerably. The relativistic evolution models were not positively confirmed, but indirectly they were and after 1963 few astronomers questioned that the universe expands in accordance with the laws of general relativity. Whether the expansion started in a big bang or not was a question outside the domain of radio cosmology.

This question was unexpectedly answered in 1965, when a group of physicists at Princeton University became aware of the puzzling background radiation at wavelength $\lambda = 7.3$ cm that had been detected the previous year. The celebrated and Nobel Prize-rewarded discovery of Arno Penzias and Robert Wilson was serendipitous and unrelated to cosmological research. The two physicists employed by the Bell Laboratories were engaged in work related to the possibility of transatlantic telecommunication, and they did not know about the earlier and at the time half-forgotten prediction of Alpher, Herman, and Gamow. It was only in the spring of 1965 that Robert Dicke, together with James Peebles, Peter Roll, and David Wilkinson, concluded that the Penzias–Wilson radiation was of cosmological origin, namely a fossil left over from a "fireball" phase in the early big-bang universe.[77] The two papers in the July 1965 issue of *Astrophysical Journal*, one by Penzias and Wilson and the other by the Princeton group, mark the real discovery of the CMB and initiated a new chapter in big-bang theory. Detection of the radiation at other wavelengths followed in 1966, when measurements were made at $\lambda = 3.2, 20.7$, and 0.26 cm. They strongly suggested that the CMB was black-body-distributed, which further strengthened the big-bang interpretation and made it almost impossible to maintain the steady-state theory of the universe.

But only almost, for none of the leading advocates of steady-state cosmology admitted right away that the new evidence—helium abundance, radio source counts, and the CMB—amounted to a refutation of their favored view of the cosmos. In a paper of 1965, Sciama insisted, perhaps with forced optimism, that "the steady-state model remains in the field, bloody but unbowed."[78] His statement referred to the situation in radio astronomy. A year later Sciama admitted that the state of the theory was worse than just bloody. It had lost so much blood, been refuted by so many reliable observations, that it could safely be declared dead.

What finally convinced Sciama were studies of the redshifts of the quasars, or originally quasi-stellar radio sources, which were discovered in 1963 by Maarten Schmidt. If the quasar redshifts were really cosmological, the objects must have been more common very far away and long ago, which squarely contradicted the PCP at the heart of classical steady-state cosmology (see also Chapter 7). Most astronomers agreed that there were compelling evidence that quasars were indeed at cosmological distances, but for a time Hoyle and a few other steady-state protagonists remained unconvinced. If the redshifts were due to cosmic expansion, a relation between the quasars' redshifts and flux densities would be expected. In 1966 the Cambridge astronomer Malcolm Longair and now also Sciama (collaborating with his student Martin Rees) argued that there was such a relation and that quasars were one more nail in the coffin of the steady-state theory.[79] Although Sciama thus abandoned the theory he had defended for more than a decade, not all steady-state advocates followed him.

Of the original Cambridge trio, Bondi and Gold quietly left cosmology under the impact of the new observations and also because they were dissatisfied with Hoyle's many attempts to revise the steady-state theory and breathe new life into it. Yet they were no less dissatisfied with the victorious big-bang theory and preferred to leave research in cosmology rather than convert to the new mainstream view of the universe such as Sciama did. With a few exceptions, after about 1964 they stopped defending the steady-state alternative. On the other hand, there is little doubt that, at least through the 1960s, they preferred this theory in spite of recognizing the successes of the big-bang theory. Bondi and Gold admitted that the epic battle between the two world views had come to an end, and that they had lost the battle, but they did it with regret.[80] The same was the case with McCrea, who without enthusiasm joined the big-bang consensus view in the late 1960s. The case of Fred Hoyle was quite different, for he never surrendered to what he considered the dogma of the big bang.

In the years after 1965 there was no shortage of alternatives to the big-bang interpretation of the CMB. Several of these short-lived hypotheses, such as one proposed by Sciama in 1966 and another by Michael Rowan-Robinson in 1974, ascribed the radiation to the integrated flux of discrete celestial sources. Hoyle later claimed that "There has never been a difficulty in the steady-state theory over the energy-density of the cosmic microwave background."[81]

What Hoyle had in mind was his favored alternative, namely that the microwaves originated in galaxies whose light was converted into heat radiation of a shape close to the blackbody spectrum. In papers of 1967 he, Narlikar and Chandra Wickramasinghe suggested that the agents for the necessary thermalization were interstellar grains of graphite. The idea was that the grains would absorb light and re-emit it in the far infrared region, and in this way produce a microwave background of temperature approximately 3 K. Hoyle and Wickramasinghe continued to develop their hypotheses well into the 1990s, hoping in this way to prove a viable alternative to the standard big-bang interpretation of the CMB. But the thermalization hypothesis attracted little interest and was generally seen as artificial and no more credible than the tired-light hypotheses denying the expansion of the universe (for these, see Section 4.7).

With the *C*-field model of 1960 Hoyle initiated a development that eventually turned the steady-state theory into versions quite different from its starting point in 1948. For example, whereas the Hubble constant had originally been taken as a true constant in conformity with the PCP, in the later versions Hoyle allowed Hubble's constant to vary in time. As he put in a lecture of 1982:

> There is no sense in which I would think it helpful to return to the concept of the universe that is strictly steady, not even on the scale H^{-1}.... My 1948 form of the steady-state theory was rather like a phonograph record stuck in a groove, playing endlessly the same phrase, while the Bondi–Gold form of the theory was like a record that plays only one note, the sort of test record one uses to check a stereo system.[82]

Hoyle and Narlikar introduced the *C*-field as a scalar quantity associated with a negative energy density that would compensate the positive energy of the created matter, and in this way they secured strict energy conservation. The PCP was now definitively abandoned and the universe pictured as having developed into an exponentially expanding state from some initial but non-singular state. Whereas Hoyle and Narlikar had hitherto followed the standard assumption of a homogeneous universe, in 1966 they abandoned the assumption and with it the idea of matter being created uniformly through space. As an alternative they explored the possibility of matter pouring into space from discrete sources of supermassive bodies. The creation processes might temporarily be caught off, with the result that the universe would follow a series of minor expansions and contractions around the general exponential expansion. They referred to this scenario, vaguely related to the modern idea of the multiverse, as a steady-state "bubble universe."[83]

Hoyle's variety of post-1965 revised steady-state models had in common an uncompromising opposition to the big-bang assumption of a universe born a finite time ago. Together with Narlikar and Geoffrey Burbidge, during the 1990s he continued to criticize what by then had been standard cosmology for two decades, and in 1993 their efforts to formulate an alternative culminated in what they called the QSSC theory, meaning quasi steady-state cosmology.[84] The universe of QSSC is eternal and creative. Although on a very long timescale the size of the universe, as given by the scale factor *R*, increases exponentially, there is superposed on the expansion oscillations of a smaller timescale. The $R(t)$ function can be written as

$$R(t) = \exp\left(\frac{t}{P}\right)\left[1 + \alpha \cos\left(\frac{2\pi}{Q}\right)\right], \tag{5.23}$$

where $P \gg Q$ and α is a constant in the range $0 < \alpha < 1$ which signifies the amplitude of the oscillations. The time parameter Q is the cyclical period and P is a measure of the exponential growth of *R* on a very long time scale (Fig. 5.8). Hoyle, Burbidge and Narlikar derived a number of consequences from their theory, one of them being that

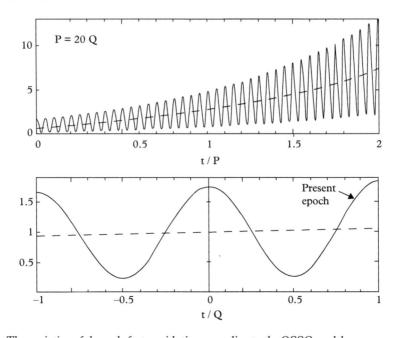

Fig. 5.8 *The variation of the scale factor with time according to the QSSC model.*

Source: F. Hoyle, G. Burbidge, and J. Narlikar, "Further astrophysical quantities expected in a quasi-steady state universe," Astronomy and Astrophysics **289** *(1994), p. 732. Reproduced with permission of ESO.*

the present mass density is almost entirely baryonic (no exotic dark matter) and another that there has never been a radiation-dominated phase of the universe.

The QSSC theory survived the death of Hoyle in 2001, but just barely so. Attempts to develop the theory continued for a decade or so after which they seem to have ceased. The theory was meant to challenge the standard big-bang theory, but only very few mainstream cosmologists took up the challenge or showed any interest in the alternative. From a sociological point of view it clearly was a failure, as indicated by the few citations to the literature on QSSC theory from scientists outside the small group defending the theory.

Although the QSSC and its preceding theories were never taken very seriously, it is worth noting that their shadows are still to be found in modern theoretical cosmology. The Hoyle–Narlikar universe of the 1960s has been described as inflationary, with inflation going on at all time. Indeed, the later versions of the steady-state theory had features in common with the inflationary models developed from the early 1980s onwards and in particular with so-called chaotic or eternal inflation models. In these models the universe as a whole is described as self-reproducing and with neither a single beginning nor an end. As two physicists noted, "while the SS [steady state] has approached the BB [big bang], the BB has also approached the SS, in the form of 'eternal inflation'."[85] Although there are scientific similarities between the post-1965 steady-state theory and the later inflation theory, the historical trajectories of the two research traditions were quite separate. Inflation theory did not grow out of or was even remotely inspired by the steady-state theory.

....

NOTES

1. Bondi (1952), p. 5.
2. Suzuki (1928). The problem of the helium–hydrogen ratio had first been investigated by Richard Tolman in a paper six years earlier.
3. Atkinson (1931a, 1931b). For historical reviews of early nuclear astrophysics, see Kragh (1996), pp. 84–102 and Shaviv (2009), pp. 275–312.
4. Lanczos (1925)
5. Lemaître (1931a). For contexts and details, see Kragh and Lambert (2007). See also Chapter 3 of the present volume.
6. See Bonolis (2017), pp. 336–9.
7. Lemaître (1931c), p. 408.
8. Lemaître (1997), a translation of the French 1933 article. For this article, see Eisenstaedt (1994).
9. For the cool reception of Lemaître's theory, see Kragh (2012b).
10. Merrill (1933), p. 28.
11. Nobel Archive, Stockholm. The nominator was the American chemist Don Merlin Yost. Lemaître was also nominated for the 1954 physics prize, but in this case only for his prediction of the expanding universe. The nominator was Alexandre Dauvillier, a French astrophysicist. See Kragh (2017c).
12. On Gamow's life and career, see Harper (2001).
13. Gamow (1940), p. 202.
14. Weizsäcker (1938), p. 644. For English translations, see Lang and Gingerich (1979), pp. 309–19 and Drieschner (2014), pp. 7–30. On Weizsäcker's contributions to nuclear stellar physics, cosmology, and cosmogony, see *Acta Historica Leopoldina*, **63** (2014), a special issue devoted to Weizsäcker's life and work.
15. For the emergence of the historical style in cosmology and its later entrenchment, see Pearce (2017).
16. Goldschmidt's research program and the development of so-called cosmochemistry are described in Kragh (2001).
17. Gamow and Fleming (1942).
18. Letter of 24 October 1945, reproduced in Kragh (1996), pp. 106–7.
19. Gamow (1946); Kragh (1996), p. 110.
20. Peebles (2014).
21. Alpher, Bethe, and Gamow (1948). Bethe contributed only nominally and without his knowledge. For the three authors and the "$\alpha\beta\gamma$" label, see Kragh (1996), p. 113.
22. *Science News Letter*, **53**, (24 April, 1948), p. 259.

23. Alpher and Herman (1948).
24. Alpher and Herman (1949), p. 1094. See also Peebles (2014) for a critical analysis of the Alpher–Herman papers of 1948–1949.
25. On the differences of Gamow's use of the CMB and the one of Alpher and Herman, see Alpher (2012), written by the son of Ralph Alpher.
26. Kragh (1996), pp. 134–5; Peebles, Page, and Partridge (2009), pp. 43–4. For the later history of CMB, see Chapter 8 in this volume.
27. Gamow (1949), p. 367. See Kragh (1996), pp. 73–9 for the timescale problem.
28. See Kragh (2014a) for details about the name "big bang" and its early history.
29. Gamow (1954), p. 63.
30. See Longair (2008), pp. 3–26 for a historical survey of theories of galaxy formation.
31. Shaviv (2009), pp. 294–9.
32. Gamow (1954), p. 62.
33. Hayashi (1950).
34. Alpher, Follin, and Herman (1953), p. 1347.
35. There is no scholarly analysis of the question, but see Kragh (1996), pp. 135–41.
36. The history of the steady-state theory is fully described in Kragh (1996) on which source the present chapter relies and in which most of the relevant publications can be found.
37. MacMillan (1925), p. 99; Kragh (1995).
38. See O'Raifeartaigh and McCann (2014).
39. Bondi and Gold (1948); Hoyle (1948).
40. McCrea (1950).
41. Bondi (1952).
42. Hoyle (1950), p. 102. For details about the term and its history, see Kragh (2014a).
43. Couderc (1952), pp. 213–4, 220.
44. For the work of McCrea and others on matter creation, see Kragh (1999).
45. For context, detail and references, see Kragh (1997).
46. Interview of 26 May 2004, quoted in Gregory (2005), p. 70.
47. For references to most quotations in this section, see Kragh (1996), pp. 219–50.
48. Ellis and Penrose (2010).
49. Bridgman (1955), p. 275, from a paper originally published in 1932.
50. *British Journal for the Philosophy of Science*, **10** (1959), pp. 55–7. The Bondi–Whitrow discussion appeared in the same journal, **4**, (1954), pp. 271–83. On Popper and cosmology, see Kragh (2013c).
51. Pauli to A. Jaffé, 3 December 1951. In Pauli (1996), p. 447
52. Einstein to J. J. Fehrer, 1952, as quoted in O'Raifeartaigh et al. (2014).
53. Douglas (1956).
54. On McVittie and his important work in cosmology, see Sànchez-Ron (2005).
55. Letter to A. Sandage of 23 September 1958, quoted in Sànchez-Ron (2005), p. 205.
56. See the proceedings of the conference, Stoops (1958).
57. Bondi et al. (1960).
58. The special issue was published as a book the following year. See Piel et al. (1956), where the quotation is on p. 73.

59. Hoyle (1955), p. 319.
60. Grünbaum (1952). For this and other references to the philosophy-cosmology literature ca. 1950–1965, see Kragh (1996).
61. Hanson (1963), p. 467.
62. See, for example, *Nature*, **196**, (1962), pp. 1015–17.
63. See, for example, Ellis, Kirchner, and Stoeger (2004).
64. See www.papalencyclicals.net. See McConnell (2002) for a concise review of the relation between religion and modern cosmology.
65. Mascall (1956), p. 158.
66. For details, see Kragh (2013b) and also Chapter 9 of this volume.
67. Hutten (1962), p. 113, who referred in particular to the recent observations of Martin Ryle that allegedly disproved the steady-state theory.
68. Gribbin (1976); Longair (2008).
69. Einstein (1945), p. 132. For aspects of the timescale problem, see Trimble (1996) and Lepeltier (2007).
70. According to Osterbrock (2001), p. 205, Baade considered cosmology "a waste of time."
71. Hoyle's argument for a 7.7 MeV resonance in C-12 has been described as an "anthropic" prediction, that is, motivated by the existence of human beings. However, this is quite wrong. For the anthropic principle, see Chapter 12, and see Kragh (2010) for a detailed analysis of Hoyle's prediction.
72. Burbidge et al. (1957).
73. Bondi (1960), p. 58.
74. Osterbrock and Rogerson (1961), p. 134. See also Osterbrock's recollections in Peebles, Page, and Partridge (2009), pp. 86–92.
75. Sandage (1970).
76. For the history of early radio cosmology, see Sullivan (1990).
77. For details, see Peebles, Page, and Partridge (2009). See also Chapter 8 in this volume. The term "fireball" was suggested by John Wheeler.
78. Sciama (1965), p. 15.
79. Longair (1966a).
80. Bondi and Gold were co-signers of the anti-big-bang letter of 2004 mentioned in Section 4.9.
81. Hoyle (1990), p. 224.
82. Hoyle (1982), p. 48. Note that Lemaître (1931a) used the phonograph metaphor when he introduced the primeval-atom hypothesis (see also Chapter 3).
83. Hoyle and Narlikar (1966).
84. For QSSC theory, see Hoyle, Burbidge, and Narlikar (2000).
85. Aguirre and Gratton (2002). For the history of inflation cosmology, see Chapter 11 in this volume.

6

Observational and astrophysical cosmology: 1940–1980

Malcolm S. Longair

6.1 Introduction

The dramatic story of the beginnings of modern observational and theoretical cosmology
has been told by Robert Smith (Chapter 2) and Matteo Realdi (Chapter 3), while
the myriad of alternative theoretical possibilities inspired by these discoveries has been
critically reviewed by Helge Kragh in Chapter 4. This burst of activity was stimulated
by the theoretical breakthroughs of Albert Einstein, Willem de Sitter, Hermann Weyl,
Alexander Friedman, George Lemaître, Howard Robertson, Arthur Walker, and Arthur
Eddington among others, while the observational foundations for the models were provided by Knut Lundmark, Vesto Slipher, Carl Wirtz, Edwin Hubble, Milton Humason,
and Fritz Zwicky. These pioneers began the conversion of cosmology from a matter of
speculative philosophy to a discipline subject to observational and experimental validation. The realization of this transformation was to take many years before Rutherford's
famous condemnation,

> Don't let me catch anyone talking about the Universe in my department,

could be put to rest.

This chapter spans the period from the Second World War to roughly 1980.[1] This
period saw the gradual emergence of the new disciplines of radio, infrared, ultraviolet,
X-ray, and γ-ray astronomy, all of which were to bring new insights into the astrophysics
of cosmology. By 1980, these fields were well established and awaited the next generation
of instruments and telescopes to consolidate the understanding of their role in cosmology.
In optical astronomy, the Hale 200-inch telescope came into full operation in the early
1950s and was to dominate optical cosmological studies until the 1970s when other
4-metre class telescopes were constructed which enabled the community to compete
with the privileged astronomers who had access to the Hale telescope. The 1970s also saw
the introduction of electronic detectors into optical astronomy and these began to replace

Longair, M. S., 'Observational and astrophysical cosmology: 1940–1980' in *The Oxford Handbook of the History of Modern
Cosmology*, edited by Kragh, H. and Longair, M. S. © Oxford University Press 2019.
DOI: 10.1093/oxfordhb/9780198817666.013.6

photographic plates, greatly increasing the sensitivities achievable by the largest optical telescopes. By the 1980s, these detectors had become the gold standard for many aspects of cosmological research, gradually replacing the old techniques used by the pioneers of modern cosmology.

Undoubtedly the most important discovery of this period was that of the cosmic microwave background (CMB) radiation in 1965 by Arno Penzias and Robert Wilson, which rightly deserves a chapter to itself by Bruce Partridge, himself one of the pioneers of the subject (Chapter 8). It was soon accepted that this radiation is the cooled remnant of the very hot early phases of the early Universe, in due course opening up new possibilities for the determination of cosmological parameters with remarkable precision, as well as providing constraints on fundamental physical processes in the Universe.

Bondi's *Cosmology* (Bondi, 1952, 1960) provides a valuable benchmark for the state of theory and observation in cosmology in the immediate post-War period. Although one of the proponents of the steady-state model of the Universe, Bondi's presentation is judicious and gives a splendid feel for the uncertainties about the whole study of cosmology at the time, as amplified by the discussion of Helge Kragh in Chapter 4. In Chapter 5, Kragh then gives a detailed account of the confrontation between the steady-state and evolutionary scenarios for the large-scale structure and evolution of the Universe. By the mid-1960s, the debate was convincingly resolved in favour of the Friedman–Lemaître evolutionary models of the Universe. By the end of the period covered in this Chapter, theory and observation had put much more flesh on the bare bones of the standard Friedman–Lemaître cosmological models so that they could be confronted with observation in new ways. The first glimpses of the concepts of gravitational lensing, cold dark matter and inflation were on the horizon.

It is not unfair to say that, after 1980, the whole discipline changed gear with outstanding applications of the new technologies for all wavebands, the beginning of the epoch of large space observatories, new ways of tackling the old cosmological problems, the application of large scale computing to the problems of astrophysical cosmology and an influx of new generations of observers, instrumentalists and theorists to capitalize upon these opportunities. How this came about and its consequences are the subjects of Chapter 10.

In contrast to the discussions of Chapters 3–5, this chapter is largely focussed upon the challenges of observational and astrophysical cosmology. As emphasized in the earlier chapters, the models constructed by the pioneers of theoretical cosmology in the contemporary sense needed to be confronted with observation before they could begin to be taken seriously as contributions to the physics of the Universe. That is the thrust of this chapter.

6.2 The legacy of the 1930s

By 1940, much of the infrastructure of modern cosmology was in place. On the observational side, the galaxies seemed to be reasonably uniformly distributed out to large distances as indicated by Hubble's counts of galaxies. The linear relation between recession velocity and distance indicated that this distribution was expanding uniformly, the Friedman–Lemaître world models providing a wide range of potential dynamics for the expanding Universe. The Robertson–Walker metric provided a very convenient tool for understanding how to relate intrinsic properties to observables for the cosmologist's preferred dynamical model of the Universe.

On the observational side, much of the most important research had been carried out with the 100-inch Hooker Telescope at the Mount Wilson Observatory. By 1926, the first application of the ideas of relativistic cosmology to the Universe of galaxies had been made. It comes as no surprise that in 1928 George Ellery Hale, Director of the Mount Wilson Observatory, began his campaign to raise funds for the construction of the Palomar 200-inch telescope. Early work on 'big-bang' cosmology—the study of the Universe of distant galaxies—needed the largest telescopes that could be built (Hale, 1928). In the great American tradition of private sponsorship of observational astrophysics, in which the USA had taken a decisive lead, Hale was successful in obtaining a grant of $6,000,000 from the Rockefeller Foundation for the telescope before the year was out.

Undoubtedly, the construction of the 200-inch telescope on a good, dark site on Mount Palomar in Southern California was Hale's masterpiece. Technologically, the 200-inch telescope stretched mirror and telescope technology to their limits and incorporated a number of important innovations which were to be built into succeeding generations of large telescopes.[2] Those of us privileged to have used the 200-inch telescope know what a tremendous feat of engineering it was. As one writer put it,

> the Hale telescope stands among all telescopes as the climax of dreadnought design,

and represented the pre-Second World War 'brute force' approach to telescope construction. It was to dominate much of observational cosmological research until the next generation of 4-metre class telescopes began to appear in the 1960s and 1970s.

6.3   Hubble's constant

A legacy of the 1930s was the problem that the age of the Universe, as estimated from the inverse of Hubble's constant H_0^{-1}, was less than the age of the Solar System. In 1952, Walter Baade announced that the value of Hubble's constant, H_0, had been overestimated because the distance to the Andromeda Nebula, M31, adopted by Hubble was about a factor of 2 too small (Baade, 1952). The cause of the discrepancy was the difference in the period–luminosity relations for Cepheid variables stars of Populations I and II. By using the same type of Cepheid variable in our own galaxy, in the Magellanic Clouds and in M31, the distance to M31 increased by a factor of 2. Consequently, Hubble's constant was reduced to 250 km s^{-1} Mpc^{-1} and H_0^{-1} was increased to 4×10^9 years, roughly the age of the Solar System.

In 1956, Humason, Mayall, and Sandage showed that the expected redshift–magnitude relation, $m = 5 \log_{10} z + \text{constant}$, is observed for galaxies selected at random,

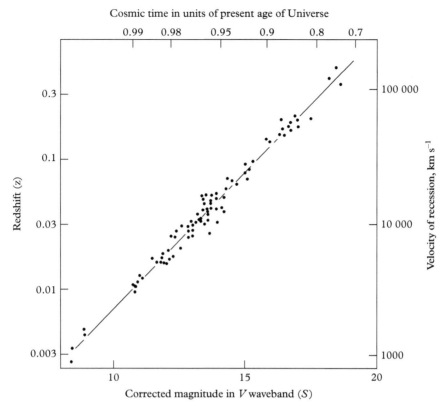

Fig. 6.1 *The redshift–V magnitude relation for the brightest galaxies in clusters (Sandage, 1968). The straight line shows the expected relation if the galaxies all have the same intrinsic luminosity, m = 5 log$_{10}$ z + constant. The sparsity of points at redshifts greater than 0.3 illustrates the difficulty of finding clusters of galaxies at large redshifts which would be suitable for cosmological tests.*

but that there is a large scatter about the mean relation because of the breadth of the luminosity function of galaxies (Humason et al., 1956). It had been known since Hubble's pioneering studies of the 1930s, however, that the brightest galaxies in clusters of galaxies define a very much tighter relation which follows precisely Hubble's law $v = H_0 r$ (Fig. 6.1). Thus, in order to estimate the value of H_0, it was only necessary to calibrate the observed relation by measuring the distance of the nearest rich cluster of galaxies, the Virgo cluster of galaxies, by techniques independent of its redshift. Humason, Mayall, and Sandage estimated the distance of the giant spiral galaxy NGC 4321, one of the brightest galaxies in the Virgo cluster, assuming that the brightest stars and nebulae in that galaxy were the same as those in M31. Hubble's constant was revised downwards again to 180 km s^{-1} Mpc^{-1}. In 1958, Sandage's best estimate of H_0 was reduced yet again from 180 to 75 km s^{-1} Mpc^{-1} (Sandage, 1958). The principal reason for this further downward revision was that what had been thought to be the brightest stars in some of the most distant galaxies studied turned out to be regions of ionized hydrogen and star clusters.

Hubble's constant H_0 appears ubiquitously in cosmological formulae and its value was the subject of considerable controversy for many years. The use of the redshift–magnitude relation for brightest cluster galaxies had the advantage that Hubble's law is defined well beyond distances at which there might have been deviations associated with the peculiar motions of clusters and superclusters of galaxies.

The traditional approach to the calibration of the relation shown in Fig. 6.1 involves a hierarchy of distance indicators to extend the local distance scale from the vicinity of the Solar System to the nearest giant cluster of galaxies, the Virgo cluster. The only direct methods of distance measurement involve stellar parallaxes and these can only be used for stars in the neighbourhood of the Sun. To extend the distance scale further, it is assumed that objects of the same intrinsic types can be identified at greater distances. Then, their relative brightnesses provide estimates of their distances.[3]

The period–luminosity relation, discovered by Henrietta Leavitt in 1912 (Leavitt, 1912), provides one of the best means of extending the distance scale from our own Galaxy to nearby galaxies, but even using the 200-inch telescope it was only possible to use this procedure to distances of about 1–2 Mpc. Other techniques were used to extend the distance scale from the neighbourhood of our Galaxy to the Virgo cluster, including the luminosity functions of globular clusters, the brightest stars in galaxies and the luminosities of Type I supernovae at maximum light. In 1977, Brent Tully and Richard Fisher discovered the relation between the absolute magnitudes of spiral galaxies and the velocity widths of their 21-cm line emission (Tully and Fisher, 1977). This relation could be determined for a number of spiral galaxies in a nearby group or cluster and then relative distances found by assuming that the same correlation between their intrinsic properties is found in more distant groups and clusters.

Extensive studies were made of correlations between various properties of elliptical galaxies, specifically their luminosities, sizes, central velocity dispersions, their abundances of heavy elements, and so on. In 1976, Sandra Faber and Robert Jackson discovered a strong correlation between luminosity, L, and central velocity dispersion σ of the form $L \propto \sigma^x$, where $x \approx 4$ (Faber and Jackson, 1976). This correlation has been studied and refined by subsequent authors, who have found values of x ranging from about 3 to 5. The significance of this relation is that, if the velocity dispersion σ is measured for an elliptical galaxy, its intrinsic luminosity can be inferred from the Faber–Jackson relation and hence its distance found.

From the 1970s until the 1990s, there was an ongoing controversy concerning the value of Hubble's constant.[4] In a long series of papers, Sandage and Gustav Tammann found values of Hubble's constant of about 50 km s^{-1} Mpc^{-1}, whereas de Vaucouleurs, Aaronson, Mould, and their collaborators found values of about 80 km s^{-1} Mpc^{-1}. The nature of the discrepancy can be appreciated from their estimates of the distance to the Virgo cluster. If its distance is 15 Mpc, the higher estimate of H_0 is found, whereas if the distance is 22 Mpc, values close to 50 km s^{-1} Mpc^{-1} are obtained. Sandage and Tammann repeatedly emphasized how sensitive the distance estimates are to observational selection effects, such as the Malmquist bias, and systematic errors.

When the Hubble Space Telescope (HST) project was approved by NASA and the US government in 1977, one of its major scientific objectives was to use its superb sensitivity for faint star-like objects to enable the light curves of Cepheid variables in the Virgo cluster to be determined precisely and so estimate the value of Hubble's constant to 10% accuracy. This programme was raised to the status of an HST Key Project in the 1990s with a guaranteed share of observing time to enable a reliable result to be obtained.

6.4   Sandage's assessment of the route to the determination of cosmological parameters

In 1961, Allan Sandage published an influential paper entitled *The Ability of the 200-inch Telescope to Discriminate between Selected World Models* in which different approaches to the determination of cosmological parameters were discussed critically (Sandage, 1961a). The observed properties of galaxies at large redshifts depend upon the geometry of the world model and upon its kinematics between the epochs of emission and reception of the radiation. The list of parameters to be determined were:

- *Hubble's constant, $H_0 = (\dot{a}/a)_{T_0}$, the present rate of expansion of the Universe;*
- *The deceleration parameter, $q_0 = -(\ddot{a}/\dot{a}^2)_{T_0}$, the present dimensionless deceleration of the Universe;*
- *The curvature of space, $\kappa = \mathcal{R}_0^{-2}$;*
- *The mean density of matter in the Universe at the present epoch, ρ_0, and its value relative to the critical density, $\rho_{\rm crit} = 3H_0^2/8\pi G$;*
- *The present age of the Universe, T_0;*
- *The cosmological constant, Λ.*

Here, a is the scale factor which describes the relative distance between objects which partake in the uniform expansion of the Universe as a function of cosmic time, κ is the curvature of the spatial geometry, and $\mathcal{R}_0$ the radius of curvature of the 3-space isotropic geometry at the present epoch. These are not independent parameters, *provided* the Friedman–Lemaître models are a correct description of the large-scale dynamics of the Universe.

$$\kappa = \mathcal{R}^{-2} = \frac{(\Omega_0 - 1) + \frac{1}{3}(\Lambda/H_0^2)}{(c/H_0)^2}; \quad q_0 = \frac{\Omega_0}{2} - \frac{1}{3}\frac{\Lambda}{H_0^2}, \tag{6.1}$$

where $\Omega_0 = \rho_0/\rho_{\rm crit}$ is the *density parameter*. If $\Lambda = 0$, there is a simple one-to-one relation between the geometries of the world models, their densities and dynamics, $q_0 = \Omega_0/2$ and $\kappa = \mathcal{R}^{-2} = (\Omega_0 - 1)/(c/H_0)^2$. Thus, if the Friedman–Lemaître models with non-zero cosmological constant are adopted, three independent parameters need to be determined, for example, H_0, q_0 and Ω_0. If the cosmological constant is zero, only two parameters need be determined, say, H_0 and q_0. The steady-state model was uniquely defined by the single parameter H_0 (Chapter 5). If all three parameters are determined independently, expression (6.1) provides a test of the general theory of relatively on the largest scales accessible to us.

Sandage concluded that the most promising route was the use of the redshift–apparent magnitude relation for the brightest galaxies in clusters for which the dispersion in absolute magnitude about the mean relation was only about $\Delta m \approx 0.3$. His best estimate for q_0 was 1 ± 0.5, but it could have ranged between 0 and 3. He warned that the analysis involves a number of important selection effects which need to be taken into account before a convincing estimate could be made. In particular, he emphasized the importance of the *Malmquist bias*, according to which intrinsically brighter objects are selected in studies which extend to a limiting apparent magnitude (Malmquist, 1920).

Sandage also noted that there was a discrepancy between the ages of the oldest globular clusters, which were estimated to be about 15×10^9 years, and the age of the Universe which, for the $q_0 = 1, \Lambda = 0$ model with $H_0 = 75$ km s^{-1} Mpc^{-1}, was $T_0 = 7.42 \times 10^9$ years. A solution to this problem would be to assume that the cosmological constant was positive which would result in a negative value for q_0. Sandage took the view that there were probably too large uncertainties in the estimates of H_0, q_0 and T_0 for this result to be taken too seriously. He was to devote an enormous effort to the determination of these basic cosmological parameters. Until the 1970s, his work dominated the field.[5] Indicative of his approach to observational cosmology during these years was the title of his paper *Cosmology: A Search for Two Numbers* (Sandage, 1970).

6.5   Gamow and primordial nucleosynthesis

During the 1930s, there were two reasons why the synthesis of the chemical elements in the early stages of the Friedman–Lemaître world models was taken seriously. First, the studies of Cecilia Payne(-Gapochkin) and Henry Norris Russell had shown that the abundances of the elements in stars were remarkably uniform, suggesting they had a common origin. The second was that the interiors of stars seemed not to be hot enough for the nucleosynthesis of the chemical elements to take place. The starting point for studies of primordial nucleosynthesis was therefore to work out the equilibrium abundances of the elements at some high temperature and assume that, if the density and temperature decreased sufficiently rapidly, these abundances would remain 'frozen' as the Universe expanded and cooled.

Detailed calculations were carried out in 1942 by Subrahmanyan Chandrasekhar and Louis Henrich who confirmed the expectation of equilibrium theory that, if the elements were in equilibrium at a high temperature, their abundances would be inversely correlated with their binding energies (Chandrasekhar and Henrich, 1942). The typical physical conditions under which this result was found involved densities of $\rho \approx 10^9$ kg m^{-3} and temperatures $T \approx 10^{10}$ K. There were, however, several gross discrepancies between their predictions and the observed abundances of the elements. The light elements, lithium, beryllium and boron, were predicted to be vastly overproduced relative to their cosmic abundances and iron underproduced, as were all the heavier elements with mass numbers greater than about 70. This result was referred to as the 'heavy element catastrophe'. It was concluded that the chemical elements could not have been synthesized at a single density and temperature. Chandrasekhar and Henrich suggested that some non-equilibrium process had to be involved.

As described by Helge Kragh in Section 5.2, Georges Lemaître proposed in 1931 that the Friedman–Lemaître models had evolved from an initial state which he termed a 'primaeval atom', consisting of vast numbers of protons, electrons and α-particles packed together at nuclear densities (Lemaître, 1931b). Such a huge 'atom' is necessarily unstable and Lemaître proposed that the process of disintegration would give rise to the formation of the chemical elements. He also suggested that the energy released in the nuclear fission processes could account for the high energies of the cosmic rays.

Lemaître's ideas provided the starting point for George Gamow's attack on the problem of the origin of the chemical elements.[6] In 1946, he investigated the synthesis of the chemical elements during the early phases of the Friedman models, extrapolating them back to very early cosmological epochs at which the densities were high enough for nucleosynthesis to take place (Gamow, 1946). He found that the timescale of the Universe was then too short to establish an equilibrium distribution of the elements. In his original proposal, the initial state consisted of a sea of neutrons and subsequent β-decays and neutron capture processes moved nuclei towards the locus of nuclear stability.

Ralph Alpher joined Gamow as a graduate student in 1946 and was given the task of working out the products of nucleosynthesis according to Gamow's prescription. Neutron capture cross-sections were available as a by-product of the nuclear physics programmes carried out during the Second World War and these showed the encouraging result that there is an inverse correlation between the relative abundances of the chemical elements and their neutron capture cross-sections. In Alpher's first calculations, it was assumed that the initial conditions consisted of a sea of free neutrons. As protons became available as a result of the β decay of the neutrons, heavier elements were synthesized by neutron capture. The nuclear reactions were assumed to begin only after the temperature had fallen below that corresponding to the binding energy of deuterium, $kT = 0.1$ MeV, and the Universe was assumed to be static. This theory was published in 1948 by Alpher, Bethe and Gamow, Bethe's name being added to complete the $\alpha\beta\gamma$ pun, and they found reasonable agreement with the observed abundances of the elements (Alpher et al., 1948). The importance of the paper was that it drew attention to the necessity of a hot, dense phase in the early Universe if the chemical elements were to be synthesized cosmologically.

In the same year, Alpher and Robert Herman began improved calculations of primordial nucleosynthesis, but now including the dynamics of the expansion of the early Universe (Alpher and Herman, 1948). At the very high temperatures at early epochs, the Universe was radiation- rather than matter-dominated and they could then work out the subsequent thermodynamic history of the Universe. They found that the temperature history of the thermal background radiation corresponded closely to the adiabatic expansion of a photon gas, $T \propto a^{-1}$, where a is the scale factor of the Universe

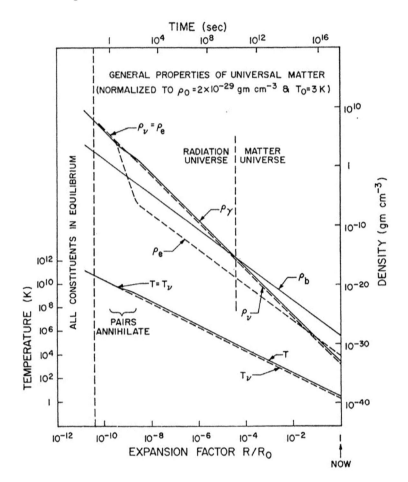

Fig. 6.2 *The thermal history of the Universe containing many of the key features described by Alpher and Herman in 1948 (Alpher and Herman, 1948). This diagram was published by Wagoner, Fowler, and Hoyle in 1967, following an earlier version by Robert Dicke and his colleagues (Dicke et al., 1965).*

(Fig. 6.2). From these results, they came to the far-reaching conclusion that a cooled remnant of these hot early phases should be present in the Universe today and estimated that the temperature of the thermal background should be about 5 K. This was the first prediction that there should exist diffuse background radiation in the centimetre and millimetre wavebands associated with what became known as the *Big Bang theory* of the evolving Universe.[7]

There was, however, a major problem with this picture which Gamow and his colleagues were well aware of—there are no stable nuclei with mass numbers 5 and 8 and hence it was difficult to understand how elements such as carbon, nitrogen, and oxygen could have been created by the addition of further protons, neutrons, or α-particles to helium nuclei. Enrico Fermi and Anthony Turkevich carried out calculations of the

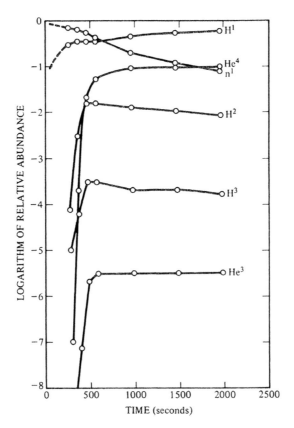

Fig. 6.3 *The evolution of the fraction (by number) of the light nuclei in a radiation-dominated Universe, according to calculations by Fermi and Turkevich and published by Alpher and Herman in 1950 (Alpher and Herman, 1950). The models began with 100% of the material in the form of neutrons.*

evolution of the abundances of the light elements including 28 nuclear reactions for elements up to mass number 7 in a radiation-dominated expanding Universe. Their results, published by Alpher and Herman in 1950 (Alpher and Herman, 1950), showed that only about one part in 10^7 of the initial mass was converted into elements heavier than helium, far less than the cosmic abundances of the heavy elements (Fig. 6.3).

In 1950, another key link in the chain was provided by Chushiro Hayashi who pointed out that, in the early phases of the Universe at temperatures only ten times greater than those at which nucleosynthesis takes place, the neutrons and protons would be maintained in thermodynamic equilibrium through the weak interactions (Hayashi, 1950):

$$e^+ + n \leftrightarrow p + \bar{\nu}_e, \qquad \nu_e + n \leftrightarrow p + e^-. \tag{6.2}$$

Furthermore, at about the same temperature, electron-positron pair production ensures a plentiful supply of positrons and electrons. The result was that, rather than assume arbitrarily that the initial conditions consisted of a sea of neutrons, the equilibrium abundances of protons, neutrons, electrons, and all the other constituents of the early Universe could be calculated exactly. In 1953, Alpher, James Follin, and Herman worked out the evolution of the proton–neutron ratio as the Universe expanded and obtained answers remarkably similar to modern calculations (Alpher et al., 1953). They left, however,

> ...for future study to re-examine the formation of the elements by thermonuclear reactions as a subsequent part of the picture developed here.

They had come very close indeed to the modern picture of the thermal and nuclear evolution of the early Universe, including the important result that, in the standard Big Bang picture, about 25% of the primordial baryonic matter by mass is converted into helium, as well as very much smaller abundances of the light elements deuterium and helium-3 (Fig. 6.3). Before this result became an established feature of astrophysical cosmology, however, steady state cosmology and the nucleosynthesis of the chemical elements in stars occupied centre stage. The discovery of the triple-α resonance by Fred Hoyle (Hoyle, 1954) led to the monumental paper by Geoffrey Burbidge, Margaret Burbidge, William Fowler, and Hoyle, always known as B^{2}FH (Burbidge et al., 1957), in which it was demonstrated how all the chemical elements from carbon onwards could be synthesized in stars—a cosmological origin was unnecessary for these elements.

Helium is one of the more difficult elements to observe astronomically because of its high excitation potential and so can only be observed in very hot stars. Donald Osterbrock and John Rogerson had shown in 1961 that the abundance of helium seemed to be remarkably uniform wherever it could be observed and corresponded to about 25% by mass (Osterbrock and Rogerson, 1961). A further important observation was reported by O'Dell in 1963 of a similar helium abundance in a planetary nebula in the old globular cluster M15 (O'Dell et al., 1964). Despite the fact that the heavy elements were deficient relative to their cosmic abundances, the helium abundance was still about 25%.

Although helium is synthesized in the central regions of stars during their long phases of evolution on the main sequence, it is most unlikely that this process could have created as much helium as 25% by mass of the baryonic matter in the Universe. Most of the luminosity of galaxies is associated with the burning of hydrogen into helium in main sequence stars and so, if the luminosity of our Galaxy had remained more of less the same throughout its lifetime, an upper limit of about 1% of the mass of the Galaxy could have been converted into helium. Furthermore, the stars move off the main sequence when only about 10% of their mass has been converted into helium, and then the helium is burned into heavier elements. It was difficult to understand why there should be a universal abundance of about 25% by mass if the helium was created in stars.

In 1964, Hoyle and Roger Tayler worked out in much more detail the formation of helium in the early phases of the Big Bang (Hoyle and Tayler, 1964). They reaffirmed the result that about 25% helium by mass is synthesized in the Big Bang, in remarkable agreement with observation and the result was essentially independent of the overall baryonic matter density in the Universe. The reason for the constancy of the cosmic helium abundance is that it is primarily determined by the thermodynamics of the early universe, rather than by the microphysics involved in the nuclear reactions. One consequence of the Big Bang model which Hoyle and Tayler did not mention explicitly in their paper was that the cooled remnant of the thermal radiation present during the very hot early phases should be detectable at centimetre and millimetre wavelengths.[8]

6.6   Radio astronomy and cosmology

Quite different directions for cosmological studies were opened up as a result of advances in the new discipline of radio astronomy. The significance of these innovations for high energy astrophysics are reviewed in much more detail in Chapter 7 on *relativistic astrophysics* and in Chapter 8 by Bruce Partridge on the *cosmic microwave background radiation*.

6.6.1   The counts of radio sources

Large-scale surveys of extragalactic radio sources began in the early 1950s. A central figure in this story was Martin Ryle who led the initiatives in radio astronomy at the Cavendish Laboratory in Cambridge. By early 1954, he was converted to the idea that most of the radio sources observed at high galactic latitudes are extragalactic objects. Ryle and Anthony Hewish designed and constructed a large four element interferometer to carry out a new survey of the sky at 81.5 MHz which, being an interferometer, would be sensitive to small angular diameter sources. The second Cambridge (2C) survey of radio sources was completed in 1954 and the first results published in the following year (Shakeshaft et al., 1955). Ryle and his colleagues found that the small diameter radio sources were uniformly distributed over the sky but that their numbers increased dramatically as the survey extended to fainter and fainter flux densities. In any uniform Euclidean model, the numbers of sources brighter than a given limiting flux density S are expected to follow the relation $N(\geq S) \propto S^{-3/2}$. In contrast, Ryle found a huge excess of faint radio sources, the slope of the source counts between 20 and 60 Jy being described by $N(\geq S) \propto S^{-3}$ (Fig. 6.4). He concluded that the only reasonable interpretation of these data was that the sources were extragalactic, that they were objects similar to the extremely luminous radio source Cygnus A and that there was a much greater comoving number density of radio sources at large distances than there are nearby. As Ryle expressed it in his 1955 Halley Lecture in Oxford (Ryle, 1955),

> This is a most remarkable and important result, but if we accept the conclusion that most of the radio stars are external to the Galaxy, and this conclusion seems hard to avoid, then there seems no way in which the observations can be explained in terms of a steady-state theory.

These remarkable conclusions came as a surprise to the astronomical community. There was enthusiasm and also some scepticism that such profound conclusions could be

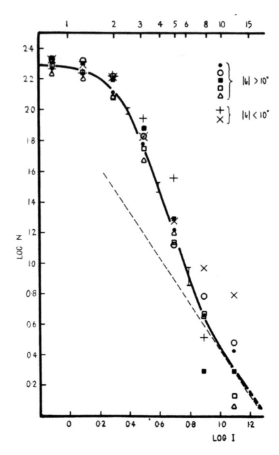

Fig. 6.4 *The integral number counts of radio sources from the 2C survey of radio sources. N is the number of radio sources brighter than flux density I, the units of I being 10^{-25} W m^{-2} sr^{-1}. The dashed line shows the 'Euclidean' number counts of radio sources. The observations show a very large excess of faint radio sources relative to the expectations of the Euclidean world model (Ryle, 1955).*

drawn from the counts of radio sources, particularly when their physical nature was not understood and only the brightest twenty or so objects had been associated with relatively nearby galaxies.

The Sydney group led by Bernard Mills carried out similar radio surveys of the southern sky at about the same time with the Mills Cross and they found that the source counts could be represented by the relation $N(\geq S) \propto S^{-1.65}$, which they argued was not significantly different from the expectation of uniform world models. In 1957 Mills and Bruce Slee stated (Mills and Slee, 1957):

> We therefore conclude that discrepancies, in the main, reflect errors in the Cambridge catalogue, and accordingly deductions of cosmological interest derived from its analysis are without foundation. An analysis of our results shows that there is no clear evidence for any effect of cosmological importance in the source counts.

The problem with the Cambridge number counts was that they extended to surface densities of radio sources such that the flux densities of the faintest sources were overestimated because of the presence of faint sources in the beam of the telescope, a phenomenon known as *confusion*. Peter Scheuer devised a statistical procedure for deriving the number counts of sources from the survey records themselves without the need to identify individual sources (Scheuer, 1957). This technique, which he referred to as the $P(D)$ technique and which has since been adopted in many other astronomical contexts, showed that the slope of the source counts was actually -1.8. The dispute reached its climax at the Paris Symposium on Radio Astronomy in 1958 but the conflicting positions were not resolved (Bracewell, 1959).[9]

The resolution of the controversy only came with the construction of the next generation of radio telescopes which had higher angular resolution and hence were less sensitive to the effects of source confusion. In the next Cambridge catalogues, the 3C catalogue (Edge et al., 1959) and, particularly, the revised 3C catalogue (Bennett, 1962), much more care was taken to eliminate the effects of confusion. The accuracy of the radio source positions also improved so that identifications could be made with fainter galaxies. These showed that Ryle's conclusions of 1955 were basically correct but the magnitude of the excess had been significantly overestimated. More radio sources were identified with distant galaxies and the optical identification programmes led to the discovery of quasars in the early 1960s (see Chap. 7). The radio source counts derived from the 4C catalogues (Pilkington and Scott, 1965; Gower et al., 1967) showed an excess over the expectations of Euclidean world models (Gower, 1966). In fact, the discrepancies with the uniform Friedman–Lemaître and steady-state models were much greater than this simple comparison suggested because the predicted radio source counts converge rapidly in all these models as soon as the source populations extended to significant redshifts (Longair, 1971; Scheuer, 1975). This point had already been forcefully made by William Davidson as soon as the first source counts from the 4C survey were published (Davidson, 1962). As he wrote,

> If the Cambridge data are accepted, the above facts, taken together with the consistent observations of isotropy and lack of clustering referred to in the Introduction, present very convincing evidence against the steady-state model. A confirmation of the Cambridge survey by an equally sensitive one elsewhere, and sustained evidence of the absence of any significant clustering of the sources, would together invalidate the steady-state theory beyond all reasonable doubt.

By the mid-1960s, the evidence was compelling that there was indeed an excess of sources at large redshifts and this was at variance with the expectations of the steady-state theory.

6.6.2   The discovery of the CMB radiation

Alpher and Herman's prediction of a cooled remnant of the hot early phases of the Universe was more or less forgotten when Gamow's theory of primordial nucleosynthesis failed to account for the creation of the chemical elements. The idea of searching for thermal radiation from the Big Bang was revived in the early 1960s by Yakov Zeldovich in Moscow and by Robert Dicke in Princeton and their colleagues.

In 1964, Andrei Doroshkevich and Igor Novikov re-analysed the physics of the Big Bang model and showed that thermal background radiation with a Planck spectrum at radiation temperature between about 1 and 10 K should be present in the Universe at the present day (Doroshkevich and Novikov, 1964). They noted that useful limits to the background radiation temperature could be obtained from the measurements of Edward Ohm of 1961 of the radio background emission at centimetre wavelengths published in the reports of the Bell Telephone Laboratories (Ohm, 1961). In fact, Ohm had discovered an excess noise temperature of 3.3 K in his experiments, but believed that this figure was within the measurement errors of the total signal detected by his antenna and receiver system, which was 22.3 K. This was one of a number of 'near-misses' which are discussed by Bruce Partridge in Section 8.3.

The very next year, in 1965, the microwave background radiation was discovered by Arno Penzias and Robert Wilson, more or less by accident. Using the same 20-foot horn reflector used by Ohm, they discovered that, wherever they pointed the telescope on the sky, there was an excess antenna temperature, which could not be accounted for by noise sources in the telescope or receiver system. Having carefully calibrated all parts of the telescope and receiver system, they found that there remained about 3.5 ± 1 K excess noise contribution (Penzias and Wilson, 1965).

At almost exactly the same time, Robert Dicke's group in Princeton was preparing exactly the same type of experiment to detect the cooled remnant of the Big Bang. It became apparent that Penzias and Wilson had discovered exactly what the Princeton physicists were searching for. Within a few months, the Princeton group had measured a background temperature of 3.0 ± 0.5 K at a wavelength of 3.2 cm, confirming the blackbody nature of the background in the Raleigh–Jeans region of the spectrum (Roll and Wilkinson, 1966).

Remarkably, there was earlier evidence for a diffuse component of millimetre radiation with this radiation temperature from the study of several faint interstellar absorption lines associated with the molecules CH, CH^+, and CN. In the case of CN, for example, absorption was observed from the first rotationally excited state of the molecule as well as the ground state. In 1941, Andrew McKellar had shown that the necessary excitation temperature to populate the first excited state was 2.3 K although the origin of the excitation was then unknown (McKellar, 1941).

Many more details of the history of observations and measurements of the CMB radiation are described in Chapter 8.

6.7   Helium problem revisited

The paper by Hoyle and Tayler in 1964 and the discovery of the cosmic background radiation in 1965 stimulated a number of detailed studies of the primordial synthesis of the light elements. The standard radiation-dominated Big Bang picture makes quite definite predictions about the abundance of the light elements, but it depends upon the dynamics of the Universe through the epochs when the neutrinos decouple from matter. Hoyle, Tayler, Fowler, Wagoner, and Peebles and the Moscow group working with Zeldovich appreciated that the synthesis of the light elements provide an important diagnostic tool for the dynamics of the Universe at the epoch of nucleosynthesis. If the Universe expanded too rapidly, the neutron–proton ratio would freeze out at a higher temperature and helium would be overproduced. This result enabled important constraints to be placed upon any variations of the gravitational constant with time, such as is found in the Brans–Dicke cosmology, as well as restricting the number of permissible neutrino species to 3, a result subsequently confirmed by experiments with the Large Electron–Positron Collider at CERN from studies of the energy widths of the decay products of the Z^0 bosons. It could also set constraints upon anisotropic cosmological models.

In 1967, Robert Wagoner, William Fowler, and Fred Hoyle repeated the analysis carried out by Hoyle and Tayler, but now using all the available cross-sections for many more nuclear interactions between light nuclei and with the knowledge that the CMB

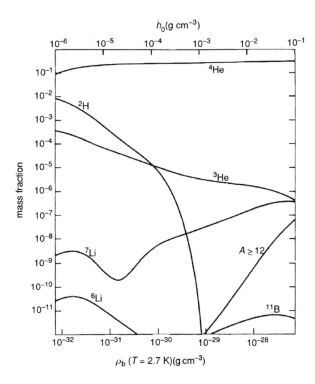

Fig. 6.5 *The synthesis of the light elements in the Big Bang according to the calculations of Wagoner, Fowler, and Hoyle, as revised by Wagoner in 1973 (Wagoner et al., 1967; Wagoner, 1973). These computations demonstrated the sensitivity of the abundances of deuterium, ^{3}He and ^{7}Li to the present baryon density of the Universe.*

radiation had radiation temperature about 2.7 K (Wagoner et al., 1967). Fowler's deep understanding of nuclear physics contributed greatly to all aspects of these computations, enabling a very detailed network of the many nuclear interactions involved in the synthesis of the light elements to be created. These calculations confirmed that about 25% of helium by mass is created by primordial nucleosynthesis and that the figure is remarkably independent of the present density of baryonic matter in the Universe today. Of particular importance was their demonstration that the abundances of other products of nucleosynthesis, deuterium, ^{3}He and ^{7}Li are sensitive to the mean baryon density in the Universe (Fig. 6.5). These elements are very difficult to synthesize in stars because they have relatively small nuclear binding energies—deuterium and ^{3}He are destroyed rather than created in stars.

Interstellar absorption lines of deuterium were discovered in the ultraviolet region of the spectrum by John Rogerson and Donald York in 1973 from observations made by the Copernicus ultraviolet satellite (Rogerson and York, 1973). An interstellar deuterium abundance of 1.5×10^{-5} by mass relative to hydrogen was found. Subsequent observations showed that the same deuterium abundance is found along the line of sight to other stars which could be observed by the Copernicus satellite (Vidal-Madjar et al., 1977). These observations enabled an upper limit to be placed upon the mean baryonic density of the Universe of 1.5×10^{-28} kg m^{-3}, corresponding to $\Omega_B h^2 \leq 10^{-2}$ where $h = H_0/100 \, \mathrm{km \, s^{-1} \, Mpc^{-1}}$. If the mean baryon density of the Universe were any greater, deuterium would be underproduced by primordial nucleosynthesis and no other way of creating deuterium astrophysically was known. This important upper limit to the baryon density in the Universe was at least an order of magnitude less than the critical cosmological density.

By the early 1970s, the overwhelming balance of opinion was that there was convincing evidence for two relics of the hot early phases of the Universe, the CMB radiation and the cosmic abundances of the light elements. Since then the standard Friedman–Lemaître models have become the preferred framework for astrophysical cosmology.

6.8   Cosmological parameters revisited

6.8.1   The deceleration parameter

The redshift–magnitude relation for the brightest galaxies in clusters showed an impressive linear relation (Fig. 6.1) (Sandage, 1968), but it only extended to redshifts $z \sim 0.5$ at which the differences between the world models are relatively small. Sandage was well aware of the many effects which needed to be considered before a convincing estimate of q_0 could be found. Some of these were straightforward, such as the need to determine the luminosities of galaxies within a given metric diameter, but others were more complex. For example, Sandage and Hardy discovered that the brightest galaxy in a cluster is more luminous, the greater the difference in magnitude between the brightest and next brightest galaxies in the cluster (Sandage and Hardy, 1973). In what they termed the *Bautz–Morgan effect*, the second and third ranked members of the cluster were intrinsically fainter than the corresponding galaxies in other clusters with less dominant first ranked galaxies. It seemed as though the brightest galaxy became brighter at the expense of the next brightest members, a phenomenon which could plausibly be attributed to the effects of galactic cannibalism (Hausman and Ostriker, 1977). Sandage adopted an empirical correction to reduce the clusters to a standard Bautz–Morgan type.

Sandage was also well aware of the need to take account of the evolution of the stellar populations of the galaxies with cosmic time. These corrections followed naturally from his work on the Hertzsprung–Russell diagrams of globular clusters of different ages which mimic the cosmic evolution of the old stellar populations of galaxies. He included evolutionary corrections in the K-correction to the absolute magnitudes of the galaxies. The significance of these corrections for galactic evolution was emphasized by the pioneering calculations of Beatrice Tinsley of the expected magnitude of evolutionary changes with cosmic epoch for a wide range of different scenarios for the evolution of the stellar and gaseous content of galaxies (Tinsley, 1976, 1980; Tinsley and Gunn, 1976).

There were, however, other worrying pieces of evidence which did not fit easily into a picture of passive evolution of the galaxies in clusters. Dramatic evidence for the evolution of galaxies in rich, regular clusters at relatively small redshifts was first described in the pioneering analyses of Harvey Butcher and Augustus Oemler (Butcher and Oemler, 1978, 1984). They found that the fraction of blue galaxies in such clusters increased from less than 5% in a nearby sample to percentages as large as 50% at redshift $z \sim 0.4$. The Butcher–Oemler effect was the subject of a great deal of study and debate, the major observational problems concerning the contamination of the cluster populations by foreground and background galaxies, as well as bias in the selection criteria for the clusters selected for observation (Dressler, 1984).

The determination of q_0 might seem to be easier if the samples of galaxies extended to larger redshifts, but it proved far from trivial to find suitable clusters at redshifts greater than 0.5. Those in which the brightest galaxies were observed often turned out to be bluer than expected. This finding reflects a basic problem with this approach to measuring the deceleration parameter—the differences between the expectations of the world models only become appreciable at large redshifts when the Universe was significantly younger than it is now. Consequently careful account has to be taken of the evolutionary changes of the objects which are assumed to have 'standard' properties. Even by the time of Sandage's review of the problem in 1993, the uncertainties in the value of q_0 had not decreased, his estimate being $q_0 = 1 \pm 1$ (Sandage, 1995).

6.8.2 The density parameter Ω_0—the dark matter problem to 1980

The critical cosmological density $\rho_{\mathrm{crit}} = 3H_0^2/8\pi G$ depends upon the value of Hubble's constant and so it is convenient to write the density parameter

$$\Omega_0 = \frac{\rho_0}{\rho_{\mathrm{crit}}} = \frac{\rho_0}{2 \times 10^{-26} h^2 \text{ kg m}^{-3}} \quad \text{where} \quad h = \frac{H_0}{100 \text{ km s}^{-1} \text{Mpc}^{-1}}, \qquad (6.3)$$

and ρ_0 is the present average mass density in the Universe. A convenient way of evaluating the average mass density of galaxies in the Universe was first to work out their average luminosity density by integrating over the luminosity function of the galaxies and then converting this into a mass density by adopting a suitable average value for the mass-to-luminosity ratio of the matter in galaxies. An analysis of this nature was first carried out by Hubble in his pioneering paper of 1926 in which he found $\rho_0 = 1.5 \times 10^{-28}$ kg m^{-3} (Hubble, 1926a). This calculation was repeated by Oort in 1958 who found that the average mass density was 3.1×10^{-28} kg m^{-3} assuming that Hubble's constant was 180 km s^{-1} Mpc^{-1} (Oort, 1958).

In 1978, James Gunn expressed the same result in terms of the mass-to-light ratio which would be needed if the Universe were to attain the critical density (Gunn, 1978). He found $(M/L)_{\mathrm{crit}} = 2600h$, very much greater than the values found in our vicinity in the plane of the Galaxy and in the visible parts of galaxies. The mass of dark matter in galaxies and clusters of galaxies, however, far exceeds that in the visible parts of galaxies. If account is taken of the dark matter, the overall mass-to-luminosity ratio attains values of $M/L \sim 100 - 150$. In well-studied rich clusters, such as the Coma cluster, the value of M/L is of the order of 250, but this value is biased towards elliptical and S0 galaxies which have three times larger values of M/L than the spiral galaxies, the latter contributing most of the light per unit volume in the Universe. These values of M/L are significantly less than the value needed to close the Universe. Gunn's best estimate of the density parameter for bound systems such as galaxies, groups and clusters of galaxies was about 0.1 and was independent of the value of h.

Through the 1970s, there was growing evidence for the presence of dark haloes about spiral galaxies from their rotation curves, the variation of the rotational speed $v_{\mathrm{rot}}(r)$ about the centre of the galaxy with distance r. In many cases, the rotational velocities in the outer regions of spiral galaxies determined from radio 21-cm HI velocity measurements are remarkably constant with increasing distance from the centre (Bosma, 1981). In the case of a spherical distribution of mass, the rotation curve is expected to follow the relation $v_{\mathrm{rot}}^2 = GM(\leq r)/r$. Hence the mass within radius r must increase linearly with distance from the centre, in contrast to the light distribution which decreases exponentially with increasing distance from the centre. For the visible parts of spiral galaxies, mean mass-to-luminosity ratios in the B waveband in the range $1 - 3$ are found. The M/L ratio must however increase to values of $M/L \approx 10 - 20 \, M_\odot/L_\odot$ in the outer regions of spiral galaxies, similar to the values found for elliptical galaxies. These data provide evidence for the presence of dark matter haloes in spiral galaxies.

There were also theoretical reasons why spiral galaxies should possess dark haloes. Ostriker and Peebles showed that, without such a halo, a differentially rotating disc of stars is subject to a bar instability (Ostriker and Peebles, 1973). Their argument has been confirmed by subsequent computer simulations and suggests that dark haloes are needed to stabilize the discs of spiral galaxies.

On scales greater than those of clusters of galaxies, estimates of the mass density in the general field can be found from the *cosmic virial theorem* (Peebles, 1976). In this procedure, the random velocities of galaxies with respect to the mean Hubble flow are compared with the varying component of the gravitational acceleration due to large scale inhomogeneities in the distribution of galaxies. As in the other methods of mass determination, the mass density is found by comparing the random kinetic energy of galaxies with their gravitational potential energy, this comparison being carried out in terms of two-point correlation functions for both the velocities and positions of galaxies selected from the general field. Application to the random velocities of field galaxies suggested that Ω_0 might be larger than 0.2 (Davis et al., 1978; Davis and Peebles, 1983).

A similar argument involved studies of the infall of galaxies into superclusters of galaxies. Galaxies in the vicinity of a supercluster are accelerated towards it, thus providing a measure of the mean density of gravitating matter within the system. The velocities induced by large-scale density perturbations depend upon the *density contrast* $\Delta\rho/\rho$ between the system studied and the mean background density. A typical formula for the infall velocity u of test particles into a density perturbation is (Gunn, 1978)

$$u \propto H_0 r \Omega_0^{0.6} \left(\frac{\Delta\rho}{\rho}\right)_0. \tag{6.4}$$

In Gunn's analysis, this method resulted in values of Ω_0 of about 0.2 to 0.3. This technique was to provide convincing values for the density parameter once large-scale surveys of the redshifts of galaxies became available in the 1990s (see Section 10.6).

6.8.3 The age of the universe, T_0

The ages of the oldest stars in globular clusters were considerably greater than that of the Solar System and could be estimated by the method pioneered by Sandage and Schwarzschild in 1952 (Sandage and Schwarzschild, 1952). This involved comparing the Hertzsprung–Russell diagrams of the oldest, metal-poor, globular clusters with the expectations of the theory of stellar evolution from the main sequence onto the giant branch. The *main sequence termination point* is particularly sensitive to the age of the cluster and, in the oldest globular clusters, it has reached a mass of about 0.9 $M_\odot$. In the most metal-poor, and presumably oldest, clusters the abundances of the elements with atomic number Z are about 150 times lower than their Solar System values. In 1970, Schwarzschild used this approach to estimate the ages of globular clusters, taking account of all the evidence available at that time. His final result was an age of the Universe of $10 \pm 4 \times 10^9$ years (Schwarzschild, 1970).

Just as Rutherford had used the relative abundances of the radioactive species to set a lower limit to the age of the Earth in 1904, so lower limits to the age of the Universe can be derived from the discipline of *nucleocosmochronology*. A secure lower limit to the age of the Universe can be derived from the abundances of long-lived radioactive species. In 1963 Edward Anders used these to determine an accurate age for the Earth of 4.6×10^9 years (Anders, 1963). Some pairs of long-lived radioactive species, such as ^{232}Th–^{238}U, ^{235}U–^{238}U, and ^{187}Re–^{187}Os can provide information about nucleosynthetic timescales before the formation of the Solar System (Schramm and Wasserburg, 1970). In Fowler's review in the early 1970s, this technique gave the age of the Universe prior to the formation of the Solar System of $6.9 \pm 2 \times 10^9$ years, resulting in a total age of the Universe of $11.7 \pm 2 \times 10^9$ years (Fowler, 1972).

Perhaps the most important result of these studies was that these independent methods now gave timescales consistent with the dynamical timescale of the Universe, although the uncertainties were too large to enable any choice of a preferred cosmological model to be made.

6.8.4   Inhomogeneous world models

Another issue which arose during the 1960s and 1970s was the impact of inhomogeneities in the distribution of matter in the Universe. In the standard application of the Robertson–Walker metric, it is assumed that matter is homogeneously distributed within the light cone subtended by a distant object at the observer. But suppose there is no matter within the light cone, as might be the case for point-like objects in the Universe. This problem was addressed by Zeldovich and Dashevsky (Dashevsky and Zeldovich, 1964; Zeldovich, 1964b) who showed that the observed intensities of objects are little affected, but the characteristic minimum in the angular diameter–redshift relation is not present if there is no matter within the light cone. As a consequence, the absence of a minimum in the angular diameter–redshift relation does not necessary imply that the Universe is of low density. These calculations were amplified in the papers by Roeder and his colleagues who provided a convenient set of relations for different fractions of the average mass density in the Universe within the past light cone from distant objects (Dyer and Roeder, 1972).

6.8.5   The status of the cosmological constant, Λ

One of the continuing dilemmas for cosmologists was the status of the cosmological constant Λ. With the discovery of the expansion of the system of galaxies, Einstein regretted the inclusion of the cosmological constant into the field equations. According to Gamow, Einstein stated that the introduction of the cosmological constant was 'the greatest blunder of my life' (Gamow, 1970). But, there was no consensus about whether or not it should be present in Einstein's field equations. Many regarded its presence as something which detracted from the simple beauty of the relation between the mass density of the Universe and its scalar curvature (see expression 6.1). But, George McVittie argued forcefully and presciently for its inclusion in the field equations (McVittie, 1965):

> Historically, the constant Λ was introduced by Einstein in the course of working out a cosmological problem.... This cosmical significance is accidental; the presence of Λ arises through a purely mathematical operation. The cosmical constant would, in fact, be present in Einstein's equations even if cosmology had never been thought of.

Inspection of the field equations of general relativity indicates that, if Λ is positive, the cosmological term may be thought of as 'the repulsive effect of a vacuum', a concept with no meaning in classical physics (Zeldovich, 1968). As Zeldovich pointed out, however, quantum electrodynamics allows such forces to exist as, for example, in the zero-point energy of the vacuum.

As is apparent from Section 6.8.1, it was difficult enough trying to determine two cosmological parameters from the available data, let alone three. Nonetheless, the cosmological constant Λ appeared regularly in the literature in response to various apparent anomalies in extragalactic research. A prime example is its application to resolve the timescale problem whenever the age of the Universe appeared to be less than the age of astronomical objects since appropriately chosen values of Λ allowed cosmological timescales to exceed H_0^{-1}. A distinctive feature of such models is a long 'coasting phase' when the Universe expands at almost constant speed for a long period. This would allow light time enough to circumnavigate the closed geometry of certain world models so that the same object could be observed in opposite directions on the sky (Solheim, 1968a,1968b). It was proposed that this phenomenon might be the origin of the excess of faint radio sources observed in the number counts of radio sources, but this possibility was discounted once account was taken of the wide dispersion in the luminosity function of the radio sources (Longair and Scheuer, 1970). At one stage there appeared to be an excess of absorption line systems in the spectra of quasars with redshifts $z \geq 2$ at $z = 1.95$ (Burbidge, 1967), but the statistical significance of the excess disappeared in due course.

The issue was not to be resolved until the late 1990s and the early 2000s when the redshift–magnitude relation for Type 1a supernovae and, even more spectacularly, the WMAP and Planck observations of the CMB radiation demonstrated unambiguously the presence of the cosmological constant in the field equations.

6.9   The large-scale distribution of galaxies

Groups and clusters of galaxies come in a wide variety of different types ranging from rich regular clusters with smooth galaxy density profiles and roughly circular appearance to irregular systems which have a ragged appearance without any prominent central concentration of galaxies. While the Palomar 48-inch Schmidt Sky Survey was being carried out during the 1950s, George Abell, one of the principal observers, classified and catalogued the rich clusters of galaxies as the survey proceeded over a period of many years. The catalogue was compiled by visual inspection of the plates according to strict selection criteria such that only the most prominent clusters were included. The *Abell Catalogue of clusters of galaxies* was published in 1958 and contains about 2,400 of the richest clusters of galaxies north of declination $-20°$ away from the Galactic plane (Abell, 1958). The survey was extended to include the southern hemisphere when the UK Schmidt Telescope survey plates were completed and a catalogue of over 4,000 rich clusters over the whole sky was prepared by Abell, Harold Corwin, and Ronald Olowin in 1989, six years after Abell's untimely death in 1983 (Abell et al., 1989). In Abell's catalogue, there is a bias towards the richest, symmetrical systems, reflecting his strict criteria.

The clustering of galaxies in fact occurs on a very wide range of physical scales from small groups containing only a few galaxies to giant clusters and superclusters of galaxies. Following Hubble's pioneering studies of the counts of galaxies in the 1930s, a major effort was made after the Second World War to define the large scale structure of the distribution of galaxies using the large-scale plates of the Lick Northern Proper Motion Surveys. These large plates had been taken with the 51-cm Carnegie double astrograph for the purpose of measuring the proper motions of stars, but they also included a great deal of information about the distribution of galaxies on the sky. Counts of galaxies on these plates were made by Donald Shane, Carl Wirtanen, and their colleagues during the 1950s and published as counts of galaxies in $1° \times 1°$ boxes for the sky north of $\delta = -23°$ (Shane and Wirtanen, 1957). Jerzy Neyman and Elizabeth Scott used correlation functions to analyse the variance of the numbers of galaxies in these cells (Neyman et al., 1954). In turn, these studies led to the use of two-point correlation functions to describe the large-scale clustering properties of galaxies. The two-point spatial correlation function $\xi(r)$ for galaxies can be written

$$N(r)\,\mathrm{d}V = N_0[1 + \xi(r)]\,\mathrm{d}V, \tag{6.5}$$

where $N(r)$ is the number density of galaxies at radial distance r from any given galaxy and $\xi(r)$ describes the excess probability of finding a galaxy at distance r over a uniform distribution N_0. Without distances, the three-dimensional function $\xi(r)$ has to be related to the angular two-point correlation function which is what is observed on the photographic plates.

One important issue was whether or not there exist preferred scales of clustering of galaxies. Abell and de Vaucouleurs had shown that there exists non-random clustering of clusters of galaxies, but it was not clear whether or not there is a continuous range of clustering (Abell, 1962; De Vaucouleurs, 1971). Tao Kiang and William Saslaw proposed that there were no preferred scales, but that clustering could occur on all scales (Kiang and Saslaw, 1969). In 1969, Hiroo Totsuji and Taro Kihara were the first to show that the galaxy correlation function $\xi(r)$ can be approximated by a power law over a wide range of scales and this approach was developed extensively by James Peebles and his colleagues in an important series of papers in the 1970s.[10] The function $\xi(r)$ can be described by a power-law function,

$$\xi(r) = \left(\frac{r}{r_0}\right)^{-\gamma}, \tag{6.6}$$

where $\gamma = 1.77$ and $r_0 = 5h^{-1}$ Mpc. This function gives a good representation of the clustering of galaxies on scales from about $10h^{-1}$ kpc to $10h^{-1}$ Mpc, but on scales $r \geq 20h^{-1}$ Mpc, the function decreases more rapidly with increasing physical size (see Chapter 10, Fig. 10.1).

The form of correlation function (6.6) is spherically symmetric about any point and so washes out a great deal of information about the structure of the clustering, but it makes the important point that clustering occurs on a very wide range of physical scales from small groups of galaxies to systems much greater than even the largest clusters of galaxies. The rich clusters are no more than the most prominent features of a continuous spectrum of clustering.

In fact, the distribution of galaxies is much more complicated than this. In the 1970s, Peebles and his colleagues re-analysed the Lick counts of galaxies using the original $10' \times 10'$ cells used by Shane and Wirtanen and demonstrated that, on scales greater than those of clusters of galaxies, the distribution of galaxies has a stringy, cellular appearance

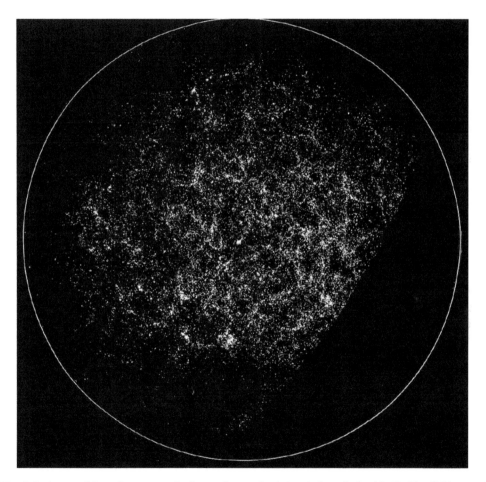

Fig. 6.6 *A map of the galaxy counts in the northern galactic hemisphere derived by Peebles, Seldner, and their colleagues from a reanalysis of the Lick counts of galaxies. The northern galactic pole is at the centre and the galactic equator is the white bounding circle. The galactic latitude is a linear function of radius from the pole. Galactic longitude increases clockwise with $l = 0°$ at the bottom of the map. The prominent 'cluster' in the centre of the image is the Coma cluster (Seldner et al., 1977).*

(Seldner et al., 1977) (Fig. 6.6). In parallel with these studies, increasing numbers of redshifts for nearby galaxies became available and these enabled the three-dimensional distribution of galaxies to be defined directly. Jaan Einasto and his colleagues emphasized the importance of large scale structures on scales much greater than those clusters of galaxies for cosmology (Jöeveer and Einasto, 1978).

6.10   The formation of structure in the Universe

The challenge for the cosmologist is not to explain the detailed features of galaxies and clusters but to explain how large-scale structures formed in the expanding Universe in the first place in the sense that, if $\delta\rho$ is the enhancement in density of some region over the average background density ρ, the *density contrast* $\delta\rho/\rho$ reached amplitude 1 from initial conditions which must have been remarkably isotropic and homogeneous. Once the initial perturbations had grown in amplitude to $\delta\rho/\rho \sim 1$, their development became non-linear and they rapidly evolved towards bound structures in which star formation and other astrophysical processes led to the formation of galaxies and clusters of galaxies as we know them today.

Taking the average matter density in the Universe today to correspond to a density parameter $\Omega_0 \sim 0.2$, the average densities of gravitationally bound systems such as galaxies and clusters of galaxies are very much greater than this value, typically being about 10^6 and 1,000 times greater than the mean background density respectively. Superclusters have mean densities a few times the background density. Since the average density of matter in the Universe ρ changes as $a^{-3} = (1+z)^3$, where a is the scale factor and z is redshift, typical galaxies must have had $\delta\rho/\rho \sim 1$ at a redshift $z \approx 100$. They could not have separated out as discrete objects at larger redshifts, or else their mean densities would be greater than those observed at the present epoch. By the same argument, clusters and superclusters could not have separated out from the expanding background at redshifts greater than $z \sim 10$ and $z \sim 1$, respectively. Therefore galaxies and larger scale structures must have separated out from the expanding Universe at redshifts very significantly less than 100, long after the epoch of recombination at $z \approx 1,000$.

6.10.1   Gravitational collapse and the formation of structure in the expanding Universe

The next step is to include small density perturbations into the Friedman–Lemaître models and study their development under gravity. This problem was solved by James Jeans in 1902 for the case of a stationary medium (Jeans, 1902). He derived the *dispersion relation* between wavenumber k and angular frequency ω, for perturbations in a medium of density ρ_0 and pressure p_0,

$$\omega^2 = c_s^2 k^2 - 4\pi G\rho_0, \tag{6.7}$$

where c_s is the speed of sound in the medium and $k = 2\pi/\lambda$, λ being the wavelength of the perturbation. Thus, for small wavelengths and large wavenumbers, the right hand side is positive and the perturbations behave as propagating sound waves. If the right hand side is negative, gravitational collapse occurs and the density perturbations grow exponentially with characteristic timescale $\tau \sim (G\rho_0)^{-1/2}$ in the limit of long wavelengths. The criterion for collapse is thus that the size of the perturbation should exceed the *Jeans' length* $\lambda_J = c_s/(G\rho_0/\pi)^{1/2}$, meaning that the gravitational force of attraction by the matter of the perturbation exceeds the pressure gradients which resist collapse.

The analysis was repeated for the case of an expanding medium in the 1930s by Georges Lemaître and by Richard Tolman for the case of spherically symmetric perturbations (Lemaître, 1933; Tolman, 1934a) and the solution for the general case was found by Evgenii Lifshitz in 1946 (Lifshitz, 1946). In the non-relativistic case, the development of a density perturbation $\Delta = \delta\rho/\rho$ is

$$\frac{d^2\Delta}{dt^2} + 2\left(\frac{\dot{a}}{a}\right)\frac{d\Delta}{dt} = \Delta(4\pi G\varrho_0 - k^2 c_s^2), \tag{6.8}$$

where k is the proper wavenumber and c_s is the speed of sound. Analysing this equation, Lifshitz found that the condition for gravitational collapse is exactly the same as the Jeans' criterion at any epoch but, crucially, the growth of the density contrast is no longer exponential but only algebraic. In the case of a matter-dominated Universe with the critical density, $\Omega_0 = 1$, the density contrast grows linearly with the scale factor a, that is,

$$\delta\rho/\rho \propto a = (1 + z)^{-1} \propto t^{2/3}. \tag{6.9}$$

A similar result is found for radiation-dominated universes, $\delta\rho/\rho \propto a^2 = (1 + z)^{-2}$. For other Friedman–Lemaître world models with $\Lambda = 0$, the growth rate (6.8) is a good approximation for redshifts $z > \Omega_0^{-1}$, but at smaller redshifts, the perturbations no longer grow. The implication of these results is that the fluctuations from which the large scale structure of the Universe formed cannot have grown from infinitesimal statistical perturbations in the number density of particles—they must have developed from perturbations of finite amplitude. For this reason, Lemaître, Tolman, and Lifshitz inferred that galaxies were not formed by gravitational collapse.

Other authors took the point of view that finite amplitude perturbations should be included in the initial conditions from which the Universe evolved and then the evolution of the perturbation spectrum with cosmic time studied in detail. During the 1960s, the Moscow school led by Yakov Zeldovich, Igor Novikov, and their colleagues and James Peebles at Princeton pioneered the study of the development of structure in the Universe. If perturbations on a particular physical scale were tracked backwards into the past, it was found that, at some large redshift, the scale of the perturbation was equal to the horizon scale of the Universe at that time, that is $r \simeq ct$, where t is the age of the Universe. In 1964, Novikov showed that, to form structures on the scales of galaxies and clusters of galaxies, density perturbations on the scale of the horizon had to have amplitude $\delta\rho/\rho \sim 10^{-4}$ in order to guarantee the formation of galaxies by the present epoch (Novikov, 1964)—these are certainly *not* infinitesimal perturbations and their origin had to be ascribed to processes occurring in the very early Universe.

6.11   The thermal history of the Universe

To determine which physical scales are unstable to gravitational collapse, we need to know the evolution of the speed of sound of the pre-galactic gas as a function of cosmic epoch. With the discovery of the CMB radiation, the thermal history of the matter and radiation content of the Universe could be worked out in detail. This history, first worked out by Alpher and Herman (1948), could now be placed on an firm observational foundation. In the simplest picture, the temperature of the background radiation changes with scale factor as $T = T_0/a = T_0(1 + z)$, exactly as in the adiabatic expansion of a photon gas, but there are a number of important refinements (Fig. 6.7).

6.11.1   The epoch of recombination and the last scattering surface

At redshift $z = 1,500$, the temperature of the background radiation was $T = T_0(1 + z) \approx 4,000$ K, at which there were sufficient photons in the Wien region of the Planck distribution to ionize all the intergalactic hydrogen. This epoch is referred to as the *epoch of recombination* since the hydrogen was fully ionized at earlier cosmic epochs. The details of the process of recombination of the primordial plasma as the Universe expands and cools are important in understanding the origin of the temperature fluctuations in the CMB radiation. These calculations were first carried out by James Peebles and by Yakov Zeldovich, Vladimir Kurt, and Rashid Sunyaev independently in the late 1960s (Peebles, 1968; Zeldovich et al., 1968). Recombinations to the ground state of hydrogen release Lyman continuum photons which can immediately reionize any neutral hydrogen atoms which have recombined. The recombination rate is determined by the rate at which Lyman-α photons are destroyed by the rare two-photon process and takes place over a finite redshift range. Detailed calculations show that the pregalactic gas was 50% ionized at a redshift $z_r \approx 1,500$. At earlier epochs, $z \approx 6,000$, helium was 50% ionized and rapidly became fully ionized before that time.

Consequently, at redshifts greater than about 1,000, the Universe became opaque to *Thomson scattering*. The intergalactic gas was essentially fully ionized at $z > 1,000$ and so the optical depth at larger redshifts is

$$\tau_T = 0.035 \frac{\Omega_B}{\Omega_0^{1/2}} h z^{3/2}. \tag{6.10}$$

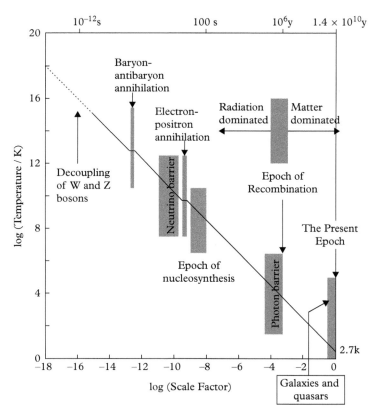

Fig. 6.7 *A summary of the thermal history of the CMB radiation, showing some of the important events at different cosmic epochs. This diagram is a simplified version of Fig. 6.2, first derived by Alpher and Herman in 1948 (Alpher and Herman, 1948; Wagoner et al., 1967).*

For reasonable values of Ω_B, Ω_0, and h, $\tau_T \gg 1$. Therefore, the Universe beyond a redshift of 1,000 is unobservable by electromagnetic radiation. There is therefore a *photon barrier* or *last scattering surface* at a redshift of 1000 beyond which we cannot obtain information directly using photons. This is the surface on which the ripples in the CMB Radiation were imprinted by matter perturbations.

6.11.2   The radiation-dominated era

If the matter and radiation were not thermally coupled, they would cool independently, the hot gas having ratio of specific heats $\gamma = 5/3$ and the radiation $\gamma = 4/3$, the corresponding variations of the temperature of the matter and radiation with scale factor a being $T_m \propto a^{-2}$ and $T_r \propto a^{-1}$, respectively. The CMB radiation provides by far the greatest contribution to the energy density of radiation in intergalactic space. Therefore, comparing the inertial mass density in the radiation and the matter as a function of redshift,

$$\frac{\rho_{\mathrm{r}}}{\rho_{\mathrm{m}}} = \frac{aT^4(z)}{\Omega_0 \rho_{\mathrm{c}}(1+z)^3 c^2} = \frac{2.48 \times 10^{-5}(1+z)}{\Omega_0 h^2}. \tag{6.11}$$

Hence, the Universe became radiation- rather than matter-dominated at redshifts $z \gg 4 \times 10^4 \Omega_0 h^2$. It would be expected that the matter would cool more rapidly than the radiation during the subsequent post-recombination era. This is not the case, however, because the matter and radiation are strongly coupled by *Compton scattering*. The optical depth of the pre-recombination plasma for Thomson scattering is so large that the small energy transfers which take place between the photons and the electrons in Compton collisions cannot be ignored. The details of this process were worked out in pioneering papers by Raymond Weymann (Weymann, 1966) and in much more detail by Zeldovich, Sunyaev, and their colleagues (Zeldovich and Sunyaev, 1969; Sunyaev and Zeldovich, 1980; Pozdnyakov et al., 1983). The analyses by Zeldovich and Sunyaev were based upon the theory of induced Compton scattering developed by Alexander Kompaneets and his colleagues which had been published in 1956, long after this remarkable classified work had been completed in 1950 (Kompaneets, 1956). These papers showed that, during the radiation-dominated epochs, the matter and radiation were maintained in very close thermal contact by Compton scattering so long as the intergalactic gas remained ionized. Since the radiation had much greater heat capacity than the matter, the matter cooled at the same rate as the radiation during the radiation-dominated epochs, $T_{\mathrm{m}} \propto a^{-1}$.

Zeldovich and Sunyaev showed how significant distortions of the black-body spectrum of the CMB could take place if the electrons were heated to a temperature greater than the radiation temperature. This might involve, for example, the dissipation of primordial sound waves or turbulence, matter–antimatter annihilation, the evaporation of primordial black holes by the Hawking mechanism, or the decay of heavy unstable leptons. If no photons were created, the spectrum of the radiation would be distorted from its Planckian form to a Bose–Einstein spectrum with a finite dimensionless chemical potential μ,

$$I_\nu = \frac{2h\nu^3}{c^2} \left[\exp\left(\frac{h\nu}{kT_{\mathrm{r}}} + \mu \right) - 1 \right]^{-1}. \tag{6.12}$$

This form of equilibrium spectrum is expected when there is a mismatch between the number of photons and the energy to be distributed among them in statistical equilibrium.

As soon as the first quasar with redshift $z > 2$ was discovered in 1965, the radio quasar 3C 9, James Gunn and Bruce Peterson and independently Peter Scheuer realized that the continuum radiation to the short wavelength side of the redshifted Lyman-α line was observable in the optical waveband and provided a sensitive test for the presence of intergalactic neutral hydrogen (Gunn and Peterson, 1965; Scheuer, 1965). The Gunn–Peterson test makes use of the fact that the cross-section for the Lyman-α transition at 121.6 nm is very large and so, when the ultraviolet continuum of distant quasars is shifted to the redshift at which it has wavelength 121.6 nm, the radiation is scattered many times so that, if sufficient neutral hydrogen is present at these redshifts, an absorption trough would be expected to the short wavelength side of the redshifted Lyman-α line.

The predicted absorption trough was searched for in those quasars with redshifts $z \geq 2$. No evidence for such a depression to the short wavelength side of the Lyman-α line was observed in 3C 9 or in any of the other large redshift quasars which were observed over the succeeding years. As larger redshift quasars were discovered, it was possible to search for a feature to the short wavelength side of the corresponding resonance line of neutral helium, He I, which has rest wavelength 58.4 nm, but no evidence of an absorption trough was observed. A typical upper limit to the number density of neutral hydrogen atoms at a redshift $z = 3$ was $N_H \leq 10^{-5}$ m^{-3}, very small indeed compared to typical cosmological baryonic densities. Therefore, if there were significant amounts of hydrogen in the intergalactic medium, it would have to be very highly ionized. Consequently, at some time between the epoch of recombination and a redshift certainly greater than 5, the intergalactic gas must have been ionized and reheated. This epoch is known as the *epoch of reionization*.

6.11.3   Earlier epochs

Extrapolating back to redshifts $z \approx 3 \times 10^8$, $T = 10^9$ K, the radiation temperature was sufficiently high for the background photons to attain γ-ray energies, $\varepsilon = kT = 100$ keV, at which the high energy photons in the Wien region of the Planck distribution were energetic enough to dissociate light nuclei such as helium and deuterium. At earlier epochs, all nuclei were dissociated into protons and neutrons. When the clocks are run forward, this was the epoch when primordial nucleosynthesis of the light elements took place (Sections 6.5 and 6.7). At redshift $z \approx 10^9$, electron–positron pair production from the thermal background radiation took place and the Universe was flooded with electron–positron pairs, one pair for every pair of photons present in the Universe now. Running the clocks forward, the electrons and positrons annihilated at about this epoch and their energy transferred to the photon field—this accounts for the little discontinuity in the temperature history when the electrons and positrons were annihilated (Fig. 6.7). At a slightly earlier epoch the opacity of the Universe for weak interactions became unity, resulting in a *neutrino barrier*, similar to the photon barrier at $z \approx 1,000$.

We can extrapolate even further back in time to $z \approx 10^{12}$ when the temperature of the background radiation was sufficiently high for baryon–antibaryon pair production to take place from the thermal background and the Universe was flooded with baryons and antibaryons, one pair for every pair of photons present in the Universe now. Again, there is a little discontinuity in the temperature history at this epoch.

These considerations lead to the *baryon-asymmetry problem*. In order to produce a matter-dominated Universe at the present epoch, there must have been a tiny asymmetry between matter and antimatter in the very early Universe. Roughly, for every 10^9

antibaryons, there must have been $10^9 + 1$ baryons. Then, the 10^9 baryons annihilated with the 10^9 antibaryons, leaving one baryon which became the Universe as we know it with the present photon-to-baryon ratio.

In 1965, Zeldovich showed that, if the Universe were completely symmetric with respect to matter and antimatter, the present day photon to baryon/antibaryon ratio would be about 10^{18} (Zeldovich, 1965), very much greater than the observed value of about 10^9. Baryon symmetric models of the Universe were proposed by Hannes Alfvén and Oskar Klein in 1962 (Alfvén and Klein, 1962) and by Omnes in 1969 (Omnès, 1969) but none of these convincingly demonstrated how the matter and antimatter could be separated in the early Universe (see also Section 4.6). The baryon asymmetry must have originated in the very early Universe. Fortunately, it was known that there is a slight asymmetry between matter and antimatter because of CP violation observed in the decays of K^0 and other elementary particles.

The process of extrapolation can be carried further and further back into the mists of the early Universe but we soon run out of physics validated experimentally in the laboratory. The boldest theorists extrapolate back to the Planck era at $t_P \sim (Gh/c^5)^{1/2} = 10^{-43}$ s, when the relevant physics was certainly very different from the physics of the Universe from redshifts $z < 10^{12}$ to the present day.

6.12   The physics of the development of small perturbations with cosmic epoch

During the 1960s and 1970s, it was generally assumed that the principal sources of inertial mass in the Universe were baryonic matter and the CMB radiation. The dark matter problem was fully appreciated, but within the limits of observational uncertainty at that time, the dark matter could well have been in some dark baryonic form. Consequently, the development of the spectrum of initial perturbations could be worked out assuming that the principal constituents of the Universe were baryonic matter and radiation. Once the variation of the speed of sound with cosmic epoch was established, the evolution of the primordial perturbation spectrum could be worked out. The speed of sound c_s can be written

$$c_s^2 = \left(\frac{\partial p}{\partial \rho} \right)_S, \tag{6.13}$$

where the subscript S means 'at constant entropy', that is, the speed of adiabatic sound waves. From the epoch when the energy densities of matter and radiation were equal to beyond the epoch of recombination, the dominant contributors to p and ρ changed dramatically. Since the matter and radiation were closely coupled throughout the pre-recombination era, the square of the sound speed can be written

$$c_s^2 = \frac{c^2}{3} \frac{4\rho_r}{4\rho_r + 3\rho_m}. \tag{6.14}$$

where ρ_r and ρ_m are the inertial mass densities in radiation and matter respectively. Thus, in the radiation-dominated era, $z \gg 4 \times 10^4 \Omega_0 h^2$, $\rho_r \gg \rho_m$ and the speed of sound tended to the relativistic sound speed, $c_s = c/\sqrt{3}$. At smaller redshifts, the sound speed decreased as the contribution of the inertial mass density of the matter became more important. Specifically, between the epoch of equality of the matter and radiation energy densities and the epoch of recombination, the pressure of the sound waves was provided by the radiation, but the inertia was due to the matter.

After the decoupling of matter and radiation soon after the epoch of recombination, the sound speed became the thermal sound speed of the matter which, because of the close coupling between the matter and the radiation, had temperature $T_r = T_m$ at redshifts $z \gtrsim 550 h^{2/5} \Omega_0^{1/5}$. Thus, at a redshift of 500, the temperature of the gas was about 1,300 K.

In the 1970s, these concepts gave rise to two alternative scenarios for the origin of structure in the Universe. In the *adiabatic* scenario, the initial perturbations were adiabatic sound waves before recombination and structure in the Universe formed by the fragmentation of the large scale structures which reached amplitude $\delta\rho/\rho \sim 1$ at relatively late epochs. A realization of this scenario was described by Doroshkevich, Sunyaev, and Zeldovich in 1974 (Doroshkevich et al., 1974). The alternative was the *isothermal model of structure formation* favoured by Peebles and his colleagues which involved fluctuations in the baryon density, whilst the temperature of the radiation-dominated plasma remained uniform. In the case of perfect gases, any pressure and density distribution in the radiation-dominated phases can be represented as the super-position of a distribution of adiabatic and isothermal perturbations. The isothermal perturbations caused no fluctuations in the background radiation temperature during the radiation-dominated era and were frozen into the radiation-dominated plasma.

6.12.1 The adiabatic model of structure formation

The variation of the *Jeans' mass*, the mass which is just stable against collapse under gravity, with cosmic epoch is of particular importance. In the *adiabatic picture* developed by Zeldovich and his colleagues, it was assumed that a spectrum of small adiabatic perturbations was set up in the early Universe and their evolution followed according to the rules developed above. Fig. 6.8(a) shows schematically how perturbations on different mass scales evolve with cosmic epoch in the standard Friedman–Lemaître models (Sunyaev and Zeldovich, 1970a). Since the speed of sound in the radiation-dominated phases was close to the speed of light, the Jeans' mass was roughly equal to the mass contained within the horizon scale $r_H \simeq ct$ during these epochs. As soon as masses on these scales came through the horizon, those perturbations with masses less than $M_J = 3.75 \times 10^{15}/(\Omega_B h^2)^2 M_\odot$ became sound waves.

The sound speed decreased after the epoch of equality of the energy densities in the matter and radiation until the epoch of recombination was approached. Then, after the decoupling of matter and radiation, the speed of sound dropped dramatically to the thermal sound speed in the baryonic matter with the result that all masses greater than

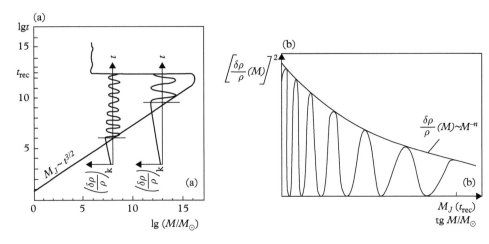

Fig. 6.8 *The 'stability diagram' of Sunyaev and Zeldovich published in 1970. (a) The region of instability is to the right of the solid line marked M_J. The two superimposed graphs illustrate the evolution of adiabatic perturbations with different masses from early times, through the times when they enter the horizon up to the epoch of recombination. (b) Perturbations corresponding to different masses arrive at the epoch of recombination with different phases, resulting in a periodic dependence of the amplitude of the perturbations upon mass (Sunyaev and Zeldovich, 1970a).*

about $M_J = 1.6 \times 10^5 (\Omega_0 h^2)^{-1/2} M_\odot \sim 10^6 M_\odot$ became unstable and began to grow in amplitude as $\delta\rho/\rho \propto (1+z)^{-1}$. It is therefore apparent why the adiabatic perturbations had to be of finite amplitude when they came through the horizon since they could only grow after the epoch of recombination and then only as $(1+z)^{-1}$. Figure 6.8(a) due to Sunyaev and Zeldovich shows schematically the oscillations of perturbations on different mass scales in the pre-recombination era.

In 1968, Joseph Silk realized that, during the pre-recombination epochs, sound waves in the radiation-dominated plasma were damped by the diffusion of radiation out of the perturbation by repeated electron scatterings (Silk, 1968). The effect of damping was to dissipate fluctuations with masses less than $M = M_D = 10^{12} (\Omega_B h^2)^{-5/4} M_\odot$ by the epoch of recombination. Thus, for adiabatic perturbations, all fine scale structure was wiped out and only objects with masses greater than those of large galaxies or clusters of galaxies survived to the epoch of recombination.

In the *adiabatic picture* developed by Zeldovich and his colleagues in the early 1970s, only large scale perturbations with masses $M \geq 10^{14} M_\odot$ survived to the epoch of recombination. During the pre-recombination era, after the perturbations came through their particle horizons, those with masses less than $M_J = 10^{16} - 10^{17} M_\odot$ were sound waves which oscillated until the epoch of recombination, when their internal pressure support vanished and the Jeans' mass dropped to $M_J \sim 10^6 M_\odot$.

Zeldovich and his colleagues also determined the amplitudes and scales of the perturbations which survived to the epoch of recombination, as illustrated in Fig. 6.8(b)

(Sunyaev and Zeldovich, 1970a). The oscillations seen in Fig. 6.8(b) result from the fact that the fluctuations which develop into bound structures at late epochs are those with large amplitudes when they came through the horizon. Figure 6.8(a) shows examples of two perturbations coming through their particle horizons and oscillating as sound waves until the epoch of recombination. The amplitude of the oscillations at the epoch of recombination depended upon the phases of oscillations of the sound waves at that time. Those oscillations which completed an integral number of oscillations would have maximum amplitude as they began to collapse under gravity after the decoupling of matter and radiation. In contrast, those oscillations which had phases such that they had zero amplitude at the decoupling epoch did not form objects at all. The mass spectrum of perturbations at the decoupling epoch is shown in Fig. 6.8(b). This spectrum of oscillations as a function of mass is referred to as the spectrum of *acoustic oscillations* and is a general prediction of theories of structure formation involving primordial adiabatic sound waves.

In the early 1970s, Zeldovich and Edward Harrison independently put together information about the spectrum of the initial fluctuations on different physical scales and showed that observed structures in the Universe could be accounted for if the mass fluctuation spectrum had the form $\delta M/M \propto M^{-2/3}$ in the very early Universe, corresponding to a power spectrum of initial fluctuations of the form $P(k) \propto k^n$ with $n = 1$ (Harrison, 1970; Zeldovich, 1972). The amplitude of the power spectrum had to be $\sim 10^{-4}$. Such a spectrum has the attractive feature that fluctuations on different mass scales had the same amplitude when they came through the horizon, in other words, it results in a fractal universe. This spectrum is known as the *Harrison–Zeldovich spectrum* of initial perturbations.

Following recombination, all the surviving perturbations grew in amplitude as $\delta\rho/\rho \propto (1+z)^{-1}$ until the epoch at which $\Omega_0 z \sim 1$. In the early 1970s, the density parameter in baryons Ω_B was known to be less than about $0.05h^{-2}$ from the constraints provided by primordial nucleosynthesis (Section 6.7) and so, even if $h = 0.5$, the perturbations would grow slowly at redshifts $z \leq 5$. In order to ensure the formation of galaxies and larger scale structures, the amplitudes of the perturbations must have attained $\delta\rho/\rho = 1$ by $z \sim 5$. This was a satisfactory result, since quasars were known to exist at redshifts greater than 2 and the number counts of quasars and radio sources indicated that these objects had flourished at these early epochs. Zeldovich and his colleagues inferred that galaxies and the large scale structure of the Universe began to form at relatively late epochs, $z \sim 3 - 5$. Since the fluctuations had attained amplitude $\delta\rho/\rho \sim 1$ at $z \sim 5$ and $\delta\rho/\rho \propto (1+z)^{-1}$, the amplitude of the density perturbations at the epoch of recombination must have been at least $\delta\rho/\rho \geq 3 \times 10^{-3}$.

6.12.2 The Zeldovich approximation

The structures which survived on the scales of clusters and superclusters of galaxies were unlikely to be perfectly spherical and, in a simple approximation, could be described by ellipsoids with three unequal axes. One of the general rules which comes out of the study of the collapse of ellipsoidal mass distributions is that collapse takes place most rapidly along the shortest axis (Lin et al., 1965). For the case of primordial density fluctuations, Zeldovich showed in 1970 how triaxial collapse could be followed into the non-linear regime (Zeldovich, 1970a).

In the *Zeldovich approximation*, the growth of perturbations into the non-linear regime is followed in Lagrangian coordinates, in other words, the motion of particles is followed in a comoving coordinate frame. If $\vec{x}$ and $\vec{r}$ are the proper and comoving position vectors of the particles of the fluid, the Zeldovich approximation can be written

$$\vec{x} = a(t)\vec{r} + b(t)\vec{p}(\vec{r}). \tag{6.15}$$

The first term on the right-hand side describes the uniform expansion of the background model and the second term the comoving deviations $\vec{p}(\vec{r})$ of the particles' positions relative to a fundamental observer located at comoving vector position $\vec{r}$. Zeldovich showed that, in the coordinate system of the principal axes of the ellipsoid, the motion of the particles in comoving coordinates is described by a 'deformation tensor' D

$$D = \begin{bmatrix} a(t) - \alpha b(t) & 0 & 0 \\ 0 & a(t) - \beta b(t) & 0 \\ 0 & 0 & a(t) - \gamma b(t) \end{bmatrix}. \tag{6.16}$$

The clever aspect of the Zeldovich approximation is that, although the constants α, β and γ vary from point to point in space depending upon the local density distribution, the functions $a(t)$ and $b(t)$ are the same for all particles. In the case of the critical model, $\Omega_0 = 1$,

$$a(t) = \frac{1}{1+z} = \left(\frac{t}{t_0}\right)^{2/3} \quad \text{and} \quad b(t) = \frac{2}{5}\frac{1}{(1+z)^2} = \frac{2}{5}\left(\frac{t}{t_0}\right)^{4/3}, \tag{6.17}$$

where $t_0 = 2/3H_0$.

For the case $\alpha > \beta > \gamma$, collapse occurs most rapidly along the x-axis and the density becomes infinite when $a(t) - \alpha b(t) = 0$. At this point, the ellipsoid has collapsed to a 'pancake' and the solution breaks down at later times. The density became large in the plane of the pancake and the infalling matter was heated to a high temperature as the matter collapsed into the pancake, a process sometimes called the 'burning of the pancakes'. Galaxies were assumed to form by fragmentation or thermal instabilities within the pancakes. In this picture, all galaxies formed late in the Universe, once the large-scale structures had collapsed. This baryonic pancake theory was developed in some detail by Zeldovich and his colleagues in the 1970s and can be thought of as a 'top–down' scenario for galaxy formation (Doroshkevich et al., 1974). Among the successes of the theory was the fact that it resulted in interconnected flattened, stringy structures, not unlike the great holes and sheets of galaxies observed in the local Universe.

6.12.3   The isothermal model of structure formation

The alternative picture, favoured by James Peebles and his colleagues at Princeton, was one in which the perturbations were not sound waves but simply *isothermal* perturbations in the pre-recombination plasma which were in pressure balance with the background radiation at the same temperature. Low mass perturbations were not damped out and so masses on all scales survived to the recombination epoch. Galaxies and clusters of galaxies then formed by the process of hierarchical clustering of low mass objects under the gravitational influence of perturbations on larger scales. Throughout the radiation-dominated era, the timescale for the expansion of the radiation-dominated Universe was much shorter than the collapse timescale and so the isothermal perturbations scarcely grew at all—the amplitude of these perturbations grew by only a factor of about 2.5 from the time they entered the horizon to the epoch of equality of matter and radiation energy densities. Subsequently, the perturbations grew according to the usual result $\delta\rho/\rho \propto a$. Just as in the adiabatic model, similar temperature fluctuations were imprinted on the CMB radiation at the epoch of recombination as the perturbations began to collapse to form bound objects. Their subsequent behaviours were, however, quite different.

In the isothermal picture masses began to collapse on all scales greater than $M = M_J \sim 10^6 M_\odot$ immediately after the epoch of recombination. This picture had the attractive feature that the first objects to form would have masses similar to those of globular clusters, which are the oldest known objects in our Galaxy. The process of galaxy and structure formation was ascribed to the *hierarchical clustering* of these small-scale structures under the influence of the power spectrum of perturbations, which extended up to the largest scales. Another attractive feature of this picture was that there would be early enrichment of the chemical abundances of the elements as a result of nucleosynthesis in the first generations of massive stars. This process could account for the fact that, even in the largest redshift quasars, the abundances of the elements were not so different from those observed locally. Many of these ideas were developed by Peebles in his important monograph *The Large-Scale Structure of the Universe* (Peebles, 1980).

6.12.4   The Press–Schechter model of hierarchical clustering

The process of structure formation by hierarchical clustering was put on a formal basis by William Press and Paul Schechter in a remarkable paper of 1974 (Press and Schechter, 1974). Their objective was to provide an analytic formalism for the process of structure formation once the density perturbations had reached amplitude $\delta\rho/\rho \sim 1$. Their analysis started from the assumptions that the power spectrum of the primordial density perturbations was of power-law form, $P(k) \propto k^n$, and that the phases of the waves were random, what are known as *Gaussian fluctuations*. When the amplitude of the perturbations reached a critical value δ_c, it was assumed that they formed bound systems with mass M. With these assumptions, they showed that the evolution of the

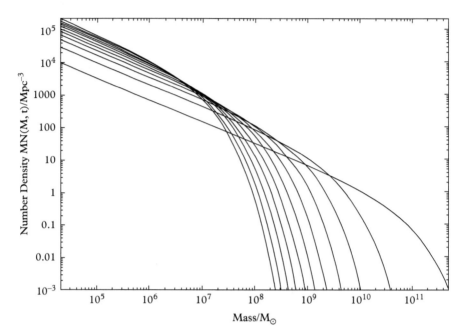

Fig. 6.9 *The variation of the form of the Press–Schechter mass function as a function of cosmic time in the Einstein–de Sitter world model $\Omega_0 = 1$, according to the expression (6.18). (Courtesy of Andrew Blain).*

spectrum of bound objects with cosmic time could be written in the remarkably simple form.[11]

$$N(M) = \frac{\bar{\varrho}}{\sqrt{\pi}}\frac{\gamma}{M^2}\left(\frac{M}{M^*}\right)^{\gamma/2}\exp\left[-\left(\frac{M}{M^*}\right)^{\gamma}\right], \qquad (6.18)$$

where $\gamma = 1 + (n/3)$ and $M^* = M^*(t_0)(t/t_0)^{4/3\gamma}$. The variation of this mass function with time is shown in Fig. 6.9. Press and Schechter were well aware of the limitations of their approach, but it turned out that their mass function and its evolution with cosmic epoch were in good agreement with more detailed analyses and with the results of subsequent supercomputer simulations. The Press–Schechter formalism has proved to be a useful tool for studying the development of galaxies and clusters of galaxies in hierarchical scenarios for galaxy formation. In contrast to the adiabatic picture, the isothermal scenario is a 'bottom–up' picture, in which galaxies and larger scale structures are assembled out of smaller objects by clustering and coalescence.

6.13   Confrontation with the observations

A key test of both models was provided by the fact that the presence of density fluctuations at the epoch of recombination should leave an imprint upon the CMB radiation. In the simplest picture, if the process of recombination were instantaneous, the adiabatic perturbations would be expected to result in temperature fluctuations $\Delta T/T = \frac{1}{3}\Delta\rho/\rho$ on the last scattering surface (Silk, 1968). In fact, the problem is much more complicated than this because the process of recombination was not instantaneous. The fluctuations which were imprinted upon the background radiation depended upon their sizes and optical depths relative to the thickness of this last scattering surface. The principal sources of temperature fluctuations on small scales were expected to be associated with first order Doppler scattering due to the collapse of the perturbations (Sunyaev and Zeldovich, 1970a). These predictions provided a challenge for the observers since the predicted amplitudes of the fluctuations in these early theories were of the order $\Delta T/T \geq 10^{-3}$–10^{-4}.

The theory of these processes for both adiabatic and isothermal baryonic perturbations was worked out by Sunyaev and Zeldovich (Sunyaev and Zeldovich, 1970a), who found the important result that, for both types of perturbation, the root-mean-square temperature fluctuations were predicted to be

$$\left\langle \left(\frac{\delta T}{T} \right)^2 \right\rangle^{1/2} = 2 \times 10^{-5} \left(\frac{M\Omega_0^{1/2}}{10^{15}\,M_\odot} \right)^{1/2} (1+z_0), \tag{6.19}$$

for masses $M \geq 10^{15}\Omega_0^{-1/2}\,M_\odot$, where z_0 is the redshift at which $\delta\varrho/\varrho = 1$.

As recounted by Partridge in Chapter 8, throughout the 1970s increasingly sensitive searches were made for fluctuations in the CMB radiation, these observations being analysed critically by him in his review of 1980 (Partridge, 1980a). His own observations had reached sensitivities of $\Delta T/T \approx 10^{-4}$ or slightly better by that time (Partridge, 1980b). Thus, by the early 1980s, the upper limits to the intensity fluctuations in the cosmic background radiation were beginning to constrain severely purely baryonic theories of structure formation.

There were further problems with both scenarios. It gradually became apparent that the dominant form of matter in the Universe was unlikely to be baryonic. The constraints from primordial nucleosynthesis of the light elements strongly suggested that the mean baryonic mass density of the Universe was about an order of magnitude less than the mean total mass density $\Omega_0 \approx 0.2$–0.3. In addition, in the early 1980s, the concept of the inflationary expansion of the early Universe, pioneered by Andrei Linde, Alan Guth, and their colleagues, caught the imagination of theorists (Guth, 1981). One of the consequences of that picture, which could resolve a number of the fundamental cosmological problems, was that the Universe should have flat spatial geometry and so, if $\Omega_\Lambda = 0$, it followed that $\Omega_0 = 1$. In this case, there was no question but that most of the mass in the Universe had to be in some non-baryonic form. Something new was needed and it was not long in coming, a story taken up in Chapter 10.

....

NOTES

1. In writing this chapter, I have made extensive use of my review *Observational Cosmology* (Longair, 1971) and my book *The Cosmic Century: A History of Astrophysics and Cosmology* (Longair, 2006). I have also used material from my book *Galaxy*

Formation (Longair, 2008), which contains much more detailed analyses of the theoretical considerations of Sections 6.9–6.13.

2. Details of some of these innovations are listed in my book *The Cosmic Century: A History of Astrophysics and Cosmology* (Longair, 2006).

3. The 'cosmological distance ladder' presented by Rowan-Robinson (1985) shows schematically the range of distances over which different classes of object can be used to estimate astronomical distances.

4. For a detailed discussion of the different approaches to the determination of Hubble's constant during the 1970s and 1980s, see Rowan-Robinson (1985); his conclusions were updated in Rowan-Robinson (1988).

5. In a survey of his contributions to observational cosmology, Sandage provided an excellent overview of the history of the determination of cosmological parameters using optical observations (Sandage, 1995). The text includes many intriguing comments and footnotes about the problems of theory and observation which faced the pioneers of the subject from the 1920s to the 1990s. His approach was that of the classical observational cosmologist and it was instructive to contrast his approach with mine in my companion set of lecture notes at the same Saas-Fee school (Longair, 1995).

6. The history of Gamow's work on the Big Bang theory is described by Alpher, R. A. and Herman, R. (1990) and by R.V. Wagoner (1990).

7. There had been references to the expansion of the standard Friedman models as an explosion or 'bang', but the term *Big Bang* entered the cosmological literature with some force following a series of BBC radio broadcasts by Fred Hoyle in the spring of 1949. Hoyle fully intended the term in a pejorative sense, in contrast to his enthusiasm for the steady-state theory which he, Bondi and Gold had developed over the preceding years (see Chapter 5). The lectures were subsequently published essentially unmodified by Hoyle in his book *The Nature of the Universe* (Hoyle, 1950). The term is not a particularly happy one, since it conjures up a somewhat misleading impression of how the standard isotropic world models are constructed, but it is now firmly embedded in the literature.

8. According to Roger Tayler (personal communication), he had included this result in his draft of their paper, but it did not appear in the published version.

9. The strong feelings aroused by the contrasting views of the proponents of the evolutionary and steady-state models are vividly described in Helge Kragh's book *Cosmology and Controversy: the Historical Development of Two Theories of the Universe* (Kragh, 1999). I was there during the period from 1963 onwards. Sadly, the conflicting views and personalities of Ryle and Hoyle and his colleagues could not be reconciled, despite the fact that at the working level their students and colleagues got on well.

10. References to the numerous papers of Peebles and his colleagues on correlation functions for galaxies can be found in Peebles' monograph *Principles of Physical Cosmology* (Peebles, 1993).

11. I have given a simple derivation of this result in my book *Galaxy Formation* (Longair, 2008).

7

Relativistic astrophysics and cosmology

Malcolm S. Longair

7.1 Introduction

The term *relativistic astrophysics* has a multitude of meanings. Often, it is taken to be synonymous with the term *high energy astrophysics*, the study of extreme objects and the radiation they emit, very often involving relativistic particles. The term includes the role of special and general relativity in understanding the physics of extreme objects in the Universe—active galactic nuclei, binary sources involving relativistic stars such as neutron stars and black holes, and, most recently, the spectacular discovery of inspiralling and coalescing black holes detected by their gravitational radiation.[1]

The emphasis in this handbook is the role which relativistic astrophysics plays in the cosmological arena. There are a number of different aspects to this role. First, relativistic objects turn out to be powerful astrophysical probes of the nearby and distant Universe. They play a central role in the discipline of *astrophysical cosmology*, meaning the huge range of astrophysical phenomena involved in understanding how the Universe has evolved from its earliest beginnings to the present day. Second, relativistic astrophysics provides key tests of relativistic theories of gravity which are fundamental to the construction of cosmological models.

7.2 Relativistic astrophysics to 1945

7.2.1 Special and general relativity—their significance for astrophysics and cosmology

With the deep insights provided by Einstein's paper of 1905, the special theory of relativity was quickly assimilated into the infrastructure of theoretical physics.[2] Relativistic length contraction and time dilation were validated by the experiments of Roy Kennedy and Edward Kennedy and Thorndike (1932).[3] Although the mass–energy

Longair, M. S., 'Relativistic astrophysics and cosmology' in *The Oxford Handbook of the History of Modern Cosmology*, edited by Kragh, H. and Longair, M. S. © Oxford University Press 2019.
DOI: 10.1093/oxfordhb/9780198817666.013.7

relation $E = mc^2$ was widely assumed to be a logical consequence of the special theory, the first direct experimental validation of the relation was provided by Cockcroft and Walton (1932). But already, Arthur Eddington (1920b) had used the relation to account for the process of energy generation in the central regions of the Sun through the conversion of four hydrogen atoms into helium, before there was any understanding of how that process could take place.

General relativity was applied by Albert Einstein (1917a) to construct the first fully self-consistent cosmological model. This story and the subsequent development of the understanding of cosmological models are described in detail by Matteo Realdi (Chapter 3) and by Helge Kragh (Chapter 4). But so far as astrophysics was concerned, general relativity was regarded as a small second-order correction to Newtonian gravity, which was generally beyond the possibilities for observation. There were undoubted triumphs, particularly in accounting for the advance of the perihelion of Mercury (Einstein, 1915) and the measurement of light deflection of stars by the Sun (Dyson et al., 1920). But for normal stars, the effects were tiny, the general relativistic parameter $2GM/rc^2$ being only 11.4×10^{-8} at the surface of the Earth and 4×10^{-6} at the surface of the Sun. Chwolson (1924) and Einstein (1936) realized that the gravitational deflection of the light from a perfectly aligned nearby star with a distant star would result in an Einstein, or Chwolson–Einstein, ring, but its diameter would be tiny and unobservable.

7.2.2   White dwarfs, supernovae, and neutron stars

With the discovery of quantum mechanics and the incorporation of Bose–Einstein and Fermi–Dirac statistics into the toolbox of the astrophysicist in the mid-1920s, relativity began to play a more significant role in stellar structure and evolution. The theory of white dwarfs was one of the first triumphs of the new quantum theory of statistical mechanics as applied to astrophysics. Fowler (1926) used these concepts to derive the equation of state of a cold degenerate electron gas. Wilhelm Anderson (1929) showed that the degenerate electrons in the centres of white dwarfs with mass roughly that of the Sun become relativistic and so a relativistic equation of state for a degenerate electron gas was needed. Anderson (1929) and Edmund Stoner (1929) realized that the change in the dependence of pressure p upon density ρ from $p \propto \rho^{5/3}$ to $p \propto \rho^{4/3}$ has the consequence that there do not exist equilibrium configurations for degenerate stars with mass greater than about the mass of the Sun.

The most famous analysis of this result was carried out by Subrahmanyan Chandrasekhar (1931), who had begun working on this problem before he arrived to take up a fellowship at Trinity College, Cambridge in 1930. According to the *virial theorem* for stable stars, the total internal energy U is one half of the total gravitational potential energy Ω_g,

$$2U = |\Omega_g| = \tfrac{1}{2} GM^2/R. \tag{7.1}$$

In the non-relativistic case, the internal energy U of the degenerate gas depends upon radius as R^{-2} and so, for low mass white dwarf stars, these have a definite radius. With increasing mass, however, the central temperature of the star increases and the electrons become relativistic. Then, the relativistic expression for the pressure of the gas has to be used and the dependence of the internal energy upon radius changes to R^{-1}. Thus, the right- and left-hand sides of (7.1) now depend upon radius in the same way. The mass of the star does not depend upon its radius. In white dwarf stars, the chemical abundances have evolved through to helium, carbon or oxygen and therefore the limiting mass for the white dwarfs, the *Chandrasekhar mass*, is

$$M_{Ch} = 1.46 \, M_{\odot}. \tag{7.2}$$

There is no equilibrium state for greater masses and the star collapses to a singularity. This conclusion was vehemently rejected by Eddington and led to the famous confrontation between Eddington and Chandrasekhar.[4]

The neutron was discovered by Chadwick in 1932. The first mention of the possibility of neutron stars appeared as the famous 'Additional Remark' to a paper by Walter Baade and Fritz Zwicky (1934a). In that year, they published two papers on the energetics of what they termed 'super-novae'. In the first paper, Baade and Zwicky (1934b) proposed that the population of novae consists of two types, the ordinary novae, which are relatively common phenomena, and the super-novae which are very rare but very energetic indeed. They identified the bright nova observed in the Andromeda Nebula in 1885 as the archetype of this class of extremely violent explosion and suggested that Tycho Brahe's nova of 1572 was another example of this class. In their second paper, they suggested that such events might be the sources of the cosmic rays. Both proposals are remarkably close to the correct answer. As an addendum to the second paper (Baade and Zwicky, 1934a), they wrote

> With all reserve we advance the view that a super-nova represents the transition of an ordinary star into a *neutron star*, consisting mainly of neutrons. Such a star may possess a very small radius and an extremely high density. As neutrons can be packed much more closely than ordinary nuclei and electrons, the 'gravitational packing' energy in a *cold* neutron star may become very large, and under certain circumstances, may far exceed the ordinary nuclear packing fractions. A neutron star would therefore represent the most stable configuration of matter as such.

Zwicky began the systematic search for supernovae in 1934, first with a $3\frac{1}{4}$-inch Wollensack lens camera mounted on the roof of the Robinson Astrophysics Laboratory of the California Institute of Technology and then, in 1936, with a wide-angle 18-inch Schmidt telescope sited at the new observatory at Palomar. The first supernova was discovered in NGC 4157 in March 1937 and the second in the dwarf spiral galaxy IC 4182 on 26 August 1937. Zwicky continued to discover about four supernovae per year in relatively nearby galaxies. With the construction of the 48-inch Schmidt telescope at Palomar in 1949, searches could be made for fainter supernovae and typically about 20 were discovered each year.

Baade and Zwicky (1938) gave the first description of the typical light curves of supernovae consisting of an initial outburst lasting for a few weeks following which the brightness decreases exponentially with a half-life of about 60 days. The light curves of the first dozen supernovae discovered by Zwicky were remarkably similar, but a few years later Minkowski (1941) discovered that there are two quite distinct types of supernovae. The primary distinction between the two types concerns differences in their spectral evolution.

- The spectra of *Type I supernovae* contain broad emission bands, which almost 30 years later were interpreted as the superposition of hundreds of lines of Fe^+ and Fe^{2+} by Robert Kirshner and Bev Oke (1975). A key feature of their spectra is the absence of hydrogen lines. The spectral and luminosity evolution of the Type I supernovae are essentially identical and they are found in all types of galaxy.

- In contrast, the *Type II supernovae* show the Balmer series of hydrogen soon after maximum light. They display a much wider range of properties than the Type II supernovae and are only found in spiral galaxies, generally within the spiral arms.

Gamow (1937, 1939) showed that a gas of neutrons could be compressed to a much higher density than a gas of nuclei and electrons and estimated the probable densities of such a stars to be about 10^{17} kg m^{-3}. The issue of the maximum mass of neutron stars was discussed by Landau (1938) and in much greater detail by Robert Oppenheimer and George Volkoff (1939). The basic physics is the same as in the case of the white dwarfs but now neutron degeneracy pressure holds up the star. Complications arise because of the necessity of using an equation of state of neutron matter at nuclear densities, as well as including the effects of general relativity. They found an upper mass limit of about $0.7 \ M_\odot$, not so different from the best modern estimates which correspond to about $2 - 3 M_\odot$.

The neutron stars are so compact that general relativity is no longer a small correction but is central to the stability of the star. Typically, for neutron stars, the general relativistic parameter $2GM/Rc^2 = 0.3$. Their radii are only about three times greater than the Schwarzschild radius R_g of a spherically symmetric black hole of the same mass. This work created some theoretical interest but little enthusiasm from the observers. The radii of typical neutron stars were expected to be about 10 km and so there was no prospect of detecting significant fluxes of thermal radiation from such tiny stars.

7.2.3   The cosmic rays

Meanwhile, a route to the study of high energy particles in the cosmos was provided by the discovery of the cosmic rays. By 1900 it was known that electroscopes discharged even if they were kept in the dark, well away from sources of natural radioactivity. Charles (C.T.R.) Wilson carried out experiments on this phenomenon on behalf of the Meteorological Council but the results were inconclusive. In his paper, he made the prophetic remark

Experiments were now carried out to test whether the production of ions in dust-free air could be explained as being due to radiation from sources outside our atmosphere, possibly radiation like Röntgen rays or like cathode rays, but of enormously greater penetrating power. (Wilson, 1901).

The concept of *cosmic radiation* had entered the literature.

The breakthrough came in 1912 and 1913 when first Victor Hess and then Werner Kolhörster made manned balloon ascents in which they measured the ionization of the atmosphere with increasing altitude. By late 1912 Hess had flown to 5 km and then by 1914 Kolhörster had made ascents to 9 km, the experiments being carried out in open balloons. They found the startling result that the average ionization increased with respect to the ionization at sea level above about 1.5 km (Fig. 7.1). The cosmic radiation entering the Earth's atmosphere was at least five times more penetrating than the γ-rays from radium C, one of the most penetrating forms of γ-rays known. Hess (1912) made the immediate inference:

> The results of the present observations seem to be most readily explained by the assumption that a radiation of very high penetrating power enters our atmosphere from above, and still produces in the lower layers a part of the ionization observed in closed vessels.

In 1929, Dmitri Skobeltsyn, working in his father's laboratory in Leningrad, noted some cloud chamber tracks which were hardly deflected at all and which looked like electrons with energies greater than 15 MeV—these were the first cloud chamber images

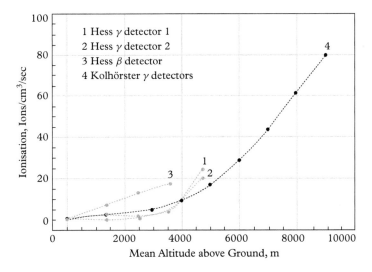

Fig. 7.1 *The variation of ionization with altitude from the pioneering observations of Hess and Kolhörster (Courtesy of Michael Walter).*

of the cosmic radiation. 1929 also saw the invention of the *Geiger–Müller detector* which enabled individual cosmic rays to be detected with precisely known arrival times. In the same year, Walther Bothe and Kolhörster performed one of the key experiments in cosmic ray physics and in the process introduced the important concept of *coincidence counting* to eliminate background events. In the crucial experiment they placed slabs of lead and then gold up to 4 cm thick between the counters and measured the decrease in the number of coincidences when the absorber was introduced. The mass absorption coefficient agreed very closely with that of the atmospheric attenuation of the cosmic radiation. The experiment strongly suggested that the cosmic radiation consisted of charged particles rather than γ-rays. The flux of these particles could also account for the observed intensity of cosmic rays at sea level. Finally, they noted that the particles would have to be very energetic because of their long ranges in matter. They estimated the energies of the particles to be about 10^9–10^{10} eV.

The experiments carried out using cloud chambers showed that showers of cosmic ray particles are often observed. The full extent of some of these *extensive air showers* was established by Pierre Auger and his colleagues (1939) from observations with a number of separated detectors. To their surprise, they found that the air showers could extend over distances greater than 100 m on the ground and consisted of the arrival of millions of ionizing particles. The particles responsible for initiating the showers must have had energies exceeding 10^{15} eV at the top of the atmosphere, direct evidence for the acceleration of charged particles to extremely high energies in extraterrestrial sources.

7.3   The opening up of the electromagnetic spectrum

Until 1945, astronomy meant optical astronomy. The commissioning of the Palomar 200-inch telescope in 1949 highlighted the dominance of the USA in observational astrophysics in the period immediately after the Second World War. The 200-inch telescope was a private telescope used more or less exclusively by a relatively small group of privileged astronomers employed at the host institutions, the Astrophysics Department of the California Institute of Technology, the Mount Wilson Observatory and the Carnegie Institute of Washington. The astronomical scene was, however, about to change with the development of new ways of tackling astrophysical and cosmological problems. There were several reasons for this major change in outlook.

- The most important of these was the *expansion of the wavebands available for astronomical observation*. A plot of the temperature of a black body against the frequency (or wavelength) at which most of the radiation is emitted is shown in Fig. 7.2(a). In the lower panel, Fig. 7.2(b), the transparency of the atmosphere to radiation as a function of frequency is presented, showing how high a telescope must be placed above the surface of the Earth for the atmosphere to become transparent. With the development of radio astronomy and the capability of placing telescopes in space, the expansion of the observable electromagnetic spectrum led to a vast increase in the range of *temperatures* accessible for astronomical study and so to a

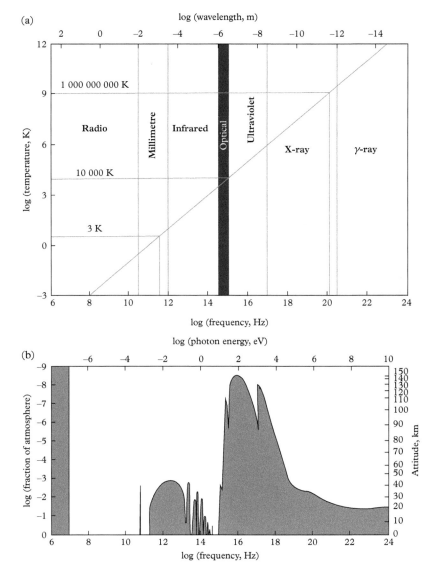

Fig. 7.2 (a) *The relation between the temperature of a black-body and the frequency ν (or wavelength λ) at which most of the energy is emitted (solid diagonal line). The frequency (or wavelength) plotted is that corresponding to the maximum of a black-body at temperature T. Convenient expressions for this relation are: $\nu_{max} = 10^{11} (T/K)\,Hz$; $\lambda_{max} T = 3 \times 10^6$ nm K. The ranges of wavelength corresponding to the different wavebands—radio, millimetre, infrared, optical, ultraviolet, X-ray, and γ-ray—are shown.*
(b) *The transparency of the atmosphere for radiation of different wavelengths. The solid line shows the height above sea level at which the atmosphere becomes transparent for radiation of different wavelengths (Giacconi et al., 1968; Longair, 1988).*

much more complete description of our physical Universe. As described by Silvia De Bianchi in Chapter 9, the competition for superiority between the USA and the USSR during the Cold War (1950–1991) resulted in accelerated opportunities for exploiting space technology for research in solar system, Galactic and extragalactic astronomy.

- Equally important was the development of *non-electromagnetic means of tackling astrophysical and cosmological problems*. *Neutrino astronomy* made spectacular contributions to astrophysics and fundamental physics. *Gravitational wave astronomy* has become a major tool for the high energy astrophysicist with the direct detections of gravitational radiation from coalescing binary black holes and binary neutron stars. *Astro-particle physics* includes the search for stable massive particles predicted by theories of elementary particles and has a key role in understanding the dark matter problem, as well as in fundamental physics for cosmological applications.

- None of these developments would have been possible without remarkable *technological developments in the design and construction of telescopes, instruments and detectors for all wavebands*. In many cases, the technologies were imported from non-astronomical disciplines and modified for the special needs of astronomical observation. The semiconductor and computer revolutions were crucial to the advance of both observation, data collection and analysis, interpretation and theory. The role of computation in astrophysics and cosmology has completely changed many aspects of research in these disciplines. Theory and observation can now be compared with a precision which would have been quite inconceivable to the pioneers of the pre-War years.

- There was a *huge increase in the volume of activity in the astronomical sciences*. At least part of the growth was associated with the influx of physicists whose research interests and expertise led them to consider astrophysical and cosmological problems. A measure of this increase in activity is provided by the membership of the International Astronomical Union (IAU). At the first General Assembly held in Rome in 1922, there were just over 200 professional astronomers from 19 adhering countries. By 1938, the numbers had risen to 550 from 26 countries. By the time of the 2012 General Assembly held in Beijing, China, the membership had risen to 9,898 from 70 adhering countries.

- *Astronomy, astrophysics, and cosmology became one of the 'big sciences'*. After the Second World War, there was a spectacular increase in investment in basic research in the USA, largely stimulated by the huge contributions which the very best research scientists had made during the period of hostilities and the realization of the enormous potential for economic growth, as well as strategic defence requirements, which the fruits of basic research could bring. In Europe, it took somewhat longer to recover from the ravages of the War but, in due course, these countries began to invest heavily in pure and applied research. The major discoveries of radio, X-ray, and γ-ray astronomy, as well as the rise of high energy astrophysics in the 1960s and 1970s, had a considerable impact upon the case for increasing investment in large astronomical facilities. The culmination of this historical progression includes the National Aeronautics and Space Administration (NASA) Great Observatories programme including the Hubble Space Telescope, the European Southern Observatory's Very Large Telescope (VLT) on Cerro Paranel in Chile, and the ALMA submillimetre aperture synthesis array, consisting of 66 high precision antennae.

All these aspects of the post-1945 explosion of investment in science contributed to the dramatic advances in relativistic astrophysics and cosmology.

7.4 Radio astronomy, extragalactic radio sources, quasars, and pulsars

7.4.1 Radio astronomy and the birth of high energy astrophysics

The expansion of the observable electromagnetic spectrum began with the announcement by Karl Jansky (1933) of the discovery of the radio emission from the Galaxy at the long wavelength of 14.6 m (20.5 MHz). This discovery was confirmed by Grote Reber, a radio engineer and enthusiastic amateur astronomer. With his home-built radio antenna and receiving system operating at a wavelength of 1.87 m (160 MHz), Reber (1940) made a radio scan along the plane of the Galaxy. Comparison of Jansky's and Reber's observations showed that the emission could not be black body radiation. Louis Henyey and Philip Keenan (1940) showed that, whilst the radiation at 1.87 m might be the bremsstrahlung of gas at 10,000 K, the intensity observed by Jansky at the longer wavelength was far too intense for this to be the emission process. The culmination of Reber's work was the publication of the first map of the radio emission from the Galaxy (Fig. 7.3) (Reber, 1944). These observations attracted little attention from professional astronomers.

The development of radar during the Second World War had two immediate consequences for radio astronomy. First, sources of radio interference which might confuse radar location had to be identified. In 1942, James Hey and his colleagues at the Army Operational Research Group in the UK discovered intense radio emission from the Sun which coincided with a period of unusually high sun-spot activity (Hey, 1946). At the end of hostilities, they began mapping the sky at 5 m wavelength and in 1946 discovered the first discrete source of radio emission, which lay in the constellation of Cygnus, the source named Cygnus A (Hey et al., 1946). The second consequence was that the extraordinary research efforts to design powerful radio transmitters and sensitive receivers for radar resulted in new technologies which were to be exploited by the pioneers of radio astronomy, all of whom came from a background in radar. The leaders of the newly-founded radio astronomy groups included Martin Ryle at Cambridge University and Bernard Lovell at Manchester University and Joseph Pawsey at Sydney. Numerous discrete sources of radio emission were discovered.

Ryle and Francis Graham Smith (1948) discovered the most powerful radio source in the Northern hemisphere, Cassiopeia A, and in the next year the Australian radio

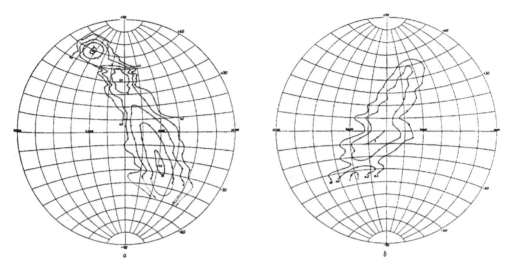

Fig. 7.3 *Reber's map of the radio emission of the Galaxy, made at a radio frequency of 160 MHz (1.87 m). The contours of radio emission are plotted in celestial coordinates and are more or less coincident with the Milky Way (Reber, 1944).*

astronomers John Bolton, Gordon Stanley, and Bruce Slee (1949) associated three of the discrete radio sources with remarkable nearby astronomical objects. One was associated with the supernova remnant known as the Crab Nebula and the others, Centaurus A and Virgo A, with the strange galaxies NGC 5128 and M87 respectively. These early surveys also established the existence of a population of discrete radio sources, some concentrated towards the plane of the Galaxy, but many lying at high galactic latitudes. There was some controversy as to whether the isotropic component of the source population was primarily associated with nearby radio stars in our own Galaxy or with distant extragalactic objects.

Distances could only be determined by finding an associated optical object and measuring its distance. Graham Smith (1951) measured interferometrically the positions of the two brightest sources in the northern sky, Cygnus A and Cassiopeia A, with an accuracy of about 1 arcmin. This led to their optical identification by Walter Baade and Rudolph Minkowski (1954) from observations with the Palomar 200-inch telescope. Cassiopeia A was associated with a young supernova remnant of a star which exploded about 250 years earlier in our own Galaxy, while Cygnus A was associated with a faint galaxy at a redshift of 0.057. There was soon a consensus that the high latitude radio sources formed an extragalactic population of sources which could be used for cosmological studies.

The nature of the Galactic radio emission and, by analogy of the discrete radio sources, was solved in the late 1940s. During the 1930s and 1940s, particle accelerators, such as cyclotrons and betatrons, were constructed in which protons or electrons moved in circular paths in a uniform magnetic field. Julian Schwinger (1946, 1949) worked out in detail the expected radiation spectrum of highly relativistic electrons spiralling in a uniform magnetic field and showed that, because of the extreme effects of aberration, the radiation is most intense at a very much higher frequency $\nu \approx \gamma^2 \nu_g$, where $\nu_g = eB/2\pi m_e$ is the non-relativistic gyrofrequency of the electron and $\gamma = (1 - v^2/c^2)^{-1/2}$ is the Lorentz factor. The 70 MeV synchrotron accelerator built at the General Electric Laboratory at Schenectady, New York had a transparent glass vacuum tube. Intense optical radiation from the accelerator was first observed in April 1947 and identified with the radiation predicted by Schwinger—it was named synchrotron radiation (Elder et al., 1947).

The first application of synchrotron radiation in an astronomical context was proposed by Hannes Alfvèn and Nicolai Herlofson (1950), who suggested that the emission of the 'radio stars' might be the synchrotron radiation of high energy electrons gyrating in magnetic fields within a 'trapping volume' of about 0.1 light year radius about the star. Karl-Otto Kiepenheuer (1950) and Vitali Ginzburg (1951) made the much better suggestion that the Galactic radio emission is the synchrotron radiation of ultra-relativistic electrons gyrating in the interstellar magnetic field. By the mid-1950s, the power-law spectrum of the Galactic radio emission and its high degree of polarization convinced everyone of the correctness of the synchrotron hypothesis.

With the identification of Cassiopeia A with a young supernova remnant, its enormous synchrotron radio luminosity provided direct evidence for the acceleration of huge fluxes of very high energy electrons in supernova remnants, an idea foreshadowed by Baade and Zwicky (1934b). Even more remarkably, the radio source Cygnus A had a radio luminosity more than a million times greater than that of our Galaxy and so had to be the source of vast quantities of relativistic material. Just as unexpected was the fact that the radio emission did not originate from the galaxy itself. Roger Jennison and Mrinal Kumar Das Gupta (1953) at Jodrell Bank used interferometric techniques to show that the radio emission originated from two huge lobes located on either side of the radio galaxy (Fig. 7.4)—enormous amounts of material had to be accelerated to relativistic energies and ejected into intergalactic space in opposite directions.

In 1959, Geoffrey Burbidge (1959) worked out the minimum amount of energy in high energy particles and magnetic fields which had to be present in the source regions to account for the radio emission. The energies proved to be enormous, in some sources corresponding to a rest mass energy of about $10^6 M_\odot$. This result marked the beginning of high energy astrophysics in its modern guise. Some astrophysical means had to be found of converting a significant fraction of the rest mass energy of a galaxy into high energy particles and magnetic fields. The problem was not, however, confined to galaxies—supernovae remnants such as Cassiopeia A were also able to convert a significant fraction of their rest mass energy into high energy particles and magnetic fields.

7.4.2 The discovery of quasars and their close relatives

The radio astronomical discoveries of the 1950s stimulated a great deal of astrophysical interest and led to major investments in the construction of radio telescope systems. The

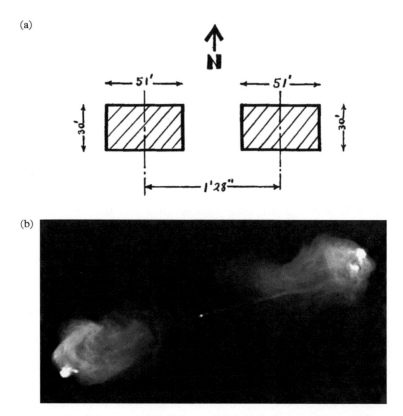

Fig. 7.4 (a) *A reconstruction of the radio structure of the radio source Cygnus A from radio interferometric observations by Jennison and Das Gupta at a frequency of 125 MHz (Jennison and Das Gupta, 1953). (b) A high angular resolution map of the same source made with the Very Large Array in the USA at a frequency of 5 GHz (Perley et al., 1984).*

radio observatories began systematic surveys of the sky in order to understand both the astrophysics of the sources of radio waves and their use as cosmological probes. Most of those which could be securely identified were found to be associated with some of the most massive and luminous galaxies known and so could be observed at cosmological distances. The largest redshift known for any galaxy was that of the radio galaxy 3C 295 at $z = 0.46$ (Minkowski, 1960).

By 1962, Thomas Matthews and Sandage (1963) had identified three of the brightest radio sources, 3C 48, 3C 196, and 3C 286, with 'stars' of an unknown type with strange optical spectra. The breakthrough came in 1962 when Cyril Hazard and his colleagues measured very precisely the position of the radio source 3C 273 by the method of lunar occultations using the recently-completed Parkes 64-m radio telescope in New South Wales, Australia (Hazard et al., 1963). 3C 273 was identified with what appeared to be a 13th magnitude star (Fig. 7.5). Maarten Schmidt (1963) immediately used the Palomar

Fig. 7.5 *The Hubble Space Telescope image of the quasar 3C 273, showing the optical jet ejected from the quasar nucleus. The faint smudges to the south of the quasar are now known to be galaxies at the same distance as the quasar (courtesy of NASA and the Space Telescope Science Institute).*

200-inch telescope to obtain its optical spectrum which contained prominent emission lines, but at unexpected wavelengths. The clue lay in the familiar pattern of lines which Schmidt realized was the Balmer series of hydrogen, but shifted to longer wavelengths with redshift $z = \Delta\lambda/\lambda_0 = 0.158$—this was certainly no ordinary star.

3C 273 was the first, and brightest, of this class of hyperactive galactic nuclei. Its optical luminosity was about 1,000 times greater than the luminosity of a galaxy such as our own (Fig. 7.5). Furthermore, searches through the Harvard plate archives revealed that the enormous luminosity of 3C 273 varied on a timescale of years (Smith and Hoffleit, 1963). Nothing like this had been observed in astronomy before. This discovery opened the floodgates for the identification of many more examples of these objects. They were termed *quasi-stellar radio sources* and within a year this term had been contracted to the word *quasar*. Within a couple of years, the redshifts of the radio quasars had reached $z = 2.012$ for the quasar 3C 9 (Schmidt, 1965), for some years the largest redshift of any astronomical object.

It soon became apparent that the quasars are among the most extreme examples of what are now termed generically *active galactic nuclei*. The first examples had been discovered in the early 1940s by Carl Seyfert (1943), who studied a number of spiral galaxies, such as NGC 1068 and NGC 4151, which possessed star-like nuclei. They possessed very intense emission lines of the Balmer series and forbidden lines such as [O II], [O III], [N II], [Ne III], [S II], and [S III]. Furthermore, the lines were very broad, the Doppler velocities required to broaden the lines being up to 8,500 km s^{-1}, quite unlike those observed in regions of ionized hydrogen in galaxies. Furthermore, the spectrum of the continuum radiation was smooth, quite unlike the spectrum of starlight. Seyfert's pioneering work was largely neglected until the 1960s.

In 1965, radio-quiet counterparts of the radio quasars were discovered by Allan Sandage (1965) and these *radio-quiet quasars* turned out to be about one hundred times more common than the radio-loud variety. The similarity between the properties of quasars and the nuclei of Seyfert galaxies was reinforced by the discovery in 1967 that both the continuum emission and the strong emission lines in Seyfert galaxies are variable (Fitch et al., 1967). It became apparent that there is a continuous sequence of high energy activity in the nuclei of galaxies, from weak nuclei such as the centre of our own Galaxy to the extreme examples of quasars in which the starlight of the galaxy is completely overwhelmed by the intense non-thermal radiation from the nucleus.

Among the more extreme examples of active galactic nuclei were the *BL-Lacertae* or *BL-Lac objects*, discovered in 1968 as extremely compact and highly variable radio sources by John McLeod and Brian Andrews (1968). Their optical spectra are smooth and featureless and display very rapid variability at optical and radio wavelengths from timescales of less than a day to weeks. Eventually, when optical spectrographs with very large dynamic ranges became available, redshifts for the host galaxies of the BL-Lac objects were measured, showing that the BL-Lac objects are indeed extragalactic objects, but that their luminosities are not as great as those of the most luminous quasars.

Once the characteristic features of active galactic nuclei were established, surveys were undertaken to find more examples. The surveys of galaxies with ultraviolet excesses carried out by Beniamin Markarian and his colleagues at the Byurakan Observatory in Armenia proved to be a rich source of Seyfert galaxies, culminating in 1981 with paper 15 of the series (Markarian, 1967; Markarian et al., 1981). About 10% of these UV-excess galaxies were Seyfert galaxies. Another rich source of Seyfert galaxies proved to be Zwicky's *Catalogue of Selected Compact Galaxies and of Post-Eruptive Galaxies* (Zwicky and Zwicky, 1971).

7.4.3   The discovery of neutron stars

The neutron stars were discovered, more or less by chance, as the parent bodies of radio pulsars by Antony Hewish and Jocelyn Bell in 1967 at Cambridge, England (Hewish et al., 1968). Hewish had pioneered the techniques of observing the scintillation or 'twinkling' of compact radio sources at long radio wavelengths due to irregularities in the plasma density along the lines of sight to the radio sources. This technique turned out to be a means of discovering radio quasars, many of which are compact radio sources and hence expected to display large radio scintillations at low frequencies.

Hewish designed a 1.8 hectare array operating at 81.5 MHz to further these studies and was awarded a grant of £17,286 by the Department of Scientific and Industrial

Research to construct it, as well as outstations for measuring the velocity of the solar wind.[5] Jocelyn Bell, Hewish's research student, discovered a strange source which seemed to consist entirely of scintillating radio signals (Fig. 7.6(a)). In November 1967, the source reappeared and was found to consist entirely of a series of pulses with a pulse period of about 1.33 s (Fig. 7.6(b)). Over the next few months three further sources were discovered with pulse periods ranging from 0.25 to almost 3 s. They were named *pulsars*.

The favoured picture for the pulsar phenomenon was described by Thomas Gold in 1968, similar in many respects to Pacini's prescient proposal of 1967, and consisted of an isolated, rotating, magnetized neutron star in which the magnetic axis of the star and its

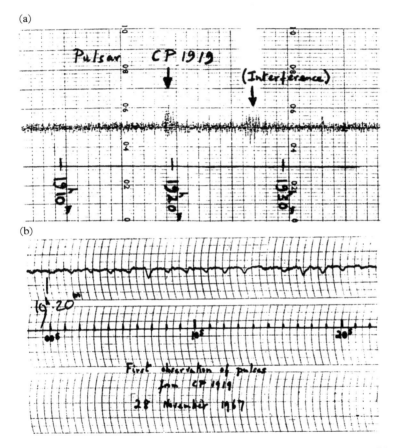

Fig. 7.6 *The discovery records of the first pulsar to be discovered, PSR 1919+21 (Hewish, 1986).*
(a) *The first record of the strange scintillating source labelled CP 1919. Note the subtle differences between the signal from the source and the neighbouring signal due to terrestrial interference.* (b) *The signals from PSR 1919+21 (top trace) observed with a shorter time-constant than the discovery record showing that the signal consists entirely of regularly spaced pulses with period 1.33 s. The lower trace shows one-second time markers.*

rotation axis are misaligned (Pacini, 1967; Gold, 1968). Two of the pulsars discovered in 1968 were of special importance. A pulsar of period 0.089 s was discovered in the young supernova remnant in the constellation of Vela (Large et al., 1968) and soon afterwards a pulsar with the very short period of 0.033 s was discovered in the centre of the Crab Nebula (Staelin and Reifenstein, 1968). Optical pulses from the Crab Nebula pulsar with precisely the same pulse period were discovered in 1969 (Cocke et al., 1969) and within three months rocket flights by the teams from the Naval Research Laboratory (Fritz et al., 1969) and the Massachusetts Institute of Technology (Rossi, 1970) had discovered X-ray pulses as well.[6] The short periods of these pulsars proved beyond a shadow of doubt that the parent bodies of the pulsars were neutron stars and their association with supernova remnants showed that neutron stars form as a result of stellar collapse.

7.5   General relativity and active galactic nuclei

7.5.1   The First Texas Symposium on Relativistic Astrophysics

Just before the discovery of quasars, Burbidge, Burbidge, and Sandage (1963) surveyed a wide range of evidence concerning activity in the nuclei of galaxies in an influential paper entitled *Evidence for the Occurrence of Violent Events in the Nuclei of Galaxies*. Their survey included a very wide range of evidence including the Seyfert galaxies, the radio galaxies and other galaxies in which explosions seemed to have occurred. These included objects such as the irregular galaxy M82 and the giant elliptical galaxy M87 which possesses the prominent optical jet first described by Heber Curtis (1918).

One of the first discussions of the implications of these remarkable discoveries was held at the First Texas Symposium on Relativistic Astrophysics held in Dallas, Texas, in 1963. For the first time, the optical and radio astronomers got together with the theoretical astrophysicists and, in particular, with the general relativists to thrash out what was known about these objects and the role which general relativity might play. Coincidentally, Hoyle and Fowler (1963) had been investigating the properties of super-massive stars with masses as great as $10^6 \, M_\odot$ but, although these might radiate enormous luminosities, they proved to be notoriously unstable.

The most important conclusions of the meeting were that the quasars must involve strong gravitational fields and that general relativity must play a central role in understanding their properties. At the closing dinner, Gold made the remark,

> Everyone is pleased: the relativists who feel they are being appreciated, who are suddenly experts in a field which they hardly knew existed; the astrophysicists for having enlarged their domain, their empire by the annexation of another subject—general relativity (Gold, 1965).

The proceedings of the First Texas Symposium contained many of the most important papers on radio sources, active galaxies and quasars published up to 1964

(Robinson et al., 1965) and undoubtedly stimulated the remarkable upsurge of interest in general relativity.

For some time, a few astronomers, the most prominent of whom were Fred Hoyle, Geoffrey Burbidge, and Halton Arp, questioned whether the redshifts of the quasars were of cosmological origin or whether they might be due to some other physical process. If they were relatively nearby objects, their luminosities would be far less extreme and their variability would be much less of a problem. Gravitational redshifts could not account for the observations, as discussed by Greenstein and Schmidt (1964), and Doppler redshifts were ruled out because of the absence of quasars with large blue shifts. Other strange phenomena were discovered, including the extreme radiation energy densities in their nuclei which would lead to the 'inverse-Compton catastrophe' (Hoyle et al., 1966). The extreme luminosities and short timescales of variability of quasars pointed inevitably to the presence of very compact luminous sources in the nuclei of galaxies and by 1967, this had been observed in the nuclei of luminous Seyfert galaxies (Fitch et al., 1967). Although widely discussed, the non-cosmological redshift hypothesis never attracted wide support—no other satisfactory explanation for the origin of the redshifts of the quasars was forthcoming.[7]

7.5.2 Advances in general relativity

In parallel with these great observational discoveries, dramatic progress was made in understanding the role of general relativity, not only in cosmology, but also in the physics of matter in strong gravitational fields (Israel, 1987). The exact solution of the Einstein's field equations for a point mass in general relativity had been discovered by Karl Schwarzschild (1916), the year after the final version of Einstein's General Theory was published. As Einstein commented on receiving Schwarzschild's paper,

> I had not expected that one could formulate the exact solution of the problem in such a simple way.

Schwarzschild derived the metric which bears his name in order to provide exact results for tests of general relativity. The *Schwarzschild metric* a point object of mass M can be written

$$ds^2 = \left(1 - \frac{2GM}{rc^2}\right) dt^2 - \frac{1}{c^2}\left[dr^2\left(1 - \frac{2GM}{rc^2}\right)^{-1} + r^2(d\theta^2 + \sin^2\theta \, d\phi^2)\right], \qquad (7.3)$$

where θ and ϕ are polar coordinates and r is a coordinate distance. Schwarzschild did not remark upon the fact that this metric contains two singularities, one at $r = 0$ and the other at coordinate distance $r = 2GM/c^2$. In fact, the second singularity is not a real physical singularity, but is associated with the particular choice of coordinate system in which the Schwarzschild metric was written, as was first demonstrated by Martin Kruskal in the mid-1950s, but only published in 1960. The singularity at $r = 0$ is, however, a real physical singularity in space-time.

If the discovery of quasars was symbolic of a turning point in modern astrophysics, the same could be said of the discovery by Roy Kerr in the same year of the *Kerr metric*, one of the most important exact solutions of Einstein's field equations (Kerr, 1963). It turned out that the Kerr solution describes the metric of space-time about a rotating black hole, a generalization of the Schwarzschild metric – the Kerr metric depends upon both the mass and angular momentum of the rotating black hole. In 1965, a further generalization of the Kerr metric was discovered by Ezra (Ted) Newman and his colleagues (1965) for the case of a system with finite electric charge by solving the combined Einstein and Maxwell field equations. Only later was it realized that the metric describes a rotating black hole with finite electric charge. In 1971, Brandon Carter showed that the only possible solutions for uncharged axisymmetric black holes were the Kerr solutions and in 1972 Stephen Hawking showed that all stationary black holes must be either static or axisymmetric so that the Kerr solutions indeed included all possible forms of black hole (Carter, 1971; Hawking, 1972).

These theorems led to important conclusions about the fate of collapsing bodies in general relativity. No matter how complex the object and its properties before collapse to a black hole, all multipole moments are radiated away in the process of formation of the black hole, leaving it with only mass, angular momentum and electric charge. This result is often referred to as the *no-hair theorem* for black holes and applies to isolated black holes. Although an isolated black hole cannot possess a magnetic dipole moment, magnetic fields can penetrate into the black hole provided they are firmly attached to the external medium. As shown by Kip Thorne and his colleagues (1986), electrodynamics in the vicinity of black holes can be precisely described by considering the surface of infinite redshift, or *event horizon*, to consist of a membrane with the resistivity of free space $Z = (\mu_0/\epsilon_0)^{1/2}$.

One of the key questions addressed by the relativists was whether or not there must be a physical singularity at the origin as a result of gravitational collapse. The problem was solved by Roger Penrose (1965) who showed quite generally that, once a surface from which light cannot escape outwards has formed, what is known as a *closed trapped surface*, there is inevitably a singularity inside that surface. The nature of the singularity is described by the Kerr metric. The same techniques were applied to the Universe as a whole by Hawking and Penrose (1969), who showed that, according to general relativity, it is inevitable that there exists a singularity at the origin of the Hot Big Bang models of the Universe, subject to some rather general physical conditions. They later extended their results to a much larger class of theories of gravity, the results of these endeavours being summarized in the book *The Large Scale Structure of Space-Time* by Hawking and George Ellis (1973). In non-technical language, applying their results to our Universe and subject to some rather general geometrical conditions, there is inevitably a singularity at the origin of the world model, provided gravity is attractive and the equation of state of the matter satisfies the energy condition $\varrho + p_i \geq 0$.

From the point of view of astrophysics, the most important results of the study of black holes, as they were named by John Wheeler (1968), concerned the behaviour of matter in the vicinity of the event horizon and the maximum amount of gravitational binding energy which could be released when matter falls into the black hole from infinity. The Schwarzschild radius, $r_g = 2GM/c^2$, of a spherically symmetric black hole is the surface of infinite redshift, meaning that radiation emitted from this surface is observed with infinite wavelength at infinity. This is the general relativistic version of the insight of John Michell (1767) that, for a massive enough star, the escape velocity exceeds the speed of light. The interpretation in general relativity is however quite different because the geometry changes on crossing the Schwarzschild radius, as can be appreciated by inspection of the metric (7.3). There is also a last stable circular orbit about a non-rotating black hole at radius $3r_g$—within this radius there are no stable circular orbits and the matter spirals inwards through the surface of infinite redshift, adding to the mass of the black hole. The maximum gravitational binding energy which can be released by infall onto a non-rotating (Schwarzschild) black hole from infinity is 5.72% of its rest mass energy.

Corresponding calculations were carried out for Kerr black holes. There is a maximum amount of angular momentum which the hole can possess $\mathscr{J}_{max} = GM^2/c$. For a maximally rotating black hole, the surface of infinite redshift shrinks to $r_g = GM/c^2$ and, in the case of corotating orbits, up to 42% of the rest mass energy of the infalling matter can be released. The rotational energy of the black hole can also be tapped, as demonstrated by Penrose (1969), amounting to a maximum of 29% of the rest mass energy of the black hole. These important results showed that the accretion of matter onto black holes is potentially an extremely powerful source of energy, very much greater than, for example, nuclear energy.

7.6   X- and γ-ray astronomy

7.6.1   The birth of X-ray astronomy

Immediately after the Second World War, those physicists and astronomers interested in ultraviolet and X-ray astronomy made the first observations from above the Earth's atmosphere.[8] The German V-2 rocket programme had made enormous strides in rocket technology during the War and many of the German scientists who had built them, led by Werner von Braun, as well as 300 box cars full of V-2 parts, were taken to the USA where they formed the core of the US Army's rocket programme. The US Army announced that these rockets would be available for scientific research.

One of the prime targets of the early rocket experiments was the ultraviolet and X-ray emission of the Sun.[9] The first successful rocket ultraviolet observations of the Sun were made in October 1946 by the group led by Richard Tousey at the Naval Research Laboratory (Baum et al., 1946). In September 1949, Herbert Friedman and his colleagues (1951) made the first successful X-ray observations of the Sun, confirming the expectation that the Sun's corona is very hot. These rocket experiments continued throughout the 1950s and elucidated many of the X-ray properties of the Sun.

The flights of Sputniks 1 and 2 in late 1957 and Yuri Gagarin's the orbital flight in 1961 came as a profound shock to the US administration who realized that the USA had fallen behind the USSR in space technology and therefore was strategically vulnerable. The US response was to set up NASA in July 1958 as a civilian organization to begin the process of catching up with the USSR. The American Science and Engineering (AS&E) group was set up in association with the Massachusetts Institute of Technology to work on military and civilian contracts.

The AS&E group led by Riccardo Giacconi developed plans for making astronomical observations in the X-ray waveband, although their theoretical calculations did not promise much success with the sensitivities available at that time. The first successful rocket flight took place in June 1962. In the five minutes of observing time above the Earth's atmosphere, Giacconi and his colleagues discovered an intense discrete source of emission in the constellation of Scorpius which became known as Sco X-1 (Fig. 7.7) (Giacconi et al., 1962). In addition, an intense background of X-rays was observed, remarkably uniformly distributed over the sky (Gursky et al., 1963). These observations were soon confirmed by other rocket flights by the AS&E group as well as by Friedman's group at NRL, which also discovered X-rays from the Crab Nebula (Bowyer et al., 1964). These were entirely unexpected discoveries.

The next decade saw a flurry of activity in which a dozen or more X-ray astronomy groups made numerous rocket flights of increasing sophistication. By 1967, more than thirty X-ray sources were known, including the detection of X-rays from a number of supernovae, the quasar 3C 273 and the radio galaxy M87. These pioneering observations provided tantalizing glimpses of the richness of the X-ray sky, but the picture was

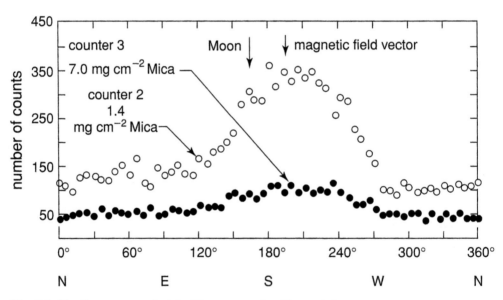

Fig. 7.7 *The discovery record of the X-ray source Sco X-1 and the X-ray background emission by Giacconi and his colleagues in a rocket flight of June 1962. The prominent source was observed by both detectors, as was the diffuse background emission (Giacconi et al., 1962).*

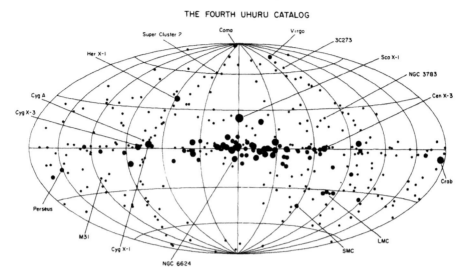

Fig. 7.8 *A map of the X-ray sky showing the objects contained in the fourth UHURU catalogue (Forman et al., 1978). The names of X-ray sources associated with well-known astronomical objects are indicated.*

confused. Some of the sources were highly variable, being present in one rocket flight and then disappearing on the next. These problems were resolved with the launch in December 1970 of the UHURU X-ray observatory, the first satellite dedicated to X-ray astronomy—it initiated the successful series of Explorer satellites sponsored by NASA. The UHURU observatory conducted the first survey of the complete X-ray sky and revealed the true nature of the X-ray population (Giacconi et al., 1971b). Some impression of the variety of sources present in the X-ray sky can be gained from the plot of the sources listed in the fourth UHURU catalogue (Fig. 7.8).

7.6.2   X-ray binaries, neutron stars, and black holes

In late 1970, the variable source Cygnus X-1 was observed by the UHURU Observatory and displayed somewhat random X-ray variability on timescales as short as 100 ms, indicating that the source region must be very compact (Oda et al., 1971; Rappaport et al., 1971). Just as remarkable were the observations of the source Centaurus X-3 (Cen X-3), first made in January 1971, which showed a clear periodicity with a pulse period of about 5 s, longer than that of any known radio pulsar. Furthermore, the pulsation period was not stable but seemed to vary with time (Giacconi et al., 1971a). In May 1971 it was found that the period of the X-ray pulsations varied sinusoidally with a period of 2.1 days, suggesting that the X-ray source was a member of a binary system. Then, on 6 May, the source disappeared, only to reappear half a day later. This pattern repeated roughly every two days. The X-ray source was being occulted by the primary star of the binary system (Schreier et al., 1972). With these clues, Wojciech Krzeminski identified the primary star with a massive blue star with the same binary period of 2.1 days as the X-ray source (Krzeminski, 1973, 1974). Soon after this discovery, another similar source was discovered (Tananbaum et al., 1972), the source Hercules X-1 (Her X-1) which had a pulse period of 1.24 s and an orbital period of 1.7 days (Fig. 7.9).

The short period of the X-ray source in Her X-1 was strong evidence that the parent body must be a neutron star, similar to those of the radio pulsars. The mechanism of energy production was immediately identified as accretion. Satio Hayakawa and Masaru Matsuoka (1964) had already proposed such a mechanism involving normal close binary systems and Shklovskii (1967) further proposed accretion onto a neutron star as the energy source for the luminous X-ray source Sco X-1. Kevin Prendergast and Geoffrey Burbidge (1968) made the important point that, in accretion from the primary star onto a compact secondary in a binary system, the accreted matter would necessarily acquire a considerable amount of angular momentum and so an *accretion disc* would form about the compact star. Accretion of matter from the primary star onto a compact neutron star is a very powerful energy source. When the effects of general relativity are taken into account, the upper limit to the energy release is 5.72% of the rest mass energy of the accreting matter onto a solar mass black hole, roughly an order of magnitude greater than can be liberated by nuclear fusion reactions.[10] The masses of the companion neutron stars could be estimated using the standard procedures of celestial mechanics. The masses of the seven binary X-ray sources for which this analysis could be carried out lay in the

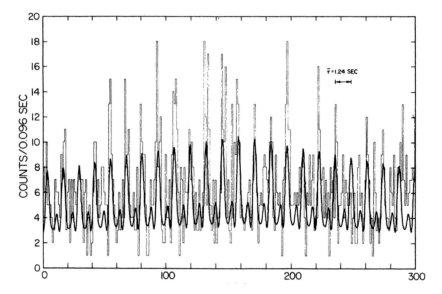

Fig. 7.9 *The discovery records of the pulsating X-ray source Her X-1. The histogram shows the number of counts observed in successive 0.096 s bins. The continuous line shows the best-fitting harmonic curve to the observations, taking account of the varying sensitivity of the telescope as it swept over the source (Tananbaum et al., 1972).*

range 1.2–1.4 $M_\odot$, consistent with the upper limit to the masses of neutron stars, which is similar to the Chandrasekhar limit for white dwarfs (Rappaport and Joss, 1983).

From these considerations, a standard picture for the pulsating binary X-ray sources was developed.[11] In the case of massive companions such as blue supergiant stars, the neutron star is embedded in a strong stellar wind and the matter within a certain radius of the neutron star is accreted onto it. In the low mass binary systems, mass transfer takes place through the process of *Roche lobe overflow*, in which the primary star fills its Roche lobe, the equipotential surface joining the two stars, and so matter attains a lower gravitational potential by collapsing to form an accretion disc about the neutron star.[12] The X-ray pulsations are attributed to accretion onto the poles of the rotating neutron star, the strong non-aligned magnetic field channelling the matter into the polar regions.

The first strong candidate for a *black hole companion* was found in the bright X-ray source Cyg X-1. In 1971, with an improved position provided by an MIT rocket flight, radio astronomers at the National Radio Astronomy Observatory in the USA and at the Westerbork Observatory in the Netherlands searched the field for a variable radio source which might be associated with Cyg X-1. These searches resulted in an accurate radio position which coincided with a 9th magnitude blue supergiant star (Braes and Miley, 1971; Hjellming and Wade, 1971). In the next year, Louise Webster and Paul Murdin (1972) and independently Thomas Bolton (1972) showed that the star was the primary star of a binary system with period 5.6 days. Assuming the mass of supergiant B star was greater than $10M_\odot$, the mass of the invisible companion had to be greater than 3 $M_\odot$, the most likely masses being 20 $M_\odot$ for the blue supergiant and 10 $M_\odot$ for the unseen companion. The latter mass far exceeded the upper limit for stability as a neutron star and so it had to be a black hole.

Over the succeeding years, many more examples of black holes in X-ray binary systems were discovered. The black hole candidates have quite different X-ray spectral and variability characteristics as compared with those in which pulsating X-rays are found and which are confidently associated with accreting neutron stars (Fig. 7.10) (Orosz, 2007; Orosz et al., 2007).

Following the launch of UHURU observatory, seven satellites with X-ray detectors were flown in the succeeding seven years. The next step was the development of imaging X-ray telescopes and this was achieved with the NASA HEAO-B satellite, named the *Einstein X-ray Observatory*. Launched in November 1978, it opened up the detailed study of the astrophysics of the classes of source so far detected with image quality of a few arcsec being achieved by the high resolution camera. Among the most significant findings was that essentially all classes of star can be X-ray emitters.

The Einstein Observatory was succeeded by the second of NASA's Great Observatories, the *Chandra X-ray Observatory*, launched in July 1999. In parallel the European Space Agency launched the *XMM-Newton X-ray Observatory* in December 1999, the acronym standing for X-ray Multi-Mirror Telescope. The primary objective of the latter mission was high sensitivity and high spectral resolution and this was achieved using a set 58 nested paraboloid-hyperboloid mirrors and high sensitivity CCD X-ray detectors.

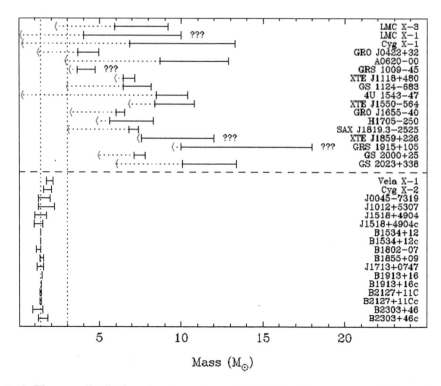

Fig. 7.10 *The mass distribution of neutron stars and black holes. Those systems which are known to possess neutron stars all have masses about $1.4M_\odot$. The masses of the black holes in X-ray binary systems for which good mass estimates are available all exceed the upper for neutron star masses of $M \approx 3M_\odot$ (Orosz, 2007; Orosz et al., 2007).*

Both missions provided astronomers with data of extraordinarily high quality for the study of all types of X-ray sources.

There remained the need to carry out surveys of the whole sky to understand in detail the nature of the population of X-ray sources. This objective was achieved by the German-led *ROSAT Observatory*, standing for *Röntgen Satellit*, launched in June 1990. After nine years in space, an all-sky catalogue of 150,000 objects was compiled, most of them belonging to an extragalactic population of faint X-ray sources.

7.6.3 γ-ray astronomy

Cosmic γ-rays were first detected in observations made by the Explorer II satellite in 1965 (Kraushaar et al., 1965), but this experiment did little more than show that there existed γ-rays which originated from beyond the Earth's atmosphere. The first significant astronomical observations were made by the third Orbiting Solar Observatory (OSO-III) launched in March 1967—γ-rays with energies $E_\gamma > 100$ MeV were detected from the general direction of the Galactic Centre (Clark et al., 1968). These were interpreted as the emission associated with the decay of neutral pions created in collisions between relativistic protons and the cold interstellar gas.

The Small Astronomical Satellite SAS-2 was launched in November 1972 and operated for only eight months but detected about 8000 γ-rays of cosmic origin (Fichtel et al., 1978). It confirmed that there is a general concentration of γ-rays towards the plane of the Galaxy. Discrete sources of γ-rays were present, two of them associated with the pulsars in the Crab and Vela supernova remnants. Evidence was also found for diffuse extragalactic γ-ray background radiation. The SAS-2 mission was followed in 1975 by the equally successful COS-B satellite launched by a European consortium, resulting in a detailed map of the Galactic plane as well as evidence for 24 discrete γ-ray sources, a number of them extreme examples of quasar phenomena (Mayer-Hasselwander et al., 1982).

In the early 1960s the Vela series of satellites were launched with γ-ray detectors on board to monitor the atmospheric nuclear test-ban treaties concluded between the USA and the USSR. To everyone's surprise, γ-ray bursts of astronomical origin were discovered, each γ-ray burst lasting typically less than a minute (Klebesadel et al., 1973). During that time, each burst was the brightest γ-ray source in the sky. The first of these had been detected by the Vela satellite in 1967, but was not reported in the open scientific literature until 1973.

The second of NASA's Great Observatories, the Compton Gamma Ray Observatory (CGRO), was successfully launched by the Space Shuttle in April 1991 and resulted in a definitive map of the whole sky in γ-rays. The map showed clearly the plane of our Galaxy and over 250 sources with photon energies greater than 100 MeV were discovered. The distribution of the γ-ray bursts was found to be isotropic over the sky and most extreme active galactic nuclei were established as the sources of intense, variable γ-ray emission.

In 1948, Patrick Blackett had realized that the speeds of the ultrarelativistic electrons and positrons were so close to the speed of light that they exceeded the local speed of light in the atmosphere $v = c/n$, where n is the refractive index of the atmosphere. As a result, the ultrarelativistic electrons and positrons should emit optical Cherenkov radiation (Blackett, 1948). This prediction was confirmed several years later by William Galbraith and John Jelley (1953, 1955). Exactly the same technique could be used to detect ultra-high energy γ-rays incident upon the top of the atmosphere.[13]

The breakthrough came with the development of imaging cameras which enabled weak γ-ray sources to be discriminated with high efficiency against the background of cosmic ray induced extensive air-showers. In 1989, the group at the Fred Lawrence Whipple Observatory used a 37-pixel camera to detect γ-rays from the Crab Nebula at the 9-sigma confidence level (Weekes et al., 1989). Two years later, with an upgraded camera with 109 pixels, the signals were detected at the 20-sigma level (Vacanti et al., 1991). These detector systems, known as atmospheric Cherenkov imaging telescopes, were successfully developed by a number of groups. Some of the most extreme active galactic nuclei were found to be sources of intense, variable ultrahigh-energy γ rays with $\varepsilon \sim 1$ TeV (Fig. 7.11).

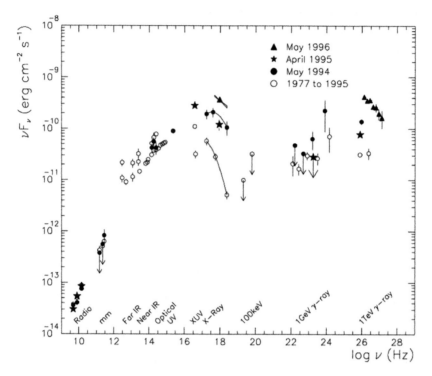

Fig. 7.11 *The energy spectrum, νF_ν of the extreme BL-Lac object Markarian 421 observed from the radio to the ultrahigh energy γ-ray waveband. The γ-ray source is highly variable. The lower energy γ-ray observations were made by the Compton Gamma-Ray Observatory and the high energy γ-ray points from observations by atmospheric Cherenkov imaging telescopes (Catanese and Weekes, 1999).*

7.7   Supermassive black holes in galactic nuclei

7.7.1   The Eddington luminosity

The case for the presence of black holes in active galactic nuclei can be appreciated by combining two results. The first is the expression for the *Eddington limiting luminosity*, L_{Edd},

$$L_{Edd} = 1.3 \times 10^{31} \, (M/M_\odot) \, \text{W}, \qquad (7.4)$$

the upper limit to the luminosity of a source of mass M—if the luminosity of the source were any greater, radiation pressure would blow the star apart or equivalently infalling matter would be blown away.

The second result is the *causality* relation, which states that variations in the intensity of a source of size r cannot be observed to take place on time scales less than r/c, the time it takes electromagnetic waves to cross the source. The smallest size which a source of mass M can have is roughly the Schwarzschild radius $r_g = 2GM/c^2$ and hence a lower limit to the timescale of variations from such a source is

$$T \geq r_g/c = 2GM/c^3 \approx 10^{-5} \, (M/M_\odot) \, \text{s}. \tag{7.5}$$

The combination of (7.4) and (7.5) can be used to illustrate how black holes can account naturally for rapid variations in the most extreme active galactic nuclei. For example, a $10^9 \, M_\odot$ black hole can have luminosity up to $\sim 10^{40}$ W and variations in this luminosity could occur on timescales as short as about 5 hours. Wandel and Mushotzky (1986) showed that, for a sample of X-ray variable quasars and Seyfert 1 and 2 galaxies, the masses inferred from the causality relation (7.5) were in reasonable agreement over a considerable range of black hole masses, $10^6 \leq M/M_\odot \leq 10^{10}$, with dynamical estimates based upon the breadth of the emission lines in their optical spectra. Typically, the luminosities amounted to about 1–10% of the Eddington limit.

7.7.2 Dynamical estimates of the masses of galactic nuclei

The simplest approach to making estimates of black hole masses is to consider test objects in bound orbits about the nucleus. Assuming a spherically symmetric mass distribution, $M(\leq r) = v^2 r/G$. A similar result is found for the velocity dispersion of stars in a star cluster or elliptical galaxy. Different approaches have been used to estimate the masses of compact objects in the nuclei of galaxies.

*M*87 (*NGC* 4486) is one of the nearest giant elliptical galaxies which displays many of the characteristics of the most extreme active galaxies. Young, Sargent, and their colleagues found that the velocity dispersion of stars increases towards the nucleus and, assuming that the velocity dispersion is isotropic, a central mass of $3 \times 10^9 \, M_\odot$ was inferred (Young et al., 1978; Sargent et al., 1978). Ford and his colleagues (1994) discovered a disc of ionized gas about the nucleus of M87 from HST imaging observations. Spectroscopy by Harms and his colleagues (1994) established that the disc is in Keplerian rotation about the nucleus, the inferred mass of the nuclear regions being $3 \times 10^9 \, M_\odot$, in agreement with the estimates of Sargent and his colleagues (1978).

*M*106 (*NGC* 4258), a nearby galaxy, is a member of a class of extragalactic maser sources known as *megamasers*, which can have luminosities up to about a million times those of galactic maser sources. Observations made with the Very Long Baseline Array (VLBA) operating at the H_2O maser wavelength of 1.3 cm by Miyoshi and his colleagues (1995) enabled the location and velocities of the H_2O masers to be determined with an angular resolution better than a milliarcsec. The mass of the central black hole was inferred to be $3.7 \times 10^7 \, M_\odot$ and the corresponding mass density greater than $4 \times 10^9 \, M_\odot \, \text{pc}^{-3}$. This figure is about 40,000 times greater than the densities of globular clusters.

The *black hole in the centre of our Galaxy* is the closest example of a supermassive black hole. The dynamical centre of our Galaxy coincides with the compact variable non-thermal radio source Sagittarius A* (Sgr A*). Diffraction limited observations at a

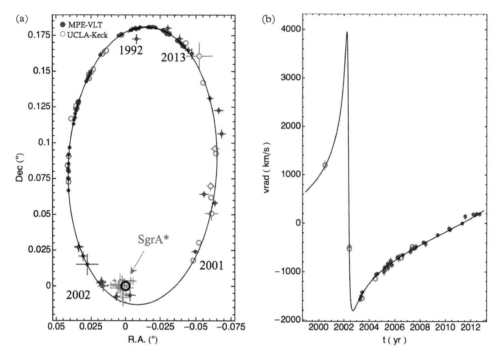

Fig. 7.12 *The orbit of the star S2 around Sgr A*, measured from 1992 to 2012. (a) The measured positions of S2 show a Keplerian ellipse. The data are from the New Technology Telescope (1992–2001), VLT (2002–2012) and the Keck Telescope (Gillessen et al., 2013). The black ellipse is the best-fitting orbit (Gillessen et al., 2009). (b) The corresponding radial-velocity data and best-fit relation.*

wavelength of 2 μm were made of the motions of stars about the galactic nucleus by Ghez and her colleagues (2000) with the Keck 10-metre telescope and by Reinhard Genzel and his colleagues with the VLT. These observations enabled them to trace the complete elliptical orbit of the star labelled S2 in Fig. 7.12 about the Sgr A* (Gillessen et al., 2013). The orbital period of S2 is 15.2 years and Sgr A* lies in one focus, as expected from Kepler's laws of planetary motion. The motions of numerous infrared stars enabled the mass distribution in the galactic nucleus to be determined. The inferred mass of the black hole is $4.31 \pm 0.06(\text{stat}) \pm 0.36(\text{syst}) \times 10^6 \, M_\odot$ and the lower limit to the mass density from observations of S2 is $10^{17} \, M_\odot \, \text{pc}^{-3}$. This is compelling evidence for the presence of a supermassive black hole in the centre of our Galaxy.

7.7.3   Black holes and spheroid masses

In their review of the determination of the masses of supermassive black holes, Kormendy and Richstone (1995) plotted the black hole masses against the absolute magnitude of the spheroids of their parent galaxies for nearby galaxies for which the mass estimates were reasonably secure. They found a correlation between these quantities, but warned of the many selection effects which might induce such a correlation.

Magorrian and his colleagues (1998) carried out a similar analysis using a larger sample of nearby galaxies and found a strong correlation. The mass estimates of the black holes and spheroid stellar masses were improved by Häring and Rix (2004) who showed that there is an almost linear relation between these quantities, $M_{\mathrm{bh}} \propto M_{\mathrm{bulge}}^{1.12 \pm 0.06}$. Typically, the black hole mass amounted to about 0.1–0.2% of the stellar mass of the bulge.

Understanding these relations is a challenging problem in the theory of galactic evolution. Apparently, the growth of supermassive black holes cannot be dissociated from the formation and evolution of the spheroids of galaxies themselves. In particular, active galactic nuclei must be involved in the feedback processes which are necessary to prevent the continued growth of giant elliptical galaxies to small redshifts.

7.8 Relativistic phenomena in galactic and extragalactic astrophysics

7.8.1 The inverse Compton catastrophe

The highly variable, non-thermal properties of active galactic nuclei posed intriguing astrophysical puzzles. As already mentioned, Hoyle, Burbidge, and Sargent (1966) showed that the quasars are susceptible to what is referred to as the *inverse Compton catastrophe*. The causality relation $r \leq c\tau$ sets an upper limit r to the size of the emitting region where τ is time scale of variability and hence a lower limit to the energy density of radiation which is assumed to be synchrotron radiation. There is then a critical brightness temperature of about 10^{12} K above which inverse Compton scattering is the dominant energy loss process and this was found to be the case in a number of sources. In his prescient paper of 1966, Martin Rees showed how these problems could be overcome if the source components moved out of the nuclear regions at relativistic speeds (Rees, 1966, 1967). Several relativistic effects contribute to the alleviation of the problems, particularly aberration effects which make the sources appear to be more luminous and larger than they are in their rest frames.

7.8.2 Radio jets and radio hot-spots

The first high resolution radio maps of double extragalactic radio sources showed that there are the 'hot-spots' towards the leading edges of the double radio structures with high energy densities in relativistic particles and magnetic fields (Fig. 7.4(b)). The synchrotron lifetimes of the ultrarelativistic electrons in the most compact hot-spots were less than the age of the source and so the electrons had to be continuously accelerated or replenished. Furthermore, the energy densities of the high energy particles and magnetic fields were too great to be confined within the hot-spots. They are temporary phenomena which need to be continuously supplied with energy from the active galactic nucleus by beams or jets of particles (Longair et al., 1973). Direct evidence for these jets was subsequently found in the radio maps of many radio galaxies and radio quasars, including Cygnus A (Fig. 7.4(b)). Variants of this model were proposed by Rees (1971) and by Scheuer (1974). The generic picture of continuous-flow models remained the preferred picture and self-similar, gas dynamical analytic models were derived by Scheuer and by Christian Kaiser and Paul Alexander (1997).

One of the difficulties associated with the presence of high energy electrons in many different astronomical environments is the fact that, as a gas of relativistic electrons and its frozen-in magnetic field expand, energy is lost adiabatically. A popular mechanism proposed by Enrico Fermi involved stochastic acceleration of particles in elastic collisions with interstellar clouds (Fermi, 1949), but this *Fermi acceleration* process was second-order in the ratio of the velocity of the clouds to the speed of light and was very slow. In 1977 and 1978, a number of independent authors showed that *first-order Fermi acceleration* in strong shock waves is a remarkably effective means of accelerating particles and creating a power-law energy spectrum (Axford et al., 1977; Krymsky, 1977; Bell, 1978; Blandford and Ostriker, 1978). The high energy particles are scattered back and forth across the shock wave and, in each crossing in either direction, the particles obtain a first-order fractional increase in energy of order v/c where v is the velocity of the shock wave. Particles are lost from the accelerating region by the bulk motion of the material behind the shock downstream. These two competing effects lead to the formation of a power-law energy spectrum of the form $N(E)\,dE \propto E^{-2}\,dE$. The attraction of this process is that it only depends upon the particles encountering a strong shock wave. In supernova remnants and at the interface between jets and the interstellar or intergalactic media, this mechanism results in acceleration of high energy particles where they are needed.

The origin of the magnetic fields in supernova remnants was solved by Stephen Gull (1975) who showed that an expanding supernova shell suffers Rayleigh–Taylor instabilities as it decelerates. The instability results in turbulent motions at the contact discontinuity between the expanding sphere of hot gas and the shocked interstellar material. The turbulent eddies can stretch and wind up the magnetic field until its energy density is roughly the same as that in the turbulent motions.

7.8.3   Superluminal radio sources

Very long baseline interferometric observations of structures of the compact radio cores of radio quasars and radio galaxies by Marshall Cohen and his colleagues provided clear evidence that on the scale of milliarcseconds these were not stationary but that the source components appeared to be moving apart at speeds exceeding the speed of light. In the case of 3C 273, for example, the fainter source component in the nucleus appeared to move a distance of about 25 light years between 1977 and 1980, implying an apparent separation velocity of about ten times the speed of light (Pearson et al., 1981, 1982). Subsequent observations showed that the phenomenon is common among radio sources with compact radio cores.

The explanation of this phenomenon had already been already described by Rees (1966, 1967). If a source component is ejected at a relativistic speed from the nucleus at a small angle to the line of sight, the component can have an apparent velocity perpendicular to the line of sight of up to γv where $\gamma = (1 - v^2/c^2)^{-1/2}$ is the Lorentz factor. In addition, the luminosity of the approaching component is enhanced by the effects of aberration and this accounts for the one-sidedness of many of the jets observed in active galactic nuclei. These superluminal motions are assumed to be associated with the jets responsible for fueling the hot-spots and extended radio structures.

In 1994, relativistic jets were found in galactic binary X-ray sources in which there is evidence for the presence of a black hole. Felix Mirabel and Luis Rodrigues (1994) observed the intense hard X-ray source GRS 1915+105 during a radio outburst and discovered that its radio components were separating from the radio core at 1.25 times the speed of light. Mirabel and Rodrigues pointed out the remarkable similarities between the radio properties of these source, which they term *microquasars*, and the radio galaxies and radio quasars (Mirabel and Rodrigues, 1998).

7.8.4   The origin of jets in active galactic nuclei

The origin of the relativistic jets observed in active nuclei proved to be a difficult problem. One possibility is that they are associated with the funnels which are formed along the rotation axes of thick accretion discs, but these are thought to be unstable configurations. Other models ascribe the jets to electromagnetic processes occurring close to the black hole itself and, in the case of the rotating black holes, there is a preferred axis along its rotation axis. In the model considered by Rees and his colleagues (Rees, 1971, 1976), electromagnetic torques are used to extract the rotational energy of a rotating black hole resulting an outflow parallel to its rotation axis.

The most promising models associate the initial collimation of the jets with the winding up of magnetic fields lines frozen into the ionized gas in the accretion disc about a rotating black hole. Richard Lovelace and his colleagues (Lovelace and Romanova, 2003) carried out analytic and numerical simulations of the force-free magnetic field structures which arise as an initial dipolar magnetic field threading the accretion disc is wound up. They found solutions in which 'magnetic bubbles' are expelled at relativistic speeds along the rotation axis of the disc. They suggested that even the most extreme relativistic jets could be accounted for by these models.

7.8.5 γ-ray sources and γ-ray bursts

Among the most luminous γ-ray sources were those associated with those radio quasars which exhibited superluminal motions (Kniffen et al., 1994). The γ-ray luminosities were so extreme and the timescales of variability so short that the energy density of γ-rays was very large, so large that the γ-rays would be degraded into electron–positron pairs through the pair-production process. The importance of this process is described by the *compactness factor* C, defined as $C = L_\gamma \sigma_T / 4\pi m_e c^4 t$, where L_γ is the γ-ray luminosity at 1 MeV, σ_T is the Thomson cross-section and t is the timescale of variability. If the compactness factor is greater than one, the γ-rays cannot escape from the source. If it is assumed that the source components are relativistically beamed however, the γ-ray luminosity is enhanced by a factor of roughly κ^5 where $\kappa = \gamma[1 + (v/c)\cos\theta]$ and θ is the angle between the axis of ejection and the line of sight. As a result, the γ-rays can survive without suffering degradation by electron–positron pair production, consistent with the observation of superluminal motion of their radio components.

The Burst and Transient Source Experiment (BATSE) of the Compton Gamma-Ray Observatory detected 2704 γ-ray bursts and established that they are uniformly distributed over the sky. The bursts were of very short duration and the angular resolution of the BATSE instrument only a few degrees so that there was not time or adequate positional accuracy to make secure identifications of the sources. In 1993, Peter Mészáros and Rees (1993) argued that, although the intense γ-ray emission only lasted a short time, the emission would remain observable at lower energies for very much longer— an *afterglow* would appear progressively in the X-ray, then optical, infrared, and radio wavebands.

In 1997, the X-ray telescope of the Italian–Dutch BeppoSAX satellite was pointed at the position of the γ-ray burst GRB 970228 within 8 hours of the event having taken place and its X-ray afterglow discovered (Costa et al., 1997). Subsequently, afterglows were observed throughout the electromagnetic spectrum. These observations enabled the burst to be associated with a very faint galaxy (Sahu et al., 1997). Within four years, over 40 γ-ray burst afterglows were detected and 30 of these could be associated with a distant host galaxy. These observations demonstrated that the γ-ray bursts are extreme events occurring in galaxies at cosmological distances, spanning essentially the same redshift range as the most distance galaxies and quasars. Thus, the GRBs provide a means of probing the large redshift Universe, particularly the epoch of reionization.

The very short timescales of variability of their energy release indicated that the sources must originate from regions less than about 100 km in size on timescales much less than a second. Hence the bursts must involve stellar mass objects and huge energy releases, typically 10^{47} J if the emission were isotropic. Furthermore, the energy density in γ-rays in the source region would be so great that, just as in the extreme γ-ray sources, the γ-ray bursts must involve extreme relativistic bulk motions.

Again, many of the results derived by Rees (1966, 1967) could be applied directly to the observed emission of the γ-ray bursts. In the simplest picture, the source region can be taken to be a *relativistic fire-ball,* meaning a relativistically expanding sphere which heats the surrounding gas and drives a relativistic shock wave into it. Relativistic beaming alleviates the problems of $\gamma\gamma$ annihilation, but does not solve the energy problem. The fireball need not, however, be isotropic. In the case in which the energy of the γ-ray burst is emitted in a narrow collimated beam, the energy requirements would be very significantly reduced to a value typical of the energy released in the core collapse of a supernova explosion. The supernova–γ-ray burst association was established with certainty for the γ-ray burst GRB 030329 observed on 29 March 2003 for which the characteristic broad lines of an extremely energetic supernova were observed within days of the event (Hjorth et al., 2003).

7.9 The environments of black holes

7.9.1 Black holes in binary systems and active galactic nuclei

Immediately following the discovery of quasars, a plethora of models appeared, all of them attempting to account for their huge variations in luminosity on short timescales. Yakov Zeldovich (1964a) in Moscow and Edwin Salpeter (1964) at Cornell independently pointed out that *accretion of matter onto black holes* is potentially a very powerful energy source. An effective way of releasing the binding energy of the infalling material is through the formation of an *accretion disc* about the black hole, just as in the case of the binary X-ray sources (Section 7.6.2). The presence of viscous forces in the disc transfers angular momentum outwards so that the matter can gradually drift inwards. At the same time, the frictional forces heat up the material of the disc and this is the means by which the matter releases its gravitational binding energy.

Donald Lynden-Bell (1969) showed how, in principle, black holes in active galactic nuclei could account for the properties of the most extreme active galactic nuclei. James Bardeen (1970) then pointed out that the black hole is likely to possess angular momentum as the infalling matter brings angular momentum with it with a correspondingly greater energy release, up to 42% of the rest mass energy of the infalling matter for a maximally corotating Kerr black hole.

The luminosities of the binary X-ray sources could be naturally accounted for by accretion onto neutron stars and, in a number of cases, these approached the Eddington limiting luminosity (equation (7.4)), indicating just how effective this energy source could be (Margon and Ostriker, 1973). Detailed models of accretion discs about black holes were published by Nikolai Shakura and Rashid Sunyaev (1973) and they showed that many of the properties of thin accretion discs are independent of the specific form of the viscosity. Balbus and Hawley (1991) then discovered a physical realization of the origin of the viscosity as a result of magnetohydrodynamic instabilities if the accretion disc is permeated by a magnetic field frozen into the ionized plasma. Accretion models for active galactic nuclei fuelled by accretion discs became the preferred model for active galactic nuclei.

7.9.2 X-ray observations of fluorescence lines in active galactic nuclei

The behaviour of matter in the strong gravitational fields close to black holes can be studied using X-ray spectral observations of active galactic nuclei. Key observations are the asymmetric iron fluorescence lines observed in the X-ray spectra of Seyfert 1 galaxies, the most convincing case being the galaxy MCG -6-30-15. There are significant variations in its X-ray flux density on the timescale of hours, indicating that the emission originates very close to the nucleus itself. The favoured picture involves a warm accretion disc about the black hole embedded in a hot X-ray emitting halo which is the source of the X-ray continuum radiation. The X-ray line is attributed to the fluorescent emission of the ions present in the accretion disc.

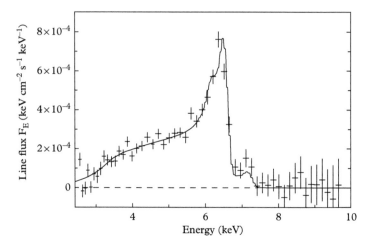

Fig. 7.13 *The broad asymmetric iron fluorescent line as observed in a long observation of the Seyfert 1 galaxy MCG -6-30-15 by the EPIC instrument of the XMM-Newton Observatory (Fabian et al., 2002). The best-fit spectrum corresponds to the disc having an angle of inclination $i = 30°$ and an inner radius $r = 2GM/c^2$.*

Matt, Fabian, and Reynolds (1997) showed that the 6.4 keV fluorescent line of iron is by far the strongest line and acts as a tracer of the velocity field in the accretion disc. In the innermost regions of the disc, special and general relativistic effects are large and strongly influence the shape of the line profile. Figure 7.13 shows the profile of the 6.4 keV line in the spectrum of MCG -6-30-15 as observed in a long spectroscopic integration by the XMM-Newton satellite (Fabian et al., 2002). There is a rather abrupt cut-off at energies greater than 6.7 keV and an extension of the line to energies less than 4 keV at the low energies. This asymmetry occurs naturally if the thin accretion disc extends inwards towards the last stable orbit about the black hole. The broadening of the line is due to the combined effect of the gravitational redshift which shifts the spectrum to lower X-ray energies and the Doppler effect of photons emitted from a source moving at an angle to the line of sight. The asymmetric profile of the 6.4 keV line has a natural interpretation if the lowest energy parts of the line originate close to the last stable orbit of an accretion disc about a massive black hole and the disc is observed more or less face-on.

7.9.3   Unified models for active galaxies

The concept of unification resulted from the realization that projection effects must be important in determining the observed properties of active galactic nuclei. The significance of projection effects in distinguishing between Seyfert 1 and Seyfert 2 galaxies was first demonstrated by Robert Antonucci and Joseph Miller (1985) in their study of the nearby Seyfert 2 galaxy NGC 1068.[14] When observed spectropolarimetrically, the polarized line emission was found to be as broad as the broad permitted lines seen in

Seyfert 1 galaxies. They interpreted these observations in terms of a model in which there is an 'obscuring torus' about the nucleus so that, when the galaxy is observed at a small angle to the axis of the torus, the nuclear regions are observed in scattered light and the characteristic Seyfert 1 broad-line regions and strong blue and UV continuum are revealed. If the axis of the torus is observed at a large angle to the line of sight, the nuclear regions are obscured and only the narrow-line regions, which are located much further out from the nucleus than the broad-line regions, are observed in direct light. The obscured central regions can, however, be observed in the light reflected from clouds outside the region of the torus, perhaps by gas and dust in the narrow-line regions.

The importance of projection effects in defining the observed properties of double extragalactic radio sources was advocated by Peter Barthel (1989, 1994). The principal arguments concerned the observation of superluminal motions in the radio quasars, suggesting that they are observed at small angles to the line of sight, the one-sided jet emission observed in the radio quasars, but not in the radio galaxies, the relative sizes of the double radio structures of the quasars and radio galaxies and so on.[15] Radio quasars are observed when the nucleus is observed within a cone of half-angle roughly 45° and a radio galaxy observed when the nuclear regions are hidden by the obscuring torus. It is natural to attribute the extreme *BL-Lac* or *blazar* phenomena to radio sources in which the relativistic jets point almost precisely along the line of sight to the observer. In a few blazars, high dynamic range radio observations enabled the underlying double radio source to be observed. Another unification scenario concerns the relation between the BL-Lac objects and radio galaxies. A good case can be made for BL-Lac objects being the relativistically beamed jets originating in the nuclei of lower luminosity radio galaxies (Urry and Padovani, 1994).

7.10   The cosmic evolution of active galaxies

7.10.1   The counts of radio sources

The early history of evidence for the strong cosmological evolution of the extragalactic radio source population was recounted in Section 6.6.1. By the mid-1960s, it was known that the majority of these sources were at large redshifts and that the source population had evolved very strongly with cosmic epoch (Davidson and Davies, 1964; Longair, 1966b). There was a dramatic increase in the comoving space density of powerful radio sources out to redshift $z = 2$, beyond which the rise could not continue because of the convergence of the number counts and the upper limit to the integrated background emission due to discrete sources.

When redshift data for complete samples of the radio quasars in the 3CR samples became available, Schmidt (1968) and Rowan-Robinson (1968) showed that these sources were concentrated towards the limits of their observable volumes. The values of $\langle V/V_{max} \rangle$ for the quasars was found to be 0.686 ± 0.042, significantly greater than the expectation for a uniform population of 0.5.[16] By the early 1990s the radio galaxies in the 3CR sample too were all identified and their redshifts measured. Their V/V_{max}

distribution and $\langle V/V_{\mathrm{max}} \rangle = 0.697 \pm 0.031$ were exactly the same as those of the radio quasars.

Over the succeeding years, radio source counts were determined over the entire radio waveband from low radio frequencies to short centimetre wavelengths (Fig. 7.14). They all show the same overall features—a steep source count at high flux densities, a plateau at intermediate flux densities and convergence at low flux densities. At the very lowest flux densities, $S \leq 10^{-3}$ Jy, the source counts flatten again.

By the mid-1990s, the changes in the source population out to redshifts $z \approx 2$ could be conveniently described by *luminosity evolution* in which the radio luminosity function of the powerful radio sources is shifted to greater radio luminosities with increasing redshift as $L(z) = L_0(1+z)^3$, whilst the normal galaxy radio luminosity function remains unchanged. The dramatic changes in the luminosity function cannot

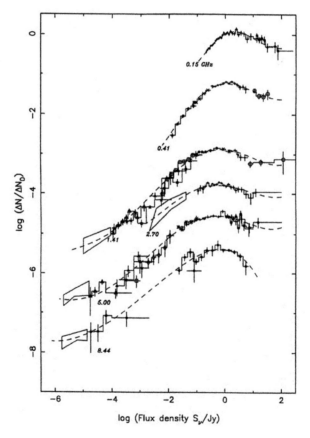

Fig. 7.14 *The differential, normalized counts of extragalactic radio sources at a wide range of frequencies throughout the radio waveband. This compilation was kindly provided by Jasper Wall (Wall, 1996). The points show the number counts derived from surveys of complete samples of radio sources. The boxes indicate extrapolations of the source counts to very low flux densities using the P(D) technique.*

continue to redshifts $z \gg 2$. In a typical model, the luminosity functions at $z = 4$ have declined from the maximum at $z \approx 2 - 3$ (Dunlop and Peacock, 1990). These results show that the radio source and associated black hole activity was at its greatest at redshifts $z \sim 2$ and that this activity decreased rather dramatically at earlier and later cosmological epochs.

7.10.2   Radio-quiet quasars

The radio-quiet quasars are about 50–100 times more common than the radio-loud variety (Sandage, 1965). The early surveys established that they also have a steep source count and values of $\langle V / V_{\mathrm{max}} \rangle$ much greater than 0.5 (Bracessi et al., 1970; Schmidt and Green, 1983), a similar evolutionary behaviour to that of the radio-loud quasars.

The construction of complete samples of optically-selected quasars was challenging because they are rare objects which must be distinguished among the very much more populous normal stars. A composite quasar spectrum spanning the wavelength range 100–600 nm is shown in Fig. 7.15, illustrating the underlying broad-band continuum spectrum, as well as the presence of strong broad emission lines, which can strongly influence their colours (Francis et al., 1991). Various features of such quasar spectra can be used to distinguish them in large scale photographic and spectroscopic surveys. These techniques include:

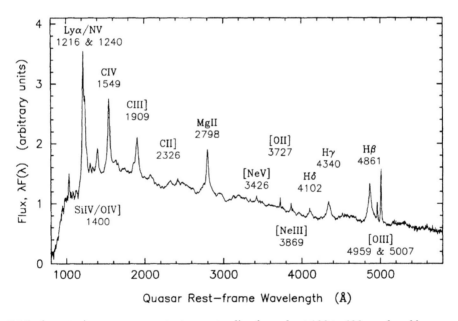

Fig. 7.15 *A composite mean quasar spectrum extending from about 100 to 600 nm, found by averaging over 700 quasars from the Large Bright Quasar Survey of Francis and his colleagues (1991). The spectrum is plotted in units of λF_λ. The spectrum contains a number of the common emission lines found in quasars superimposed upon a strong non-thermal continuum. The flux scale is arbitrary.*

- *Ultraviolet excesses* The UV-optical continuum spectra of quasars show a strong ultraviolet excess as compared with normal stars. This technique was notably successful in discovering quasars with redshifts $z \leq 2.2$, for example, by Schmidt and Green (1983) who derived a complete sample of 114 bright radio-quiet quasars over about a quarter of the whole sky. The same technique was also exploited in the 2dF Quasar Redshift Survey carried out at the Anglo-Australian Telescope.

- An extension of this technique involved the use of multicolour photometry. David Koo and Richard Kron (1982) used (U, J, F, N) photometry to find radio quiet quasars with ultraviolet excesses to B = 23 and found the first evidence for the convergence of the counts of radio-quiet quasars at faint magnitudes. Stephen Warren and his colleagues found the first quasar with redshift $z > 4$. The largest sample of quasars to date has been found from images taken through the *ugriz* filters as part of the Sloan Digital Sky Survey (SDSS) (Richards et al., 2006).

- The extension of this colour selection technique to even larger redshifts has been carried out using SDSS data by Fan and his colleagues (2001, 2004). They discovered nine quasars with redshifts $z > 5.7$, the largest redshift being 6.28.

- Another approach is to make use of the fact that the Lyman-α and CIV emission lines are always very strong in the spectra of quasars and are superimposed upon a roughly power-law continuum energy distribution. Osmer (1982) pioneered the use of a dispersion prism, or grating, in conjunction with a wide-field telescope to discover quasars with redshifts $z > 2$, at which these lines are redshifted into the optical waveband.

The changes of the optical luminosity function of optically selected quasars with cosmic epoch was derived by Boyle and his colleagues (2000) out to redshift $z \sim 2.3$ using a complete sample of 6,000 quasars from the AAT 2dF Quasar Redshift survey (Fig. 7.16). At zero redshift, the luminosity function of the overall quasar population joins smoothly on to that of the Seyfert galaxies. Over the redshift range $0 \leq z \leq 2.3$, the changes in the luminosity function can be described by 'luminosity evolution', of almost exactly the same form as that needed to account for the evolution of the radio luminosity function for luminous extragalactic radio sources, namely, $L(z) \propto (1+z)^\beta$ out to $z = 2$ with $\beta = 3.5$.

Systematic surveys were undertaken to determine the very large redshift evolution of the radio-quiet quasar population by a number of authors (Schmidt et al., 1995; Fan et al., 2001, 2004). In the SDSS observations, the five colour selection technique enabled quasars up to $z = 5$ to be studied. The comoving number density of luminous quasars peaks between redshifts 2 and 3 and then converges rapidly at larger redshifts (Richards et al., 2006), in agreement with earlier studies (Fig. 7.17). This figure also includes the results of Fan and his colleagues, who used the *i*-band drop-out technique to discover quasars with redshifts greater than $z = 5.7$. These observations showed that the decline in the comoving space density of luminous quasars continues out to redshift $z = 6$.

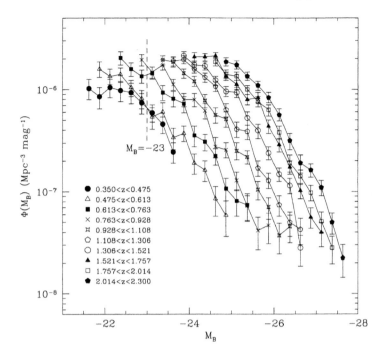

Fig. 7.16 *The evolution of the optical luminosity function for 6,000 optically selected quasars in the redshift range* $0.35 \leq z \leq 2.3$ *observed in the 2dF Quasar Redshift survey carried out at the Anglo-Australian Telescope (Boyle et al., 2000).*

7.10.3   Counts of extragalactic X-ray sources and the X-ray background

The first deep surveys of the X-ray sky were carried out by the Einstein X-ray Observatory in 1979 (Giacconi et al., 1979), one of the prime motivations being to understand the origin of the X-ray background radiation discovered in Giacconi's pioneering rocket flight in 1962. They found that about 26% of the background intensity could be attributed to discrete X-ray sources, most of which were associated with active galaxies and quasars (Hasinger et al., 1993). Definitive evidence for cosmological evolution of the population of X-ray sources was derived from the deep surveys carried out by the ROSAT satellite. A survey of the whole sky in the X-ray energy band 0.1 to 2.4 keV resulted in a catalogue of about 60,000 sources and information about their X-ray spectra in four X-ray 'colours'.

The resulting differential, normalized X-ray source counts derived by Günther Hasinger and his colleagues bears a strong resemblance to the differential counts of radio sources (Hasinger et al., 1993). At high X-ray flux densities, $S > 3 \times 10^{-14}$ erg s^{-1} cm^{-1}, the differential counts have slope $(\beta + 1) = 2.72 \pm 0.27$ and below this flux density the counts converge with source count slope $(\beta + 1) = 1.94 \pm 0.19$. The identifications

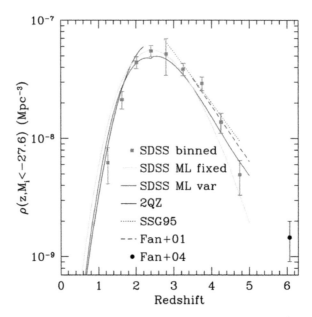

Fig. 7.17 *The integrated i-band luminosity function for quasars more luminous than $M_i = -27.6$ (Richards et al., 2006). The solid black line terminating at $z \approx 2$ is from the 2dF Quasar Redshift survey (Boyle et al., 2000). The dashed and dotted lines are from Fan and Schmidt and their colleagues (Fan et al., 2001; Schmidt et al., 1995) respectively. The point at $z \sim 6$ is the comoving space density estimated by Fan and his colleagues (Fan et al., 2004).*

of sources in the medium deep survey were consistent with a picture in which the X-ray sources follow the same type of cosmological evolutionary behaviour as the radio galaxies, radio quasars and optically selected quasars.

The next major advances were made by the Chandra and XMM-Newton X-ray Observatories. Deep, wide-field surveys with these telescopes enabled number counts of X-ray sources to be determined to faint X-ray flux densities throughout the 0.5–10 keV wavebands (Brandt and Hasinger, 2005). Major efforts were made to identify and obtain optical spectra for the X-ray sources in the Chandra surveys resulting in estimates of the evolution of the comoving space density of sources in the 0.5–2 keV and 2–10 keV wavebands. Both show the characteristic feature of the evolution of the population of active galactic nuclei—the most luminous sources show the strongest cosmological evolution out to redshifts $z \approx 2$ while the less luminous sources show much less dramatic evolution.

The X-ray spectral types change significantly with X-ray energy. Essentially all the background intensity in the waveband 0.5–2 keV is associated with active galactic nuclei in which the characteristic broad lines and continuum are observed, meaning that the nuclear regions of the active galaxies are observed directly. At hard X-ray energies ($\varepsilon \sim 10$ keV), at least 50% and probably more of the background is associated with the discrete sources already detected, but their spectral mix is somewhat different. The most important difference is that there is a major population of sources which are strongly absorbed at soft X-ray energies by photoelectric absorption, strongly suppressing the soft X-ray flux density and so hardening the X-ray spectra. The inferred column densities causing the absorption can be up to 10^{24} cm^{-2} and greater. For column depths greater than 10^{24} cm^{-2}, the effects of Comptonization cause major distortions of the X-ray source spectra whatever the initial input photon spectrum (Gilli et al., 2007). The absorption is presumed to be associated with the presence of obscuring tori about the active galactic nuclei.

At energies greater than 10 keV, the X-ray background spectrum continues to be remarkably flat up to about 30 keV beyond which it changes slope rather dramatically and continues more or less as a power-law to γ-ray energies. Comptonized sources are assumed to make a significant contribution to the background spectrum at energies greater than $\varepsilon = 10$ keV.

7.10.4 Relation to galaxy evolution in the 'thermal' wavebands

In parallel with the major advances in understanding active galactic nuclei, black holes and their evolution with cosmic epoch, enormous strides were made in understanding the astrophysics of galaxies, clusters and the intergalactic medium which provide the environments in which relativistic astrophysics takes place. These topics are dealt with extensively in Section 10.16. A few examples include the following:

- *Counts of galaxies* (Section 10.16.1) In the infrared K waveband (2.2 μm), the counts follow reasonably closely the expectations of uniform world models. In contrast, in the R, B, and U wavebands, there is a large excess of faint galaxies, particularly in the shortest wavelength bands. In the ultra-deep field (UDF) images taken with the Hubble Space Telescope, much of the excess is associated with irregular/peculiar/merger systems. In general, the faint blue objects are about an order of magnitude less luminous and smaller in physical size than nearly galaxies.

- *Mid- and far-infrared Universe* (Section 10.16.2) The IRAS Satellite surveys between 12.5 and 100 μm showed that there were more faint IRAS galaxies than expected, in the same sense as the counts of radio sources, X-ray sources, and quasars. More recently, number counts in these wavebands have been made by the Spitzer Space Telescope at 24, 70, and 160 μm and all show a clear excess of faint sources, as well as a rather dramatic cut-off at low flux densities. The counts are dominated by starburst activity (Babbedge et al., 2006).

- *Submillimetre number counts* (Section 10.16.3) The number counts greatly exceed the numbers of sources expected on the basis of the local 60 μm luminosity function of luminous far-infrared galaxies determined by the IRAS survey. The number counts of submillimetre galaxies must converge rapidly just below 0.1 mJy or the submillimetre background intensity would be exceeded.

The significance of these and other observations are discussed in Section 10.16.

7.10.5   The formation of supermassive black holes

Possible routes to the formation of supermassive black holes in active galactic nuclei are summarized in the diagram presented by Rees (1984) in his review of the physics of the black holes in the nuclei of galaxies (Fig. 7.18). As Rees emphasized, these are plausible scenarios but each involves astrophysical process which are poorly understood. For example, star formation is still quite poorly understood and empirical relations have to be used to relate the star formation rate to the ambient gas density. The liberation of gas from the debris of the star formation and of supernova explosions certainly contribute

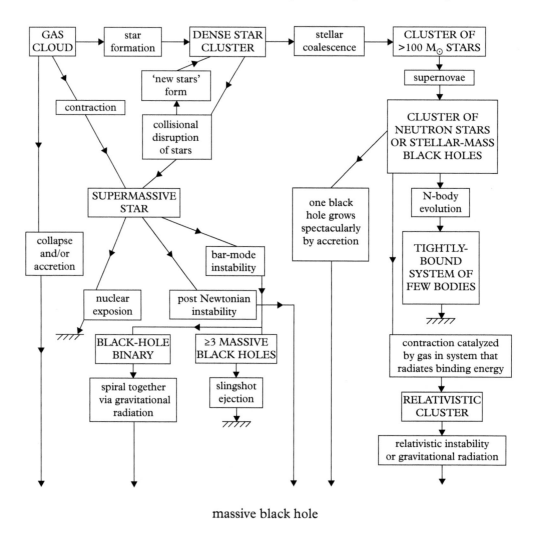

massive black hole

Fig. 7.18 *The 'Rees diagram' illustrating various possible pathways to the formation of supermassive massive black holes in the nuclei of galaxies (Rees, 1978).*

to fuelling the central regions about the black hole but quantifying these processes is a real challenge. The rates of coalescence of binary compact stars is not well known.

There is no doubt, however, that the above processes take place and can contribute to the growth of the massive black holes in active galactic nuclei. The accretion of matter onto black holes in the nuclei of galaxies undoubtedly takes place. Plentiful gas is available in the vicinity of the nuclear regions to fuel the black holes. The coalescence of black holes has now been observed through the spectacular results of the LIGO experiments. It is still too early to claim that the sequence of events is understood. How these processes can be modelled is discussed in more detail in Section 10.18 in the context of semi-empirical models of galaxy formation and evolution.

7.11   Gravitational waves

7.11.1   Radio pulsars and tests of general relativity

The radio pulsars turned out to provide some of the very best tests of general relativity. Observations by Joseph Taylor and his colleagues made with the Arecibo radio telescope demonstrated that the arrival times of the radio pulses from pulsars are the among the most stable clocks available to us (Taylor, 1992). Among the most important systems from the point of view of testing general relativity are the binary pulsars in which both stars are neutron stars in a close binary orbit. The first of these, the pulsar PSR 1913+16, was discovered by Russell Hulse and Taylor in 1974 (Hulse and Taylor, 1975) and it has been observed with very precise timing since then. The system has a binary period of only 7.75 hours and the orbital eccentricity is large, $e = 0.617$. Various parameters of the binary orbit can be measured very precisely and these provide estimates of different quantities which involve the masses of the two neutron stars, M_1 and M_2, in different ways. Assuming general relativity is the correct theory of gravity, the mass estimates of the two neutron stars are the most accurate values for any stars, besides the Sun, $M_1 = 1.4411 \pm 0.0007 \, M_\odot$ and $M_2 = 1.3873 \pm 0.0007 \, M_\odot$.

Knowing the masses of the neutron stars and their orbits, the rate of loss of rotational energy by gravitational radiation can be predicted precisely. The change of orbital phase of the system PSR 1913+16 has been observed since its discovery in 1974 and the observed changes agree precisely with the predictions of general relativity (Fig. 7.19) (Will, 2006). This result enabled a wide range of alternative theories of gravity to be excluded. For example, the gravitational waves derived from standard general relativity are quadrupolar in nature and any theory which, say, predicted the emission of dipole or scalar gravitational radiation can be excluded.[17]

The same techniques of accurate pulsar timing can also be used to determine whether or not there is any evidence for the gravitational constant varying with time (Taylor, 1992). For a range of plausible equations of state, the limits to $\dot{G}/G$ are less than about 10^{-11} year^{-1}. Thus, there can have been little change in the value of the gravitational constant over cosmological timescales (see Chapter 4).

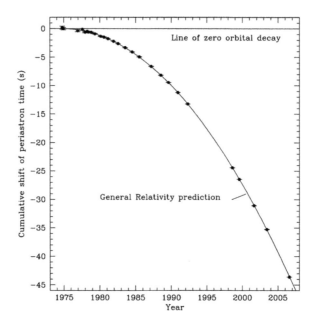

Fig. 7.19 *The change of orbital phase as a function of time for the binary neutron star system PSR 1913+16 compared with the expected change due to gravitational radiation energy loss by the binary system (solid lines) (Taylor, 1992; Will, 2006). The change in angular momentum due to gravitational radiation loss is* $-\mathrm{d}\Omega/\mathrm{d}t \propto \Omega^5$, *as shown by the solid line.*

7.11.2   The search for gravitational waves

Gravitational waves were predicted by Einstein (1916c), although there was considerable debate about the reality of the phenomenon for a number of years. The experimental search for the waves was begun by Joseph Weber in a pioneering set of experiments carried out in the 1960s. He published his proposals for their detection in his book *General Relativity and Gravitational Waves* (Weber, 1961) and constructed an aluminium bar detector to measure gravitational radiation from cosmic sources (Weber, 1966). He estimated that he could measure strains as small as 10^{-16}, the fractional change in the dimensions of the detector caused by the passage of gravitational waves. His first published results caused a sensation when he claimed to have found a positive detection of gravitational waves by correlating the signals from two gravitational wave detectors separated by a distance of 1,000 km at the University of Maryland and the Argonne National Laboratory (Weber, 1969). In a subsequent paper, he reported that the signal originated from the general direction of the galactic centre (Weber, 1970). These results were received with considerable scepticism by the astronomical community since the reported fluxes far exceeded what even the most optimistic relativist would have predicted for the flux of gravitational waves originating anywhere in the

Galaxy. The positive effect of Weber's experiments was that a major effort was made by experimentalists to check his results and, in the end, his results could not be reproduced.

The challenge to the experimental community was important in stimulating interest in how the extremely tiny strains expected from strong sources of gravitational waves could be detected. The outcome was the approval of a number of major national and international projects in order to detect the elusive gravitational waves. The LIGO

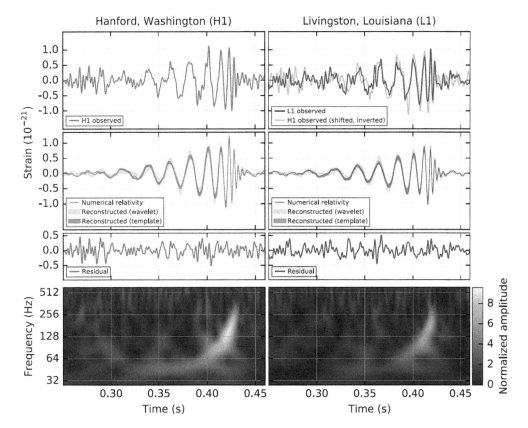

Fig. 7.20 *The gravitational-wave pulse GW150914 observed by the LIGO Hanford (H1, left column panels) and Livingston (L1, right column panels) detectors. Times are shown relative to September 14, 2015 at 09:50:45 UTC. Top row, left: H1 strain. Top row, right: L1 strain. GW150914 arrived first at L1 and $6.9^{+0.5}_{-0.4}$ ms later at H1; for a visual comparison, the H1 data are also shown, shifted in time by this amount and inverted. Second row: Gravitational-wave strain projected onto each detector in the 35–350 Hz band. Solid lines show a numerical relativity waveform for a system with parameters consistent with those recovered from GW150914 confirmed to 99.9% confidence level. Shaded areas show 90% credible regions for two independent waveform reconstructions. Third row: Residuals after subtracting the filtered numerical relativity waveform from the filtered detector time series. Bottom row: A time–frequency representation of the strain data, showing the signal frequency increasing over time, the well-known chirp pulse signature of inspiralling and coalescence of a black hole binary system (Abbott et al., 2016).*

project, an acronym for Laser Interferometer Gravitational-Wave Observatory, was initiated by Kip Thorne and Rainer Weiss in 1984 and approval given by the US National Science Foundation in 1994.[18] The project in its present incarnation was begun in the early 2000s and consists of two identical interferometers each with a 4-km baseline located at Livingston, Louisiana and Hanford near Richland, Washington. The long development programme gradually increased the sensitivity of the detector systems until by 2015, a strain sensitivity of about 10^{-23} was achieved.

Just prior to the first production run with the upgraded detector system, the two detectors observed simultaneously a transient gravitational-wave signal on 14 September 2015. The signal swept the frequency range from 35 to 250 Hz with a peak gravitational-wave strain of 1.0×10^{-21}. The waveform was a perfect match to the characteristic signature predicted by general relativity for the inspiral and merger of a pair of black holes and the subsequent ring-down to create a single black hole (Fig. 7.20). The precisely known shape of the 'gravitational chirp' enabled the cosmological redshift of $z = 0.09^{+0.04}_{-0.03}$ to be measured. The initial black hole masses were $36^{+5}_{-4} M_\odot$ and $29^{+4}_{-4} M_\odot$, and the final black hole mass is $62^{+4}_{-4} M_\odot$, with $3.0^{+0.5}_{-0.5} M_\odot c^2$ being radiated as gravitational waves. These observations demonstrate unambiguously the existence of binary stellar-mass black hole systems—it was the first direct detection of gravitational waves and the first observation of a binary black hole merger. Since then, other examples of the gravitational waves from coalescing black holes have been discovered, as well as the coalescence of two neutron stars (Abbott et al., 2017).

. . . .

NOTES

1. Many more details of the astrophysics and radiation processes involved in a wide range of these topics is contained in my text *High Energy Astrophysics*, 3rd edition (Longair, 2011).

2. The reception histories of special and general relativity are well documented, for example by Stachel (1995). Brief accounts of the events leading to the discovery of these theories are contained in my book *Theoretical Concepts in Physics* (Longair, 2003).

3. Dates in brackets following the investigators' names refer to the publication cited in the bibliography. Where the date of the work was in an earlier year, that is indicated in the text.

4. I have given a few more details of the controversy in my book *The Cosmic Century* (Longair, 2006).

5. The history of the of pulsars is described by Hewish (1986). A delightful description of the discovery of pulsars is given by Jocelyn Bell-Burnell (1983).

6. A survey of observations and the physics of pulsars, including references to many original papers, is given by Lyne and Graham Smith (1998).

7. The various issues involved are summarized in the book *The Redshift Controversy* (Field et al., 1974). A history and early summary of what was known about quasars in 1966 is provided by the book by the Burbidges entitled *Quasi-stellar Objects* (Burbidge and Burbidge, 1967).

8. The history of X-ray astronomy is told by Tucker and Giacconi in *The X-ray Universe* (Tucker and Giacconi, 1985).

9. For the early history of ultraviolet observations of the Sun, see *Sun and Earth* (Friedman, 1986).

10. I have given a proof of this result in Section 17.10 of my book *Theoretical Concepts in Physics* (Longair, 2003).

11. For details of the physics of accreting binary systems, see Shapiro. and Teukolsky (1983).

12. Many more details of these processes are contained in the book *Accretion Power in Astrophysics*, third edition, by Frank, King, and Raine, (2002).

13. Reviews of the history of ultra-high energy γ-ray astronomy are given by Aharonian and Akerlof (1997) and Catanese and Weekes (1999).

14. The Seyfert galaxies have compact luminous nuclei. They are divided into two classes depending on the breadths of their emission lines. The Type 2 Seyfert galaxies possess narrow strong emission lines with velocity widths up to about 1,000 km s^{-1}. The Type 1 Seyferts contain narrow lines as well, but there are also strong broad wings to these emission lines, with velocity widths up to about 10,000 km s^{-1}. The natural interpretation of this distinction is that the broad lines regions originate close to the nucleus and the narrow lines from much further out. This model is supported by observations of reverberation mapping of the lines (for more details, see my book *High Energy Astrophysics* (Longair, 2011)).

15. More details of these radio source unification scenarios, including the distinction between the Fanaroff–Riley classes, are given in my book *High Energy Astrophysics* (Longair, 2011).

16. $\langle V/V_{\max} \rangle$ is the average value of the ratio of the volume within which the source is observed to the total volume within which the source could have been observed at the radio and optical flux density limits of the complete sample.

17. Hulse and Taylor were awarded the 1993 Nobel Prize in Physics 'for the discovery of a new type of pulsar, a discovery that has opened up new possibilities for the study of gravitation'.

18. A very readable account of black holes, gravitational waves, and their significance for physics and astronomy is contained in Kip Thorne's book *Black Holes and Time Warps: Einstein's Outrageous Legacy* (Thorne, 1994).

8

The cosmic microwave background: from discovery to precision cosmology

R. Bruce Partridge

8.1 Introductory remarks

This chapter chronicles the discovery of a $\sim$3 K microwave background by Arno Penzias and Robert Wilson, shows how its cosmic origin was established, and then assesses the impact of detailed studies of this cosmic microwave background (CMB) on modern cosmology.

The 1965 discovery itself has been treated numerous times. In addition, we have the great benefit that virtually all the parties involved in the discovery and its interpretation could and did tell their stories: these are assembled in *Finding the Big Bang* (Peebles et al., 2009), henceforth abbreviated as FBB. Ralph Alpher and Robert Herman (2001) in *Genesis of the Big Bang* describe their earlier investigations, together with George Gamow, of the Hot Big Bang. Other accounts of the history of this field are listed in the references at the end of this chapter.

This chapter adopts a slightly more personal tone than others in this volume. This is in part because I was involved in some of the measurements analyzed here. Readers should be aware of the potential for bias in my presentation, particularly in Section 8.4. As an observer, rather than a theorist, one bias that I freely acknowledge is the focus on observations of the $\sim$3 K radiation, rather than theoretical analyses of that cosmic background. Chapter 6 by Malcolm Longair will help restore the balance.

The decision by the editors of this volume to separate out one strand of modern cosmology can clearly be justified. The CMB is a complex topic and has an intricate history. On a deeper level, there is the striking fact that CMB observations by themselves, with only minimal support from other astrophysical evidence, are able to set values for all the important cosmological parameters of the now-accepted standard model for the evolution of the Universe. However well justified, this editorial decision does place a

Partridge, R. B., 'The cosmic microwave background: from discovery to precision cosmology' in *The Oxford Handbook of the History of Modern Cosmology*, edited by Kragh, H. and Longair, M. S. © Oxford University Press 2019.
DOI: 10.1093/oxfordhb/9780198817666.013.8

burden on readers to integrate the results discussed here with those presented in other chapters, particularly Chapters 6, 10, and 11. Other contributors to this volume and I have tried to ease that burden by embedding frequent cross-references in our respective chapters.

In this chapter, the term "Hot Big Bang" is used generically and somewhat loosely to refer to any cosmological model that has a hot, dense phase early in its history. Since high density and high temperature, $T \approx 10^{10}$ K, are the only conditions needed to generate a CMB, the details of any prior evolution are glossed over until the last section of this chapter. Specifically, I introduce the issue of a very early so-called inflationary phase of exponential expansion only late and briefly in Section 8.6.[1] Second, I avoid the term *cosmic* microwave background and the abbreviation CMB in Sections 8.2–8.4, since salient parts of the story presented here are the tests to show that the microwave background discovered in 1965 is in fact *cosmic in origin* and a remnant of the early Hot Big Bang.

Section 8.2 presents the discovery as it is usually described. Section 8.3 demonstrates that there were strong hints of a cosmic background before 1965 and Section 8.4 treats efforts to prove that the microwave background found in 1965 is indeed the cooled relic of a Hot Big Bang. Section 8.5 describes advances made up to and including the results of NASA's COBE mission. The long final Section 8.6 assesses the impact of CMB observations, ground-based as well as from space, on modern cosmology.

8.2   The discovery of a microwave background: the story as it is usually told

The story begins at the Bell Telephone Laboratories in Holmdel, New Jersey. There, two recently hired radio astronomers were attempting to characterize a large horn antenna coupled to a sensitive microwave detector operating at 7 cm wavelength. The antenna had been constructed to receive signals from the first communications satellite, Echo, and was re-fitted to work at a shorter wavelength for both radio astronomy and satellite communication with Telstar, Echo's successor. The signals expected were weak, so it was important to understand competing sources of emission that entered the antenna when it was pointed at the sky. The two scientists pursuing this study, Arno Penzias and Robert Wilson, soon encountered a problem. They could not explain all the microwave radiation entering the antenna; indeed, more than half the measured signal from the sky could not be explained. The discovery of this extra signal, and the steps Penzias and Wilson took to isolate it, are elegantly described in their 1978 Nobel lectures and their recollections are gathered in FBB.

They had detected an unknown source of "excess antenna temperature." To microwave engineers and radio astronomers (many of whom emerged from an engineering community), both incoming signals and instrumental noise can be quantified in terms of an equivalent noise temperature. Despite all the careful steps Penzias and Wilson had taken to track down potential instrumental or other sources of the "excess," it appeared that a signal equivalent to the emission of a 3.5 ± 1 K black body was entering the antenna when pointed anywhere at the sky. They had a significant signal, one in need of explanation.

Meanwhile, at Princeton University, some 50 km from Bell Laboratories, Robert Dicke and his younger colleagues were constructing an argument for the existence of a specific astronomical signal, thermal radiation filling the Universe. Dicke's argument for the existence of thermal radiation was quite different from, and entirely independent of, the detailed arguments for a Hot Big Bang advanced less than twenty years before by George Gamow and his colleagues, principally Ralph Alpher and Robert Herman, as discussed in detail in Chapter 4. Gamow and colleagues built heat into their model in order to create the heavy elements, that is, those more massive than hydrogen. Dicke's cyclic model, in contrast, requires a hot dense phase to dissociate, and thus *remove* heavy elements. Recall the context in which Dicke and his "Gravity Group" were working in the mid-1960s. Cosmology, a relatively peripheral branch of science, was deeply immersed in the debate between the Steady State theory and the Hot Big Bang model. Dicke had in mind a cyclic model, one that preserves some positive features of both.[2] His cosmological model, never published in detail, calls for an endless series of expansions and subsequent contractions of the Universe. It thus allows for the expansion and evolution of the particular cycle we inhabit, yet retains, in the long term, a sort of steady state, with an unending number of cycles. Assuming our cycle of expansion is typical, however, stars form and generate heavy elements such as helium, carbon, and iron in each cycle. To prevent a buildup of such heavy elements cycle after cycle, these elements must be broken up at the end of each cycle as the Universe contracts and the density increases. Temperatures of order 10^{10} K are required for the photo-dissociation of heavy nuclei. Thus the dense phase, when each cycle of contraction ends and the next cycle of expansion begins, must be hot, just as in the model put forward by Gamow and his colleagues. Each new cycle of expansion starts hot, dense and, in terms of its elemental contents, simple: just neutrons, protons and electrons. Although the underlying cosmological model is different, each expanding phase of Dicke's cyclic Universe had the same early Hot Big Bang as the model of Gamow and his colleagues. The Princeton group referred to this hot initial state as the "primeval fireball."

In either model, as the Universe expands, the temperature drops—but not to zero. Dicke was a master experimentalist and so his "Gravity Group," was not content just to predict the cosmic thermal radiation, but also set out to *find* it. Dicke suggested a microwave search for the heat, and encouraged two members of the Gravity Group, Peter Roll and Dave Wilkinson, to take up the challenge of building a microwave device to measure the heat from the Big Bang. The results and the importance of this experiment form the core of Section 8.4.1 of this chapter.

As preparations for the experiment went forward, other members of the Gravity Group were following up the consequences of a primeval fireball, in effect duplicating much of the work of Gamow and his colleagues. Jim Peebles, in particular, was investigating, or re-investigating, nucleosynthesis, the generation of heavy elements, especially helium, in the primeval fireball, as well as the consequent constraints on the present radiation temperature. In addition, Gravity Group members were presenting their findings in scientific talks. Peebles gave one such talk at Johns Hopkins University in February 1965. In the audience was Ken Turner, then at the Department of Terrestrial Magnetism (DTM) of the Carnegie Institution. Turner in turn described the Princeton work to his colleague, Bernie Burke, then also at DTM. Sometime later, Penzias called Burke on an unrelated matter and at the end of that conversation mentioned the vexing "excess temperature." Burke passed on what he had heard of the Princeton work, and suggested that Penzias call Dicke.

That call interrupted a lunchtime meeting of Dicke's Gravity Group, which included Roll and Wilkinson. Dicke listened, asked some questions, then turned to his colleagues and said, "Well, boys, we've been scooped." The Princeton group had indeed been scooped. The remnants of the Hot Big Bang, or the primeval fireball, had been found first elsewhere by Penzias and Wilson.

Or had they? Is this a clear-cut case of scientific discovery? The standard view is that Penzias and Wilson discovered the CMB, and that the CMB rapidly became a pillar of modern cosmology. That is the brief textbook version, but in my view the story is in fact more complicated.[3] There were observational hints of a cosmic background before the Bell Laboratories work. Equally, there were doubts about the cosmic origin of the "excess temperature" well after the discovery had been announced in 1965.

It is useful to start by reviewing the two letters to *The Astrophysical Journal* in which the discovery was reported (Penzias and Wilson, 1965) and interpreted (Dicke et al., 1965). The very short letter by Penzias and Wilson clearly describes the "excess antenna temperature" they found and their painstaking attempts to rule out an instrumental origin of that signal. Penzias and Wilson made a clear decision not to ascribe an astronomical origin to the signal. To quote Wilson, from his contribution in FBB: "…after all, the measurement might be correct even if the cosmology turned out not to be!" On the other hand, in a note added in proof, they used another radio astronomical measurement to show that their "excess" was unlikely to be synchrotron radiation and so some sort of astronomical origin of the "excess" was evidently in their minds. They also noted that precursor work at Bell Laboratories, particularly by Ohm (1961), was not inconsistent with their "excess antenna temperature" measurement.[4]

Dicke and his colleagues, in their longer letter, provided a cosmological interpretation, "if we tentatively assume the measurements of Penzias and Wilson." The Princeton group went on to point out some problems reconciling a cosmic background of 3.5 K with many of the existing cosmological models. At issue were the observed helium abundance and the constraints that a relic temperature as low as 3.5 K, combined with an estimated helium abundance of $\sim 25\%$ by mass, put on the density of matter. That issue was not fully resolved until some years later, with the addition of dark matter to the cosmic mix. The discussion of helium abundance, incidentally, refers directly to the work of Alpher, Gamow, and Herman. There is more to both these seminal letters than merely "an unexplained signal" and "a straightforward explanation."

Despite the cautious tone of both letters, the observational results were secure and statistically significant: there clearly was some astronomical signal. Equally important, the letter by Dicke and colleagues offered a complete and consistent explanation of the observations. Both elements, I suggest, were required to convert a puzzle into a discovery. The need for a confluence of a significant *measurement* and a convincing *explanation* will be all the clearer when we look at the many observational hints of a microwave background that emerged before 1965, only to be forgotten, left unanalyzed, or misinterpreted. Equally, as we have seen in Chapter 6, much of the theoretical work of Alpher, Herman, Gamow, and their colleagues faded from view because there was no strong observational evidence for a Hot Big Bang or, to be more precise, no observational evidence that *required* a hot initial phase. In the struggle between the Big Bang and Steady State models, there were certainly observations, such as the counts of radio sources by Martin Ryle and colleagues that, though not without dispute, seemed to favor an evolving cosmology over a steady-state model. Likewise, measurements by Allan Sandage among others of the rate of change of cosmic expansion indicated deceleration under gravity, as expected in a Big Bang model, rather than an accelerated expansion as required in Steady State theory.[5] But these results cast doubt on the Steady State model rather than offering any direct support for a *hot* initial phase.

An interesting footnote to the discovery is that it was first publicly announced not in *The Astrophysical Journal (ApJ)* but in *The New York Times*. The two letters were submitted on May 7 and May 13, 1965. A week later, the *Times* broke the story, well before that issue of *Astrophysical Journal* was published. I suspect that more astronomers and physicists learned of the discovery through the newspaper than the journal. In any case, if a specific date is to be assigned to the announcement of "excess temperature," and its possible explanation as a cosmic relic, it is May 21, 1965.

8.3   Before 1965: early hints and missed opportunities

Penzias and Wilson had no idea they were making a discovery that would help turn cosmology into a precision science. Equally, neither the Bell Laboratories team nor Dicke and his colleagues had any recollection that the papers of Gamow and his colleagues hinted at a detectable, residual signal of the Hot Big Bang. There is, for instance, a 1949 paper by Alpher and Herman that specifically cites a cosmic temperature of 5 K (Alpher and Herman, 1949). On the one hand, as both Alpher and Herman subsequently pointed out, the initial lack of full recognition of their earlier work by both Penzias and Wilson and the Gravity Group is surprising.[6] While a central motivation of the Hot Big Bang model, the creation of heavy elements, had been undermined, the Big Bang model was still part of the discourse in cosmology, both professional and public.[7] On the other hand, it is perhaps not so surprising after all that Penzias and Wilson did not have the earlier work in mind. Both were radio astronomers, working in an engineering environment, not cosmologists. In addition, cosmology in the 1960s barely existed as a field of science. The field instead was largely mathematical; for instance, in Bondi's brief textbook *Cosmology*, only a dozen pages are devoted to astronomical observations, and there is no discussion of a Hot Big Bang model beyond one brief mention on page 120 (Bondi, 1960). The passage of time may also have played a role. The last scientific paper I can locate by Gamow and his colleagues that treated the Hot Big Bang was published in 1957, well before the Bell Laboratories experiments started and Dicke's interests turned to cosmology. It is worth adding that Dicke also managed to forget one of his own earlier results, specifically an upper limit of 20 K he and colleagues (Dicke et al., 1946) established on any astronomical microwave background!

8.3.1   Precursor observations of the CMB

Ironically, just as there were prescient theories without apparent observational backing, there were precursor observations that lacked, or escaped, theoretical understanding. Many are summarized in Chapter 3 of FBB; a sample is examined here. These were measurements that might have been interpreted as evidence for a Hot Big Bang, but were not.

The most famous are the studies by Adams (1941) and McKellar (1941) of the absorption lines in the optical spectra of stars produced by CN molecules lying along the line of sight. Using high-resolution spectra obtained by Adams, McKellar found that absorption lines were caused not just by CN in its ground state, but also by CN molecules in the first excited rotational ($K = 1$) state. Some mechanism is required to populate this excited state. That could in principle be either collisions or thermal radiation. If the latter were the case, as seems reasonable in low-density interstellar space, then, as McKellar (1941) pointed out, a thermal bath of microwave radiation with $T = 2.3$ K would be required to explain the optical observations. But no one connected the work of this stellar spectroscopist to the later predictions of Alpher, Gamow and Herman until after 1965. Instead, in his careful analysis of the observations of Herzberg (1950) summarized the results in this way: "From the intensity ratio of the lines with $K = 0$ and $K = 1$, a rotational temperature of 2.3 K follows, which has of course only a very restricted meaning." Whatever "a very restricted meaning" meant to Herzberg, it certainly did not mean evidence for the Hot Big Bang. Note that the observational results implying a 2.3 K background appeared before any of the theoretical investigations of the Hot Big Bang by Gamow and his colleagues. They missed this clue. Perhaps the mental leap from molecular spectroscopy to cosmology was too great. And then World War II intervened.

There were, however, more direct radio astronomical measurements that could have detected (or in some cases probably did detect) the microwave background. For instance, Emile Le Roux, working at the Nancy radio telescope in France, conducted a survey of the radio sky at 33 cm wavelength, and reported an apparently isotropic background of 3 ± 2 K (Le Roux, 1956). The reported value was not corrected for atmospheric (or Galactic) emission; in addition, it required a large correction for emission from the ground into side-lobes of the antenna employed. That correction in turn assumed both perfect knowledge of the side-lobes and that the ground was at a fixed temperature of 285 K. Neither assumption is well justified.

At roughly the same time that Penzias and Wilson were puzzling over their "excess temperature," a group of radio astronomers in Toronto were encountering a problem with long wavelength observations of Galactic emission.[8] Adding a frequency-independent, celestial signal of 3–5 K would have helped resolve their measurements with their expectations for synchrotron emission.

A more pertinent case is the set of observations at 3.2 cm made in the then Soviet Union by Tigran Shmaonov (1957). At 3.2 cm, synchrotron radiation from both the Galaxy and extragalactic sources is much weaker, and indeed is nowhere near able to produce the sky temperature of 4 ± 3 K he observed. I suspect the CMB was indeed the main contributor to his 4 K signal; on the other hand, a 1σ detection is very weak evidence, unlike the $> 3\sigma$ result of Penzias and Wilson.

The prime example of a missed opportunity was the work of Ohm (1961) carried out at Penzias and Wilson's own institution and indeed as part of the Echo project! This was a study of the microwave intensity of the sky at 13 cm. Just as Penzias and Wilson would later do, Ohm attempted to account for all contributions to the measured temperature of the sky, including emission from the Earth's atmosphere. The measured temperature, however, exceeded the estimated value by ~ 3 K, but with an uncertainty of ~ 3 K. Another signal, not from the atmosphere, or pick-up from the ground or the Galaxy, was entering the receiver. Ohm had come across an "excess antenna temperature" four years before another such excess was found by his Bell Labs colleagues.

This last finding also got lost, at least in the United States. It was picked up in the Soviet Union by members of the brilliant group gathered in Moscow around Ya. B. Zeldovich, a group who were together to make so many contributions to modern cosmology and astrophysics. In one of the great and sad ironies in the field, Andrei Doroshkevich and Igor Novikov may have misinterpreted Ohm's reported residual of ~ 3 K as a measure of atmospheric emission.[9] They thus missed the hint that Ohm had seen an *additional*, astronomical background. Indeed, their mentor, Zeldovich, took the Ohm results to set an upper limit of ~ 1 K on any cosmic background, in keeping with the cold big bang model he then favored. Doroshkevich and Novikov (1964) did, however, point out the need for further observations to assess "the correctness of the Gamow theory." Even more prescient is one of the figures from their 1964 paper, a comparison of a thermal spectrum drawn for $T = 1$ K to the spectrum of background emission expected from all radio galaxies. A year before the Bell Laboratories discovery, Doroshkevich and Novikov had demonstrated that a thermal spectrum would dwarf emission from galaxies at centimeter wavelengths.

There was no lack of hints, both theoretical and observational. What it took for a real "discovery" was the confluence of a clearly significant signal, the "excess" of 3.5 ± 1 K, and a convincing cosmological explanation of that signal.

8.4   1965–1970: a background, but is it cosmic?

The recognition of the discovery, and of its eventual importance to cosmology, however, was neither instantaneous nor unanimous. The status of the "excess noise" reported by Penzias and Wilson was far from settled in 1965. True, Dicke and his colleagues had advanced a convincing explanation, but not all cosmologists were convinced. There was a microwave background, but was it truly cosmic in origin? Other astronomical explanations for microwave emission were promptly and seriously presented, or at least discussed. These included a solar system origin, a Galactic origin, emission from intergalactic plasma, and the summed emission of many extragalactic radio sources, to name some of the competing explanations of the Bell Laboratories' results. Among the flurry of such papers were those by Kaufman (1965), Pariiskii (1968), Hoyle and Wickramasinghe (1967), Hazard and Salpeter (1969), Wolfe and Burbidge (1969), and Layzer and Hively (1973). Readers of this book will recognize many of these names for other notable contributions to astronomy and cosmology. Clearly, the cosmic origin of the microwave background was not immediately accepted by all those active in the nascent field of cosmology. Alternative explanations were offered and needed to be taken seriously.

To demonstrate that the microwave background was truly cosmic in origin, it had to pass two observational tests. The first was the spectrum of the radiation. Radiation in a hot, dense early phase of expansion would have a thermal or black-body spectrum. Furthermore, it had been known since the work of Tolman (1934b) that in the expanding Universe, while the temperature of the radiation decreases adiabatically, it maintains its thermal spectrum. If the "excess temperature" was indeed a cool relic of the Hot Big Bang, it should have a black-body spectrum. In contrast, most known astronomical sources of microwave emission have very different spectra.

The second test involved the isotropy of the background: was it essentially the same temperature in all directions, as most of us expected for a remnant of the Hot Big Bang? Given the Solar System's significant offset from the Galactic center, emission of any sort from the Milky Way Galaxy, in contrast, would be far from isotropic. The same argument holds for potential Solar System sources as well. Penzias and Wilson had indeed recognized the importance of this isotropy test. They describe their "excess" as "within the limits of our observations, isotropic, unpolarized, and free from seasonal variations." Although not specified in their 1965 letter, their observations limited any departure from isotropy to 10% or less.

Note that these two tests were not needed to confirm the Bell Laboratories *observations*; they were crucial instead in determining whether the Gravity Group's *interpretation* of those observations was correct. Not surprisingly, therefore, the Princeton group played an active role in both tests.

The remainder of this section explains both how these tests were carried out and the results.[10] We will see that the tests of a cosmic origin were easily passed. Most, but not all, early observations of the microwave background were consistent with a black-body spectrum and with isotropy, in line with a truly cosmic origin. The few observations that appeared inconsistent with the cosmic explanation are briefly reviewed in Section 8.4.4. These turned out to be erroneous. By about 1970, most cosmologists had been convinced that the remnant of a Hot Big Bang had been found.

8.4.1   Testing the spectrum

Both Gamow and his colleagues, and later Dicke, had implicitly assumed that matter and radiation in a Hot Big Bang would be in thermal equilibrium. Thus the radiation would have a black-body spectrum in the early phases of expansion. Subsequent investigations have shown this to be the case unless extra energy is added to the radiation well after its temperature has dropped below 10^{10} K.[11]

As already noted, Tolman (1934b) showed that an initially black-body spectrum is maintained as the Universe expands—Dicke demonstrated the same result based on a simpler argument that the expansion of the Universe can be treated as adiabatic. While the black-body spectral shape is maintained, the temperature drops in a precisely predictable way as the Universe expands, varying as the inverse of the scale factor $a(t)$. It follows from Stefan's Law that the energy density in radiation varies as $a^{-4}(t)$. On the other hand, the density of matter varies during the expansion as $a^{-3}(t)$. Hence the relation between temperature and density in an expanding universe is known exactly. Indeed, Alpher, Herman, and Gamow made use of this relation to estimate the current radiation temperature to be a few kelvin, and the brief explanatory letter of Dicke et al. (1965) shows these relations (Fig. 8.1). For future reference, note that the different dependences of matter and radiation density on $a(t)$ ensure that the latter dominated at small enough $a(t)$, that is, early in the history of the Universe. Thus, the Hot Big Bang model predicts a thermal, black-body background of a few kelvin at the present epoch. The Bell Laboratories' work established that the actual temperature is close to 3.5 K.

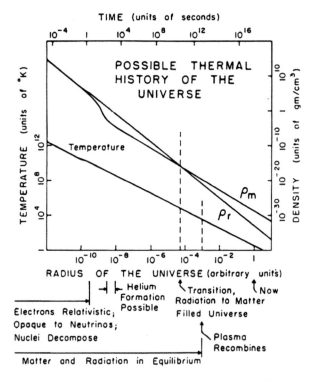

Fig. 8.1 *The evolution of matter* (ρ_{m}) *and radiation* (ρ_{r}) *density as the Universe expands. Note that radiation dominated for the first* 10^{10}–10^{11} *s. The smooth decrease in T is also shown; the tiny "kink" at t ~ 1 s results from the addition of photons from electron-positron annihilation (Dicke et al., 1965).*

As it happens, that radiation temperature of a few kelvin and the corresponding energy density of $\sim 10^{-12}$ erg cm^{-3} is close to the energy density of starlight in our Galaxy. Arthur Eddington (1926b), for instance, calculated that the energy density of Galactic starlight was equivalent to that of 3.18 K thermal radiation. The coincidence with Penzias and Wilson's 3.5 K was quickly noted. Could the Bell Laboratories signal be Galactic, not cosmic, in origin?

The same kind of argument can be made for the summed emission of all galaxies; it must have some non-zero value. Since galaxies do not fill space entirely, however, the energy density of extragalactic starlight alone would be well below the value estimated by Eddington for our Galaxy. On the other hand, astronomers were well aware in the 1960s that some galaxies pour out radio waves and other forms of electromagnetic radiation, some with luminosities exceeding their optical emission; could the observed emission of such galaxies extend to shorter, microwave, wavelengths and explain the "excess temperature" found at Bell Laboratories?

The spectrum of the microwave background provided a test. Both starlight and radio emission from galaxies have spectra that differ strongly from a black body. In the centimeter range, for instance, synchrotron radiation dominates the output of the Galaxy and most extragalactic radio sources. The synchrotron emission of known astronomical sources, including our Galaxy, has a spectrum that decreases with frequency as ν^{α}, with the spectral index α typically in the range -0.5 to -1.0. On the other hand, 3.5 K black-body emission has a ν^2 spectrum, increasing rapidly with frequency, at centimeter wavelengths. These are sharply different (Fig. 8.2) as Doroshkevich and Novikov (1964) had already pointed out. A single measurement, of course, does not determine a spectrum. Measurements of the microwave background at other wavelengths, different from 7 cm, were obviously required. Providentially, the experiment undertaken by Roll and Wilkinson quickly and critically met that need. At the time of the fateful telephone call described in the previous section, Roll and Wilkinson were already at work on a microwave receiver designed to search for a thermal background. Crucially, their instrument operated at a different wavelength, 3 cm not 7 cm. Like Penzias and Wilson's apparatus, it compared the temperature of the sky to that of a black-body absorber/emitter held at a known, low temperature, 4.2 K, the boiling point of liquid helium. The Princeton instrument, however, had a more straightforward design, one better suited to controlling some sources of systematic error that had required such careful attention by Penzias and Wilson.

In 1966, Roll and Wilkinson published their results: at a wavelength of 3 cm, the temperature of the microwave background was measured to be 3.0 ± 0.5 K, in excellent agreement with the value found by Penzias and Wilson (Fig. 8.2). To drive home the importance of the Princeton measurement, consider what Roll and Wilkinson would have measured had an alternative interpretation of the "excess temperature" been true. Had Penzias and Wilson been observing synchrotron emission at $\lambda = 7$ cm, from either our Milky Way Galaxy or extragalactic sources, the predicted temperature at 3 cm would have been ~ 0.3 K not 3 K. The huge discrepancy is clear from Fig. 8.2. The results of Roll and Wilkinson were unambiguous: they agreed with Penzias and Wilson, and the microwave background had a spectrum consistent with a black-body shape. The cosmic

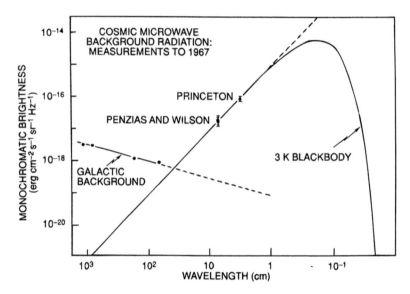

Fig. 8.2 *The first two measurements of the microwave background, both consistent with a 3 K black-body spectrum. The dashed line above the peak of the black-body curve is a gray-body spectrum discussed later (Roll and Wilkinson, 1967). To me, fitting a 3 K black-body curve to just two points is one of the most daring extrapolations in all of science—and beautifully confirmed by later measurements.*

explanation of Dicke and colleagues had survived its first test. In addition, the temperature of the background had been refined by Roll and Wilkinson (1967) to 3.0 ± 0.5 K.

In parallel, several astronomers, Field and Hitchcock (1966), Thaddeus and Clauser (1966), and Shklovski (1966), prompted by the announcement of the Bell Laboratories' results, recalled the CN absorption line measurements discussed in Section 8.3. While the temperature derived from the CN absorption line ratio was apparently a bit low at 2.3 K, it was consistent within the errors with 3 K. Field and Hitchcock (1966) also reanalyzed the optical data, finding $2.7 < T < 3.4$ K. From 7 cm to 0.26 cm, all measurements were consistent with a thermal spectrum of ~ 3 K, strongly inconsistent with the spectra expected from other known astronomical sources.

Adding to the support for a thermal spectrum, both Howell and Shakeshaft (1966) and Penzias and Wilson (1967), using or modifying existing apparatus, measured the temperature of the sky at 21 cm, extending observations to longer wavelengths. At 21 cm, a correction was needed for the synchrotron emission of the Galaxy. Once that correction had been made, these teams found residual backgrounds of $T = 2.8 \pm 0.6$ K and 3.2 ± 1.0 K, respectively. Howell and Shakeshaft (1967) pushed the observations to even longer wavelength, obtaining $T = 3.7 \pm 1.2$ K in the wavelength range 49–73 cm. Radio astronomers in the Soviet Union soon confirmed these observations. There were additional measurements made in the interval 1965–1970, most of them consistent with the value of 3 ± 0.5 K found by Roll and Wilkinson.[12]

There were a few discrepant measurements, treated separately in Section 8.4.4. These few cases aside, over a wavelength range of nearly 300, many diverse measurements were broadly consistent with a thermal, black-body spectrum of cosmic origin. Any alternative, non-cosmic explanation required increasingly complicated and ad hoc mechanisms to match the spectral observations. Some of these, including a novel suggestion by Narlikar and Wickramasinghe (1967), are analyzed in my summary of the observational situation in the late 1960s (Partridge, 1969).

One reasonable opening for an alternative explanation remained available, however. The majority of the measurements just listed were also consistent with a so-called gray-body spectrum[13], which has the same shape as a black-body spectrum, but with the intensity scaled down by a constant factor at all wavelengths. For instance, a gray-body background with a spectral shape consistent with $T = 30$ K, but scaled down by a factor 10, would appear to have an equivalent temperature of 3 K at centimeter wavelengths, matching the Bell Laboratories and Princeton results. What astronomical mechanism could produce such a spectrum was not clear, but this alternative explanation needed to be ruled out to demonstrate that the observed, microwave background was truly black-body and thus cosmic. Figure 8.2 shows how to test the gray-body hypothesis by extending observations to shorter wavelengths. A true black-body, or Planckian, spectrum at temperature T has a maximum at a wavelength $0.5/T$ cm, or roughly a few millimeters for $T = 3$ K (see Fig. 8.2). A 30 K gray-body spectrum, on the other hand, would peak at much shorter wavelengths and thus maintain a λ^{-2} or ν^2 Rayleigh–Jeans dependence to mm wavelengths, as shown by the upper dashed line in Fig. 8.2. Could the spectral curvature expected of a true 3 K black-body spectrum be detected at short wavelengths, ruling out the gray-body hypothesis?

That was the goal of 1967–1968 observing campaigns by the Princeton group (Stokes et al., 1967; Wilkinson, 1967; Boynton et al., 1968). Among the plans for the 1967 campaign were working at high altitude to reduce emission from the Earth's atmosphere and improving the coupling to the liquid-helium cooled absorber/emitter which had dominated the error budget of the Roll and Wilkinson experiment. Finally, to ensure a cleaner intercomparison of measurements made at different wavelengths, we designed three devices with very similar properties, all of which could be coupled to the same cold absorber. These operated at 3.2, 1.58, and 0.86 cm. The high, dry site at 12,500 ft in the White Mountains of California and better instrument design substantially reduced the experimental error to roughly 0.2 K at all three wavelengths. The results clearly favored a black-body spectrum over a gray-body spectrum, but the latter was excluded only at the 2.5σ level. To quote Boynton et al. (1968), which describes the follow-up campaign, the 1967 results had "not yet demonstrated conclusively the spectral curvature expected." The next and crucial step was to extend the observations to still shorter wavelengths. That was undertaken by Boynton et al. (1968), using the same techniques, but with a radiometer operating at 0.33 cm (Fig. 8.3). The team made their measurements in winter, at a site at 11,330 ft altitude in Colorado, thus freezing out some of the atmospheric water vapor. Their results excluded the gray-body hypothesis at a much higher significance level, $> 4\sigma$. Confirming evidence was soon supplied, again at 0.33 cm, by Millea et al. (1971). By 1971, the gray-body alternative had been ruled out conclusively.

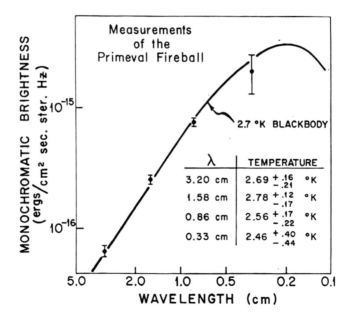

Fig. 8.3 *Spectral measurements by the Princeton group from 1967–1968, showing the departure from a pure λ^{-2} law, and agreement with a black-body spectrum. (Boynton et al., 1968)*

The Princeton campaign also refined the value of the present temperature of the cosmic background to 2.7 ± 0.2 K.[14] By 1968, the temperature of the remnants of the Hot Big Bang, T, had been determined to better than 10% accuracy, making it then the most precisely determined of the crucial parameters of cosmology. Furthermore, this result proved to be consistent with all subsequent observations and the currently accepted value of 2.7255 ± 0.0006 K (Fixsen, 2009).[15]

8.4.2   Testing isotropy

We turn now to the second test: isotropy. While there is no requirement that the Universe on a large scale be either homogeneous or isotropic, it was generally assumed to be both. Homogeneity, and, one can argue, isotropy, are enshrined in the Cosmological Principle: the notion that, on a large scale, the Universe is the same everywhere and to all observers.[16] If the Universe is isotropic and homogeneous, so too is the relic heat of the Big Bang. Certainly, if the 3 K microwave background is truly cosmic in origin, it ought not to show any strong dependence on Galactic or ecliptic coordinates. Was the microwave background detected by Penzias and Wilson indeed the same in all directions? This was the crux of the second test.

Measurements of the isotropy of the microwave background are intrinsically easier than measurements of its absolute temperature; to test isotropy, only comparative measurements are needed of the temperature at various points on the sky. Many of the systematic errors that could bias an absolute measurement cancel out in a comparative measurement. Even the relatively crude receivers available in the 1960s were capable of recording differences in temperature of a few millikelvin (mK), that is $\sim 0.1\%$ of the temperature of the background. To reach this level of sensitivity, however, required more attention to competing sources of microwave emission, such as the emission of the Galaxy, radiation from the ground that could leak into the apparatus, and gray-body emission from the Earth's atmosphere[17]. Keep in mind that the last two sources of competing emission were at roughly 300 K and 3–10 K, and the aim was to detect differences in temperature of order 0.001 K.

The first intentional project to measure or set upper limits to the anisotropy of the microwave background provides useful illustrations of some of these issues. In 1965, Dave Wilkinson and I set out to modify the 3 cm receiver used by Roll and Wilkinson to turn it into a device to make comparative measurements of a long strip of the sky; an "isotropometer," a name that has fortunately faded from the field. To reduce some sources of systematic error, we kept the instrument fixed and unmoving, with the main horn antenna pointed up at an elevation of 41 deg. As the Earth rotated, the instrument scanned a circle parallel to the celestial equator every 24 hours. Measurements of the temperature of the background along that circle were rapidly and continuously compared to values from another, smaller antenna pointed at the North Celestial Pole (NCP), a fixed point on the sky, which at Princeton also had an elevation of 41 deg. Observations of the NCP thus provided a fixed standard against which to compare the measurements along the circular scan. Since both measurements were made through the same thickness of the Earth's atmosphere, emission from the atmosphere canceled out to first order. To control possible instrumental variations, we also arranged for the main antenna to view the NCP every few minutes, by raising a reflecting sheet in front of the antenna. Screens were used to prevent radiation from the hot surface of the Earth reaching the antenna. We also took data for an entire year, so that diurnal or solar effects would average out to first order as the Sun drifted in Right Ascension.

The crucial finding of the "isotropometer" was that there was no significant difference in temperature along the scanned circle at a level of a few tenths of a percent or less (Partridge and Wilkinson, 1967) (Fig. 8.4). This limit improved Penzias and Wilson's upper limit on departures from isotropy by nearly two orders of magnitude. In particular, although we did detect a slight increase in temperature when the instrument viewed the plane of the Milky Way, it was less than 1% of the background temperature—the "excess temperature" was clearly not entirely Galactic in origin. Nor did we find any evidence that the temperature of the microwave background varied with ecliptic coordinates. The "excess temperature" did not originate in the Solar System. The Gravity Group's explanation for the 3 K background had passed another test.

One further possibility remained. Could the microwave background be the summed emission of many extragalactic sources, such as quasars, discovered just a few years earlier? This emission would be isotropic on a large scale. Such suggestions were put forward by Sciama (1966), Gold and Pacini (1968), Pariiskii (1968), and Layzer (1968) and more fully developed by Wolfe and Burbidge (1969). While the spectral observations just discussed certainly presented problems for such an alternative explanation, it was by

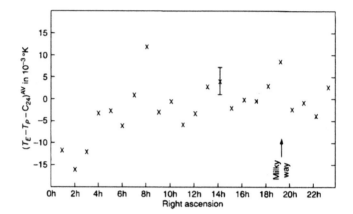

Fig. 8.4 *Measurements of fluctuations in the sky temperature along a circle at −8° declination made by the Princeton "isotropometer" (from Partridge and Wilkinson, 1967). As the representative error bar suggests, the scatter is consistent with noise, mainly atmospheric emission. The results allowed only an upper limit to be set on CMB anisotropy.*

no means fanciful. We have subsequently learned that much of the X-ray and the far infrared (IR) backgrounds is contributed by discrete sources. To test this possibility in the case of the microwave background, searches for small scale fluctuations in the "excess temperature" were needed. If emission from discrete sources were responsible for the background, we would expect small statistical variations in the measured temperature as the number of sources varied randomly across the sky. Two radio astronomers at Stanford, Edward (Ned) Conklin and Ron Bracewell (1967), had shown that anisotropies on 13′ scale were less than 0.008 K in amplitude. These results were confirmed by Epstein (1967) at a much shorter wavelength, though with a looser upper limit. The Stanford upper limit on small scale fluctuations was too tight to fit comfortably in an explanation based on emission from galaxies: the background was too smooth (Smith and Partridge, 1970).[18]

The second test of the cosmic origin of the "excess temperature" had also been quite comprehensively passed: it was remarkably isotropic. By 1970, it had become entirely appropriate to refer to the 2.7 K background as the *cosmic* microwave background (CMB), and we will do so from now on. The existence of a CMB in turn established that the Universe had a history, and that its present properties are very different from those early in its history.

8.4.3   Small amplitude anisotropies

The ability of observers to detect anisotropies at the ∼ 0.1% level heightened interest in processes that could induce small amplitude anisotropies in an initially isotropic CMB. These fall into two different classes. The first possibility is large angular scale anisotropy induced either by the motion of the observer or by anisotropic expansion. The other concerns much smaller scale variations in temperature caused by the interaction of the CMB photons with inhomogeneous clumps of matter while they are on the way to our detectors. Note that the starting assumption in each case is an initially isotropic background. I defer a discussion of what is in fact the most prominent and interesting source of fluctuations in the CMB, primordial density perturbations present on the surface of last scattering, to Section 8.5.4.

Dipole anisotropy

Motion of the observer with respect to the rest of the Universe induces a Doppler shift in radiation,[19] including the CMB. To first order, this produces anisotropy in the measured temperature of the CMB with an amplitude $\delta T/T = (v/c)\cos\theta$, where v is the speed, and θ is the angle between the direction of motion and the direction of observation (Peebles and Wilkinson, 1968; Bracewell and Conklin, 1968). In the direction of our motion, a slight excess in the temperature is seen, and a slight decrease in the opposite direction, producing a dipole pattern of amplitude δT in CMB maps.

One goal of the isotropometer experiment at Princeton was to detect the dipole signal expected from the motion of the Solar System in our Milky Way Galaxy. In the mid-1960s, the velocity of the Solar System in the Milky Way had a reasonably well-established direction, and an estimated magnitude of 200–250 km s^{-1}. This is less than 0.1% of the speed of light and so the required sensitivity for a clear detection was of order one millikelvin (mK). For the observing scheme used at Princeton, a dipole signal on the sky produces a modulation in the measured temperature with a 24 hour period. Neither that instrument, nor an improved version installed in 1967 at a better site in Yuma, Arizona, produced results of sufficient accuracy to detect the Earth's motion. The Princeton group was able only to set upper limits to a dipole modulation, $\delta T < 4$ mK, a 2σ upper limit deduced from results cited by Partridge (1969). We also set comparable limits on modulation with a 12 hour period, a signal expected if the CMB had a quadrupole moment. These experiments were not limited by sensitivity, but by fluctuating emission from the Earth's atmosphere. The main culprit was lumpy water vapor, that is, clouds. Despite the fact that Yuma was selected as the least cloudy place in the USA, atmospheric fluctuations remained a problem. In hindsight, a better choice would have been to work at a high altitude site, as indeed the Princeton group did for the spectral measurements described above.

The next year, Conklin and Bracewell took advantage of the same high altitude site at 12,500 feet at the White Mountain Research Station in California to mount another search for the dipole signal, direct evidence of the Earth's motion. Their instrument was more symmetrical, employing two similar microwave horns that scanned two points along a circle in sky separated by roughly 60 deg. They thus made differential measurements along a circle at $\sim$ 32 deg declination, rather than comparing measurements with a fixed point on the sky. The more symmetrical design and a better site enabled Conklin (1969) to detect a dipole distortion in the CMB, and thus to determine the "Velocity of the Earth with Respect to the Cosmic Background Radiation," to quote the title of this important *Nature* article.

This dipole signal was in fact another confirmation of the cosmic origin of the CMB. If thermal radiation fills the Universe, and we observe it from a moving platform like the Solar System, we must see the dipole anisotropy induced by the Doppler effect. That is just what Conklin reported in 1969, and elaborated in his 1972 PhD thesis (Conklin, 1972). There was, however, an apparent problem with these results: the motion he reported lay almost opposite in direction to the then-accepted velocity of the Solar System in the rotating Galaxy as determined from observations of stellar motions. Did this discrepancy reflect an error in the measurements? Daringly, and correctly as it turned out, Conklin argued that the apparent discrepancy was in fact evidence for the motion of the entire Milky Way Galaxy relative to the rest of the Universe. What we observe is the vector sum of the Earth's velocity in the rotating Galaxy and the velocity of the Galaxy itself. Conklin and Bracewell's measurements indicated that the latter dominated. The implied magnitude of this "peculiar velocity," to use the usual astronomical term, was large, at several hundred km s^{-1}. At the time, such a large peculiar velocity was generally unexpected. The guarded reception of the Stanford results owed in part to this discrepancy, and in part to the large and uncertain correction for foreground synchrotron emission needed to extract the small Doppler signal, as shown in Fig. 8.5.

The Stanford experiment clearly benefited from a better site with reduced atmospheric emission. Could even better results be obtained from an instrument lofted to the top of the atmosphere by a balloon? Several such attempts were made, two early attempts involving Dave Wilkinson (Henry, 1971; Corey and Wilkinson, 1976). As Paul Henry ruefully points out in his contribution to FBB, straylight emission from the Moon and then instrumental problems kept him from a clear detection (Henry, 1971). Brian Corey (1978) was more successful, flying instruments operating at several wavelengths

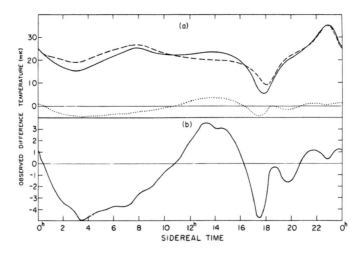

Fig. 8.5 *Top: The solid line represents the smoothed measurements by Conklin, and the dashed line his estimate of the Galactic contribution. The difference, shown amplified in the lower panel, is the evidence for the dipole anisotropy caused by the motion of the Solar System (Conklin, 1972).*

in order to control better the Galactic emission. The direction of future attempts to find the dipole signal had been set: work from high altitude or even space.

Quadrupole and related signals

The equations of general relativity that govern the expansion of the Universe are consistent with anisotropic as well as fully isotropic expansion and with large scale rotation (and shear). In the simplest case, consider a Universe expanding more rapidly along the *x*-axis than along the two orthogonal axes. Faster expansion implies more rapid cooling of the CMB; hence the measured temperature is lower along the *x*-axis. The resulting anisotropy has a quadrupolar distribution, oblate like a tangerine. Other, more complicated, geometries are possible which could induce dipole, quadrupole, and even more complex patterns (Kantowski and Sachs, 1966; Novikov, 1968; Misner, 1968). A primordial, large-scale, magnetic field could also induce quadrupole isotropy (Thorne, 1967). So too could large-scale rotation. The rough upper limits on a quadrupolar signal set by Partridge and Wilkinson (1967) were used by Hawking (1969) to establish stringent limits on cosmic rotation. Interest in these anisotropic models faded, and I will not consider them further.[20]

Smaller scale anisotropies caused by interaction with intervening matter

Theorists quickly realized that microwave photons streaming towards us would interact with matter lying along their path, and that these interactions could leave imprints on the CMB. Specifically, such interactions could produce small changes in the observed temperature, that is, fluctuations or small scale anisotropies. Two classes of interaction were recognized and evaluated in the 1960's: purely gravitational interactions and scattering by free electrons.

The first paper on gravitational interactions, by Rainer Sachs and Arthur Wolfe (1967), examined the effect on photon trajectories of perturbations in a uniform Robertson–Walker metric along the line of sight. As the photons propagate through an expanding perturbation small changes in the observed redshift is induced, and hence in the temperature of the CMB. Such fluctuations integrated over all perturbations along the line of sight are referred to as the "integrated Sachs–Wolfe" or ISW effect. As Sachs and Wolfe presciently pointed out, however, ISW temperature fluctuations could easily be swamped by any *intrinsic* temperature variations on the surface of last scattering. Finally, Sachs and Wolfe made a rough estimate of $\delta T/T$, finding a value of ~ 0.005 for a set of extreme assumptions. So large a signal was quickly ruled out by the observations, but the basic argument of this foundational paper was correct.

The presence of density inhomogeneities along the line of sight also causes small deflections of the photon trajectories, referred to as weak gravitational lensing. The relevance of weak lensing to CMB studies was first analyzed by Blanchard and Schneider (1987). Gravitational lensing of an entirely isotropic background has no effect—lensing affects only the anisotropies in the background, subtly modifying the pattern of any intrinsic anisotropies.

As we will see in the subsequent section, both these gravitationally-induced effects are swamped by the larger amplitude and much more interesting intrinsic fluctuations in the CMB. Indeed, it took decades to produce clear observational evidence for the ISW effect (Boughn and Crittenden, 2004). Lensing of intrinsic anisotropies in the CMB was detected by Smith et al. (2007) and Hirata et al. (2008).

The interaction between microwave photons and neutral atoms and molecules is generally very small: intergalactic space is truly transparent. On the other hand, the scattering cross section is much higher for free electrons. If the kinetic temperature of the free electrons greatly exceeds 3 K, scattering adds energy to the photons by the inverse Compton effect. This process was noted by Weymann (1966) and studied in more detail by Yakov B. Zeldovich and Rashid Sunyaev (1968). A global distribution of hot electrons will affect the overall CMB spectrum.[21]

A local assembly of hot electrons, on the other hand, induces a perturbation only in that part of the sky. One astronomical locale for large numbers of hot, free electrons is the plasma in clusters of galaxies (Sunyaev and Zeldovich, 1972). Some CMB photons passing through clusters receive a small boost in energy, and hence frequency. The net effect on the spectrum of the CMB is to induce a small shift of an initially black-body spectrum to higher frequencies, as shown in Fig. 8.6. Surprisingly, as the figure demonstrates, the result is to lower the observed CMB temperature for frequencies

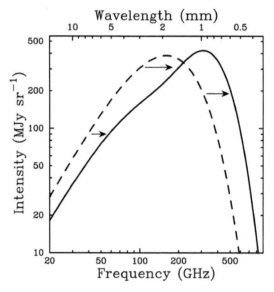

Fig. 8.6 *Illustrating the shift to higher frequencies caused by the Sunyaev–Zeldovich (SZ) effect. The magnitude of the effect is strongly exaggerated. Note that the SZ effect results in a lower intensity at wavelengths > 1.4 mm. (Adopted from an original in Sunyaev and Zeldovich (1980).) The slight "wiggle" in the shifted spectrum at wavelengths ~ 2–3 mm is present in the original drawing but is an oft-repeated artifact.*

below a critical value of $\nu_c = 217$ GHz, corresponding to 1.4 mm wavelength. The amplitude of the small temperature decrement in the direction of clusters depends directly on the product of the number density and temperature of the free electrons. Early X-ray observations of the hot plasma in clusters allowed a rough estimate of the decrement at frequencies below 217 GHz: ~ 1 mK. This was of course an enticing target for observers, and programs were promptly launched in the Soviet Union, the UK and the USA to detect the Sunyaev–Zeldovich (SZ) decrement in the direction of known clusters of galaxies. Early claims of detection were made by Yuri Pariiskii (1973), but the first truly convincing results were by Birkinshaw et al. (1984). Since the amplitude of the SZ signal is almost independent of redshift or distance, SZ searches offer a means of finding clusters at all redshifts; the first detection of previously unknown clusters of galaxies using the SZ effect was reported by the South Pole Telescope team in Staniszewski et al. (2009). The early measurements are reviewed by John Carlstrom and his colleagues (2002), who have led the study of SZ signals for several decades. Despite their importance to both astrophysics and cosmology, I will not consider SZ observations of clusters further here.

8.4.4 Inconsistent measurements

Sections 8.4.1 and 8.4.2 ended with ringing claims that the cosmic origin of the microwave background had been firmly established. There were, however, a few early experimental results which cast doubt on that conclusion. This section describes two of these, with a focus on an innovative measurement of the spectrum of the background carried out above the Earth's atmosphere by a team from Cornell and the Naval Research Laboratory (Shivanandan et al., 1968). This measurement was important for four reasons. It extended spectral measurements to much shorter wavelengths, 0.4–1.3 mm, where the contrast between gray-body and black-body spectra is strong. To reach these short wavelengths, a new detector technology was needed: this was the first CMB experiment to employ a bolometric detector. At such short wavelengths, the Earth's atmosphere is far from transparent; hence the experiment was lofted above the atmosphere by a rocket. Only the high sensitivity and wide bandwidth of the NRL/Cornell detector made possible a measurement in the short time the instrument was above the atmosphere. Finally, these results caused a considerable stir because they were in much better agreement with a gray-body spectrum than a ~ 3 K black-body spectrum. Was something odd going on at and beyond the peak of the putative 3 K spectrum? To answer this question, additional measurements at short wavelengths were needed. There was no hope of making them from the surface of the Earth, even at high altitude. An alternative to a short rocket flight was available: lifting an instrument above much of the atmosphere using a large balloon. Just this approach was adopted by Rai Weiss and his colleague at MIT, Dirk Muehlner. They employed a cryogenically-cooled, solid-state detector with a smaller bandwidth than the NRL/Cornell device. Their experiment was carried above most of the Earth's atmosphere (and an even larger fraction of the atmospheric water vapor) by a large balloon. The results, reported in 1970, cast serious doubt on the NRL/Cornell measurements, albeit over a smaller wavelength range (Muehlner and Weiss, 1970). On the other hand, the MIT measurement of the spectrum still fell above the level expected for a true black body.

In retrospect, it is clear that the apparent gray-body radiation detected in the rocket experiment came from rocket exhaust and other debris that traveled along with the experimental package after it was released from the rocket. The smaller "excess" seen in the balloon experiment has a more complex explanation. A possible spectral peculiarity at short wavelengths remained a niggling problem, even as further balloon and rocket-borne experiments were carried out, until the spectacular results of the COBE mission were announced in 1990 (see Section 8.5.3). Both of the pioneering experiments just listed, and the problems that led to false results, are discussed by the men who led them in FBB. Martin Harwit concludes his reminiscences by reminding us, "Sometimes it may be better to try difficult experiments and fail, than not to try at all." No observer will disagree.

8.5   Developments up to 1990 and the COBE results

While the cosmic origin of the microwave background had largely been accepted by 1970, the full power of CMB observations to reform and refine our understanding of the Universe took time to realize. This section briefly summarizes some of the developments leading up to the stunning results of the COBE mission, launched in 1989. While there is no attempt to treat these developments in strict chronological order, this section begins with a lull in CMB studies in the 1970s.[22]

8.5.1   A pause in the 1970s

In the late 1960s, in addition to the results on the temperature and isotropy of the CMB, a number of important papers on the physical and astronomical implications of the CMB appeared. Among the topics treated were: detailed calculations of nucleosynthesis in the initial minutes of the Hot Big Bang, including the production of both deuterium and helium (Peebles, 1966; Wagoner et al., 1967); fluctuations in the CMB temperature introduced by interactions with matter along the line of sight (Section 8.4.3); polarization of the CMB (Rees, 1968); and a precise estimate of when the CMB photons were last strongly scattered by matter. The last interaction, dominated by Thomson scattering of the CMB photons by free electrons, occurred a few hundred thousand years after the Big Bang, when the temperature was high enough to keep matter ionized. As the temperature dropped further, free electrons disappeared into neutral atoms, and the scattering of photons dropped sharply (Section 6.11). In 1968, Peebles and, independently, Zeldovich, Vladimir Kurt, and Sunyaev calculated both the epoch and redshift of this surface of last scattering, $z \sim 1,000$.

After their last scattering, the CMB photons are free to travel directly to our instruments. Thus when we study the CMB, we are observing this moment of last scattering, very early in the history of the evolving Universe. Maps of the distribution of the CMB across the sky are images of this surface of last scattering. If the density on this surface is inhomogeneous, we can expect small temperature fluctuations on that surface as well.[23] This crucial source of anisotropies was first investigated by Sunyaev and Zeldovich (1970a) and by Peebles and Yu (1970).[24]

These important theoretical papers were the first freshets of what was to become a flood of results. Surprisingly, however, there was a lull before that full flood began. In the 1970s, there was a nearly decade-long pause in CMB studies.[25] On the observational side, there is a plausible explanation: all the "easy" experiments had been done. Existing instruments, such as the Bell Laboratories horn antenna, or the Stanford radio telescope, had been pushed to their limits. Likewise, the first generation of specially designed instruments, such as those used by the Princeton group on White Mountain, had run their course. New techniques, new detectors, and better sites were needed.

Why interest in the cosmological consequences of a 3 K background leveled off is less clear. As noted in Chapter 6, during the 1970s cosmology as a field of research was making a transition from mathematical models to observation-based science. Relatively few theorists were at first interested in the *physics* of cosmology. There were a few clear exceptions. One is Peebles (1971), whose book is tellingly titled *Physical Cosmology*. Another is *Relativistic Astrophysics* by Zeldovich and Novikov, first published in Russian in 1967 but only in 1983 in English, which treats much of the physics laid out in this chapter (Zeldovich and Novikov, 1983). Also contributing to the "lull," I suspect, was the wealth of other recent astronomical discoveries engaging theorists: quasars, pulsars, gravitational lensing, and the evolution of galaxies. One salient example: in the 1970s, Rashid Sunyaev turned from cosmology to the study of X-ray binary systems and accretion disks.

IAU Symposium 63 (Longair, 1974), with the informative title "Confrontation of Cosmological Theories with Observational Data," provides a useful contemporary summary of the field in the early 1970s. This meeting, held in Krakow in 1973, brought together cosmologists from both the West and the Soviet bloc as well as theorists and observers. It clearly shows how far we had to go to convert cosmology into a precision science.

By the 1980s, the pace in CMB studies was picking up, as was true in other areas of the newly-emerging science of cosmology. Interest in the CMB was also fanned by anticipation of the results from COBE, a NASA mission that promised unrivalled spectral measurements and far more precise limits on, or even detection of, large scale anisotropies in the CMB.

8.5.2   Observational hurdles and the means to overcome them

If the "easy" experiments had been done by 1970, what made improved and more sensitive observations "hard"? We start with difficulties CMB observers faced, then turn to means to mitigate them.

Observational hurdles

First, there were instrumental hurdles, features of the experimental design that introduced either noise or bias into measurements of the CMB. For measurements of the CMB spectrum, raw sensitivity was not a problem. In rough numbers, 1960-era receivers could reach a few hundredths of a kelvin, a $\sim 1\%$ measurement, in an hour of observation. For anisotropy searches, however, where the goal was 1 mK or below, sensitivity was indeed a problem. Better, less noisy, detectors were needed. The "off-the-shelf" microwave receivers used in the first few experiments were not adequate. The stability of instruments was also an issue for both spectral and anisotropy observations. The Princeton "isotropometer", for instance, depended on the instrument remaining stable through a full cycle as the reflector was raised and lowered. Better observational techniques were needed. A third issue was stray radiation, especially from the hot ground, reaching the detector. Microwave instruments used in CMB research do not have sharp, well-defined beams. There are also subsidiary diffraction maxima, or "side lobes," that allow off-axis radiation to enter the antenna and reach the detector. This stray radiation plagued several early CMB measurements.

Angular resolution was also an issue. Diffraction limits the angular resolution of an antenna of diameter D to approximately λ/D radians. To achieve resolution of 1 deg at $\lambda = 3$ cm, for instance, a 2-m antenna is needed. This is an awkward size, larger than the specially designed instruments, such as those of the Princeton group, and smaller than conventional radio telescopes, for example, the 18 m dish employed by Conklin and Bracewell. For spectral measurements, resolution is unimportant, thus permitting a compact design. Not so for the search for CMB anisotropy on various scales—resolution does matter. Pushing to higher resolution carries the cost of larger antennae, and the additional cost of more integrating time to survey a given solid angle of the sky.

The observers also had to contend with sources of foreground radiation, as had Penzias and Wilson. These include emission from our Milky Way Galaxy and gray-body emission from the Earth's atmosphere. The spectrum of Galactic emission is complex at cm and mm wavelengths, since synchrotron, free-free, and dust emission can all contribute. For spectral observations at wavelengths shorter than a few cm and made away from the Galactic plane, the Galactic contribution is far smaller than the ~ 3 K CMB signal, and hence not a major problem. At longer wavelengths, corrections are required and depend on both the spectrum and the distribution of the Galactic emission. For anisotropy searches, on the other hand, Galactic emission matters a great deal. The signals sought are smaller and Galactic emission is both patchy and strongly anisotropic. To avoid the noise and bias introduced by Galactic emission, the only recourses are to make observations well away from the Galactic plane and to select an observing frequency where the Galactic emission is a minimum. The best wavelengths are roughly 0.2–1 cm (30–150 GHz). Note that the first generation of "easy" anisotropy experiments were all carried out at longer wavelengths, with the exception of those by Eugene Epstein (1967) and Stephen Boughn and his colleagues (1971).

For ground-based experiments, the choice of observing wavelength is further restricted by emission from the Earth's atmosphere. Atmospheric gases also weakly absorb microwave radiation, but the gray-body emission from a ~ 270 K atmosphere is the larger problem. The main culprits are water vapor, which has strong and broad emitting/absorbing spectral lines at 22 and 180 GHz, and oxygen, with lines at 60 and 120 GHz. For observations at wavelengths $\lambda > 2$ cm, say, atmospheric absorption is of order 1%, and thus contributes several K to any measurement made from the Earth's surface. At shorter wavelengths, the atmospheric emission is stronger, and observers need to work in atmospheric "windows" between H_2O and O_2 lines. Windows at ~ 30, 90, 150, and 220 GHz, corresponding to wavelengths $\lambda = 10$, 3.3, 2, and ~ 1.4 mm, are frequently exploited, but even in these windows, atmospheric emission exceeds the CMB signal. Emission from water vapor is a particular curse because it is not homogeneously distributed.

Leaping the hurdles

In the decades following the discovery of the CMB, observers worked to mitigate the effects listed above by minimizing them when possible, and then by measuring and correcting for any residual effects. That two-step process continues up to the latest CMB observations.

Atmospheric emission, for instance, can be minimized by working at high altitude and at a sensible frequency. The amount of water vapor drops off rapidly with altitude and it helps further to work at a cold site. Many CMB observers sought out mountain tops and arctic sites. They also took more care in measuring and correcting for residual atmospheric emission.[26] Clearly, working above the atmosphere, or most of the atmosphere, was the ultimate goal. That meant stratospheric balloon flights (see Section 8.4.4; see also Woody and Richards, 1979; Peterson et al., 1985), rockets (Shivanandan et al., 1968; Gush, 1981; Matsumoto et al., 1988), high flying aircraft (Smoot et al., 1977), and ultimately satellite platforms. Both rockets and high altitude flights suffered from restricted observing time, amounting to a few minutes above the atmosphere for sub-orbital rocket flights, and from the difficulty of ensuring both stability and adequate shielding from stray light. Experiments lofted by balloons could be larger and better shielded; in addition, a typical balloon flight allowed many hours of observation above most of the Earth's atmosphere. Balloon flights, however, often failed. Carefully designed experiments ended up in the ocean, the Brazilian jungle (Lubin et al., 1985) or in free fall from stratospheric altitudes (see Weiss in FBB). For all these flight modes, the drive for altitude inexorably increased the complexity and cost of CMB observations. Teams, not individuals, were required to mount such larger-scale experiments.

Also adding to the cost and complexity was better and more intentional design of the apparatus, including receivers. More attention was paid to sources of systematic error as well as sensitivity. One example is the need to shield the detectors from stray light, emission from the ~ 300 K surface of the Earth diffracted into the receiver. CMB experiments began to sprout elaborate ground shields. In addition, horn antennae were designed to reduce the amplitude of side lobes.[27] Faster switching reduced the effect of instabilities in the instrument.

The most valuable and lasting technological advance in CMB studies occurred when astronomers familiar with IR techniques entered the field. The first few years of CMB studies were dominated by radio astronomers and engineers. They naturally employed techniques they were used to, such as heterodyne receivers and Dicke-switched radiometers. These techniques had to be pushed to shorter wavelengths with increased sensitivity, and they were. IR techniques, brought to CMB studies by astronomers and physicists such as Harwit, Francesco Melchiorri, Paul Richards, and Weiss, transformed the field. IR astronomers use bolometers: small solid-state devices that register the intensity of incoming radiation independent of its wavelength. For instance, incident radiation, of whatever wavelength, slightly raises the temperature of a small device, and this temperature change can be detected electronically. Bolometric devices soon proved to be far more sensitive at short wavelengths than radio-astronomical detectors, and IR astronomers worked both to make the bolometers more sensitive and to extend their response to longer wavelengths.

One contributing element in the greater sensitivity of bolometers is their wide bandwidth. That very property, however, raises a question—how can one restrict the response to a desired range of wavelengths? This is accomplished using filters, but care needs to be taken to avoid thermal emission from the filters themselves. Hence they are frequently cooled. The bolometers themselves also had to be cryogenically cooled to realize their full sensitivity. The IR astronomers also brought their expertise in cryogenics to CMB observations, and cooling detectors to reduce instrument noise became the norm even for heterodyne devices.

Larger antennae and/or shorter wavelengths were introduced to improve the angular resolution of CMB anisotropy searches. The drive to larger apertures was pushed furthest by the use of interferometers, arrays of telescopes for which the effective resolution is set by the size of the array.[28] One early attempt was made by Martin et al. (1980), but better results were soon produced by Fomalont et al. (1988) and Subrahmanyan et al. (1993). To some degree, interferometry also averages out atmospheric fluctuations and mitigates some other sources of systematic error. Hence the use of interferometric techniques was extended to smaller arrays in order to reduce errors rather than to obtain higher resolution (Timbie and Wilkinson, 1990). Higher resolution, however obtained, carries a cost. The apparatus is larger or more complex, especially in the case of multi-antenna arrays.

Many of the early developments just described are examined critically and in greater detail in a review article on CMB observations by Weiss (1980).

8.5.3   Selected observational advances before COBE (1970–1990)

As CMB observers overcame the hurdles just described, the pace of observations picked up. In this section, I present a selection of the experiments leading up to the launch of NASA's COBE mission in late 1989, with an emphasis on those that had the most impact on the field or those that introduced new observational techniques.

Improved measurements of the spectrum of the CMB

Many groups tried to improve measurements of the CMB spectrum at short wavelengths, near the peak of a ~ 3 K black-body spectrum. An early and important result emerged from the balloon-borne experiment of Dave Woody and Paul Richards (1979). Richards and his colleagues at Berkeley were to make important advances in the design of bolometers; the bolometer arrays used in current CMB experiments from the *Planck* mission to the South Pole Telescope are his "children." The Berkeley balloon flight was successful, and a spectrum spanning 0.25–6 mm was obtained. There was, however, a niggling feature in the spectrum, an excess around 1.6 mm wavelength. In addition, the measured temperature $T = 2.96$ K was high compared to CN and ground-based measurements. At about the same time, a rocket experiment by Herb Gush and his colleagues at the University of British Columbia (Gush, 1981) re-measured the short-wavelength spectrum of the CMB, again using a scanning interferometer. Despite the short duration of the rocket flight, this measurement was more precise than the Berkeley experiment. They did not find the puzzling spectral feature, but they did find an excess at a different wavelength. Uncertainties about the CMB spectrum at short wavelengths lingered.

One more, pre-COBE, attempt to sort out the sub-millimeter spectrum was made by Jeff Peterson, Richards, and Thomas Timusk (1985), again using a balloon to lift their experiment to high altitudes. It employed a bolometric detector and a set of filters to define five spectral bands in the 1–4 mm range, shown in the lower half of Fig. 8.7. Their results did not show the various spectral excesses and features seen by earlier programs, and the temperature they measured was fully consistent with the most recent CN results available to them (Fig. 8.7). The short-wavelength problem had been solved.

Or had it? Richards was also working with a Japanese group, headed by Toshio Matsumoto, to fly a carefully calibrated experiment on a sounding rocket. Their results once again appeared to show a short-wavelength excess (Matsumoto et al., 1988).[29]

At the same time, astronomers working at optical wavelengths were improving measurements of the interstellar CN absorption lines.[30] Some of these projects were also able to detect faint traces of absorption by CN in its second excited state, yielding

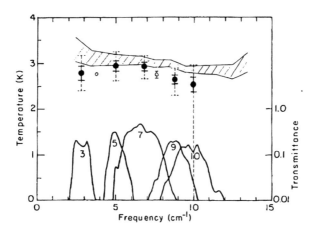

Fig. 8.7 *A summary of some pre-COBE spectral measurements at short wavelengths (Peterson et al., 1985). The results of Peterson and his colleagues are shown by filled circles with band passes shown below and the contemporary CN measurements by small open circles. The broad hatched band traces the earlier results of Woody and Richards (1979).*

values for T at 1.3 mm as well as 2.6 mm (Fig. 8.7). Kogut et al. (1990b) used absorption lines of a different interstellar molecule, H_2CO, to probe the background temperature at 2.1 mm. Atomic and molecular lines in sources at high redshift were also used to confirm the relationship $T(z) = T_0(1 + z)$.[31] As the CN results became more precise, they agreed better and better with a black-body spectrum, and cast increasing doubt on the apparent excesses at short wavelength. The issue was settled with iron-clad finality by the results from the FIRAS instrument flown on COBE.

Observers were well aware that COBE would make spectral measurements only at wavelengths shorter than 1 cm. As a complement to these anticipated results, more precise measures of the CMB spectrum at longer wavelengths were sought. A large international consortium, led by Reno Mandolesi, Giorgio Sironi, and George Smoot, set out to make such measurements, returning to White Mountain in 1982 and 1983. This was in effect a scaled-up version of the Princeton 1967-8 experiments, but with more instruments, some of them tuned to longer wavelengths. Technical improvements included an improved cold comparison source and more consistent monitoring of atmospheric emission. Six wavelengths were probed from 12 cm down to 0.33 cm, ensuring some overlap with COBE's planned wavelength range at the lower end. Results were consistent with a black-body spectrum (Smoot et al., 1985). There was no evident trend in T with wavelength. This work was continued, mainly by the Berkeley and Milan groups, for several more years.[32]

The elegant experiment of David Johnson and Wilkinson (1987) also deserves mention. Their balloon-borne instrument operated at 1.2 cm. At a flight altitude exceeding 25 km, the residual atmospheric emission was only 2 ± 2 mK, permitting a high precision determination of T. The final value was $T = 2.783 \pm 0.025$ K, precise but not in particularly good agreement with other results, or the later findings of COBE.

Tighter upper limits on anisotropy of the CMB

In the two decades leading up to COBE, numerous efforts were made to find fluctuations in the CMB, or to set more restrictive upper limits on the amplitude $\delta T/T$ of such fluctuations. For much of that period, observations were made on a scatter-shot basis, without much regard for theoretical guidance on which angular scale to emphasize. Observers simply tried to drive $\delta T/T$ down on a variety of angular scales, determined primarily by the frequency and by the aperture of the antenna they employed. Many of the techniques outlined in Section 8.5.2 were brought to bear to increase sensitivity and control sources of systematic error.

At least by 1970 there was a tentative goal. Several early theory papers made approximate predictions of the expected amplitude of CMB anisotropies, Sachs and Wolfe (1967), Peebles and Yu (1970), and Zeldovich (1970b) prominent among them. These estimations of $\delta T/T$ were based on the then reasonable assumption that baryons made up most of the density of the Universe. On this assumption, $\delta T/T \sim 10^{-4}$ or more, corresponding to $\delta T \sim 0.3$ mK, was expected. This was just in the grasp of the improved observational techniques becoming available in the 1980s. The hunt was on.

Observers had already learned that atmospheric instability made sensitive observation on large angular scales very difficult. Consequently, searches for anisotropy on degree scales and above were mostly balloon-borne; rocket flights were too short to yield the necessary sensitivity. One instance is the work of Fabbri and his colleagues (1980)—the claimed detection of quadrupole anisotropy in that paper, however, was not borne out by later observations. A pioneer satellite experiment, Relikt 1, was launched by the Soviet Union in 1983. Like its successor, COBE, it mapped the entire sky and robustly detected the dipole signal. Relikt 1 also set a stringent limit on the quadrupole anisotropy of 0.2 mK (Strukov and Skulachev, 1984).

On smaller angular scales, atmospheric fluctuations are less pronounced, so ground-based observations are possible. A salient example is the careful work by Juan Uson and Wilkinson (1984) using the large Green Bank antenna. Their upper limit of roughly 0.03 mK on scales of ~ 5 arcmin was the most stringent at the time. Five years later, the upper limit was further tightened by Readhead et al. (1989). These upper limits to anisotropies fell well below the expected level ($\delta T \sim 0.3$ mK) for baryon fluctuations. Figure 8.8 summarizes the many pre-COBE attempts to find or set limits on anisotropies in the CMB.

8.5.4 Advances in theory

Malcolm Longair argues in Chapters 6 and 10 that cosmology as a field of science "changed gears" in the 1980s. That was largely true for CMB studies as well. Prominent theorists such as Peebles, Joe Silk, and Zeldovich and their colleagues made major contributions to our understanding of the CMB and its relevance to cosmology. The

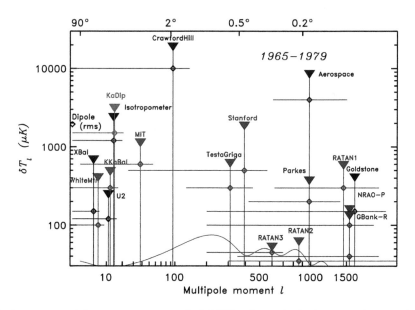

Fig. 8.8 *A summary of the many attempts to find CMB anisotropies in the years 1965–1979. All save the dipole measurement are upper limits; the now-accepted level of fluctuations is shown by the solid line. (From FBB, which includes citations.)*

arrival of a new wave of young researchers—Dick Bond, George Efstathiou, Alan Guth, and many others—also enriched the field. The treatment here of advances in theory in the years before COBE focusses on theoretical developments with close ties to observations of the CMB.[33] What physical insights might help guide experimental projects? What cosmological information could be wrung from increasingly sensitive CMB observations?

At first blush, Alan Guth's introduction of inflation (1981), the subject of Chapter 11, might seem far removed from either question. An inflationary period, however, was introduced in part to solve the problem of the large-scale isotropy of the CMB. How did two regions of the Universe large enough not to have been causally connected at the epoch of last scattering "know" to have the same temperature? Inflation provided a brief and much earlier, but adequate, period of causal connection to explain the observed isotropy. A period of rapid expansion also offered an explanation of the large size, or more precisely, the small geometrical curvature, of the Universe. Inflation also predicted the existence of density fluctuations on astronomical scales, as discussed in Chapter 11. These inhomogeneities are quantum fluctuations, increased dramatically in scale by inflation. Inflation even predicted, or at least strongly favored, a particular spectrum for these fluctuations, meaning how the density contrast $\delta\rho/\rho$ varied with mass or linear scale (Harrison, 1970; Zeldovich and Novikov, 1970). That so-called Harrison–Zeldovich spectrum leads to fluctuations of the same amplitude on all angular scales when they came through their cosmological horizons. It also sets one of the fundamental parameters of cosmology, the scalar index of the perturbations n_s, at unity. The basic concept of inflation has spawned a vast number of variants, as outlined in Chapter 11, including many different descriptions of how inflation ends. These in turn predict slightly different, but always small, departures from the exact, scale-independent, Harrison–Zeldovich spectrum: n_s falls a bit below 1.000 and so fluctuations on small scales have a slightly smaller amplitude than those on large scales.

Primary (intrinsic) fluctuations in the CMB

Inflation provides an explanation for density inhomogeneities on astronomically relevant scales at the epoch of last scattering. These in turn imprint temperature fluctuations in the CMB on a wide range of angular scales on the surface of last scattering (Sunyaev and Zeldovich, 1970a; Peebles and Yu, 1970). A map of the distribution of the CMB intensity across the sky is in effect a "baby picture" of the Universe, showing how matter was distributed at the particular, early moment when the CMB photons last scattered. These fluctuations are *intrinsic* to the CMB, generated before or at the epoch of last scattering, and not those imposed later by interaction with intervening matter. The growing understanding of the origin and properties of these intrinsic, or *primary*, fluctuations is, to my mind, the major theoretical advance made in the years between the discovery of the CMB and the launch of COBE a quarter of a century later.

While inflation specified a *spectrum* for density inhomogeneities, it did not, and could not, provide any firm estimate of the overall *amplitude* of these fluctuations A_s. Estimates of the magnitude of $\delta\rho/\rho$ and $\delta T/T$ came from quite different physical insights. First of all, assume that all the structure visible in the Universe today, from galaxies to clusters to even larger systems, was formed by the action of gravity alone. Both detailed simulations and observations have shown that this assumption is a strikingly good one for the early, linear evolution of such systems. Gravity acts to increase $\delta\rho/\rho$; regions of over-density contract and thus increase their density contrast.[34] Evgenii Lifshitz showed in 1946 that $\delta\rho/\rho$ grows linearly with decreasing redshift as $(z+1)^{-1}$ as long as matter dominates the rate of expansion—the relation for the earlier radiation-dominated phase is $\delta\rho/\rho \propto (z+1)^{-2}$.

We know that matter in the present Universe at $z=0$ is inhomogeneously distributed. On length scales smaller than about 8 Mpc, $\delta\rho/\rho$ now exceeds unity[35]. Assuming matter dominated between last scattering and the present, we can use Lifshitz's simple rule to see that the amplitude of density fluctuations at $z=1000$ had to be of order 0.001 or greater on these scales. Next, if these fluctuations were adiabatic, that is, if both radiation and baryonic matter were similarly compressed,[36] then $\delta T/T = \frac{1}{3}\delta\rho/\rho$. This connection is far from exact, as we will see, but gives an order of magnitude estimate of $\delta T/T$, and was so used until better calculations emerged. These arguments suggest an amplitude $\delta T/T \sim (1-3) \times 10^{-4}$, that is δT would be expected to be a few tenths of a mK. In fact, the calculated amplitude $\delta T/T$ is lower, as explained below. The important point, however, is that it *can* be calculated.

Another crucial theoretical advance after the lull of the 1970s was the recognition that physical processes can alter the initial flat $n_s \sim 1$ spectrum of fluctuations emerging from inflation, at least on length scales smaller than the causal horizon, ct. These processes affect the spectrum of density fluctuations before or at the epoch of last scattering, but only on scales $< ct$. The time that is relevant is the epoch of last scattering, now known to be 380,000 years.

The "physical processes" just referred to are in principle quite simple, but in fact quite intricate.[37] Here, I skip most of the details, and instead emphasize the end results. As already noted, gravity tends to amplify the density contrast $\delta\rho/\rho$ of regions that happen to be over-dense. Countering gravity is pressure, both thermal and radiation pressure. The struggle between these two produces oscillations in the density of the plasma-radiation mixture: sound waves.[38] As the important early work of Sunyaev and Zeldovich (1970a) showed, sound waves with certain physical scales are favored; on these scales, $\delta\rho/\rho$ and hence $\delta T/T$ reach maxima. A rough estimate of the linear scale of the largest of these harmonic oscillations can be obtained from $R_s = c_s t$, where c_s is the sound speed and t the epoch of last scattering. In turn, c_s depends on the physical properties of the early plasma, including the ratio of the density of baryons to that of the CMB. To some degree, the epoch of last scattering also depends on cosmological parameters. In any case, density fluctuations of scale R_s and its harmonics, $R_s/2, R_s/3, \ldots$, are prominent on the surface of last scattering. These in turn produce temperature fluctuations. The connection between $\delta\rho/\rho$ and $\delta T/T$ is more complicated than the simple estimate $\delta T/T = \frac{1}{3}\delta\rho/\rho$ used earlier. Sachs and Wolfe (1967), for instance, pointed out that overdensities perturb the metric, affecting the redshift. The consequence is that overdense regions actually appear cooler than average, not hotter, as expected from the naïve argument that $\delta T/T = \frac{1}{3}\delta\rho/\rho$. It remains true, however, that the surface of last scattering is marked by regions of slightly higher temperature and regions of slightly lower temperature, and these have characteristic linear scales that depend only on the mix of baryons, photons and dark matter at the time of last scattering, through the sound speed. There are discernible features in the "baby picture" of the Universe provided by the CMB (see the frontispiece).

If these fluctuations are adiabatic, as expected if they originate from very early quantum fluctuations, both photons and baryons are equally compressed.[39] Photons, however, are free to diffuse out of overdense regions, thus reducing the density contrast in radiation. This diffusion of photons, first recognized by Silk in 1967, and hence known as *Silk damping*, further reduces the amplitude of temperature fluctuations for a given level of $\delta\rho/\rho$ (Silk, 1968). Since photons can more easily diffuse out of small scale fluctuations than large ones, Silk damping washes out the amplitude of temperature fluctuations more and more as their linear scale decreases. Both the resonant oscillations and the damping caused by diffusion of photons, of course, can operate only on scales smaller than the causal horizon at the epoch of last scattering. On larger scales, physical processes had no time to operate, and we see the initial, nearly flat spectrum of primordial fluctuations.

The baryonic matter is ionized and so the photons scatter from the free electrons. This creates a drag on the electrons and the positively charged ions follow thanks to the Coulomb forces between them. While the density of radiation dominates, this drag acts to decrease $\delta\rho/\rho$ in the baryonic matter. That is not the case for the dark matter, however, since photons do not interact with it, except through gravity. Thus $\delta\rho/\rho$ in baryonic matter can and does fall below the density contrast in dark matter. Note that all these processes tend to reduce, as well as complicate, the level of fluctuations $\delta T/T$ expected on the surface of last scattering, as noted by Sunyaev and Zeldovich (1970a). It is important to add, however, that all of these processes are well understood and linear; hence they can be modeled well once we know the mixture of photons, baryons, and dark matter at the epoch of last scattering.

The power spectrum of CMB fluctuations

It is a challenge to present all the information on $\delta\rho/\rho$ compactly and clearly. Correlation functions offer one choice and were used by the COBE DMR team to present their first results (Smoot et al., 1992). Another compact and more informative means of characterizing fluctuations as a function of their angular scale is the power spectrum, which shows the fluctuation power as a function of spatial frequency, ℓ, where $\ell \sim 180\,\deg/\theta$ and θ is expressed in degrees. Provided the temperature fluctuations are statistically random, the power spectrum contains all the statistical information contained in a two-dimensional map of the CMB (such as Fig. 8.12 or the frontispiece of this volume).

We begin by noting that the complete distribution of CMB temperature across the sky can be fully described in terms of the spherical harmonics, $Y_{\ell m}$:

$$T(x) = \sum_{\ell=2}^{\ell_{max}} \sum_{m=-\ell}^{\ell} a_{\ell m} Y_{\ell m}(x). \tag{8.1}$$

The coefficients $a_{\ell m}$ have the units of temperature. For Gaussian random fluctuations, the fluctuation power is defined by $\ell(\ell+1)C_\ell/(2\pi)$, where C_ℓ is a weighted average over all m of the coefficients $a_{\ell m}$:

$$C_\ell = \frac{1}{2\ell+1} \sum_{m=-\ell}^{\ell} |a_{\ell m}|^2. \qquad (8.2)$$

Since C_ℓ is expressed as power,[40] not amplitude, the units are K^2 or more usually $(\mu K)^2$. This representation was introduced into theory papers in the 1980s by Bond and Efstathiou (1987). It has now become the standard way of presenting both calculations and measurements of CMB fluctuations.

Given the various physical processes sketched in this subsection, we can construct an instructive cartoon version of the power spectrum (Fig. 8.9). At low ℓ, corresponding to large angular scales, the spectrum of the fluctuations takes the unaltered Harrison–Zeldovich form, with a slight slope caused by the departure of n_s from exactly unity. At angular scales smaller than the causal horizon at last scattering, the resonant oscillations produce peaks and troughs in $\delta\rho/\rho$ and hence fluctuations $\delta T/T$; since power rather than amplitude is plotted, however, these both produce peaks in the power spectrum. Silk damping decreases the amplitude of the resonant fluctuations as θ decreases or ℓ increases. These combine to produce the cartoon shown in Fig. 8.9. In this cartoon, the slight departure of n_s from exactly 1.000 is greatly exaggerated for clarity. The vertical dashed line indicates the angular scale corresponding to the causal horizon on the last scattering surface, $\theta \sim 2\,\mathrm{deg}$ assuming a flat, Euclidean geometry. The first resonant peak

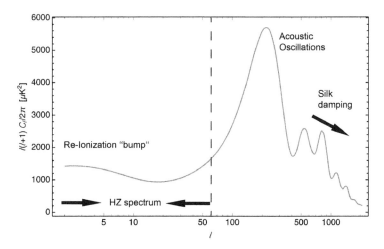

Fig. 8.9 *A cartoon version of the power spectrum, showing the approximately flat (Harrison–Zeldovich) spectrum at large scales, and the oscillations at $\ell > 100$ or so. These in turn are more strongly damped at still smaller scales (higher ℓ).*

appears at a somewhat smaller scale (larger ℓ), since its physical scale is determined by the speed of sound in the mix of matter and radiation rather than the speed of light. Recall also that the overall amplitude of the power spectrum can be substantially smaller than that naïvely expected from the simple relationship $\delta T/T = \frac{1}{3}\,\delta\rho/\rho$ because of photon drag on the baryons and other effects.

We need to make two adjustments to this cartoon version, both affecting the power spectrum at low ℓ. The first is cosmic variance. This refers to the uncertainty in predicting the power spectrum from a given cosmological model. A given model can produce many realizations of the sky, but we have only one sky to observe. A model produces values of C_ℓ that are ensemble averages over the $a_{\ell m}$. But the sky represents just a single realization, which could differ from the average. The amplitude of this unavoidable cosmic uncertainty is given by $\Delta C_\ell = (2/(2\ell+1)^{1/2})C_\ell$, so that it affects the power spectrum only at low ℓ. Cosmic variance for one particular model is indicated by the shaded area in Fig. 8.13.[41]

The second adjustment involves an important transition in the baryonic contents of the Universe that we have not yet considered.[42] At redshifts of order 10, the baryonic matter was re-ionized, presumably by the UV radiation streaming out from the first generation of stars or galaxies. From the results of the Gunn–Peterson test (1965), we know that the intergalactic baryonic matter is now highly ionized. While the details of how and when that re-ionization takes place are still not completely settled, re-ionization reintroduces free electrons, which then scatter the CMB photons. Since the Universe is far less dense at $z \sim 10$ than it was at $z \sim 1{,}000$, this scattering is weak, but it still acts to damp the amplitude of CMB fluctuations. Current estimates for the scattering optical depth, τ, range from 0.05 to 0.09, so that the effect on the power spectrum is small, a reduction by a factor $e^{-2\tau}$, or $\sim 10-20\%$. Since the scattering is another physical process, it can affect only scales smaller than ct, where t now refers to the epoch corresponding to $z \sim 10$. Thus the power spectrum is unaffected at very low ℓ (large scales) but slightly diminished at $\ell > 10$ or so. It is the high-ℓ spectrum that is affected by re-ionization, but the detectable "fingerprint" of re-ionization is the apparent, slight upward bump in the power spectrum at low ℓ, shown in exaggerated form in the cartoon Fig. 8.9. Detailed models such as that shown in Fig. 8.13 properly include this effect.

Let us list some of the cosmological parameters invoked to construct the power spectrum, whether it be the cartoon version of Fig. 8.9 or the precise curve shown in Fig. 8.13. A_s determines the amplitude of the power spectrum, and n_s determines the overall slope of the initial spectrum of fluctuations, quite small if n_s is close to unity. The scattering of CMB photons caused by re-ionization also affects the overall amplitude at $\ell > 10$ or so. The amplitude and the physical scale of the resonant peaks R_s and its harmonics depend on the properties of the matter–radiation mix before and at the epoch of last scattering, including the density of baryons, Ω_b. In particular, the ratio of the amplitudes of odd numbered peaks, those corresponding to compressions or over-densities, to even numbered peaks, corresponding to under-densities, is sharply dependent on Ω_b.[43] The angular scale at which we detect the resonant peaks depends on the angular diameter distance, which in turn depends on both H_0 and the geometric curvature of space. To construct an accurate power spectrum, we need values for all these parameters. Conversely, once we have an observed power spectrum, we can determine the values of these same parameters which produce a power spectrum best matching the observations. Herein lies the power of CMB observations for the determination of cosmological parameters and the history of the Universe.

By the 1980s, increasingly stringent upper limits on $\delta T / T$ came into tension with the predictions of cosmological models which contained only baryonic matter. Low values of $\delta T / T$ implied low values for the amplitude of density fluctuations on the surface of last scattering, too low to serve as the "seeds" of the structure we see in the Universe today. Dark matter broke this tension, since larger values of $\delta\rho/\rho$ were allowed for dark matter; these fluctuations seeded all later structures. This indeed became one of several arguments for the existence of substantial amounts of dark matter.[44]

8.5.5 COBE and some precursors

The Cosmic Background Explorer (COBE) space mission first detected fluctuations in the CMB, albeit only at low ℓ. Plans for COBE were developed late in the 1970s, while many of the theoretical advances just discussed were still in gestation. Several groups applied to NASA, with different goals in mind, sometimes overlapping.[45] Two paramount goals emerged: a refined measurement of the CMB spectrum across the peak of a ~ 3 K black body, and a search for large-scale anisotropies in the CMB. Measurements were also proposed at sub-millimeter wavelengths to search for a potential far IR background. In the end, under the firm, effective oversight of Nancy Boggess, NASA decided to merge all the proposed experiments and proposing groups. Thus COBE emerged with three instruments: the Far Infrared Absolute Spectrophotometer (FIRAS) to measure the CMB spectrum, the Differential Microwave Radiometers (DMR) to measure its isotropy, and the Diffuse Infrared Background Experiment (DIRBE) to find and characterize the cosmic IR background (Boggess et al., 1992). At the time, some CMB researchers, both those in the COBE team and those outside it, considered this merger odd and potentially unhelpful. As it turned out, Boggess and NASA were right: the three instruments could work together and share expensive elements such as the cryostat.

The DMR experiment (Smoot et al., 1990) employed three pairs of matched horn antennas, a design that goes back to Conklin and Bracewell. Each pair made comparative measurements of the sky as the satellite rotated; these were subsequently unfolded to produce maps of the CMB. Each pair of antennas operated at a different wavelength, 0.33 cm, 0.57 cm, and 0.95 cm, to provide a means of separating CMB fluctuations from Galactic signals since their wavelength dependence is quite different. FIRAS, like several balloon and rocket experiments preceding it, employed optical/IR technology, including a cooled Michelson interferometer to scan the CMB spectrum (Mather et al., 1990). Two sources were compared simultaneously, the sky and a carefully designed blackbody source whose temperature could be modified in a narrow range around 2.7 K. When the temperature of the internal black–body exactly matched that of the sky, the signals canceled; any temperature difference caused a departure from that null. This design permitted a very precise determination of the CMB temperature. Changing the length of the arms of the interferometer permitted the FIRAS team to sample easily the radiation temperature T at different wavelengths with the same, stable instrument.

There is no room or need here to describe the various vicissitudes faced by the COBE team (see Mather and Boslough, 1996). These delayed the launch until 1989, but all three instruments worked to perfection, and all three produced results that profoundly affected and advanced cosmology.

The first results to be made public were the initial FIRAS observation of the CMB spectrum at wavelengths of 1 to 0.05 cm. The preliminary CMB spectrum shown by John Mather at the 1990 winter meeting of the American Astronomical Society was based on just nine minutes of observation! Those stunning results resulted in a rare but well-deserved standing ovation from the audience of astronomers and, a decade later, in an equally well-deserved Nobel Prize for Mather. There was not a single hint of any departure from a black-body curve in the range of wavelengths observed (Mather et al., 1990). The final FIRAS spectrum is shown in Fig. 8.10. Furthermore, the measured temperature, later refined to 2.7255 ± 0.0006 K by the COBE team (Fixsen, 2009), agreed with the painfully accumulated ground-based and CN measurements, but of course was far more precise.

An even bigger impact, particularly on the public, was made in 1992 when the DMR results were announced by the COBE team. For the first time, after a quarter century of searching for them, fluctuations in the temperature of the CMB had been detected (Smoot et al., 1992). The resolution of the DMR was limited to several degrees by the small size of the antennae. Thus the anisotropies detected were the large-scale primordial perturbations arising from early quantum fluctuations inflated to astronomical scales, not

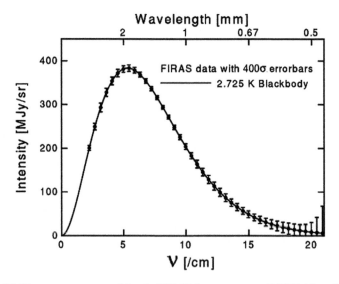

Fig. 8.10 *The CMB spectrum measured by the FIRAS instrument on COBE. Note that the error bars are 400σ; 1σ errors would be invisible. (From the COBE Science Working Group.)*

those affected by later physical processes. At first, this detection was only statistical, in effect extra fluctuations on top of instrumental noise. It took further analysis, and more years of observation, to produce robust maps showing the CMB fluctuations (Bennett et al., 1996).

Results from the third instrument, DIRBE, emerged more gradually. DIRBE measurements of the far- and mid-IR radiation filling the Universe established that $\sim 1/2$ the energy poured out by stars was re-processed by dust absorption and re-emission into IR photons.[46]

COBE had a long gestation period. Consequently, there were opportunities to try some of the same measurements planned for the mission. One salient instance was the improved spectral measurement made by Herb Gush, Mark Halpern, and Ed Wishnow in Canada (Gush et al., 1990). They too employed an interferometric technique, but their experiment relied on a suborbital rocket flight, launched just after the FIRAS announcement. In a mere 5 minutes of observation, they were able to measure $T_0 = 2.736 \pm 0.017$ K, in excellent agreement with the COBE FIRAS measurement. Another is the Relikt 1 satellite mentioned earlier, which had the goal of measuring or setting limits on very large angular scale anisotropies.[47]

Other teams were pressing hard on anisotropy limits on smaller scales. In retrospect, some of these experiments probably did detect CMB fluctuations, but generally at

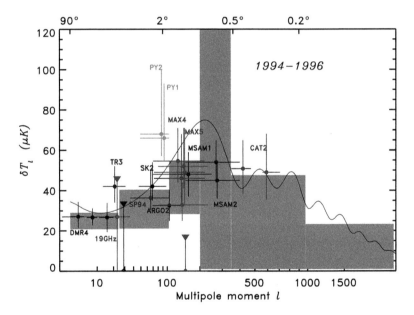

Fig. 8.11 *A summary of the many attempts to find CMB anisotropies in the years 1994 to 1996, following the announcement of the COBE results. The now-accepted level of fluctuations is shown by the solid line. Note the difference in the vertical scales in Figs. 8.8 and 8.11, reflecting the great improvement in experimental technique through the 1980s and 1990s. (From FBB, which includes citations.)*

Fig. 8.12 *The map of CMB temperature made by the WMAP satellite. The entire sky is shown, with the plane of the Milky Way Galaxy centered horizontally. The map is based on nine years of WMAP observations. (Courtesy of NASA). A later, higher resolution version of the CMB map, obtained by the Planck Team, appears in the frontispiece of this volume.*

too low a significance level to allow the teams to claim detections. These precursor experiments did, however, pave the wave for a rapid burst of experiments that verified the DMR results and, crucially, extended them to smaller angular scales, where the interesting features of the power spectrum lie. The first four decades of long history of CMB anisotropy searches are summarized in Chapter 5 of FBB. A summary of the situation not long after the 4-year COBE results were announced is shown in Fig. 8.11 and can be compared with the results shown in Fig. 8.8.

8.6   The CMB and the standard model of cosmology

In the preceding two sections, we have seen how the CMB developed into a pillar—perhaps the central pillar—of modern cosmology. This final section jumps ahead to the present to assess the role CMB studies play in today's "precision cosmology." The progress made since COBE is too vast to treat step-by-step, advance by advance. Instead, the goal in this section is to present the best observational results available in late 2017, and to summarize the wealth of cosmological conclusions that can be drawn from them. In many cases, the "best observational results" are those obtained by the European Space Agency's *Planck* mission, so this section will rely heavily on the newest *Planck* results and values of cosmological parameters determined from them. This approach unfortunately minimizes the exceptionally valuable measurements made earlier by a crucial NASA mission, the Wilkinson Microwave Anisotropy Probe. WMAP, appropriately named to honor Dave Wilkinson following his untimely death, was the first project to determine the power spectrum of CMB anisotropies in full detail (Bennett et al., 2003; Hinshaw et al., 2003). It also mapped CMB fluctuations with higher resolution and sensitivity than COBE (Fig. 8.12), showed $n_s < 1.000$ (Spergel et al., 2003) and produced the first measurement of the optical depth to re-ionization (Kogut et al., 2003, later revised by Spergel et al., 2007). While *Planck* has improved these results, WMAP opened the era of precision cosmology.

The focus on *Planck* also does not do justice to the important contributions made by ground-based programs, including the South Pole Telescope, POLARBEAR, and the Atacama Cosmology Telescope, in particular by extending the power spectrum to higher ℓ. These higher resolution studies[48] reveal more of the acoustic peaks and the Silk damping tail, facilitate CMB lensing measurements and have detected hundreds of clusters of galaxies through the SZ effect.

While the CMB is indeed central to modern cosmology, a wealth of other observations also contribute to our understanding of the properties and evolution of the Universe. These also buttress conclusions drawn from studies of the CMB alone. It is this interconnected web of evidence that bolsters our confidence in the now standard cosmological model, despite the complexities listed in Chapter 10. This mutual reinforcement of CMB and other cosmological observations has made cosmology the precision science it is today, and can best be appreciated by reading this chapter along with Chapters 6, 10, and 11.

Section 8.6 begins with an analysis of what CMB measurements alone can tell us. In Section 8.6.3, the cosmological information imprinted in polarized CMB fluctuations is discussed. Section 8.6.4 demonstrates briefly how complementary astronomical observations enrich and strengthen results derived from the CMB.

8.6.1 What a precise value of T_0 brings to cosmology

Of the fundamental parameters of modern cosmology, T_0, the present value of the CMB temperature, is known to the highest precision: $T_0 = 2.7255 \pm 0.0006$ K (Fixsen, 2009). Furthermore, we can project that value into the distant past using the exact and simple relation $T(t) = T_0/a(t) = T_0(1+z)$, where $a(t)$ is the scale factor at any earlier epoch t and z is the corresponding redshift. For instance, at a redshift $1 + z = 1,000$, roughly when the CMB photons last interacted with matter, $T_0 = 2,725.5$ K, so there were $N_{ph} = 4.1 \times 10^{11}$ photons cm^{-3}. That number density exceeds by a factor greater than 10^9 the number density of baryons N_b in the Universe at that epoch, and this ratio N_{ph}/N_b remains fixed as the Universe expands. As discussed in Chapter 6, this ratio strongly affects nucleosynthesis at a much earlier epoch. The mass fraction of deuterium, in particular, increases sharply as this ratio rises, that is, as N_b decreases. Since we know N_{ph} with precision, the deuterium abundance becomes a direct measure of the density of baryons N_b. The observed deuterium abundance implies $\Omega_b = 0.0500 \pm 0.0015$ (Cooke et al., 2014). The $\sim 3\%$ precision is limited solely by uncertainties in nuclear reaction rates and the astronomical measurements of deuterium, not the tiny uncertainty in N_{ph}.

High energy cosmic rays passing through the sea of CMB photons interact with them, the primary process being photo-pion production. Since we know N_{ph}, we can calculate the mean free path of the highest energy cosmic rays with precision limited only by uncertainties in the photo-pion cross sections.[49]

Limits on spectral distortions

As explained earlier, the spectrum of the CMB is expected to be precisely Planckian or black-body. The FIRAS spectrum (Fig. 8.10) shows no evidence for any departure from an exact black-body spectrum at wavelengths < 1 cm. There are, however, mechanisms that could introduce very small spectral distortions, especially at longer wavelengths.[50] The first is the SZ effect, already noted in Section 8.4.3. The distortion in the overall spectrum of the CMB caused by scattering from a uniformly distributed background of hot electrons is described by the parameter y which is proportional to the pressure of the background electrons. At wavelengths > 1.4 mm, this y distortion decreases the measured T_0, as shown in Fig. 8.6, and increases T_0 at shorter wavelengths. There is a second possibility: suppose that conditions in the early Universe were adequate to produce only *kinetic* equilibrium between the matter and the CMB photons, rather than complete *thermal* equilibrium. The photon number then differs from Planckian, and is instead a Bose–Einstein distribution, given by equation (6.12) with the chemical potential μ slightly greater than zero. The spectral distortions appear only when $h\nu/kT$ is less than or of order μ, that is, at long wavelengths. There are also processes which can replenish the photons at long wavelengths.[51] Adding in these processes, there would be a dip in the CMB spectrum over a range of wavelengths in the Rayleigh–Jeans region of the spectrum. If μ is not large, Burigana et al. (1991) show that the dip, or μ distortion, is centered at a wavelength of roughly 30 cm.

The COBE FIRAS measurements were able to establish strong limits on both y and μ (Fixsen et al., 1996): $y < 1.5 \times 10^{-5}$ and $\mu < 9 \times 10^{-5}$. Given that the FIRAS measurements did not extend beyond 1 cm, it was important to push spectral measurements to longer wavelengths, where the μ distortion, in particular, might be more apparent. Some of the early ground-based efforts to make spectral measurements at longer wavelengths were mentioned in Section 8.4.1. There was also a later balloon experiment, ARCADE, that extended observations to 10 cm (Singal et al., 2011; Fixsen et al., 2011). In the end, ARCADE was not able to improve on the FIRAS limit on μ, in part because of a competing and unexplained residual signal at ~ 10 cm wavelength, probably of astronomical origin.[52] Nevertheless, with the possible exception of the unexplained ARCADE excess at 10 cm, there is currently no evidence for any departure from a perfect black-body spectrum.

8.6.2   The power spectrum of CMB anisotropies and what it tells us

The details of the power spectrum of CMB fluctuations depend sensitively on the values of a small set of cosmological parameters (Section 8.5.4). The importance of CMB anisotropy measurements lies in our ability to exploit this dependence to go in the opposite direction, from a power spectrum derived from CMB maps to the set of cosmological parameters that best matches the observations. The process is simple in principle. Construct a power spectrum from a CMB map along with uncertainties in that power spectrum. Then compare it to a network of model power spectra constructed

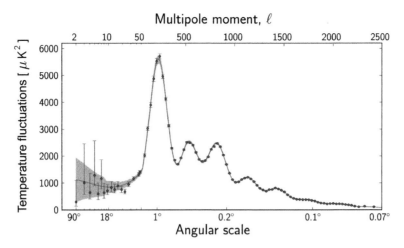

Fig. 8.13 *Points and experimental error bars show the CMB power spectrum measured by* Planck, *adapted from Planck Collaboration (2016b). The map used to generate the power spectrum is shown as the frontispiece to this volume. The continuous curve is a ΛCDM model generated from the best-fit set of cosmic parameters. The shaded area at large scales shows cosmic variance in the model. (Courtesy of the European Space Agency.)*

with varying values of the cosmological parameters to select the set of parameters which most closely fits the measured power spectrum.

The astonishing result of this process is that six and only six parameters are necessary to fit all the intricate details of the power spectrum, as shown in Fig. 8.13. Six parameters alone specify the overall amplitude of the power spectrum, its slight tilt, the location of the peaks, and the onset of damping at high ℓ. Furthermore, these six parameters alone are sufficient to select a single, best fit, cosmological model. There are clearly far more features in the power spectrum than six, so the fit is highly constrained. That in turn implies that we may add other cosmic or physical parameters to the mix and determine their values as well. The six parameters needed are:

1. $100\,\theta_{\mathrm{MC}}$. θ_{MC} is defined as the angle in radians subtended by the sound horizon, $R_{\mathrm{s}} = c_{\mathrm{s}}t$ on the last scattering surface. That angle is $R_{\mathrm{s}}/D_{\mathrm{A}}$ for small angles, where D_{A} is the angular diameter distance to the last scattering surface. θ_{MC} fixes the angular scale of the first peak in the power spectrum. Note that D_{A} in turn depends on both the overall curvature of space and H_0.

2. The density of baryonic matter. This is usually expressed as $\Omega_b h^2$ using the notation Ω for the ratio of the density of a component of the cosmic mix, in this case baryons, to the critical density (Chapter 6). h is the ratio of the observed, present value of Hubble's constant to the fiducial value 100 km s^{-1} Mpc^{-1}.

3. The density of dark matter, Ω_{c}, assumed to be cold dark matter, as defined in Chapter 11.

4. n_s, the slope of the spectrum of primordial density perturbations.
5. The optical depth, τ, from Thomson scattering at and after the epoch of re-ionization at $z \sim 10$.
6. The overall amplitude of the CMB power spectrum, in μK^2. Since this is measured primarily at medium and large ℓ, where scattering at and after re-ionization damps the signal, we take account of that small effect by expressing the initial amplitude as $A_s = A(\text{measured}) \, e^{2\tau}$. Since A_s is small, the quantity quoted to compare with the observations is generally $10^9 A_s \, e^{-2\tau}$ with units of μK^2.

These six parameters suffice to define a standard lambda cold dark matter (ΛCDM) model.

The standard model depends on a set of assumptions as well as these six parameters. One is the universality and constancy of physical laws; another is the assumption that General Relativity is the correct theory of gravity. As the name suggests, the ΛCDM standard model also assumes that the density of matter in the Universe is dominated by dark matter (see Chapter 11), and that the dark matter is cold, meaning non-relativistic for most of the history of the Universe. The main contributor to the overall cosmic density, however, is dark energy, generally assumed to be a cosmological constant, Λ. Finally, the use of the power spectrum to represent CMB fluctuations rests on the assumption that the statistics of these fluctuations are random and Gaussian.

Two other assumptions deserve more careful consideration. The first is that the space curvature of the Universe is negligibly small (so that the rules of flat, Euclidean geometry apply). The action of inflation (Chapter 11) in reducing curvature by something like $60 - 80$ orders of magnitude makes this assumption reasonable. If the curvature is not negligible, however, angular diameter distances are changed, and the connection between physical lengths on the surface of last scattering and the observed angles subtended by them is altered. That change directly affects θ_{MC}. The second assumption is that the density fluctuations are adiabatic, as predicted by most models of inflation. There is an alternative: isothermal, or isocurvature, fluctuations.[53] For these, while the matter density varies from place to place, the radiation density (or total density, including both matter and radiation), stays constant, as the names suggest. For a given level of density inhomogeneity, $\delta\rho/\rho$, the resulting value of $\delta T/T$ is thus far smaller for such fluctuations.

Given the constraining power of the CMB power spectrum, the same game can be played with these assumptions as with the six parameters. Permit one of these assumptions to break, and compare the resulting model power spectrum with the observations. Does the resulting model power spectrum fit the observations better? The answer so far is a clear "no," as we will see in the following subsection. There is no evidence from the CMB of any required adjustment to any of these basic assumptions.

Planck values of the six cosmological parameters

The most recent comprehensive determination of the six parameters of ΛCDM cosmology was by the *Planck* team, based on all the assumptions just enumerated, including flat geometry (Planck Collaboration, 2016b). At the time this chapter was written, updated results were expected from the *Planck* team within a few months. Little change in any of the parameters is anticipated with the possible exception of τ, the reionization optical depth. From this paper:

1. $100\,\theta_{MC} = 1.04086 \pm 0.00048$ radians.
2. The density of baryons: $\Omega_b h^2 = 0.02222 \pm 0.00023$ (this parameter, like θ_{MC}, is measured to 1% precision); for *Planck's* value of H_0, $\Omega_b = 0.049$.
3. The density of Cold Dark Matter: $\Omega_c = 0.265 \pm 0.005$.
4. $n_s = 0.9652 \pm 0.0062$, as expected just less than unity.
5. $\tau = 0.078 \pm 0.019$.
6. $10^9 A_s e^{-2\tau} = 1.881 \pm 0.014 \ \mu K^2$ (again <1% precision).

Note that the *Planck* team reports 95% confidence levels (2σ) in the "parameters" paper (Planck Collaboration, 2016b), and those 2σ errors are cited above. Thus the more conventional 1σ uncertainty on Ω_c for instance is ~ 0.0025 or $\sim 1\%$. In addition, the Planck Collaboration (2016b) reports many values of several of these parameters, depending on exactly which *Planck* data and ancillary data are included. The figures shown above are based only on the *Planck* temperature power spectrum, and are taken from column 6 in Table 1 of that paper. They do not include small shifts and improvements in the values introduced by either *Planck* polarization data or other astrophysical data, samples of which are treated separately below.

These results, derived from the CMB temperature power spectrum alone, establish a particular ΛCDM model with impressive precision. The *Planck* results can be used to set constraints on other quantities of cosmological interest as well. The value of τ, the optical depth to re-ionization for instance, translates to a value for the redshift of the epoch of re-ionization, $z \sim 9$. If we assume flat geometry, the total mass-energy density must equal the critical density, so $\Omega_{tot} = 1.00$. But the power spectrum provides values for two of the three components of Ω_{tot}, so we can find the density of dark energy by simple subtraction: $\Omega_E = 1 - \Omega_b - \Omega_c = 0.686$. The CMB, like the supernova results described in Chapters 10 and 11, thus shows that dark energy dominates the dynamics of the expanding, accelerating Universe. The good agreement between this value of Ω_E and the independent value from supernova observations in turn provides observational support for flat geometry. Finally, measurements of CMB anisotropies can determine a value for H_0 independent of all steps in the distance ladder, as discussed in the next subsection and again in Section 8.6.4.

Planck constraints on assumptions underlying the standard model and other physical quantities

As already noted, the constraining power of the CMB power spectrum is very large and so we may test the assumptions underlying the standard model, or add parameters to the six just listed. A few instances are summarized here.[54] One example is the assumption that dark energy is an unvarying cosmological constant (Chapter 11). That assumption can be tested by adopting a simple model allowing for secular change: $w(t) = w_0 + [1 - a(t)]w_a$, where w specifies the equation of state p/ρ. The equation of state for a pure cosmological constant is $w_a = 0$ and $w(t) = w_0 = -1$. The current best limit on w_a is -0.4 ± 0.8, with $w_0 = -1.02 \pm 0.1$, both consistent with a cosmological constant (Planck Collaboration, 2016b).

Testing the assumption that the statistics of CMB fluctuations are purely Gaussian is a bit more complicated (Planck Collaboration, 2016d), but there is no evidence for any significant departure from Gaussian statistics. Likewise, there is no evidence for isocurvature fluctuations (Planck Collaboration, 2016b). Two general predictions of inflation, adiabatic density fluctuations with Gaussian statistics, are thus fully consistent with these observational results.

The constraining power of CMB observations extends beyond cosmology. We can use the early Universe as a laboratory to test atomic and particle physics as well. The production of helium in the first few minutes of cosmic history has already been discussed. The agreement between the observed helium abundance of 0.2465 ± 0.0097 by mass (Aver et al., 2013) and that predicted, first from primordial nucleosynthesis and second from details of the CMB power spectrum (0.24665 ± 0.0002; Planck Collaboration, 2016b), tells us that we have nuclear reaction rates right, and more fundamentally that General Relativity was as valid at $t = 3$ min as it is now. In perhaps the first use of cosmology to constrain fundamental physics, Steigman, Schramm, and Gunn (1977) extended that analysis by using the observed helium abundance to show that the number of lepton families was $N < 5$. Well in advance of accelerator measurements, Yang et al. (1984) refined this analysis to show that preferred number was $N = 3$. The most recent CMB results from *Planck* (Planck Collaboration, 2016b) constrain N to a precision of ± 0.3.[55]

The CMB power spectrum can also be used to set constraints on the mass of neutrinos, or more precisely on the sum of masses of the three neutrino species.[56] In this case, the best results come from observations made on small angular scales, where the diffusion of neutrinos can most visibly wash out fluctuations. One recent result (Couchot et al., 2017) is $\sum m_\nu < 0.17$ eV. This limit lies far below another limit on neutrino masses that can be set by analyzing the contribution of neutrinos to Ω, a limit first explored by Gershtein and Zeldovich (1966). The richness of neutrino physics lies beyond the scope of this paper,[57] but it is worthy of note that neutrino oscillations already establish a lower limit of ~ 0.06 eV on the sum of masses, only a factor of 3 below the current CMB upper limit. An actual, CMB-based, measurement of $\sum m_\nu$ is within sight.

As a final and informative example, consider the assumption that the geometry of the Universe is Euclidean. If it is not, and space is curved, then θ_{MC} will not match the Euclidean value. The same is true for the angular scale of the peaks in the CMB power spectrum. We know both the physical dimension of the fluctuations producing the peaks and the distance to the surface of last scattering, but the angle subtended by the former will be larger than the flat-space value if space is positively curved, and less if space is negatively curved. As Fig. 8.13 demonstrates, the first peak of the observed power spectrum falls almost exactly at ~ 0.9 deg, the value expected for Euclidean geometry.

We cannot, however, stop there. The angular diameter distance to the surface of last scattering also depends on the value of Hubble's constant. Thus we cannot test the geometry without assuming a value of H_0. This is one example of the degeneracy of cosmological parameters: change in one parameter can be compensated by a change in another. Increasing H_0, for instance, can keep the angular scale of the first peak of the temperature power spectrum at ~ 0.9 deg even if the geometry is negatively curved. The Planck Collaboration (2016b) treats this problem by assigning a wide prior to H_0, then finding the resulting value of curvature consistent with the data. The result slightly favors weak positive curvature at the $\sim 2\sigma$ level.[58] We will see below that adding other data pulls the value of curvature back to zero.

This curvature–H_0 degeneracy (Bond et al., 1997; Zaldarriaga et al., 1997) also explains why H_0 is not listed among the six fundamental parameters of the standard ΛCDM cosmology. The CMB power spectrum can yield a value of H_0, but only if a value for space curvature is assumed. For flat space, as expected from inflation, the *Planck* results give $H_0 = 67.3 \pm 1.0$ km sec^{-1} Mpc^{-1}. We will see in Section 8.6.4 that this value is in tension with measurements derived from supernovae (as discussed in Chapter 10), the one serious disagreement between CMB results and those from other astronomical observations.

The curvature–H_0 degeneracy is not the only one. For instance, two of the six parameters listed above are at least partially degenerate: τ and n_s. The value of the optical depth τ affects the relative amplitude of undamped large scale CMB anisotropies and the slightly damped fluctuations on smaller scales. And n_s affects that ratio as well by introducing an overall tilt into the power spectrum. To break these degeneracies we need additional observational input—the power spectrum of fluctuations in the CMB temperature alone is not able to do so. These additional observations can come either from the CMB itself, specifically its polarization, or from other astronomical observations. These are treated in Sections 8.6.3 and 8.6.4, respectively.

8.6.3   What the power spectrum of polarized anisotropies tells us

Small amounts of polarization can be induced in the CMB as a result of Thomson scattering by free electrons, if the incident radiation field is anisotropic. In particular, a quadrupole moment induces linear polarization, as first noted by Rees (1968).[59] Since polarization is produced by Thomson scattering, it is induced only at the epoch of last scattering, and more weakly at and after the epoch of reionization when free electrons are present.

E-mode polarization

While the $\ell = 2$ quadrupole moment in the CMB can introduce large scale polarization (Rees, 1968; Basko and Polnarev, 1980; Negroponte and Silk, 1980), the most interesting polarization patterns are those resulting from local temperature fluctuations (Hu and White, 1997). In the vicinity of fluctuations of higher and lower temperature, free

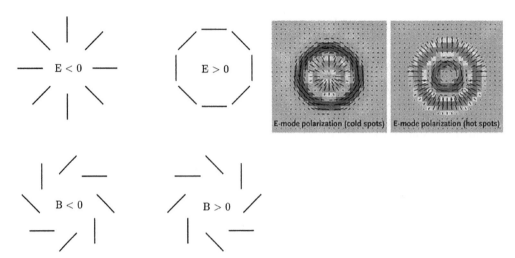

Fig. 8.14 *(Top left) Polarization patterns produced by regions of higher and lower temperature. Note that the two patterns are orthogonal, but both have radial symmetry. These are referred to as E-mode polarization. The top right panel shows actual data. The* Planck *maps of polarization are stacked at the locations of cold and hot fluctuations. (Bottom left) The B-mode patterns produced by tensor fluctuations discussed in Section 8.6.3* B-mode polarization *(ESA and the Planck Collaboration).*

electrons encounter slightly anisotropic radiation fields, including a quadrupole moment. The scattered radiation will thus be partially polarized. The two orthogonal states of polarization produced by regions of higher or lower temperature are shown in Fig. 8.14. For scalar density fluctuations, these so-called E-modes have radial symmetry, as shown.

This polarization is local, present only in the vicinity of temperature fluctuations. There is thus a strong link between temperature and polarization fluctuations. Consequently, the power spectrum of E-mode fluctuations can be directly and uniquely calculated from the power spectrum of temperature fluctuations (Crittenden et al., 1995). That is also true of the power spectrum of the correlation between temperature and polarization, known as the TE correlation.

Polarization of the CMB was first detected by the Degree Angular Scale Interferometer (DASI) experiment (Kovac et al., 2002) at the expected level. Subsequent observations by ground-based experiments and *Planck* have confirmed and greatly improved these early results. Figure 8.15 shows the power spectrum derived from *Planck* measurements of the E-mode polarization and the TE correlation, compared with the theoretical predictions derived entirely from the temperature power spectrum. No free parameters are employed in this fit and yet the agreement is excellent.

Careful instrumental design is required to measure CMB polarization. On the other hand, many sources of foreground emission are not strongly polarized, and hence there is less contamination of EE and TE measurements. As a consequence, polarization power spectra, such as those shown in Fig. 8.15, can now set independent constraints

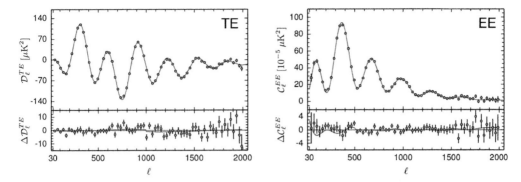

Fig. 8.15 Planck *power spectra of polarized E-mode fluctuations and the TE correlation (points) compared with theoretical power spectra derived from the same cosmological model shown in Fig. 8.13, with no adjustments or free parameters. The small departures from the model are shown in the two lower boxes (Planck Collaboration, 2016b).*

on cosmological parameters competitive with those derived from CMB temperature fluctuations.

B-mode polarization

The radially symmetric E-mode patterns are induced by scalar fluctuations in density and hence temperature, the only class of fluctuation we have considered so far. We now add the possibility of *tensor* fluctuations, perturbations in the metric of space-time, or gravitational waves. As shown in Chapter 11, these are expected in inflationary models. While quantum fluctuations in density grow into scalar perturbations, quantum fluctuations in the metric grow into tensor fluctuations.

Tensor fluctuations induce some E-mode polarization. More importantly, they also produce a different mode of polarization, the B-modes, which are divergence-free (Kamionkowski et al., 1997; Seljak and Zaldarriaga, 1997). Figure 8.14 (lower) shows the two orthogonal B-modes. The absence of radially symmetry shows that B-modes cannot be induced by scalar fluctuations. Since the existence of tensor modes is an integral prediction of inflation, the detection of B-mode polarization provides a fundamental test of inflation itself.

The amplitude of the B-mode signal compared to E-modes is described by r, the ratio of the amplitude of tensor modes to the amplitude of scalar modes. A value for that ratio, in turn, can be calculated for each of the many models of inflation. It follows that once the B-mode polarization has been detected, we will have a measure of r, and can then eliminate models that are inconsistent with that value. Even upper limits on r can rule out some classes of inflationary models. Furthermore, the value of r is a measure of the energy scale of inflation: $r = (E/1.9 \times 10^{16} \text{ GeV})^4$. Thus detection of B-mode polarization would yield clues to both the very early Universe at time-scales of order 10^{-35} s and extremely high energy physics.

It is not surprising, therefore, that many groups are actively pursuing the B-modes. Unfortunately, the signal is very weak and there are competing signals. One, lensing, is well understood. As the polarized CMB streams towards us, it is gravitationally lensed by intervening lumps of matter. Gravitational lenses are highly distorting, and can covert local E-mode polarization into B-modes.[60] The angular distribution of this lensed B-mode signal mimics the power spectrum of the E-modes and is identified in Fig. 8.16— it can be seen that it dominates the intrinsic B mode signal at small scales, $\ell \geq 100$. These lensed B-modes were first detected by the South Pole Telescope (Hanson et al., 2013), but of course they provide no information about primordial tensor fluctuations. Indeed, to find the tensor B-modes, it may be necessary to remove the competing, lensing signal from maps of CMB polarization. This process is called "de-lensing," and it makes use of our increasingly complete knowledge of the large scale distribution of matter which causes the lensing. The *Planck* Team, for instance, has published a map of the lensing mass (Planck Collaboration, 2016c). The statistical properties of such maps can be used to subtract at least part of the lensed B-mode signal.

The second competing signal is polarized emission from the Galaxy, indicated schematically in Fig. 8.16. That emission, both synchrotron radiation and dust re-emission, is strong and chaotic in orientation. In certain locations, it can mimic a B-mode pattern. Indeed, it is now believed that dust emission was responsible for the apparent detection of a B-mode signal claimed by the BICEP2/Keck Team (BICEP2 Collaboration et al., 2014).[61] To detect weak cosmic B-modes in the presence of

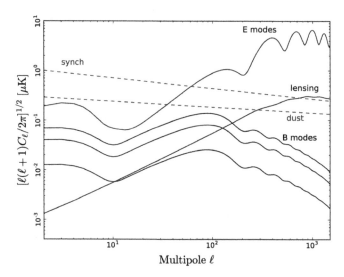

Fig. 8.16 *Predicted power spectra for the E-modes (upper curve) and the cosmological B-modes for three different values of the tensor-scalar ratio, r. Also shown are the lensing B-modes, which dominate the more interesting cosmological signal at $\ell \gtrsim 100$. Note the compressed vertical scale; B-modes are an order of magnitude weaker than E-modes. The dashed lines show estimates of foreground noise, specifically synchrotron radiation and dust emission. (Courtesy of Marco Bersanelli and Davide Maino.)*

polarized dust foregrounds will require some of the techniques described in Section 8.5.2, including multi-frequency observations. Attention to foregrounds is an important part of planning for future CMB experiments.[62]

The current best limits on B-mode signals, and hence r, come from new work by the BICEP2 & Keck Array and Planck Collaborations et al. (2016): $r < 0.07$. Given the emphasis on B-mode searches, it is likely that this limit will be lowered. Reaching a precision equivalent to $r \sim 0.01$, corresponding to energy scale of $\sim 6 \times 10^{15}$ GeV for inflation, seems possible if competition from other signals can be controlled.

Recall from Section 8.5.4 that models of inflation predict, and indeed require, n_s to be slightly different from unity. Each of the many variants of inflation predict both a value for n_s and a value for r. Hence a plot of these two parameters provides a useful diagnostic. Figure 8.17 is such a plot from the Planck Collaboration (2016e), and shows that a substantial swathe of inflationary models is already excluded at some level by the CMB observations. It is equally clear that many models remain viable—hence the continued interest in better constraints on r.

Polarization and degeneracies

Observations of polarized fluctuations are useful to break the degeneracies between some of the cosmic parameters, or at least sharpen the constraints set by the CMB temperature measurements alone. The clearest example is that observations of E-mode polarization

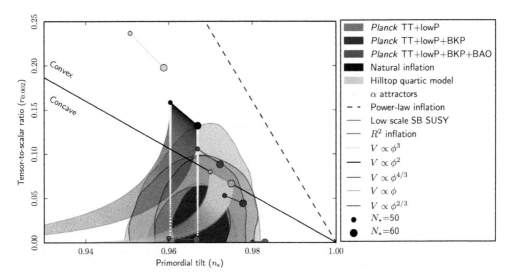

Fig. 8.17 *Two parameters from inflationary models, r and n_s, are plotted for a variety of models. The various ellipses show the constraints set by* Planck *alone and by* Planck *plus other cosmological data, such as baryon acoustic oscillations (BAO) and the BICEP2/Keck–Planck results (BKP). The results clearly exclude some models. See* Planck Collaboration (2016e) *for further details on constraints placed on inflation. Note that more recent BICEP2/Keck results have further reduced the upper limit on r to 0.07.*

at large angular scales provide a separate measurement of the re-ionization optical depth, thus breaking the degeneracy between τ and n_s cited in Section 8.6.2. Including low-ℓ polarization data allowed the *Planck* team to refine the value of τ cited above to 0.060 ± 0.009 and n_s to 0.9619 ± 0.0045 (Planck Collaboration, 2016f). Adding in polarization data also sharply improves the constraint on isocurvature fluctuations.[63]

Lensing has been mentioned as an impediment in B-mode searches. Lensing of the CMB, on the other hand, is beneficial in breaking degeneracies between cosmological parameters (Planck Collaboration, 2016c). It offers, for instance, an independent means of measuring A_s. Combining lensing with the temperature power spectrum helps break the H_0-curvature degeneracy, thus enabling a calculation of H_0 based solely on CMB observations, with no need for the distance ladder discussed in Chapters 6 and 10. Adding lensing changes H_0 slightly to 67.8 in the usual units (Planck Collaboration, 2016b, Table 4). Rapidly improving CMB lensing results on small scales will also tighten constraints on the masses of neutrinos.

8.6.4   Cosmic concordance and some tensions

While the CMB alone is able to determine the parameters of a standard ΛCDM model, other astrophysical and cosmological observations fully deserve all the attention and resources focused on them. First, ancillary measurements, for instance values of H_0 and of baryon acoustic oscillations (BAO),[64] play an important role in breaking degeneracies in the CMB results and thus in sharpening constraints on some cosmological parameters. Far more important, however, is the broad agreement between the CMB results and those derived from other, entirely independent, measurements. It is this cosmic concordance that gives us confidence that we have correctly modeled the properties and history of the Universe.

The concordance between the values for the mass fraction of helium derived from nucleosyntheses and from the CMB power spectrum on the one hand, and the observed helium abundance on the other, has already been mentioned. Studies of Type Ia supernovae, employed as standard candles, first showed the need for dark energy and established a value for Ω_E. That in turn matches the value derived from the CMB. Very different physical processes are involved and very different observational techniques employed; and yet the results agree. With the exception of a few observational "tensions" listed below, all the current cosmological measurements discussed in this chapter and Chapter 10 are in acceptable agreement. Modern "precision cosmology" does not rest solely on the CMB.

Selected supporting observations

Just which ancillary observations are needed or useful to supplement CMB results? There are two general groups. The first are physical constants or measurements, often laboratory-based, used in the analysis of the physical processes that generate or influence $\delta T/T$ fluctuations on the surface of last scattering. Among these are the half-life of the free neutron, various atomic scattering cross sections and atomic transition probabilities such as the 2S to 1S two-photon transition probability for hydrogen. Since these constants or measurements are needed to construct a model power spectrum, they can in turn be tested by CMB observations.[65]

More central to this volume are other *cosmological* observations, many described in Chap. 10. Supernova observations not only furnish a value for Ω_E, they also supply a precise value for H_0. Counts of clusters of galaxies as a function of redshift, often made using the Sunyaev–Zeldovich effect, constrain the growth of large scale structure. There is certainly an argument for including all such cosmological observations on an equal footing. The Planck Collaboration (2016b), however, adopted a very conservative stance on the inclusion of ancillary observations. It combines *Planck* CMB data with a limited set of other results, most prominently the BAO, to resolve degeneracies and sharpen the CMB-only results. The stated reasons for favoring BAO results are that they are a linear phenomenon, avoiding the complexities of non-linear theory, and that BAO measurements are geometrical. It happens also that they are nicely orthogonal to some CMB measurements in the sense that lines of degeneracy in the two data sets are roughly at right angles. Consider, for instance, the degeneracy between H_0 and angular diameter distance, or geometrical curvature. The CMB data favor lower values of H_0 if the angular diameter distance is adjusted upwards. The BAO data,[66] in contrast, require a lower angular diameter distance for lower values of H_0. Combining the two data sets pins down both H_0 and space curvature.

Remaining tensions

Even the BAO results are not in perfect agreement with the CMB. For instance, recent BAO measurements at high redshifts, admittedly difficult, favor a slightly lower value of H_0 than either the CMB values or those from low redshift BAO measurements (Aubourg et al., 2015).[67]

Since the *Planck* team first released its tables of cosmological parameters, there has been a much-discussed, and much-debated tension between the value of H_0 favored by *Planck* and some other CMB data on the one hand and the precision measurements of H_0 derived from supernovae on the other (Riess et al., 2011, 2016). The Planck Collaboration (2016b) cites $H_0 = 67.3 \pm 1.0$ km s^{-1} Mpc^{-1}; Adam Riess and his colleagues (2016) cite 73.24 ± 1.74 km s^{-1} Mpc^{-1}. The numbers differ by less than 10%, not the factor of two that enflamed cosmological debates in the 1970's; but 10% is several standard deviations in either value. If we want our cosmology to be truly "precision," we need to take the difference seriously.

As we saw earlier, the CMB value is model dependent; equally, the supernova value depends on a number of intricate astrophysical details and corrections. Even fiddling with the model-dependence or astrophysical details, however, does not bring the values from the two techniques into alignment (Freedman, 2017). They remain in tension. Adding to the turbulence, some CMB measurements, specifically those reported by the South Pole Telescope team, agree better with the high H_0 favored by the supernova results, giving $H_0 = 71.3 \pm 2.1$ km s^{-1} Mpc^{-1} (Story et al., 2013; Henning et al., 2018). The *Planck* side of the story is fully laid out by the Planck Collaboration (2016b) in

Section 5 of their paper. The other side of the story is elaborated by Riess et al. (2016) and by Bernal et al. (2016).

Are we back to the situation in the 1970s and 1980s where there were manifest tensions between different measurements of the CMB spectrum? We do not yet know. And that, it seems to me, is an appropriate note on which to end this chapter. There are still results we do not understand, and decisive measurements to be made. Perhaps new physics lurks in the difference: *Planck* measures H_0 effectively at $z \sim 1000$, supernovae at $z \sim 1$. The history of the CMB, and of cosmology more generally, is far from a closed book. That cheers me as an observer. There is still a lot of work to do.

Acknowledgments

Sections of this chapter, and in some cases, the entire chapter were read by Paul Boynton, Malcolm Longair, Jim Peebles, Virginia Trimble, and Martin White, who provided many useful comments and corrections. The remaining errors and biases are my own. A Haverford student, Gerrit Farren, provided much-needed technical and TeX assistance.

Further reading

First, I list a few general works treating the history of CMB studies and further details of the physics. Efforts to demonstrate that the microwave background discovered in 1965 was indeed cosmic are described in Chapter 2 of *3K: The Cosmic Microwave Background Radiation* (Partridge, 1995). Wilkinson and Peebles (1983) summarize both the early observations and some of the missed opportunities. A less formal history is available in *The Afterglow of Creation* by Marcus Chown (1993). *The First Three Minutes* by Steven Weinberg (1977) elegantly describes the earliest moments of the Universe.

While the field is advancing so rapidly that any summation of the impact of the CMB on cosmology is likely to be dated quickly, there are useful reviews and summaries. Among these are articles for the public such "The Cosmic Symphony" (Hu and White, 2004) and a fine pedagogical website maintained by Wayne Hu (http://background. uchicago.edu/). For readers desiring more technical detail, there is a review of the field supplied by Hu and Dodelson (2002); the volume by Ruth Durrer (2008), *The Cosmic Microwave Background*, largely devoted to the theory of the CMB; a less formal treatment, and a useful introduction to CMB studies in Russia, in *The Physics of the Cosmic Microwave Background* by Naselsky ct al. (2006); and two recent articles summarizing both the results of the European Space Agency's *Planck* CMB mission and the impact of these results on cosmology (Planck Collaboration, 2016a, 2016b). Finally, a summary of the goals and promise of future CMB experiments is provided by the CMB-S4 Science Book (arXiv:1610.02743).

....

NOTES

1. For more details of inflationary models of the early universe, see Chapter 11.
2. See Chapter 4 for a discussion of these ideas.
3. See Wilkinson and Peebles (1983) for their views.
4. See Section 8.3.
5. Both are discussed in more detail in Chapter 6, which also provides relevant references.
6. See their joint paper (Alpher and Herman, 1988), or their 2001 book *Genesis of the Big Bang*.
7. See Chapter 5 and also Gamow's paper in *Scientific American* (Gamow, 1956).
8. See Jasper Wall's contribution to FBB.
9. Atmospheric emission is indeed about 2–3 K, but Ohm had already corrected for that.
10. This ground is covered less formally in my contribution to FBB.
11. See the paragraphs entitled *Limits on spectral distortions* in Section 8.6.1.
12. These are analyzed in detail in Chapter 4 of my Partridge 1995 book *3K: The Cosmic Microwave Background* (hereafter abbreviated as *3K*), and also in Chapter 5 of FBB.
13. See Section 8.4.4 for a discussion of one early experimental result that heightened interest in gray-body spectra.
14. Note that the published tables in Stokes et al. (1967) and Wilkinson (1967) are jumbled. The values of T in Fig. 8.3 (from Boynton et al., 1968) are correct.
15. Those interested in further details of these early spectral measurements can consult *3K*, Chapters 2 and 4 for a technical analysis, and FBB for the recollections of those making the observations.
16. The cosmological principle, and extensions of it, are treated in more depth in Chapter 5.
17. See Chapter 4 of *3K* for technical details.
18. But see also Wolfe and Burbidge (1969) for a more extreme case.
19. More precisely, the peculiar velocity of the Earth or Solar System relative to the smoothly expanding comoving coordinate system of the Robertson–Walker metric.
20. For details, consult Barrow et al. (1985) and references therein.
21. See the paragraphs *Limits on spectral distortions* in Section 8.6.1.
22. A fuller account of 40 years of progress, from roughly 1970 to 2009, is contained in Chapter 5 of FBB. The emphasis there, as here, is on observations.
23. These so-called "primary fluctuations," as opposed to the "secondary" fluctuations discussed in Section 8.4.3, are treated in detail in Section 8.5.4.
24. Note the many parallel efforts by Peebles and the group in Moscow gathered around Ya B. Zeldovich. Was this apparently duplicate effort unfortunate? It did cause some friction (see recollections by, *inter alia*, Malcolm Longair and Rashid Sunyaev in FBB). I would argue that in the end it was valuable. Peebles and the Zeldovich group

approached these theoretical issues in different ways, yet emerged with generally concordant results. That agreement bolstered confidence in the conclusions. In any case, some level of overlap and duplication was probably inevitable given restrictions on the flow of information between the Soviet Union and the West in the 1960s. US journals barely trickled in to Russia; few Western scientists could read Russian, and hence needed to wait for English translations of the main Russian journals.

25. See Chapter 2 of *3K*, in particular Fig. 2.8.
26. This can be achieved by tipping the beam to different zenith angles: atmospheric emission varies as the secant of the zenith angle. See, for instance, Partridge et al., 1984.
27. See, for example, Gorenstein et al. (1978).
28. See *3K*, Chapter 3.
29. These balloon and rocket measurements and possible problems with them are analyzed in more detail in Chapter 4 of *3K*.
30. See, for instance, Crane et al. (1986, 1989), and Meyer et al. (1989).
31. This is in effect another test of the cosmic origin of the CMB. See Noterdaeme et al. (2011) for a summary of the observational results.
32. See, for example, Bersanelli et al. (1989), Kogut et al. (1990a), and Sironi et al. (1990). These pre-COBE measurements are also analyzed in detail in Chapter 4 of *3K*.
33. Hence I skip over an important technical advances: for example, deciding on the proper gauge to employ when treating density perturbations; see Bardeen (1980).
34. For details, see Chapter 6, Section 6.10.
35. See Chapter 6 or, for more detail, Peebles' book, *The Large-Scale Structure of the Universe* (1980).
36. Note the restriction to baryonic matter; it turns out to be quite important.
37. The calculations are presented briefly in Chapter 6, Section 6.12; see also Hu and Dodelson (2002) for a full review.
38. Hence the title of a helpful *Scientific American* article on the subject by Hu and White: "The Cosmic Symphony" (2004).
39. Chapter 6, Section 6.12.1.
40. There are several conventions for representing fluctuation power. The usual one is to plot $C_\ell \ell(\ell+1)/2\pi$, often written as D_ℓ, versus ℓ. When the aim is to present details of the power spectrum at high ℓ, however, $D_\ell[\ell(\ell+1)]$ is often plotted instead. When a larger dynamic range of power is needed, $D_\ell^{1/2}$ in μK is plotted.
41. If only a fraction of the sky f_{sky} is mapped, the cosmic variance increases as $1/f_{sky}$. $f_{sky} < 1$ also introduces correlations between the ℓs.
42. See Chapter 10, Section 10.19.
43. For details, see Hu and Dodelson (2002). Baryons add inertia, hence reducing the relative amplitude of even-numbered peaks corresponding to regions of lower density; see Chapter 10, Section 10.19.
44. See Peebles (1982) for an early formulation of the key role of dark matter in suppressing the predicted large amplitude temperature fluctuations in the CMB.

45. For an overview of the gradual and sometimes turbulent genesis of the mission, see *The Very First Light*, by Mather and Boslough (1996).
46. Important as it is, this topic lies outside the scope of this chapter; see Hauser and Dwek (2001) for a review.
47. See Chapter 9, especially Section 9.3.2 and Fig. 9.3.
48. For example, Henning et al. (2018), POLARBEAR Collaboration et al. (2017) and references therein, and Louis et al. (2017).
49. For early papers on this topic, see Greisen (1966) and Stecker (1968).
50. See the papers by Sunyaev and Zeldovich (1970b), Sunyaev and Zeldovich (1980), Danese and de Zotti (1982), and Burigana et al. (1991).
51. I omit here a description of processes which can return the spectrum to black-body form at long wavelengths. See the references in endnote 50 and Zeldovich et al. (1972).
52. For details, see Seiffert et al. (2011).
53. See Section 6.12.1 of Chapter 6.
54. For details, see the very helpful treatments in Planck Collaboration (2014, 2016b).
55. Recall that these are 95% confidence level constraints, in keeping with the practice adopted by the *Planck* team in their presentation of parameters.
56. See Bashinsky and Seljak (2004) for details and a description of earlier work.
57. See Abazajian et al. (2011).
58. A measure of the impact of curvature can be specified by assigning an equivalent density to space curvature; the "weak positive curvature" is equivalent to $\Omega_K = -0.05 \pm 0.05$.
59. See also the useful pedagogical treatment by Wayne Hu: http://background.uchicago.edu/index.html
60. See Blanchard and Schneider (1987).
61. See Mortonson and Seljak (2014) and particularly BICEP2 and Keck Array Collaborations et al. (2015) for a re-analysis which gave instead an upper limit $r < 0.13$ at 95% confidence.
62. See, for instance, the CMB-S4 Science Book (arXiv:1610.02743).
63. See figure 24 of Planck Collaboration (2016b).
64. See Section 10.12 of Chapter 10.
65. See Section 8.6.2 for neutrino properties, and Planck Collaboration (2016b) for other constraints on atomic physics.
66. See, for example, the BOSS CMASS results of Anderson et al. (2014).
67. Readers desiring more detailed explanations and references to papers describing this work should consult Section 5.2 of Planck Collaboration (2016b).

9

Space science and technological progress: testing theories of relativistic gravity and cosmology during the Cold War

Silvia De Bianchi

9.1 Introduction: general relativity, cosmology, and the space race

In the second half of the twentieth century, astrophysics and cosmology made unprecedented advances, not only in gaining new information about galaxies, black holes, and the constitution of our universe, but also in increasing the capability for testing the general theory of relativity (GR) in various ways. The great achievements in applying and testing GR was not just the result of technological developments but also of the specific historical circumstances that accelerated the process. The context of the Cold War, indeed, worked as a powerful vehicle for technological development in both the USA and the USSR, leading to a competition that informed this remarkable attempt to gain knowledge about the universe and astrophysical objects in entirely new ways.

In this chapter, the focus is upon technological developments that both United States of America and the Soviet Union promoted in radio astronomy and space science. These were translated into specific scientific campaigns aimed at gaining knowledge of the universe, testing relativistic theories of gravity and providing new technology to be employed in both space and on Earth.[1] To frame the development of observational cosmology within the Cold War context also means considering the role played by military technologies and facilities that were at the heart of a number of missions and successful observations. In addition, greater mobility of scientists was fostered by the reorganization of the academic system in the West in the post-War period.

Blum and his colleagues (2016) have identified factors that led to advances in the GR in the second half of the twentieth century. Not only the fundamental role played by theoretical physics in World War II, but also its contribution to the global

De Bianchi, S., 'Space science and technological progress: testing theories of relativistic gravity and cosmology during the Cold War' in *The Oxford Handbook of the History of Modern Cosmology*, edited by Kragh, H. and Longair, M. S. © Oxford University Press 2019. DOI: 10.1093/oxfordhb/9780198817666.013.9

arms race during the Cold War were key factors. The development of cosmology as a science benefitted greatly as a result of the substantial increases in funding and of the influx of talented individuals working in the field of theoretical physics.[2]

In the mid-1950s, several research centres promoting the study of general relativity flourished and increased their activities in various universities worldwide. These were typically small research groups of young researchers generally led by a principal investigator. In the USA, examples could be found at Syracuse University with Peter Bergmann, at Princeton University with John A. Wheeler and at the Institute of Field Physics of the University of North Carolina run by Bryce DeWitt and Cécile DeWitt-Morette. In Europe there were the Paris groups led by André Lichnerowicz and Marie-Antoinette Tonnelat, respectively, the King's College London team with Hermann Bondi, the University of Warsaw with Leopold Infeld, the University of Hamburg with Pascual Jordan, and the Institute for Pure Mathematics of the German Academy of Sciences at Berlin under Achilles Papapetrou. According to Blum et al. (2016), the majority of these research centres profited from the boom in theoretical physics in the postwar period that led to a substantial increase in the number of new PhD students.

The increased scientific productivity of these centres contributed to the establishment of a stable long-term post-doctoral education programmes. The growing number of young PhD students in theoretical physics, however, could not easily and quickly be absorbed by the academic system, resulting in the increased need to foster the mobility of these young researchers (Kaiser, 2005). These changes in the academic organization of research thus facilitated the transfer of knowledge and know-how among the research centres. As shown in the following sections, the transfer of knowledge and know-how in the field of theoretical astrophysics was a fundamental feature of collaboration between the two blocs during the Cold War, especially from mid-1960s onwards. However, observational cosmology and the testing of general relativity relied on technological facilities borrowed from the military, resulting in a dichotomy between competition and collaboration for the USA and the USSR.

The literature tends to stress the collaborative aspects of relations between the USSR and the USA in the field of astrophysics and space science, especially in the 1970s with the Apollo–Soyuz joint mission. Taking a closer look into the history of modern cosmology and by focusing on the history of testing of relativistic theories of gravity, we can however appreciate the complexities of USSR/USA relations. The converse is also true—to understand the major achievements of the last 70 years, which consolidated observational and astrophysical cosmology as a science, account has to be taken of the history of technology during the Cold War and we should frame it within the context of testing general relativity. This accounts for the big changes of approach that the community of scientists witnessed in order to achieve their scientific goals.

As Longair (1971) suggested, it is impressive to observe the change of perspective with respect to cosmology as presented in the classic textbook *Cosmology* by Bondi (1952), for instance, as compared with the way the subject had to be presented in 1970. New types of observation and theoretical analyses added important clues about how the universe must have evolved astrophysically. The detection of new classes of sources of radiation came about thanks to military technology developed in the 1940s and 1950s, while the new organization of academia and the realization of space and deep-space research, as well as the increase of funding deployed by the two blocs to support the space race, led to a major change of perspective for modern cosmology.

Within this 'big picture' one can appreciate how, from the 1960s onwards, the relationship of collaboration and competition between the two blocs gave rise to interesting dynamics that we can might be called 'a virtuous circle' through which not only was general relativity tested, but also knowledge gained about the phenomenology of the universe.

9.2   The origin of the testing of relativistic theories of gravity in the USA

What do we mean by a programme of testing relativistic theories of gravity? Three tests of general relativity were first proposed by Einstein himself—the advance of the perihelion of Mercury's orbit, the deflection of light by the Sun and the gravitational redshift of light.[3] Measurements of the gravitational redshift of light, for instance, were attempted between 1925, beginning with W.S. Adams, and 1954.[4] However, before 1961, no official programme of testing general relativity was promoted or discussed by the scientific community and the United States government. The first draft of a strategy to test general relativity was presented in 1961, soon after the creation of the National Aeronautical and Space Administration (NASA), the principal promoter of this initiative being Howard P. Robertson.[5] In the post-war period, he worked at various institutions, including Princeton University and Caltech from 1947 to 1961.

Robertson promoted the *NASA Conference on Experimental Tests of Theories of Relativity* held at Stanford University on 20–21 July 1961. The event included thirty-five participants from the most prestigious institutions in the USA, such as Princeton University, MIT, Harvard, Stanford, Caltech, Syracuse University, NASA, the Aerospace Corporation, the Goddard Space Flight Center, and General Motors.[6] This conference was responsible for creating new elements within the NASA programme dedicated to the testing of relativistic theories of gravity and promoting the advance of cosmology at a theoretical and experimental level. The range of potential observational and experimental projects which were to lead to future milestones in the testing of theories of gravity included post-Newtonian tests of gravity, gravitational lensing, light travel time delay testing, tests of the equivalence principle, frame-dragging tests, binary pulsars, and gravitational wave detection. The scientific justification for this ambitious programme was the fact that, apart from the advance of the perihelion of Mercury's orbit, no other accurate test of Einstein's theory was available in the early 1960s.

In his introductory remarks, Robertson reviewed the experimental basis of general relativity and underscored the point that the gravitational redshift could not be considered a full test of relativistic theories of gravity. He also noticed that the deflection of light by the Sun had not been measured with great precision and that only the advance of the perihelion of Mercury's orbit provided an accurate test of Einstein's theory.[7] According to the report on the conference (Schiff, 1961), Robertson suggested that tests of the special theory of relativity could best be performed on the surface of the earth, and that rockets would be more useful for tests of the GR and alternative theories of gravity.

The participants discussed only six pre-circulated papers. A first presentation given by R.V. Pound (Harvard) highlighted the results of his precision terrestrial experiments to measure the gravitational redshift of γ-rays using the Mössbauer effect. There was general agreement with Robertson's conclusion that a satellite effort was not worth the effort to refine this experiment. R.H. Dicke (Princeton) discussed proposals for modified theories of gravitation and stressed the importance of null experiments, such as the Eötvös experiments performed at Princeton on the equivalence of gravitational and inertial masses. He also suggested testing the prediction of Dirac's cosmological theory of a variable gravitational constant and Jordan's proposal of replacing the Newtonian gravitational constant by a scalar field that would depend on the proximity of matter.[8] The third contribution by J. Siry (Goddard Space Flight Center) described the minitrack radio and optical methods for tracking satellites. C.W. Sherwin (Aerospace Corporation) suggested in the discussion that the motion of a satellite be slaved to a freely falling test object so that the latter is always at the centre of the satellite. In principle, this test object could be free of all forces except that arising from the gravitational field of the earth, moon, and so on. The satellite would then be constrained to follow its motion, and at the same time this arrangement would protect it from environmental disturbances such as radiation and atmospheric gases. The trajectory of such a slaved satellite would be truly representative of the gravitational field, and could provide very high-quality information from which the mass multipole moments of the earth could be computed.[9] Then L.I. Schiff (Stanford) described the predictions of Einstein's theory with regard to the motion of the spin axis of a gyroscope that is either at rest in an earth-bound laboratory, or in a free-fall orbit about the earth. The discussion was devoted mainly to two possible satellite-gyroscope experiments. W.A. Little (Stanford) described a proposal for a system that consisted of a superconducting sphere supported stably in a static magnetic field. A different kind of extreme-precision gyroscope was described by A. Nordsieck (University of Illinois and General Motors). Finally, the last paper was presented by J. Weber (University of Maryland) on the detection and production of gravitational waves. He first discussed naturally occurring detectors such as the earth and the moon and then focused on laboratory detectors of gravitational waves. Weber, Dicke, and P.G. Bergmann (Syracuse University) discussed the measurability of the various components of the Riemann tensor and the need for an invariant formulation of the results of particular experiments. Bergmann also commented on the maximum luminosity of gravitational radiation expected from binary-star systems. He expressed his concern that solutions of Einstein's field equations for gravitational radiation were far from complete, so that if a measurement could be made, it would have theoretical significance. In closing, Robertson summarized those parts of the conference that were of greatest interest to NASA. There was general agreement with his conclusion that some or all of the three types of gyroscope precession experiments (superconductive, electric, slaved satellite) were promising enough to warrant encouragement by NASA.

This conference was highly relevant for the history of modern cosmology, not only because NASA wanted to supply vehicles that could be used for scientific experiments, but also because this meeting marked a milestone that indissolubly linked the development of modern cosmology with technology employed in the space race programmes of the two competing superpowers. To give an impression of the relevance of this meeting, even Thomas Kuhn mentioned it in *The Structure of Scientific Revolutions*.[10] According to Kuhn, this conference represented the attempt to normalize the GR as a paradigm. For him, the success of a paradigm 'is at the start largely a promise of success discoverable in selected and still incomplete examples'.[11] Normal science consists in the actualization of that promise, an actualization achieved by extending the knowledge of those facts that the paradigm displays as particularly revealing, by increasing the extent of the match between those facts and the paradigm's predictions and by further articulation of the paradigm itself.[12] The case of the gravitational shift of Mossbauer radiation, for instance, was described as a dormant field that was awakened by the 1961 NASA conference. Kuhn ascribes the attempt of this conference to normalize science insofar as it pointed to facts that can be compared directly with predictions from the paradigm theory.[13] Kuhn's interpretation suggested that the 1961 conference was a fundamental step towards the acquisition of successful science and then to use of these achievements to compete with USSR by exerting hegemony in the field of theoretical and experimental physics. Even if one puts aside Kuhn's interpretation, it is clear that the NASA conference prompted a select number of excellent theoretical and experimental physicists to bring forward ambitious proposals to test general relativity and to gain knowledge about the universe in new ways. Throughout the 1960s and 1970s, other tests were proposed and realized, for example, using the measurement of gravitational lensing and undertaking tests of the equivalence principle. The latter in particular was at the centre of the experiments performed by both blocs by lunar laser ranging (LLR) experiments, to be discussed below.

It should be mentioned that attitudes changed dramatically towards cosmology as a science in the USA, the USSR and world-wide as a result of the discovery of the cosmic microwave background (CMB) radiation by Penzias and Wilson in 1965 (see Chapter 8). Many factors—technological, experimental, observational, and theoretical—played a role in making cosmological studies an integral part of the frontiers of physical enquiry and so the links between cosmology and relativistic gravity continued to be strengthened over succeeding years.

9.3   Cosmology and testing the theory of relativity in the USSR

The USSR had the technology to proceed with the testing of theories of relativistic gravity and in the 1950s and 1960s experiments were also proposed to detect gravitational waves.[14] They had know-how and capability in space science that enabled them to compete with the American scientists and engineers.[15] There was a similar event to the NASA conference in the USSR, the first *Soviet Gravitation Conference* held at the School of Physics of the Moscow State University from 27–30 June 1961.[16] In contrast to the NASA conference at which only 6 papers were discussed, the programme of the conference in Moscow included 83 papers presented by Soviet authorities in the fields of gravitation, relativity, and quantum theory. The objective of this conference was to establish the 'science of gravitation' both theoretically and experimentally.

Two sessions were devoted to experiments and cosmology with the following contents. The fifth session of the conference focused on the possible use of satellites and space rockets to verify the physical foundations of both the special and general theories of relativity. The scientists reporting on their research in this session were allowed to talk from 20 to 40 minutes, whereas the theoretical physicists in previous sessions had only been allocated from 5 to 15 minutes for their contributions, meaning that experimental activities dominated the discussion. The Soviet institutions participating in this fifth session were the Physics Institute of the Academy of Sciences, the Moscow State University, the Joint Nuclear Research Institute at Dubna, and the Astronomical Observatory of Pulkovo. The first speaker of the session was V.L. Ginzburg of the Physics Institute in Moscow who discussed in detail Schiff's 1960 proposal to measure the relativistic precession of the rotational axis of a satellite-borne or earth-based gyroscope (see Section 9.2).[17] In fact, the USSR could compete with the USA in military facilities and technology but could not count on the same investments in enhancing the potentialities and the application of their technology in a flexible way. The result was some delay as compared with the USA in implementing activities in other scientific, non-military sectors. They were however perfectly aware of the strategy pursued by the Americans in tests of relativity and attempted to compete with them.

The sixth session was even more intriguing, because it mirrored the profound dissatisfaction of the Soviet scientific community with existing cosmological theory, in particular, the rejection of the conception of the universe being homogeneous and isotropic. The session was divided into five main sub-sections. The first dealt with criticisms of homogeneous and isotropic models of the universe. The second focused on the 'gravitational paradox', namely on the apparent indeterminacy of the gravitational force exerted upon a mass by the infinite number of other masses in an infinitely large universe. A solution by Kipper was discussed with a very similar outcome to that of Milne's concept of a duality of the definition of time.[18] The third sub-section included the discussion of 'metagalactic' formation and the origin of galactic nuclei. This section preceded one on the 'ageing' of the universe in which Belokon presented his proposal of an explosive expansion of the universe that slowed down the rate of annihilation of the universe that he believed to have originally balanced energy densities and pressures. Finally, the last sub-section was devoted to neutrino astrophysics and on how a neutrino–antineutrino background could affect the earliest stages of the universe.[19] The latter was a prominent sector of research, not only theoretically but also experimentally because of the presence of Bruno Pontecorvo in Dubna.

With a few exceptions, however, there was relatively little interaction between the radiophysicists and the more traditional Soviet community of astrophysicists.[20] Many different Soviet laboratories were engaged in radio astronomy, and there was often significant competition between them both for academic recognition and for resources. Soviet observational radio astronomy depended on the Scientific Council for Radio Astronomy of the USSR Academy of Sciences, which was for many years under the leadership of Vladimir Kotel'nikov. Unlike the laboratories directly dependent on space missions, such as those included in the facilities of Dubna, observational radio astronomers had greater difficulties in supporting their science at the beginning of the 1960s. In what follows, the main achievements in radio astronomy and cosmology are analysed to provide a picture of the development of observational cosmology in the Soviet Union.

9.3.1   Radio astronomy facilities in the USSR

In the mid-1950s, the first generation of Soviet radio telescopes and radio interferometers operated at a wide range of wavelengths (from 4 m to 2 cm). They were used to carry out pioneering series of radio observations of the quiescent and perturbed Sun, the Moon and the brightest discrete cosmic radio sources. Theoretical studies of the physics of the Sun, planets, interstellar medium, cosmic rays, supernova remnants, galactic nuclei, extragalactic objects, and the expanding universe were carried out at the Sternberg Astronomical Institute of Moscow State University (GAISH), the Institute of Applied Mathematics of the USSR Academy of Sciences (IPM), the Byurakan Astronomical Observatory of the Academy of Sciences of the Armenian SSR (BAO), the Lebedev Physical Institute of the USSR Academy of Sciences (FIAN), and the Radio Physical Research Institute in Gorkii (NIRFI).[21] From the mid-1950s the development of Soviet radio astronomy was centred on the construction of large radio telescope facilities operating in various wavelength ranges—metre, centimetre, and millimetre. Among the most important of these was the Large Pulkovo Radio Telescope, the fully steerable 22-m radio telescope of FIAN in Pushchino and the 22-m radio telescope of the Crimean Astrophysical Observatory (CrAO). Furthermore, USSR radio astronomers could make use of the fixed 31-m radio telescope of FIAN in the Crimea, the DKR-1000 wide-band cross radio telescope of FIAN, the large radio interferometer of the BAO, and the composite eight-element antenna of the Deep Space Communications Centre in the Crimea. Major contributions to astrophysics included investigations of the fluxes and polarizations of discrete radio sources, such as the Crab Nebula (FIAN, GAO, GAISH, BAO), as well as campaigns that ultimately led to the discovery of the variability of extragalactic radio sources (GAISH).

In the mid-1960s Soviet radio astronomy entered a further period of expansion in scope. Radio recombination lines of excited hydrogen were discovered thanks to a collaboration between GAISH, FIAN, and GAO. The fundamentally new technique of very long baseline interferometry (VLBI) was proposed (FIAN, GAISH, GAO) and various institutions (GAISH, FIAN, NIRFI, the Space Research Institute, BAO, CrAO, GAO) became involved in collaboration with NASA and the Australian government in 1976 (see Section 9.5). In the 1970s studies of pulsars (FIAN, IRE, IKI) led to the discovery of new sources as well as a number of important properties in the structures of the pulses themselves. In the early 1980s the construction of the 70-m radio telescope of the Deep Space Communications Centre began in the Crimea, thereby enabling investigations at wavelengths as short as 8 mm. The role of some of these Soviet facilities and the discoveries made with them are put into the context of the Cold War in the following sections.

9.3.2   How CMB discovery changed cosmology in USSR

The proposal to undertake tests of theories of relativistic gravity in the Soviet Union was fully accepted in early 1970s. Its close relationship to space and military programmes during the Cold War accounts for why the USSR did not detect the CMB before the USA. In simple terms, military technology and radio astronomical methods of studying the surface of Moon and planets through manned and unmanned missions gave an extraordinary boost to these areas of astrophysics. But in addition, the environments within which the Soviet scientists and American scientists undertook their research programmes were rather different and certainly impacted the ability of the Soviet scientists to capitalize upon their technical expertise.

The restricted mobility of Soviet experimentalists during the Cold War played a significant role. Because of their lack of contact with Western radio astronomers, the Soviet scientists were unable to appreciate the technical advances of the 1960s and 1970s which led to the construction of large arrays of modest sized dish antenna, such as the Cambridge 5-km telescope, Westerbork Array, the Very Large Array (VLA), and the Australia Telescope Compact Array. There were issues related to phase stability inherent in multi-element interferometer arrays. But perhaps most crucially, the lack of access to state-of-the-art computing facilities necessary to analyse multi-element interferometer data meant that the Soviet radio astronomers concentrated on large one-dimensional filled aperture fixed or phased arrays, such as the Pulkovo antenna designed by Khaikin, and later the RATAN-600 antenna. But neither of these made a significant impact on radio astronomy outside the Soviet Union. The 22-m steerable radio telescope located in Pushchino, near Moscow, and its twin even more accurate, version built in Crimea on the shores of the Black Sea were, however, the first radio telescopes of their size to operate at millimetre wavelengths. Both the Pushchino and Crimean antennas were used for some of the earliest radio observations at millimetre wavelengths, particularly of the planets, but their impact was limited by the poor sensitivity of the receivers available to the Soviet radio astronomers. The Crimean antenna was used in 1969 for the first VLBI observations between the USSR and the USA, and later, it was used together with the European VLBI network. The technology developed in the USSR was such that they could compete in these areas and also collaborate with what had been developed in Western Countries, but this occurred well after 1965.

Before Penzias and Wilson's milestone discovery of the CMB, A. Doroshkevich and I. Novikov published a paper entitled 'Mean density of radiation in the metagalaxy and certain problems in relativistic cosmology' in the Academy of Sciences of the USSR *Doklady* in 1964. Their basic idea involved making certain assumptions concerning the evolution of galaxies in the past and taking into account the redshift of the radiation from distant galaxies to calculate the intensity of the background radiation from galaxies at the present epoch for each wavelength. Among their considerations were galaxies as powerful emitters of radio waves at metre and decimeter wavelengths. Gas and dust were also considered as sources of background radiation. The upshot was that, if the universe had been hot at some point, the primordial radiation had to be added to the background radiation spectrum and this they accomplished in their paper. The wavelengths of the primordial background radiation lay in the centimetre and millimetre wavebands and so fell within the region of the spectrum where the contribution of galaxies is practically zero. Thus, the CMB in this wavelength range should exceed the emission of known sources of radio emission by a factor of tens of thousands, or even millions. It should be therefore observable. What is remarkable is that Penzias and Wilson made their discovery quite independently of knowledge of the paper by Doroshkevich and Novikov. In his 1978 Nobel Lecture, Penzias wrote that Doroshkevich and Novikov's (1964) 'remarkable paper not only points out the spectrum of the relict radiation as a black-body microwave phenomenon, but also explicitly focuses upon the Bell Laboratories twenty-foot horn reflector at Crawford Hill as the best available instrument for its detection! Having found the appropriate reference,[22] they misread its results and concluded that the radiation predicted by the "Gamow theory" was contradicted by the reported measurement'.[23]

Therefore, both Soviet and American scientists identified correctly the means for detecting the cosmic background radiation. Without entering into the details of the well-known history of the CMB discovery,[24] it is intriguing to understand why the Soviet scientists recommended the use of a facility belonging to their competitors. The Bell Laboratory facilities were known to Doroshkevich and Novikov thanks to the paper published by Edward Ohm in 1961.[25] It is interesting to note that the Bell Laboratory's radiometer prepared by Penzias and Wilson was a slightly modified version of Ohm's receiver. In 1961, Edward Ohm had constructed a radiometer to receive microwave signals from the Echo balloon (see Fig. 9.1), a reflector of radio and television signals launched by a rocket in 1960. Ohm found an excess radiation temperature in observations at 11 cm wavelength of 3.3 K and this was noticed by Doroshkevich and Novikov who attributed it to the CMB predicted by Gamow's theory. The horn antenna, originally designed for communications through the Echo balloon with the Bell Company's Telstar satellite became available in 1963.[26]

Even if Soviet radio astronomy was technologically sophisticated, their communication system was underdeveloped as compared with the American system that largely had the public sector investing in telecommunications. This may explain why and Novikov thought that the American facilities might have been able to verify quickly Gamow's theory. It was only in 1965 that the Soviet Union launched the Molniia satellite communications system linking Moscow to remote towns and military installations in the northern parts of the country. The Molniia system, the world's first domestic satellite communications network, relayed radio and television broadcasts from Moscow. The system also transmitted signals to spacecraft in the Soyuz, Salyut and other space programmes. The Molniia system employed several satellites and several ground stations. Therefore, one of the factors to be taken into account when reconstructing the history

Fig. **9.1** *Static inflation test of Echo satellite in Weeksville, NC. ID: EL-1996-00052 Other ID: L61-4623 Credits: NASA.*

of the CMB and the priority of the US discovery is that the Soviet technology employed was not primarily 'tuned' on the civil sector, but rather to military applications, at least until 1965. This of course does not mean that the USSR did not invest in radio and telecommunication for civil purposes but just that there was a delay as compared with the USA, but this resulted in higher costs in terms of gaining 'firsts' in testing theories of relativity and cosmology. That this gap was easily filled is testified by the fact that in early 1970s the USSR was at the same level as the USA in the technology for testing relativity, for instance, in competing in laser lunar ranging (LLR) experiments.

As was the case in Western Countries, experimental radio astronomy in the Soviet Union largely grew out of wartime radar research programmes. However, unlike in Europe, Australia, and the United States where post-war research was implemented primarily in universities and in civilian research laboratories, any research in the USSR with potential military applications, such as radio and radar astronomy, remained largely within military-oriented and tightly controlled laboratories. As such, publication in the open literature was restricted, and when published, important experimental details were usually omitted and so the results were often suspect or ignored by the Western scientific community. Thus, although starting in 1958, many of the most important Soviet journals were translated into English, for the most part Soviet observational radio astronomy had little impact outside of the Soviet Union. By 1971 only 2% of Russian scientific papers were published outside the country.[27]

The bar on travel was very serious and had a negative impact on Soviet research. In the field of astronomy and astrophysics the International Astronomical Union and other international organizations, which were recognized by the USSR, were crucial.

For instance, the first major IAU Symposium which brought together the best of the US, Western, and Soviet cosmologists at the 1973 Krakow meeting in Poland represented an opportunity to overcome such barriers. This was an important meeting that gathered together the giants of general relativity and cosmology, such as Wheeler, Thorne, Zeldovich, Sunyaev, and Novikov.[28]

In contrast, the theoretical work of Soviet scientists such as Iosif Shklovskii, Solomon Pikel'ner, Vitaly Ginzburg, Yakov Zeldovich in Moscow and Victor Ambartsumian in Armenia, and later their students, including Nikolai Kardashev, Igor Novikov, and Vyacheslav Slysh, at the Sternberg Astronomical Institute, Yuri Pariiskii in Leningrad, and Rashid Sunyaev at the Institute of Applied Mathematics, was widely recognized and considerably influenced both theoretical thinking as well as motivating new observational radio astronomy programmes in the United States, Australia, and Europe. This explains why soon after the discovery of the CMB, Soviet astrophysics and cosmology became major drivers of cutting-edge cosmological research. By the mid-1960s, Zeldovich had assembled a number of excellent scientists at the Institute of Applied Mathematics in Moscow.[29] In 1974, his group moved to the Space Research Institute of the Academy of Science, also known as the *Institut Kosmicheskikh Issledovanii (IKI)* in Moscow which was much more outwardly-facing.

Zeldovich was a key member of the Soviet atomic bomb project. In 1946, he was named head of the theoretical division at Arzamas-16, the most important Soviet nuclear research centre where, with other prominent scientists and mathematicians, he led the study and analysis of the highest plasma temperatures that could be reached in a hydrogen bomb explosion[30]. He also engaged Landau and Gelfand in their efforts to solve the Kompaneets equation.[31] Thanks to these studies on the hydrogen bomb, it was found in the 1950s that the equation was useless for military purposes and its study was declassified in 1955.[32] It is worth noticing that the USSR sought to develop extremely powerful bombs and tested the massive Tsar Bomb in 1961, which had a yield of 50 megatons of TNT. In contrast, the United States focused its efforts on producing smaller, more effective nuclear weapons that could be deployed on its medium-range ballistic missiles (MRBM) and inter-continental ballistic missiles (ICBM).[33] However, precisely this competition in the nuclear weapons programme enabled the Soviet scientists, and Zeldovich in particular, to transfer the knowledge gained in that sector to pure and theoretical research after 1961. It comes as no surprise that in 1961 it was possible to organize the first Soviet conference on testing gravitation and the theory of relativity, precisely because it was not necessary to invest more theoretical resources on the hydrogen bomb and classified studies, and also because there was progressively more tolerance of the application of relativity to cosmology.[34]

Even if it was not built for this purpose, IKI was meant to play a public role once Leonid Brezhnev recognized that it would be very costly for the Soviet Union to build up its own expertise in every sphere of scientific and industrial endeavour: better to share distinct areas of expertise with socialist and Western countries.[35] Cooperation was more economical than using foreign currency to buy in and copy Western technology. This led to more collaboration, circulation of knowledge within and outside the USSR, and exchange of information with Western Countries.[36]

In Moscow and in the rest of the World, IKI was associated to the name of Zeldovich, who led the Department of Theoretical Astrophysics. Zeldovich was the major driving force behind the major USSR scientific contributions to cosmology from 1966 to 1974. His contributions to the theory of the hot universe were of particular importance. In the late 1960s, he and Sunyaev showed that, due to energy released in the early stages of the evolution of the universe, the spectrum of the CMB radiation would be slightly distorted from that of a black body. Between 1969 and 1974 they, together with Illarionov, studied perturbations of the spectrum expected for various energy loss processes in detail. They showed that there should exists two main types of spectral distortions arising in the early universe. Therefore, they concluded that analyses of the spectrum of the CMB could reveal the epoch for specific types of energy loss.

Among their more striking predictions was the Sunyaev–Zeldovich (SZ) effect which results in distortions of the spectrum of the CMB radiation through inverse Compton scattering by hot electrons in clusters of galaxies—the low energy CMB photons receive on average an energy boost in each collision with the hot electrons in the intracluster medium (Fig. 9.2). Current research is focused on modelling how the effect is produced in the intracluster plasma in galaxy clusters, and on using the effect to estimate Hubble's constant. This also involves separating the different components in the angular average

Fig. 9.2 *The image shows the two galaxy clusters as seen at optical wavelengths with ground-based telescopes and through the Sunyaev–Zeldovich effect (diffuse emission) with the Planck satellite. The pair of clusters is located about a billion light-years from Earth, and the gas bridge extends approximately 10 million light-years between them. Credits: ESA Planck Collaboration; optical image: STScI Digitized Sky Survey.*

statistics of fluctuations in the background. The use of surveys of clusters detected by the SZ effect for the determination of cosmological parameters has been demonstrated by Barbosa et al. (1996). Measurements of the thermal SZ effect have now been obtained from the Atacama Large Millimeter Array (ALMA) and other facilities. From its first detection in 1983 to the recent past, particular milestones include:

1983   Cambridge Radio Astronomy Group and Owens Valley Radio Observatory first detect the SZ effect from clusters of galaxies.

1993   The Ryle Telescope images the SZ effect for the first time in a cluster of galaxies.

2003   The WMAP spacecraft maps the CMB over the whole sky with some sensitivity to the SZ effect.

2005   The Atacama Pathfinder Experiment—Sunyaev–Zeldovich camera saw first light and shortly after began pointed observations of galaxy clusters.

2005   The Arcminute Microkelvin Imager and the Sunyaev–Zeldovich Array begin surveys for very high redshift clusters of galaxies using the SZ effect.

2007   The South Pole Telescope (SPT) saw first light on 16 February 2007, and began science observations in March of that same year.

2007   The Atacama Cosmology Telescope (ACT) saw first light on 8 June.

2008   The SPT discovers for the first time clusters of galaxies by the SZ effect.

2009   The Planck spacecraft, launched on 14 May 2009, carries out a full sky SZ effect survey of galaxy clusters.

2012   The ACT performs the first statistical detection of the kinematic SZ effect.

2012   The first detection of the kinematic SZ effect is observed in MACS J0717.5+3745, confirmed in 2013.

A completely different type of distortion of the CMB spectrum is due to the recombination of hydrogen in the universe. In 1968, Zeldovich, Kurt, and Sunyaev showed that radiation associated with two-photon transitions and Lyman-alpha emission lead to the appearance of bands in the submillimetre part of the spectrum. Calculation of the recombination for a non-equilibrium CMB spectrum carried out in 1983 by Sunyaev and others, showed that the recombination lines of hydrogen and helium were accessible to observations using radio telescopes with the sensitivity available at that time. At a redshift $z = 1,500$ the background radiation attained a temperature $T = T_0(1 + z) \approx 4,000$ K at which there were sufficient photons in the Wien region of the Planck distribution to ionize all the intergalactic hydrogen. This epoch is referred to as the *epoch of recombination* since the hydrogen was fully ionized at earlier cosmic epochs. The process of recombination of the primordial plasma is important in understanding the origin of temperature fluctuations in the CMB radiation. The recombination calculations were first carried out by James Peebles, Yakov Zeldovich, Vladimir Kurt, and Rashid Sunyaev in the late 1960s.[37]

Another major experimental contribution to CMB studies during the period 1980 to 1991 was the map of the CMB produced by Strukov and Skulachev. Prognoz 9, launched on 1 July 1983 included the Relikt-1 experiment to investigate the anisotropy of the CMB at 37 GHz, using a Dicke-type modulation radiometer. The entire sky was observed in 6 months with an angular resolution of 5.5 degree and temperature resolution of 0.6 mK. A map of most of the sky at 37 GHz was made available (Fig. 9.3). They mapped the CMB dipole over the sky for the first time with the Relikt-83 experiment.[38] This experiment was prepared by the Space Research Institute of the USSR Academy of Sciences and supervised by Igor Strukov, who published their results and their implication for cosmology with Skulachev and Klypin in 1988,[39] one year before COBE started operating (for further details see Chapter 8).

By this date, the Soviets could no longer compete with the capabilities in radio astronomy and space science, unless they were prepared to invest heavily in updating the technology. In radio astronomy, for example, pioneering aperture synthesis radio telescopes were constructed in Cambridge (the One-Mile Telescope and the 5-km Ryle Telescope) and these led the research field until the advent of the VLA in the USA. Aperture synthesis imaging was first developed at radio wavelengths by Martin Ryle and coworkers from the Radio Astronomy Group at Cambridge University. Martin Ryle and Antony Hewish jointly received a Nobel Prize for this and other contributions to the development of radio astronomy. During the late 1960s and early 1970s, computers, such as the Titan machine in Cambridge, became capable of handling the computationally-intensive Fourier transform inversions which were central to the success of the One-Mile

Fig. 9.3 *Map of the sky at 37 GHz from the Relikt-83 experiment. The dipole component associated with the Earth's motion through the CMB has been removed. Zero longitude is on the right. (Reprinted from 'The anisotropy of the microwave background: Space experiment "relict"', 17, 8, I.A. Strukov, D.P. Skulachev, A.A. Klypin, Copyright 1988, with permission from Elsevier).*

and Ryle (5-km) telescopes. The technique was later developed and used in VLBI to obtain baselines of thousands of kilometres. Aperture synthesis is also used in synthetic aperture radar and in optical telescopes.[40] Suitable images could also be made with a relatively sparse and irregular sets of baselines, especially by using non-linear deconvolution algorithms such as maximum entropy methods. Furthermore, the alternative synthesis imaging produced a shift in emphasis from trying to synthesize the complete aperture (allowing image reconstruction by Fourier transform) to trying to synthesize the image from available data using computationally expensive algorithms. The Soviets could not compete in this area because they did not have access to the computer power needed to make aperture synthesis maps.

9.3.3   Some USSR firsts

The USSR boosted investment in its telecommunication systems and research into computation and cybernetics, as exemplified by the facts that they sent robots to the Moon well before the USA and obtained a richer number of samples of the Moon's surface than the USA in the early 1970s. The USSR could also count on powerful radio astronomy facilities by the end of the 1960s. In 1969, the Radio Astronomy Section of the Pulkovo Observatory became the Leningrad Branch of the SAO which served as the lead organization, directing all the work associated with the planning and construction of the RATAN-600. The telescope was intended to be a general-purpose instrument, a counterweight to the tendency toward specialization, while at the same time surpassing existing telescopes in resolving power, sensitivity both in flux density and brightness temperature, and in the accuracy with which the celestial coordinates of objects could be determined.[41]

Another important facility was based in Pushchino,[42] which still hosts the Radio Astronomy Observatory of the AstroSpace Center, Lebedev Physical Institute, Russian Academy of Sciences (LPI PRAO ASC).[43] LPI scientists had organized permanent observation stations in both the Crimea and the Moscow regions by 1966. The Crimean research station at Katsiveli became the largest of them. On both sites, RT-22 fully steerable 22-m parabolic radio telescopes were installed. The LPI RT-22 radio telescope was a unique instrument among the world's single dishes thanks to its record-breaking high angular resolution.

The wide-band cross-type radio telescope (DKR-1000) was another large radio telescope installed at the Pushchino observatory. It consisted of two parabolic cylinders 1km long and 40 m wide, extending from east to west and from north to south. The DKR-1000 differed from the best foreign competitors, which operated in a fixed and relatively narrow frequency range, in that it allowed simultaneous observations at any wavelength in the range from 2.5 to 10 m. Because of this capability and its high sensitivity, the instrument even now provides a unique facility for a variety of research projects. Equally fruitfully, RT-22-based research concerned the physical characteristics of the lunar surface layers and the conditions in the atmospheres and on the surfaces of the planets. In 1962, Kuz'min summarized the data obtained with RT-22 in one of the first catalogues of centimetre radio sources. The high frequency break in the spectrum of radio galaxy Cygnus-A found in the same series of observations enabled the first experimental estimates of the age of this class of objects to be obtained. Finally, it was with the LPI RT-22 that Sorochenko and his co-workers first observed hydrogen radio recombination lines in the radio waveband in 1964. Investigations of gaseous nebulae in the hydrogen, carbon, and helium radio recombination lines were the main research priorities of the RT-22 team for many years.

The first important studies carried out with the DKR-1000 in the mid-1960s included observations of a large sample of radio sources at frequencies 38, 60, and 86 MHz that enabled the dependence of the spectral index on flux density for extragalactic radio sources to be determined and also investigations of circumsolar and interplanetary plasma based on scintillations of compact radio sources. In the late 1960s, simultaneous observations were made with the DKR-1000 east-west arm and smaller radio telescopes. Following the discovery of pulsars, reported in early 1968, it became clear that DKR-1000 was ideally matched to studying their radio emission. The first Pushchino pulsar, PP0943, was discovered in the same year. Many important data on the spectral characteristics of individual and averaged pulses of pulsars, the frequency dependence of polarization of their radio emission and the changes of pulse profiles with frequency were obtained in the following few years. A super-dispersive pulse delay system at low frequencies enabled the deformation of magnetic dipole field lines close to the light cylinder region about the central neutron star to be detected.

The use of the DKR-1000 north-south arm was much more restricted because operation of such a sophisticated phased array over a wide frequency range was a difficult technical problem. However, for the observations requiring long integration times, the use of the antenna with a wide beam proved valuable. Specifically, low-frequency carbon radio recombination lines in the spectrum of the Cassiopeia A supernova remnant were detected with the north–south arm of the radio telescope. The discovery of interplanetary scintillations of compact radio sources and the subsequent discovery of pulsars demonstrated that the confusion effect due to the limited resolving power of the telescope was not essential for some important observational programmes. In such cases, it is possible to use large filled aperture radio telescopes and there was no need for cross-type antennas or other complex telescopes.

For this reason, Vitkevich decided in the late 1960s to build one more radio telescope in Pushchino, now known as the Large Phased Array (BSA FIAN). It took relatively little time and money to construct and commission the BSA, which comprised over 16,000 dipoles and covered an area of 7.2 hectares. The BSA has a record sensitivity in the metre wavelength range. Observations with BSA FIAN at wavelengths around 3 m began in 1974. The high sensitivity of BSA allowed daily observations of interplanetary scintillations of about 150 compact radio sources to be made. Such observations for several hundred extragalactic radio sources of different classes revealed the degree of their compactness, the physical conditions in the central regions of different types of galaxies and the distribution of compact radio sources in the Universe. The data obtained with BSA FIAN were used to derive the radio luminosity function of clusters of galaxies and to determine the peculiarities of the radio emission from those clusters rich in hot intergalactic gas. Observation of the Andromeda Nebula, our neighbouring massive spiral galaxy, showed that it has an extended radio halo responsible for most of its low frequency radiation. As expected, the most interesting results of BSA studies pertain to pulsars. The unique high sensitivity of this radio telescope has enabled the majority of pulsars in the northern sky to be observed. These observations and the data obtained with other radio telescopes were used to compile a catalogue of averaged pulse profiles at different frequencies for more than 150 pulsars.

9.4   Military technology and testing general relativity

Many of the most important discoveries in twentieth-century observational cosmology were the result of a complex network of collaboration between the USA and Western Countries, but the dynamics of competition with the USSR also played a role. The desire to catch up with the Soviet scientists and technologists and overtake their achievements was a powerful stimulus in the history of space science. This is evident from the very beginning, when the Soviets launched the first satellite Sputnik, triggering the reaction of the USA that they had to develop competitive technology and to create a strategy and organization to mastermind the competition. This led to the formation of NASA and the huge investment in the Apollo programme.

It would however be misleading to believe that the USA simply reacted to the great advances in Soviet space science and engineering. The Americans had their own agenda and the structures which led to the creation of NASA were already present well before the Sputnik launch. This was already evident from the important Rand Corporation report led by Lyman Spitzer, which set out a roadmap for the future NASA space science programme. In 1946, Spitzer wrote the report for the RAND Corporation entitled 'Astronomical Advantages of an Extra-Terrestrial Observatory', in which he explored the advantages of a space-based telescope of 5–15 m in diameter. At the time of his proposal, the Palomar 5-m telescope was still under construction. Spitzer suggested that the best reason to build a space-based telescope would be to discover new phenomena, and perhaps to modify our concepts of space and time.[44]

After the Soviets launched Sputnik, NASA was created and began experiments with a series of orbiting astronomical observatories (OAOs) and small satellites. In the 1960s, however, the Apollo lunar landing project absorbed most of NASA's resources and orbiting telescopes were given low priority. After the successful lunar landing program, NASA began a modest study project for a large space telescope (LST). By 1977 Spitzer and the astronomer John Bahcall had brought about the necessary political support for the project, by convincing colleagues, congressmen, and NASA headquarters that an optical–ultraviolet space telescope would represent one of the most important advances in observational astronomy for the coming decades. Once the telescope and its instruments were completed, the telescope would make some of the most precise and deepest observations of the universe ever recorded. The Hubble Space Telescope (HST) was finally launched in April 1990, after a long delay following the explosion of the Space Shuttle Challenger in 1986. When the first images were received, there was a terrible disappointment. The images were blurred as a result of an error in the optical design. Within a few years, a set of corrective optics was installed during a special Space Shuttle mission which enabled the LST to realize fully its goals. Since then, the HST, as the LST was renamed, has contributed enormously to our understanding of the large-scale structure of the universe, of stellar evolution, and of the dynamics of galaxies in an unprecedented way. It has produced firm evidence for black holes and provided images of objects and phenomena not even imagined in the 1940s. Lyman Spitzer was right when he predicted that the space telescope would profoundly affect our understanding of the workings of the universe.

It is interesting to contrast the approaches of the USSR and the USA in obtaining access to space. The USA opted for the Shuttle programme as the successor to the Apollo project. It undoubtedly had major successes including the launch of the four great observatories, the HST, the Compton Gamma-ray Observatory, the Spitzer (SIRTF) Observatory, and the Chandra X-ray mission. These missions all had goals that strongly contributed to cosmology and they were planned and began their construction during the Cold War period with technical connections with the military sector.

Radio astronomy was not the only field in which USSR and USA competed. Military research and technology also opened a new path in the application of radar to investigate the cosmos. The past 70 years have seen a unique capability for research and expanding scientific knowledge of the Solar System by using radar technology to conduct planetary astronomy. This technology involves aiming a carefully controlled radio signal at a planet, or some other Solar System target, such as a planetary satellite, an asteroid, or a ring system, detecting its echo and analysing the information that the echo carries. This capability has contributed to the scientific knowledge of the Solar System in two fundamental ways. Most directly, planetary radars can create images of the target surfaces which are otherwise hidden from sight and can furnish other kinds of information about target surface features. Radar also can provide highly accurate measurements of a target's rotational and orbital motions. Such measurements are invaluable for the navigation of Solar System exploratory spacecraft, a principal activity of NASA since its inception in 1958.[45] Focusing on radar astronomy enables us to highlight strengths and weaknesses of the two blocs.

The development of radar astronomy was originally linked to the need to intercept missiles. The USA designed their early warning system to detect the Semyorka intercontinental ballistic missiles produced from 1957 to 1961. In the same period the USA produced the SM-65 Atlas missiles (1957–1959) that were to be intercepted by USSR. Upgraded facilities, increased transceiver power, and improved apparatus increased observational capabilities. Radar techniques could also be used to test general relativity by observing Mercury and providing a refined value for the astronomical unit. As mentioned above, radar images provide information about the shapes and surface properties of solid bodies, which cannot be obtained by other ground-based techniques. In this sense the application of radar technology to astrophysics had a direct and important impact on the study of the planets of the solar system and for all future manned missions. Furthermore, radar directly measures the distances to objects and how fast they are changing. This aspect also proved of major importance for the development of space science and for astronomical and astrophysical studies.

The two major sites in the two blocs in which radar astronomy was developed were the Lincoln Laboratory and the Goldstone Deep Space Communications Complex in the United States and the Pluton complex in the USSR—they directly competed in military and scientific technology. The latter was a system of deep space communications and planetary radars built in the Deep-Space Communication Centre near Yevpatoria in 1960 and consisted of at least three antennas. The Pluton complex supported all the Soviet space programmes until 1978, when the Yevpatoria RT-70 radio telescope was built. Then the Pluton complex became a backup system for the RT-70. Until 1966, the Pluton complex was the world's highest capacity deep space communication system prior to the Goldstone Deep Space Communication Complex in the USA. Among its achievements were the radar detection of the planet Venus in 1961, of the planet Mercury in June 1962 and of the planet Mars in February 1963. In September–October 1963 it made the radar detection of the planet Jupiter. Finally, Pluton also represented an important technological achievement for radio astronomy and astrophysics. In October and November 1960, for instance, an observation programme was carried out at the Crimean radio astronomy station of Lebedev Physical Institute to study and determine the characteristics of the radiation of the Crab Nebula at a wavelength of 21 cm with greater accuracy than before.[46]

A remarkable way in which the American scientists exploited radar technology and applied it to test general relativity was through the Shapiro delay. In the 1960s, Shapiro realized that there was an accurate way of testing general relativity by measuring the time it would take a radar pulse to travel to and from a planet and how this would be affected if the pulse passed close to the Sun. Radar signals passing near a massive object take slightly longer to travel to a target and back than it would be if the mass of the object were not present. Because of the curvature of space-time about the Sun, a radar pulse would not reach Mars along a straight path but along a slightly curved trajectory. The Shapiro, or gravitational time delay, effect is the fourth classic solar system tests of general relativity. In an article entitled *Fourth test of general relativity*, Shapiro stated:

> Because, according to the general theory, the speed of a light wave depends on the strength of the gravitational potential along its path, these time delays should thereby be increased by almost 2×10^{-4} s when the radar pulses pass near the Sun. Such a change, equivalent to 60 km in distance, could now be measured over the required path length to within about 5–10% with presently obtainable equipment.[47]

The value of the time delay can be evaluated using the Schwarzschild solution of the Einstein field equations.

An important improved test of the Shapiro delay was performed in 1977 and was directly connected with the development of the US space programme.[48] As described above, for signals propagating between the earth and a planet near superior conjunction there was a predicted increase in this round trip (Shapiro, 1964). This prediction has been tested several times, first by means of radar signals reflected from the planets Mercury and Venus,[49] and later by radio tracking of the Mariner 6, 7, and 9 space-craft.[50] The results of these tests were consistent with general relativity to within the estimated errors, which ranged from 5% for the radar experiment to 2% for Mariner 9 mission.

The Viking spacecraft offered the opportunity to increase the accuracy of the time delay test. The Viking experiment overcame the limitations of previous experiments in two ways—measurement and interpretation were significantly more accurate. The ranging transponders on the Viking spacecraft enabled measurements of the round-trip group delays of signals propagating between the earth and the spacecraft to be made with an uncertainty of only about 10 ns. This level of accuracy, by itself, represented an improvement over that attained during previous interplanetary ranging experiments. To interpret these accurate delay measurements as a test of general relativity, Shapiro emphasized that one must be able to determine, with at least comparable accuracy to all other non-relativistic contributions to the delays. These contributions came from two sources: (1) the orbits of the spacecraft and of the tracking stations on the earth and (2) dispersion in the solar corona, which increases the group delays significantly for signal paths that pass near the Sun. The Viking spacecraft enabled the contributions from these two sources to be determined accurately, because the two landers were securely placed on another planet and the two orbiters possessed ranging transponders transmitting in the band frequencies of interest. In the most accurate previous time delay test of general relativity both these features were lacking.[51]

9.5   From competition to collaboration and beyond

At least until 1961 the USSR invested more in manned space missions than in testing theories of relativity and cosmology, whereas in the US during the first 15 years of the Cold War it was the other way round. At the end of the 1960s the USSR invested more in robotics (Lunar modules 1970–1973) and testing relativity as part of the LLR programme (performed in the same period as those in the USA), radio astronomy, radar astronomy, and the detection of gravitational waves. The crucial years 1972–1973 opened up the possibility of collaboration between the USA and the USSR and, in so doing, opened up a 'new era' in space science. At the Moscow Summit in May 1972, Nixon and Brezhnev signed the agreement for the Apollo–Soyuz Test Project (ASTP).

Generally, the focus is on attributing the collaboration to the process of détente,[52] but the scenario becomes somewhat more complex when it is realized that the collaboration on the Apollo–Soyuz test was also accompanied by competing tests in the space programmes of the two blocs. Even if the USSR and the USA apparently started to collaborate fully in sharing technology to travel above the atmosphere, to make rendezvous possible and to reproduce optimal pressurized conditions, yet the competition increased from 1972 to 1975 in the experiments performed in space, as well as in testing relativity—for example, tracking technology, clean signals, Schiff's proposal of Gravity Probe B, and so on. This can be appreciated not only by considering the LLR experiments performed between 1970 and 1973, but also from the fact that still today there are classified documents about the internal structure of the Soyuz 7K-T

(1973–present) because the Soyuz 7K-T/A9 is still employed for military purposes and, in particular, because the USSR did not want to reveal difference between the Soyuz 7K-T (1973–1981) and Soyuz T (1976–1986) used in the 1975 tests. In other words, the collaboration certainly mirrored the willingness of the two actors to give a new direction to their diplomatic and commercial relations, but it did not decrease military competition and, more interestingly from the present perspective, it did not prevent them from competing in the area of testing relativity.

US and Soviet scientists designed and planned the joint mission which resulted in the first rendezvous and docking of their separate spacecraft in earth orbit on 17 July 1975. This joint effort strove to loosen tensions between the two superpowers and to pursue shared interests in the fields of science, health and space exploration. The project exposed a small group of American and Soviet engineers and scientists to the culture, lifestyles, and values of their competitor. The mission gave a political meaning to the process of collaboration, even if it also displayed several aspects of technological and scientific competition, both with respect to targets and experiments.

In the mid-1970s Soviet the astrophysics and space science programmes appeared to be extremely strong. On the one hand, the role played by European Countries, such as France, strongly favoured Soviet scientists in the dichotomy between cooperating with the USA and USSR. We have evidence, indeed, for the collaboration between France and USSR (1970–1972) at various levels, such as radio astronomy, D1 E Proton missions (1972–1977), and exchange of know-how.[53]

On the other hand, the new era opened up by the Nixon administration led to an important collaboration between the two blocs in VLBI. On 28 April and 6 May 1976, radio telescopes on the two continents were linked together to perform observations of H_2O maser sources with an angular resolution of less than 0.0001". Among the radio sources observed were W49N and W51. The study was useful in understanding the physical processes and the conditions involve in creating cosmic masers. The VLBI experiments included four radio telescopes located on different continents linked to form a radio interferometer (Fig. 9.4).

The longest baseline was about 0.94 of the Earth's diameter. Observations were made in the 1.35-cm-water-vapour radio line, and the instrument had the maximum attainable resolution from the ground. The four antennas included in the first global radio telescope were: the 22-m antenna of the CrAO at Simeiz, the 26-m antenna at the Maryland Point Observatory of the US Naval Research Laboratory (NRL), the 40-m antenna of the Owens Valley Radio Observatory (OVRO) in California, and the 64-m NASA antenna at Tidbinbilla, Australia. This collaboration gave the USA the chance to use the latter facility which was a specialized instrument for long-range space communications.[54]

The realization of collaboration in VLBI was made possible thanks to an agreement between the Australian government and USSR in 1975.[55] In 1977 the Australian Minister for Science informed the Australian parliament about the ongoing activities and collaboration between the two countries.[56] Not only the collaboration in VLBI was on the table but also the LLR programme involving the National Mapping Division, Department of National Resources, and NASA. Ten Russian scientists visited

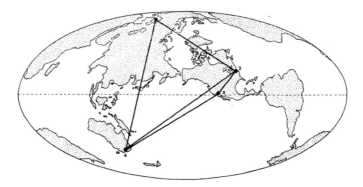

Fig. 9.4 *The locations of the four radio telescopes forming the very long baseline interferometer: Simeiz in Crimea, Maryland Point, Big Pine California, and Tidbinbilla Australia. Reproduced from Batchelor, R. et al. (1976). First global radio telescope,* Soviet Astronomy Letters, *2, No. 5 181, with the permission of AIP Publishing.*

Australia and seven Australian scientists visited the USSR for to engage in research. The agreement between Australia and the USSR involved not only VLBI studies, but also the exchange of information and know-how in entomological studies, mineralogy and geochemistry.[57] D.L. Jauncey of the CSIRO Division of Radio Physics visited USSR to discuss the results of the Australia/USSR/US VLBI experiment, while V.I. Kostenko of the Institute of Space Research visited Australia. The collaboration also included research on cosmology and high-energy astrophysics. Y.P. Uyasov and V.M. Malofeev (Lebedev Institute of Physics of the Academy of Sciences) visited Australia from 18 November to 12 December 1976 to conduct pulsar observations using the radio telescope of the University of Tasmania.

Thus, the collaboration in VLBI between Australia, the USA and the USSR was backed by the exchange of information concerning strategic fields for the Soviet and Australian economies and agriculture. The détente certainly allowed the realization of previously underdeveloped collaborations and fostered the construction of a unique experimental community studying the universe. However, there are still traces of competition between the USA and USSR in the 1970s in their respective space programmes, an interesting aspect being that involving another actor, France, on the stage.

According to Krige et al. (2013), the USA did not encourage technological innovation of the French space programme, because of proliferation risks.[58] The USA adopted strategies to encourage the constitution of a European Community with a relatively low-profile space programme. France, however, wanted to achieve supremacy in the Western European Space Programme and its collaboration with the Soviets in terms of know-how exchange should be understood as part of the attempt that France made to create greater independence from the USA and take a powerful role in European space research and technology. This strategy resulted in the relativity tests performed as part of the LLR experiments carried out in the 1990s.

9.6 Lunar laser ranging (LLR)

An example of France's strategy of collaborating with the USSR without American interference concerned the experiments and missions carried out in the 1970s.[59] The France–USSR collaboration was intended to strengthen competition in the market of the two partners in various sectors—scientific, military, medical, and commercial—by fostering laser technology and be competitive with the USA.

The competing LLR programmes in the USA and USSR paralleled those of the Apollo (1961–1972) and Luna (1959–1976) missions, respectively. The USA LLR experiments ran from 1969 and were included in the Apollo 11, 14, and 15 missions. USSR activities involved experiments on the Luna 17 and 21 missions. These produced sensitive tests of gravitational physics: the equivalence principle (EP), the accurate determination of the parameters of the parametrized post-Newtonian formalism, limits to the time variation of the gravitational constant, geodetic precession, frame-dragging, and the gravitational inverse square law.

LLR has been a pillar for testing general relativity over the previous four decades as well as dramatically increasing our understanding of the geophysics, geodesy and dynamics of the Earth and Moon. The French built arrays for the Soviet Lunokhod rovers which continue to be useful targets for laser ranging and have provided the most stringent tests of the strong equivalence principle and the time variation of Newton's gravitational constant.

As mentioned in Sections 9.1 and 9.2, Einstein's GR enjoyed its first major success in 1915 by accounting for the anomalous perihelion advance of Mercury's orbit with no adjustable theoretical parameters. Soon after, Eddington's 1919 observations of the displacement of star positions close to the Sun during a solar eclipse confirmed the doubling of the deflection angles predicted by GR as compared with Newtonian and EP arguments. After these beginnings, the GR has continued to be verified to ever-higher levels of accuracy. Microwave ranging to the Viking landers on Mars yielded $\sim 0.2\%$ accuracy through the Shapiro time-delay technique. Spacecraft and planetary radar observations reached an accuracy of $\sim 0.15\%$. LLR has provided verification of GR, improving the accuracy to $\sim 0.05\%$ through precision measurements of the lunar orbit.[60] The astrometric observation of the deflection of quasar positions with respect to the Sun performed with VLBI over the whole sky improved the accuracy of the GR tests to $\sim 0.045\%$.[61]

How does LLR work? LLR accurately measures the time of flight of a laser pulse fired from an observatory on Earth, bounced off a corner cube retroreflector on the Moon, and returned to the observatory (Fig. 9.5).[62] In essence, LLR tests the EP by comparing the free-fall accelerations of the Earth and Moon toward the Sun. If the EP were violated, the lunar orbit would be displaced along the Earth–Sun line producing a range signature having a 29.53 day synodic period, different from the lunar orbit period of 27 days. Since the first LLR tests of the EP were published in 1976, the precision of the test has increased by two orders-of-magnitude.[63]

The testing of general relativity through LLR experiments was only made possible thanks to the Apollo programme and the competition with the Soviets. On 20 July 1969,

Fig. 9.5 *Goddard's laser ranging facility directing a laser toward the LRO spacecraft in orbit around the moon (white disk). The Moon has been deliberately over-exposed to show the laser. Credits: Tom Zagwodzki/Goddard Space Flight Center.*

humans landed on the Moon for the first time and they carried with them the technology necessary to test the EP. On 21 July 1969 the Apollo 11 astronauts deployed the first laser ranging retroreflector (LRRR or LR3). Within a month, return photons were successfully observed at several observatories. Two more retroreflector arrays were placed on the Moon during the Apollo 14 and 15 missions. In addition, two French arrays were secured to the Russian Lunokhod rovers carried on Luna landers. These five retroreflector arrays remain visible today,[64] and the laser ranging data collected over the past 40 years has dramatically continued to increase our understanding of gravitational physics, as well as Earth and Moon geophysics, geodesy, and dynamics. They remain the only operating lunar experiments from the Apollo era and, more remarkably, only recently has ground station technology advanced to the point where uncertainties associated with the lunar arrays are limiting the precision of the range measurements.

The main difference between the Apollo and Luna LLR missions is that the Soviets decided to send fully automated self-moving modules (Lunokhod 1 and 2, see Fig. 9.6) and therefore could collect more data than the American static reflectors.

On 17 November 1970, the Soviet Luna 17 spacecraft landed the first roving remote-controlled robot on the moon. Known as Lunokhod 1, it weighed just under 2,000 pounds and was designed to operate for 90 days guided by a five-person team on planet Earth at the Deep Space Center near Moscow, USSR. Lunokhod 1 actually toured the

Fig. 9.6 *The module Lunokhod 1 with the French LLR reflector. Credits: NASA.*

lunar Mare Imbrium (Sea of Rains) for 11 months in one of the greatest successes of the Soviet lunar exploration programme.

Therefore, in the early 1970s with the support of French laser technology, the USSR demonstrated the capacity to compete with the USA in testing general relativity in an entirely new way and contributed substantially to such studies in the following decade, thereby strengthening the underlying physics of modern cosmology, as was also the case with the Cassini mission. After the collapse of the Soviet Union and the end of the Cold War, the science element of the Soviet space programme collapsed and funding for X-ray and gamma-ray missions stopped for some time. The collaboration between the UK and the Soviet programme in X-ray imaging also ground to a halt, despite major investments by the UK in X-ray instrumentation. But the Soviets were still able to provide launchers and of course provided key facilities for servicing the International Space Station (ISS).

9.7   Concluding remarks

As we have seen, the dichotomy of collaboration versus competition during the Cold War (1950–1991) can be used to cast a different perspective on the history of modern cosmology. In particular, connecting the history of testing the theories of relativistic gravity with the development of space science and its political context leads us to identify important characteristics of the development of modern cosmology. It also explains why fields immediately related to military research and sites, as in the case of radio astronomy, successfully reached their scientific goals in a relatively short time in both blocs. The same cannot be said for fields such as those involving the search for gravitational waves, because the military impact was low and the facilities and technology implementation costs were extremely high.

The birth of the new field of 'relativistic astrophysics' since the 1950s, particularly from the beginning of the 1960s onwards, can be directly connected to the development of space programmes and the dynamics of collaboration/competition between the two blocs. The need for an explanation of nature of quasars (1963), pulsars (1967), neutron stars (1971), and binary pulsars (1974), and the indirect evidence of gravitational waves, led to the wider introduction of general relativity into astronomy.[65] Moreover, with the prediction and observation of the cosmic background radiation, cosmology no longer remained an academic subfield of general relativity but became part of mainstream relativistic astrophysics and astrophysical cosmology. The dynamics of competition and collaboration involved actors from the two blocs and improved technological trade-offs as well as the empowerment of space science collaboration in various technologies. Testing relativity was still an area of competition in the 1970s. In this sense we can use the history of testing general relativity as a case study to reconstruct the dynamics of the competition between the USSR and the USA. It can also be used as a model for the connection between the development of cosmology and the technology and know-how of the space programme. But, most importantly, the history of technology and space science throws a fresh light on the history of general relativity and cosmology, linking them to the history of human endeavour and to politics.

Acknowledgements

The research leading to this article has been made possible thanks to the Ramón y Cajal programme (RYC-2015-17289) and the Mineco project HAR2014-57776.

....

NOTES

1. For an account of the ideology behind the disputes on cosmology in the early years of the Cold War, see Chapter 8.
2. See Blum et al. (2016).
3. Kenyon (1990), Chap. 8
4. See Holberg (2010) for the measurement of the gravitational redshift of Sirius B.
5. Robertson is not only known for his major contributions on physical cosmology and the uncertainty principle. He was involved during World War II in the National Defense Research Committee (NDRC) and the Office of Scientific Research and Development (OSRD). He served as Technical Consultant to the Secretary of War, the OSRD Liaison Officer in London, and the Chief of the Scientific Intelligence Advisory Section at Supreme Headquarters Allied Expeditionary Force. He was Chairman of the Defense Science Board from 1956 to 1961 and a member of the

President's Science Advisory Committee (PSAC) from 1957 to 1961. He had deep knowledge of the military facilities and technology used by the US army that could have been redirected for civil and research purposes.

6. See Schiff (1961), p. 42.
7. See Schiff (1961), pp. 340–3.
8. In the discussion, W. A. Fowler (California Institute of Technology) and Heckmann underlined how no conclusions could be reached by comparing the theoretical parameters of cosmological models with existing cosmological observations (Schiff, 1961).
9. The concept of a slaved satellite was proposed in 1959 by G.E. Pugh. In a companion paper, J. Mitchell (NASA) described the capabilities of existing and anticipated satellite vehicles (Schiff, 1961).
10. Kuhn (1970), p. 26, footnote.
11. Kuhn (1970), pp. 23–4.
12. Kuhn (1970), p. 24.
13. Kuhn (1970), p. 26.
14. For early proposals to detect and study gravitational waves, see Gertsenshtein (1962), Gerstenshtein and Pustovoit (1963), and Braginskii and Gertsenshtein (1967). These were presented at the 1961 conference held in Moscow, as reported by Garbell (1963), pp. 14–15. For a brief history of the topic, see Amaldi (1989).
15. This can be observed in particular in the engineering studies for Y. Gagarin's mission employing the Vostok 1 re-entry module (Vostok 3KA, 1961). H. Julian Allen and A.J. Eggers Jr. wrote a classified report concerning the feasibility of re-entry in 1953 that was published in 1958, see Vincenti et al. (2007). However, the Soviet engineers were able to translate the concept into practice before the US team.
16. See Garbell (1963).
17. See Garbell (1963), pp. 54–9.
18. See Garbell (1963), pp. 61.
19. See Garbell (1963), pp. 61–2.
20. See Braude et al. (2012).
21. The important results obtained by the Soviet radio astronomers included the discovery of the linear polarization of the diffuse cosmic radiation, the detection of circular polarization of the radio emission of active regions on the Sun and the detection and study of the spectra of thermal and non-thermal sources at centimetre wavelengths thanks to the joint efforts of FIAN, NIRFI, GAISH. See Braude et al. (2012).
22. Ohm (1961).
23. Penzias (1979).
24. For further details, see Chapter 8, Kragh (1999a), pp. 343–7, Naselsky et al. (2006), pp. 3-ff, and Peebles et al. (2009).
25. See Ohm (1961).
26. See Kragh (1996), p. 346.
27. See Harvey and Zakutnyaya (2011), p. 485.

28. See Longair (1974).
29. See Peebles et al. (2009), pp. 138ff. Also see Gindilis (2012), pp. 110ff.
30. See Peebles et al. (2009), p. 122.
31. The Kompaneets equation is a form of a Fokker–Planck equation that first attracted attention in military studies and then in cosmology and astrophysics as it describes the evolution of the spectrum of radiation exposed to the frequency shifts of Compton scattering by electrons in a hot plasma.
32. See Kompaneets (1956).
33. See Central Intelligence Agency, National Intelligence Estimate 11-2A-62, "Soviet Atomic Energy Program", (16 May 1962). https://www.cia.gov/library/readingroom/docs/DOC_0000843187.pdf. Accessed online on 20/03/2018.
34. For the debates on cosmology as being opposed to dialectic materialism, see Kragh (1996) and Peebles et al. (2009).
35. See Harvey and Zakutnyaya (2011), p. 485.
36. Sunyaev describes aspects of these developments in Peebles et al. (2009), pp. 116–7.
37. See Longair (2006), pp. 394ff. See also Peebles (1968), Zeldovich et al. (1968). The latter paper was originally submitted on 27 December 1967 and published in *Zh. Eksp. Teor. Fiz.* 55, 278–286 (July 1968).
38. The discovery of anisotropy by the RELIKT-1 spacecraft was first reported officially in January 1992 at the All-Moscow Astronomy Seminar held at the Sternberg Astronomical Institute. The Nobel Prize in Physics for 2006 was awarded to a team of American scientists, who observed the same large-scale anisotropy, but also crucially detected the small-scale fluctuations associated with the formation of large-scale structures in the universe on 23 April 1992, based on data taken by the COBE spacecraft. As a follow-up to RELIKT-1, it was decided in 1986 to study the anisotropy of the CMB as part of the Relikt-2 project with greatly increased sensitivity. The spacecraft was scheduled to be launched in 1993–1994, but the launch never took place because of the break-up of the Soviet Union and lack of funding.
39. See Strukov et al. (1988). For further developments, see Strukov et al. (1993).
40. The name aperture synthesis derives from the fact that originally scientists thought that they had to carry out measurements at essentially every baseline and orientation out to some maximum spacing. Therefore, a fully-sampled Fourier transform formally contains the information exactly equivalent to the image from a conventional telescope with an aperture diameter equal to the maximum baseline.
41. The RATAN-600 was the world's first multi-element-reflector radio telescope without any structure linking the surface elements. The functions normally performed by such a structure are instead, executed by exploiting terrestrial rotation. RATAN-600 has the ability to shift its operating frequency quickly, thereby allowing the telescope to scan a source over a range of frequencies in a very short time. This enables radio frequency observations in the frequency band ranging from 610 MHz to 30 GHz, making it especially useful for discovering variable radio sources.
42. Dagkesamanskii (2009).

43. See the USSR Council of Ministers decree no. 2006-p (11 April 1956) allowing 'the USSR Academy of Sciences to build premises for the LPI Radio Astronomy Station and to establish a radio telescope'.

44. Spitzer described two major advantages of a space telescope. First, the atmosphere absorbs almost all the ultraviolet and most of the infrared light from space before it reaches the ground. Observations at such wavelengths must necessarily be made above the atmosphere. Second, the turbulence of the atmosphere blurs the images of celestial bodies. If astronomers could put a telescope in orbit, they would finally realize the full power of the telescopes they had been using for hundreds of years. Images of such clarity and precision had never been obtained.

45. The development of radar astronomy and the use of planetary radar astronomy provides a means for understanding the larger context of the planning and execution of big science in the USA. For an interesting case study on planetary radar astronomy and testing relativistic theories of gravity, see Wilson and Kaiser (2014).

46. See Udaltsov (1963). Originally published in Russian in 1962.

47. See Shapiro (1964) and Shapiro et al. (1968).

48. See Reasenberg et al. (1979).

49. See Shapiro (1958) and Shapiro et al. (1971).

50. See Anderson et al. (1975), (1978) and Reasenberg et al. (1979).

51. Shapiro et al. (1977).

52. The 'Agreement Concerning Cooperation in the Exploration and Use of Outer Space for Peaceful Purposes', signed by Nixon and Soviet Premier Alexsey Kosygin in May 1972, was only one issue that the US and Soviet Union dealt with in Moscow that spring. The summit was the result of years of negotiations with the Soviet Union, and the outcome of the meeting became a feather in the administration's cap. Nixon and Brezhnev signed three significant agreements: SALT I, an ABM treaty, and finally a document called "Basic Principles of US–Soviet Relations." SALT I, which continues to overshadow the space agreement in historical studies about Nixon's foreign policy, failed as an arms control treaty.

53. In the USA the rendezvous-and-dock scenarios and timelines were all tailored to a specific mission. No attempt was made to standardize them. Consequently, the US approach was very labor-intensive and expensive, requiring extensive crew training and system redundancy to ensure mission success. In contrast, the Soviets pursued a course of approach-rendezvous-and-dock (AR&D) that was primarily automated with standardized operations—the flight crew was relegated to override and monitoring functions (Polites, 1998). Their efforts in AR&D date back to October 1967 when they joined two unmanned Cosmos spacecraft in orbit. The AR&D system that they had developed and refined over the years has been used repeatedly for docking the unmanned Progress and the manned Soyuz vehicles to the Mir space station. Interestingly, the hardware in this system is similar to that employed in the manual rendezvous and docking systems flown by the USA. It includes guidance digital computers, IMU-type inertial sensors (that is, rate gyros and accelerometers), optical devices for inertial sensor alignment, rendezvous radar, and TV cameras. The rendezvous radar system is called Kurs.

54. See Batchelor et al. (1976). The Canberra Deep Space Communication Complex (CDSCC) is an Earth station in Australia that is located Tidbinbilla in Paddy's River. The complex is part of the Deep Space Network run by NASA Jet Propulsion Laboratory (JPL). It is commonly referred to as the Tidbinbilla Deep Space Tracking Station and was officially opened on 19 March 1965. During the mid-1960s NASA built three tracking stations in the Australian Capital Territory. The Tidbinbilla Tracking Station is the only NASA tracking station in Australia still in operation. During the Apollo programme, Tidbinbilla was used for tracking the Apollo Lunar Module. The Orroral Valley Tracking Station was opened in May 1965. Its role was orbiting satellite support, although it also supported the Apollo–Soyuz Test Project in 1975. It was closed in 1985.

55. The Australia/USSR Scientific and Technical Cooperation Agreement was signed on 15 January 1975.

56. Minister for Science Australian government on Tuesday, 19 April 1977: http://parlinfo.aph.gov.au/parlInfo/genpdf/hansard80/hansardr80/1977-04-19/0285/hansard_frag.pdf;fileType=application%2Fpdf, pp. 3–4 accessed 30/11/2017.

57. Minister for Science Australian government on Tuesday, 19 April 1977: http://parlinfo.aph.gov.au/parlInfo/genpdf/hansard80/hansardr80/1977-04-19/0285/hansard_frag.pdf;fileType=application%2Fpdf, pp. 3–4 accessed 30/11/2017.

58. See Krige et al. (2013), pp. 3–20.

59. For the collaboration between Intercosmos (USSR) and CNES (France), see Moroz (2001).

60. See Nordtvedt 1968, 1991, 1998; Dickey et al. (1994); Williams et al. (1996), Anderson and Williams (2001), and Williams et al. (2001).

61. See Robertson et al. (1991), Lebach et al. (1995), and Shapiro et al. (2004).

62. For a general review of LLR see Williams et al. (2004).

63. See Merkowitz (2010).

64. See Murphy et al. (2011).

65. For a concise summary of the golden age of general relativity, see Goenner (2017).

10

Observational and astrophysical cosmology: 1980–2018

Malcolm S. Longair

10.1 From acts of faith to precision cosmology

The year 1980 marked a turning point in astrophysical cosmology and the determination of cosmological parameters. The pioneering efforts up till that year demonstrated just how challenging it was to obtain reliable values for the various cosmological parameters. New cosmological facts had emerged, pride of place going to the discovery in 1965 by Penzias and Wilson of the cosmic microwave background (CMB) radiation and the subsequent compelling evidence for its Planckian spectrum and its remarkable isotropy (Chapter 8). In addition, there was unambiguous evidence for the evolution with cosmic epoch of the populations of active galaxies, specifically, the extragalactic radio sources and the quasars. Observations of these sources extended the upper redshift limit of the observable Universe of discrete objects to redshifts z well beyond $z = 1$ and so epochs could now be probed when the Universe was significantly younger than it is today, albeit for rather exotic types of galaxy.

1980 was also a critical year for astronomical technology. The monopoly of the Palomar-Hale 200-inch telescope was challenged by a new generation of 4-m telescopes on excellent astronomical sites which were in full operation. These included the Mayall 4-m telescope at Kitt Peak (1973), the 3.9-m Anglo–Australian Telescope at Siding Spring Mountain in Australia (1974), the 4-m Blanco Telescope at the Cerro Tololo International Observatory in Chile (1976), the ESO 3.6-m telescope at La Silla, Chile (1977), and the Canada–France–Hawaii 3.6-m telescope on Mauna Kea, Hawaii (1979). For all these telescopes, the astrophysics of distant galaxies and related cosmological studies were priority programmes, while the number of observers involved in frontier cosmological problems increased dramatically.

But there was much more. The electronic semiconductor revolution resulted in huge gains in the sensitivities of imagers and spectrographs and in the quality of the digital data produced by them. In the mid-1970s, digital spectrographs such as the Boksenberg image photon counting system revolutionized optical spectroscopy and in 1977 charged coupled devices (CCDs) were selected as the imagers for the wide field camera of the

Longair, M. S., 'Observational and astrophysical cosmology: 1980–2018' in *The Oxford Handbook of the History of Modern Cosmology*, edited by Kragh, H. and Longair, M. S. © Oxford University Press 2019.
DOI: 10.1093/oxfordhb/9780198817666.013.10

Hubble Space Telescope (HST). The CCDs had quantum efficiencies of about 70% compared with about 1–2% for the best optical plates. In addition in the mid-1980s, the first infrared arrays operating in the 1–5 μm waveband became available to astronomers. The impact of these new technologies was immediate and the quantity and quality of astronomical data increased enormously.

Surveys were of special importance for cosmology and the astronomical technologists came up with larger and larger array cameras and multi-object spectrographs which enabled the spectra of hundreds of spectra of faint galaxies to be obtained in a single night's observing. The HST and the Sloan Digital Sky Survey (SDSS) produced vast amounts of data of cosmological relevance which changed dramatically the face of astrophysical cosmology and brought the epochs when galaxies first formed within range.

From the perspective of theory, 1980 also saw the last gasp of cosmological models in which the gravitating matter in the Universe could be assumed to be entirely baryonic. Calculations of the products of primordial nucleosynthesis showed that the mean baryonic density in the Universe was about an order of magnitude less than that observed in galaxies and clusters on the large scale. These problems were compounded by the limits to spatial fluctuations in the CMB radiation which were in serious conflict with models in which all the gravitating matter in the Universe was assumed to be baryonic. Some form, or forms, of dark matter were needed and this was formalized in the early 1980s. These developments stimulated the interests of particle physicists since the dark matter might be in some as yet unknown form of elementary particle, for example, the lightest supersymmetric particle or the axions. The beginning of the dramatic growth of *astroparticle physics* can be traced to this era.

With the exponential increase in computing power through these decades, computational astrophysics proved its enormous value in understanding quantitatively the structures which would form if the gravitating matter were in various hypothetical forms. The culmination of these endeavours was a remarkable synthesis between deductions from the power spectra of the fluctuations in the CMB radiation and ground-based observations of Type Ia supernovae which provided compelling evidence for the necessity of cold dark matter in the Universe at large and the presence of the cosmological constant in Einstein's field equations of general relativity.

The discussion of these developments is considered in two parts. The first concerns the continuing efforts to determine cosmological parameters. The most powerful evidence is undoubtedly derived from the spectacular results of the WMAP and Planck experiments, as described in detail by Bruce Partridge (Chapter 8). All the other observational evidence had, however, to be consistent with these results. Even one seriously discrepant estimate of one of the parameters would cast doubt upon the whole scheme of things. In the second part, the remarkable advances in understanding the origin of galaxies and large-scale structures in the Universe are reviewed—these are strongly dependent upon the global properties of the cosmological models.

The title of this introduction reflects the fact that, in the 'classical' approach to the estimation of cosmological parameters, it is an act of faith that the standard 'candles' and 'rigid rods' do not change with cosmic epoch. Galaxies, for example, are complex systems, the evolution of which is determined by non-linear processes, many of which are poorly understood. Another example is the nature of the Type 1a supernova explosions which are used to estimate the acceleration of the Universe—the astrophysics of these events is not fully understood. In contrast, the observations of the CMB radiation and of the very large-scale distribution of matter in the Universe involve linear physics for which the computations can be carried with considerable confidence. As a result, the principal cosmological parameters have been determined to better than 1%. In 1980, that would have seemed a quite unattainable goal.

Part 1. Cosmological parameters

In this part, the emphasis is upon the 'standard' routes to the determination of cosmological parameters, as listed in Section 6.4. These include enhancements of those discussed by Sandage in his review of 1961, as well as quite different astrophysical approaches. In the second part, other cosmological parameters will be added to the list, many of them discussed in Chapter 8.

10.2   The isotropy and homogeneity of the Universe

The remarkable isotropy of the CMB radiation showed that, on the large scale, the Universe is the same in all directions to better than one part in 10^5 (Chapter 8). With the availability of large surveys of galaxies during the 1990s, it became possible to study the *homogeneity* of the Universe as well as its isotropy. The Universe of galaxies is highly inhomogeneous, consisting of structures from the scale of isolated galaxies, through groups and clusters of galaxies to superclusters and giant voids in the distribution of galaxies (Fig. 6.6). The simplest approach is to use *two-point correlation functions* defined by

$$N(\theta)\,d\Omega = n_g[1 + w(\theta)]\,d\Omega, \tag{10.1}$$

in which $w(\theta)$ describes the excess probability of finding a galaxy at an angular distance θ from any given galaxy, $d\Omega$ is the element of solid angle, and n_g is a suitable average surface density of galaxies. $w(\theta)$ contains information about the clustering properties of galaxies to a given limiting apparent magnitude and can be measured with some precision from large statistical surveys of galaxies such as the Cambridge APM surveys which contained over two million galaxies.

The homogeneity of the distribution of galaxies can be studied by measuring the angular two-point correlation function as a function of increasing apparent magnitude. If the galaxies are sampled from a homogeneous, but clustered, distribution, the angular two-point correlation function scales with increasing limiting distance D as

$$w(\theta, D) = \frac{D_0}{D} w_0 \left(\theta \frac{D}{D_0} \right), \tag{10.2}$$

where the function $w_0(\theta)$ has been determined to distance D_0. The observed angular two-point correlation function scales exactly as expected if the distribution of the galaxies displayed the same degree of spatial correlation throughout the local Universe out to $z \sim 0.2$, as seen in Fig. 10.1 (Maddox et al., 1990). Excellent agreement was found for a larger sample of galaxies from the SDSS which extended to apparent magnitude $r^* = 23$ (Connolly et al., 2002; Scranton et al., 2002). The mean redshift in the magnitude interval $21 \le r^* \le 22$ is 0.43.

Peebles (1993) emphasized the importance of these results for cosmology. As he expressed it:

> ... the correlation function analyses have yielded a new and positive test of the assumption that the galaxy space distribution is a stationary (statistically homogeneous) random process.

Figure 10.1 also shows that the correlation function for galaxies is smooth—clustering is found on all angular scales with no prominent features on the scales of clusters or

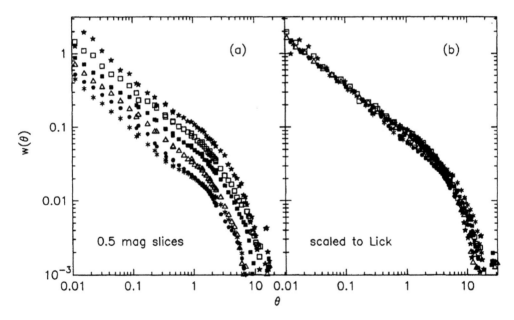

Fig. 10.1 *The two-point correlation function for galaxies over a wide range of angular scales. (a) The correlation functions for galaxies derived from the APM surveys at increasing limiting apparent magnitudes in the range $17.5 < m < 20.5$. The correlation functions are displayed in intervals of 0.5 magnitudes. (b) The two-point correlation functions scaled to the correlation function derived from the Lick counts of galaxies using the expression (10.2) (Maddox et al., 1990).*

superclusters of galaxies. The correlation function $w(\theta)$ can be characterized by a power law of the form $w(\theta) \propto \theta^{-(0.7-0.8)}$ with a cut-off on large angular scales, $\theta \gtrsim 1°$.

10.3   Hubble's constant

10.3.1   The HST Key Project

During the 1980s and 1990s, a major effort was made to resolve the differences between the estimates of Hubble's constant, much of it stimulated by the capability of the HST to measure Cepheid variable stars in the Virgo cluster of galaxies. When the HST project was approved in 1977, one of its major scientific objectives was to use its superb sensitivity for faint star-like objects to enable the light curves of Cepheid variables in the Virgo cluster to be determined precisely with the aim of estimating the value of Hubble's constant to 10% accuracy. This programme became an HST Key Project with a guaranteed large share of observing time.

The Key Project team, led by Wendy Freedman, carried out an outstanding programme of observations and analysis of these data. The team used not only the HST data, but also all the other distance measurement techniques to ensure internal self-consistency of the distance estimates. These included an improved determination of the local distance scale in our own Galaxy from the parallax programmes of the *Hipparcos* astrometric satellite, the calibration of the local Cepheid distance scale, improved application of the Tully–Fisher and Faber–Jackson relations, and so on. The result was that by the late 1990s the distances of many nearby galaxies were known very much more precisely than previously. By 2000, there was relatively little disagreement about the distances of those galaxies which had been studied out to the distance of the Virgo cluster. The remaining differences concerned how the data were to be analysed once the distances were known, in particular, in eliminating systematic errors and biases in the observed samples of galaxies. The final result of the project published in 2001 was $H_0 = 70 \pm 7$ km s^{-1} Mpc^{-1}, where the errors are 1-σ errors (Freedman et al., 2001).

10.3.2   The value of Hubble's constant by independent routes

A number of *physical methods* of measuring H_0 were based upon measuring a physical dimension l of a distant object, independent of its redshift, and its angular size θ, so that an angular diameter distance $D_A = l/\theta$ could be found at a known redshift z. An example of this technique was the combination of IUE observations of the time-variability of the emission lines from the supernova SN 1987A in the Large Magellanic Cloud (LMC) with HST observations of the emission-line ring observed about the site of the explosion to measure the physical size of the ring (Panagia et al., 1991). The distance found for the LMC was as accurate as that found by the traditional procedures.

Another promising method, suggested originally by Walter Baade in 1926 and modified by Adriaan Wesselink in 1947, involves measuring the properties of an expanding stellar photosphere (Baade, 1926; Wesselink, 1947). The velocity of expansion is measured from the Doppler shifts of the spectral lines and the increase in physical size estimated from the change in luminosity and temperature of the photosphere. The *Baade–Wesselink method* was first applied to supernovae by Branch and Patchett (1973) and by Kirshner and Kwan (1974). It was successfully applied to the supernovae SN 1987A in the LMC, resulting in a distance consistent with other precise distance measurement techniques (Eastman and Kirshner, 1989). Extending the Baade–Wesselink technique to 10 Type II supernovae with distances ranging from 50 kpc to 120 Mpc, Brian Schmidt and his colleagues found a value of H_0 of 60 ± 10 km s^{-1} Mpc^{-1} (Schmidt et al., 1992).

Another promising approach involves the use of the very hot gaseous atmospheres within clusters of galaxies, the properties of which can be measured from their X-ray spectra and from the Sunyaev–Zeldovich decrement in the CMB radiation due to inverse Compton scattering (Gunn, 1978). The X-ray surface brightness depends upon the electron density N_e and the electron temperature T_e through the relation $I_\nu \propto \int N_e^2 T_e^{-1/2} dl$, the electron temperature T_e being found from the shape of the bremsstrahlung spectrum. The Sunyaev–Zeldovich decrement is proportional to the Compton optical depth $y = \int (kT_e/m_e c^2) \sigma_T N_e \, dl \propto \int N_e T_e \, dl$. Thus, the physical properties of the hot gas cloud are over-determined and the physical dimensions of the X-ray emitting volume can be found. Steven Myers and his colleagues estimated a value of $H_0 = 54 \pm 14$ km s^{-1} Mpc^{-1} from detailed studies of the Abell clusters A478, 2142 and 2256 (Myers et al., 1997).

Another physical method of measuring H_0 is to use gravitational lensing of distant objects by intervening galaxies or clusters. The first gravitationally lensed quasar 0957+561 was discovered by Walsh et al. (1979). The gravitational deflection of the light from the quasar by the intervening galaxy splits its image into a number of separate components. If the background quasar is variable, a time delay is observed between the variability of the different images because of the different path lengths from the quasar to the observer on Earth. A time-delay of 418 days was measured for the two components of the double quasar 0957+561, enabling physical scales at the lensing galaxy to be determined. A value of Hubble's constant of $H_0 = 64 \pm 13$ km s^{-1} Mpc^{-1} at the 95% confidence level was found.

The estimates of Hubble's constant found by these physical methods are consistent with the value found by Freedman and her colleagues and with that derived from the power spectrum of the fluctuations in the CMB radiation (Chapter 8).

10.4 Acceleration of the Universe

The use of supernovae of Type 1a to extend the apparent magnitude–redshift relation to redshifts $z > 0.5$ has a number of attractive features. Firstly, it is found empirically that these supernovae have a very small dispersion in absolute luminosity at maximum light (Branch and Tammann, 1992). This dispersion can be further reduced if account is taken of the correlation between the maximum luminosity of Type 1a supernovae and the duration of the initial outburst. This *luminosity–width relation* is in the sense that the supernovae with slower decline rates from maximum light are more luminous than those which decline more rapidly. Secondly, there are astrophysical reasons why these objects are likely to be good standard candles. The Type Ia supernovae result from the explosion of white dwarfs which are members of binary systems—mass is accreted onto the surface of the white dwarf from the other member of the binary. Although the precise mechanism which initiates the explosion has not been established, the preferred picture is that the mass accreted onto the white dwarf raises the temperature of the surface layers to such a high value that nuclear burning is initiated and a deflagration front propagates into the interior of the star, causing the explosion.

In 1995 Ariel Goobar and Saul Perlmetter discussed the feasibility of observing Type 1a supernova out to redshift $z \approx 1$ in order to estimate the values of the density parameter Ω_0 and the cosmological constant (Goobar and Perlmutter, 1995). In 1996, they and their colleagues described the first results of systematic searches for Type 1a supernovae at redshifts $z \sim 0.5$ (Perlmutter et al., 1996). Deep images of selected fields were taken during one period of new moon and the field imaged in precisely the same way during the next new moon. Any supernovae which appeared between the first and second epoch observations were quickly identified and observed photometrically and spectroscopically over the succeeding weeks.

Perlmutter and his colleagues discovered 27 supernovae of Type 1a between redshifts 0.4 and 0.6 in three campaigns in 1995 and 1996 (Perlmutter et al., 1996, 1997). These data demonstrated convincingly the effects of cosmological time dilation by comparing the light curves of Type 1a supernovae at redshifts $z \sim 0.4-0.6$ with those at the present epoch (Goldhaber et al., 1996). The luminosity–width corrected intrinsic spread in the luminosities of these supernovae was only 0.21 magnitudes.

Observations with the HST were used to discover Type 1a supernovae at redshifts greater than $z = 0.8$. In two independent programmes, Peter Garnavich, Perlmutter and their colleagues discovered the Type 1a supernovae SN1997ck at redshift $z = 0.97$ and SN1997ap at redshift $z = 0.83$ respectively (Garnavich et al. 1998, Perlmutter et al. 1998).

The redshift–apparent magnitude relation presented by Perlmutter is shown in Fig. 10.2, similar to that found by Garnavich and his colleagues. Both groups found independently that the data favoured cosmological models in which the cosmological constant Λ is non-zero. Both groups continued to discover Type 1a supernovae at very large redshifts (Knop et al., 2003; Tonry et al., 2003). The significance of these results is displayed in a diagram in which the density parameter of matter Ω_0 is plotted against the value of the cosmological constant, written as a density parameter in the vacuum fields $\Omega_\Lambda = \Lambda/3H_0^2$—if the global geometry of the Universe were flat, $\Omega_0 + \Omega_\Lambda = 1$. The results of the Supernova Cosmology Project are shown in Fig. 10.3.

There are various ways of interpreting Fig. 10.3, particularly when taken in conjunction with independent evidence on the mean mass density of the Universe and the evidence from the spectrum of fluctuations in the CMB radiation. The most conservative approach is to note that the matter density in the Universe must be greater than 0, all the data being consistent with values of $\Omega_0 \approx 0.25-0.3$. Hence, the cosmological constant must be non-zero, $\Omega_\Lambda \sim 0.7$.

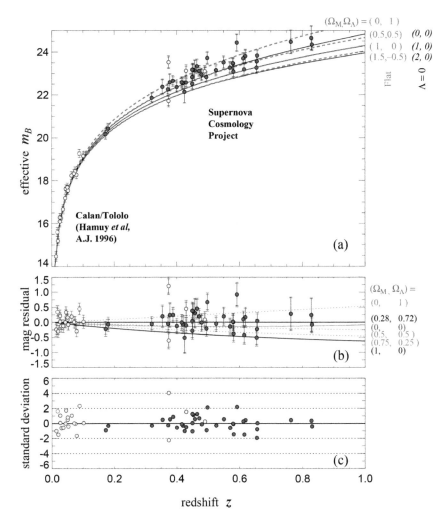

Fig. 10.2 *The apparent magnitude–redshift relation for supernovae of Type 1a. All the magnitudes plotted include the appropriate K-corrections and have been corrected for the width–luminosity relation. The theoretical lines in the top diagram are for: $(\Omega_0, \Omega_\Lambda) = (0,0); (1,0); (2,0)$ from top to bottom—solid lines; $(\Omega_0, \Omega_\Lambda) = (0,1); (0.5,0.5); (1,0); (1.5, -0.5)$ from top to bottom—dotted lines. (Perlmutter et al., 1999).*

10.5   The age of the Universe

Globular clusters contain among the oldest stars in our Galaxy. The features of their Hertzsprung–Russell diagrams which are particularly sensitive to their ages is the *main sequence termination point*. In the oldest globular clusters, the main sequence termination

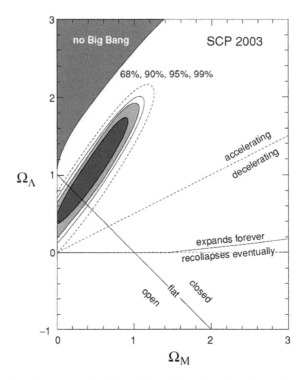

Fig. 10.3 *The 68%, 90%, 95%, and 99% confidence limits for the values of Ω_0 and Ω_Λ determined by the Supernova Cosmology Project. The condition $\Omega_0 + \Omega_\Lambda = 1$ is shown on the diagram by the line labelled 'flat' (Knop et al., 2003).*

point has reached a mass of about 0.9 $M_\odot$ by the present epoch and, in the most metal-poor and presumably oldest clusters, the abundances of the elements with atomic number $Z \geq 3$ are about 150 times lower than their Solar System values. These features make the determination of stellar ages more tractable than might be imagined. Low-mass, metal-poor stars have radiative cores and so are unaffected by the convective mixing of unprocessed material from their envelopes into their cores (Bolte, 1997). Furthermore, the corrections to the perfect gas law equation of state are relatively small throughout most solar mass stars. The surface temperatures of these stars are also high enough for molecules to be rare in their atmospheres, simplifying the conversion of their effective temperatures into predicted colours. Chaboyer (1998) demonstrated that the absolute magnitude of the main sequence termination point is the best indicator of the age of the cluster.

An example of this technique is its application to the old globular cluster 47 Tucanae. Its heavy element abundance is only 20% of the Solar abundance and the age of the cluster is estimated to be between $(12-14) \times 10^9$ years (Hesser et al., 1989).

André Maeder (1994) reported evidence that the ages of the oldest globular clusters are about 16×10^9 years and similar results were reported by Sandage (1995). Bolte (1997) argued that the ages of the oldest globular clusters were

$$T_0 = 15 \pm 2.4 \text{ (stat)} \, {}^{+4}_{-1} \text{ (syst) Gy.} \tag{10.3}$$

The first results of the *Hipparcos* astrometric survey concerning the local distance scale in our Galaxy were announced in 1997, increasing its value by about 10% (Feast and Catchpole, 1997). Consequently the stars in globular clusters were more luminous and their main sequence lifetimes were reduced. In Chaboyer's (1998) review, the ages of globular clusters were estimated to be $T_0 = (11.5 \pm 1.3)$ Gy.

Constraints on the age of the Galaxy can be obtained from estimates of the cooling times of white dwarfs. According to Chaboyer, these provide a firm lower limit of 8 Gy. The numbers of white dwarfs observed in the vicinity of the Solar System enable an estimate of $\left(9.5^{+1.1}_{-0.8}\right)$ Gy to be made for the age of the disc of our Galaxy (Oswalt et al., 1996).

In 1904, Rutherford had used the relative abundances of the radioactive species to set a lower limit to the age of the Earth and similarly lower limits to the age of the Universe can be derived from *nucleocosmochronology* (see Section 6.8.3). An example of the application of this technique was carried out by Sneden et al. (1992) for the ultra-metal-poor K giant star CS 22892-052 in which the iron abundance is 1000 times less than the solar value. A number of species never previously observed in such metal-poor stars were detected, for example, Tb (terbium, $Z = 65$), Ho (holmium, $Z = 67$), Tm (thulium, $Z = 69$), Hf (hafnium, $Z = 72$), and Os (osmium, $Z = 76$), as well a single line of Th (thorium, $Z = 90$). The thorium abundance is significant smaller than its scaled solar system abundance and so the star must have been formed much earlier than the Solar System. A lower limit to the age of CS 22892-052 of $(15.2 \pm 3.7) \times 10^9$ years was found.

A conservative lower bound to the cosmological timescale can be found by assuming that all the elements were formed promptly at the beginning of the Universe. From this line of reasoning, David Schramm found a lower limit to the age of the Galaxy of 9.6×10^9 years (Schramm, 1997). His best estimates of the age of the Galaxy are somewhat model-dependent, but typical ages of about $(12-14) \times 10^9$ years are found (Cowan et al., 1991).

In summary, all the astrophysical estimates of the age of the Universe are in agreement within the statistical errors with the value of 13.799 ± 0.038 Gy derived from analysis of the power spectrum of the CMB radiation (Chapter 8).

10.6   Mass density of the Universe

Throughout the 1980s and 1990s, the value of the density parameter Ω_0 remained a thorny issue. The use of the mean mass-to-luminosity ratios for large-scale systems such as clusters of galaxies was studied extensively by Bahcall (2000). She analysed the many different approaches which can be taken to derive values of M/L for clusters of galaxies – cluster mass-to-light ratios, the baryon fraction in clusters and studies of cluster evolution. These all found the same consistent result that the mass density of the universe corresponds to $\Omega_0 \approx 0.25$ and furthermore that the mass approximately traces the light on large scales. These results reflected the generally accepted view that, if mass densities are determined for bound systems, the total mass density of the Universe is about a factor of 4 less than the critical cosmological density.

As discussed in Section 6.8.2, the infall of galaxies into superclusters of galaxies can provide an estimate of Ω_0 on scales greater than those of clusters of galaxies. In the 1990s, complete samples of IRAS galaxies became available and they were used to define the local density and velocity fields. A mean density close to the critical density $\Omega_0 = 1$ was found (Lynden-Bell et al., 1988). Subsequently, Avishai Dekel and his colleagues devised numerical procedures for deriving the distribution of mass in the local Universe entirely from the measured velocities and distances of complete samples of nearby galaxies, the objective being to produce a three-dimensional map of velocity deviations from the mean Hubble flow (Hudson et al., 1995). Despite using only velocities and distances, and *not* their number densities, many of the familiar features of the local Universe are recovered—the Virgo supercluster and the 'Great Attractor' as well as voids in the mean mass distribution were recovered. These procedures tended to produce somewhat larger values of Ω_0, Dekel estimating that the density parameter is greater than 0.3 at the 95% confidence level.

The issue of the total amount of dark matter present in the Universe was the subject of heated debate throughout the 1990s.[1] The upshot was that there was agreement that the value of Ω_0 is greater than 0.1 and a value of 0.2 to 0.3 would probably be consistent with most of the data, the only discrepancy being the somewhat larger values favoured by Dekel.

As discussed in Section 6.8.2, large-scale density perturbations induce potential motions with the result that galaxies are expected to be observed 'falling into' large-scale density perturbations and so the projected velocity component along the line of sight differs from that associated with its cosmological redshift. For large scales, on which the linear relation (6.4) is valid, Kaiser (1987) showed how this redshift bias could be estimated and eliminated from the inferred two-point correlation functions.

The 2dF data set was used by Peacock and his collaborators to measure the magnitude of the various redshift distortions and so make an independent estimate of the value of Ω_0 on scales greater than those of clusters of galaxies (Peacock et al., 2001). Two-dimensional correlation functions in the radial (π) and transverse (σ) directions are shown as a two-dimensional plot in Fig. 10.4 for a sample of 141,000 galaxies from the 2dF Galaxy Redshift Survey. The stretching of the correlation function along the central vertical axis is due to the velocity dispersion of galaxies in groups and clusters. If there were no infall of galaxies into large-scale structures, the contours away from the vertical axis would be circular, meaning that the correlated structures have the same dimensions in the radial and transverse directions. In fact, the two-dimensional correlation function

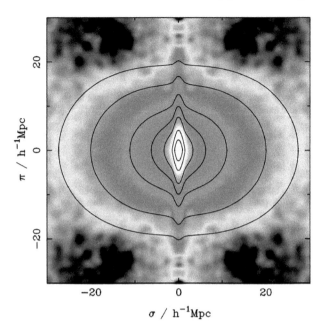

Fig. 10.4 *The two-dimensional correlation function for galaxies selected from the 2dF Galaxy Redshift Survey, $\xi(\sigma,\pi)$, plotted as a function of the inferred transverse (σ) and radial (π) pair separation. To illustrate deviations from circular symmetry, the data from the first quadrant have been repeated with reflection in both axes. This plot shows clearly the redshift distortions associated with the 'fingers of God' elongations along the central vertical axis and the coherent Kaiser flattening of the correlation function in the radial direction at large radii (Peacock et al., 2001).*

is flattened in the radial direction, associated with the infall of galaxies on either side of the cluster in the radial direction. Peacock and his colleagues concluded that

$$\Omega_0^{0.6}/b = 0.43 \pm 0.07. \tag{10.4}$$

The analysis of Verde and her colleagues showed that for galaxies in general the bias parameter b is very close to unity (Verde et al., 2002) and so the dark matter density parameter can be found, $\Omega_0 = 0.25 \pm 0.06$. This is among the most powerful routes to the value of Ω_0 because it refers to the distribution of gravitating mass on very large scales, typically $\sim 20h^{-1}$ Mpc, on which the perturbations are still in the linear regime.

When taken in conjunction with the results derived from the power spectrum of fluctuations in the CMB radiation discussed in Chapter 8, the consensus view is that the best estimate of the overall density parameter for the Universe is $\Omega_0 \approx 0.25-0.3$. Consequently, most of the gravitating mass cannot be in the form of baryonic matter, which is strongly constrained by the production of the light elements in the early stages of the Big Bang, as discussed in the next section.

10.7   The abundances of the light elements and the baryonic density parameter $\Omega_{\rm B}$

As discussed in Sections 6.5 and 6.7, the abundances of the light elements—deuterium, helium-4, helium-3, and lithum-7—created by primordial nucleosynthesis are of the greatest importance for astrophysical cosmology. The predicted abundances of these elements as a function of the present baryon-to-photon ratio is shown in Fig. 10.5 (Steigman, 2004, 2006). The challenge is to estimate the light element abundances in systems which have not been 'contaminated' by astrophysical processes in stars and the interstellar medium.

Helium-4 (^{4}He) Helium can only be observed in hot stars and in regions of ionized hydrogen. It is synthesized in the course of stellar evolution and so its abundance should be estimated in systems which are uncontaminated by stellar nucleosynthesis. This has be achieved using its abundance in low-metallicity, extragalactic H II regions. Steigman adopts a mean value for the primordial helium abundance of $Y_{\rm p} = 0.238 \pm 0.005$, where the errors encompass the small residual differences between independent estimates.

Deuterium (D) The abundance of deuterium is crucial cosmologically because of its strong dependence upon the present baryon density (Fig. 10.5). The local interstellar abundance of deuterium in our Galaxy has been well determined by observations of the resonance absorption lines of deuterium in interstellar clouds by the Copernicus satellite and by the HST (Linsky et al., 1994). The values found, $({\rm D/H}) = (1.5 \pm 0.2) \times 10^{-5}$, is a secure lower limit to the primordial deuterium abundance since deuterium is destroyed when it is circulated through the hot central interiors of stars. This value is not so different from the deuterium abundance observed in the solar atmosphere. The deuterium abundance has also been determined in the Lyman-α absorbers observed in the spectra of high redshift quasars. Quasars with redshifts $z \geq 2.5$ have sufficiently large redshifts that the deuterium Lyman resonance lines are redshifted into the region of the optical spectrum accessible to ground-based high-resolution spectroscopy. Steigman's preferred value of (D/H) was $(2.6 \pm 0.4) \times 10^{-5}$.

Helium-3 (^{3}He) is observed in the oldest meteorites, the carbonaceous chondrites, with an abundance $(^3{\rm He/H}) = 1.4 \pm 0.4 \times 10^{-5}$. This value is taken to be representative of the ^{3}He abundance about 4.6×10^9 years ago when the solar system formed. ^{3}He has also been observed at radio wavelengths through the equivalent of the 21-cm hyperfine line of neutral hydrogen and the abundances in interstellar clouds lie in the range $[^3{\rm He/H}] = (1.2 \text{ to } 15) \times 10^{-5}$. ^{3}He is destroyed inside stars but it is a more robust isotope than deuterium.

Lithium-7 (^{7}Li) ^{7}Li is also a fragile element and so its primordial abundance can be depleted by circulation through the hot inner regions of stars. It can also be enhanced by spallation collisions between cosmic ray protons and nuclei and the cold interstellar gas and by cosmic ray nucleosynthesis interactions in the interstellar medium. Therefore, the ^{7}Li abundance as a function of metallicity should reach a 'plateau' in metal-poor stars. In 1982, Spite and Spite made the first estimates of the ^{7}Li abundance for metal-poor halo stars and found that it converged at low metallicities (Spite and Spite, 1982). Steigman's

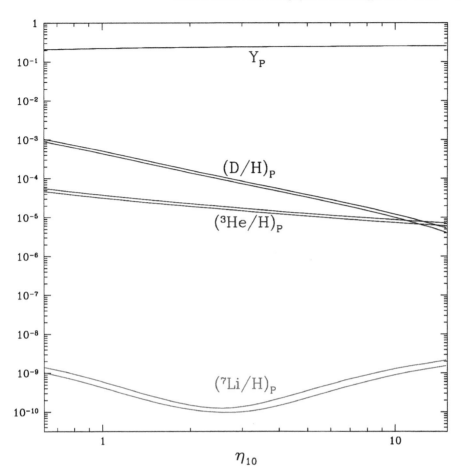

Fig. 10.5 *The predicted primordial abundances of the light elements as a function of the present baryon-to-photon ratio in the form $\eta = 10^{10} n_B/n_\gamma = 274 \Omega h^2$. Y_p is the abundance of helium by mass, whereas the abundances for D, ^3He, and ^7Li are plotted as ratios by number relative to hydrogen. The widths of the bands reflect the theoretical uncertainties in the predictions (Steigman, 2004).*

compilation of ^{7}Li abundances shows convincingly the 'Spite plateau' at metallicities [Fe/H] ≤ -2. For larger values of [Fe/H], a much wider spread in the lithium abundance is observed. Steigman adopted a ^{7}Li abundance of $[^7\text{Li}] = 12 + \log(\text{Li/H}) = 2.1 \pm 0.1$.

Steigman advocated the use of the deuterium abundance as the most sensitive probe of the baryon density parameter Ω_B. Adopting this best fitting value, the predicted ^{4}He, ^{3}He, and ^{7}Li abundances could be compared with observation. For (D/H) = $(2.6 \pm 0.4) \times 10^{-5}$,

$$\Omega_B h^2 = 0.022^{+0.003}_{-0.002}. \tag{10.5}$$

The predicted value of $(^3\text{He}/\text{H})$ is $(1.0 \pm 0.1) \times 10^{-5}$ is then also in excellent agreement with the best-estimate of the primordial ^3He abundance. For ^4He and ^7Li, the abundances are in agreement within the 2-sigma confidence levels. This value of $\Omega_B h^2$ is in remarkable agreement with the entirely independent estimate derived from analysis of the WMAP and Planck observations (Chapter 8).

Steigman showed that, leaving the number of neutrino species N_ν as a free parameter, the best-fitting value is $N_\nu = 2.3$, less than the known number of neutrino species $N_\nu = 3$, but this discrepancy is only at about the 1.5-sigma level. More than three neutrino species is, however, excluded at a high degree of significance. This limit to N_ν was derived *before* the number was measured from the energy width of the decay products of the Z^0 boson by the Large Electron–Positron collider (LEP) at CERN (Opal Collaboration, 1990).

Thus, the primordial abundances of the light elements are in remarkable agreement with the predictions of the standard model of Big Bang nucleosynthesis. Adopting a reference value of Hubble's constant $h = 0.7$, the density parameter in baryons is $\Omega_B = 0.045$. Hence, *there is insufficient baryonic matter to close the Universe*. Furthermore, since we have argued that the overall density parameter $\Omega_0 \approx 0.3$, it follows that the dark matter cannot be in the form of baryonic matter.

10.8 Ω_Λ and the statistics of gravitational lenses

One way of testing models with finite values of Ω_Λ is to make use of the fact that the volume enclosed by a given redshift z increases as Ω_Λ increases. The statistics of gravitationally lensed images by intervening galaxies therefore provides a test of models with finite Ω_Λ (Fukugita et al., 1992). Assuming that the population of lensing galaxies can be represented by identical isothermal spheres with constant comoving space density N_0, the expression for the Einstein radius θ_E, within which strong distortions of the image of a background object are expected, is

$$\theta_E = \frac{4\pi \langle v_\parallel^2 \rangle}{c^2} \frac{D_{LS}}{D_S},$$

where $\langle v_\parallel^2 \rangle$ is the mean square velocity dispersion along the line of sight of the particles which make up each isothermal sphere, D_S is the angular diameter distance of the background quasar and D_{LS} the angular diameter distance from the lens to the background source. The probability of lensing of a source at redshift z_S is then

$$p(z_S) = A N_0 \int_0^{z_S} \left(\frac{D_L D_{LS}}{D_S} \right)^2 \frac{c(1+z)^2 \, dz}{H_0[(1+z)^2(\Omega_0 z + 1) - \Omega_\Lambda z(z+2)]^{1/2}}. \tag{10.6}$$

Carroll et al. (1992) normalized this integral to the probability of lensing in the case of the Einstein–de Sitter model, $\Omega_0 = 1, \Omega_\Lambda = 0$. The resulting probability of strong lensing as a function of Ω_0 and Ω_Λ is plotted in Fig. 10.6, the contours showing the relative lensing probabilities for a quasar at a typical redshift $z_S = 2$. If $\Omega_\Lambda = 0$, as represented

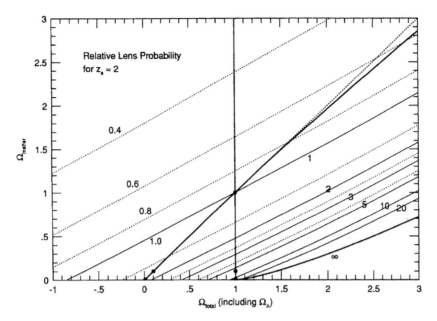

Fig. 10.6 *The probability of observing strong gravitational lensing relative to that of the critical Einstein–de Sitter model, $\Omega_0 = 1, \Omega_\Lambda = 0$ for a quasar at redshift $z_S = 2$ (Carroll et al., 1992). The contours show the relative probabilities derived from the integral (8.30) and are presented in the same format as in Fig. 7.7.*

by the solid diagonal line, there are only small changes in the probability of lensing. For low values of Ω_0, the probability increases by about a factor of 2 as compared with the $\Omega_0 = 1$ model. In contrast, the probability of lensing is very strongly dependent upon the value of Ω_Λ. For example, for the flat world models with $\Omega_0 + \Omega_\Lambda = 1$, the probability of lensing increases by almost a factor of ten as Ω_0 changes from 1 to 0.1.

To obtain limits to the value of Ω_Λ from the frequency and properties of gravitational lenses in complete samples of quasars and radio galaxies, modelling of the lens and background source populations needs to be carried out. The amplification of the brightness of the images as well as the detectability of the distorted structures need to be included in the computations (Carroll et al., 1992; Kochanek, 1996).

The largest survey to date designed specifically to address this problem has been the Cosmic Lens All Sky Survey (CLASS) in which a very large sample of flat spectrum radio sources were imaged by the Very Large Array (VLA), the Very Long Baseline Array (VLBA), and the MERLIN long-baseline interferometer. Thirteen sources were found to be multiply-imaged out of a sample of 8958 radio sources (Chae et al., 2002). More recently, the CLASS collaboration has reported the point-source lensing rate to be one per 690 ± 190 targets (Mitchell et al., 2005). The CLASS collaboration found that the observed fraction of multiply-lensed sources was consistent with flat world models, $\Omega_0 + \Omega_\Lambda = 1$, in which density parameter in the matter Ω_0 was

$$\Omega_0 = 0.31 \, {}^{+0.27}_{-0.14} \, (68\%) \, {}^{+0.12}_{-0.10} \, \text{(systematic)}. \tag{10.7}$$

Alternatively, for a flat universe with an equation of state for the dark energy of the form $p = w\varrho c^2$, they found an upper limit to w,

$$w < -0.55 \, {}^{+0.18}_{-0.11} \, (68\%), \tag{10.8}$$

consistent with the standard value for the cosmological constant $w = -1$ (Chae et al., 2002).

10.9   Summary

The upshot of this discussion is that the many routes to the determination of cosmological parameters are in good agreement with the rather precise values derived from the power spectrum of the CMB radiation as discussed in Chapter 8. There is, however, no room for complacency. The confrontation between theory and observation must continue with increasing precision, not only in the search for new physics, but also to enable limits to other cosmological parameters which may make their presence known at high levels of sensitivity. Examples include the impact of the finite neutrino rest mass and the search for the polarization signature of gravitational waves in the CMB radiation (see Chapter 8).

Part 2. Structure and galaxy formation

The theory of the origin of structure in a universe composed of baryonic matter reached crisis point in 1980 (Section 6.13). There had to be dark matter present in order to circumvent the upper limits to the temperature fluctuations in the CMB radiation and to account for the total masses of galaxies and clusters, which were about an order of magnitude greater than the amounts present as baryonic matter. Almost immediately, two scenarios were developed involving *massive neutrinos* and what became generically known as *cold dark matter*.

10.10   Dark matter and galaxy formation

10.10.1   Hot dark matter—neutrinos with finite rest mass

The first potential solution appeared in 1980 when Valentin Lyubimov and his collaborators (1980) reported that the electron neutrino had a finite rest mass of about 30 eV. Semion Gershtein and Zeldovich (1966) had noted that relic neutrinos of finite rest mass could make an appreciable contribution to the mass density of the Universe.

Later, Györgi (George) Marx and Alexander Szalay (Marx and Szalay, 1972; Szalay and Marx, 1976) considered the role of neutrinos of finite rest mass as candidates for the dark matter, as well as studying their role in galaxy formation. The intriguing aspect of Lyubimov's result was that, if the relic neutrinos had this rest mass, the Universe would be closed, $\Omega_\nu = 1$. Zeldovich and his colleagues developed a new version of the adiabatic model in which the Universe was dominated by neutrinos with finite rest mass (Doroshkevich et al., 1980a,b; Zeldovich and Sunyaev, 1980).

Neutrinos of rest mass energy 30 eV would have been highly relativistic during the epoch of nucleosynthesis and so none of the predictions of the standard Big Bang were changed. The differences appeared at later epochs when the neutrinos became non-relativistic. Prior to this epoch the energy densities in the photons and neutrinos were roughly the same and so, when the neutrinos become non-relativistic, their inertial masses no longer decreased as the Universe expanded, unlike the photons, and so the Universe became matter-dominated. The neutrino fluctuations began to grow under gravity as soon as they became non-relativistic but, since the neutrinos are ultra-weakly interacting, they streamed freely out of the perturbations and these small-scale perturbations were damped out. The amplitudes of the sound waves in the matter and radiation remained at roughly the same level they had when they came through the horizon. After recombination, the baryonic matter collapsed into the surviving larger amplitude neutrino fluctuations.

Only neutrino perturbations on the very largest scale with masses $\geq 10^{16} M_\odot$ survived to the epoch of recombination. Thus, the largest scale structures formed first and smaller scale structures formed by the process of fragmentation. This model had the advantage of reducing very significantly the expected amplitude of the fluctuations in the CMB radiation since the fluctuations in the baryonic matter were of low amplitude during the critical phases when the background photons were last scattered. This scenario for galaxy formation became known as the *hot dark matter* picture of galaxy formation since the neutrinos were highly relativistic when they decoupled from thermal equilibrium.

There were, however, concerns about this picture. Perhaps most seriously, there were reservations about the experiments which claimed to have measured the rest mass of the electron neutrino and the results could not be reproduced—the upper limit to the rest mass of the electron neutrino is now about $m_\nu \leq 3$ eV (Weinheimer, 2001). In addition, constraints could be set to the mass of the neutrinos if they were to constitute the dark matter in galaxies, groups and clusters of galaxies. Tremaine and Gunn (1979) showed that, while 30 eV neutrinos could bind clusters and the haloes of giant galaxies, those needed to bind dwarf galaxies would have to have masses much greater than 30 eV.

10.10.2 Cold dark matter and structure formation

Once it was appreciated that non-baryonic dark matter had to be taken really seriously, many possible candidates were proposed by the particle physicists. Examples included the axions, supersymmetric particles such as the gravitino or photino and ultraweakly interacting neutrino-like particles, all of which might be relics of the very early Universe.[2] From roughly 1980 onwards, the particle physicists began to take the early Universe really seriously as a laboratory for particle physics at energies which could not conceivably be achieved in terrestrial laboratories.

According to Peebles (1993), Richard Bond introduced the term *cold dark matter* in 1982 to encompass many of these exotic types of particle suggested by the particle physicists. The matter is 'cold' in the sense that these particles decoupled from the thermal background after they had become non-relativistic. In the same year, Peebles (1982) demonstrated how the presence of such particles could reduce the amplitude of the predicted fluctuations in the CMB radiation to levels consistent with the observational upper limits.

A change of terminology was also introduced about this time. In the purely baryonic picture, the perturbations in the early Universe could be decomposed into isothermal and adiabatic modes. Now, another independent component, the cold dark matter, was added to the mix. In the three-component case, the decomposition can be made into similar modes, but the names 'isothermal' and 'adiabatic' were scarcely appropriate for fluids containing collisionless dark matter particles. The types of *metric perturbation* were now referred to as *curvature* and *isocurvature* modes.

- The *curvature modes* are similar to the adiabatic modes in that, during the radiation-dominated era and before the epoch of equality of mass densities in the matter and radiation, the amplitudes of the perturbations in the radiation, the baryonic matter and the dark matter were similar and driven by gravitational potential perturbations in the metric. The relation between the density contrasts in the various mass energies $\delta\varrho/\varrho$ is similar to the adiabatic law:

$$\frac{1}{3}\frac{\delta\varrho_B}{\varrho_B} = \frac{1}{3}\frac{\delta\varrho_D}{\varrho_D} = \frac{1}{4}\frac{\delta\varrho_{rad}}{\varrho_{rad}} = \frac{1}{4}\frac{\delta\varrho_\nu}{\varrho_\nu} = \left(\frac{1}{4}\frac{\delta\varrho_C}{\varrho_C}\right), \qquad (10.9)$$

  where the subscripts are B = baryons, D = cold dark matter, rad = radiation, and ν = neutrinos. The final term in large brackets is the total density contrast which is determined by the curvature perturbations, C. Thus, there were variations in the local mass-energy density from point to point in the Universe, resulting in local perturbations in the curvature of space. The above relation is known as the *adiabatic condition* and the perturbations are often referred to as *adiabatic curvature perturbations*.

- In the *isocurvature modes*, the total mass-energy density is constant at a particular epoch everywhere and so there are no perturbations of the spatial curvature of the background model, but there are fluctuations in the mass-energy density of each of the four above components from point to point in the Universe.

Any general distribution of perturbations can be decomposed into the sum of curvature and isocurvature modes. The adiabatic curvature modes have received the most attention because they are generated rather naturally in the inflationary model of the early Universe.

The cold dark matter scenario was similar in many ways to the isothermal model (Davis et al., 1992). Since the matter was very cold, the perturbations were not damped by free streaming. Fluctuations on all scales survived and so, when the pre-recombination Universe became matter dominated, these perturbations begin to grow, decoupled from the matter and radiation. As in the hot dark matter scenario, after the epoch of recombination, the baryonic matter collapsed into the growing potential wells in the dark matter.

10.10.3 The evolution of the primordial perturbation spectrum—transfer functions

A number of effects significantly modify the initial perturbation spectrum which was presumed to have originated in the very early Universe. These can be described by the *transfer function* $T(k)$ which shows how the shape of the initial power spectrum $\Delta_k(z)$ in the dark matter is modified by different physical processes through the relation

$$\Delta_k(z=0) = T(k)f(z)\,\Delta_k(z). \qquad (10.10)$$

$\Delta_k(z=0)$ is the power spectrum at the present epoch and $f(z) \propto a \propto t^{2/3}$ is the linear growth factor between the scale factor at redshift z and the present epoch. The wavenumbers k are comoving wavenumbers and so they follow the evolution of a perturbation of a particular dark matter mass. The expected forms of the function $T(k)$ can provide important information about cosmological parameters.

Adiabatic cold dark matter

The initial power spectrum of the curvature perturbations is assumed to be power-law form,

$$P(k) = |\Delta_k|^2 \propto k^n, \qquad (10.11)$$

recalling that the Harrison–Zeldovich spectrum has $n = 1$.[3] If the perturbations came through the horizon during the radiation-dominated phase, the dark matter perturbations were gravitationally coupled to the radiation-dominated plasma and their amplitudes were stabilized by the Mészáros effect. Thus, as soon as the perturbations came through the horizon, the perturbations ceased to grow until the epoch of equality of matter and radiation energy densities. After that time, all the perturbations grew as $\Delta_k \propto a$ until very late epochs. Thus, between crossing their particle horizons at scale factor a_H and the epoch of equality a_{eq}, the amplitudes of the perturbations were damped by a factor $(a_H/a_{eq})^2$ relative to the unmodified spectrum.

The transfer function in Fig. 10.7 shows that there is a large change in slope of the predicted power spectrum at the wavenumber k_{eq}, or the mass M_{eq}, corresponding to the wavenumbers and masses of the horizon scale at the epoch of equality of matter and radiation energy densities. Representative values of these quantities are given in Table 10.1 for two reference world models. These examples show that the location of

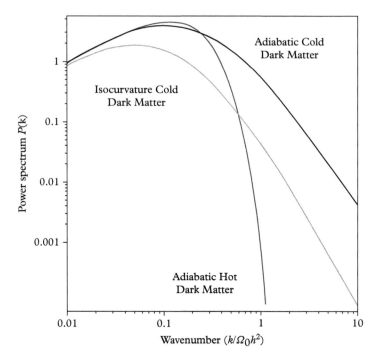

Fig. 10.7 *The predicted power spectra $P(k)$ for different models of structure formation (Bardeen et al., 1986; Peacock, 2000). For the cold dark matter models, a Harrison–Zeldovich power spectrum, $n = 1$ has been assumed. In the case of the isocurvature model, the value $n = -3$ has been adopted. The scaling has been chosen so that the power spectra are the same at small wavenumbers, that is, on very large physical scales. In both cases, the wavenumbers are in Mpc^{-1}.*

Table 10.1 *Properties of adiabatic cold dark matter perturbations which enter the particle horizon at the epoch of equality of matter and radiation energy densities.*

	$\Omega_0 = 0.3$	$\Omega_0 = 1$
World model	$\Omega_\Lambda = 0.7$	$\Omega_\Lambda = 0$
	$h = 0.7$	$h = 0.7$
z_{eq}	3,530	11,760
t_{eq}	47,500 years	4,277 years
Comoving horizon scale		
$r_{eq} = 2ct_{eq}/a_{eq}$	100 Mpc	26 Mpc
$M_{eq} = (\pi/6)r_{eq}^3\varrho_0$	$2.3 \times 10^{16}\,M_\odot$	$1.2 \times 10^{15}\,M_\odot$

the maximum of the power spectrum is sensitive to the parameters of the cosmological model. The turn-over in the spectrum is expected to occur on scales of the order of, or greater than, those of clusters of galaxies.

Notice that primordial perturbations on all scales and masses survive into the post-recombination era in the cold dark matter picture.

Adiabatic hot dark matter

In the case of the adiabatic hot dark matter model with massive neutrinos, small-scale perturbations are damped by the free-streaming of neutrinos as soon as they come through the horizon during the radiation-dominated era. Notice the exponential cut-off in the power spectrum in Fig. 10.7. Again the power spectrum has a maximum on scales greater than those of clusters of galaxies.

Isocurvature cold dark matter

The isocurvature modes behave quite differently from the adiabatic modes discussed above. When the perturbations were on superhorizon scales, any perturbation can be decomposed into curvature and isocurvature modes. Therefore, although most current discussion of structure formation has concentrated upon the adiabatic curvature modes, the isocurvature modes may play a role. For these modes, the curvature must take the same value everywhere at a given epoch and so any perturbation in the dark matter must be compensated by the opposite perturbation in the radiation. These perturbations can be thought of as fluctuations in the local equation of state. The power spectrum is expected to turn over at somewhat smaller wavenumbers than in the case of the adiabatic cold dark matter model.

10.10.4 Computational simulation of large-scale structure formation

By early 1980s, the power of digital computers had developed to such an extent that simulations of the non-linear evolution of the various scenarios for galaxy and large-scale structure formation could be carried out meaningfully. Figure 10.8 shows a sample of the results of early computer simulations of the hot and cold dark matter models carried out by Marc Davis, Carlos Frenk, and their colleagues (Frenk, 1986). These numerical simulations represented the state of the art in simulations of non-linear gravitational clustering in the mid-1980s, the *N*-body codes involving periodic boundary conditions and 32,768 particles.

There were problems with both dark matter models.

- In the *hot dark matter* picture, elongated and flattened structures were formed very effectively. in fact, too effectively in creating highly clustered structures (Fig. 10.8(b))—the observed Universe is not as highly structured as this (Fig. 10.8(c)). The baryonic matter formed 'pancakes' within the large neutrino haloes and their evolution was similar to that in the adiabatic baryonic picture. The instabilities and fragmentation of the baryonic material in the pancakes, which

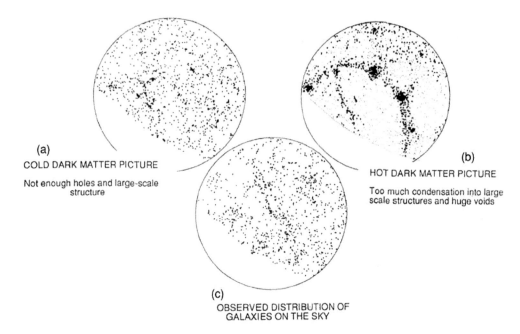

(a)
COLD DARK MATTER PICTURE

Not enough holes and large-scale
structure

(b)
HOT DARK MATTER PICTURE

Too much condensation into large
scale structures and huge voids

(c)
OBSERVED DISTRIBUTION OF
GALAXIES ON THE SKY

Fig. 10.8 *Simulations of the expectations of* (a) *the cold dark matter with* $\Omega_0 = 0.2$ *and* $\Omega_\Lambda = 0$ *and* (b) *the hot dark matter model with* $\Omega_0 = 1$ *and* $\Omega_\Lambda = 0$ *for the origin of the large-scale structure of the Universe (Frenk, 1986).* (c) *These simulations can be compared with the large-scale distribution of galaxies observed in the Harvard–Smithsonian Center for Astrophysics Survey of Galaxies. The unbiased cold dark matter model does not produce sufficient large-scale structure in the form of voids and filaments of galaxies, whereas the unbiased hot dark matter model produces too much clustering.*

resulted in the formation of galaxies and smaller structures, took place at late epochs and so it was difficult to account for the early formation of stars and the subsequent early enrichment of the heavy element abundances in the interstellar media of distant galaxies and quasars.

- In the *cold dark matter* picture, large-scale structures such as galaxies and clusters of galaxies were assembled by the process of hierarchical clustering of lower mass objects. Figure 10.8(a) shows that large-scale structures indeed developed but were not as pronounced on the large scale as is observed in the local Universe. There were, however, significant successes for the cold dark matter picture. In particular, it could account for the observed two-point correlation function of galaxies $\xi(r) \propto r^{-1.8}$ over a wide range of physical scales as a result of non-linear interactions once the perturbations had developed in amplitude to $\Delta_k > 1$. The cold dark matter picture was favoured by many investigators, but it was not without its problems. For example, in realizations of the cold dark matter model with $\Omega_0 = 1$, the velocity dispersion of galaxies chosen at random from the field was found to be too large (Efstathiou, 1990).

In both cases, the match to observation could be improved if it was assumed that the galaxies provide a biased view of the large-scale distribution of mass in the Universe (see Section 10.10.6).

10.10.5   Non-linear development of the density perturbations

The power-law form of the two-point correlation function $\xi(r) = (r/r_0)^{-1.8}$ on scales much larger than the characteristic length scale r_0 means that the perturbations are still in the linear stage of development and so provide directly information about the form of the processed initial power spectrum. On scales $r \leq r_0$, the perturbations are non-linear. An important insight was provided by Hamilton et al. (1991) who showed how it is possible to relate the observed spectrum of perturbations in the non-linear regime, $\xi(r) \geq 1$, to the processed initial spectrum in the linear regime.

The density perturbations behave like little closed universes which reach maximum size at some epoch, known as the 'turnround' epoch, after which they collapses to form bound structures. Bound structures satisfying the virial theorem are formed when the perturbations have collapsed to half the dimensions at the turnround epoch. By the time the virialized structures had formed, the density contrast had reached values greater than 100.

Hamilton and his colleagues showed that the evolution from the linear to the non-linear regime follows closely a self-similar solution which could be found from the numerical computations of Efstathiou et al. (1988). Figure 10.9(a) shows how the amplitude of the spatial two-point correlation function changes between the linear and non-linear regimes. Numerical simulations also showed that the assumption of virialization when the perturbations had collapsed by a factor two in radius from the turnround epoch was a good one, the best-fit value corresponding to a factor of 1.8.

This formula can be used in conjunction with the expression for the evolution of the power spectrum to work out the evolution of the two-point correlation function from its processed initial form to the present day. The results presented by Hamilton and his colleagues are shown in Fig. 10.9(b). This diagram shows rather convincingly the evolution of the processed initial power spectrum as a function of redshift. The horizontal line at $\xi = 1$ divides the linear from the non-linear evolution of the power spectrum.

Hamilton and his colleagues then showed how the procedure could be inverted in order to derive the processed initial mass function from the observed two-point correlation functions for different samples of galaxies. The procedures were extended by Peacock and Dodds (1994) for a wider range of world models with $\Omega_\Lambda = 0$.

10.10.6   Biasing

The generic term *biasing* means the preferential formation of galaxies in certain regions of space rather than in others. Part of the motivation behind the introduction of biasing was to improve the agreement between the predictions of the models and the observed distribution of galaxies.

(a)   (b)

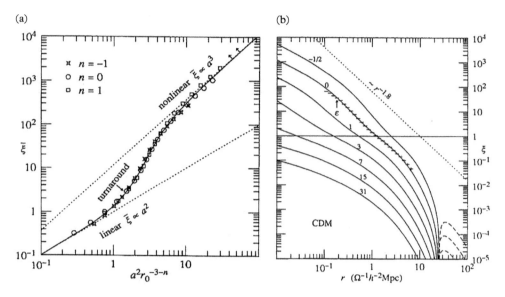

Fig. 10.9 (a) *The variation of the spatial two-point correlation function with the square of the scale factor as the perturbations evolve from linear to non-linear amplitudes.* (b) *The evolution of the spatial two-point correlation function as a function of redshift. The function has been normalized to result in a correlation function which resembles the observed two-point correlation function for galaxies which has slope* −1.8. *Non-linear clustering effects, as represented by the function shown in* (a), *are responsible for steepening the processed initial power spectrum (Hamilton et al., 1991).*

Many possible biasing and anti-biasing mechanisms were described by Dekel and Rees (Dekel, 1986; Dekel and Rees, 1987) who discussed how these can be tested by observations of the nearby Universe. Some biasing mechanisms come about rather naturally. For example, Kaiser (1984) realized that inherent in the description of the power spectrum of the perturbations is the assumption that the fluctuations are Gaussian with variance $\langle \Delta^2 \rangle$. Thus, the probability of encountering a density contrast Δ at some point in space is proportional to $\exp(-\Delta^2/\langle \Delta^2 \rangle)$. Kaiser argued that galaxies are most likely to form in the highest peaks of the density distribution. This process could, for example, account for the fact that the clusters of galaxies are much more strongly clustered than galaxies in general. This scenario was worked out in detail by Peacock and Heavens (1985) and by Bardeen et al. (1986).

Another example of biasing involves galactic explosions which sweep gas out of the galaxy. This process could result in positive or negative biasing. A violent explosion can remove gas from the vicinity of the galaxy and heat it to too high a temperature for further star formation to occur. On the other hand, the swept-up gas may be highly compressed at the interface between the hot expanding 'sphere' and the intergalactic gas. In the case of galactic supernova remnants, star formation can be stimulated by the passage of a strong shock wave and the same process might operate on a galactic scale (Ostriker and Cowie, 1981).

The availability of large statistical samples of galaxies such as the 2dF Galaxy Redshift Survey enabled bias parameters to be determined for galaxies in general and for those of different luminosities, types and colours. Verde et al. (2002) showed how the 2dF redshift survey can be used to estimate the bias to second order in the bias parameter b. The density contrast in the galaxies and the dark matter are written to second order as

$$\Delta_{\text{gal}} = b_1 \Delta_{\text{D}} + b_2 \Delta_{\text{D}}^2. \tag{10.12}$$

They found that the overall linear bias parameter b_1 is close to unity, specifically, $b_1 = 1.04 \pm 0.11$, and the non-linear bias parameter consistent with zero, $b_2 = -0.054 \pm 0.08$ on scales between roughly $5h^{-1}$ and $30h^{-1}$ Mpc.

Although the distribution of galaxies overall is unbiased on large scales, this does not exclude the possibility that there is bias on small scales or for different classes of galaxies, which is inevitably the case in, for example, accounting for the morphology–clustering relation described by Dressler (1980). The analyses by Norberg et al. (2001, 2002) demonstrated how the bias parameter depend upon the luminosities and spectral types of the galaxies. There is a clear variation of the bias parameter b with absolute luminosity in the sense that the most luminous galaxies are more strongly correlated than galaxies in general.

10.11 The Jeans' instability in the presence of cold dark matter

A key aspect of the cold dark matter scenario is how the fluctuations in the baryonic plasma were suppressed through the recombination era, the epoch at which the temperature fluctuations were imprinted on the CMB radiation. The development of the gravitational instability in the dark and baryonic matter in the expanding Universe, when the internal pressure can be neglected, can be written

$$\ddot{\Delta} + 2\,(\dot{a}/a)\,\dot{\Delta} = A\varrho\Delta, \tag{10.13}$$

where $\Delta = \delta\rho/\rho$ and $A = 4\pi G$ in the matter-dominated cases. Writing separately the equations for the evolution of the density contrasts in the baryons (B) and the dark matter (D) as Δ_{B} and Δ_{D} respectively, the following coupled equations need to be solved:

$$\ddot{\Delta}_{\text{B}} + 2\,(\dot{a}/a)\,\dot{\Delta}_{\text{B}} = A\varrho_{\text{B}}\Delta_{\text{B}} + A\varrho_{\text{D}}\Delta_{\text{D}}, \tag{10.14}$$

$$\ddot{\Delta}_{\text{D}} + 2\,(\dot{a}/a)\,\dot{\Delta}_{\text{D}} = A\varrho_{\text{B}}\Delta_{\text{B}} + A\varrho_{\text{D}}\Delta_{\text{D}}, \tag{10.15}$$

where the gravitational driving terms on the right-hand side are the same for both components. As an illustration, consider the case in which the dark matter has $\Omega_0 = 1$ and the baryon density and its perturbations are taken to be negligible compared with those of the dark matter. Then, the perturbations in the dark matter grow as $\Delta_{\text{D}} = Ba$ where B is a constant. It follows that

$$\Delta_B = \Delta_D \left[1 - (z/z_0) \right]. \tag{10.16}$$

The amplitudes of the perturbations in the baryons grow rapidly to the same amplitude as that of the dark matter perturbations. To express this result more crudely, the baryons fall into the dark matter perturbations and, within a factor of two in redshift, have already grown in amplitude to half that of the dark matter perturbations. The evolution of the density perturbations in the dark matter, the baryonic matter and the radiation for a mass $M = 10^{15} M_\odot$ is shown in Fig. 10.10(b). This behaviour can be contrasted with the evolution of density perturbations in the purely baryonic picture (Fig. 10.10(a)) (Coles and Lucchin, 1995).

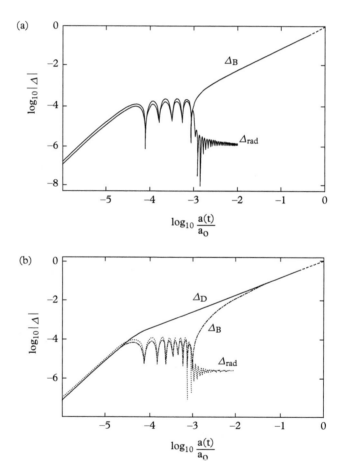

Fig. 10.10 *Illustrating the evolution of density perturbations in: (a) baryonic matter Δ_B and radiation Δ_{rad} in the standard baryonic adiabatic model. (b) The baryonic matter Δ_B, the radiation Δ_{rad} and the dark matter Δ_D according to the cold dark matter scenario. In both cases, the mass of the perturbation is $M \sim 10^{15} M_\odot$ (Coles and Lucchin, 1995).*

10.12   Variations on a theme of cold dark matter

Although the 'standard' cold dark matter picture with $\Omega_0 = 1$, $\Omega_\Lambda = 0$ and $n = 1$ had some success in reproducing a number of features of observed structures in the Universe on large-scales, the consensus of opinion among cosmologists was that it was not good enough. These concerns were sufficiently worrying for Davis et al. (1992) to entitle their *Nature* review paper *The End of Cold Dark Matter?*. The outcome of these studies was the development of a number of variants of the 'standard' picture.[4]

The problem facing the cosmologists is illustrated in Fig. 10.11, which compares the predictions of various models of structure formation with estimates of the processed initial power spectrum of galaxies by Peacock and Dodds (1994). The shape of the power spectra can be recognized from Fig. 10.7. The best fit to the processed initial power spectrum was provided by an open cosmology with $\Omega_0 \sim 0.2$ if the cosmological constant

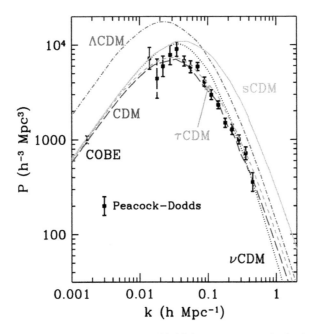

Fig. 10.11 *Some examples of the predicted processed initial power spectra of galaxies for different models of structure formation (Dodelson et al., 1996). The models shown involve the standard cold dark matter (sCDM), open cold dark matter (CDM), cold dark matter with a finite cosmological constant (ΛCDM), cold dark matter with decaying neutrinos (τCDM), and an alternative neutrino dark matter model (νCDM) described by Dodelson and his colleagues. The models are compared with the power spectrum of galaxies derived by Peacock and Dodds (1994) and the normalization at small wavenumbers derived from the COBE observations of the temperature fluctuations in the CMB radiation. The ΛCDM model has been shifted upwards for the sake of clarity.*

was zero. The 'standard' cold dark matter model with $\Omega_0 = 1$ resulted in too much power at large wavenumbers.

The origin of the problem can be traced to the physics which determines the location of the maximum in the power spectrum, namely the wavenumber which corresponds to the horizon scale at the epoch of equality between the matter and radiation content of the Universe (Section 6.11.2). The epoch of equality occurred at a redshift

$$z_{eq} = 3c^2 \Omega_0 H_0^2 / 8\pi G \chi a T_0^4, \qquad (10.17)$$

while the particle horizon was $r_H = 2ct_{eq}$. Putting these together, Hu et al. (1997) found the comoving wavenumber corresponding to the maximum of the power spectrum was

$$k_{eq} = (2\Omega_0 H_0^2 z_{eq}/c^2)^{1/2} = 7.3 \times 10^{-2} \Omega_0 h^2 \, \text{Mpc}^{-1}, \qquad (10.18)$$

where the temperature of the CMB radiation has been assumed to be 2.728 K. A value of $\Omega_0 = 1$ results in the maximum of the power spectrum being shifted to larger wavenumbers than the observed maximum and so to excess power on small scales. The aim of the computational cosmologists was to find ways of moving the maximum, which depends upon $\Omega_0 h^2$, to smaller wavenumbers. This was achieved in the open CDM model with $\Omega_0 = 0.2$ and $h = 0.7$ and in the ΛCDM model which also has $\Omega_0 = 0.2$ but with $\Omega_0 + \Omega_\Lambda = 1$.

A wide range of models was studied by Kauffmann et al. (1999), the initial perturbation spectrum being chosen to reproduce the same large-scale structure at the present epoch. They all make quite different predictions, however, about their evolution at earlier epochs. A few examples illustrate the range of possibilities they considered.

Cold dark matter with a finite cosmological constant (ΛCDM).
This model with $\Omega_0 \approx 0.25 - 0.3$, $\Omega_\Lambda \approx 0.7 - 0.75$, $n \approx 1$ was to become the industry standard for the formation of large-scale structure once it was established that the cosmological constant had the value $\Omega_\Lambda \approx 0.7$.

Open cold dark matter (OCDM).
The OCDM model with $\Omega_0 \approx 0.2$, $\Omega_\Lambda = 0$ has a similar evolutionary history to the standard ΛCDM model. because there is not a great deal of difference in the dynamics of the underlying models as compared with the OCDM model until late epochs.

Standard cold dark matter (SCDM).
The standard CDM model with $\Omega_0 = 1$, $\Omega_\Lambda = 0$ could only be made consistent with the present observed large-scale structure by adopting a low value of Hubble's constant, $h = 0.5$ so that a lower value of Ωh^2 is obtained, and a value of the cluster density parameter $\sigma_8 = 0.5$, lower than the present best estimates $\sigma_8 = 0.9$. Reducing σ_8 corresponds to increasing the bias parameter. There is very much more rapid evolution of the large-scale structure over the redshift range $3 > z > 0$ than in the ΛCDM or open models.

Cold dark matter with decaying neutrinos (τCDM).
In this scenario, the objective was to enhance the radiation to matter energy densities so that the epoch of equality of matter and radiation energy densities was shifted to lower redshifts as compared with the sCDM model, mimicking the case of the OCDM picture. The trick was to suppose that there existed a massive neutrino-like particles which decayed into relativistic forms of matter after the epoch of nucleosynthesis (Bond and Efstathiou, 1991; McNally and Peacock, 1995).

Besides these four models, many other possibilities were considered. For example, a *mixed dark matter* (HCDM) model in which the dark matter was a mixture of hot and cold dark matter. In the *tilted cold dark matter* model, the power spectrum of the initial fluctuations departed from the standard value $n = 1$. In the *broken scale invariant cold dark matter* model, the initial power spectrum changed slope at the appropriate wavenumber to reduce the power on small scales.

10.13 The acoustic peaks in the power spectrum of galaxies

Although the baryons constitute only about 20% of the total mass density, they leave a perceptible imprint upon the galaxy power spectrum. The expected effects were illustrated by the analyses of Eisenstein and Hu (1998) who considered the transfer functions and power spectra for models with mixed baryons and dark matter.

Considering first the case of baryon-only models, the upper panels of Fig. 10.12 show the predicted transfer functions $T(k)$ for an open model with $\Omega_0 = 0.2$ and the critical model $\Omega_0 = 1$, in both cases with $\Omega_\Lambda = 0$. For comparison, these diagrams also show as dotted lines the cold dark matter transfer function with no baryons present. Notice the differences in the location of the wavenumber k_{eq} associated with the horizon scale at the epoch of equality which is proportional to Ω_0. The transfer functions for purely baryonic matter result in very pronounced oscillations in the power spectrum of the distribution of galaxies at the present epoch.

The lower pair of diagrams in Fig. 10.12 show the results of computations for mixed baryonic and cold dark matter models for $\Omega_0 = 0.2$ and the critical model $\Omega_0 = 1$ with $\Omega_\Lambda = 0$. In both cases, there are equal amounts of baryonic and dark matter. From the analysis of Section 10.10, the perturbations in the baryons had amplitudes much smaller than those in the cold dark matter immediately after recombination, but the acoustic oscillations were present in the baryonic component. After recombination, the perturbations in the baryons were amplified by the gravitational influence of the perturbations in the dark matter, as described in Section 10.11.

10.13.1 The 2dF Galaxy Redshift Survey

The results of the analysis by Cole et al. (2005) of 221,414 galaxies, all with measured redshifts, from the 2dF data set showed that the overall shape of the power spectrum is similar to the form shown in Fig. 10.10 for the cold dark matter model but modified to take account of the baryons. They found significant 'ripples' which were interpreted as the acoustic peaks in the power spectrum.

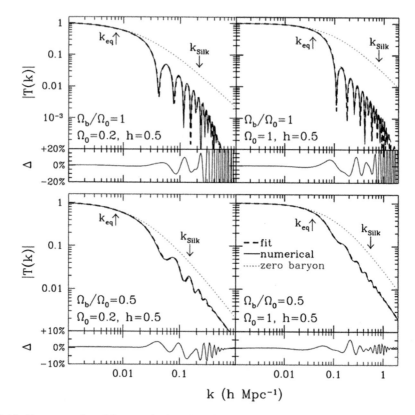

Fig. 10.12 *Four examples of the transfer functions for models of structure formation with baryons only (top pair of diagrams) and with mixed cold and baryonic models (bottom pair of diagrams) (Eisenstein and Hu, 1998). The numerical results are shown as solid lines and fitting functions by dashed lines. The lower small boxes in each diagram show the percentage residuals to their fitting functions, which are always less than 10%.*

In interpreting the power spectrum, the 2dF team compared the observed spectrum with the expectations of a range of cold dark matter models with different amounts of baryonic matter present. They found evidence for the first and second peaks of the baryon power spectrum at wavenumbers 0.06 and 0.12, respectively, corresponding to about $100h^{-1}$ and $60h^{-1}$ Mpc. Good agreement with the observations was obtained for the set of cosmological parameters listed in Table 10.2. The derived value of the density parameter was $\Omega_0 = 0.231 \pm 0.021$.

10.13.2 The Sloan Digital Sky Survey

The power spectrum derived from a sample of 46,748 luminous red galaxies in the redshift range 0.16−0.47 from the SDSS derived by Hogg et al. (2005) is shown in Fig. 10.13. The mean redshift of the sample was about 0.3 compared with 0.1 for the complete 2dF sample. Although the statistics are smaller than in the 2dF sample, the restriction to luminous galaxies meant that better statistics were achieved over larger volumes, particularly in the crucial $50h^{-1}$ to $200h^{-1}$ Mpc range of scales.

The two-point correlation function is presented in Fig. 10.13 in the form $s^2\xi(s)$ where s is the separation of the galaxies in order to highlight the curvature of the power spectrum on small physical scales. The clear feature in the power spectrum at physical scale $100h^{-1}$ Mpc corresponds to the first acoustic peak in the power spectrum of primordial fluctuations, in good agreement with that inferred from the 2dF Galaxy Redshift Survey.

From the overall shape of the correlation function, the matter density was found to correspond to $\Omega_0 = 0.273 \pm 0.025$, if it is assumed that the dark energy is associated with the cosmological constant and the global geometry of the Universe is flat. If the scale of

Table 10.2 *Cosmological parameters derived from analysis of the 2dF Galaxy Redshift Survey.*

Power spectrum spectral index	$n = 1$	assumed
Hubble's constant	$h = 0.72$	assumed
Neutrino masses	$m_\nu = 0$	assumed
Overall density parameter	$\Omega_0 h = 0.168 \pm 0.016$	derived
Baryon fraction	$\Omega_B/\Omega_0 = 0.185 \pm 0.046$	derived

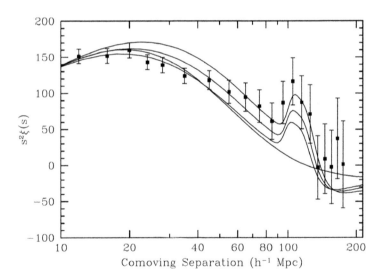

Fig. 10.13 *The large-scale redshift–space correlation function of the Sloan Digital Sky Survey luminous red galaxy sample plotted as the correlation function times s^2. This presentation was chosen to show the curvature of the power spectrum at small physical scales. The models have $\Omega_0 h^2 = 0.12$ (top), 0.13 (middle), and 0.14 (bottom), all with $\Omega_B h^2 = 0.024$ and $n = 0.98$. The smooth line through the data with no acoustic peak is a pure Cold Dark Matter model with $\Omega_0 h^2 = 0.105$ (Eisenstein et al., 2005).*

the acoustic peak is included in the estimates, the constraint on the spatial curvature was found to be $\Omega_\kappa = -0.010 \pm 0.009$. Notice that these conclusion are independent of the information derived from analyses of the fluctuations in the CMB radiation.

Both the Sloan and 2dF teams recognized the central importance of the discovery of baryon oscillations in the power spectrum of galaxies. In conjunction with the perturbations observed in the CMB radiation discussed in Chapter 8, the observations are compelling evidence that the structures we observe in the Universe today resulted from the gravitational collapse of small amplitude perturbations originating in the early Universe.

10.14   The post-recombination era

The astrophysics of the post-recombination era from $z \approx 1,500$ to the present epoch involves the non-linear evolution of perturbations in the dark and baryonic matter, as well as a host of astrophysical phenomena associated with the formation and death of stars and the associated feedback mechanisms. In addition, the formation and evolution of the supermassive black holes in the nuclei of massive galaxies needs to be incorporated into the story.

The *post-recombination era* can be divided into two phases. The first corresponds to the epochs between the epoch of recombination, when the temperature of the thermal background radiation was about 4000 K, and the *epoch of reionization* when the intergalactic gas was reionized. This era is often referred to as the *dark ages* since it is very difficult to study observationally.

Gravitational collapse resulted in the formation of growing gravitational wells in the dark matter into which the baryonic matter accumulated until the first stars were formed–that epoch is often referred to as the *cosmic dawn*. The first massive stars emitted intense ultraviolet radiation which heated and reionized the diffuse neutral hydrogen. The key to unlocking what took place at this transitional epoch is the observation of neutral hydrogen through the reionization era. Neutral hydrogen can be observed at a radio wavelength of 21 cm through a highly forbidden hyperfine transition and, by observing it at low radio frequencies, its highly redshifted radio emission should be observable. The reionization process must have occurred at some epoch in the redshift interval $30 > z > 6$, but, despite considerable efforts on the part of the radio astronomers, the cosmic signal has proved elusive. The reionization history of the Universe is one of the major challenges for observational cosmology.

The second phase corresponds to the redshift interval $0 < z < 6-7$, what may be termed the *observable Universe of galaxies* and is the range of redshifts over which luminous galaxies and quasars have now been observed. There is plentiful evidence for the dramatic evolution of the populations of galaxies and quasars over this redshift interval (Section 10.16). The intrinsically non-linear nature of the processes involved has been the subject of large-scale computer simulations which provide clues to the underlying astrophysics (Section 10.17).

10.14.1 The non-linear collapse of density perturbations

The development of dark matter perturbations was followed into the non-linear regime following the insights of Hamilton et al. (1991) described in Section 10.10.5. Their argument can be developed to provide further insight into the early formation history of galaxies and clusters of galaxies.

One calculation, which can be carried out exactly, is the collapse of a uniform spherical density perturbation in an otherwise uniform Universe, a model sometimes referred to as *spherical top-hat collapse*. The dynamics of such a region are precisely the same as those of a closed Universe with $\Omega_0 > 1$. The perturbation therefore reached maximum size, which is referred to as the 'turnround' scale-factor a_{max} at time t_{max}, and then collapsed to infinite density in twice that time. For illustration, we take the background model to be the critical model, $\Omega_0 = 1, \Omega_\Lambda = 0$, to which all the models tend at early epochs. It is straightforward to show that, at t_{max}, the density contrast $\rho_{max}/\rho_0 = 5.5$. According to this idealized mode, the perturbation then collapsed to infinite density at twice the time it took to reach maximum expansion. Bound structures must, however, have formed before this ultimate collapse could take place.

Because of the presence of dark matter density perturbations and the tidal effects between them, the perturbations fragmented into sub-units which then came to a dynamical equilibrium under the influence of large-scale gravitational potential gradients, the process of *violent relaxation* described by Lynden-Bell (1967). He showed that the system rapidly evolved towards an equilibrium configuration in which all the masses attained the same velocity distribution satisfying the virial theorem, in which the internal kinetic energy of the system is half its (negative) gravitational potential energy. By collapsing by a factor of two in radius from its maximum radius of expansion, this condition is satisfied. Once the kinetic energy was randomized by violent relaxation, the virial theorem condition for dynamical equilibrium was satisfied. The density of the perturbation increased by a further factor of 8, while the background density continued to decrease. The net result is that, when the collapsing cloud became a bound virialized object, its density was about 150 times the background density at that time.

These estimates were confirmed by N-body computer simulations of structure formation in the expanding Universe. According to Coles and Lucchin (1995), the system was virialized after a few crossing times, when the density contrast would be closer to 400 . The general rule was found that discrete objects such as galaxies and clusters of galaxies only became distinct gravitationally bound objects when their densities were at least 100 times the background density.

These simple calculations can be used to estimate the redshift at which systems of different mass were virialized in the post-recombination era:

$$(1 + z_{vir}) \leq 0.47 \left(\frac{v}{100\,\text{km s}^{-1}} \right)^2 \left(\frac{M}{10^{12}\,M_\odot} \right)^{-2/3} (\Omega_0 h^2)^{-1/3}. \tag{10.19}$$

Adopting reference values $\Omega_0 = 0.3, h = 0.7$, for galaxies having $v \sim 300$ km s^{-1} and $M \sim 10^{12} M_\odot$, the redshift of formation must have been less than about 7. For clusters of galaxies for which $v \sim 1000$ km s^{-1} and $M \sim 10^{15} M_\odot$, the redshift of formation cannot have been much greater than 1. This is consistent with the observation that the density contrast of 100 is not so different from the ratio of the mean density of clusters of galaxies to typical cosmological densities at the present day.

10.14.2   The Zeldovich approximation

The top-hat model applies for precisely spherically symmetric perturbations. The next approximation is to assume that they are ellipsoidal with three unequal principal axes. The study of ellipsoidal mass distributions shows that collapse takes place most rapidly along the shortest axis (Lin et al., 1965). For the case of primordial density fluctuations, Zeldovich (1970a) showed how the collapse could be followed into the non-linear regime in this more general case.

In the *Zeldovich approximation*, the development of perturbations into the non-linear regime is followed in Lagrangian coordinates. If $\vec{x}$ and $\vec{r}$ are the proper and comoving position vectors of the particles of the fluid, the Zeldovich approximation can be written

$$\vec{x} = a(t)\vec{r} + b(t)\vec{p}(\vec{r}). \tag{10.20}$$

The first term on the right-hand side describes the uniform expansion of the background model and the second term the comoving deviations $\vec{p}(\vec{r})$ of the particles' positions relative to a fundamental observer located at comoving vector position $\vec{r}$. Zeldovich showed that, in the coordinate system of the principal axes of the ellipsoid, the motion of the particles in comoving coordinates is described by a 'deformation tensor' D

$$D = \begin{bmatrix} a(t) - \alpha b(t) & 0 & 0 \\ 0 & a(t) - \beta b(t) & 0 \\ 0 & 0 & a(t) - \gamma b(t) \end{bmatrix}. \tag{10.21}$$

The clever aspect of the Zeldovich approximation is that, although the constants α, β, and γ vary from point to point in space, the functions $a(t)$ and $b(t)$ are the same for all particles. In the case of the critical model, $\Omega_0 = 1$,

$$a(t) = (1+z)^{-1} = (t/t_0)^{2/3} \quad \text{and} \quad b(t) = \tfrac{2}{5}(1+z)^{-2} = \tfrac{2}{5}(t/t_0)^{4/3}, \tag{10.22}$$

where $t_0 = 2/3H_0$.

For the case $\alpha > \beta > \gamma$, collapse occurs most rapidly along the x-axis and the density becomes infinite when $a(t) - \alpha b(t) = 0$. At this point, the ellipsoid has collapsed to a 'pancake' and the solution breaks down at later times. The cold dark matter particles move purely under gravity and have no internal pressure. Consequently, collapse of the ellipsoids into pancakes does not give rise to strong shock waves, which occur when baryonic matter is included in the simulations.

The results of numerical N-body simulations showed that the Zeldovich approximation is remarkably effective in describing the non-linear evolution of large-scale structures

(a)   (b)

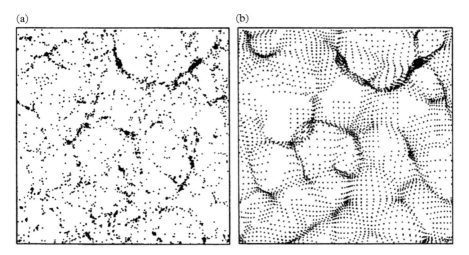

Fig. 10.14 *A comparison between the formation of large-scale structure according to (a) N-body simulations, and (b) the Zeldovich approximation, which began with the same initial conditions (Coles et al., 1993; Coles and Lucchin, 1995). The agreement between the two approaches is remarkably good.*

up to the point at which caustics are formed. An comparison between the results of adopting the Zeldovich approximation and of N-body simulations by Coles et al. (1993, 1995) is shown in Fig. 10.14. The Zeldovich approximation was accurate quantitatively and gives insight into the results of the N-body simulations.

10.14.3   The Press–Schechter formalism revisited

Press and Schechter were well aware of the limitations of the *Press–Schechter model* (Monaco, 1998) and a number of their assumptions turned out to be incorrect—for example, they assumed that all the matter in the Universe is baryonic and has the critical cosmological density (Schechter, 2002). Monaco (1999) remarked that:

> There is a simple, effective and wrong way to describe the cosmological mass function. Wrong, of course, does not refer to the results, but to the whole procedure.

Monaco described numerous reasons why the calculation should not work. Nonetheless, the theory predicted that the mass spectrum of clusters is of power-law form with an exponential cut-off at high masses, in reasonable agreement with the observed mass function of clusters (Schechter, 2002). In the process of achieving this, however, it was implicitly assumed that smaller mass structures were wiped out as they coalesced to form more massive structures. Massive N-body simulations, the *Millennium Simulation*, showed however that there is disruption of the coalescing dark matter haloes, but a significant fraction survives and can populate the haloes of giant clusters (Springel et al., 2005).

More recent large-scale simulations for the standard ΛCDM model show that the Press–Schechter mass function is a remarkably effective description of the form and evolution of the mass function of dark matter halos with cosmic epoch. Figure 10.15 shows the results of the Millennium Simulation which involved following the evolution of more than 10^{10} particles. $\bar{\rho}$ is the mean density of the background model. The solid lines show parametric fits to the simulations and the faint dotted lines the predicted Press–Schechter function at redshifts $z = 10.07$ and $z = 0$. The success of the Press–Schechter formalism is apparent. Springel and his colleagues caution that at the high mass end where the 'exponential' cut-off occurs, the Press–Schechter function underpredicts the mass function by up to an order of magnitude.

The Press–Schechter formalism is a useful tool for studying the development of galaxies and clusters of galaxies in hierarchical scenarios of galaxy formation. As an example of the use of the function, Efstathiou (1995) matched the Press–Schechter mass function to the results of N-body simulations of the development of galaxies and clusters for the standard $\Omega_0 = 1$ cold dark matter model. He then illustrated how the comoving number density of dark matter haloes with masses greater than a given value,

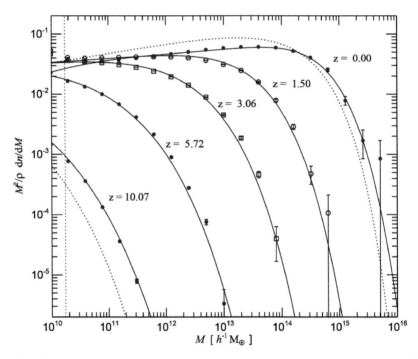

Fig. 10.15 *Differential halo number density as a function of mass and redshift from the Millennium Simulation (Springel et al., 2005). The function n(M, z) is the comoving number density of halos with masses less than M. What is plotted is the differential halo multiplicity function in the form $(M^2/\bar{\rho})\, dn/dM$, where $\bar{\rho}$ is the mean density of the Universe. The solid lines represent an analytic fitting function, while the dotted lines show the Press–Schechter function at $z = 10.07$ and $z = 0$.*

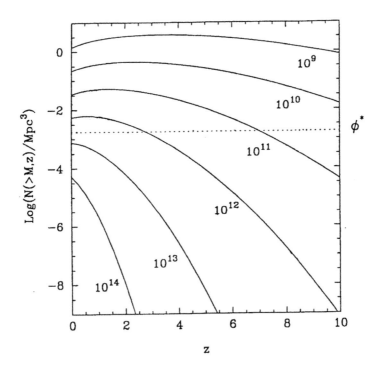

Fig. 10.16 *The evolution of the comoving number density of dark matter haloes with masses greater than M as a function of redshift for a standard cold dark matter model with $\Omega_0 = 1$. The curves have been derived using the Press–Schechter form of evolution of the mass spectrum which is a close fit to the results of N-body simulations. The dotted line labelled ϕ^* shows the present number density of L^* galaxies (Efstathiou, 1995).*

$N(\geq M, z)$ changes with redshift (Fig. 10.16) (Efstathiou and Rees, 1988; Efstathiou, 1995). The present number density of L^* galaxies is shown as a dotted line. Although derived for the $\Omega_0 = 1, \Omega_\Lambda = 0$ model the results are similar for the preferred ΛCDM model. Figure 10.16 demonstrates vividly how the most massive systems form rather late in the Universe, galaxies with masses $M \sim 10^{12} M_\odot$ only appearing in substantial numbers at redshifts $z \leq 4$.

10.15   The role of dissipation

By *dissipative processes* we mean that baryonic matter can lose energy by a myriad of radiative processes, resulting in the loss of thermal energy from the system. These processes enable the baryonic component of galaxies to lose internal energy and so shrink within the dark matter haloes.

The role of dissipative processes in galaxy formation was described by Rees and Ostriker (1977), who considered the cooling of a primordial plasma with the primaeval abundances of hydrogen and helium. Silk and Wyse (1993) subsequently included cooling by heavy elements at different levels of enrichment relative to the primordial values into their cooling curves. In the absence of heavy metals, the dominant loss mechanism at high temperatures, $T > 10^6$ K, is thermal bremsstrahlung, the energy loss rate being proportional to $N^2 T^{1/2}$. At lower temperatures, the main loss mechanisms are free–bound and bound–bound transitions of hydrogen at $T \approx 10^4$ K and of ionized helium at $T \approx 10^5$ K. As the abundance of the heavy elements increases, the overall energy loss rate can be more than an order of magnitude greater than that of the primordial plasma at temperatures $T \leq 10^6$.

For a fully ionized plasma, the cooling time is defined as the time it takes the plasma to radiate away its thermal energy

$$t_{\text{cool}} = \frac{E}{|dE/dt|} = \frac{3NkT}{N^2 \Lambda(T)}, \tag{10.23}$$

where N is the number density of hydrogen ions. This timescale can be compared with that for gravitational collapse, $t_{\text{dyn}} \approx (G\rho)^{-1/2} \propto N^{-1/2}$. The significance of these timescales is best appreciated by inspecting the locus of the equality $t_{\text{cool}} = t_{\text{dyn}}$ in a temperature–number density diagram (Fig. 10.17). Inside this locus, the cooling time is shorter than the collapse time and so it is expected that dissipative processes are more important than dynamical processes in determining the behaviour of the baryonic matter. Also shown in Fig. 10.17 are lines of constant mass, as well as the loci corresponding to

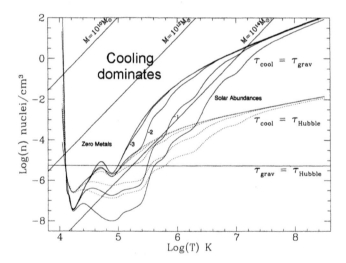

Fig. 10.17 *A number density–temperature diagram showing the locus defined by the condition that the collapse time of a region t_{dyn} should be equal to the cooling time of the plasma by radiation t_{cool} for different abundances of the heavy elements (Silk and Wyse, 1993). Also shown are lines of constant mass, a cooling time of 10^{10} years (dotted lines), and the density at which the perturbations are of such low density that they do not collapse in the age of the Universe.*

the radiation loss time being equal to the age of the Universe, and to the perturbations having such low density that they do not collapse gravitationally in 10^{10} years. The masses which lie within the critical locus, and which can cool in 10^{10} years, is $10^6 \leq M/M_\odot \leq 10^{12}$. The fact that the masses lie naturally in the range of observed galaxy masses suggests that their properties are not only determined by the initial fluctuation spectrum, but also by dissipative astrophysical processes.

Blumenthal et al. (1984) plotted the observed location of galaxies on a temperature–number density diagram similar to Fig. 10.17, but using velocity dispersion in place of temperature, that is, $\frac{1}{2}kT_{\text{eff}} \approx \frac{1}{2}mv^2$. The irregular galaxies fell well within the cooling locus and the spirals, S0, and elliptical galaxies all lay close to the critical line. On the other hand, the clusters of galaxies lay outside the cooling locus. Thus, cooling is expected to be an important factor in the formation of galaxies.

In the analysis of Rees and Ostriker, it was assumed that all the mass of galaxies was in the form of baryonic matter. White and Rees (1978) studied the more realistic case in which the galaxies and clusters are dominated by dark matter, the baryonic matter constituting only about 20% of the total mass. The dark haloes provided the potential wells into which the baryonic matter could collapse and cool.

10.16   Evolution of galaxies and the global star formation rate

Enormous strides were made in understanding the evolution of the stellar and gaseous components of galaxies through the period 1980–2018. The number counts of galaxies in different 'thermal' wavebands provided key evidence on their overall evolution with cosmic epoch.

10.16.1   The evolution of galaxies

With the development of sensitive semiconductor detector arrays for the optical and near infrared wavebands and the construction of large space and ground-based telescopes, the high redshift Universe was opened up for the study of the evolution of galaxies. The counts of galaxies presented by Metcalfe et al. (1996) provide a good impression of the overall counts of galaxies in the B (440 nm), I (800 nm), and K (2.2 μm) wavebands (Fig. 10.18), including deep number counts in the Hubble Deep Field. The lines labelled 'No evoln.' show the expectations of uniform world models and include appropriate K-corrections for the types of galaxy observed in bright galaxy samples. These studies were extended to even fainter magnitudes through observations of the Hubble Ultra Deep Field, which made use of the wider field capabilities of the Advanced Camera for Surveys (Beckwith et al., 2006). This project resulted in the deepest image ever taken of the sky and involved 400 orbits of observing time.

In the infrared K waveband (2.2 μm), the counts follow reasonably closely the expectations of uniform world models. In contrast, in the R, B, and U wavebands, there is a large excess of faint galaxies at $B > 23$. The nature of the excess of blue galaxies

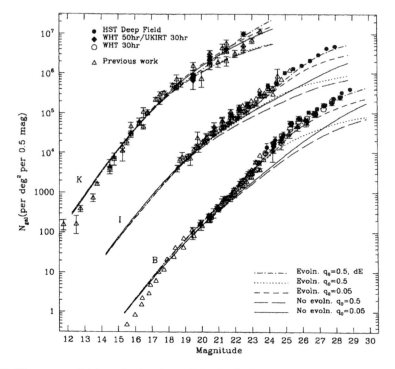

Fig. 10.18 *The counts of faint galaxies observed in the B, I, and K wavebands compared with the expectations of various uniform world models, as well as other models in which various forms of the evolution of the luminosity function of galaxies with redshift are assumed (Metcalfe et al., 1996). The galaxy counts follow closely the expectations of uniform world models at magnitudes less than about 22, but there is an excess of galaxies in the B and I wavebands at fainter magnitudes.*

was elucidated by studies with the HST by Abraham et al. (1996) who first showed that the objects classified as irregular/peculiar/merger systems are responsible for the blue galaxy excess relative to their numbers in bright galaxy samples. Cowie et al. (1996) confirmed that there is little change in the K-band luminosity function out to redshifts $z \approx 1$, indicating that most of their old stellar populations were already in place by a redshift of 1.

Many of the faint blue objects appearing in the Hubble Ultra Deep Field bear little resemblance to the classical forms of galaxy. This is particularly true if attention is restricted to the sample of galaxies at large redshifts $z \geq 3.5$ through the observation of 'drop-outs' through the various filter bands. These are interpreted as star-forming galaxies in which the Lyman cut-off is redshifted into the optical waveband. These galaxies are much smaller and much more irregular than those observed in the nearby Universe, in general agreement with the expectations of hierarchical clustering models of galaxy formation.

10.16.2   Mid- and far-infrared Universe

The IRAS satellite carried out the first essentially complete sky survey of the infrared sky in those wavebands between 12.5 and 100 μm which are inaccessible from the ground. The differential number counts showed that there were more faint IRAS galaxies than expected, in the same sense as the counts of radio sources, X-ray sources, and quasars, although the range of redshifts sampled was very much smaller than those of the latter objects (Oliver et al., 1992). More recently, number counts in these wavebands have been made by the NASA Spitzer Space Telescope, an 85-cm cryogenic telescope with three cooled scientific instruments for imaging and spectroscopy in the 3.6–160 μm waveband. These instruments resulted in number counts at 24, 70, and 160 μm and all show a clear excess of faint sources, as well as a rather dramatic cut-off at low flux densities.

The mid- and far-infrared counts are dominated by starburst activity, far outnumbering those in which the heating of the dust grains is due to the ultraviolet emission of embedded active galactic nuclei (Babbedge et al., 2006). The phenomenological model of Lagache et al. (2003, 2004) provided a simple picture for the evolutionary behaviour of the starburst population in the mid-infrared to submillimetre wavebands. The population consists of two components, a population of normal galaxies which evolves passively with cosmic epoch and a strongly evolving starburst component which consists of all the highest luminosity far-infrared sources, $L \geq 3 \times 10^{11} M_\odot$. The evolution of the luminous star-forming galaxies can be described by luminosity evolution of the form $L \propto (1+z)^3$ from $z = 0$ to $z = 1.5$, beyond which the luminosity function remains constant out to redshift $z = 5$.

10.16.3   Submillimetre number counts

The predicted number counts of galaxies which are intense dust emitters in the submillimetre waveband, $0.1 \leq \lambda \leq 1$ mm are quite different from those at shorter wavelengths. The maximum intensity at 50–100 μm is associated with heated dust grains with temperatures of about 30–60 K. To the long-wavelength side of this maximum, the dust clouds are optically thin and the spectra of their thermal emission is typically $S_\nu \propto \nu^{3-4}$. The luminosities of these far-infrared sources can exceed those in the optical waveband.

Because of the very large negative K-corrections in these wavebands, once galaxies of a given luminosity are observed at $z \sim 1$, the whole of the redshift range out to $z \sim 5-10$ becomes observable at roughly the same flux density (Blain and Longair, 1993, 1996)—there is a corresponding dramatic increase in the number counts of sources at this flux density. Counts in the submillimetre waveband made with the SCUBA array detector on the James Clerk Maxwell Telescope, resulted in the discovery of a large population of submillimetre sources (Smail et al., 1997). Over the succeeding years a major effort was made to determine their number counts (Fig. 10.19) (Cowie et al., 2002). These number counts greatly exceed the numbers of sources expected on the basis of the local 60 μm luminosity function of luminous far-infrared galaxies determined by the IRAS survey.

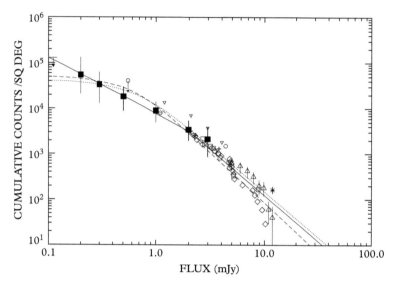

Fig. 10.19 *The counts of submillimetre sources at 850μm (Cowie et al., 2002). The filled squares with error bars have been derived from an analysis of submillimetre sources observed in the vicinity of clusters of galaxies which enhances their flux densities because of gravitational lensing. The points with flux densities greater than about 2 Jy are derived from blank field surveys. The dotted and dashed lines show parametric fits to the counts which converge to the total extragalactic background emission observed in the submillimetre waveband.*

10.16.4   The global star formation rate

The observations discussed in Sections 10.16.1–10.16.3 are related to the global rate of formation of stars as a function of cosmic epoch. Lilly and Cowie (1987) and Cowie et al. (1988) first showed how the rate of formation of stars and heavy elements could be inferred from the flat blue continuum spectra of star-forming galaxies and that these estimates are remarkably independent of the choice of cosmological model. A prolonged burst of star formation has a flat intensity spectrum at wavelengths longer than the Lyman limit at 91.2 nm. This distinctive form of spectrum has been used in different ways to find star forming galaxies and so estimate the rate of star formation out to large redshifts. It was shown that there was an order of magnitude decrease in the ultraviolet ionizing radiation between redshifts $z \sim 1$ and the present epoch (Madau et al., 1996).

Steidel and his colleagues extended the multicolour technique for finding star-forming galaxies to redshifts $z > 3$ (Steidel, 1998; Steidel et al., 1999) by searching for star-forming galaxies in which the Lyman-limit is redshifted into the optical waveband, the objects referred to as *Lyman drop-out galaxies*. Observations of the Hubble Deep and Ultra Deep Fields provided very precise photometry in four wavebands spanning the wavelength range $300 < \lambda < 900$ nm as well as high resolution optical images which enabled the morphologies of these galaxies to be studied (Beckwith et al., 2006; Bouwens et al., 2006). Of the 10,040 objects appearing in the catalogue in the i_{775} waveband, there were 504 B_{435}-dropouts, 204 V_{204} dropouts, and 54 i_{775}-dropouts, corresponding to star-forming galaxies with redshifts in the ranges $3.5 \leq z \leq 4.7$, $4.6 \leq z \leq 5.7$, and $5.7 \leq z \leq 7.4$, respectively. The Subaru Deep Surveys made use of the wide-field capability of the Japanese 8-m Subaru telescope to show that, if attention is restricted to luminous far ultraviolet galaxies, there is a factor of ten decrease in the far ultraviolet luminosity density between redshifts $z \sim 3$ and $z \sim 6$ (Shimasaku et al., 2005).

There are other routes to determining the star formation history of the Universe using the reprocessed radiation from the star formation activity. These include the bright hydrogen and oxygen emission lines from the surrounding gas, enhanced emission at mid-, far-infrared and submillimetre wavelengths from heated dust and radio emission from relativistic electrons accelerated by the shock waves produced by the supernovae explosions of short-lived massive stars.

A problem with optical determinations of the cosmic star formation rate is that the statistics are strongly influenced by the effects of dust extinction. The solution is to make observations in the far-infrared and submillimetre wavebands in which star-forming regions are intense dust emitters, the interstellar extinction is very small and the total millimetre/submillimetre luminosity of a galaxy provides a measure of the star formation rate. Dunlop surveyed the many biases and corrections which need to be applied to these and found a consistent solution which can accommodate all the observations relevant to the determination of the global rates of star formation (Dunlop, 2011, 2013). Figure 10.20(a) shows the rapid decrease in the global star formation rate from redshift $z = 1$ to the present epoch, as well as its build-up from large redshifts to $z \sim 1$. This can also be converted into a diagram showing the global build-up of the stellar content of galaxies (Fig. 10.20(b)). Much of the conversion of the diffuse gas into stars took place a relatively late epochs.

Despite the evidence of the global star formation rate, the most massive galaxies were already 'old' at redshifts of about 2 and so must have formed the bulk of their stellar

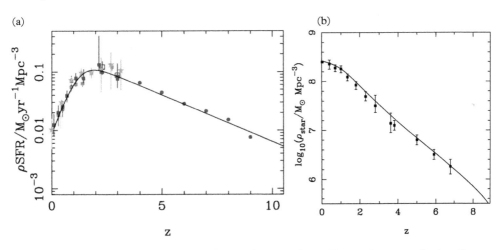

Fig. 10.20 *(a) The star formation rate density as a function of redshift, once the many selection effects are taken into account. (b) The build-up of stellar populations of galaxies as a function of redshift (Dunlop, 2011, 2013).*

populations at larger redshifts. These galaxies have more or less the same maximum mass over the redshift interval $0 < z < 2$, contrary to what might have been expected if they had increased in mass and luminosity through the favoured process of hierarchical clustering.

This is illustrated by the plot of stellar mass against redshift for the Gemini Deep Deep and K20 surveys (McCarthy, 2006). There was little evidence for any increase in the baryonic masses of the galaxies with deceasing redshift. Similar results were reported by Cimatti et al. (2004) from their analyses of the properties of four very red galaxies with redshifts in the range $1.6 \leq z \leq 1.9$ from the K20 sample. The spectroscopic and morphological properties of these galaxies indicated that they were 'old, fully assembled, spheroidal galaxies' with masses $M > 10^{11} M_\odot$. Most of the active star formation at $z \sim 1-2$ must have taken place in less massive galaxies. This phenomenon, that the massive galaxies formed their stellar populations at early epochs and star formation continues for lower mass galaxies at much later epochs, is referred to as 'down-sizing'.

10.17   Putting it all together

As discussed by Bruce Partridge in Chapter 8, the concordance ΛCDM model has had remarkable success in accounting for the observed features of the Universe on large scales. It is assumed that the creation of a fully-fledged theory of the formation of galaxies as we know them is associated with astrophysical processes in which baryonic matter evolved within dark matter haloes. This necessarily involves processes such as the dependence of the star formation rate upon the properties of the interstellar medium, the role of supernova explosions in sweeping matter away from star-forming regions and out of galaxies, the role of collisions between galaxies, the formation and growth of black holes in the nuclei of galaxies and so on. These processes provide *feedback mechanisms* which can promote or suppress the formation of new stars and structures—galaxy formation is necessarily highly non-linear.

Computational astrophysics has made enormous contributions to understanding the origin and evolution of cosmic structures. As an example of what is now feasible, the *Millennium Simulation* for the growth of dark matter perturbations under gravity involved following the evolution of over 10^{10} particles from a redshift $z = 127$ to the present epoch within a cubical region of comoving dimension $500h^{-1}$ Mpc (Springel et al., 2005). The next step has been to build on the success of these programmes to address the astrophysical issues involved in the formation of stars and of supermassive black holes in galactic nuclei.

This is the domain of *semi-analytic models of galaxy formation* in which rules need to be set up to describe how the baryonic matter cools, forms stars and completes the full life-cycle of stellar evolution. In the words of Volker Springel and his colleagues:

> The term 'semi-analytic' conveys the notion that while in this approach the physics is parameterised in terms of simple analytic models, following the dark matter merger trees over time can only be carried out numerically. Semi-analytic models are hence best viewed as simplified simulations of the galaxy formation process.

This approach can be considered to be 'experimental computational astrophysics' in the sense that there are many possible realizations of these models and the aim is to constrain the necessary and essential physics on a more or less trial and error basis when the modellers run out of secure astrophysics. The list of necessary ingredients for the models include the following.

- *Radiative cooling and star formation.* Cold gas accumulates in the central regions of the dark matter haloes and is identified with the interstellar medium of the protogalaxy. The cool gas settles into a disc supported by rotation. Once the surface gas density exceeds the critical density found from observational studies such as those of Kennicutt (1989), star formation is assumed to take place with an efficiency of 10% on the dynamical time-scale of the disc. The parameters describing these processes are chosen to reproduce the phenomenological laws of star formation and the observed gas fractions in galaxies at small redshifts.

- The most massive stars have short lifetimes and explode as *supernovae*. Guided by observation, it is assumed that supernovae can blow gas out of star-forming galaxies and that the rate of mass ejection is proportional to the total mass of stars formed.

- The *morphologies of the galaxies* are parameterized by their bulge-to-disc ratios which are correlated with their Hubble types. Any surviving cold gas collapses to the nuclear regions of the galaxy where it gives rise to a nuclear starburst. The parameterization of the latter process is derived from systematic studies of hydrodynamical simulations of galaxy collisions (Mihos and Hernquist, 1994, 1996).

- The models should reproduce the observed spectra of galaxies at different stages of their evolution. Stellar population synthesis codes enable predictions to be made of the *spectrophotometric properties of galaxies* (Bruzual and Charlot, 2003). The modelling should include the recycling of processed stellar material through the interstellar gas, both to the hot and cold phases—the next generation of stars is formed in the cold phases. Dust extinction needs to be included and this depends upon the enrichment of the interstellar gas with heavy elements and their condensation into dust grains.

- It is assumed that the *growth of black holes* takes place as a result of galaxy mergers during which cold gas is dragged into the central regions under the influence of tidal forces. This process creates a nuclear starburst as well as providing mass which is accreted into the black hole. The parameterisation of this process is chosen so that the observed relation between the mass of the central black hole and the mass of the bulge is obtained. The accretion process results in the huge luminosities of the quasars, so long as the supply of fuel lasts. The powerful relativistic jets originating in active galactic nuclei have a profound influence upon the surrounding interstellar and intergalactic media. The effect of this heating is to inhibit the further accretion of baryonic mass onto the galaxy. This process is adopted by the modellers to account for the fact that massive galaxies do not grow indefinitely and so provides a cut-off at the high mass end of the luminosity function.

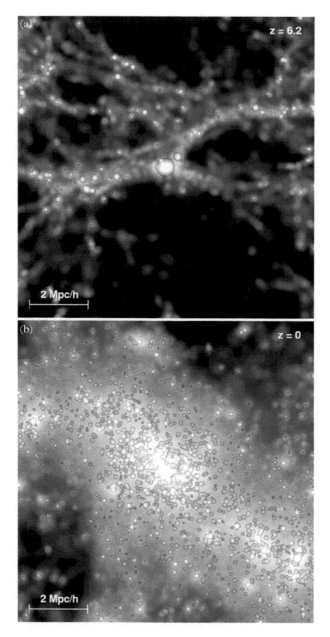

Fig. 10.21 (a) *The environment of a 'first quasar candidate' in the* Millennium Simulation *at redshift* $z = 6.2$ *and* (b) *at the present epoch in a cube of comoving dimension* $10h^{-1}$ *Mpc. The galaxies of the semi-analytic model are shown as circles overlayed on a grey-scale image of the dark matter density distribution. The volume of the sphere representing each galaxy is proportional to its stellar mass. At* $z = 6.2$, *all the galaxies are blue because of ongoing star formation, whereas many of the galaxies which have been accreted into the rich cluster at* $z = 0$ *have evolved into red galaxies (Springel et al., 2005).*

This list is indicative of the types of essential input physics which the semi-analytic modellers have to grapple with in order to account for the observed properties of galaxies. Having incorporated the baryonic physics into the *Millennium Simulations*, the two-point correlation function for galaxies provides a much better fit to the observations than that of the dark matter haloes on their own. The models can account for the two-point correlation function for galaxies of different luminosities and colours (Springel et al., 2005). The models also provide clear predictions about the evolution of the mass function of clusters of galaxies as a function of cosmic epoch.

The models can account for the formation of quasars at redshifts as large as $z = 6$. Figure 10.21(a) shows the region of a simulation containing a massive galaxy with a supermassive black hole at its centre at redshift $z = 6.2$, as well as the cellular large-scale structures about it. Although the high density peaks necessary for this to come about are rare, they must occur statistically and lead to the early formation of a galaxy massive enough to host a supermassive black hole. The same region can be tracked to the present epoch when it turns out that the region has evolved into a rich cluster of galaxies with a massive galaxy at its centre (Fig. 10.21(b)) (Springel et al., 2005).

The importance of the semi-analytic approach is that it provides a powerful new tool for testing hypotheses about the physical processes involved in galaxy and supermassive black hole formation. The simulations also provided strong motivations for future programmes of observation and theoretical analysis.

. . . .

NOTES

1. Some flavour of the points of contention among the experts in the field can be found in the proceedings of the 1996 Princeton meeting *Critical Dialogues in Cosmology*, particularly the debate between David Burstein and Dekel (Dekel et al., 1997).
2. The discussion by Kolb and Turner is recommended for the enthusiast who wishes to obtain insight into the motivation for the introduction of these particles into theories of elementary particles (Kolb and Turner, 1990). If these particles exist, they would have important astrophysical consequences which are discussed by Kolb and Turner. A gentle introduction to the physics of these possibilities is included in my book *Galaxy Formation* (Longair, 2008).
3. The analysis is most conveniently carried out in Fourier space and so the various correlation functions for galaxies have to be related to the corresponding power spectra. An introduction to how this is done is described in my book *Galaxy Formation* (Longair, 2008).
4. A good impression of the status of the subject during the 1990s is contained in *Critical Dialogues in Cosmology* (Turok, 1997).

11

Inflation, dark matter, and dark energy

Malcolm S. Longair and Chris Smeenk

11.1 Introduction

The story of how we arrived at the present picture of the structure and evolution of the Universe has concentrated largely upon observation, interpretation, and the judicious application of theory through Chapters 6–10. The developments in astrophysical and geometrical cosmology represent quite extraordinary progress in understanding the origins and evolution of our Universe and its contents. The contrast between the apparently insuperable problems of determining precise values of cosmological parameters up till the 1990s and the present era of *precision cosmology* in the first decades of the twenty-first century is startling. But these achievements also resulted in a significant change of perspective in that they involved the introduction of new aspects of physics into cosmology, largely as a result of the increased confidence in favour of the now-standard ΛCDM model. These in turn led to a better understanding of the energy budget of the Universe and a much clearer understanding of the early Universe.

We now need to pull all these strands together to address the major issues of cosmology as a science and pave the way for the considerations of Chapters 12 and 13, which review potential future directions for contemporary cosmology.[1] The steps towards the realization that dark matter and dark energy are essential components of the physical content of our universe will be briefly reviewed (Section 11.2). Then, the major physical problems which have to be addressed by observers and theorists are discussed (Section 11.3). In Section 11.4, a brief pedagogical interlude will help bring some of the issues into clearer focus. This leads to a critical discussion of the inflationary paradigm for the very early history of our Universe in Section 11.5 and subsequent sections.

Longair, M. S., Smeenk, C., 'Inflation, dark matter, and dark energy' in *The Oxford Handbook of the History of Modern Cosmology*, edited by Kragh, H. and Longair, M. S. © Oxford University Press 2019.
DOI: 10.1093/oxfordhb/9780198817666.013.11

11.2   Dark matter and dark energy

11.2.1   Dark matter

In the early days of astrophysical cosmology, there were many reasons why there should be dark matter in the Universe made out of familiar baryonic material—low mass stars, dead stars, interstellar and intergalactic gas, dust, and so on. If the matter did not radiate in the optical waveband, it was invisible. The subsequent story breaks naturally into two parts—first establishing the amount of dark matter present in the Universe and then determining whether or not it is baryonic. This endeavour was to require the full power of the information-gathering capacities of the new post-War astronomies, the advent of new technologies and associated astronomical facilities, and advances in interpretation and theory (Section 7.3). Among key astrophysical steps along the way were the following:

- Oort's pioneering determination of the mass density in the plane of the Galaxy from the velocity dispersion of stars perpendicular to the plane of the Galaxy ($0.092 \, \mathrm{M_\odot}$ $\mathrm{pc^{-3}}$) showed that gravitationally there was more mass present than the sum of the masses of all types of star in our vicinity ($0.038 \, \mathrm{M_\odot} \, \mathrm{pc^{-3}}$) (Oort, 1932).

- Zwicky's remarkable pioneering demonstration of the enormous mass-to-light ratios of clusters of galaxies, as determined by application of the virial theorem to the velocity dispersion of galaxies in the Coma cluster, brought vividly to light just how much dark matter there had to be in these systems (Zwicky, 1933, 1937).[2]

- Once powerful long-slit optical spectroscopic facilities became available, the rotation curves of galaxies could be traced well beyond their central regions and flat rotation curves were observed by Vera Rubin and her colleagues (Rubin et al., 1980) (Fig. 11.1). At the same time, studies of spiral galaxies using the 21-cm line

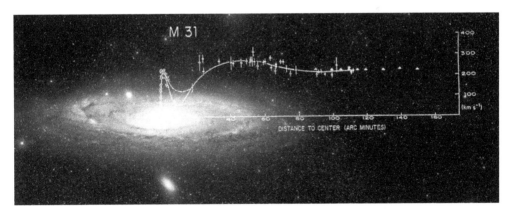

Fig. 11.1 *The rotation curve for the nearby giant spiral galaxy M31, showing the flat rotation curve extending well beyond the optical image of the galaxy thanks to observations of the velocities of interstellar neutral hydrogen by Morton Roberts and his colleagues (courtesy of the late Dr. Vera Rubin).*

of neutral hydrogen determined the rotation curves to much greater distances from their centres than optical observations and showed that flat rotation curves are the norm, rather than the exception (Bosma, 1981).[3]

- Theoretical studies of the stability of the mass distributions in disk galaxies by Miller et al. (1970) and Hohl (1971) found that these were unstable. Ostriker and Peebles (1973) showed that the presence of dark matter haloes could stabilize disc galaxies.

- X-ray imaging of clusters of galaxies enabled the total mass distribution within the cluster gravitational potential to be determined and there was found to be much more mass present than could be attributed to galaxies, based on their average mass-to-luminosity ratios (Fabricant et al., 1980; Böhringer, 1994).

- The low mass-to-luminosity ratios for the visible parts of galaxies were consistent with the low baryonic mass density inferred from primordial nucleosynthesis (Sections 6.7 and 10.7). While the parameters could be stretched to explain the dark matter in clusters by baryonic matter, it was at the verges of plausibility by the 1980s.

- Finally, the low limits to the fluctuations in the cosmic microwave background (CMB) radiation forced cosmologists to take non-baryonic dark matter really seriously in the early 1980s in order to account for the formation of structure in the Universe by the present epoch, while depressing the predicted level of fluctuations in the CMB radiation below the observational upper limits (Section 6.13).

Thus, from the early 1980s onwards, non-baryonic dark matter had to be taken seriously. From the point of view of the origin of cosmic structure, models were developed to study the astrophysical implications of different forms of dark matter candidates, for example, hot versus cold dark matter, top-down versus bottom-up approaches to structure formation and so on (Section 10.10). The observational and experimental challenges now shifted to developing more detailed models to understand the nature of the dark matter, either in terms of specific classes of astrophysical objects, or by following up clues from particle physics.

11.2.2   Constraining dark matter candidates

Baryonic dark matter

By *baryonic matter*, we mean ordinary matter composed of protons, neutrons and electrons and for convenience we include the black holes in this discussion. Certain forms of baryonic matter are very difficult to detect because they are very weak emitters of electromagnetic radiation. Important examples include stars with masses $M \lesssim 0.08M_{\odot}$, in which the central temperatures are not hot enough to burn hydrogen into helium—they are known collectively as *brown dwarfs*. They have no internal energy source and so the source of their luminosity is the thermal energy with which they were endowed at birth. There could be a small contribution from deuterium burning, but even this is not possible for stars with masses $M \leq 0.01M_{\odot}$. Brown dwarfs are normally classified as inert stars with masses in the range $0.08 \geq M \geq 0.01 M_\odot$. Below that mass, they are normally referred to as planets, $0.01 M_\odot$ corresponding to ten times the mass of Jupiter.

Until relatively recently, brown dwarfs were very difficult to detect. The situation changed dramatically with a number of technical advances in optical and infrared astronomy. The 2MASS infrared sky survey, conducted at a wavelength of 2 μm, discovered many cool brown dwarfs. The NICMOS infrared camera on the Hubble Space Telescope (HST) discovered numerous brown dwarfs in nearby star clusters. The same techniques of high precision optical spectroscopy, which has been spectacularly successful in discovering extrasolar system planets, was also used to discover a number of brown dwarfs orbiting normal stars. Although the brown dwarfs are estimated to be about twice as common as stars with masses $M \geq 0.08 M_\odot$, they contribute very little to the mass density in baryonic matter as compared with normal stars because of their low masses. The consensus of opinion is that brown dwarfs could only make a very small contribution to the dark matter problem.

Black holes are potential candidates for the dark matter. The supermassive black holes in the nuclei of galaxies have masses which are typically only about 0.1% of the mass of the bulges of their host galaxies and so they contribute negligibly to the mass density of the Universe. There might, however, be an intergalactic population of massive black holes. Limits to their number density can be set in certain mass ranges from studies of the numbers of gravitationally-lensed galaxies observed in large samples of extragalactic radio sources. In their VLA survey of a very large sample of extragalactic radio sources, designed specifically to search for gravitationally lensed structures, Hewitt and her colleagues set limits to the number density of massive black holes with masses in the range $10^{10} \leq M \leq 10^{12} M_\odot$. They found that the numbers corresponded to $\Omega_{BH} \ll 1$ (Hewitt et al., 1987). The same technique using very long baseline interferometry (VLBI) can be used to study the mass density of lower mass black holes by searching for the gravitationally lensed images on an angular scale of a milliarcsecond, corresponding to masses in the range $10^6 \leq M \leq 10^8 M_\odot$ (Kassiola et al., 1991). Wilkinson et al. (2001) searched a sample of 300 compact radio sources for examples of multiple gravitationally lensed images but none were found. The upper limit to the cosmological mass density of intergalactic supermassive compact objects in the mass range $10^6 \leq M \leq 10^8 M_\odot$ corresponded to less than 1% of the critical cosmological density.

Another possibility raised by Mészáros (1975) was that the dark matter might consist of black holes of mass roughly 1 $M_\odot$. It cannot be excluded that the dark matter might consist of a very large population of very low mass black holes, but these would have to be produced by a rather special initial perturbation spectrum in the very early Universe before the epoch of nucleosynthesis. The fact that black holes of mass less than about 10^{12} kg evaporate by Hawking radiation on a cosmological timescale sets a firm lower limit to the possible masses of mini-black holes which could contribute to the dark matter at the present epoch (Hawking, 1975).

An impressive approach to setting limits to the contribution which discrete low mass objects, collectively known as MAssive Compact Halo Objects, or MACHOs, could make to the dark matter in the halo of our own Galaxy, has been the search for gravitational microlensing signatures of such objects as they pass in front of background stars. The

MACHOs include low mass stars, white dwarfs, brown dwarfs, planets, and black holes. These events are very rare and so very large numbers of background stars have to be monitored. The beauty of this technique is that it is sensitive to MACHOs with a very wide range of masses, from 10^{-7} to 100 $M_\odot$, and so the contributions of a very wide range of candidates for the dark matter can be constrained. In addition, the expected light curve of such gravitational lensing events has a characteristic form which is independent of wavelength. The timescale for the brightening is roughly the time it takes the MACHO to cross the Einstein radius of the dark deflector. Two large projects, the MACHO and the EROS projects, have made systematic surveys over a number of years to search for these events. The MACHO project, which ran from 1992 to 1999 used stars in the Magellanic Clouds and in the galactic bulge as background stars and millions stars were monitored regularly (Alcock et al., 1993b). The first example of a microlensing event was discovered in October 1993 (Fig. 11.2), the mass of the invisible lensing object being estimated to lie in the range $0.03 < M < 0.5\ M_\odot$ (Alcock et al., 1993a).

By the end of the MACHO project, many lensing events had been observed, including over 100 in the direction towards the galactic bulge, about three times more than expected. In addition, 13 definite and four possible events were observed in the direction

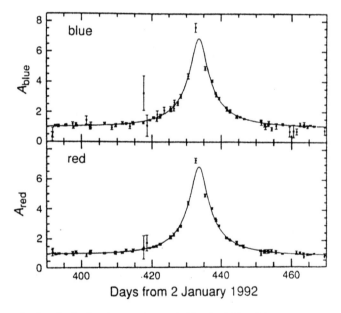

Fig. 11.2 *The gravitational microlensing event recorded by the MACHO project in February and March 1993. The horizontal axis shows the date in days measured from day zero on 2 January 1992. The vertical axis shows the amplification of the brightness of the lensed star relative to the unlensed intensity in blue and red wavebands. The solid lines show the expected variations of brightness of a lensed star with time. The same characteristic light curve is observed in both wavebands, as expected for a gravitational microlensing event (Alcock et al., 1993b).*

of the Large Magellanic Cloud (Alcock et al., 2000). The numbers are significantly greater than the 2–4 detections expected from known types of star. The technique does not provide distances and masses for individual objects, but, interpreted as a galactic halo population, the best statistical estimates suggest that the mean mass of these MACHOs is between 0.15 and $0.9\,M_\odot$. The statistics are consistent with MACHOs making up about 20% of the necessary halo mass, the 95% confidence limits being 8–50%. Somewhat fewer microlensing events were detected in the EROS project which found that less than 25% of the mass of the standard dark matter halo could consist of dark objects with masses in the range 2×10^{-7} to $1\ M_\odot$ at the 95% confidence level (Afonso et al., 2003). The most likely candidates for the detected MACHOs would appear to be white dwarfs, which would have to be produced in large numbers in the early evolution of the Galaxy, but other more exotic possibilities cannot be excluded. The consensus view is that MACHOs alone cannot account for all the dark matter in the halo of our Galaxy and so some form of non-baryonic matter must make up the difference.

As discussed in Sections 6.7 and 10.7, a strong limit to the total amount of baryonic matter in the Universe is provided by considerations of primordial nucleosynthesis. A consequence of that success story is that the primordial abundances of the light elements, particularly of deuterium and helium-3, are sensitive tracers of the mean baryon density of the Universe. Steigman found a best estimate of the mean baryon density of the Universe of $\Omega_B h^2 = (0.0223 \pm 0.002)$ (Steigman, 2006). Adopting $h = 0.7$, the density parameter in baryonic matter is $\Omega_B = 0.0455$, compared with a mean density of matter in the Universe of $\Omega_0 \approx 0.25$ (see Section 10.7). Thus, ordinary baryonic matter is only about one tenth of the total mass density of the Universe, most of which must therefore be in some non-baryonic form.

11.2.3   Non-baryonic dark matter

The dark matter may consist of the types of particle predicted by theories of elementary particles but not yet detected experimentally. Three of the most popular possibilities are described briefly in the following paragraphs.

Axions

The smallest mass candidates are the *axions* which were invented by particle theorists in order to 'save quantum chromodynamics from strong CP violation'. If these particles exist, they would have important astrophysical consequences (Kolb and Turner, 1990). If the axions were produced in thermal equilibrium, they would have unacceptably large masses, which would result in conflict with observations of the Sun and the supernova SN1987A. Specifically, if the mass of the axion were greater than 1 eV, the rate of loss of energy by the emission of axions would exceed the rate at which energy is generated by nuclear reactions in the Sun and so its centre would need to be hotter, resulting in a shorter age than is acceptable and greater emission of high energy neutrinos. There is, however, another non-equilibrium route by which the axions could be created in the early Universe. If they exist, they must have been created when the thermal temperature of the Universe was about 10^{12} K but they were out of equilibrium and never acquired thermal velocities—they remained 'cold'. Their rest mass energies are expected to lie in the range 10^{-2}–10^{-5} eV. The role of such particles in cosmology and galaxy formation are discussed by Efstathiou (1990) and Kolb and Turner (1990).

Neutrinos with finite rest mass

A second possibility is that the three known types of neutrino have finite rest masses. Laboratory tritium β-decay experiments have provided an upper limit to the rest mass of the electron antineutrino of $m_\nu \leq 2$ eV (Weinheimer, 2001). This measurement does not exclude the possibility that the two other types of neutrino, the μ and τ neutrinos, could have greater masses. However, the discovery of neutrino oscillations has provided a measurement of the mass difference between the μ and τ neutrinos of $\Delta m_\nu^2 \sim 3 \times 10^{-3}$ (Eguchi et al., 2003; Aliu et al., 2005). Thus, although their masses are not measured directly, they probably have masses of the order of 0.1 eV. This can be compared with the typical neutrino rest mass needed to attain the critical cosmological density of about 10–20 eV.[4]

WIMPs

A third possibility is that the dark matter is in some form of *weakly interacting massive particle*, or WIMP. This might be the gravitino, the supersymmetric partner of the graviton, or the photino, the supersymmetric partner of the photon, or some form of as yet unknown massive neutrino-like particle. In particular, the dark matter might be in the form of the lightest supersymmetric particle which is expected to be stable.

There must, however, be a suppression mechanism to avoid the problem that, if the WIMPs were as common as the photons and neutrinos, the masses cannot be greater than about 30 eV. The physics of this process is described by Kolb and Turner (1990).[5] According to particle theorists, almost all theories of physics beyond the standard model involve the existence of new particles at the TeV scale because of the symmetries which have to be introduced to avoid proton decay and violations of the precision tests of the electro-weak theory. These considerations lead to the expectation of new particles at the weak energy scale.

An example of the type of experiment which could demonstrate the presence of new particles has been carried out at the LHCb and CMS experiments at CERN. The cross-section of the extremely rare decay of the B_s meson into two muons has been measured. The observed branching fraction for this process compared with the predictions of the standard model provides a means of searching for physics beyond the standard model. The measurements were statistically compatible with standard model predictions and so allow stringent constraints to be placed on theories beyond the standard model. This experiment, involving the simplest of the routes to the detection of supersymmetric particles, gave a null result but this does not rule out the importance of this type of search for supersymmetric particles since there are other ways in which they could be involved in particle decays (CMS and LHCb Collaborations et al., 2015).

11.2.4 Astrophysical and experimental limits

Useful astrophysical limits can be set to the number densities of different types of neutrino-like particles in the outer regions of giant galaxies and in clusters of galaxies. The WIMPs and massive neutrinos are collisionless fermions and therefore there are constraints on the phase space density of these particles, which translate into a lower limit to their masses (Tremaine and Gunn, 1979).

Being fermions, neutrino-like particles are subject to the Pauli exclusion principle according to which there is a maximum number of particle states in phase space for a given momentum $p_{\max}$. It is a straightforward calculation to show that the resulting lower bound to the mass of the neutrino is:

$$m_\nu \geq \frac{1.5}{(N_\nu \sigma_3 R_{\mathrm{Mpc}}^2)^{1/4}} \, \mathrm{eV}, \tag{11.1}$$

where the velocity dispersion σ_3 is measured in units of 10^3 km s^{-1} and R is measured in Mpc.

In clusters of galaxies, typical values are $\sigma = 1000$ km s^{-1} and $R = 1$ Mpc. If there is only one neutrino species, $N_\nu = 1$, we find $m_\nu \geq 1.5$ eV. If there were six neutrino species, namely, electron, muon, tau neutrinos and their antiparticles, $N_\nu = 6$ and then $m_\nu \geq 0.9$ eV. For giant galaxies, for which $\sigma = 300$ km s^{-1} and $R = 10$ kpc, $m_\nu \geq 20$ eV if $N_\nu = 1$ and $m_\nu \geq 13$ eV if $N_\nu = 6$. For small galaxies, for which $\sigma = 100$ km s^{-1} and $R = 1$ kpc, the corresponding figures are $m_\nu \geq 80$ eV and $m_\nu \geq 50$ eV, respectively. Thus, particles with rest masses $m_\nu \sim 1$ eV could bind clusters of galaxies but they could not bind the haloes of giant or small galaxies.

The search for evidence for different types of dark matter particles has developed into one of the major areas of *astroparticle physics*. An important class of experiments involves the search for weakly interacting particles with masses $m \geq 1$ GeV, which could make up the dark halo of our Galaxy. In order to form a bound dark halo about our Galaxy, the particles would have to have velocity dispersion $\langle v^2 \rangle^{1/2} \sim 230$ km s^{-1} and their total mass is known. Therefore, the number of WIMPs passing through a terrestrial laboratory each day is a straightforward calculation. When these massive particles interact with the sensitive volume of a detector, the collision results in the transfer of momentum to the nuclei of the atoms of the material of the detector and this recoil can be measured in various ways. The challenge is to detect the very small number of events expected because of the very small cross-section for the interaction of WIMPs with the nuclei of atoms. A typical estimate is that less than one WIMP per day would be detectable by 1 kilogram of detector material. There should be an annual modulation of the dark matter signal as a result of the Earth's motion through the galactic halo population of dark matter particles.

A good example of the quality of the data now available is provided by the results of the Super Cryogenic Dark Matter Search (SuperCDMS) at the Soudan Laboratory.

With an exposure of 1690 kg days, only a single candidate event was observed, consistent with the expected background in the detector. The upper limit to the spin-independent WIMP-nucleon cross-section is $(1.4 \pm 1.0) \times 10^{44}$ cm^2 at 46 GeV c^{-2}. These results are the strongest limits to date for WIMP-germanium-nucleus interactions for masses greater than 12 GeV c^{-2} (SuperCDMS Collaboration et al., 2017).

Alternatives to dark matter have been proposed, including modifying Newtonian dynamics (MOND models). But the constraints and the undoubted successes of the standard picture based on standard general relativity sets a very high bar for alternatives to dark matter and will not be discussed further here.

11.2.5   The dark energy

The compelling evidence for the finite value of the cosmological constant Λ has been reviewed in Sections 8.6.1 and 10.4–10.8. The evidence for an accelerating universe from the redshift–magnitude relation for Type 1A supernovae and, even more compellingly in our view of these authors, from the many different aspects of analysing the properties of the power spectrum of the fluctuations in the CMB radiation, is unambiguous. It is particularly impressive that, using the scalar power spectrum of the fluctuations, their polarization power spectra, and the large-scale power spectrum of the dark matter derived from the *Planck* observations themselves, the six parameter family of the best-fit model can be derived without recourse to any other observations.[6] The independence of this result from all the other estimates is striking.

But there is more to it than that. The ΛCDM model solves the problem of creating the large-scale structure of the distribution of dark matter in a simple and elegant manner without the need to patch it up essentially arbitrarily with astrophysical phenomena, which is necessary in the other viable models.[7]

If dark matter is a hard problem, the dark energy is very, very hard. The contrast between the dark matter and the dark energy is striking. The estimates of the amount of dark matter depend on Newtonian gravity in domains in which we can have a great deal of confidence that it is the appropriate limit of general relativity. The dark matter is acted upon by gravity in the usual way, whereas the dark energy term in Einstein's equations does not depend upon the mass distribution, as can be seen from the expression for the variation of the scale factor a with cosmic time.

$$\ddot{a} = -\frac{4\pi G}{3}a\left(\varrho + \frac{3p}{c^2}\right) + \tfrac{1}{3}\Lambda a. \tag{11.2}$$

The Λ term provides a uniform background against which the evolution of the contents of the Universe unfold. It only makes its presence known on the largest scales observable at the present epoch and becomes of decreasing importance at large redshifts.

There is also the issue of on which side of (11.2) the cosmological constant term should appear. The Einstein field equations are written by Matteo Realdi in his equation (3.3) above as follows:

$$G_{\mu\nu} - \tfrac{1}{2}g_{\mu\nu}G - \Lambda g_{\mu\nu} = -\kappa T_{\mu\nu}. \tag{11.3}$$

The left-hand side of this equation describes the geometry of space-time as described by $G_{\mu\nu}$ while the stress-energy tensor $T_{\mu\nu}$ appears on the right-hand side. Is the Λ-term part of the intrinsic geometry of the universe, in which case it should appear on the left-hand side, or is it a source term for the gravitational field in which case it should appear on the right-hand side of (11.3)? These are hard questions to answer observationally, but some aspects of them are feasible. For example, if the dark energy term were to change with cosmic epoch, that would imply that it is a physical field. Experiments such as the *Euclid* experiment of the European Space Agency and the *WFIRST* mission of NASA aim to tackle that very issue.

11.3   The big problems

The concordance model discussed in Sections 8.6.1 and 10.4 is undoubtedly a remarkable triumph, but, like all good theories, it raises as many problems as it solves. The picture is incomplete in the sense that, within the context of the standard Friedman world models, the initial conditions have to be put in by hand in order to create the Universe as we observe it today. How did these initial conditions arise? Let us review these basic problems.

11.3.1   The horizon problem

The horizon problem, clearly recognized by Dicke (1961) is the question 'Why is the Universe so isotropic?' At earlier cosmological epochs, the particle horizon $r \sim ct$ encompassed less and less mass and so the scale over which particles could be causally connected was smaller and smaller. We can illustrate this by working out how far light could have travelled along the last scattering surface at $z \sim 1,000$ since the Big Bang. In matter-dominated models, this distance is $r = 3ct$, corresponding to an angle $\theta_{\mathrm{H}} \approx 2°$ on the sky. Thus, regions of the sky separated by greater angular distances could not have been in causal communication. Why then is the CMB radiation so isotropic? How did causally separated regions 'know' that they had to have the same temperature to better than one part in 10^5?

11.3.2   The flatness problem

Why is the Universe geometrically flat, $\Omega_\kappa = 0$? The flatness problem was also recognized by Dicke in 1961 and reiterated by Dicke and Peebles in 1979 for standard world models with $\Omega_\Lambda = 0$ (Dicke, 1961; Dicke and Peebles, 1979). In its original version, the problem arises from the fact that, according to the standard world models, if the Universe were set up with a value of the density parameter differing even slightly from the critical value $\Omega = 1$, it would diverge very rapidly from this value at later epochs. If the Universe has density parameter Ω_0 today, at redshift z, $\Omega(z)$ would have been given by

$$\left[1 - \frac{1}{\Omega(z)} \right] = f(z) \left[1 - \frac{1}{\Omega_0} \right], \tag{11.4}$$

where $f(z) = (1 + z)^{-1}$ for the matter-dominated era and $f(z) \propto (1 + z)^{-2}$ during the radiation dominated era. Thus, since $\Omega_0 \sim 1$ at the present epoch, it must have been extremely close to the critical value in the remote past. Alternatively, if $\Omega(z)$ had departed from $\Omega(z) = 1$ at a very large redshift, Ω_0 would be very far from $\Omega_0 = 1$ today. Thus, the only 'stable' value of Ω_0 is $\Omega_0 = 1$. There is nothing in the standard world models that would lead us to prefer any particular value of Ω_0. This is sometimes referred to as a *fine-tuning problem*.[8]

When Dicke described the horizon problem, the value of the overall density parameter was poorly known, but his argument was still compelling. Now we know that the value of the spacial curvature parameter is $\Omega_\kappa = \Omega_D + \Omega_B + \Omega_\Lambda - 1 = 0.00 \pm 0.01$ (Planck Collaboration, 2016b). The Universe really is geometrically flat—there is no hiding place.

11.3.3   The baryon-asymmetry problem

The baryon-asymmetry problem arises from the fact that the photon-to-baryon ratio today is

$$\frac{N_\gamma}{N_B} = \frac{4 \times 10^7}{\Omega_B h^2} = 1.6 \times 10^9, \tag{11.5}$$

where Ω_B is the density parameter in baryons and the values of Ω_B and h have been taken from Table 10.2. If photons are neither created nor destroyed, this ratio is conserved as the Universe expands. At temperature $T \approx 10^{10}$ K, electron–positron pair production takes place from the photon field. At a correspondingly higher temperature, baryon–antibaryon pair production takes place with the result that there must have been a very small asymmetry in the baryon-antibaryon ratio in the very early Universe if we are to end up with the correct photon-to-baryon ratio at the present day. At these very early epochs, there must have been roughly $10^9 + 1$ baryons for every 10^9 antibaryons to guarantee the observed ratio at the present epoch. If the Universe had been symmetric with respect to matter and antimatter, the photon-to-baryon ratio would now be about 10^{18}, in gross contradiction with the observed value (Zeldovich, 1965). Therefore, there must be some mechanism in the early Universe which results in a slight asymmetry between matter and antimatter. Fortunately, we know that spontaneous symmetry breaking results in a slight imbalance between various classes of mesons and so there is hope that this can be explained by 'standard' particle physics, but the precise mechanism has not been identified.

11.3.4   The primordial fluctuation problem

What was the origin of the density fluctuations from which galaxies and large-scale structures formed? According to the analyses of Section 6.10.1, the amplitudes of the density perturbations when they came through the horizon had to be of finite amplitude, $\delta\varrho/\varrho \sim 10^{-4}$, on a very wide range of mass scales. Such density perturbations could not have originated as statistical fluctuations in the numbers of particles on, say, the scales of superclusters of galaxies. As discussed in the previous chapter, this problem led pioneers such as Lemaître, Tolman, and Lifshitz to conclude that galaxies could not have formed by gravitational collapse. Others, such as Zeldovich, Peebles, and their colleagues, pressed ahead and assumed that such fluctuations had their origin in the very early universe and followed up the consequences of that assumption. There must have been some physical mechanism which generated finite amplitude perturbations with power-spectrum close to $P(k) \propto k$ in the early Universe.

11.3.5   The values of the cosmological parameters

The horizon and flatness problems were recognized before compelling evidence was found for the finite value of the cosmological constant. The concordance values for the cosmological parameters create their own problems. The density parameters in the dark matter and the dark energy are of the same order of magnitude at the present epoch but the matter density evolves with redshift as $(1 + z)^3$, while the dark energy density is unchanging with cosmic epoch. Why then do we live at an epoch when they have more or less the same values?

The tortuous history of the cosmological constant was recounted in Sections 6.8.5 and 10.4. A key insight resulted from the introduction of Higgs fields into the theory of weak interactions (Higgs, 1964). The Higgs fields are *scalar* fields, which have negative pressure equations of state, $p = -\varrho c^2$.[9] The theoretical value of ϱ_Λ can be estimated from quantum field theory and is found to be $\varrho_v = 10^{95}$ kg m^{-3}, about 10^{120} times greater than the value of ϱ_Λ at the present epoch, which corresponds to $\varrho_\Lambda \approx 10^{-27}$ kg m^{-3} (Carroll et al., 1992).[10] This is usually regarded as quite a problem.

As if these problems were not serious enough, they are compounded by the fact that the nature of the dark matter and the dark energy are unknown. Thus, one of the consequences of precision cosmology is the remarkable result that we do not understand the nature of about 95% of the material which drives the large-scale dynamics of the Universe. The concordance values for the cosmological parameters listed in Section 10.6.2 really are extraordinary—many of our colleagues regard them as crazy. Rather than being causes for despair, however, these problems should be seen as the great challenges for the astrophysicists and cosmologists of the twenty-first century. It is not too far-fetched to see an analogy with Bohr's theory of the hydrogen atom, which was an uncomfortable mix of classical and primitive quantum ideas, but which was ultimately to lead to completely new and deep insights with the development of quantum mechanics (Longair, 2013).

11.3.6   The way ahead

In the standard Friedman models, the problems are solved by assuming that the Universe was endowed with appropriate initial conditions in its very early phases. To put it crudely, we get out at the end what we put in at the beginning. In a truly physical picture of our Universe, we should do better than this.

There are five possible approaches to solving these problems (Longair, 1997):

- That is just how the Universe is—the initial conditions were set up that way.
- There are only certain classes of Universe in which 'intelligent' life could have evolved. The Universe has to have the appropriate initial conditions and the fundamental constants of nature should not be too different from their measured values or else there would be no chance of life forming as we know it. This approach involves the *anthropic cosmological principle* according to which, in an extreme version, it is asserted that the Universe is as it is because we are here to observe it.
- The inflationary scenario for the early Universe. This topic is taken up in Section 11.5 and subsequent sections.
- Seek clues from particle physics and extrapolate that understanding beyond what has been confirmed by experiment to the earliest phases of the Universe.
- Something else we have not yet thought of. We can think of this in terms of what Donald Rumsfeld called the 'unknown unknowns—the ones we don't know we don't know'.[11] This would certainly involve new physical concepts.

Let us consider aspects of these approaches.

11.3.7 The limits of observation

Even the first, somewhat defeatist, approach might be the only way forward if it turned out to be just too difficult to disentangle convincingly the physics responsible for setting up the initial conditions from which our Universe evolved. In 1970, McCrea considered the fundamental limitations involved in asking questions about the very early Universe, his conclusion being that we can obtain less and less information the further back in time one asks questions about the early Universe (McCrea, 1970). A modern version of this argument would be framed in terms of the limitations imposed by the existence of a last scattering surface for electromagnetic radiation at $z \approx 1,000$ and those imposed on the accuracy of observations of the CMB radiation and the large-scale structure of the Universe because of their cosmic variances.

In the case of the CMB radiation, the observations made by the *Planck* experiment are already cosmic variance limited for multipoles $l \leq 1,500$—we will never be able to learn much more than we know already about the form of the scalar power spectrum on these scales. In these studies, the search for new physics will depend upon the discovering discrepancies between the standard concordance model and future observations. The optimists would argue that the advances will come through extending our technological capabilities so that new classes of observation become cosmic variance limited. For example, the detection of primordial gravitational waves through their polarization signature at small multipoles in the CMB radiation, the nature of dark matter particles and the nature of the vacuum energy are the cutting edge of fundamental issues for astrophysical cosmology. These approaches will be accompanied by discoveries in particle physics with the coming generations of ultra-high energy particle experiments.

It is also salutary to recall that the range of particle energies which have been explored by the most powerful particle accelerators is about 200 GeV, corresponding to a cosmological epoch of about 1 microsecond from the Big Bang. This seems very modest compared with the Planck era which occurred at $t \sim 10^{-43}$ s. Is there really no new physics to be discovered between these epochs?

It is folly to attempt to predict what will be discovered over the coming years, but we might run out of luck. How would we then be able to check that the theoretical ideas proposed to account for the properties of the very early Universe are correct? Can we do better than boot-strapped self-consistency? The great achievement of modern observational and theoretical cosmology has been that we have made enormous strides in defining a convincing framework for astrophysical cosmology through precise observation and the basic problems identified above can now be addressed as areas of genuine scientific enquiry.

11.3.8   The anthropic cosmological principle

There is certainly some truth in the fact that our ability to ask questions about the origin of the Universe says something about the sort of Universe we live in. The cosmological principle asserts that we do not live at any special location in the Universe, and yet we are certainly privileged in being able to make this statement at all. In this line of reasoning, there are only certain types of Universe in which life as we know it could have formed. For example, the stars must live long enough for there to be time for biological life to form and evolve into sentient beings. This line of reasoning is embodied in the *anthropic cosmological principle*, first expounded by Carter (1974) and dealt with *in extenso* in the books by Barrow and Tipler (1986) and Gribbin and Rees (1989). Part of the problem stems from the fact that we have only one Universe to study—we cannot go out and investigate other Universes to see if they have evolved in the same way as ours. There are a number of versions of the principle, some of them stronger than others. In extreme interpretations, it leads to statements such as the strong form of the principle enunciated by Wheeler (1977),

> Observers are necessary to bring the Universe into being.

It is a matter of taste how seriously one wishes to take this line of reasoning. To many cosmologists, it is not particularly appealing because it suggests that it will never be possible to find physical reasons for the initial conditions from which the Universe evolved, or for the values of the fundamental constants of nature. But some of these problems are really hard. Weinberg, for example, found it such a puzzle that the vacuum energy density Ω_Λ is so very much smaller than the values expected according to current theories of elementary particles, that he invoked anthropic reasoning to account for its smallness (Weinberg, 1989, 1997). Another manifestation of this type of reasoning is to invoke the range of possible initial conditions which might come out of the picture of chaotic or eternal inflation (Linde, 1983) and argue that, if there were at least 10^{120} of them, then we live in one of the few which has the right conditions for life to develop as we know it. We leave it to the reader how seriously these ideas should be taken, having first read Chapters 12 and 13. Some of us prefer to regard the anthropic cosmological principle as the very last resort if all other physical approaches fail.

11.4 A pedagogical interlude—distances and times in cosmology

First, let us summarize the various times and distances used in the study of the early universe.[12] Some of the terminology used in the subsequent discussion may seem somewhat non-intuitive and so this short pedagogical interlude is intended to help the non-expert appreciate the importance of the physics which follows.

Comoving radial distance coordinate: In order to define a self-consistent distance at a specific epoch t, we projected the proper distances along our past light cone to that reference epoch which we take to be the present epoch t_0. In terms of cosmic time and scale factor a, the comoving radial distance coordinate r is then defined to be

$$r = \int_t^{t_0} \frac{c\,dt}{a} = \int_a^1 \frac{c\,da}{a\dot{a}}. \tag{11.6}$$

Proper radial distance coordinate: The same problem arises in defining a proper distance at an earlier cosmological epoch. We *define* the proper radial distance r_{prop} to be the comoving radial distance coordinate projected back to the epoch t. From (11.6), we find

$$r_{\text{prop}} = a\int_t^{t_0} \frac{c\,dt}{a} = a\int_a^1 \frac{c\,da}{a\dot{a}}. \tag{11.7}$$

Particle horizon: The particle horizon r_{H} is defined as the maximum proper distance over which there can be causal communication at the epoch t:

$$r_{\text{H}} = a\int_0^t \frac{c\,dt}{a} = a\int_0^a \frac{c\,da}{a\dot{a}}. \tag{11.8}$$

Radius of the Hubble sphere: The Hubble radius is the proper radial distance of causal contact *at a particular epoch*. It is the distance at which the velocity in the velocity–distance relation at that epoch is equal to the speed of light. This Hubble sphere has proper radius

$$r_{\text{HS}} = \frac{c}{H(z)} = \frac{ac}{\dot{a}}. \tag{11.9}$$

This is the maximum distance over which causal astrophysical phenomena can take place at the epoch t.

Event horizon: The event horizon r_E is defined as the greatest proper radial distance an object can have if it is ever to be observable by an observer who observes the Universe at cosmic time t_1:

$$r_E = a \int_{t_1}^{t_{\max}} \frac{c\,\mathrm{d}t}{a(t)} = a \int_{a_1}^{a_{\max}} \frac{c\,\mathrm{d}a}{a\dot{a}}. \tag{11.10}$$

The presence of an event horizon reflects the space-time structure of the universe in the infinite future.

Cosmic time: Cosmic time t is defined to be time measured by a fundamental observer who reads time on a standard clock:

$$t = \int_0^t \mathrm{d}t = \int_0^a \frac{\mathrm{d}a}{\dot{a}}. \tag{11.11}$$

Conformal time: The conformal time is found by projecting time intervals along the past light cone to the present epoch, using the cosmological time dilation relation. There are similarities to the definition of comoving radial distance coordinate:

$$\mathrm{d}t_{\mathrm{conf}} = \mathrm{d}\tau = \frac{\mathrm{d}t}{a}. \tag{11.12}$$

Thus, according to the cosmological time dilation formula, the interval of conformal time is what would be measured by a fundamental observer observing distant events at the present epoch t_0. At any epoch, the conformal time has value

$$\tau = \int_0^t \frac{\mathrm{d}t}{a} = \int_0^a \frac{\mathrm{d}a}{a\dot{a}}. \tag{11.13}$$

11.4.1 The past light cone

This topic requires a little care because of the way in which the standard models are set up in order to satisfy the requirements of isotropy and homogeneity. Because of these, Hubble's linear relation $v = H_0 r$ applies at the present epoch *to recessions speeds which exceed the speed of light*. Consider the proper distance between two fundamental observers at some epoch t,

$$r_{\mathrm{prop}} = a(t)r, \tag{11.14}$$

where r is comoving radial distance. Differentiating with respect to cosmic time,

$$\frac{\mathrm{d}r_{\mathrm{prop}}}{\mathrm{d}t} = \dot{a}r + a\frac{\mathrm{d}r}{\mathrm{d}t}. \tag{11.15}$$

440 Malcolm S. Longair and Chris Smeenk

The first term on the right-hand side represents the motion of the substratum and, at the present epoch, becomes $H_0 r$. The second term on the right-hand side of (11.15) corresponds to the velocity of peculiar motions in the local rest frame at r, since it corresponds to changes of the comoving radial distance coordinate. The element of proper radial distance is $a \, dr$ and so, if we consider a light wave travelling along our past light cone towards the observer at the origin, we find

$$v_{\text{tot}} = \dot{a}r - c. \tag{11.16}$$

This key result defines the propagation of light from the source to the observer in space-time diagrams for the expanding Universe.

We can now plot the trajectories of light rays from their source to the observer at t_0. The proper distance from the observer at $r = 0$ to the past light cone r_{PLC} is

$$r_{\text{PLC}} = \int_0^t v_{\text{tot}} \, dt = \int_0^a \frac{v_{\text{tot}} \, da}{\dot{a}}. \tag{11.17}$$

Notice that, initially the light rays from distant objects are propagating away from the observer – this is because the local isotropic cosmological rest frame is moving away from the observer at $r = 0$ at a speed greater than that of light. The light waves are propagated to the observer at the present epoch through local inertial frames which expand with progressively smaller velocities until they cross the *Hubble sphere* at which the recession velocity of the local frame of reference is the speed of light. Note that r_{HS} is a proper radial distance. From this epoch onwards, propagation is towards the observer until, as $t \to t_0$, the speed of propagation towards the observer is the speed of light.

It is simplest to illustrate how the various scales change with time in specific examples of standard cosmological models. We consider first the critical world model and then our reference Λ model. It is convenient to present these space-time diagrams with time measured in units of H_0^{-1} and distance in units of c/H_0. The diagrams shown in Figs. 11.3 and 11.4 follow the attractive presentation by Davis and Lineweaver, but the time axis has been truncated at the present cosmological epoch (Davis and Lineweaver, 2004).

The critical world model $\Omega_0 = 1, \Omega_\Lambda = 0$

Two different versions of the space-time diagram for the critical world model are shown in Fig. 11.3(a) and (b). The world lines of galaxies having redshifts 0.5, 1, 2, and 3 are shown. As expected, in Fig. 11.3(a) the world lines of galaxies follow the relation $r \propto t^{2/3}$. When plotted against comoving radial distance coordinate in Figs. 11.3(b), these become vertical lines. Using the conformal time coordinate, the Hubble sphere and particle horizon, as well as the past light cone, become straight lines. There is no event horizon in this model. The initial singularity is now stretched out to become the abscissa of Fig. 11.3(b).

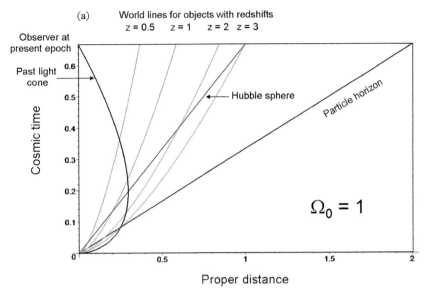

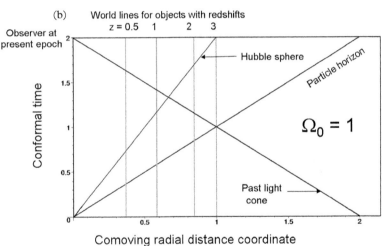

Fig. 11.3 *Space-time diagrams for the critical cosmological model, $\Omega_0 = 1, \Omega_\Lambda = 0$. The times and distances are measured in units of H_0^{-1} and c/H_0, respectively. (a) This diagram is plotted in terms of cosmic time and proper distance. (b) The same space-time diagram plotted in terms of conformal time and comoving radial distance coordinate.*

The reference world model $\Omega_0 = 0.3, \Omega_\Lambda = 0.7$

Taking $\Omega_0 = 0.3$ and $\Omega_\Lambda = 0.7$, the rate of change of the scale factor with cosmic time in units in which $c = 1$ and $H_0 = 1$ is

$$\dot{a} = \left[\frac{0.3}{a} + 0.7(a^2 - 1) \right]^{1/2}. \tag{11.18}$$

The diagrams shown in Fig. 11.4(a) and (b) have many of the same general features as Fig. 11.3(a) and (b), but there are key differences, the most significant being associated with the dominance of the dark energy term at late epochs.

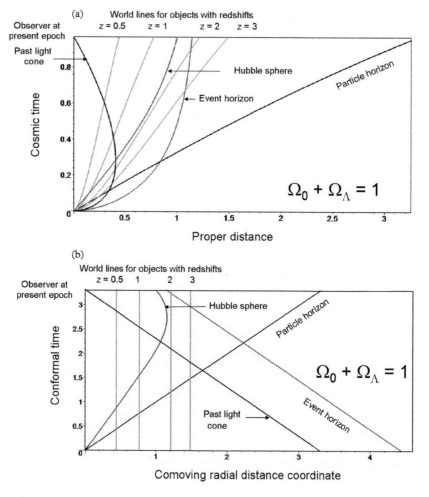

Fig. 11.4 *Space-time diagrams for the reference cosmological model, $\Omega_0 = 0.3, \Omega_\Lambda = 0.7$. The times and distances are measured in units of H_0^{-1} and c/H_0, respectively (Davis and Lineweaver, 2004).*

- Note that the cosmic timescale is stretched out relative to the critical model.
- The world lines of galaxies begin to diverge at the present epoch as the repulsive effect of the dark energy dominates over the attractive force of gravity.
- The Hubble sphere begins to converge to a proper distance of 1.12 in units of c/H_0. The reason for this is that the expansion rate becomes exponential in the future while Hubble's constant tends to a constant value of $\Omega_\Lambda^{1/2}$.
- Unlike the critical model, there is an event horizon in the reference model. The reason is that, although the geometry is flat, the exponential expansion drives galaxies beyond distances at which there could be causal communication with an observer at epoch t. It can be seen from Fig. 11.4(a) that the event horizon tends towards the same asymptotic value of 1.12 in proper distance units as the Hubble sphere. To demonstrate this, we need to evaluate the integral

$$r_{\mathrm{E}} = a \int_a^\infty \frac{\mathrm{d}a}{\left[0.3a + 0.7(a^4 - a^2)\right]^{1/2}}. \tag{11.19}$$

For large values of a, terms other than that in a^4 under the square root in the denominator can be neglected and the integral becomes $1/0.7^{1/2} = 1.12$, as found above for the Hubble sphere. In Fig. 11.4(b), the comoving distance coordinates of the Hubble sphere and the event horizon tend to zero as $t \to \infty$ because, for example, (11.8) has to be divided by a to convert it to a comoving distance and $a \to \infty$. This shrinking of the Hubble sphere is the origin of the statement that ultimately we will end up 'alone in the Universe'.

The papers by Davis and Lineweaver (2004) and Ellis and Rothman (1993) repay close study. The remarkable Appendix B of the former paper indicates how even some of the most distinguished cosmologists and astrophysicists can lead the unwary newcomer to the subject astray.

11.5   The inflationary Universe—historical background

The most important conceptual development for studies of the very early Universe can be dated to about 1980 and the proposal by Guth of the *inflationary model* for the very early Universe (Guth, 1981). Guth fully acknowledged that there had been earlier suggestions foreshadowing his proposal.[13] Zeldovich had noted in 1968 that there is a physical interpretation of the cosmological constant Λ in terms of the zero-point fluctuations in a vacuum (Zeldovich, 1968). Linde in 1974 and Bludman and Ruderman in 1977 had shown that the scalar Higgs fields of particle physics have similar properties to those which would result in a positive cosmological constant (Linde, 1974b; Bludman and Ruderman, 1977).[14] In 1975, Gurevich noted that an early initial vacuum-dominated phase would provide a 'cause of cosmological expansion' (Gurevich, 1975), this solution having later to be joined onto the standard Friedman–Lemaître solutions.

Starobinsky (1980), a member of Zeldovich's group of astrophysicists/cosmologists, found a class of cosmological solutions which indeed did just that, starting with a de Sitter phase and ultimately ending up as Freidman–Lemaître models—he noted that the exponential de Sitter expansion could lead to a solution of the singularity problem by extrapolating the de Sitter solutions back to $t \to -\infty$. He also predicted that gravitational waves would be generated during the de Sitter phase at potentially measureable levels. Commenting on this work, Zeldovich also noted that the exponential expansion would eliminate the horizon problem.

Guth realized that, if there were an early phase of exponential expansion of the Universe, this could solve the horizon problem and drive the Universe towards a flat spatial geometry, thus solving the flatness problem, simultaneously. The great merit of Guth's insights was that they made the issues of the physics of the early Universe accessible to the community of cosmologists and spurred an explosion of interest in developing genuine physical theories of the very early Universe by particle theorists.

Suppose the scale factor, a, increased exponentially with time as $a \propto e^{t/T}$. Such exponentially expanding models were found in some of the earliest solutions of the Friedman equations, in the guise of empty de Sitter models driven by what is now termed the vacuum energy density Ω_Λ (Lanczos, 1922). Consider a tiny region of the early Universe expanding under the influence of the exponential expansion. Particles within the region were initially very close together and in causal communication with each other. Before the inflationary expansion began, the region had physical scale less than the particle horizon, and so there was time for it to attain a uniform, homogeneous state. The region then expanded exponentially so that neighbouring points in the substratum were driven to such large distances that they could no longer communicate by light signals—the causally-connected regions were swept beyond their particle horizons by the inflationary expansion. At the end of the inflationary epoch, the Universe transformed into the standard radiation-dominated Universe and the inflated region continued to expand as $a \propto t^{1/2}$.

Let us demonstrate to order of magnitude how the argument runs. The timescale 10^{-34} s is taken to be the characteristic e-folding time for the exponential expansion. Over the interval from 10^{-34} s to 10^{-32} s, the radius of curvature of the Universe increased exponentially by a factor of about $e^{100} \approx 10^{43}$. The horizon scale at the beginning of this period was only $r \approx ct \approx 3 \times 10^{-26}$ m and this was inflated to a dimension of 3×10^{17} m by the end of the inflationary era. This dimension then scaled as $t^{1/2}$, as in the standard radiation-dominated Universe so that the region would have expanded to a size of $\sim 3 \times 10^{42}$ m by the present day—this dimension far exceeds the present particle horizon $r \approx cT_0$ of the Universe, which is about 10^{26} m. Thus, our present Universe would have arisen from a tiny region in the very early Universe which was much smaller than the horizon scale at that time. This guaranteed that our present Universe would be isotropic on the large scale, resolving the horizon problem. At the end of the inflationary era, there was an enormous release of energy associated with the 'latent heat' of the phase transition and this reheated the Universe to a very high temperature indeed.[15]

The exponential expansion also had the effect of straightening out the geometry of the early Universe, however complicated it may have been to begin with. Suppose the tiny region of the early Universe had some complex geometry. The radius of curvature of the geometry $R_c(t)$ scales as $R_c(t) \propto a(t)$, and so it is inflated to dimensions vastly greater than the present size of the Universe, driving the geometry of the inflated region towards flat Euclidean geometry, $\Omega_\kappa = 0$, and consequently the Universe must have $\Omega_0 + \Omega_\Lambda = 1$. It is important that these two aspects of the case for the inflationary picture can be made independently of a detailed understanding of the physics of the inflation. There is also considerable freedom about the exact time when the inflationary expansion could have occurred, provided there are sufficient e-folding times to isotropize our observable Universe and flatten its geometry.

The problem with this realization was that it predicted 'bubbles' of true vacuum embedded in the false vacuum, with the result that huge inhomogeneities were predicted. Another concern was that an excessive number of monopoles were created during the grand unified theory (GUT) phase transition. Kibble (1976) showed that, when this phase transition took place, topological defects are expected to be created, including point defects (monopoles), line defects (cosmic strings) and sheet defects (domain walls). Kibble also showed that one monopole is created for each correlation scale at that epoch. Since that scale cannot be greater than the particle horizon at the GUT phase transition, it is expected that huge numbers of monopoles are created. According to the simplest picture of the GUT phase transition, the mass density in these monopoles in the standard Big Bang picture would vastly exceed $\Omega_0 = 1$ at the present epoch (Kolb and Turner, 1990).

In Guth's original inflationary scenario, the exponential expansion was associated with the spontaneous symmetry breaking of grand unified theories of elementary particles at very high energies through a first-order phase transition, only about 10^{-34} s after the Big Bang, commonly referred to as the GUT era. The Universe was initially in a symmetric state, referred to as a false vacuum state, at a very high temperature before the inflationary phase took place. As the temperature fell, spontaneous symmetry breaking took place through the process of barrier penetration from the false vacuum state and the Universe attained a lower energy state, the true vacuum. At the end of this period of exponential expansion, the phase transition took place, releasing a huge amount of energy.

The model was revised in 1982 by Linde and by Albrecht and Steinhardt who proposed instead that, rather than through the process of barrier penetration, the transition took place through a second-order phase transition which did not result in the formation of 'bubbles' and so excessive inhomogeneities (Linde, 1982, 1983; Albrecht and Steinhardt, 1982a). This picture, often referred to as *new inflation*, also eliminated the monopole problem since the likelihood of even one being present in the observable Universe was very small.

11.6   New inflation and the Nuffield workshop

By the spring of 1982 several groups were at work fleshing out the details of the new inflationary scenario: Turner and Kolb at the University of Chicago and Fermilab, Steinhardt and Albrecht at the University of Pennsylvania, Guth at MIT, Linde and his collaborators in Moscow, Laurence Abbott at Brandeis, Hawking and others at Cambridge, and John Barrow at Sussex. With notable exceptions, such as Hawking and Barrow, nearly everyone in this research community came from a background in particle physics. They all met in Cambridge at a workshop sponsored by the Nuffield Foundation to hammer out the developing issues in the physics of the early Universe.[16]

Nearly half the lectures at the Nuffield workshop were devoted to inflation. One important focus of the conference was the calculation of density perturbations produced during an inflationary era. Steinhardt, Starobinsky, Hawking, Turner, Lukash, and Guth had all realized that this was a 'calculable problem', the answer being an estimate of the magnitude of the density perturbations, measured by the dimensionless density contrast $\Delta = \delta\rho/\rho$, produced during inflation. In this intense period of calculation and critical discussion, the particle physicists adopted grand unified theories of particle physics as the basis for their calculations, particular attention being paid to Higgs fields which had just the right equation of state to drive inflation. Preliminary calculations of this magnitude disagreed by an astounding 12 orders of magnitude: Hawking found $\Delta \approx 10^{-4}$, whereas Steinhardt and Turner (1984) initially estimated a magnitude of 10^{-16}. After three weeks of effort, the various groups working on the problem had converged on an answer.

Mukhanov and Chibisov (1981) had argued that a de Sitter phase could generate perturbations by 'stretching' the zero-point fluctuations of quantum fields to significant scales. This idea, carried out quite independently of Guth's work, would become the basis for the generation of seed perturbations in inflationary cosmology. Prior to the workshop, Hawking had circulated a preprint which argued that initial inhomogeneities in the scalar field ϕ would result in inflation beginning at slightly different times in different regions; the inhomogeneities reflect the different 'departure times' of the scalar field. Hawking's preprint claimed that this resulted in a scale-invariant spectrum of adiabatic perturbations with $\Delta \approx 10^{-4}$, exactly what was needed in accounts of structure formation.

But others did not trust Hawking's method. At the heart of the debate was the 'gauge problem', reflecting the fact that a 'perturbed space-time' cannot be uniquely decomposed into a background space-time plus perturbations. Slicing the space-time along different surfaces of constant time leads to different magnitudes for the density perturbations. The perturbations 'disappear', for example, by slicing along surfaces of constant density. In practice, almost all studies of structure formation used a particular choice of gauge, generally the synchronous gauge, but this leads to difficulties in interpreting perturbations with length scales greater than the Hubble radius. Length scales 'blow up' during inflation since they scale as $R(t) \propto e^{Ht}$, but the Hubble radius remains fixed since H is approximately constant during the slow roll phase of inflation.[17] For this reason it is especially tricky to calculate the evolution of physical perturbations using a gauge-dependent formalism.

Hawking and Guth pursued refinements of Hawking's approach during the Nuffield workshop, the centrepiece of these calculations being the 'time delay' function characterizing the start of the scalar field's slow roll down the effective potential. This 'time delay' function can be related to the two-point correlation function characterizing fluctuations in ϕ prior to inflation, and it is also related to the spectrum of density perturbations, since these are assumed to arise as a result of the differences in the time at which inflation ends (see Section 11.7.3).

Steinhardt and Turner then enlisted James Bardeen's assistance in developing a third approach; he had recently formulated a fully *gauge invariant formulation* for the study of density perturbations on all scales (Bardeen, 1980). Using Bardeen's formalism, the three aimed to give a full account of the behaviour of different modes of the field ϕ as these evolved through the inflationary phase and up to recombination. The physical origin of the spectrum was traced to the qualitative change in behaviour as perturbation modes expand past the Hubble radius: they 'freeze out' as they cross the horizon, and leave an imprint that depends on the details of the model under consideration. Despite the conflicting assumptions and other differences, the participants of the Nuffield workshop gave increasing credibility to these results because of the rough agreement between the three different approaches.

The key results were that inflation leads naturally to an almost Harrison–Zeldovich spectrum of density fluctuations and these have Gaussian phases (Bardeen et al., 1983). But reducing the magnitude of these perturbations to satisfy observational constraints required an unnatural choice of coupling constants. In particular, the self-coupling for the Higgs field apparently needed to be of the order of 10^{-8}, in contrast to the 'natural' value which would be of the order of 1.

The Higgs model was not successful but it was clear how to develop a 'newer inflation' model. Bardeen, Steinhardt, and Turner suggested that the effective potential for a scalar field in a supersymmetric theory, rather than the Higgs field of a GUT, would have the appropriate properties to drive inflation. Finding a particular particle physics candidate for the scalar field driving inflation would provide an important independent line of evidence. The Nuffield workshop marked the start of this new approach, as the focus shifted to implementing inflation successfully, rather than starting with a candidate for the field driving inflation derived from particle physics. The introduction of an 'inflaton' field, a scalar field custom-made to produce an inflationary stage, roughly a year later illustrates this methodological shift.

Following the demise of the minimal GUT models, there was an ongoing effort to implement inflation within new models provided by particle physics. Following the Nuffield workshop, inflation turned into a 'paradigm without a theory', to borrow Turner's phrase, as cosmologists developed a wide variety of models bearing a loose family resemblance. The models share the basic idea that the early universe passed through an inflationary phase, but differ on the nature of the "inflaton" field (or fields) and the form of the effective potential $V(\phi)$. Keith Olive's review of the first decade of inflation cnded by bemoaning the ongoing failure of any of these models to renew the strong connection with particle physics achieved in old and new inflation:

> A glaring problem, in my opinion, is our lack of being able to fully integrate inflation into a unification scheme or any scheme having to do with our fundamental understanding of particle physics. . . . An inflaton as an inflaton and nothing else can only be viewed as a toy, not a theory.[18]

5-dimensional assisted inflation	extended open inflation	late-time mild inflation	pre-Big-Bang inflation
anisotropic brane inflation	extended warm inflation	low-scale inflation	primary inflation
anomaly-induced inflation	extra dimensional inflation	low-scale supetrgravity inflation	primordial inflation
assisted inflation	F-term inflation	M-theory inflation	quasi-open inflation
assisted chaotic inflation	F-term hybrid inflation	mass inflation	quintessential inflation
boundary inflation	false vacuum inflation	massive chaotic inflation	R-invariant topological inflation
brane inflation	false vacuum chaotic inflation	moduli inflation	rapid asymmetric inflation
brane-assisted inflation	fast-roll inflation	multi-scalar inflation	running inflation
brane gas inflation	first order inflation	multiple inflation	scalar-tensor gravity inflation
brane-antibrane inflation	gauged inflation	multiple-field slow-roll inflation	scalar-tensor stochastic inflation
braneworld inflation	generalized inflation	multiple-stage inflation	Seiberg-Witten inflation
Brans-Dicke chaotic inflation	generalized assisted inflation	natural inflation	single-bubble open inflation
Brans-Dicke inflation	generalised slow-roll inflation	natural Chaotic inflation	spinodal inflation
bulky brane inflation	gravity driven inflation	natural double inflation	stable starobinsky-type inflation
chaotic hybrid inflation	Hagedorn inflation	natural supergravity inflation	steady-state eternal inflation
chaotic inflation	higher-curvature inflation	new inflation	steep inflation
chaotic new inflation	hybrid inflation	next-to-minimal supersymmetric	stochastic inflation
D-brane inflation	hyperextended inflation	hybrid inflation	string-forming open inflation
D-term inflation	induced gravity inflation	non-commutative inflation	successful D-term inflation
dilaton-driven inflation	induced gravity open inflation	non-slow-roll inflation	supergravity inflation
dilaton-driven brane inflation	intermediate inflation	nonminimal chaotic inflation	supernatural inflation
double inflation	inverted hybrid inflation	old inflation	superstring inflation
double D-term inflation	isocurvature inflation	open hybrid inflation	supersymmetric hybrid inflation
dual inflation	K inflation	open inflation	supersymmetric inflation
dynamical inflation	kinetic inflation	oscillating inflation	supersymmetric topological inflation
dynamical SUSY inflation	lambda inflation	polynomial chaotic inflation	supersymmetric new inflation
eternal inflation	large field inflation	polynomial hybrid inflation	synergistic warm inflation
extended inflation	late D-term inflation	power-law inflation	TeV-scale hybrid inflation

Fig. 11.5 *Paul Shellard's table showing the proliferation of inflationary models from an archive search (Shellard, 2003).*

Many different versions of the inflationary picture of the early Universe emerged, an amusing table of over 100 possibilities being presented by Shellard (2003) and shown in Fig. 11.5.

As a result, there is not a genuine physical theory of the inflationary Universe, but its basic concepts resolve some of the problems listed in Section 11.3. What it also does, and which gives it considerable appeal, is to suggest an origin for the spectrum of initial density perturbations as quantum fluctuations on the scale of the particle horizon.

11.7   The origin of the spectrum of primordial perturbations

As Liddle and Lyth (2000) have written,

> Although introduced to resolve problems associated with the initial conditions needed for the Big Bang cosmology, inflation's lasting prominence is owed to a property discovered soon after its introduction: It provides a possible explanation for the initial inhomogeneities in the Universe that are believed to have led to all the structures we see, from the earliest objects formed to the clustering of galaxies to the observed irregularities in the microwave background.

The theory also makes predictions about the spectrum of primordial gravitational waves which are accessible to experimental validation.[19] The enormous impact of particle theorists taking these cosmological problems really seriously has enlarged, yet again, the domain of astrophysical cosmology. For the 'cosmologist in the street', the theory of inflation does not make for particularly easy reading, because the reader should be comfortable with many aspects of theoretical physics which lie outside the standard tools of the observational cosmologist—ladder operators, quantum field theory, zero point fluctuations in quantum fields, all of these applied within the context of curved space-times. Developing the theory of the quantum origin of density perturbations in detail cannot be carried out with modest effort. There is no question, however, that these remarkable developments are at the cutting edge of cosmological research and have the potential to reveal new physics.

Let us list some of the clues about the formulation of a successful theory.[20]

The equation of state. We know from analyses of the physical significance of the cosmological constant Λ that exponential growth of the scale factor is found if the dark energy has a negative pressure equation of state $p = -\varrho c^2$. More generally, exponential growth of the scale factor is found provided the strong energy condition is violated, that is, if $p < -\frac{1}{3}\varrho c^2$. To be effective in the very early Universe, the mass density of the scalar field has to be vastly greater than the value of Ω_Λ we measure today.

The duration of the inflationary phase. In the example of the inflationary expansion given above, we arbitrarily assumed that 100 e-folding times would take place during the inflationary expansion. A more careful calculation shows that there must have been at least 60 e-folding times and these took place in the very early Universe, much earlier than those which have been explored experimentally by particle physics experiments. It is customary to assume that inflation began not long after the Planck era, but there is quite a bit of room for manoeuvre.

The shrinking Hubble sphere. There is a natural way of understanding how fluctuations can be generated from processes in the very early Universe. It is helpful to revisit the conformal diagrams for world models discussed in Section 11.4, in particular, Fig. 11.4(b). Recall that these diagrams are exact in the sense that the comoving radial distance coordinate and conformal time are worked out for the reference model with $\Omega_0 = 0.3$ and $\Omega_\Lambda = 0.7$. The effect of using conformal coordinates is to stretch out time in the past and shrink it into the future. Notice that, because of the use of linear scales in the ordinate, the radiation-dominated phase of the standard Big Bang is scarcely visible.

In Fig. 11.6(a), there are two additions to Fig. 11.4(b). The redshift of 1,000 is shown corresponding to the last scattering surface of the CMB radiation. The intersection with our past light cone is shown and then a past light cone from the last scattering surface to the singularity at conformal time $\tau = 0$ is shown as a shaded triangle. This is another way of demonstrating the *horizon problem*—the region of causal contact is very small compared with moving an angle of 180° over the sky which would correspond to twice the distance between the origin and the comoving radial distance coordinate at 3.09.

Let us now add the inflationary era to Fig. 11.6(a). It is useful to regard the end of the inflation era as the zero of time for the standard Big Bang and then to extend the diagram back to negative conformal times. In other words, we shift the zero of conformal time very slightly to, say, 10^{-32} s and then we can extend the light cones back through the entire inflationary era (Fig. 11.6(b)). This construction provides another way of

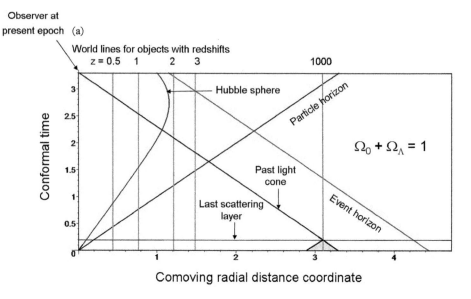

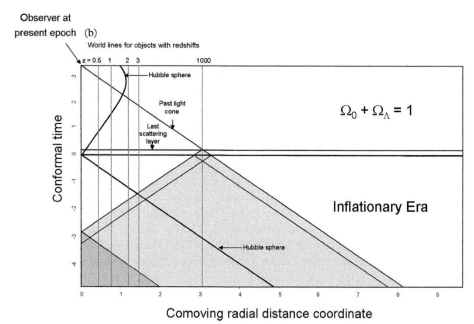

Fig. 11.6 *(a) A repeat of conformal diagram Fig. 11.4(b) in which conformal time is plotted against comoving radial distance coordinate. Now, the last scattering surface at the epoch of recombination is shown as well as the past light cone from the point at which our past light cone intersects the last scattering surface. (b) An extended conformal diagram now showing the inflationary era. The time coordinate is set to zero at the end of the inflationary era and evolution of the Hubble sphere and the past light cone at recombination extrapolated back to the inflationary era.*

understanding how the inflationary picture resolves the causality problem. The light cones have unit slope in the conformal diagram and so we draw light cones from the ends of the element of comoving radial distance at $\tau = 0$ from the last scattering surface. Projecting far enough back in time, the light cones from opposite directions on the sky overlap, represented by the dark grey shaded area in Fig. 11.6(b). This is the region of causal contact in the very early Universe.

There is, however, an even better way of understanding what is going on. We distinguished between the Hubble sphere and the particle horizon in Section 11.4—now this distinction becomes important. The particle horizon is defined as the maximum distance over which causal contact could have been made from the time of the singularity to a given epoch. In other words, it is not just what happened at a particular epoch which is important, but the history along the past light cone. Writing the exponential inflationary expansion of the scale factor as $a = a_0 \exp[H(t - t_i)]$, where a_0 is the scale factor when the inflationary expansion began at t_i, $r_{HS} = c/H$ and the comoving Hubble sphere has radius $r_{HS}(com) = c/(Ha)$. Since H is a constant throughout most of the inflationary era, it follows that the comoving Hubble sphere *decreases* as the inflationary expansion proceeds.

We now need to join this evolution of the comoving Hubble sphere onto its behaviour after the end of inflation, that is, join it onto Fig. 11.6(a). The expression for conformal time during the inflationary era is

$$\tau = \int \frac{da}{a\dot{a}},$$
(11.20)

and so, integrating and using the expression for $r_{HS}(com)$, we find

$$\tau = \text{constant} - \frac{r_{HS}(com)}{c}.$$
(11.21)

This solution for $r_{HS}(com)$ is joined on to the standard result at the end of the inflationary epoch, as illustrated in Fig. 11.6(b). The complete evolution of the Hubble sphere is indicated by the heavy line labelled 'Hubble sphere' in that diagram.

Figure 11.6(b) illustrates very beautifully how the inflationary paradigm solves the horizon problem. It will be noticed that the point at which the Hubble sphere crosses the comoving radial distance coordinate of the last scattering surface, exactly corresponds to the time when the past light cones from opposite directions on the sky touch at conformal time -3. This is not a coincidence—they are different ways of stating that opposite regions of the CMB were in causal contact at conformal time $t = -3$.

But we learn a lot more. Because any object preserves its comoving radial distance coordinate for all time, as represented by the vertical lines in Fig. 11.6, it can be seen that, in the early Universe, objects lie within the Hubble sphere, but during the inflationary expansion, they pass through it and remain outside it for the rest of the inflationary expansion. Only when the Universe transforms back into the standard Friedman model does the Hubble sphere begin to expand again and objects can then 're-enter the horizon'. Consider, for example, the region of the Universe out to redshift $z = 0.5$ which corresponds to one of the comoving coordinate lines in Fig. 11.6(b). It remained within the Hubble sphere during the inflationary era until conformal time $\tau = -0.4$ after which it was outside the horizon. It then re-entered the Hubble sphere at conformal time $\tau = 0.8$. This behaviour occurs for all scales and masses of interest in understanding the origin of structure in the present Universe.

Since causal connection is no longer possible on scales greater than the Hubble sphere, it follows that objects 'freeze out' when they pass through the Hubble sphere during the inflationary era, but they come back in again and regain causal contact when they re-cross the Hubble sphere. This is one of the key ideas behind the idea that the perturbations from which galaxies formed were created in the early Universe, froze out on crossing the Hubble sphere and then grew again on re-entering it at conformal times $\tau > 0$.

Notice that, at the present epoch, we are entering a phase of evolution of the Universe when the comoving Hubble sphere about us has begun to shrink again. This can be seen in the upper part of Fig. 11.6(b) and is entirely due to the fact that the dark energy is now dominating the expansion and its dynamics are precisely those of another exponential expansion. In fact, the Hubble sphere tends asymptotically to the line labelled 'event horizon' in Fig. 11.6(a).

11.7.1 Scalar fields

As Baumann (2007) noted, there are three equivalent conditions necessary to produce an inflationary expansion in the early Universe:

- the decreasing of the Hubble sphere during the early expansion of the Universe;
- an accelerated expansion;
- violation of the strong energy condition, meaning, $p < -\varrho c^2/3$.

How can this be achieved physically? To quote Baumann's words, written before the discovery of the Higgs boson in 2012:

> Answer: scalar field with special dynamics! Although no fundamental scalar field has yet been detected in experiments, there are fortunately plenty of such fields in theories beyond the standard model of particle physics. In fact, in string theory for example there are numerous scalar fields (moduli), but it proves very challenging to find just one with the right characteristics to serve as an inflaton candidate.

The results of calculations of the properties of the scalar field $\phi(t)$, which is assumed to be homogeneous at a given epoch, are as follows. There are a kinetic energy $\dot{\phi}^2/2$ and a potential energy, or self-interaction energy, $V(\phi)$ associated with the field. Putting these through the machinery of field theory results in expressions for the density and pressure of the scalar field:

$$\varrho_\phi = \frac{1}{2}\dot{\phi}^2 + V(\phi); \qquad p_\phi = \frac{1}{2}\dot{\phi}^2 - V(\phi). \tag{11.22}$$

Clearly the scalar field can result in a negative pressure equation of state, provided the potential energy of the field is very much greater than its kinetic energy. In the limit in which the kinetic energy is neglected, we obtain the equation of state $p = -\varrho c^2$, where the c^2, which is set equal to one by professional field theorists, has been restored.

To find the time evolution of the scalar field, we combine (11.22) with the Einstein field equations with the results:

$$H^2 = \frac{1}{3}\left(\frac{1}{2}\dot{\phi}^2 + V(\phi)\right); \qquad \ddot{\phi} + 3H\dot{\phi} + V(\phi)_{,\phi} = 0, \tag{11.23}$$

where $V(\phi)_{,\phi}$ means the derivative of $V(\phi)$ with respect to ϕ. Thus, to obtain the inflationary expansion over many e-folding times, the kinetic energy term must be very small compared with the potential energy and the potential energy term must be very slowly varying with time. This is formalized by requiring the two *slow-roll parameters* $\epsilon(\phi)$ and $\eta(\phi)$ to be very small during the inflationary expansion.[21] These parameters set constraints upon the dependence of the potential energy function upon the field ϕ and are formally written:

$$\epsilon(\phi) \equiv \frac{1}{2}\left(\frac{V_{,\phi}}{V}\right)^2; \quad \eta(\phi) \equiv \frac{V_{,\phi\phi}}{V} \quad \text{with} \quad \epsilon(\phi), |\eta(\phi)| \ll 1, \tag{11.24}$$

where $V(\phi)_{,\phi\phi}$ means the second derivative of $V(\phi)$ with respect to ϕ. Under these conditions, we obtain what we need for inflation, namely,

$$H^2 = \frac{1}{3}V(\phi) = \text{constant} \quad \text{and} \quad a(t) \propto e^{Ht}. \tag{11.25}$$

At this stage, it may appear that we have not really made much progress since we have adjusted the theory of the scalar field to produce what we know we need. The bonus comes when we consider fluctuations in the scalar field and their role in the formation of the spectrum of primordial perturbations.

11.7.2   The Quantized harmonic oscillator

The key result can be illustrated by the elementary quantum mechanics of a harmonic oscillator. The solutions of Schrödinger's equation for a harmonic potential have quantized energy levels and wave functions

$$E = \left(n + \frac{1}{2}\right)\hbar\omega; \qquad \psi_n = H_n(\xi)\exp\left(-\frac{1}{2}\xi^2\right), \tag{11.26}$$

where $H_n(\xi)$ is the Hermite polynomial of order n and $\xi = \sqrt{\beta}x$. For the simple harmonic oscillator, $\beta^2 = am/\hbar^2$, where a is the constant in the expression for the harmonic potential $V = \frac{1}{2}ax^2$ and m is the reduced mass of the oscillator.

We are interested in fluctuations about the zero-point energy, that is, the stationary state with $n = 0$. The zero-point energy and Hermite polynomial of order $n = 0$ are

$$E = \tfrac{1}{2}\hbar\omega \quad \text{and} \quad H_0(\xi) = \text{constant.} \tag{11.27}$$

The first expression is the well-known result that the oscillator has to have finite kinetic energy in the ground state. It is straightforward to work out the variance of the position coordinate x of the oscillator,[22]

$$\langle x^2 \rangle = \frac{\hbar}{2\omega m}. \tag{11.28}$$

These are the fluctuations which must necessarily accompany the zero-point energy of the vacuum fields. This elementary calculation sweeps an enormous number of technical issues under the carpet. Baumann's clear presentation of the proper calculation can be warmly recommended. It is reassuring that his final answer agrees exactly with the above results for the one-dimensional harmonic oscillator.

11.7.3   The spectrum of fluctuations in the scalar field

We need only one more equation—the expression for the evolution of the vacuum fluctuations in the inflationary expansion. The inflaton field is decomposed into a uniform homogeneous background and a perturbed component $\delta\phi$ which is the analogue of the deviation x of the zero point oscillations of the harmonic oscillator. Baumann outlines the derivation of this equation, warning of the numerous technical complexities which have to be dealt with. In Bertschinger's review of the physics of inflation, he deals with these issues and finds the following equation:

$$\delta\ddot{\phi}_k + 3\left(\frac{\dot{a}}{a}\right)\delta\dot{\phi}_k + (k_c^2 c_s^2 - 2\kappa)\delta\phi_k = 0, \tag{11.29}$$

where κ is the curvature of space at the present epoch (Bertschinger, 1996). This has a familiar form which can be understood by comparing it with (6.8) for the evolution of density perturbations in the Friedman models

$$\frac{\mathrm{d}^2\Delta}{\mathrm{d}t^2} + 2\left(\frac{\dot{a}}{a}\right)\frac{\mathrm{d}\Delta}{\mathrm{d}t} = \Delta(4\pi G\varrho_0 - k^2 c_s^2), \tag{11.30}$$

where k is the proper wavenumber and c_s is the speed of sound.[23]

Since we are interested in flat space solutions, $\kappa = 0$. Furthermore, for matter with equation of state $p = -\varrho c^2$, the speed of sound is the speed of light, which according to

Baumann's conventions is set equal to unity, and so we obtain an equation of the form (11.28). A big advantage of Baumann's proper derivation of (11.28) is that it can be applied on superhorizon scales as well as for those within the horizon, thanks to the use of Bardeen's gauge-invariant formulation of the perturbation equation.

We recognize that (11.28) is the equation of motion for a damped harmonic oscillator. If the 'damping term' $3H\delta\dot{\phi}_k$ is set equal to zero, we find harmonic oscillations, just as in the case of the Jeans' analysis of Section 11.3. On the other hand, for scales much greater than the radius of the Hubble sphere, $\Lambda \gg c/H$, an order of magnitude calculation shows that the damping term dominates and the velocity $\delta\dot{\phi}_k$ tends exponentially to zero, corresponding to the 'freezing' of the fluctuations on superhorizon scales.

Both x and $\delta\phi_k$ have zero point fluctuations in the ground state. In the case of the harmonic oscillator, we found $\langle x^2 \rangle \propto \omega^{-1}$. In exactly the same way, we expect the fluctuations in $\delta\phi_k$ to be inversely proportional to the 'angular frequency' in (11.29), that is,

$$\langle (\delta\phi_k)^2 \rangle \propto \frac{1}{k/a} \propto \lambda, \tag{11.31}$$

where λ is the proper wavelength. Integrating over wavenumber, we find the important result

$$\langle (\delta\phi)^2 \rangle \propto H^2. \tag{11.32}$$

At the end of the inflationary expansion, the scalar field is assumed to decay into the types of particles which dominate our Universe at the present epoch, releasing a vast amount of energy which reheats the contents of the Universe to a very high temperature. The final step in the calculation is to relate the fluctuations $\delta\phi$ to the density perturbations in the highly relativistic plasma in the post-inflation era. In the simplest picture, we can think of this transition as occurring abruptly between the era when $p = -\varrho c^2$ and the scale factor increases exponentially with time, as in the de Sitter metric, to that in which the standard relativistic equation of state $p = \frac{1}{3}\varrho c^2$ applies with associated variation of the inertial mass density with cosmic time $\varrho \propto H^2 \propto t^{-2}$ (see (9.7)). Guth and Pi (1982) used the time-delay formalism which enables the density perturbation to be related to the inflation parameters (see Section 11.6.1). The end results is

$$\frac{\delta\varrho}{\varrho} \propto \frac{H_*^2}{\dot{\phi}_*}. \tag{11.33}$$

where H_* and ϕ_* are their values when the proper radius of the perturbation is equal to the Hubble radius.

This order of magnitude calculation illustrates how quantum fluctuations in the scalar field ϕ can result in density fluctuations in the matter which all have more or less the same amplitude when they passed through the horizon in the very early Universe. They then remained frozen in until they re-entered the horizon very much later in the radiation-dominated era, as illustrated in Fig. 11.6(b).

This schematic calculation is only intended to illustrate why the inflationary paradigm is taken so seriously by theorists. It results remarkably naturally in the Harrison–Zeldovich spectrum for the spectrum of primordial perturbations.

In the full theory, the values of the small parameters ϵ and η defined by (11.23) cannot be neglected and they have important consequences for the spectrum of the perturbations and the existence of primordial gravitational waves. Specifically, the spectral index of the perturbations on entering the horizon is predicted to be

$$n_S - 1 = 2\eta - 6\epsilon. \tag{11.34}$$

Furthermore, tensor perturbations, corresponding to gravitational waves, are also expected to be excited during the inflationary era. Quantum fluctuations generate quadrupole perturbations and these result in a similar almost scale-invariant power spectrum of perturbations. Their spectral index is predicted to be

$$n_T - 1 = -2\epsilon, \tag{11.35}$$

where scale invariance corresponds to $n_T = 1$. The tensor-to-scalar ratio is defined as

$$r = \frac{\Delta_T^2}{\Delta_S^2} = 16\epsilon, \tag{11.36}$$

where Δ_T^2 and Δ_S^2 are the power spectra of tensor and scalar perturbations respectively.

These results illustrate why the deviations of the spectral index of the observed perturbations from the value $n_S = 1$ are so important. The fact that the best fit value $n_S = 0.961^{+0.018}_{-0.019}$ is slightly, but significantly, less than one suggests that there may well be a background of primordial gravitational waves. The detection of a background of gravitational waves is really a very great observational challenge, but they provide a remarkably direct link to processes which may have occurred during the inflationary epoch. To many cosmologists, this would be the 'smoking gun' which sets the seal on the inflationary model of the early Universe.

Whilst the above calculation is a considerable triumph for the inflationary scenario, we should remember that there is as yet no physical realization of the scalar field. Although the scale-invariant spectrum is a remarkable prediction, the amplitude of the perturbation spectrum is model dependent. There are literally hundreds of possible inflationary models depending upon the particular choice of the inflationary potential. We should also not neglect the possibility that there are other sources of perturbations which could have resulted from various types of topological defect, such as cosmic strings, domain walls, textures and so on (Shellard, 2003). Granted all these caveats, the startling success of the inflationary model in accounting for the observed spectrum of fluctuations in the CMB radiation has made it the model of choice for studies of the early Universe.

11.8 Topological defects

Throughout the 1980s and 1990s the most important alternative account of the origins of structure was based on topological defects. These ideas were first studied in the 1970s prior to the introduction of the concepts of inflation, as a general feature of spontaneous symmetry-breaking phase transitions in the early universe. Several theorists took up the challenge of understanding whether defects formed in the early universe could produce the appropriate seeds for structure formation.[24]

Starting in the early 1970s the ideas of spontaneous symmetry breaking were applied to cosmology. Extrapolating the Friedman–Lemaître models back to very early epochs, the early universe reaches arbitrarily high temperatures at early times. Kirzhnits (1972) suggested that symmetries in particle physics would be restored at sufficiently high temperatures, by analogy with symmetry restoration in condensed matter systems. Further calculations of symmetry restoration in the standard model of particle physics supported the idea that as the universe cooled it passed through a series of phase transitions that broke the symmetries between various interactions. Many symmetry breaking phase transitions in condensed matter systems lead to the formation of topological defects, such as vortices in liquid helium and so it is natural to expect that defects also arise in early universe phase transitions.

In a seminal paper, Kibble (1976) argued that topological defects would be produced as a result of the horizon structure of the early universe. Given that the correlation length of the order parameter is bounded by the horizon distance, the phase transition produces domains in which the order parameter takes on different values determined by random fluctuations. The implication is that there must be a 'defect', namely a region of space in which the fields cannot reach the vacuum state, and instead remain trapped in a state of higher energy. The nature of these regions of higher energy is fixed by the structure of the manifold. In the case of a non-simply connected vacuum manifold, the phase transition leads to two-dimensional defects called 'cosmic strings'. There are several other possibilities. A phase transition breaking a discrete symmetry leads to regions in which the order parameter takes on discrete values separated by domain walls, which are three-dimensional surfaces in space-time. If the vacuum manifold has non-contractible two-spheres rather than circles, then the phase transition produces point-like defects, such as magnetic monopoles; for non-contractible three-spheres the corresponding zero-dimensional defects are called 'textures', event-like defects that do not have a stable localized core.

Early studies showed that domain walls and some types of monopoles had disastrous consequences, conflicting with observational constraints by several orders of magnitude (see, for example, Zeldovich et al., 1975; Zeldovich and Khlopov, 1978; Guth and Tye, 1980). However, other types of defects—in particular, cosmic strings—were more plausible candidates for the seeds of structure formation. The defects are inherently stable regions of higher energy density, whose scale is set by the energy scale of the phase transition. The defects have an important impact on the dynamical evolution of the system following the phase transition, and in particular it is plausible that they provide seeds that are subsequently enhanced by gravitational instability, as described by linear perturbation theory. These theories passed an important initial test in that they lead to an approximately scale-invariant Harrison–Zeldovich spectrum of perturbations, compatible with the first generation of CMB radiation observations and the general picture of structure formation described above. However, there are important general differences between the inflationary account and that provided by topological defects, and these were clarified by a substantial research effort throughout the 1980s and 1990s.

To determine whether topological defects suffice as the primary mechanism for producing seeds for structure formation, researchers had to tackle two challenging problems. The first was to describe the phase transition itself and determine the nature of the defects produced with sufficient quantitative detail to determine the consequences for the later stages of evolution. Second, one had to describe the subsequent evolution of the network of defects left over following the phase transition over a wide range of dynamical scales. Throughout the 1980s, for example, the general picture of how strings seeded galaxy formation changed considerably in light of numerical simulations establishing details regarding the size of typical closed loops of strings and the behaviour of open strings. These two problems are exacerbated by uncertainty regarding the relevant fundamental physics. The details of the phase transitions depend on specific features of the physics—specifically concerning proposed extensions of the standard model.

Despite these difficulties, by about 1997 there was a consensus regarding the generic consequences of structure formation through defects and the contrast with the consequences of inflation. Perturbations produced in defect theories 'decohere', as first noted by Albrecht et al. (1996), in the sense that fluctuations at all wavenumbers are not in phase. This is a consequence of the non-linear evolution of the source term, which leads to mixing of perturbations across different modes. The perturbations are also non-Gaussian due to the correlations that this mixing produces between perturbations. Finally, defects generate scalar, vector, and tensor perturbations of roughly equal magnitude.

The most striking contrast with the inflationary theories is that inflation leads to phase coherence of the perturbations because the dynamics leads to synchronization of the Fourier modes with the consequent prediction of Doppler peaks. The position of the first peak also differs between the inflationary and topological defect models, with defect models generally predicting a primary peak at a larger multipole moment ($l \geq 300$) than inflation ($l \approx 200$) (Fig. 11.7). Observational results starting in the late 1990s and culminating in the WMAP and Planck results provided decisive support for inflation with respect to both of these features.

In addition to the physical contrast between the mechanisms for structure formation, there are important methodological contrasts between the two approaches. First, despite uncertainty regarding the detailed physics of the phase transitions, the account of structure formation via defects is sufficiently constrained by general theoretical principles to produce specific observational signatures. Physicists working on defects often highlighted this rigidity as a virtue of the theory, characterizing it as 'falsifiable' in a Popperian sense. Second, accounts based on topological defects do not address the problems related to initial conditions highlighted by Guth. Those who accepted Guth's approach to fine-tuning and initial conditions could still use defects, however. Inflation could still be

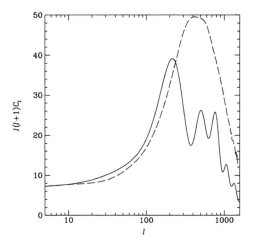

Fig. 11.7 *This figure (from Albrecht et al., 1996) shows the predicted angular power spectrum of temperature fluctuations in the cosmic background radiation for a particular model of cosmic strings (dashed line), and a generic inflationary model (solid line).*

invoked to solve the problems related to initial conditions (see, for example, Vilenkin and Shellard, 2000), as long as inflation set the stage for subsequent phase transitions that would produce appropriate topological defects.

11.9 Baryogenesis

A key contribution of particle physics to studies of the early Universe concerns the baryon-asymmetry problem, a subject referred to as *baryogenesis*. In a prescient paper of 1967, Sakharov enunciated the three conditions necessary to account for the baryon–antibaryon asymmetry of the Universe (Sakharov, 1967). *Sakharov's rules* for the creation of non-zero baryon number from an initially baryon symmetric state are:

- *Baryon number* must be violated;
- C (charge conjugation) and CP (charge conjugation combined with parity) must be violated;
- The asymmetry must be created under *non-equilibrium conditions*.

The reasons for these rules can be readily appreciated from simple arguments (Kolb and Turner, 1990). Concerning the first rule, it is evident that, if the baryon asymmetry developed from a symmetric high temperature state, baryon number must have been violated at some stage—otherwise, the baryon asymmetry would have to be built into the model from the very beginning. The second rule is necessary in order to ensure that a net baryon number is created, even in the presence of interactions which violate baryon conservation. The third rule is necessary because baryons and antibaryons have the same mass and so, thermodynamically, they would have the same abundances in thermodynamic equilibrium, despite the violation of baryon number and C and CP invariance.

There is evidence that all three rules can be satisfied in the early Universe from a combination of theoretical ideas and experimental evidence from particle physics. Thus, baryon number violation is a generic feature of GUTs which unify the strong and electroweak interactions—the same process is responsible for the predicted instability of the proton. C and CP violation have been observed in the decay of the neutral K^0 and $\bar{K}^0$ mesons. The K^0 meson should decay symmetrically into equal numbers of particles and antiparticles but, in fact, there is a slight preference for matter over antimatter, at the level of 10^{-3}, very much greater than the degree of asymmetry necessary for baryogenesis, $\sim 10^{-8}$. The need for departure from thermal equilibrium follows from the same type of reasoning which led to the primordial synthesis of the light elements. As in that case, so long as the timescales of the interactions which maintained the various constituents in thermal equilibrium were less than the expansion timescale, the number densities of particles and antiparticles of the same mass would be the same. In thermodynamic equilibrium, the number densities of different species did not depend upon the cross-sections for the interactions which maintain the equilibrium. It is only after decoupling, when non-equilibrium abundances were established, that the number densities depended upon the specific values of the cross-sections for the production of different species.

In a typical baryogenesis scenario, the asymmetry is associated with some very massive boson and its antiparticle, $X, \bar{X}$, which are involved in the unification of the strong and electroweak forces and which can decay into final states which have different baryon numbers. Kolb and Turner provided a clear description of the principles by which the observed baryon asymmetry can be generated at about the epoch of grand unification or soon afterwards, when the very massive bosons can no longer be maintained in equilibrium (Kolb and Turner, 1990). Although the principles of the calculations are well defined, the details are not understood, partly because the energies at which they are likely to be important are not attainable in laboratory experiments, and partly because predicted effects, such as the decay of the proton, have not been observed. Thus, although there is no definitive evidence that this line of reasoning is secure, well-understood physical processes of the type necessary for the creation of the baryon-antibaryon asymmetry exist. The importance of these studies goes well beyond their immediate significance for astrophysical cosmology. As Kolb and Turner remark,

> ... in the absence of direct evidence for proton decay, baryogenesis may provide the strongest, albeit indirect, evidence for some kind of unification of the quarks and the leptons.

11.10   The Planck era

Enormous progress has been made in understanding the types of physical process necessary to resolve the basic problems of cosmology, but it is not clear how independent evidence for them can be found. The methodological problem with these ideas is that they are based upon extrapolations to energies vastly exceeding those which can be tested in terrestrial laboratories. Cosmology and particle physics come together in the early Universe and they boot-strap their way to a self-consistent solution. This may be the best that we can hope for but it would be preferable to have independent constraints upon the theories.

A representation of the evolution of the Universe from the Planck era to the present day is shown in Fig. 11.8. The *Planck era* is that time in the very remote past when the energy densities were so great that a quantum theory of gravity is needed. On dimensional grounds, this era must have occurred when the Universe was only about $t_{Pl} \sim (hG/c^5)^{1/2} \sim 10^{-43}$ s old. Despite enormous efforts on the part of theorists, there is no quantum theory of gravity and so we can only speculate about the physics of these extraordinary eras.

Being drawn on a logarithmic scale, Fig. 11.8 encompasses the evolution of the whole of the Universe, from the Planck area at $t \sim 10^{-43}$ s to the present age of the Universe which is about 4×10^{17} s or 13.6×10^9 years old. Halfway up the diagram, from the time when the Universe was only about a millisecond old, to the present epoch, we can be confident that the Big Bang scenario is the most convincing framework for astrophysical cosmology.

At times earlier than about 1 millisecond, we quickly run out of known physics. This has not discouraged theorists from making bold extrapolations across the huge gap

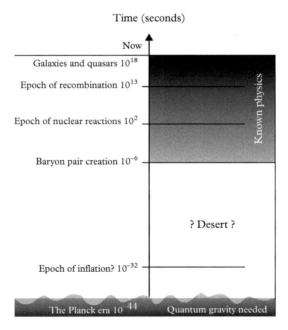

Fig. 11.8 *A schematic diagram illustrating the evolution of the Universe from the Planck era to the present time. The shaded area on the right of the diagram indicates the regions of known physics.*

from 10^{-3} s to 10^{-43} s using current understanding of particle physics and concepts from string theories. Some impression of the types of thinking involved in these studies can be found in the ideas expounded in the excellent volume *The Future of Theoretical Physics and Cosmology*, celebrating the 60th birthday of Stephen Hawking (Gibbons et al., 2003). Maybe many of these ideas will turn out to be correct, but there must be some concern that some fundamentally new physics will emerge at higher and higher energies before we reach the GUT era at $t \sim 10^{-36}$ s and the Planck era at $t \sim 10^{-43}$ s. This is why the particle physics experiments being carried out at the Large Hadron Collider at CERN are of such importance for astrophysics and cosmology. The discovery of the Higgs boson was a real triumph (Aad et al., 2012), providing essential support for our understanding of the standard model of particle physics. In addition, there is the possibility of discovering new types of particles, such as the lightest supersymmetric particle or new massive ultra-weakly interacting particles, as the accessible range of particle energies increases from about 100 GeV to 1 TeV. These experiments should provide clues to the nature of physics beyond the standard model of particle physics and will undoubtedly feed back into understanding of the physics of the early Universe.

It is certain that at some stage a quantum theory of gravity is needed which may help resolve the problems of singularities in the early Universe. The singularity theorems of Penrose and Hawking show that, according to classical theories of gravity under very general conditions, there is inevitably a physical singularity at the origin of the Big Bang, that is, as $t \to 0$, the energy density of the Universe tends to infinity. However, it is not clear that the actual Universe satisfies the various energy conditions required by the singularity theorems, particularly if the negative pressure equation of state $p = -\varrho c^2$ holds true in the very early Universe. All these considerations show that new physics is needed if we are to develop a convincing physical picture of the very early Universe.

· ·

NOTES

1. The contents of the book *Galaxy Formation* (Longair, 2008) by one of us (MSL) has been used extensively in preparing this chapter. It may be consulted for more of the technical details of the observations and the theoretical background.

2. The method Zwicky used to estimate the total mass of the cluster had been derived by Eddington in 1916 to estimate the masses of star clusters. Eddington derived the *virial theorem* which relates the total internal kinetic energy T of the stars or galaxies in a cluster to the total gravitational potential energy, $|U|$, assuming the system to be in a state of statistical equilibrium under gravity (Eddington, 1916). The total mass of the cluster can be found from the virial theorem to be $M \approx 2R_{cl}\langle v^2 \rangle / G$. Zwicky measured the velocity dispersion $\langle v^2 \rangle$ of the galaxies in the Coma cluster and found that there was much more mass in the cluster than could be attributed to the visible masses of galaxies. In solar units of $M_\odot / L_\odot$, the ratio of mass-to-optical luminosity of a galaxy such as our own is about 3, whereas for the Coma cluster

the ratio was found to be about 500. In other words, there must be about 100 times more dark, or hidden, matter as compared with visible matter in the cluster.

3. The significance of these flat rotation curves can be appreciated from the following argument. For simplicity, assume that the distribution of mass in the galaxy is spherically symmetric, so that we can write the mass within radius r as $M(\leq r)$. According to Gauss's law for gravity, we can then find the radial acceleration at radius r by placing the mass within radius r, $M(\leq r)$, at the centre of the galaxy. Then, equating the centripetal acceleration at radius r to the gravitational acceleration, we find $M(\leq r) = v_{\text{rot}}^2(r)\, r/G$. If the rotation curve of the spiral galaxy is flat, $v_{\text{rot}} = $ constant, $M(\leq r) \propto r$ and so the mass within radius r increases linearly with distance from the centre. This contrasts dramatically with the distribution of light in the discs, bulges and haloes of spiral galaxies which decrease exponentially with increasing distance from the centre.

4. See the discussion of Section 10.10.1.

5. An outline of the physics involved is given in Section 13.3 of *Galaxy Formation* (Longair, 2008).

6. See Chapter 8.

7. The reasons for this are illustrated in Section 11.4.2 of *Galaxy Formation* (Longair, 2008). See also Fig. 14.10 of that text.

8. Note that this is only one aspect of fine-tuning in order to ensure that there are observers capable of asking these questions in the Universe. See Chapter 13.

9. More exactly, the equation of state for a Higgs field takes this form if derivative terms are negligible and the effective potential is displaced from its true minima. The stress-energy tensor for the scalar field takes the form:

$$ T_{ab} = \nabla_a\phi\nabla_b\phi - \frac{1}{2}g_{ab}\left(g^{cd}\nabla_c\nabla_d\phi - V(\phi)\right). \qquad (11.37) $$

If the first two terms are negligible, we find $T_{ab} \approx \frac{1}{2}g_{ab}V(\phi)$, which is the equation of state discussed in the main text. See also Section 11.7.

10. This calculation of the theoretical value of the cosmological constant was first carried out by Wolfgang Pauli in the 1930s, but he did not take the result seriously. See, for example, Rugh and Zinkernagel (2000).

11. Donald Rumsfeld was President George W. Bush's United States Secretary of Defense and played a central role in the planning of the United States' response to the September 11 attacks, which included two wars, one in Afghanistan and one in Iraq.

12. For a more detailed discussion of these topics, see Section 12.2 of *Galaxy Formation* (Longair, 2008).

13. These earlier works, particularly the work of the Soviet theorists Sakharov, Zeldovich, Starobinsky, and their colleagues are surveyed in detail by Smeenk (2005).

14. A popular account of the history of the development of ideas about the inflation picture of the early Universe is contained in Guth's book *The Inflationary Universe: The Quest for a New Theory of Cosmic Origins* (Guth, 1997). The pedagogical

review by Lineweaver can also be recommended. He adopts a somewhat sceptical attitude to the concept of inflation and our ability to test inflationary models through confrontation with observations (Lineweaver, 2005).

15. In fact, the situation is somewhat more complex than this simple picture. Although the matter and radiation in the very early universe would have been homogenized on the small-scale, the matter and energy densities of everything other than the inflaton field are rapidly diluted during inflationary expansion—pre-existing matter and radiation are dynamically irrelevant after the first few e-folds. What explains the uniformity of different regions is instead the fact that the inflaton field decays and reheats the universe in the same fashion in different regions. Why then do the temperatures of the CMB in different patches of the sky agree? The inflaton field had a homogeneous state over some region, triggered an inflationary state that proceeded in the same way throughout this region, and then decayed into other types of matter and energy at the end of inflation. But, inflation does not guarantee that the outcome of inflation is a smooth universe by itself – it instead magnifies any non-uniformities that exist at much shorter length scales to cosmologically relevant scales (see the discussion of Section 11.7). The simple picture has to be combined with an assumption, which has generally been agreed to be quite plausible, about the small-scale non-uniformities in a pre-inflationary patch.

16. This section draws upon material contained in Smeenk (2005) and Smeenk (2018). See Vilenkin and Shellard (2000) for a masterful overview of this line of research, which includes more detailed discussion of the historical development of the field and references to original papers.

17. See Section 11.7.1.

18. Olive (1990), pp. 389.

19. There are now several recommendable books on this subject (Liddle and Lyth, 2000; Dodelson, 2003; Mukhanov, 2005).

20. The pedagogical exposition by Baumann (2007) is a very helpful guide.

21. See also note 9.

22. See Longair (2008), Section 20.5.5.

23. Equation (11.29) is often referred to as the Mukhanov–Sasaki equation for the evolution of linearized perturbations on sub- and superhorizon scales during the inflationary era.

24. This section draws upon material contained in Smeenk (2005) and Smeenk (2018).

12

Stranger things: multiverse, string cosmology, physical eschatology

Milan M. Ćirković

In each age of the world distinguished by high activity there will be found at its culmination some profound cosmological outlook, implicitly accepted, impressing its own type upon the current springs of action…In each period there is a general form of the forms of thought; and, like the air we breathe, such a form is so translucent, and so pervading, and so seemingly necessary, that only by extreme effort can we become aware of it.

Alfred North Whitehead (1933)

12.1 Introduction

This chapter will focus on the interplay between two major structural developments:

- Solidification of the inflationary cosmology, especially after the discovery of the accelerating universe dominated by dark energy in 1998.
- Diversification and increased engagement of cosmology with other fields and disciplines, be it on methodological (e.g., with computational science), phenomenological (with black hole physics, astrobiology, and so on), or conceptual (string theory or some strands of analytic metaphysics) level.

These two complementary developments offer a framework for gauging the extent of the "great transformation" of the cosmological field: from the historical *search for three numbers* (H_0, Ω_m, and Λ) to the wild variety of challenges we face once we have accepted the new standards. What local perturbations and changes led to the formation and evolution of cosmic structure as we see it? What is the origin of the initial conditions of the universe? What does its long-term future look like? In what sense could we talk about the universal cosmic evolution, encompassing not only obvious physical processes, but also chemical, planetological, and even biological evolution? These and other similar

Ćirković, M. M., 'Stranger things: multiverse, string cosmology, physical eschatology' in *The Oxford Handbook of the History of Modern Cosmology*, edited by Kragh, H. and Longair, M. S. © Oxford University Press 2019.
DOI: 10.1093/oxfordhb/9780198817666.013.12

questions, hitherto being mostly a province of metaphysics, theology, and fiction, have in this period become part and parcel of the new cosmological discourse, which put the model described by those three numbers into the background. We have seen, in Chapter 10, how deviations from the cosmological principle of large-scale homogeneity and isotropy have become the subject of new generation of theories of structure formation based on different forms of dark matter (HDM, CDM, ΛCDM, etc.). Various classical alternative cosmologies—with nuances in the meaning of "alternative"—have been analyzed in detail in Chapter 4. Here, we try to list some of the new and still-emerging topics of the period following the emergence of eternal or chaotic inflation in mid-1980s, and especially in the post-1998 period in which the dynamical dominance of dark energy has become universally (even if sometimes grudgingly) accepted.

Half-jocularly, this post-ΛCDM era has occasionally been dubbed the "post-modern cosmology," and without giving any particular credit to philosophical postmodernism,[1] there is a grain of truth in it: we have found ourselves in a new land, devoid of all the old and familiar landmarks and signposts. Not only the contents, but the very structure, form, and, epistemology of the cosmological narrative has undergone substantial changes and can only be justly described as being in the state of uncertainty and flux. Observers and critics of these trends talk about "epistemic shift," "infinite turn," or "excessive extrapolation," if not outright controversy and crisis.[2] Contemporary papers in cosmology are not only containing new research results, but often offer novel epistemological and methodological recipes, as well as considerations of problem situations which would just a few decades ago have been labelled as pure science fiction, like properties of stars in universes governed by different laws of nature, the number of microscopic degrees of freedom within *an observer*, advanced civilizations moving house to another universe via wormhole, or recipes how to make new universes in a laboratory![3] Accounting for all this "craziness" is a heavy task for both expert and a lay observer, and it is even heavier for the historian of science, especially since the ground is still shifting and the major tectonic motions are quite likely. Therefore, what follows should be taken as preliminary and incomplete more than any other chapter in this *Handbook*.

As an important indicator of the prevailing climate, new and highly respected journals have been founded, occasionally reflecting in the very title the multidisciplinary nature of the endeavor (e.g., *Journal of Cosmology and Astroparticle Physics*). Even more suggestive is the fact that prestigious journals in ostensibly unrelated fields tend to publish papers with cosmology (or specific cosmological terms) among keywords.[4] Philosophy, which has had a kind of "special relationship" with cosmology since antiquity, has dramatically increased its presence in cosmological research both in terms of relevance of its concerns, as is the number of philosophers (by primary vocation) publishing in cosmology and vice versa.[5] Also, changes introduced by the age of information have transformed cosmological practices of publication and dissemination of research results: not only has it become acceptable and uncontroversial to cite papers appearing only on the preprint servers such as https://arxiv.org/ or http://philsci-archive.pitt.edu/, but in contrast to some of the earlier epochs (as covered in detail in Chapter 4 above), internet publishing and self-publishing on online archives or even private blogs has gained in respectability. If not equal to the standard peer-reviewed publishing yet, it has at least gained a modicum of influence and importance. This process parallels another controversial feature of the first decades of the twenty-first century, namely the ascent of open-access presses and publishing houses. While this is not the place for delving upon these transformational trends, we note that cosmology as a deeply human enterprise—in Whitehead's time as well as in ours—cannot escape (and can even benefit from) cultural streams and mores of its time.

Due to all these circumstances and due to the fact that the subjects of this chapter are all rapidly changing, developing and self-transforming matters, it is quite difficult to offer anything but the fleeting, sketchy, blurred snapshot of a stormy landscape. Such a perspective differs dramatically—but necessarily—from the well-defined and established accounts of most of the previous history of physical cosmology. The jagged nature of contemporary cosmological thought thus justifies a somewhat different approach, although the one which has been tried and tested in other contexts.

Following the instructive and inspiring lead of Gerald Holton, we can recognize several important *themes* within the framework of this "postmodern" cosmological narrative.[6] Arguably the newest and the most controversial is the convergence, from several directions, onto the concept of the multiverse as the highest-level ontological structure. While having its origin in the multiple co-existent *kosmoi* of the pre-Socratic era of Greek philosophy (especially in connection to the ancient atomists, Leucippus, Democritus, Epicurus, and Lucretius, as well as somewhat controversial in the case of Milesians, in particular Anaximandros[7]), the concept first appeared in the modern scientific context in the course of the debate between Boltzmann and Zermelo in 1895, as briefly recounted in Chapter 1, Section 1.4. The modern resurgence of the concept itself (and the appropriation of the hitherto philosophical/literary term "multiverse") is in itself a signpost of the major turn. In more than one sense, to be elaborated in Sections 12.2 and 12.4 below, the multiverse represents a triumph of diversity over the monist approaches of the previous epochs of the cosmological narrative. Not only is the multiverse an embodiment of physical diversity, but its explanatory power stems directly from the epistemic diversity of approaches leading to the concept itself.

The second thematic circle, necessarily intersecting the first, is the one centered on the *observer*, and in particular the rather novel notion of the *physics of the observer*. The old-fashioned Baconian notion of an entirely implicit observer, inhabiting—at best!—some fixed, Archimedean point outside of spacetime and physical processes has already been abandoned long before the advent of "postmodern" cosmology.[8] Although both relativity theory and especially quantum mechanics as practiced by Bohr and his pupils contributed much to that abandonment, the project of reevaluation of the place of the observer within the naturalistic universe has started only recently, however, and in distinctly cosmological milieu. Questions previously belonging to purely speculative tradition of philosophy (and even there rather peripherally) have turned out to be critical and quantitative. What are the necessary physical conditions for the evolution of observers? How to compare cosmological theories implying vastly different number of observers in the universe? Does that matter as long as sets of observers are isolated at all times? What degrees of freedom (complexity, information depth, etc.) are required for a physical account of observership? Obviously, *human* observers as we know them are a topic for many sciences other than cosmology, in particular evolutionary biology and cognitive sciences, which in turn rely on ostensibly "deeper" accounts of magisteria such as molecular biology, biochemistry, astrobiology, physics of information, etc. Therefore, to address these and many other similar questions, requires strongly multi-disciplinary research activities; this is a necessity now, not a luxury as it has been hitherto regarded.

Thus, the third and final theme we shall repeatedly encounter is not only the increased multidisciplinary nature of the cosmological research, but also something which could be, for the lack of better term, called *cross-fertilization* of cosmology and a widening circle of other disciplines, from already established connection with particle physics, via computer science and astrobiology, to modern-day analytic metaphysics, historiography, theology, and even pop-culture. This thematic circle goes further and beyond the type of interdisciplinary work practiced in the aftermath of the emergence of the Big Bang paradigm in 1970s and 1980s. In some areas, it is explicitly acknowledged as such—most notably in the so-called "Big History" movement which aspires to incorporate multiple timescales, from cosmological to political and cultural, into a single historical narrative.[9] Very similar openness could be found in and around physics of information or the so-called "information paradigm" not only in physics, but in theoretical biology, astrobiology, applied mathematics, and especially modern computer science, where cosmological ideas and input are becoming welcome and explicitly so.[10] In other areas and contexts it is still to a large degree *sub rosa*, due to pressures exerted by tradition, administrative circles, funding conditions, intellectual inertia, religious and ideological influences, and other factors of interest for sociologists of science and social epistemologists. Emerging sub-disciplines, such as Bayesian cosmology, string cosmology and physical eschatology, to be briefly discussed below, are also prime items within this thematic circle, being inherently interdisciplinary and involving researchers of heterogeneous profiles and interests integrated within the "grand flow" of cosmological thought. As we discuss several such examples, justification is gathered for an important aspect to be elaborated in the concluding section of this chapter, that contemporary cosmology has acquired the key features of a true intellectual synthesis, appropriate to the third millennium.

12.2   New inflationary models and the (re)emergence of multiverse

> I myself feel impelled to the fancy—without daring to call it more—that there does exist a limitless succession of Universes …
>
> Edgar Allan Poe, *Eureka: A Prose Poem* (1848)

In the late 1970s, the concept of cosmological inflation was "in the air." If ever we could argue for *Zeitgeist* to prop ideas whose "time has come," the confluence of ideas leading to the inflationary paradigm could serve as a case study. As discussed in Chapter 11, the solution for problems arising from joining classical cosmology based on general relativity with quantum field theories of particle physics, in particular the so-called grand unified theories (GUTs), had been found by several researchers almost simultaneously at the very beginning of 1980s.[11] Any form of inflation, it turned out quickly, makes the universe much larger than the observable part. Sir Martin Rees put it succinctly:[12]

> An astonishing concept has entered the mainstream of cosmological thought: Physical reality could be hugely more extensive than the patch of space and time traditionally called "the universe." A further Copernican demotion may loom ahead.

However, it was not the inflationary *mechanism* which provoked—and to a large degree still provokes—most controversy among researchers and lay audiences alike. It is a particular consequence of the most successful and *generic* type of inflation which usually goes under the name of eternal of chaotic inflation.[13]

In the true "eye of the storm" was Andrei Linde, the main figure of the new inflationary/multiverse revolution. Andrei Dmitriyevich Linde (born 1948) is a Russian-American theoretical physicist, working for the last quarter of century as the Harald Trap Friis Professor of Physics at Stanford University. Linde was born in Moscow (then the Soviet Union) and received his Bachelor of Science degree from Moscow State University. In 1975, at—for Soviet academic programs and habits—an extremely young age of 27, Linde was awarded a Ph.D. from the Lebedev Physical Institute in Moscow. He worked at CERN since 1989 and moved to the USA in 1990. Since then he was awarded many of the highest prizes for his pioneering work on inflation (Dirac Medal, Gruber Prize, Kavli Prize, etc.), but as in the case of most other revolutionaries in the history of science, it is perhaps too early to gauge the true significance of his "Copernican turn."

Linde himself describes the role of inflation in cosmology in the following way:[14]

> It took some time before we got used to the idea that the large size, flatness, isotropy and uniformity of the universe should not be dismissed as trivial facts of life. Instead of that, they should be considered as experimental data requiring an explanation, which was provided with the invention of inflation.

While still a graduate student at the Lebedev Institute, Linde published his first results on how the energy density of a hypothetical cosmological scalar field could play the role of Einstein's classical cosmological constant Λ, as well as how vacuum phase transitions could provide energy for (re)heating up the universe after the inflationary phase.[15] Therefore, he was well-positioned to show, soon after Alan Guth's ground-breaking paper on "old" inflation that the reheating via bubble coalescence suggested by Guth does not work. The outcome was "new inflation," formulated by Albrecht and Steinhardt on the basis of Linde's work in 1982–83. New inflation resolved the reheating problem, but even more importantly from the historical point of view, opened up completely new vistas for cosmological work by incorporating quantum fluctuations of the inflaton field which drives the exponential expansion.[16] They play two crucial roles—one by creating the "seed" for subsequent structure formation, which has been an unexpected boon brought by inflation (especially since it predicts a specific and observationally testable power spectrum of the initial perturbations).

The other crucial role was even more unexpected. Quantum fluctuations in the inflaton field(s) produce changes in the rate of expansion. Some regions expand slower and others expand faster. Those regions with a higher rate of inflation expand faster and dominate the entire manifold, despite the natural tendency of inflation to end in other regions. This allows inflation to continue forever *somewhere*, and there is no necessity to fine-tune the potential of the scalar field in order to enable well-timed end of inflation ("graceful exit"). Linde's 1983 paper has thus been a watershed: "This suggests that inflation is not a peculiar phenomenon which is desirable for a number of well-known reasons... but that it is a natural and may be even inevitable consequence of the chaotic initial conditions in the very early universe."[17] Three years later, the overall picture has become even clearer and more dramatic:[18]

> As a result, our universe at present should contain an exponentially large number of mini-universes with all possible types of compactification and in all possible (metastable) vacuum states consistent with the existence of the earlier stage of inflation. If our universe would consist of one domain only (as it was believed several years ago), it would be necessary to understand why Nature has chosen just this one type of compactification, just this type of symmetry breaking, etc. At present it seems absolutely improbable that all domains contained in our exponentially large universe are of the same type. On the contrary, all types of mini-universes in which inflation is possible should be produced during the expansion of the universe, and it is unreasonable to expect that our domain is the only possible one or the best one.

The consequence was obvious: since inflation happens *somewhere* at any given time, instead of "bubble universes" all nucleating and coalescing at about the same time, we have *coexistence* of many such bubbles with the regions where inflation proceeds. The fact that inflation ended in our causally connected domain ("the universe") about 13.7 billion years ago does not tell us much about the wider whole: the ensemble of all such causally connected cosmological domains, plus the de Sitter inflationary space—or, as it came to be called, the *inflationary multiverse*.

The word "multiverse" was coined by the philosopher and psychologist William James about 1895, but with an entirely different meaning.[19] About the same time, the first scientific multiverse appeared in research publications in the polemic between Boltzmann and Zermelo, although it was not labelled as such, of course.[20] In the modern meaning of the ensemble of cosmological domains, it has been introduced and popularized by the famous British author Michael Moorcock, first in his novel *The Sundered Worlds*, originally published in 1966.[21] In a large part of Moorcock's fictional opus, his "Eternal Champion" series, the multiverse represents the central plot device, giving the framework for what fiction writers have always been doing: investigating similar plotlines in different fictional universes.[22] This notion of *variability* of individual cosmological domains ("universes") is accepted in modern physical cosmological discourse. In recent years, the multiverse has become the topic of many book-length popular expositions.[23]

One of the most unexpected—even bizarre—developments in contemporary cosmology has been a return to a form of stationary universe, or indeed *stationary multiverse*. If the eternal inflation has no beginning or end, then the multiverse as a whole is in a stationary state: the rate of "bubble universe" nucleation is constant averaged over sufficiently large comoving volume. Inflation occurring outside of the bubbles even has a de Sitter expansion scale factor, just like in the classical steady state cosmology (see Chapter 5)! Of course, each bubble universe, including the observable region, is still changing and evolving, but the multiverse as a whole is stationary. The precondition for this is, of course, that the inflation is eternal into the past as it is apparently into the future. While Linde and his collaborators continue to argue for this option, there are many opposing voices, including the one of Alan Guth.[24] The matter is still an open and active area of research. The very fact of posing and discussing this question testifies how *physics of the multiverse* has actually become an entirely legitimate topic for the cosmological discourse. Even Boltzmann's original anthropic hypothesis for the origin of thermodynamical asymmetry of our world can be remade in modern fashion using the inflationary multiverse.[25]

A different kind of multiverse has appeared as a consequence of the attempts at ultimate unification of GUTs with gravity. In the period under consideration, by far the most promising candidate is a set of ideas going under the name of *string theory*, or in the post-1995 era, *M-theory*. While some specific ideas of string cosmology will be discussed in Section 12.5, here we need to mention the most important prediction of string theory: what is variously called *string theory landscape, anthropic landscape of string theory*, or *string multiverse*.

Since string theory needs to produce the low-energy physics we are accustomed to in the limit of small energy densities, it is of paramount importance to find its vacuum solutions. In quantum field theories, the possible vacuums are labelled by the vacuum expectation values of scalar fields, as Lorentz invariance prohibits nonzero vacuum expectation values of any higher spin fields. These vacuum expectation values can take any value for which the potential function is a minimum. On the other hand, we need to label both global minima ("true vacuums") and local minima ("false vacuums"). False vacuums are metastable in principle, but in reality they could be arbitrarily long lived, for instance longer lived than the Hubble time.[26] It turns out that the number of false vacuums in string theory is huge, of the order of 10^{500}; the reason for such a hyper-astronomically large number is that the number of possible compactifications of fields in 11 dimensions is a huge combinatorial problem.[27] At a first glance, this is still not a problem for the theory, since many other accepted physical theories have multiple solutions (even if the number is not so extreme). In such situations, a researcher usually appeals to specific boundary conditions to select the "true" solution for the problem at hand; the explanation of these boundary conditions is outsourced, so to speak, to a more general theory. This would not, however, work in string theory, since it is *by definition* supposed to be the most general physical theory—where could boundary condition come from in the naturalistic view of the world?[28] *Our* vacuum certainly is one out of 10^{500} vacuums, but there are no grounds whatsoever to assign it a preferred ontological status. On the contrary, all such vacuums should be ontologically equal–each representing a different choice of the six-dimensional Calabi–Yau manifolds and generalized magnetic flux compactifications. Each would represent a different low-energy physics and a different cosmological domain (or a set of cosmological domains, see Tegmark's classification in the next section) ruled by such low-energy physics. Thus, it is our most inclusive home: the string theory multiverse.

The collection of papers edited by Bernard Carr titled *Universe or Multiverse?*, published in 2007, is a symbolic milestone, representing a codification of this new view.[29] Even skeptics toward the multiverse which are included in the volume have accepted it as a legitimate field of scientific discourse, in sharp contrast to viewing the multiverse as "metaphysical monstrosity" which predominated until a couple of decades ago. In spite of the question mark in the title, almost all contributors agreed (even if hesitatingly) that the multiverse is the only rational choice in face of our best scientific theories of the moment. Even those who are specifically skeptical toward inflation or string theory have come up with their own multiverse versions, notably Lee Smolin (see Section 12.4). The majority opinion has been encapsulated by words of Sir Martin Rees:[30]

> Some might regard other universes—regions of space and time that we cannot observe (perhaps even in principle and not just in practice)—as being in the province of metaphysics rather than physics. Science is an experimental or observational enterprise, and it is natural to be troubled by invocations of something unobservable. But I think other universes (in this sense) already lie within the proper purview of science. It is not absurd or meaningless to ask "Do unobservable universes exist?," even though no quick answer is likely to be forthcoming. The question plainly cannot be settled by direct observation, but relevant evidence can be sought, which could lead to an answer.

For whatever exact reasons, the multiverse seems here to stay and it has brought boon to the philosophy of cosmology. Among many examples we shall consider below are Tegmark's mathematical multiverse, Gerig's "demographic" multiverse, and Smolin's self-reproducing black hole multiverse.

To some observers, the multiverse represents a return to one of the old highlights of metaphysics: the principle of plenitude (or one of many other names it has been called, such as "the Lucretian argument," "Augustine's principle," logical completeness, etc.).[31] However, today's multiverse is much more tangible than those vague philosophical precursors—and even more than those well-elaborated schemes of modern analytical philosophy, like David Lewis's modal realism, which posits that all possible worlds are as real as the actual world.[32] Modern cosmological multiverse is often compared to modal realism, but the differences are significant (and acknowledged by Lewis himself); in particular, modal realism seems to imply much more universes than either cosmological inflation or string theory. There might be a relationship of modal realism with Tegmark's mathematical universe hypothesis, to be discussed in Section 12.4. Before that, we need to consider the main source of tangibility of the physical multiverse.

12.3 Observation selection effects and habitability

> If the doors of perception were cleansed, everything would appear to man as it is, infinite.
> William Blake, *The Marriage of Heaven and Hell* (ca. 1790).

In the period under consideration (roughly 1980–today), the question of the habitability of the universe for intelligent observers like us (or possible extraterrestrial observers or even future Earth-originating artificial intelligence) came to the fore of cosmological thinking. Although present since early speculations of Weyl, Eddington, Dirac, and others, the topic clearly came into focus with the controversial exposition of Brandon Carter in September 1973 at the IAU meeting in Krakow.[33] For most people and sources, Carter's study is the ultimate origin of anthropic reasoning, although it is clear that it existed since ancient Ionian philosophy and even in its physical cosmological edition it followed upon the ideas of Eddington, Dirac, and Dicke.[34]

So, historically, anthropic reasoning *has* indeed emerged from cosmological considerations; it went much farther than this circumscribed domain in the 40+ years since Carter's seminal work, however. And there is really no reason why it should not have, since (as Carter himself understood well,[35] but what subsequent critics *and* supporters often failed to understand and appreciate) the anthropic reasoning deals with *observation-selection effects*, and has no teleological meaning whatsoever. It simply applies whenever the number or other properties of observers come into play and influence particular scientific result or claim. This dysteleological nature of the anthropic reasoning has been best explicated in the seminal monograph by the Oxford philosopher and futurist Nick Bostrom.[36] In the period we consider here, more serious and formal construals of anthropic principles have emerged. In fact, the label "principle" itself has been fading for some quite time and it is more and more substituted by "anthropic reasoning."

Fine-tunings of various constants of physics and cosmological parameters of our universe have been known for quite some time.[37] They are usually expressed in the form of counterfactuals: if the parameter A was outside of a small interval around its actual value, the universe would not be habitable for intelligent observers. This is *an empirical matter*, with ample support from both observations and theoretical models of counterfactual physical systems.[38] Our universe is much more complex than most universes even with the same laws but different values of the parameters of those laws. In particular, it has a complex astrophysics, including galaxies and long-lived stars, and a complex chemistry, including carbon chemistry. These necessary conditions for life as we know it are present in our universe because of its complexity, which in turn is made possible by the special values of the parameters. For example, we observe that the cosmological density of matter at the present time ($\Omega_{m0} \approx 0.3$) and the dark energy density ($\Omega_{\Lambda 0} \approx 0.7$) are of the same order of magnitude. There is no causal explanation of that observation; it is a *prima facie* coincidence. However, a study clearly demonstrates that the distribution of ages of habitable planets is such that observers are most probably located in the cosmological epoch in which these two values are close, $\Omega_{m0} \approx \Omega_{\Lambda 0}$. In other words, the fine-tuning "coincidence" is an observational selection effect.[39]

Study and search for explanations of these and many other fine tunings constitute the bread-and-butter of both physical and philosophical research on the multiverse, for the obvious reason that under naturalism, it is very difficult to account for empirical fine-tunings without the multiverse. In one sense, it is a small part of the application of the Bayesian methods in cosmology, which has been flourishing for quite some time in other subfield, in particular in study of the large-scale structure and cosmic microwave background (CMB) anisotropies.[40] Alternatives to the multiverse explanation include the design hypothesis (which is mostly understood in the religious context of supernaturalism, although naturalistic design hypotheses are also possible, usually going under the label of "simulation hypotheses"[41]) and the rejection of the question itself (or claiming that the fine-tunings are, in philosophical parlance, *brute facts*). Since the alternatives are unsatisfactory on multiple counts, many researchers have adopted the view that the multiverse explanations at least offer some hope for a naturalistic explanatory account.

The multiverse hypothesis implies that we are living in a universe described by particular habitable combination of natural laws and physical constants; but our universe is certainly not unique in that respect. The idea of the "Archipelago of Habitability" was introduced by the Swedish-American cosmologist Max Tegmark in his intriguing 1998 paper as a set of regions in the parameter space describing those parts of the multiverse which are hospitable to life and intelligent observers of any kind.[42] Physically, the Archipelago is part of the abstract space defined by whatever physics determines the structure of the multiverse. The island is a set of parameters describing habitable universes that are close in parameter space (Fig. 12.1); whether we can specify the meaning of "closeness" beyond simple intuition depends crucially on the structure of the multiverse itself. By definition, there is at least one island in the Archipelago— our home island. If the multiverse contains arbitrarily small variations in the constants of nature or cosmological parameters or even the mathematical shape of physical laws, then it is reasonable to conclude that the Archipelago is as *dense* as the remainder of the multiverse. For instance, it is clear that the change in the coupling constants of fundamental forces of one part in 10^{20} in comparison to those actually existing in our universe (not those actually measured, since we are unable to measure them with such precision yet!), will not change anything in the habitability status of our universe. If such minuscule variations are actually realized within the multiverse (for instance, through eternal inflation), a universe otherwise described by identical laws and parameters to ours also belongs to our home island.

As mentioned above, Steven Weinberg made the prediction of the existence of a small positive cosmological constant in the anthropic manner more than a decade before its actual observational discovery.[43] Since the discovery was so unexpected in many quarters, Weinberg's success set the stage for many subsequent developments. Irrespective of whether one regards Weinberg's result as a prediction of something which has been occasionally invoked for 60 years or something truly new, this theoretical success opened the way for dramatic expansion of anthropic arguments in cosmology. In the current cottage-industry of multiverse predictions in theoretical physics and cosmology,

Fig. 12.1 *Islands in the "Archipelago of Habitability:" a parameter X can have other values than the observed X_0, but most of those regions are "below sea-level," that is, non-habitable. The reasons for the lack of habitability are just plausible examples. (Adapted from Ćirković (2012). Courtesy of Slobodan Popović Bagi and the Cambridge University Press.)*

a special role is reserved for the equation giving a probability $p(X)$ that some observer anywhere in the multiverse measures a feature X:[44]

$$p(X) = \frac{\sum_n \sigma_n(X) V_n \rho_n^{\text{obs}}}{\sum_n V_n \rho_n^{\text{obs}}}, \tag{12.1}$$

where the index n labels all vacuum states or all different universes in a differently constructed multiverse. In current versions of M-theory there is a finite number of such states, although it is huge, but in principle it could be infinite. The latter case poses some interesting problems in the theory of probability, but in general will not preclude the usage of (12.1) with appropriate weightings. V_n is the spacetime volume belonging to the universe n, ρ_n^{obs} is the density of observers in the same universe, and the indicator function is given as

$$\sigma_n = \begin{cases} 1, & \text{if universe } n \text{ has property } X \\ 0, & \text{otherwise} \end{cases}. \tag{12.2}$$

In principle, V_n is calculable from our understanding of cosmological physics, although in weird enough universes it might be impossible to calculate in practice. It is also likely to be infinite in some or most of the universes, so some appropriate weighting procedure is certainly necessary. But, of course, the biggest uncertainty comes from the quantity ρ_n^{obs}, the density of intelligent observers. It is usually assumed to be proportional to the density of galaxies,

$$\rho_n^{\text{obs}} \propto \langle \rho_{\text{gal}} \rangle_n, \tag{12.3}$$

the latter being the main feature of the structure of our own universe. Two kinds of uncertainty accompany this assumption:

i. We assume that the density of galaxies is a well-defined function of the constants of nature (including coupling constants of all low-energy interactions) and a number of cosmological parameters, usually thought about as the initial conditions of a particular universe. However, even if we fix the low-energy physics of a universe—as we do in standard astrophysical cosmology—we still fail to unequivocally derive the density of galaxies without additional assumptions dealing with the process of structure formation. In other words, our understanding of the astrophysics of galaxy formation is still insufficient for giving anything similar to the accurate prediction of the number of galaxies per unit comoving volume.

ii. There are "internal" uncertainties as to the degree galaxies are necessary as preconditions for life. Even if we discard nit-picking about what exactly can be regarded as a galaxy (some objects, especially in the low-mass range, keep defying simple classification schemes), we are still left with controversial issues such as: Which types of galaxies are conductive to the evolution of observers? How to treat possibly extremely non-uniform temporal distribution of habitable sites in different types of galaxies? Is it physically possible—and it is certainly conceivable—that a stellar population with habitable planets arises independently of galaxies? How are the details of galactic dynamics on large scale influence formation of habitable planets in any given locale? etc. All these questions are extremely difficult to tackle at present, since we know little about the astrophysical preconditions for observership under the known laws of physics—not to mention the vast generalization which answering to these questions in the general multiverse context requires. Study of familiar astrophysical objects, like stars and planets, under different low-energy physical laws is part of this novel research program.[45]

Obviously, more sophisticated research on the conditions for habitability, that is, in astrobiology, is needed here; one promising way to go is usage of the cosmological simulations (*N*-body and hydrodynamical ones) to assess habitability of simulated galaxies at various epochs. Until recently, resolution of even the highest cosmological simulation was unsatisfactory for such an involved task, but in the last several years things have finally moved in the realm of the feasible.[46] That is but one—albeit prominent—example of the newly forged and as yet mainly unacknowledged partnership of cosmology and astrobiology.

As described by Leonard Susskind, the usual complaint (which is repeated verbatim by critics of astrobiology in general) is that anthropic arguments deal with only the kind of life we know about, with the implication that some other forms of life will require different chemical, physical and cosmological prerequisites. As correctly emphasized:[47]

> "How do we know that life can't exist in radically different environments?"...I think that correct response to this criticism is that there is a hidden assumption that is an integral part of the anthropic principle, namely: the existence of life is extremely delicate and requires very exceptional conditions. This is not something that I can prove. It is simply part of the hypothesis that gives the anthropic principle its explanatory power. Perhaps we should turn the argument upside down and say that the success of Weinberg's prediction supports the hypothesis that robust intelligent life requires galaxies, or at least stars and planets...The populated landscape plays the same role for physics and cosmology as Darwinian evolution does for the life sciences.

So we have it all here: although Susskind does not elaborate, there is clearly a job for astrobiological research *within the framework of cosmology*.

Using the multiverse in this manner prompted a beginning of what could be potentially a great partnership for the twenty-first century cosmology (similar to what partnership with nuclear/particle physics was for the twentieth century cosmological discourse): the partnership with life sciences, in particular astrobiology. A prototype of such new interdisciplinary provocation of cosmology in life sciences is the study of distinguished Russian-American computational biologist Eugene V. Koonin on the relationship of the cosmological model of eternal inflation to the puzzle of abiogenesis on Earth.[48] In a thought-provoking study, Koonin concludes that only the multiverse of eternal inflation guarantees that the highly improbable steps related to life's origin will inevitably occur by chance. Therefore, problematic issues like "irreducible complexity" or unproven ribozyme-catalyzed RNA replication could be completely sidestepped—*somewhere* in the multiverse abiogenesis could proceed by chance, and we need just to apply anthropic (self-) selection to conclude that one of these places is called Earth! Irrespective of specifics of Koonin's scenario, one thing is new here. As he writes: "The plausibility of different models for the origin of life on earth directly depends on the adopted cosmological scenario."

Finally, it is important to disperse another item of confusion, often repeated in the popular science and media: while the anthropic reasoning has historically emerged in the context of cosmology, it is in no way limited to applications in cosmology or fundamental physics. It simply applies whenever the number or other properties of observers come into play and influence a particular scientific result or claim. Historical roots of the anthropic reasoning have been gradually decoupled from its epistemological and research utility. Therefore, in stark contrast to the prevailing belief, anthropic arguments have an important tradition, and a role to play *outside* highly speculative fields of string theory and particle/quantum cosmology.[49] Some of the examples include planetary sciences, astrobiology, SETI studies, risk analysis, strategic studies, safety engineering, and even a famous "financial proof" of the impossibility of time travel.[50]

12.4   Multiverse in other relevant contexts

> How odd it is that anyone should not see that all observation must be for or against some view if it is to be of any service!
>
> Charles Darwin, letter to Henry Fawcett (1861).[51]

Classification of multiverse hypotheses introduced by Tegmark in his popular *Scientific American* article in 2003 has already turned out to be immensely useful tool for thinking about these radical—and proliferating—ideas.[52] Tegmark considers four tiers or levels of the multiverse, dependent upon their basic ontology:

- **Level I: the universe beyond our horizon.** A straightforward extension of open and flat Friedman models, whose infinite spatial sections have already presented us with a host of philosophical problem which are often misaddressed to the higher levels of the multiverse. Classical cosmological principle has always implied that our Hubble volume is not special.[53]
- **Level II: different bubble universes with varying constants.** This is the truly revolutionary contribution of the inflationary paradigm, as discussed above. String theory landscape also belongs to this level.
- **Level III: Everett's branches of the quantum-mechanical multiverse.** The many-worlds interpretation of quantum mechanics implies that there is a huge number of branches of the universal wave function of the universe. This is a consequence of consistent application of the unitarity of quantum processes, without non-unitary "jumps" such as the "collapse" or "reduction" of wave function.
- **Level IV: the multiverse of all mathematical structures.** All universes described by different mathematical structures are equally real within this level multiverse.

The first two levels have been discussed elsewhere in this chapter. Level III, although undoubtedly of extremely high importance, does not strictly belong to the discourse of cosmology.[54] On the other hand, Tegmark's Level IV is something new and is in a sense both the most extreme and logical completion of the taxonomy.

In several papers and a book in the period from 1998 to the present, Tegmark advances a radical idea that there is no ontological difference between physical and mathematical structures.[55] Scientists have for quite a long time struggled with Wigner's "unreasonable effectiveness of mathematics in the natural sciences," as well as the related issue why some mathematical structures are seemingly privileged to explain physical structures (e.g., Hilbert's space of quantum states), while others have no such applications. Tegmark simply cuts the Gordian knot by asserting that *all* mathematical structures are realized *somewhere*, so there is no distinction between physical and mathematical reality. Of course, the fact that we perceive only *some* physically realized mathematical structures testifies upon the strength of observation-selection effects: as "self-aware substructures," we observe only those structures supporting sufficient complexity, stability, and other physical pre-requisites for observership. In other parts of this radical Platonist multiverse, other mathematical structures are realized without observers around to notice that.

It is an understatement that Tegmark's hypothesis has been and remains controversial. Although Tegmark has argued that it has observable consequences, most critics have disagreed.[56] Without going into extremely complex issues of possible computable measures in such a large multiverse, it is important to state that the Level IV mathematical multiverse is arguably the most extreme position within the corpus of "postmodern cosmology." As such, it is met with agnosticism by many proponents of other multiverse schemes—but it has undoubtedly "stirred the pot" and brought further exposure and visibility to cosmological issues.

As an illustration of both a problem with the new default thinking and a daring reach of the "postmodern cosmology" we might wish to briefly consider the infamous problem of *Boltzmann brains*. It was first explicitly posed in the "disturbing" 2002 study of Dyson, Kleban, and Susskind as a consequence of the post-1998 ΛCDM standard cosmology.[57] Such a cosmology leads asymptotically to a future-eternal de Sitter spacetime. In the eschatological limit of future eternity (to be discussed in more detail in Section 12.6), it is equivalent to a thermal state with an extremely low characteristic temperature of $\sim 10^{-29}$ K (roughly corresponding to thermal radiation with a wavelength of the size of our cosmological event horizon[58]). At first glance, such a low temperature coupled with the extreme dilution of matter at such late epochs is unlikely to yield anything interesting, but in fact, like any thermal system, such a late vacuum will *fluctuate* at a nonzero rate per unit spacetime volume. And, out of such fluctuations, across large eschatological time spans, complicated systems, for example, intelligent observers, can arise, again at a nonzero rate. Since de Sitter spacetime is eternal, such "freak observers" (= Boltzmann brains) will be infinite in number (in fact, their production rate will exponentially increase with time) and will thus completely overwhelm "normal" observers, like us, who emerged through physical, chemical and biological evolution. Why, then, per anthropic reasoning, are *we* not such freak observers?[59]

The Boltzmann brains problem is a close relative to other anthropic puzzles, like the Doomsday argument or Olum's problem, perhaps a bit more unpleasant being put on stronger footing by our seemingly good understanding of the de Sitter space.[60] An overview of the literature shows that the threat has been taken seriously. To give just a few examples of possible responses, Don Page argues that the simplest answer to the puzzle is that the future is finite and that our universe will vanish before Boltzmann brains start to dominate the total observers' census in the entire spatiotemporal volume.[61] (This is connected to the possibility of "Big Rip", on which more in Section 12.6.) On a more optimistic note—at least from the selfish point of view of us evolved observers!—Carlip shows that the time variation of fundamental constants can lead to a small measure of freak observers, and Vilenkin demonstrates that the same effect is achieved by the bubble nucleation in a de Sitter vacuum under some fairly reasonable assumptions.[62] Linde tackled the problem in a detailed manner in 2007, and concluded that the problem hinges on the correct choice of probabilistic measure in the multiverse context; with particular choices, the problem does not arise.[63] (Parenthetically, Linde's well-intentioned humor, characteristic for much of the cosmological discourse discussed in this chapter is exemplified by his caveat [p. 23]: "That is, of course, if BBs [Boltzmann brains] are not aggressive and do not participate in any experiments that could trigger a transition to a different vacuum. We presume that this is only a prerogative of honest folk like ourselves.")

It is important to note that the problem of Boltzmann brains appears even in a single universe provided that they asymptotically expanded in the de Sitter manner, as we have all observational reasons to believe. It is not necessary, therefore, to pose it in the multiverse context, although multiverse could help us *resolve* it, as per Vilenkin's and Linde's approach. A rather uncontroversial aspect of anthropic reasoning is that we need to suppress the measure of freak observers in the entire multiverse in order to explain our evolution and our ordered observations (and, clearly, Page's recipe will not work for a multitude of universes). Although this problem will require further work, the good news is that there are ways out, such as those shown by studies of Carlip, Vilenkin, and Linde. As the titan of science after whom those freaks were named put it in another context: "It may be objected that the above is nothing more than a series of imperfectly proved hypotheses. But granting its improbability, it suffices that this explanation is not impossible. For then I have shown that the problem is not insoluble, and nature will have found a better solution than mine."[64] The same sentiment underlies much of the recent cosmology.

Of course, in a sufficiently large universe (not to mention the multiverse), there are likely to be Boltzmann brains of all kinds, including the deluded ones thinking that they are you or me having exactly this conversation. While this may raise some interesting issues in ontology and the theory of identity, it is difficult to see why anyone would consider it paradoxical or detrimental for anthropic reasoning *per se*. We have already known for quite some time that the prediction of the classic Friedman open and flat universes (without dark energy) is that there are infinitely many galaxies, which contain infinitely many copies of Earth, this book, you and me.[65] If the existence of doubles does not disturb us too much—and nobody has argued against classical relativistic cosmology on that basis—why would a phenomenologically weirder, but essentially similar situation with Boltzmann brains cause us any sleepless night? Many contemporary cosmologists actually hold a somewhat surprising philosophical sentiment, succinctly expressed by Yasunori Nomura: "It is intriguing that such a basic fact that we observe an ordered, comprehensible world may deeply rely on the structure of quantum gravity at the fundamental level."[66]

Conservative reaction is, unsurprisingly, quite strong. Some of it is based upon the unreflective repetition of older views, such as the one of the philosopher of science John Earman who in his well-publicized—and largely outdated—critique of the anthropic reasoning wrote:[67]

> But to be legitimate, the anthropic reasoning must be backed by substantive reasons for believing in the required worlds-within-worlds structure ... Neither classical general relativity nor quantum mechanics provide any firm grounds for taking worlds-within-worlds models seriously, and while various speculative versions of the new inflationary cosmology may eventually provide such ground, the verdict is at present very much in doubt.

Note that not only is the application of the anthropic reasoning strictly circumscribed in Earman's account, but specific advances in cosmology are required as *prerequisites* for anthropic explanations. The situation has dramatically changed in comparison to the late 1980s, which is often overlooked by the philosophical opposition to the multiverse.[68]

Perhaps the most active and vocal sceptic toward the multiverse is South African cosmologist, philosopher, and human rights activist George Francis Rayner Ellis. Another student of Dennis Sciama and an early collaborator with Stephen Hawking on the seminal *Large Scale Structure of Space-Time*, Ellis has, with collaborators, written a series of studies elaborating both scientific and philosophical opposition to the concept.[69] They have been responded in return by Bernard Carr, Rees, and others—the debate is still going on at full speed at the time of writing.

The oft-repeated charge that the multiverse ideas are not testable has been actually been refuted in many cases. There *are* observable effects which are predicted by the both leading multiverse conceptions, eternal inflation and the string theory landscape. Apart from things like large-scale homogeneity and isotropy, which have been interpreted, as we have seen, as evidence for the generic (= eternal) inflation by Linde and others, and apart from the empirical matter of fine-tunings which require an explanatory mechanism, among such relics we may count cross-correlation spectra of the large-scale structure and CMB,[70] remnants of inflationary bubbles' collisions,[71] primordial gravitational wave relics,[72] and even astrophysical parameters related to structure formation such as the primordial density contrast and the virial density of dark matter.[73] This comes on the top of all those improbable coincidences and fine-tunings which are best treated when *embedded* into a larger set, here the multiverse.

To what extent could even our existence as present offer a probabilistic argument for a large world (i.e., the multiverse)? Economist and complexity theorist Austin Gerig has suggested several intriguing arguments to that effect.[74] It is not trivial to determine the corresponding bias:[75]

> Although it is relatively straightforward to demonstrate that a typical observer in a large universe will exist in a world biased towards their existence, it is perhaps not as easy for an observer to determine that the bias exists. The observer would need a correct model of cosmology, whereas often cosmological models are formulated by extrapolating from an observer's surroundings.

This might account for the fact that we discovered the multiverse so relatively late. Gerig goes further to derive a demographic argument that the slowing of the growth of the population of only observers we know so far (humans) noticed in the last decades is probabilistically consistent with the sampling of the sets of observers from a multiverse, and inconsistent with the single-universe expectations.

Perhaps the best illustration that the multiverse is here to stay is the fact that, although the concept itself is most often associated with either eternal inflation or the string theory landscape, one of the most vocal and vehement critics of both ideas, the American theoretical physicist Lee Smolin has proposed—as a solution to the fine-tuning problem—an alternative theory of the multiverse! Dubbed cosmological natural selection

(CNS), it first appeared in a research paper in 1992, and five years later it was expanded in a popular book *Life of the Cosmos*.[76]

CNS is based upon the hypothesis—in itself highly controversial—that the endpoint of gravitational collapse into a black hole is actually not the classical singularity, but a bounce creating a new universe. While it is not entirely implausible—given our poor understanding of quantum gravity—this hypothesis is entirely unsupported by independent *a priori* evidence; Smolin finds it is not that important as long as interesting *a posteriori* evidence could be provided, in form of an explanatory account of fine tuning. He theorizes that these descendant universes will be likely to have similar fundamental physical parameters to the parent universe, but that these parameters (and perhaps the laws derived from them), may be slightly altered in some stochastic fashion during the replication process. Each universe therefore potentially gives rise to as many new universes as it has black holes. In such a setup, the variations of the laws and constants will behave in an optimizing, quasi-teleological manner to create universes which maximize the production of black holes. According to Smolin, universes with many (smaller) black holes will enable higher complexity than the universes with only a few (large) black holes, and therefore habitability and the emergence of intelligent life and observers are a byproduct by the increase in black-hole fecundity. It is exactly this parallel of CNS with the Darwinian theory of biological evolution, coupled with the prospect of offering an explanation for the cosmological rise of complexity, which attracted significant attention.[77]

Many critics have pointed out weaknesses in Smolin's theory from both physical and biological points of view: universe formation in gravitational collapse is quite speculative, variations in constants are also unwarranted, star formation in our observed universe does not seem to maximize black hole production, CNS lacks adequate notion of selection pressure to be a true Darwinian theory.[78] Even a somewhat tongue-in-cheek criticism—or an attempt at a *reductio*—has been made: advanced civilizations, extraterrestrials or future human, should be still much more efficient in artificially creating black holes, so we would expect to see the universe filled with black hole-producing supercivilizations in obvious contrast with reality![79] All in all, while CNS has been tremendously inspirational and influential in the course of the last quarter of century, it has been largely abandoned as a research program. The wider point, however, seems confirmed: the multiverse is an inescapable part of any and all attempts to understand empirical fine tunings in a naturalistic manner.

12.5   The emergence of string cosmology

> Prepare your work outside; get everything ready for yourself in the field, and after that build your house.
>
> *Proverbs* 24:27.

The best candidate for the much-sought unified field theory—also jovially dubbed "The Theory of Everything"—at present is a set of ideas usually going under the title of string theory (or M-theory, after the unified form of various string theories, as presented by Edward Witten in 1995).[80] In an extremely simplified view, the point-like particles of conventional particle physics are replaced by one-dimensional objects called strings. String theory describes how strings oscillate, propagate through space, and interact with each other. In a given version of string theory, there is only one kind of string, which may look like a small loop or segment of an ordinary string, and it can vibrate in different ways. On distance scales larger than the string scale, a string will look just like an ordinary particle, with its mass, charge, and other properties determined by the vibrational state of the string. In this way, all of the different elementary particles—including the all-important graviton—may be viewed as vibrating strings.

Historically, several versions of string theory have emerged: Type I, Type IIA, Type IIB, and two so-called heterotic string theories (denoted by their groups of symmetry as SO(32) and E8×E8). The different theories allow different types of strings, and the particles that arise at low energies exhibit different symmetries. For example, the type I theory includes both open strings (segments with endpoints) and closed strings (which form closed loops), while types IIA and IIB include only closed strings. As shown by Witten, each of these five string theories arises as a special limiting case of M-theory.[81]

String theory—in the most inclusive sense—has become, for good or bad, bread-and-butter of contemporary fundamental physics. Considering the fact that most of string-related phenomena are linked to energies many orders of magnitude higher than anything accessible in the laboratory, it has been only expected that string theorists turn to cosmology as the only "natural laboratory" for testing specific predictions of the theory. Hence the emergence of string cosmology as a well-defined field in 1990s.[82] We have already discussed the "string theory landscape" of Susskind and others, which offers a natural and reasonable realization of the multiverse; in this section the more general emerging framework for application of string theory will be briefly covered.

The motivation has actually been somewhat more complex. Inflation, as discussed in Chapter 11 and above, provides a solution to problems such as flatness and horizon problems, as well as the overproduction of monopoles, all within the framework of the quantum field theory and general relativity. However, in doing so, it occludes the preceding epochs, in particular those which are consequences of the physics of the ultraviolet completion of the QFT+GR framework. String theory plays the role of the main (only?) contender for the ultraviolet completion—therefore, it encounters a "trouble" with inflation.

However, cosmology cannot avoid questions about the initial conditions. Since the days of Friedman and Lemaître, spacelike singularity ("Big Bang") at the origin of classical cosmological narrative troubled researchers and lay audiences alike. Behavior of spacelike singularities is still very much an open question, but some newer work indicates that the new degrees of freedom, introduced by string theory play the key role and could plausibly *resolve* the singularity under special conditions. If these results could be generalized (and, of course, string theory in the form of M-theory or some other similar garb continues to hold sway in our attempts to unify gravitation with other fundamental forces), this would mean solving of probably the oldest and the most controversial feature of relativistic cosmologies since the times of Friedman. No mean feat by any measure!

There is another strong motivation for attempts to employ string theory in theoretical cosmology: a suggestive analogy between black hole event horizon and the cosmological event horizon. As is rather well-known, a great breakthrough for string theory was made in 1996, when Andrew Strominger and Cumrun Vafa were able to show how microscopic degrees of freedom within the theory allow for the derivation of the famous Bekenstein–Hawking formula for the entropy of black holes.[83] Many cosmologists and string theorists hope that a similar approach could be fruitful for the cosmological event horizon, which possesses its own, Gibbons–Hawking entropy. While the hope has not been realized thus far, the entire program is appealing in view of the acknowledged capacity of string theory to resolve other problems related to black hole horizons, such as the black hole evaporation.[84]

As alluded to above, considerations of the pre-Big Bang physics have become possible within string theory. A general scenario, due mainly to two Italian theoreticians, Maurizio Gasperini and Gabriele Veneziano, and developed since early 1990s goes roughly as follows. Due to duality symmetries of string theory, we may envision conventional Big Bang as a kind of "Big Bounce," a transitory phase from the previous, gravitationally unstable evolutionary regime. String-size effects and loop corrections have been calculated to enable such bounce for a plausible range of different initial conditions. An initial unstable state of small curvature collapses into high-curvature, strongly interacting intermediate state which simulates the classical Big Bang singularity without being truly singular. Instead, copious quantum production of particles occurs in this state, which causes bouncing into the expanding phase of decreasing curvature and heating. Gasperini and Veneziano have shown that under some assumptions, a viable spectrum of density perturbations can be produced by such models and they emphasize that there are potentially observable relics in form of contributions to the primordial mode of CMB polarization, as well as the gravitational wave background.[85]

When we take a look from the "other side of the street," so to speak, we perceive that the love story of cosmology and nuclear/particle physics could well be repeated in the cosmology-string theory relationship. Since the times of Gamow, it has been repeatedly realized that the early universe provides a unique natural laboratory in which tests of concrete high-energy physics sectors could be performed. Energies characteristic of nuclear physics have been probed by primordial nucleosynthesis, those characteristic of the standard model of particle physics have been—in a somewhat more relaxed sense—probed by the inflation. Now, string theory deals with phenomena manifested at higher energies still, available only in the pre-inflationary phase (or, what amounts to essentially the same thing, pre-Big Bang physics).[86] It is conceivable that imprints of these early processes are detectable indirectly, through their impact on inflation and some of its consequences, as well as some non-inflationary sources of perturbation. In the words of Liam McAllister and Eva Silverstein, "signals of string theory will be seen, if they are seen at all, in the sky."[87] While these models are still considered an "alternative" to the standard picture (or the slow-roll chaotic/eternal inflation), the continued dominance and expansion of the string theory in fundamental physics, as well as recent successes in gravitational-wave astrophysics, have led to increased interest in them.

Possible observable relics which could serve as indicators of string physics include primordial gravitational waves, cross-correlations of the large-scale structure with the cosmic background temperature anisotropies and polarization, as well as the first ("seed") galactic magnetic fields.[88] A major problem is that there is no single model of inflation (as discussed in detail in Chap. 11), and features of a particular model could mask or emulate the effects of pre-inflationary physics. Generic inflation provides a single source of perturbations; therefore, any additional observationally detected source would be an indication of non-inflationary physics and could be compared with predictions from string theory. Alternative inflationary models which introduce degeneracy in predictions are, for example, multi-field inflations involving Brans–Dicke terms coupling the inflaton with some other field; these models seem disfavored by the current data, but the situation is far from clear at the time of writing. Supporters of string theory should be particularly wary of confirmation biases, especially taking into account the 2014 affair with the false claim of detection of primordial gravitational waves by the BICEP-2 project, which has been subsequently withdrawn.[89] Everything here is extremely tentative and uncertain—but this should not discourage future work in the area, which is quite likely to be one of the most active sectors of the cosmological work in the twenty-first century.

12.6 Physical eschatology: "cosmology of the future" in both senses

[Eddington] told me once, with evident pleasure, that the expanding universe would shortly become too large for a dictator, since messages sent out with the velocity of light would never reach its more distant portions.

Bertrand Russell[90]

The bright sun was extinguish'd, and the stars
Did wander darkling in the eternal space,
Rayless, and pathless...

Lord Byron, *Manfred* (1816)

Quite natural—in more than one sense!—a counterpart to physical cosmology emerged as a new discipline in the last quarter of the twentieth century under the title of *physical eschatology*. The word eschatology (*éschato* = last) was used originally in an exclusively religious light, as "any system of religious doctrines concerning last or final matters, as death, judgment, or an afterlife" and "the branch of theology dealing with such matters" (*Random House Webster*). The physical sciences slowly encroached upon this field, however, and in the last ca. 40 years a respectable astrophysical discipline arose as a consequence both of the improvement of our empirical knowledge of the universe and of the explosive advances in the theoretical techniques of modelling and prediction. Sir Martin Rees first employed the word "eschatology" in an astrophysical context in the title of his pioneering article of 1969, and Fred C. Adams and Gregory Laughlin used the term "physical eschatology" to denote the field in 1997.[91] As described in Chapter 1, considerations of cosmic entropy and the impending heat death of the universe have long history and are correctly considered as precursors of the modern physical eschatological discourse.

Physical eschatology, therefore, deals with the future evolution of astrophysical objects, including the universe itself. It immediately follows that a large chunk of physical eschatology is irrelevant for the cosmological discourse *sensu stricto*. One can engage in physical eschatology, as many researchers have indeed done, by investigating the future of Earth (and Earth's biosphere), the Solar System, and even our Galaxy, without reference—at least in a superficial sense—to the universe as a whole (not to mention the multiverse!).[92] One could compare this to the temporally inverted discourse of *cosmogony*, which is often defined as dealing with the origin and formation of various objects and levels of structure. This is misleading, however, since there are indeed essentially cosmological preconditions for any kind of physical eschatology, most notably the existence of an open, infinite future (or else, depending on the cosmological model), but also the notion of the constancy and universality of physical laws, which is clearly necessary for any predictive endeavor. To give just an obvious example, there are processes which would certainly impact the future of *all* astrophysical objects, stars, planets, our Earth, etc. in the fullness of time, like the proton decay occurring on timescales of $\sim 10^{34}$ years, but which are irrelevant if the cosmological future is not open, ending in the Big Crunch or the Big Rip (see below) in $\sim 10^{11}$ years. There are other, more subtle limitations, having to do with uniformities of fundamental constants, laws, and causal inferences. So the cosmological part of physical eschatology is indeed necessary for all but the nearest-future events and processes.

In order to adequately describe immense future timescales, dwarfing everything we have grown accustomed to in classical past-oriented cosmology, Adams and Laughlin have introduced a useful notation in the form of *cosmological decades*. A cosmological decade η is defined simply as:

$$\eta = \log_{10}\left(\frac{t}{1 \text{ year}}\right),$$
(12.4)

where t is the conventional cosmic time, measured from the Big Bang. Although it might seem a trivial exercise, the labelling of epochs by cosmological decades is in fact quite useful tool for intuiting the great size of the cosmological future (and hence the increased relevance of physical eschatology as a nascent field of study). We are currently living in the cosmological decade roughly $\eta \approx 10$, and everything what has been happening since the Big Bang is squeezed into these first ten cosmological decades (for example, CMB photons as observed today originated in the cosmological decade $\eta = 5.6$). In contrast, many cosmological decades will elapse before the universe dramatically changes its appearance, which is clear even on the basis of the quite rough calculations of Adams and Laughlin.

For instance, some of the signposts of the future evolution are assigned to particular cosmological decades as follows. Current epoch of star formation in galaxies and stellar evolution of individual stars will end about cosmological decade $\eta \approx 12$–14. From that time on, the stellar component of galaxies will be in form of inert stellar remnants, and the galaxies will become dark, with only extremely rare flickers of supernovae of Type Ia, even rarer binary neutron star mergers or black holes swallowing another stellar remnant or unbound planet. Galaxies themselves will persist as well-defined entities up to the cosmological decade $\eta \approx 19$; about that epoch all objects in galaxies will be either ejected from the galactic gravitational well by the slingshot mechanism or will be swallowed by central supermassive black holes. Depending on which GUT turns out to be correct, at some point between cosmological decades 32 and 49, protons will decay, leading to the end of baryonic matter as we know it. This will cause evaporation of even isolated cold white dwarfs ("black dwarfs") and cold neutron stars. At some point, only black holes and incredibly rare and huge positronium atoms (bound systems of electron and positron) will remain. Black holes themselves will gradually decay through slow emission of Hawking radiation; the supermassive ones, having absorbed large fraction of mass of their long-gone galaxies, will decay by the cosmological decade $\eta \approx 98$–100. Whether anything else can happen in those remote and strange future epochs depends on the exotic physical processes, like vacuum tunnelling or topological instabilities.

This picture presupposes future temporal infinity, as in open or flat classical Friedman models. For most of the history of physical cosmology, the basic dilemma presented by the evolutionary narrative has been whether the universe will expand forever or the gravitational pull of matter fields will be strong enough to halt the expansion and turn it into contraction towards the "Big Crunch." In the older literature, one can find equality between eternal expansion and topological openness and, conversely, between recollapse and topological closeness. However, as elaborated by Lawrence Krauss and Michael Turner, dark energy introduces a degeneracy into the cosmological future, which indicates that even a topologically closed universe ($\Omega > 1$) can expand forever in the presence of, say, a positive cosmological constant.[93] Conversely, a topologically open universe can recollapse into the Big Crunch if the dark energy is attractive (e.g., a negative cosmological constant). However, since all observations suggest a *repulsive* form of dark energy, this option is of rather academic interest. Therefore, we almost certainly live in an ever-expanding cosmological domain ("universe"). Of course, observational cosmology long ago suggested similar conclusions on the long-term future of the universe, since all surveys of gravitating matter fell short of the critical density for recollapse. One of the first interesting results of physical eschatology has been a refutation of more than century-old idea of the "heat death" of the universe, confirming the early intuition of Pierre Duhem that entropy in the cosmological context only can approach its maximum value asymptotically.[94]

One intriguing consequence of the cosmological revolution of 1998 and the discovery of dark energy-dominated universe has been another "return of a static universe" (compare with Linde's view of stationary multiverse discussed above!) in the context of our cosmological *future*. Krauss and Scherrer demonstrate that in the distant future our descendants or any other intelligent observers present will have a difficult time to derive *any* cosmological knowledge whatsoever.[95] The sequence of events is rather instructive: as galaxies cross the cosmological event horizon, eventually all sources not gravitationally bound to our Galaxy will vanish from the view of our descendants. On the other hand, objects gravitationally bound to us (= objects in the Local Group) will eventually coalesce into a single system, the process which begins with the impending merger of the Milky Way and M31.[96] The emerging "supergalaxy" will be something similar to a giant elliptical galaxy, as numerical simulations show. All visible sources, therefore, will be decoupled from the Hubble expansion and any local observer will essentially see a single stellar system *without any observational evidence of the expansion of the universe!* Interestingly enough, the situation will be somewhat similar to the classical Kapteyn universe popular about the beginning of the twentieth century, before Hubble's discoveries (see Chapter 2). It can be shown that other possible cosmological clues (CMB, dark energy, relics of primordial nucleosynthesis, etc.) will be rendered undetectable or useless for making any cosmological inferences at that remote future epoch.

A curious conclusion seems inescapable here: for the very first time in its history, physical cosmology is able *to predict its own ending* as a meaningful scientific discipline! The ending is essentially independent of the quality of observational or theoretical work, or indeed, on any property of the observers whatsoever—it is in a sense an ontological, not an epistemological occurrence. As a form of consolation, one might note that in the preceding epochs, as Krauss and Glenn Starkman smugly noted, "funding priorities for cosmological observations will become exponentially more important as time goes on."[97]

As a final topic for this rough overview of physical eschatology, an alternative cosmological future has emerged in recent years due to elaborations of the concept of dark energy. Consider the generic equation of state of a cosmological fluid (in $c = 1$ units):

$$p = w\rho. \tag{12.5}$$

Traditionally, the equation of state parameter w has been set to be $w = -1$ for vacuum energy or cosmological constant. However, apart from some reasons of elegance, there is no compelling reason for this choice. Any negative value of w would cause accelerated expansion, as discovered in 1998, but their dynamic evolution drastically differs in three cases: $0 > w > -1$, $w = -1$, and $w < -1$. Only in the classical case of cosmological constant $w = -1$, the vacuum energy density ρ_Λ stays constant throughout. If the equation of state parameter is larger than the critical value of $w = -1$, while still negative in order to provide for the accelerated expansion ($0 > w > -1$), the vacuum energy density ρ_Λ decreases. This process proceeds, supposedly, by creating various quanta out of the vacuum energy, which decrease the magnitude of the vacuum energy; this case corresponds to the *decaying dark energy*. On the other hand, for $w < -1$, ρ_Λ increases without limit with the passage of cosmic time, opening up some deeply interesting (and disturbing) physical eschatological consequences.

The case $w < -1$, corresponding to a "super-negative" equation of state, has been dubbed *phantom energy* by its discoverer, American theoretical physicist Robert Caldwell of Dartmouth College.[98] Since the vacuum energy density diverges in models with phantom energy, it eventually overcomes any other energy density anywhere, including the one keeping together bound systems, like galaxies, planetary systems, solid bodies, atoms and nuclei, and even individual baryons. In other words, all matter of the universe, from stars and galaxies to atoms and subatomic particles is progressively torn apart by the accelerated expansion of the universe at a *finite time* in the future. This singular event, which represents the end of matter—and by virtue of general relativity the end of spacetime—has been called the *Big Rip*. It represents a conceptual opposite to the classical Big Crunch singularity in the recollapsing Friedman models: instead of being annihilated by diverging density in the Big Crunch singularity in the finite future, all matter is dispersed to nothing by the diverging expansion rate in the Big Rip singularity in the finite future. The timescale for the Big Rip obviously depends on the true value of the equation of state parameter w; we are left with, roughly:[99]

$$t_{\mathrm{BR}} - t_0 \approx \frac{2}{3|1 + w|H_0\sqrt{1 - \Omega_{\mathrm{m}}}}, \tag{12.6}$$

where t_{BR} is the epoch of the Big Rip, and other symbols have their standard cosmological meaning. For a fiducial value of $w = -1.5$, and the new standard values for Hubble's constant ($H_0 = 70\,\mathrm{km\ s^{-1}\ Mpc^{-1}}$) and the cosmological matter density fraction ($\Omega_{\mathrm{m}} = 0.3$), Caldwell with collaborators calculated that the Big Rip lies about 22 billion years in our future. In this scenario (admittedly extreme, since w is likelier to be much closer to -1 according to the CMB data, see below), although the future lifetime of the universe is longer than its past so far, the Big Rip is still a more pressing concern than either the heat death of the universe, or the recollapse into Big Crunch in closed cosmological models popular in, say, 1980s.

How real is the danger of future Big Rip? So far it has not been considered very much, for several reasons, probably the most efficient being predilection of researchers for interpretation of dark energy as $w = -1$ cosmological constant. However, this type of intellectual inertia could be seriously misleading, as was similar predilection for the Einstein–de Sitter $\Omega = \Omega_{\mathrm{m}} = 1$ universe in years before 1998. As mentioned above, Don Page has in fact suggested that the Big Rip (or any other version of global doomsday) actually helps explaining why we are not Boltzmann brains ourselves! More serious reasons to doubt the Big Rip include developments of models which, although containing phantom energy, do not end in the universal dissolution of matter.[100] Some of these are wide generalizations, which delineate the regions of parameter space in which phantom energy leads "only" to asymptotic de Sitter expansion, as opposed to the Big Rip singularity. In any case, only future observations will be able to precisely determine the value of w. At present, the best observations are highly inconclusive, being in principle consistent with all the three options and leading most cosmologists to prefer $w = -1$ case. For example, the constraint from observations of distant X-ray clusters by the Chandra X-ray telescope gives $w = -1.14 \pm 0.31$.[101] Observations of the CMB anisotropies by the Planck satellite give the best-fit value as $w = -1.13^{+0.13}_{-0.10}$.[102] While this apparently supports phantom energy, one should keep in mind that the uncertainties quoted are almost certainly underestimates. Only future concordance observations will be able to settle this question decisively.[103]

12.7   Rocky road ahead: cosmology's transition into a "postmodern" epoch

You can't go home again.

Thomas Wolfe

Beside the "Great Confluence" of particle physics and (astro)physical cosmology which dominated the last quarter of the twentieth century, we are witnessing many smaller confluences, where disciplinary barriers become thin and unstable, and researchers with quite heterogeneous profiles and backgrounds participate in the grand cosmological enterprise. We have already mentioned the interaction with astrobiology in relationship to the multiverse and fine-tuning themes. There are other such examples—important branches of this flourishing tree which we have not had remotely enough space to adequately consider.

In a bigger picture, there are more and more questions which *necessarily* require multidisciplinary work and expertise. Those on the topic of habitability of the universe/multiverse or the future evolution of the observable universe are excellent examples in this regard. They are parts of a wider trend, which has gradually obtained somewhat systematic and canonical status.

The impact of cosmology on pop-culture, although not specifically researched, seems to have significantly grown (at least from a subjective point of view and on the anecdotal evidence). It is exactly in the period under consideration here, from mid-1980s to mid-2010s that terms such as "singularity," "Big Bang," "parallel universe," or "black hole" have become household terms, to a large degree due to successful TV programs of both popular-science and fictional nature. More people than ever before have been exposed to at least some elements of cosmological thinking.

The first half a century of physical cosmology has amply illustrated the tenet that philosophy plays important role in the development of a scientific discipline only in its early, formative stages.[104] This canonical view suggests that as long as the empirical database of a discipline is small and its methods rudimentary and rough, philosophical preferences of distinguished researchers and their schools exert a significant impact on the course of research. Later, when our insight progresses and our grasp of theory improves, those philosophical inputs fade and become less relevant. At least this is valid for scientists in the narrow sense; among philosophers some of the early concerns may remain relevant for much longer.

Even a cursory look at the themes discussed in this chapter suggests that this view is inapplicable to the "postmodern cosmology" and its concerns. We can conclude either: (i) that we have entered a new disciplinary field in which philosophy is again relevant, since the field is still immature, or (ii) that the model is at least sometimes wrong. At present, it does not seem clear which of the two options is likelier in the case of cosmology, and in particular its "postmodern" sector. In any case, this debate is likely to be with us for quite some time. Ironically, not only the multiverse could have a fractal structure, as predicted by most models of the chaotic inflation, but our cosmological understanding might in a way be fractal itself. The multiverse seems to be getting out of the closet and becoming a legitimate, though not widely accepted topic of research and discussions—similarly to what the classical ("hot") Big Bang was roughly between the times of Lemaître and 1965.

Considerations of Chapter 4, especially Section 4.9, apply here with a vengeance. Ironically, one might argue that, from a conservative viewpoint, cosmology almost became "normal science" again. Previously, during most of the twentieth century, cosmology has been criticized by conservative sceptics among astronomers (e.g., Dingle and Disney) and philosophers (Toulmin, Scriven, and Bunge) as being too dashing and reliant on speculations. Nowadays, things are being repeated "on a higher tier", so to speak. The twenty-first-century cosmology, painted by the widest strokes, cannot escape similar criticism, with its vacuum manifolds, braneworlds, tachyonic inflation models, Boltzmann brains, phantom energy, and the elusive multiverse itself. The opposition to this contemporary meta-paradigm exists and is vocal from time to time in both science and philosophy.[105] This novel type of scrutiny and criticism should be distinguished from the "distribution tail" of the older, twentieth-century school of criticism, where ideas of Wickramasinghe, the Burbidges, Arp, Narlikar, and possibly a few other authors going back to Sir Fred Hoyle and the classical steady-state theory properly belong; these have been covered in detail in Chapter 4. The latter criticism has been essentially a rear-guard action on the part of overt and covert fans of the steady state, sad for "ugly facts murdering a beautiful theory." The modern criticism, on the contrary, is not directed any more against the hot Big Bang as such (which won the battle in any case), but against its extensions such as inflationary paradigm. While it has not been studied sufficiently by philosophers and sociologists of science so far, one can discern several new elements. Apart from the old adage about the lack of testability, it seems that the new criticism embraces stronger realist views than it has been the case with most of the modern physics; after all, a tradition of theoretical freedom to speculate has been regarded as a strength, rather than a weakness, of physical science in the bigger part of the twentieth century.

The defenders would claim that the critics' views betray an essentialist view of the science itself: the tacit assumption that there is a single, universal, and only legitimate way to do science, valid for all times. Undoubtedly, such a view would have surprised a Galileo, a Darwin, or a Pasteur—didn't they carve their place in history by showing that previous ways of doing science (Aristotelian, typological, or spontaneous-generation, respectively) were not sound? In this sense, the critics of "postmodern cosmology" come across as profoundly Whiggish: it is *their* epoch, the second half of the twentieth century (say), which created the only correct way of doing science and anything else is either pseudoscience or a passing fad. Needless to say, such a self-centered attitude is not only unsupported by the real history, but it is hard to see how could it ever be supported; what could count as evidence in the first place?

Similar issues bothered the forefather of the multiverse, Ludwig Boltzmann, although in an entirely different context: his famous polemic against Ernst Mach about the reality of atoms.[106] While atoms were completely unobservable at the time of the debate—just as other universes are now—it was both rational and fruitful to accept their reality, in spite of the protests of Mach and others about flights of metaphysical fancy or atomism as being unscientific.

Instead, it is only reasonable to embrace the changing nature and structure of science itself. The change is evolutionary and creative, not revolutionary or destructive. In the celebrated image of the historian Arnold Toynbee, wheels are turning around, making the chariot move forward. The picture might as well be applicable to the discourse of physical cosmology: while particular problems pointed out by the critics are re-emerging time and again in differing contexts, the overall image is the one of indubitable progress. And while there certainly are many open questions related to this part of the overall narrative of physical cosmology—questions to occupy historians, philosophers, and sociologists of science for years and decades to come—there are more reasons still for Popper's famous statement: "All science is cosmology, I believe."[107]

Acknowledgements

Critical comments of both Editors have been enormously useful in improving upon a previous version of this chapter. The author uses this opportunity to thank Zoran Knežević, Marko Stalevski, Srdja Janković, Slobodan Perović, Maša Lakićević, and Anders Sandberg for their helpful suggestions and comments. Dušan Pavlović has kindly helped with several hard-to-find references. The author acknowledges financial support from the Ministry of Education, Science and Technological Development of the Republic of Serbia through the projects ON176021 and ON179048.

....

NOTES

1. Whose intellectual—and often moral—bankruptcy has been amply demonstrated by Koertge (1998) and Sokal and Bricmont (1998), and essentially admitted by Latour (2004), among others.
2. See, for instance, Kragh (2009a, 2014b, 2017b), Křížek and Somer (2016), López-Corredoira (2017), and Tavakol and Gironi (2017).
3. Farhi et al. (1990), Ansoldi and Guendelman (2007), Chavanis (2007), Adams (2008), Linde and Vanchurin (2010), Garriga et al. (2000), and Olson (2015).
4. An almost perfect example is Koonin (2007), which will be discussed in some detail. See also examples such as Trotta (2007), Russell et al. (2013), and Gardner and Conlon (2013). The last one is also a fine example of a research paper in cosmology written collaboratively by a biologist and a physicist—something which has become less and less of an exception during the period under consideration.

5. For example, Oppy (2001), Olum (2004), Knobe et al. (2006), Crawford (2013), and Azhar (2014).
6. For example, Holton Holton (1988, 1993).
7. See Gregory (2007) for a modern treatment and a good entry point into the sizeable and often highly polemical literature on the subject.
8. cf. Price (1996).
9. For different perspectives on this large and controversial project, see Chaisson (1997, 2001), Christian (2004), Nazaretyan (2005), and Last (2017).
10. Koonin, *op. cit.*, but see also Treumann (1993), Lloyd (2000, 2002), Gleiser (2010), and Lineweaver et al. (2013).
11. Starobinsky (1979), Kazanas (1980), Guth (1981), and Sato (1981).
12. Rees (2013), p. 59.
13. Albrecht and Steinhardt (1982b), Vilenkin (1983), and Linde (1986).
14. Linde (2017), p. 6.
15. For example, Linde (1974a).
16. See Chapter 11 for more on various inflaton fields.
17. Linde (1983), p. 180. See also Vilenkin (1983).
18. Linde (1986), p. 399.
19. James (1895), p. 10.
20. See Chapter 1, Section 1.4. Also see Boltzmann (1895) and Steckline (1983).
21. Moorcock (1995).
22. For a critical overview, see Gardiner (2014).
23. For a necessarily subjective choice, see Rees (1997, 2001), Vilenkin (2006), Susskind (2006), Hawking and Mlodinow (2010), and Manly Manly (2011).
24. Linde et al. (1994) and Guth (2007).
25. Ćirković (2003), partially in response to the philosophical criticism of Price (1996).
26. Opening historically all-important question whether we could expect future phase transition in "our" observed vacuum as well; cf. Turner and Wilczek (1982) and considerations of Section 12.6.
27. Susskind (2003, 2006) and Bousso and Polchinski (2004).
28. Of course, proponents of supernaturalism might have an easy answer: the deity presumably chose the "right" universe out of all possibilities. This would lose the main argument for the multiverse ideas in the first place, the fine-tuning problem, as we shall see in the next section.
29. Carr (2007).
30. Carr (2007), p. 61.
31. Lovejoy (1936) and Mash (1993).
32. Lewis (1986).
33. Published as Carter (1974); see also Ellis (2011).
34. A detailed, although hardly unbiased, historical survey up to mid-1980s can be found in Barrow and Tipler (1986).
35. For example, Carter (1983, 1993).
36. Bostrom (2002).

37. The best recent review of fundamental constants of particle physics and cosmology is given in Tegmark et al. (2006).

38. Carr and Rees (1979), Barrow and Tipler (1986), Hogan (2000), Carroll (2006), Barnes (2012), and Barnes and Lewis (2017).

39. Lineweaver and Egan (2007).

40. For a gentle introduction, see Hobson et al. (2013).

41. Bostrom (2003).

42. Tegmark (1998).

43. Weinberg (1987).

44. See Carroll (2006) for a fine review.

45. For example, Chavanis (2007), Adams (2008), and Adams et al. (2015). For a popular exposition of this research program, see Jenkins and Perez (2010).

46. Vukotić et al. (2016) and Forgan et al. (2017).

47. Susskind (2006), p. 356.

48. Koonin (2007).

49. See Ćirković (2016) for a list of relevant examples from other fields.

50. Reinganum (1986).

51. Letters of Charles Darwin (http://charles-darwin.classic-literature.co.uk/, last accessed March 11, 2018).

52. Tegmark (2003).

53. Ellis (1979).

54. See Bousso and Susskind (2012), however, for a maximally inclusive view of the Level III multiverse—in perhaps the most poetic and literary sophisticated paper ever published by *Physical Review*.

55. Tegmark (1998, 2008).

56. Schmidhuber (2002), Vilenkin (2006), and Porpora (2013).

57. Dyson et al. (2002).

58. Often called the Gibbons–Hawking radiation, after the classical study of Gibbons and Hawking (1977).

59. Subsequent elaborations of the Boltzmann brains problem can be found in Bousso and Freivogel (2007), De Simone et al. (2010), and Nomura (2015). Multiple studies of this topic manifest somewhat jovial tone; a fine example is Boddy and Carroll (2013).

60. Gott (1993), Leslie (1996), Olum (2004), and Richmond (2006).

61. Page (2006).

62. Carlip (2007) and Vilenkin (2007).

63. Linde (2007). Similar solution is elaborated in De Simone et al. (2010). See, however, Page (2008a) for an opposing view. For issues with measures in the multiverse see also Ellis et al. (2004), Page (2008b), Wilczek (2013), and Azhar (2014).

64. Boltzmann (1895), p. 484.

65. Ellis (1979).

66. Nomura (2015), p. 518.

67. Earman (1987), p. 316. Here, one should note that what Earman calls "worlds-within-worlds" is what we call the multiverse.

68. For example, Maynard-Smith and Szathmáry (1996), Pagels (1998), Manson and Thrush (2003), and Draper et al. (2007). See, however, Maudlin (2007) for significantly more nuanced view of the role of metaphysics in contemporary physical science.

69. Ellis et al. (2004), Ellis (2011), and Ellis and Silk (2014).

70. Hannestad and Mersini-Houghton (2005).

71. Aguirre et al. (2007b) and Dahlen (2010).

72. Tegmark (2005).

73. Bousso et al. (2009).

74. Gerig et al. (2013) and Gerig (2014).

75. Gerig (2014), p. 3.

76. Smolin (1992, 1997).

77. Dennett (1995), Smolin (2004), and Gardner and Conlon (2013).

78. Rothman and Ellis (1993), Vilenkin (006b), and Price (2017).

79. Harrison (1995) and Vaas (2012). See also Crane (2010).

80. For a good introduction, see Becker et al. (2006).

81. For history of string theory, see Rickles (2014), as well as the introductory chapters of Becker et al. (2006).

82. Some of the reviews are Gasperini and Veneziano (2003), McAllister and Silverstein (2008), Burgess et al. (2013), Calcagni (2017), and Antoniadis and Cotsakis (2017).

83. Strominger and Vafa (1996).

84. For example, Bowick et al. (1987) and Gasperini and Veneziano (2003).

85. Gasperini and Maharana (2007); Gasperini and Veneziano (2003, 2015).

86. Gasperini and Veneziano (1993, 2003) and Veneziano (1998).

87. McAllister and Silverstein (2008), p. 568.

88. Brustein et al. (1995), Gasperini et al. (1995), Hannestad and Mersini-Houghton (2005), and Mersini-Houghton (2006).

89. For example, Flauger et al. (2014) and Ade (2015).

90. According to Douglas (1956), pp. 122–3.

91. Rees (1969), Dyson (1979), Adams and Laughlin (1997). For an overview of the relevant literature and some of the methodology up to 2003, see Ćirković (2003). For popular introductions, see Davies (1994) and Adams and Laughlin (2016).

92. Examples of very good research in this domain include Caldeira and Kasting (1992), Laughlin et al. (1997), and Schröder and Connon Smith (2008). It is not necessary to emphasize that, insofar as a sector of physical eschatology dealing with the future of the terrestrial biosphere is impacted by present or near-future human activities, the subject matter possesses a very practical value. In this sense, physical eschatology is interesting from the point of view of moral philosophy as well; cf. Leslie (1996) and Kahane (2014).

93. Krauss and Turner (1999).

94. Frautschi (1982); cf. Avelino et al. (2001). See also Chapter 1.
95. Krauss and Scherrer (2007).
96. Van Der Marel et al. (2012).
97. Krauss and Starkman (2000), p. 23.
98. Caldwell (2002). The influence of George Lucas, whose epic Star Wars: Episode I—The Phantom Menace, opened just a few years earlier, remains conjectural.
99. Caldwell et al. (2003).
100. González-Díaz (2003), Štefancic (2004), Bouhmadi-López and Madrid (2005), and Astashenok et al. (2012).
101. Allen et al. (2008).
102. Planck Collaboration (2014).
103. Jiménez et al. (2016).
104. Kragh (1996).
105. Trotta and Starkman (2006), Moffat (2011), Moffat (2014), and McCoy (2015).
106. For example, Cercignani (1998).
107. Popper (1959), preface, p. xviii.

13

Philosophical aspects of cosmology

Chris Smeenk

13.1 Introduction

Throughout the last century, cosmologists have revisited, time and again, basic questions regarding the appropriate aims of cosmology and how best to achieve them. These debates reflect a tension between the cosmologists' ambitions to provide a scientific account of the structure, evolution, and origins of the universe, and methodological limitations imposed by cosmology's unusual object of study. To the extent that cosmology proceeds by observational study of a unique object, the universe-as-a-whole, the understanding of method relevant to other areas of physics—which depend on experimental manipulation, access to a collection of similar systems, or a combination of both—does not straightforwardly apply. Progress in cosmology has depended in part on further insights into scientific inquiry itself, to clarify how cosmology should proceed despite this contrast.

Prior to the twentieth century, many philosophers held, like Kant, that, because they cannot emulate the methods used in other fields to achieve secure knowledge, cosmologists can never reach their ambitious goals. The unambiguous progress in cosmology, in the century following Einstein's introduction of the first relativistic cosmological model in 1917, shows that the skeptics were mistaken. Yet this progress has been accompanied by ongoing debates regarding the proper aims and methods of cosmology. Historically these debates can be roughly divided into two periods. The first period featured intense debates about what qualifies as a satisfactory cosmological theory, at a time when there was not a consensus view and the sparse evidence available left many theoretical avenues open for exploration. Debates among proponents of different views often appealed to philosophical considerations. Advocates of the steady-state theory, in particular, argued that their approach was the only legitimate way to pursue cosmology scientifically. By the late 1960s, many of the avenues explored earlier had reached dead ends. Following the widespread acceptance of the hot big bang model, the focus shifted to questions raised by this model and extensions of it. For example, is a purely scientific account of the origins of the universe possible—and if so, how does a "theory" of origins compare to other scientific theories? Can we explain features of the observed universe if we treat it as part of a much larger "multiverse," and if they succeed, to what extent do such explanations

Smeenk, C., 'Philosophical aspects of cosmology', in *The Oxford Handbook of the History of Modern Cosmology*, edited by Kragh, H. and Longair, M. S. © Oxford University Press 2019.
DOI: 10.1093/oxfordhb/9780198817666.013.13

justify accepting the multiverse? At the end of a century of relativistic cosmology, there is a remarkable juxtaposition between the rich, diverse evidence for cosmology's standard model, and strident debates about such foundational questions.

The debates in both periods reveal implicit views about philosophy of science—regarding what constitutes a properly scientific theory, what counts as compelling evidence, and what kinds of explanatory demands a theory should meet. Cosmologists have occasionally engaged in explicit arguments about these issues. But more often positions on these issues are implicit in cosmologists' arguments and choices regarding what lines of research to pursue. One aim of this chapter is to clearly state positions on four central issues. Providing an adequate characterization of what is at stake in these debates will hopefully set the stage for the more challenging project of assessing and defending various positions. I will not pursue that project here, focusing instead on providing an overview of the debates and their role in the historical development of cosmology.

This chapter is organized thematically rather than chronologically, with each section considering one of four inter-related issues: (i) the uniqueness of the universe and its methodological implications; (ii) the underdetermination of theory by evidence; (iii) theories of the origins of the universe; and (iv) anthropic reasoning and multiverse theories. The main shortcoming of this structure is that it threatens to downplay the historical evolution of the field. This is not a great loss since philosophy of cosmology has primarily been driven by research trends within cosmology itself, rather than following its own internal dynamics. As a result, I have emphasized how conceptions of the appropriate aims of cosmology inquiry have shifted in response to the historical evolution of the field, characterized briefly here and in much more detail in other contributions to this volume. Finally, there are several topics that I do not have the space to discuss. Those seeking a more systematic overview of the philosophy of cosmology should consult, in particular, Ellis (2007).[1]

13.2   Uniqueness of cosmology

Cosmologists have had impressive success in extrapolating local physics, applying the laws governing the fundamental forces to domains increasingly far removed from those where they were originally established. This approach treats cosmology as, in Bondi's (1960) words, "the largest workshop in which we may assemble equipment, the elements of which are entirely composed of terrestrially verified laws of physics" (p. 4). On this view, cosmology proceeds by extrapolating the laws of local physics, discovered by studying sub-systems of the universe, to much larger scales. Einstein's seminal paper (Einstein, 1917a) showed that it was indeed possible to construct a consistent cosmological model based on general relativity. The subsequent development of relativistic cosmological models is, in part, based on assembling further "equipment" crafted in other areas of physics.

This approach has been challenged repeatedly on the grounds that the distinctive subject matter of cosmology requires something other than physics as usual.[2] Suppose that the proper subject matter of cosmology is the whole universe, or more precisely a maximally extended, connected spacetime of which the observed universe is a part.[3] Then cosmology would require a distinctively global view, not limited to the study of large-scale structure of the observed universe. In relativistic cosmology, it is at least possible to define global properties of mathematical models that represent the whole universe. But it is less clear how to develop and justify a physical theory of the whole universe and its global properties, because familiar distinctions from other theories do not apply.

The universe can be neither compared to an ensemble of other similar objects, as in other observational sciences, nor manipulated experimentally. By contrast, the physics of projectile motion, for example, describes a space of dynamically allowed trajectories, including (approximations to) actual trajectories as well as possible trajectories that could be realized with different initial conditions. In cosmology, "The distinction between impossible and possible, but "accidentally" not realized states, becomes absurd when we have to deal with something as fundamentally unique as the universe" (Bondi, 1948, p. 106). On the one hand, a distinction between laws and initial conditions seems at least superfluous, if not absurd, in describing a unique object—such as a single trajectory, or the whole universe. On the other hand, dispensing with theory entirely would leave us without the tools needed to formulate a description. Critics of the standard approach agree that it is a mistake to treat possibility and necessity in cosmology just as in other areas of physics. But there is little consensus regarding what this implies for the aim and structure of cosmological theories.

One line of thought holds that cosmology requires something more than local laws to offer satisfactory explanations. Cosmological theories constructed as assemblages of local physical laws are too liberal: they allow many possible cosmological models, most of which bear little resemblance to the observed universe. Cosmology, regarded as merely an extrapolation of local physics, fails to explain why the actual universe must have the properties that it does. Instead many properties follow from initial conditions rather than the laws; they apparently hold merely contingently, as a result of "accidental realization," not as a matter of necessity. To close this explanatory gap, on this line of thought, a theory dealing with the universe-as-a-whole must introduce laws that rule out many of the models allowed by extrapolations of local physics. Such a theory would be able to explain why the universe has to have the features it does.

Critics of the standard approach have also challenged the idea that it will identify the correct laws.[4] The laws of local physics, they argue, are established based on the study of particular subsystems, typically regarded as isolated, or without external influences— in effect as "island universes." Treating subsystems as completely isolated in this sense is, however, at best an approximation in theories with universal interactions. In the case of gravity, for example, it is impossible to entirely "screen off" the effects of distant bodies. Consider a Newtonian treatment of the solar system. The equations of motion derived for the solar system treated as an island universe differ from those that follow from treating it as a part of a larger system, such as the Milky Way. These differences are negligible, of course, given the distance from the Sun to nearby stars. Treating the system as isolated, even if it is a good approximation, is potentially misleading. The full physical description of a larger system may include interactions among subsystems that is effaced by such an approximation. Using the laws discovered for a subsystem, regarded as an island universe, as the basis for extrapolation excludes such interactions. Of particular concern for cosmology, could local physics be coupled to large-scale properties that vary on cosmological scales? If so, we cannot accurately identify the laws relevant to cosmology via extrapolation from local laws.

Reflections along these lines have inspired very different proposals for how cosmology should proceed. Arguments about the nature of the field were a recurring theme in the roughly first half century of relativistic cosmology. The steady-state theory, in particular, put these methodological concerns front and center. Its proponents argued that certain principles must be accepted in order to make cosmology properly scientific.[5] Bondi (1948) took the inability to draw a contrast between accidental and law-like features to have "obscure" implications, but for Bondi and Gold (1948) it inspired a new theory based on the "perfect cosmological principle." Local laws cannot, in general, be reliably extrapolated because of their possible dependence on large-scale properties of the universe. There is no obstacle to constructing models of the whole universe, however, if the perfect cosmological principle holds: it requires that the universe is stationary in time, with unchanging large scale properties. Bondi and Gold derived several consequences of this principle, leading to what came to be called the steady-state theory.[6] This theory had a substantial impact on research in cosmology for the next 15 years.

Recently the theoretical physicist Lee Smolin has defended a view that is, in some respects, the opposite of the steady-state theory, despite a similar stance on the distinctive nature of cosmology (Smolin, 2015; Unger and Smolin, 2015). Although this line of work has not had the broad impact of the steady-state theory (at least, not yet), it illustrates the persistence of debates regarding how cosmology should be pursued. According to Smolin, the standard approach mistakenly applies what he calls the "Newtonian paradigm," appropriate for the study of subsystems, to the universe as a whole. Insofar as laws apply to subsystems, they are approximate because they leave out interactions with other subsystems; yet if a law encompasses the entire universe, it applies to a single case, and is no longer a law. Smolin proposes to resolve this dilemma by introducing a distinctive understanding of laws: the laws of nature evolve with respect to ("real") global time. Far from being a threat, Smolin sees evolving laws as the key to answering basic questions about why the universe has the properties that it does.

Both proposals demand a great deal of cosmological theories. Smolin endorses a version of Leibniz's principle of sufficient reason: a satisfactory cosmological theory must explain why the universe has the properties it does, and treating our universe as merely one part of a multiverse, discussed in Section 13.6, does not suffice. Advocates of the steady-state theory had similar commitments:

> [We must] find some way of eliminating the need for an initial condition to be specified. Only then will the universe be subject to the rule of theory.... [A cosmological theory]

should imply that the universe contains no accidental features whatsoever. This provides us with a criterion for assessing the validity of rival theories. We believe this criterion to be so compelling that the theory of the universe which best conforms us to it is almost certain to be right. (Sciama, 1959, pp. 166–7).

Sciama would himself soon abandon the steady-state theory, based on the more compelling criterion of empirical adequacy. But Sciama's position that cosmological theories should not leave so much room for contingency, by allowing a variety of possibilities, still retains adherents.

The idea that cosmological theories must provide such explanations has been as controversial as it is persistent. In response to the steady-state theory, the philosopher Milton Munitz criticized such rationalist demands (Munitz, 1952) and offered an alternative account of the kind of explanations that cosmology should pursue. Rather than trying to show why things must be as they are, cosmologists should aim, on Munitz's view, to provide a coherent description of the structure of the observed universe and its evolution, along with an understanding of how the part of the universe we can see fits into the whole universe (Munitz, 1962).[7] More recently, Ellis (2007) has noted that cosmology may pursue explanatory aims like those in historical sciences, such as paleontology and evolutionary biology. Developing a historical reconsctruction has proceeded successfully in these areas, despite limitations similar to those faced by cosmologists: an inability to manipulate or experiment with the system under study, or compare it to an ensemble of similar systems. This success is obtained without subjecting the past to "the rule of theory:" explanations in historical sciences typically depend on assumptions regarding earlier states, but are not rejected as incomplete or unsatisfactory as a result. The demand to go beyond this and give an explanation of why the earlier state had to obtain, as a matter of necessity, reflects the physicists' interest in discovering laws. But this methodological orientation is not necessary if the aim of cosmology is limited, as with the other historical sciences, to developing and justifying a particular historical reconstruction.

In addition to this deflationary response to the demand for explanations, many cosmologists would object to the very first step above: why should cosmology be defined as the science of the whole universe? Rovelli and Vidotto (2014), for example, reject this in no uncertain terms:

> Cosmology is *not* the study of the totality of the things in the universe. It is the study of a few very large-scale degrees of freedom (p. 215, original emphasis).

Many cosmologists draw a similar contrast, declaring that physical cosmology consists in the proposal and observational testing of cosmological models assembled from local physics. This is not a minor terminological point: insofar as cosmology is the science of large scale structures and their dynamical evolution, there is no need to develop a distinctive methodology compared to other physical theories. (As we will see in Section 13.4, however, methodological questions arise again regarding attempts to give a "theory" of origins, or of the initial state.)

The calls for an alternative methodology have stemmed in part from doubts about the viability of a physics-as-usual approach. When the steady-state theory was first proposed, expanding universe models faced significant empirical problems that have since been resolved, such as the "age crisis" and the lack of a plausible account of structure formation.[8] Cosmology's subsequent track record of building successful models based on extrapolating local physics undercuts the appeal of alternative methodologies. As the example of Smolin illustrates, however, it is still possible to regard cosmology as succeeding in spite of confusion about foundational issues; on Smolin's view, cosmology is in a state of crisis, as reflected in widespread acceptance of the multiverse idea (discussed in Section 13.6). The fact that the standard methodology has succeeded reflects an empirical fact: namely, that any interplay between physics at global and local scales is sufficiently weak that it does not hinder the bottom-up construction of successful cosmological models. Persistent failure to develop a satisfactory cosmological theory might lead us to reconsider whether this is the case. It is easy to conceive of worlds in which this approach to cosmology would not be productive. For example, if gravity at solar system scales depended directly on large-scale properties, then we could not straightforwardly apply general relativity to distant regions. It would be challenging to pursue cosmology in such a world, given the difficulty of obtaining evidence regarding such functional relationships. We appear to live instead in a world in which cosmologists can thrive, and build up a successful account of the universe by extrapolating local physics. This deflationary response to the concerns of Bondi, Smolin, and others treats the success of the standard approach as itself contingent on the nature of our universe.

13.3   Underdetermination

Scientists inevitably pursue questions that evidence available at a given time cannot answer. Uncertainty is often transient, resolved with the next step in a research program, but in some cases there are permanent obstacles to obtaining decisive evidence. The extent to which evidence can settle theoretical questions—the "underdetermination of theory by evidence"—is a central theme in philosophy of science. Given its aim to describe the universe and its evolution at large scales, and the limited evidence available, it would not be surprising for transient and permanent underdetermination to be ubiquitous in cosmology.

Cosmologists' expectations regarding whether evidence can settle the central questions of their field have shifted profoundly over the last century. At mid-century cosmology was regarded as closer to mathematics, or even philosophy, than empirical science. Whitrow argued for this position in a debate with Bondi in 1954 (Whitrow and Bondi, 1954); Bondi, by contrast, was more optimistic that empirical evidence would resolve foundational disputes in the field, and emphasized that the steady-state theory at least made definite predictions. A decade later, Trautman expressed a common view in the epilogue of his Brandeis lectures on general relativity: "…it is not worthwhile to work in theoretical cosmology at the present time. I think it would be better to sit and wait for the astronomers to get more data on the motion and distribution of distant galaxies" (Trautman, 1965). Trautman's expectation that the wait would not be long was correct. Source counts derived from the 4C survey, available shortly thereafter, showed an increase in sources at high redshift incompatible with the sharp predictions made by the steady-state theory emphasized by Bondi. The serendipitous discovery of the cosmic microwave background (CMB) by Penzias and Wilson (1965) was even more significant. It encouraged cosmologists to take extrapolations of the big bang models to early times, and the application of nuclear physics to that regime, seriously (as emphasized by Weinberg, 1977). Furthermore, it provided a target for precision measurements that depends on relatively well understood physics rather than relying on galaxies and other complicated systems as standard candles. Precision measurements of the CMB have been combined with other lines of evidence, leading to consistent estimations of the parameter values in the standard model of cosmology. Cosmologists now have sufficient confidence in this model to assert the existence of new types of matter, and other additions to the standard model of particle physics, based on their cosmological effects.

This shift reflects an effective response to underdetermination, despite severe limitations on the available evidence. Here I will assess different aspects of underdetermination, to elucidate the reasons for this shift and the current limits to what evidence can establish.

13.3.1   Horizons

A particularly clear observational limit follows from the finitude of the speed of light. Physical signals moving at or below the speed of light can reach us only from a limited region of spacetime. If we represent our location as point p in a relativistic spacetime, the in principle accessible region consists of the past light cone (and its interior) at p, also called the causal past, $\mathcal{J}^-(p)$.[9] The limits to observational access are often described instead in terms of the existence of horizons, of different types (Rindler, 1956).[10] Horizons measure the maximum distance from which a signal emitted at a specified time t_e can reach an observer. The visual horizon takes t_e to be the decoupling time, prior to which the universe is opaque to electromagnetic radiation; whereas the particle horizon is defined as the limit $t_e \to 0$. Objects at positions separated by distances greater than the particle horizon have non-overlapping past light cones. Event horizons, by contrast, are defined as the boundary of the causal past in the limit as $t_0 \to \infty$—the limit of what could be seen by an "immortal" observer.

Discussions of horizons go back to the advent of relativistic cosmology, when Einstein took the existence of an event horizon in de Sitter's solution as a reason to doubt its physical viability. Rindler's work was prompted by a similar debate regarding the steady-state theory (Whitrow, 1953), which represented spacetime geometry with part of the de Sitter solution. The existence of horizons in a cosmological model has various counterintuitive consequences, but does not provide sufficient reason to reject a model outright. There is no *a priori* reason to demand that the universe can, in principle, be exhaustively surveyed by observers within it. But to what extent does the existence of horizons limit cosmologists' ability to answer central questions?

McCrea (1960, 1962) addressed this issue explicitly, arguing that observations in cosmology have an inherent uncertainty that increases with redshift (z). The argument starts from the legitimate observation that, except in very unusual cases, information available at a point p cannot be used to predict physical properties at a distinct point p'. Because the past light cones of any distinct points, $\mathcal{J}^-(p)$ and $\mathcal{J}^-(p')$, do not completely overlap, an observer at p will not be able to determine some of the physical processes that can affect the physical state at p'.[11] Any prediction thus requires some assumptions regarding what lies beyond $\mathcal{J}^-(p)$—for example, that there is no "source-free radiation" propagating to p', or that the physical properties of $\mathcal{J}^-(p')$ should resemble those of $\mathcal{J}^-(p)$ in some specific respects. General relativity does not impose such constraints, so they must come from some other source. As we will see shortly, the cosmological principle could play this role, but this is a substantial further assumption—that McCrea did not accept. He gave an obscure argument that the uncertainty associated with predictions increases with redshift (linearly with z), because the amount of relevant information decreases. The consequences McCrea claimed to find for the central cosmological question of his day are clear: this uncertainty allegedly undercut efforts to discriminate between the steady-state theory and evolutionary cosmological models.

This pessimistic conclusion was overstated, but the characterization of the epistemic predicament faced by cosmologists is apt: to what extent can we answer cosmological questions based on observations confined to $\mathcal{J}^-(p)$? Considering an idealized data set can help to distinguish between limitations that arise from the finite speed of light and from other sources. (This is not to deny that various other sources of uncertainty are far more important to observational cosmology.) To that end, we will imagine cosmologists inhabiting a universe filled with "standard objects" whose properties are fully understood (including luminosity, mass, shape, and so on, and the evolution of these properties with cosmic time), and which are targeted by sophisticated, well-funded observational programs. This idealization eliminates the uncertainty associated with "acts of faith" required in real observational cosmology, in using galaxies and other complicated systems as standard objects (see Longair's discussion in Chapter 10 of this volume). This approach further assumes that classical general relativity holds, but does not impose other constraints on the background spacetime geometry. Given access to this data set, what questions could cosmologists then answer?

Ellis et al. (1985) proved that an appropriate idealized data set of this kind is sufficient, granting that general relativity holds, to determine the spacetime geometry and distribution of matter on the past light cone $C^-(p)$ (out to the maximum redshift at which this ideal set can be observed).[12] For the ideal data set, observations can directly determine the area (or luminosity) distance of the sources, and the distortion of distant images determines lensing effects. These observations directly constrain the spacetime geometry of the past light cone $C^-(p)$. Although the ideal objects can be used effectively as "tracers" to determine the spacetime geometry, further substantive modeling assumptions are required to make claims regarding the distribution of matter and energy. (In particular, modeling assumptions are needed to resolve various degeneracies, and disambiguate the effects of different kinds of matter and energy, including dark matter and dark energy, on the ideal data set.) Observers do not have access to anything like the ideal data set, obviously, and in practice there are substantial obstacles to determining the spacetime geometry in this fashion because of uncertainty regarding the nature of the standard objects and their evolution with cosmic time.[13]

There are exotic cases in which cosmologists would be able to determine *global* properties of spacetime, and not only the spacetime geometry of $\mathscr{J}^-(p)$, using such an ideal data set.[14] In some cosmological models, well-situated observers could see the entire universe. These models are baroque variations on Einstein's closed universe: space at a given cosmic time is finite and without boundary, with an intricate topology. A common feature of these models is that observers can see multiple "ghost images" of a single object, given that light can reach an observer along many different paths through the spacetime. If the maximum spatial length in every direction is shorter than the visual horizon, observers would be able to "see around the universe," and (in principle) fully determine its spacetime geometry.[15]

Aside from these exotic cases, the finite speed of light poses a fundamental obstacle to determining global properties of spacetime empirically. To what extent does a single observer's window on the universe, or even a collection of such views, fix the overall structure of a cosmological model? This question, initially posed by physicists studying the causal structure of relativistic spacetimes, was taken up by the philosophers Glymour and Malament in the late 1970s, and revisited by Manchak in the 2000s. The Malament–Manchak formulation of the problem starts by defining what it means for two spacetimes to be "observationally indistinguishable," and then asks what global properties must be shared by indistinguishable spacetimes. Two spacetimes are indistinguishable if and only if for every observer p in the first spacetime, there is a "copy" of their $I^-(p)$ in the second spacetime.[16] Given this definition, there is a clear procedure for settling the general question: for a given global property $\mathcal{G}$, and an initial spacetime that has the property, is it possible to construct an indistinguishable counterpart that lacks $\mathcal{G}$?[17] Malament (1977) and Manchak (2009) give a series of clever constructions establishing that this is so for nearly all global properties. Since, by construction, the full data available to any observer is compatible with either counterpart, the spacetime geometry of $I^-(p)$ does not suffice to establish that $\mathcal{G}$ holds. The only properties that are guaranteed to hold for an indistinguishable counterpart are those that can be established based on the chronological past of a single point.

These results do not pose a challenge as stark as that suggested by McCrea: they do not show that observational evidence will necessarily fall short in answering central theoretical questions. Although cosmologists often speculate about the global structure of the universe on enormous scales, well beyond our past light cone, these claims play no direct role in evidential reasoning in cosmology. What does play a crucial role is the spacetime geometry of $\mathscr{J}^-(p)$ itself, which we will turn to in the next section. Furthermore, the choice at hand is between cosmological *models* rather than *theories*, since we have assumed throughout that classical general relativity holds.

The extent to which these results support, nonetheless, an interesting claim of "model underdetermination" depends on one's view of inductive inference. More straightforward cases of induction can be described in similar terms: evidence that a particular regularity obtained in the past is compatible with a model according to which it continues to hold, but it is also compatible with a "counterpart" in which the regularity fails to hold at some point in the future. Only adherents of a very strict empiricism hold that we have no reason to prefer a model in which observed uniformities can be projected to new cases. Accounts of induction, or ampliative inference, seek to provide some justification for choosing among models that are all logically compatible with the available evidence. The challenge in this case is to clarify what justifies accepting one spacetime over its indistinguishable counterparts (Earman, 2009; Norton, 2011; Butterfield, 2014).

13.3.2   The cosmological principle

The discussion above contrasts with a more typical way of using evidence, namely to choose an optimal cosmological model from a restricted set of solutions. Rather than considering the full space of solutions to general relativity, one restricts attention to a space of symmetric cosmological models (or models sharing some other global properties). Sandage's observational program (Sandage, 1961a, 1970), for example, was devoted to determining the values of two parameters sufficient to determine the "best-fit" expanding universe model. (These parameters were Hubble's constant H_0 and the deceleration parameter q_0, which fix a model provided that $\Lambda = 0$.) The expanding universe models discovered by Friedman and Lemaître (hereafter, FL models) are *isotropic* (there are no geometrically preferred spatial directions) and *homogeneous* (at a given moment of cosmic time every spatial point "looks the same").[18] Due to this symmetry, the models have a particularly simple structure: spacetime can be decomposed into three dimensional hypersurfaces of constant cosmic time $\Sigma(t)$ (topologically, $\Sigma \times \mathbb{R}$), and the field equations of general relativity reduce to a pair of ordinary differential equations. These equations fix the behavior of the scale factor $R(t)$ given the equations of state for the different types of matter present. (The scale factor represents the spatial distance in Σ between observers moving along time-like geodesics, as a function of cosmic time.) Generalizing from Sandage's two numbers, the contribution of various types of matter can be characterized in terms of dimensionless density parameters. On this approach, observations are used to determine the "best fit" model and fix the relevant parameter values.

Even this drastically oversimplified sketch of model choice suffices to illustrate the significance of the symmetry principle assumed at the outset. The empiricist approach considered in the previous section did not impose any global constraints on the space of models. Imposing homogeneity and isotropy, by contrast, is extremely restrictive and enables evidence from a limited region to determine global properties. An observer can take their observations to reveal properties of, not just the spacetime region within the past light cone, but other regions related by the symmetry. They are then entitled to claims about global structure. Other symmetry principles, or stipulations that lead to a limited space of viable models, also allow local-to-global inferences. The Copernican principle is sometimes formulated as the requirement that we do not occupy a "privileged position" in the universe. (Excessive modesty may, however, lead us astray, if we fail to acknowledge that there are various ways in which our location is privileged in the sense of being suitable for life—a topic we will return to in Section 13.5) More generally, we can require that there are no "special locations" in the universe: no point p is distinguished from other points q by any spacetime symmetries, or lack thereof.

There have been four quite distinctive ways that cosmologists have tried to justify imposing such a global symmetry, or restricting consideration to a subset of possible models on similar grounds (see also Beisbart, 2009). Milne (1933)'s influential discussion introduced the "cosmological principle" as the requirement that the universe must appear to be the same to all observers. Homogeneity and isotropy were sufficient to secure this form of equality. Milne regarded the cosmological principle as the most important axiom of a deductive system leading to a distinctive theory, kinematical relativity, rather than a claim in need of empirical support. Requiring that the principle would hold addressed the lack of predictivity for which he faulted relativistic cosmology. Bondi and Gold were influenced by this line of thought, and took a similar stance toward their proposed modification, the perfect cosmological principle. Yet after the fall from favor of the steady-state theory, few cosmologists have treated the cosmological principle as an *a priori* principle needed to formulate a cosmological theory.

In early discussions of the FL models, it was far more common to treat their high degree of symmetry as a useful simplification that should not be taken too seriously. Tolman emphasized the mathematical value of the simple FL models, but did not take observations to provide strong justification for accepting the cosmological principle. Tolman concluded with a warning that:[19]

> [W]e must be careful not to substitute the comfortable certainties of some simple mathematical model in place of the great complexities of the actual universe (Tolman, 1934b, p. 487).

The qualitative match between the FL models and observations, such as Hubble's measurements of the redshift-distance relation, encouraged the thought that the study of these models was not a purely mathematical exercise. Yet cosmologists were wary of accepting features of these models that depended on these symmetries holding exactly. Einstein and Tolman, for example, regarded early singularity theorems (which showed that curvature blows up as $t \to 0$ in the FL models) as artifacts of physically unreasonable idealizations.[20] The same reasons would undermine trust in extrapolations of the FL models to the early universe.

The situation changed dramatically with the discovery of the CMB in 1965: its isotropy provided much stronger evidence that the FL models apply, even to the very early universe and at the largest possible scales. But many cosmologists were puzzled rather than exhilarated by the success of the FL models; if anything, the models seemed to work *too* well. Why should the universe be so symmetric? The isotropy of the universe went from being a simplifying assumption to a target for physical explanations; as Misner (1968) put it,

> [The isotropy of the CMB] surely deserves a better explanation than is provided by the postulate that the Universe, from the beginning, was remarkably symmetric (p. 431).

Misner's "chaotic cosmology" program aimed to explain isotropy as the consequence of dynamical effects in the early universe (specifically, damping of anisotropies due to neutrino viscosity).[21] Although this particular proposal was ultimately unsuccessful, it suggested a new "philosophy for big bang cosmology:"

> The universe must start with a big bang (or bangs) and, almost independently of any special initial conditions, it must have a particular chemical composition, it must exhibit a Hubble expansion, it must be isotropic if it is homogeneous, and we do expect it to be homogeneous (McCrea, 1970, p. 22).

(We should expect the universe to be homogeneous due to the Copernican principle.) Misner and McCrea both clearly prefer a theory that is "indifferent" to the exact features of the initial state, with dynamics that drive a large range of possible initial states to converge to properties compatible with the observed universe (such as the specific chemical composition expected as a result of big bang nucleosynthesis, and isotropy as a consequence of Misner's proposal). This conception of what constitutes a successful early universe theory was more durable than the specific proposals McCrea discussed (see also McMullin, 1993). A decade later, Guth (1981) made a persuasive case that a different dynamical mechanism in the early universe, driving a stage of inflationary expansion, should be explored further because it promised to be successful in just this sense.

The final historical shift in the status of the cosmological principle has occurred within the last two decades: it has been scrutinized more carefully due to its essential role in precision cosmology. Insofar as the universe is well-approximated by an FL model, observations of the redshift-distance relation for standard objects can be used to constrain the density parameters characterizing different types of matter. In particular, observations of this type using supernovae (type Ia) as a standard candle led to the remarkable discovery that the expansion of the universe is accelerating (see Chapter 11 for further discussion). For this to be true in an FL-model (that is, for $\ddot{R}(t) > 0$), there must be some source that has approximately the same dynamical effects as a positive cosmological constant (often referred to as "dark energy"). This inference depends crucially not just on the physics linking $R(t)$ to different types of matter and energy, but on the accuracy of the FL-models as a description of spacetime geometry at the relevant scales. The accuracy of the FL models has shifted from being a target of physical explanation to an essential ingredient of an observational program.

One way of clarifying the status of the FL models is to carry out something like the observational cosmology program described above, with observations of the CMB as a crucial addition to the "ideal data set". Observers can use the CMB to put upper bounds on the isotropy of the universe from one point (once their proper motion with respect to the CMB is taken into account). If the Copernican principle holds, then the observed isotropy holds generally rather than as a special property of this point—suggesting that the FL geometry holds. Cosmologists can now do much better than invoking the Copernican principle. There are several generalizations of the seminal result due to Ehlers et al. (1968), which establish what different types of observations—such as that of an isotropic radiation field, by an observer moving along a geodesic—imply regarding spacetime geometry. Furthermore, observations of the CMB can provide indirect evidence regarding whether the universe appears to be isotropic from the vantage point of distant galaxies (Goodman, 1995). One type of indirect evidence takes advantage of the fact that the CMB has a black-body spectrum. Roughly put, scattering of CMB photons due to the Sunyaev–Zeldovich effect will lead to distortions in the spectrum if the CMB is anisotropic to the scatterer; if the CMB is isotropic, the spectrum will retain its black-body shape with a shift in temperature. Granted that the CMB has a black-body spectrum at decoupling, an observer who observes that the CMB has a black-body spectrum can then infer that it is also isotropic with respect to distant regions where the CMB photons are scattered. This line of argument, in conjunction with some other types of indirect evidence, can be used to make a direct empirical case that the universe is well-approximated by an FL model (Clarkson, 2012).

This raises a further question regarding the status of the FL models: how does their uniformity at large scales relate to the manifest lack of uniformity at smaller scales? This question was posed in the early days of relativistic cosmology, with the construction of models representing the gravitational field of stars (or other local systems) embedded within an expanding universe model (Einstein and Straus, 1945) or combined in a lattice (Lindquist and Wheeler, 1957). The huge density contrasts at galactic scales and smaller are certainly not "small perturbations" away from an FL model. What, then, does it mean to say that these models "approximate" the actual universe? If the FL models are taken to describe an "averaged" or smoothed out matter density, at some suitably large scale, there is a natural further question: do the FL dynamics correctly describe the dynamical evolution of this "smoothed out" distribution? Ellis (1984) and Ellis and Stoeger (1987) argued that smoothing out a general relativistic solution does not also lead to a solution—that is, the averaged spacetime geometry and stress-energy tensor do not satisfy the field equations. The field equations can be satisfied if an "effective stress energy" tensor representing the back-reaction effect of the inhomogeneities is included. The degree to which the FL models are a good approximation can then be quantified in terms of the size of these additional terms.[22]

13.3.3 Status of the ΛCDM model

A distinctive underdetermination challenge arises in considering the "best fit" model of some phenomena, based on a background physical theory. To what extent does success—that is, finding parameter values in the model consistent with a body of data—justify confidence in the accuracy of the theory? Perhaps the model succeeds due to its flexibility, by introducing new degrees of freedom that can be carefully tuned to reproduce the data without accurately representing the relevant physics. How can we discriminate among models that are all compatible with the data, but based on different underlying physics? (Such models may give similar descriptions of the phenomena being studied, but have different implications for other phenomena.) Here I will briefly assess the extent to which the success of the ΛCDM model, in fitting a wide array of cosmological observations with a small number of parameters, meets this underdetermination challenge.

One promising line of response to this challenge, formulated in general terms, is to demand, first, that there are multiple, independent ways of determining the parameters of the model, and, second, that the theory can be consistently applied in light of systematic improvements in measurement precision. The first demand exploits the theory's unification of diverse phenomena to "overdetermine" the parameters appearing in the model (see, e.g., Norton, 2000). In his defense of atomism, for example, Perrin (1923) emphasized the agreement among several strikingly diverse ways of determining Avogadro's number N, drawing on phenomena ranging from Brownian motion to the sky's color. The strength of this reply depends on the extent to which the phenomena probe the underlying theoretical assumptions in distinct ways, and whether there is an alternative hypothesis that also accounts for the agreement. As the number of methods used to determine a parameter increases, the probability that agreement among them can be attributed to chance, or to systematic errors, decreases. Turning to the second point, do increasingly precise measurements lead to refinements of the underlying model, or to anomalies? In several historical cases, a theory has guided the development of models that do meet steadily improving standards of precision, without setting aside core principles. Furthermore, the refined models often incorporate further details that can be independently checked (see, e.g. Smith, 2014). Success in these two respects provides a strong response to the underdetermination challenge. If the underlying physics were false, it would be a coincidence for the multiple ways of measuring model parameters to agree, and it would be unlikely that increasingly precise measurements would lead to further discoveries rather than anomalies. Any rival theory should be expected to agree with a theory that is successful in this sense, at least as an approximation within the relevant domain.

Turning back to cosmology, the current standard model of cosmology, the ΛCDM model, fits an impressive array of cosmological data with a small number of parameters. These include the density parameters characterizing the abundance of different types of matter, each of which can be measured by a variety of different types of observations.[23] There are two different questions that this model raises regarding the physics used in its construction: to what extent does the success of the model support, first, extrapolations of well-tested local physics, and second, novel physics tested only through cosmological applications? The model assumes the validity of extrapolating general relativity, for example, to length scales roughly 14 orders of magnitude greater than those where the theory is subject to high precision tests. There are also several aspects of the model based on novel physics that cannot be independently tested through terrestrial experiments.

The case in favor of the standard model has been strengthened considerably through precision measurements of the cosmological parameters. Many of the methods of measuring these parameters at the time Sandage formulated his observational program required a variety of astrophysical assumptions, regarding, for example, the use of galaxies as standard objects to determine spacetime geometry. Systematic uncertainties of more recent measurements of these parameters are easier to control, insofar as they rely on very well-understood physics. The CMB, in particular, provides powerful constraints on cosmological parameters due to our confidence in our physical description of recombination and of the subsequent propagation of the CMB photons. The consistent determination of these parameter values from many different types of observations supports an overdetermination argument much like Perrin's. In Perrin's case, accepting the atomic hypothesis implied that many different phenomena indirectly measure the scale of the atomic constituents of matter; any measurement incompatible with the others would cast doubt on the hypothesis. Similarly, the ΛCDM model leads to systematic connections between a diverse array of observable features of the universe. Peebles (2005), for example, enumerates 13 distinct ways of measuring the overall matter density Ω_0 at large scales: several distinct techniques based on using galaxies as mass tracers; weak lensing; cluster mass functions; the mass fluctuation power function; and so on. Accepting the ΛCDM model comes with an obligation to resolve any discrepancies among the various measurements of the basic parameters appearing in the model. While there are still open questions regarding discrepancies in some of these parameter measurements, the overall agreement among different measurements of the parameters appearing in the ΛCDM model provides strong evidence in its favor. (See Longair's Chapter 10 in this volume for an overview of measurements of the cosmological parameters.)

The strength of this case depends in part on whether the agreement among different measurements would be expected to hold even if the theories used in constructing the model were false. Perrin, for example, argued that the relationships between N and observable magnitudes, derived in the variety of cases he considered based on the atomic hypothesis, do not hold according to competing theories. Obviously, the evidence is not as decisive if several alternative theories imply that similar relationships hold. The strength of the evidence thus reflects how distinctive the theory under consideration is, as compared to the space of competing theories. This assessment is more challenging for aspects of the standard model that employ novel physics, due to the greater uncertainty regarding the space of viable alternatives.

Ellis' (2007) idea of a "physics horizon" helps to clarify the status of different parts of the standard model. As with other horizons, the physics horizon marks the limit of what is accessible; in this case, it is the physical regime accessible to terrestrial experiments and non-cosmological observations. This is not nearly as sharply defined as the horizons discussed above, as it reflects an assessment of what experiments or observations are feasible, leading to a rough division in terms of relevant energy or length scales. The qualifier "non-cosmological" is also admittedly vague, but it is intended to allow for observations, such as those of the solar system, that do not depend on a background cosmological model. For a specific theory it is possible to be more precise; for example, Baker et al. (2015) describe the regimes of parameter space characterizing gravitational theories that can be probed by different types of observations (solar system tests, gravitational waves, etc.). Several components of the standard model—such as dark matter, dark energy, and inflation—currently lie beyond the physics horizon in this sense. Although each proposal is based on plausible extensions of well established physical theories, currently the only way to evaluate these ideas is through their implications for cosmology. Insofar as they extend beyond the physics horizon, making a strong case in favor of these proposals based on multiple independent measurements, or developing more detailed models in response to systematic improvements in precision measurements, is particularly challenging.

The physics community has taken on this challenge because there are few other opportunities to test several of the most intriguing aspects of new ideas in fundamental physics. The Soviet cosmologist Yakov Zeldovich called the early universe the "poor man's accelerator," because relatively cheap observations of the early universe may reveal features of high-energy physics well beyond the reach of even the most expensive earth-bound accelerators. For many aspects of fundamental physics, in particular quantum gravity, cosmology provides the best testing ground for competing ideas. To what extent can cosmological observations replace other kinds of tests, such as accelerator experiments, in providing evidence for theories?

A brief discussion of three different cases of new physics incorporated in the ΛCDM model illustrates the challenge to providing evidence of comparable strength to that achieved in other areas of physics. For the last several decades, cosmological models have included a substantial contribution to the total matter density from non-baryonic dark matter. Dark matter was first proposed to account for the dynamical behavior of galaxy clusters and galaxies, which could not be explained with only the observed luminous matter. Dark matter plays a crucial role in accounts of structure formation, as it provides the scaffolding necessary for baryonic matter to clump, without conflicting with the uniformity of the CMB.[24] These inferences to the existence of dark matter, as well as many others, rely on gravitational physics. Obviously, it is possible that these observations reveal a flaw in our understanding of gravity rather than the presence of a new type of matter. There have been sustained efforts to clarify what form a modified gravity theory would have to take to account all of the relevant observations entirely (or mostly) without dark matter. In this context direct experimental detection of dark matter would make a decisive contribution. Several research groups aim to find dark matter particles through direct interactions with a solid-state detector, mediated by the weak force. A positive outcome of such experiments would provide evidence of the existence of dark matter that does not depend upon gravitational theory.[25]

This kind of decisive evidence is precisely what is in short supply for theories that extend beyond the "physics horizon." There is reason to hope that this situation is only temporary in the case of dark matter, but the prospects of providing independent evidence regarding the nature of "dark energy" are much worse. Cosmological models began to incorporate "dark energy" (in the sense of a non-zero cosmological constant Λ) in the early 1990s, as an essential ingredient in accounts of structure formation. By the mid-1990s, there was growing evidence in favor of a value $\Omega_\Lambda \approx 0.7$, from several different lines of evidence (Ostriker and Steinhardt, 1995). The case in favor of a non-zero Λ was strengthened considerably, and gained much more widespread attention, due to observations of the redshift-distance relation of supernovae published in 1998 (see, for example, Frieman et al., 2008, for a summary of this line of work). These observations indicated that, granting the applicability of an FL model, the expansion of the universe is accelerating. In an FL model, accelerated expansion must be driven by a contribution that has the same dynamical effects as a non-zero Λ. Just as in the case of dark matter, there is of course the possibility that the various observations taken to motivate the introduction of dark energy instead indicate either a mistake in our description of gravity, or that the FL models do not apply. Both possibilities have been the focus of sustained research efforts (see, for example, Uzan, 2010). Unlike dark matter, however, the properties of dark energy ensure that any attempt at non-cosmological detection would be futile: the energy density is so small, and uniform, that any local experimental study of its properties is practically impossible.

Turning to the third case, inflationary cosmology originally promised a powerful unification of particle physics and cosmology. The earliest inflationary models explored the consequences of specific scalar fields introduced in particle physics (the Higgs field proposed in studies of the strong interactions). Yet theory soon shifted to treating the scalar field responsible for inflation as the "inflaton" field, leaving its relationship to particle physics unresolved, and the promise of unification unfulfilled (Zinkernagel, 2002). If the properties of the inflaton field are unconstrained, inflationary cosmology is extremely flexible: it is possible to reverse engineer an inflationary model that yields any chosen evolutionary history of the early universe.[26] Specific models of inflation, insofar as they specify the features of the field or fields driving inflation and its initial state, do have predictive content. In principle, cosmological observations could determine some of the properties of the inflaton field and so select among them (Martin et al., 2013). This could in principle then have implications for a variety of other experiments or observations. In practice, however, the features of the inflaton field in most viable models of inflation guarantee that it cannot be detected in other experimentally accessible regimes. The predictive content of inflation is further weakened if it leads to an inflationary multiverse, as discussed below.

The physics horizon poses a challenge because one particularly powerful type of evidence—direct experimental detection or observation, with no dependence on cosmological assumptions—is unavailable for the physics relevant in the very early universe, or at extremely large length scales. Yet this does not imply that competing theories, such as dark matter vs. modified gravity, should be given equal credence. The case in favor of dark matter draws on diverse phenomena, and it has been difficult to produce a compelling modified theory of gravity, consistent with general relativity, that captures the full range of phenomena as an alternative to dark matter. Cosmology typically demands a more intricate assessment of background assumptions, and the degree of independence of different tests, in evaluating proposed extensions of local physics.

13.4   Origins of the universe

One profound shift marks a clear contrast between the first half-century of relativistic and the second. The idea that, at the largest observable scales, the universe does not evolve over time was no longer viable as of roughly 1965, as various observations effectively ruled out the steady-state theory. New theoretical arguments showed that the singularity known to be present in the FL models could not be dismissed as an artifact of idealizations, absent in more realistic models. As a result of these developments, cosmologists had to take seriously the prospect that time has a beginning, and to ask whether it is possible to formulate a scientific theory governing the "origin of the universe," and if so what form such a theory might take.

Philosophers have been wary of proposing theories of origins in the aftermath of the incisive critiques of cosmological arguments for the existence of God due to Hume and Kant. Hume, in particular, argued that an understanding of causal relationships in ordinary circumstances does not illuminate the "causes" relevant to the origin of the universe. Rephrasing Hume, should we expect a "theory of origins" to have anything like the structure of other physical theories? What is the appropriate explanatory target for such a theory, and how does the explanation proceed? The concerns raised in Section 13.2 seem particularly pressing here. Some cosmologists have sought to avoid an "origin" entirely. Hoyle (1975), for example, effectively demands that a good cosmological theory does not include a singular origin. Others regard the singularity as indicating the limits of applicability of classical general relativity, rather than an actual singularity; a theory of quantum gravity may lead to a fundamentally different picture. After a brief review of the arguments in favor of taking an initial singularity seriously, I will outline and assess the options for a theory of origins that have been explored.

13.4.1 Singularities

Contemporary cosmology at least has a clear target for a theory of origins: the best-fit FL model describes the universe as having expanded and evolved over ≈ 13.7 billion years. This "age of the universe" is the total proper time elapsed that would be measured by a clock moving along the worldline of a fundamental observer (moving along a geodesic), from the "origin" until now. Singularities are signaled by the existence of inextendible geodesics with bounded length. Extrapolating backwards from the present, an inextendible geodesic reaches an "edge" beyond which it cannot be extended; the finite age of the universe is the temporal distance to this "edge." This does *not* imply that there is a "first moment," just as there can be an open interval of the real number line of a specified length without a "first point."

Theorems establishing the existence of a singularity in the FL models (e.g., Tolman, 1934b) follow from the Raychaudhuri equation, which describes the evolution of a set of nearby worldlines, such as those making up a small ball of dust. It takes on the following simple form in the FL models due to their symmetry (Ellis, 1971a):

$$3\frac{\ddot{R}}{R} = -4\pi G(\rho + 3p) + \Lambda, \tag{13.1}$$

where R is the scale factor and $\ddot{R}$ is its second derivative with respect to cosmic time. A small ball of dust (with $R(t)$ measuring the distance between nearby trajectories) changes volume as a function of time, in response to the mass-energy. (More generally, there can also be a volume-preserving distortion (shear) and rotation of the ball.) Given that the universe is currently expanding, (13.1) implies that the expansion began at some finite time in the past. As this "big bang" is approached, the energy density and curvature increase without bound provided $\rho + p > 0$ (which guarantees that $\rho \to \infty$ as $R \to 0$). As $R(t)$ decreases, the energy density and pressure both increase, and they both appear with the same sign on the right hand side of (13.1)—which illustrates the instability of gravitational collapse.

Obviously the symmetries of the FL models do not hold exactly in the actual universe, and it was essential to see whether the presence of singularities was robust to relaxing these idealizations. The singularity theorems proved in the 1960s (see, in particular, Hawking and Ellis, 1973) established that singularities still arise given much weaker, and more physically well-motivated, assumptions. The singularity theorems apply to a much broader class of models, many of which lack a uniquely defined "cosmic time." In these models there is not a cosmic time with a natural physical meaning, as in the FL models. The theorems still establish that the universe is finite to the past, in the sense that there is a maximum length for inextendible geodesics.

The singularity theorems plausibly apply to the observed universe, within the domain of applicability of general relativity. There are various related theorems differing in detail, but one common ingredient is an assumption that there is sufficient matter and energy present to guarantee that our past light cone refocuses.[27] The energy density of the CMB alone is sufficient to justify this assumption. The theorems also require an energy condition: a restriction on the types of matter present in the model, guaranteeing that gravity leads to focusing of nearby geodesics. (In Section 13.1, this is the case if $\rho > 0$ and $\Lambda = 0$; it is possible to avoid a singularity with a non-zero cosmological constant, for example, since it appears with the opposite sign as ordinary matter, counteracting this focusing effect.) Finally, the theorems require assumptions regarding the global causal structure of the model. In light of the discussion of underdetermination above, justifying such global claims based on the observed universe requires acceptance of a general principle, such as the Copernican principle.

There are two limitations regarding what we can learn about the origins of the universe based on the singularity theorems. First, although these results establish the existence of an initial singularity, they do not reveal much about its structure. The spacetime structure near a "generic" initial singularity has not yet been fully characterized. Partial results have been established for restricted classes of solutions; for example, numerical simulations and a number of theorems support the BKL (Belinsky–Khalatnikov–Lifshitz) conjecture, which holds that isotropic, inhomogeneous models exhibit a complicated form of chaotic, oscillatory behavior.[28] Second, classical general relativity does not include quantum effects, which are expected to be relevant as the singularity is approached. Crucial assumptions of the singularity theorems may not hold once quantum effects are taken into account. The standard energy conditions do not hold for quantum fields, which can have negative energy densities. This opens up the possibility that a model including quantum fields may exhibit a "bounce." More fundamentally, general relativity's classical spacetime description may fail to approximate the description provided by a full theory of quantum gravity. There are several accounts of the early universe that avoid the initial singularity due to quantum gravity effects.

13.4.2   Fine-tuning and the initial state

The singularity theorems establish that, insofar as classical general relativity applies, cosmological models must be supplemented by a theory of origins. Although there is not a "first moment," such a theory might be expected to account for the structure of the "initial state" understood, roughly, as specified at the boundary of the domain of applicability of general relativity. (The precise limits of an existing theory are often clarified once the successor theory is in hand; given uncertainty about quantum gravity, the appropriate initial state is not well understood.) The features of this initial state are fixed by extrapolating backwards from current observations.

The understanding of the initial state that came into focus in the decade following the discovery of the CMB was extremely puzzling. Several of its properties were identified as possible targets of explanation for theories of the early universe, including (but not limited to) the following:

- *Matter–antimatter asymmetry*: Evidence accumulated through the 1970s that the local asymmetry extends to cosmological scales; what explains why the initial state was baryon-dominated?

- *Uniformity*: The isotropy of the CMB indicate that distant regions of the universe have uniform physical properties. This is puzzling because the FL models have a finite particle horizon distance, much smaller than the scales at which we observe the CMB.[29] As a result, the distant regions were apparently in some sort of "pre-established harmony"—sharing the same physical features from the initial state onwards, without physical interactions. Misner (1968) argued that postulating such a symmetry did not explain it.

- *Flatness*: An FL model close to the "flat" model, with nearly critical density at some specified early time is driven rapidly away from critical density under FL dynamics if $\Lambda = 0$ and $\rho + 3p > 0$. Given later observations, the initial state has to be *very* close to the flat model (or, equivalently, *very* close to critical density, $\Omega = 1$) at very early times.[30]

- *Perturbations*: The standard model requires seeds for the formation of structures such as galaxies. These take the form of density perturbations that are coherent on large scales and have a specific amplitude, constrained by observations. It is challenging to explain both properties dynamically. In the standard FL models, the perturbations have to be coherent on scales much larger than the Hubble radius at early times.[31]

On a more phenomenological approach, the gravitational degrees of freedom of the initial state could simply be chosen to fit with later observations, but many proposed "theories of initial conditions" aim to account for these features based on new physical principles. The theory of inflation, in particular, aims to explain the last three features.

These features of the initial state were taken to be appropriate explanatory targets because they seem to reflect "fine-tuning." The existence of such fine-tuning is taken to be problematic, due to a puzzling conflict between two ways of thinking about "contingent" aspects of physical theories, such as the specific values of fundamental constants. The various coupling constants appearing in the standard model of particle physics are evaluated experimentally, and cannot be derived from first principles. Similarly, various aspects of standard cosmological models follow from properties apparently set arbitrarily in the choice of an initial state. The densities of different kinds of matter, the spectrum of initial perturbations, and the current value of Hubble's constant, for example, can be determined via measurements, but there is no expectation that they can be derived from the underlying theory.

Yet other observed features of the universe, such as the existence of life, seem to depend extremely sensitively on these contingent features. There is a small literature devoted to assessing the impact of changing the values of the coupling constants in the Standard Model, or of the parameters defining the ΛCDM model.[32] These results suggest that something very close to the current set of values for the fundamental constant are necessary to support the existence of complex structures at a variety of scales, a plausible precondition for the existence of life.

Features of our theories that appear entirely contingent, from the point of view of physics, are necessary to account for the complexity of the observed universe and the very possibility of life. Shouldn't something as fundamental as the complexity of the universe be explained by the *laws* or *basic principles* of the theory, and not left to brute facts regarding the values of various constants? The unease develops into serious discomfort if the specific values of the constants are taken to be extremely unlikely: how could the values of all these constants be *just right*, by sheer coincidence?

In many familiar cases, our past experience is a good guide to when an apparent coincidence calls for further explanation. As Hume emphasized, however, intuitive assessments from everyday life of whether a given event is likely, or requires a further explanation, do not extend to cosmology. Recent formulations of fine-tuning arguments often introduce probabilistic considerations. The constants are "fine-tuned," meaning that the observed values are "improbable" in some sense. Introducing a well-defined probability over the constants would provide a response to Hume: rather than extrapolating our intuitions, we would be drawing on the formal machinery of our physical theories to identify fine-tuning. Promising though this line of argument may be, there is not an obvious way to define physical probabilities over the values of different constants, or over other features of the laws. There is nothing like the structure used to justify physical probabilities in other contexts, such as equilibrium statistical mechanics.[33]

One response to fine-tuning essentially rejects these arguments as so much mysterymongering, perhaps following Hume's lead.[34] What exactly is the problem? This question can be raised at a general or more specific level. Quite generally, the various features that are allegedly finely-tuned have to take on *some* value or other, and without a well-justified assignment of probabilities there is nothing demanding a further explanation. (Even if probabilities can be justifiably introduced, why should we demand that *all* "low probability" events or outcomes be explained?) A different line of thought to the same conclusion holds that fine-tuning problems reveal that dynamical explanations have limited scope. A full explanation of the regularities of the observed world must also appeal to initial and boundary conditions, possibly including features of the initial state. Specific fine-tuning problems have also been criticized for a failure to acknowledge salient aspects of the physics. The statement of the flatness problem, for example, highlights an aspect of FL dynamics (roughly, that all FL models approach the flat model as $R \to 0$) and then claims it is problematic. Rather than highlighting a distinctive type of fine-tuning, this seems to boil down—as with the horizon problem—to reflecting surprise that the FL models work as well as they do.[35]

Three other responses take fine-tuning as identifying a legitimate problem that needs to be addressed:

- *Designer*: Newton famously argued, for example, that the stability of the solar system provides evidence of providential design. For the hypothesized Designer to be supported by fine-tuning evidence, we require some way of specifying what kind of universe the Designer is likely to create; only such a specific design hypothesis, based in some theory of the nature of the Designer, can offer an explanation of fine-tuning.

- *New physics*: The fine-tuning can be eliminated by modifying physical theory in a variety of ways: altering the dynamical laws, introducing new constraints on the space of physical possibilities (or possible values of the constants of nature), etc.

- *Multiverse*: Fine-tuning is explained as a result of selection, from among a large space of possible universes (or multiverse).

The second response is the topic of the next section, whereas the multiverse is discussed in Section 13.6.

13.4.3   Theories of initial conditions

There are three main approaches to theories of the initial state, all of which have been pursued by cosmologists since the late 1960s in different forms. Expectations for what a theory of initial conditions should achieve have been shaped, in particular, by inflationary cosmology. Inflation provided a natural account of three of the otherwise puzzling features of the initial state emphasized in the previous section. Prior to inflation, these features were regarded as "enigmas" (Dicke and Peebles, 1979), but after inflation, accounting for these features has served as an eligibility requirement for any proposed theory of the early universe.

The first approach aims to reduce dependence on special initial conditions by introducing a phase of attractor dynamics. This phase of dynamical evolution "washes away" the traces of earlier states, in the sense that a probability distribution assigned over initial states converges towards an equilibrium distribution. Misner (1968) introduced a version of this approach (his "chaotic cosmology program"), proposing that free-streaming neutrinos could isotropize an initially anisotropic state. Inflationary cosmology was initially motivated by a similar idea: a "generic" or "random" initial state at the Planck time would be expected to be "chaotic," far from a flat FL model. During an inflationary stage, arbitrary initial states are claimed to converge towards a state with the three features described above.

The second approach regards the initial state as extremely special rather than generic. Penrose, in particular, has argued that the initial state must be very special to explain time's arrow; the usual approaches fail to take seriously the fact that gravitational degrees of freedom are not excited in the early universe like the others (Penrose, 2016). Penrose (1979) treats the second law as arising from a law-like constraint on the initial state of the universe, requiring that it has low entropy. Rather than introducing a subsequent stage of dynamical evolution that erases the imprint of the initial state, we should aim to formulate a "theory of initial conditions" that accounts for its special features. Penrose's conjecture is that the Weyl curvature tensor approaches zero as the initial singularity is approached; his hypothesis is explicitly time asymmetric, and implies that the early universe approaches an FL solution (but there is no mechanism to account for the perturbations needed to seed structure formation). In connection with the discussion in Section 13.2, this proposal introduces a law applicable only to the universe's initial state, and the questions about how to test such a global law have some force.

A third approach rejects the framework accepted by the other two proposals, and regards the "initial state" as a misnomer. This rejection can take two forms: either the initial state is instead a "branch point" where our pocket universe separated off, in some sense, from a larger multiverse, or it is regarded as the end point of a previous contraction phase as well as the effective starting point of the observed expanding phase. Both proposals then aim to explain features of the (misnamed) initial state based on this embedding in a larger spacetime. The main challenge facing cyclic universe proposals is in reconciling proposed explanations with a physical understanding of how the singularity is resolved in a theory of quantum gravity.[36] I will return to questions regarding the explanatory power of multiverse proposals in Section 13.6.

A dynamical approach, even if it is successful in describing a phase of the universe's evolution, arguably does not offer a complete solution to the problem of initial conditions: it collapses into one of the other two approaches. For example, an inflationary stage can only begin in a region of spacetime if the inflaton field and the geometry are uniform over a sufficiently large region, such that the stress-energy tensor is dominated by the potential term (implying that the derivative terms are small) and the gravitational entropy is small. There are other model-dependent constraints on the initial state of the inflaton field. One way to respond is to adopt Penrose's point of view, namely that this reflects the need to choose a special initial state, or to derive one from a previous expansion phase. The majority of those working in inflationary cosmology instead appeal to the third approach: rather than treating inflation as an addition to standard big-bang evolution in a single universe, we should treat the observed universe as part of a multiverse, discussed below.[37]

Finally, it is worth highlighting a number of conceptual pitfalls regarding what would count as an adequate "explanation" of the origins of the universe. Take the "initial state" defined at the earliest time when extrapolations based on the FL models and classical general relativity can be trusted. This "initial state" would then be the output of an earlier phase of evolution governed by a theory of quantum gravity. Although the fundamental concepts of such a theory remain obscure, the form of explanation is at least familiar:

the aim would be to show how a "classical spacetime" with certain properties emerges from a regime described in terms of different concepts. Ultimate questions about the origin of the universe must then be reformulated in terms of the concepts of quantum gravity. Cosmologists sometimes pursue, however, a more ambitious target: to explain the creation of the universe "from nothing" (see, for example, Isham and Butterfield, 2000, for an overview). The target is the true initial state, not just the boundary of applicability of classical general relativity. The origins are supposedly then explained without positing an earlier phase of evolution; supposedly this can be achieved, for example, by treating the origin of the universe as a fluctuation away from a vacuum state. Yet obviously a vacuum state is not nothing: it exists in a spacetime, and has a variety of non-trivial properties. The proposed explanation still takes the form of showing how earlier physical conditions evolve into something like what we observe; it does not directly address the metaphysical question of why there is something rather than nothing.

13.5   Anthropic reasoning

Scientific theories are usually expected to provide an objective description of a world that exists independently of our presence. Cosmology explains the structure and evolution of the universe at enormous scales. Surely our presence is entirely irrelevant to what transpires at such scales, and in the distant past? Against the backdrop of these plausible expectations, cosmologists' willingness to explain features of the universe based on our presence is particularly striking. There has been no shortage of philosophers who reject this conception of the aim of science and, like Kant, give the human subject a more central and active role. In cosmology the status of "anthropic principles," which state that our nature as observers should be taken into account in evaluating evidence, or in explaining various features of the universe, have been a reliable source of controversy. Some cosmologists dismiss discussions of the "a word" out of hand, while for others progress in some areas of cosmology requires revising basic principles of scientific methodology to handle anthropic reasoning properly.

These quite different responses can be partially explained by the fact that discussions of anthropics typically blur together several distinct ideas, leading to a confusing muddle. At least in some cases, I expect that those arguing for and against "anthropics" are talking past one another.[38] Here I will isolate and evaluate three of the central proposals in these debates: first, that selection effects need to be taken into account in evaluating evidence in cosmology; second, that cosmological theories can be assessed in terms of "anthropic predictions;" and third, that apparent fine-tuning of some feature X can be explained by showing that X is a necessary condition for our existence. The first two proposals will be the focus of this section, and I will turn to the third in connection with the multiverse in the next section.

The importance of selection effects was prominently illustrated by Dicke's (1961) reply to a speculative cosmological proposal (Dirac, 1937). Dirac noted that the age of the universe expressed in terms of fundamental constants in atomic physics is an extremely large number (roughly 10^{39}), which coincides with other large, dimensionless numbers defined in terms of fundamental constants. He proposed that the large numbers vary to maintain this order of magnitude agreement, taking the agreement to reflect some underlying law rather than a mere coincidence. This implies (among other things) that the gravitational "constant" G varies as a function of cosmic time. Dicke (1961) pointed out a quite different reason for Dirac's coincidence to hold. If the coincidence were found to hold at a randomly chosen cosmic time, then we would have some evidence in favor of Dirac's hypothesis. But our observations take place at a quite specific cosmic epoch. Creatures like us, made of carbon produced in an earlier generation of red giants, sustained by a main sequence star, can only exist within a restricted interval of cosmic times. Dicke argued that Dirac's coincidence holds for observations made within this interval, regardless of whether Dirac's speculative hypothesis holds. Eddington gave a characteristically vivid illustration of this mistake. It is the same mistake as that made by a fisherman who concludes that there are no small fish in a pond, based on the day's catch—while forgetting that small fish can wriggle through the gaps in his net.

Given this example of "anthropic" reasoning, it is hard to see what would generate controversy (see also Earman, 1987; Roush, 2003). Any account of evidential reasoning must acknowledge the importance of selections effects and take them into consideration appropriately. Recognizing a previously unnoticed selection effect often leads to re-evaluating some body of evidence. There are important questions regarding how to handle different types of selection effects, but these are hardly confined to cosmology (see, for example, Neal, 2006; Trotta, 2008). Perhaps the controversy is limited to whether this type of argument qualifies as "anthropic," since a detailed characterization of what is required for human beings (or "observers") plays no role.

Recent discussions of anthropic reasoning clearly take it to involve more than careful attention to selection effects. Weinberg (2007), for example, celebrates the acceptance of anthropic reasoning as progress in how theories are evaluated, comparable to the progress achieved in twentieth-century physics due to the appreciation of symmetries. Weinberg's defense focuses on a successful case of what I will call an "anthropic prediction." Such predictions lead to a probability distribution over the value of one or more fundamental parameters, representing the expected value to be measured by a "typical observer." The probative value of such "predictions," and how they fit into a more general account of methodology, are matters of ongoing controversy.

The most famous example of such an anthropic prediction is Weinberg's (1987) prediction that Λ should have a small, non-zero value.[39] One part of Weinberg's argument is similar to Dicke's: he argued that there are anthropic bounds on Λ, due to its impact on structure formation. The existence of large, gravitationally bound structures such as galaxies is only possible if Λ falls within certain bounds. Weinberg went a step further than Dicke, and considered what value of Λ a "typical observer" should see. He assumed that observers occupy different locations within a multiverse, and that the value of Λ varies across different regions. (Note that this is all Weinberg needs to assume regarding the multiverse; he mentions several different proposals for generating a multiverse for which this assumption plausibly holds.) Weinberg further argues that the prior probability assigned to different values of Λ should be uniform within the anthropic bounds. Typical observers should expect to see a value close to the mean of the anthropic bounds, leading to Weinberg's prediction for Λ.

There are two immediate questions regarding this proposal:[40] how is the class of "observers" defined, and what justifies taking ourselves to be "typical" members of this class? This is an instance of the well-known "reference class" problem in probability theory. The assignment of probabilities to events requires specifying how they are grouped together, or choosing a set of "reference classes".[41] Obviously, what is typical with respect to one reference class will not be typical with respect to another (compare, for example, "conscious observers" with "carbon-based life").

The principle of indifference is usually taken to imply that we should assign equal probabilities to outcomes of a probabilistic process if we have no reasons to favor some of the outcomes. Essential to Weinberg's argument is an appeal to the principle of indifference, applied to a class of observers.[42] We should calculate what we expect to observe, that is, as if we are a "random choice" among all possible observers.[43] As a general point, information regarding how some evidence claim E is obtained is essential in determining what we can infer from it; if E is obtained as a "random sample," we are entitled to a number of further conclusions. What justifies the further assumption that we are random?

The indifference principle has been thoroughly criticized as a justification for probability in other contexts; what justifies its use in this case? Bostrom (2002) argues that indifference-style reasoning is necessary to respond to the problem of "freak observers." As Bostrom formulates it, the problem is that in an infinite universe, *any* observation O is true for *some* observer (even if only for an observer who has fluctuated into existence from the vacuum). His response is that we should evaluate theories based not on the claim that *some observer* sees O, but on an indexical claim: that is, *we* make the observation O. He assumes that we are a "random" choice among the class of possible observers. If we grant the assumption, then we can assign low probability to the observations of the "freak" observers, and recover the evidential value of O. Setting aside any qualms about the details of this argument, at best it establishes what is needed in order to make sense of anthropic predictions in an infinite universe. But this kind of conditional claim will do little to persuade a skeptic who doubts the value of these arguments, and the appeal to indifference.

Skeptics have also claimed that the arguments employed in making anthropic predictions lead to absurd consequences when applied to other cases (Norton, 2010). The Doomsday Argument, for example, claims to reach a striking conclusion about the future of the human species without any empirical input (see, for example, Leslie, 1992; Gott, 1993; Bostrom, 2002). Suppose that we are "typical" humans, in the sense of having a birth rank that is randomly selected among the collection of all humans that have ever lived. We should then expect that there are nearly as many humans before and after us in overall birth rank. For this to be true, given current rates of population growth, there must be a catastrophic drop in the human population ("Doomsday") in the near future. Some commentators are willing to bite the bullet, and accept that purely probabilistic reasoning has led to such a substantive prediction with almost no empirical input. Those who wish to avoid this conclusion, rather than endorsing it, need to provide a more refined version of the principles governing such inferences.

13.6   Multiverse

Collins and Hawking (1973) (in)famously answered the question posed in the title of their paper ("Why is the Universe isotropic?"), as follows: "because we are here." Readers who had worked through their proofs of theorems regarding the growth of anisotropies in homogeneous solutions may have been surprised (or frustrated) with this answer. The appeal to anthropic considerations was motivated by these results, however: although anisotropies grow in most of the models they considered, in one case there is an open set of initial data such that the models tend to become increasingly isotropic. If galaxies would only be expected to exist in this subset of models, then we should not be surprised to observe that the universe is isotropic.

More schematically, this style of argument has three basic elements. The first is to postulate a what is now usually called a "multiverse:" an ensemble of universes, over which some property of interest varies. Second, the "anthropic subset" of this ensemble is picked out based on a property taken to be a necessary condition for the existence of creatures like us. This is often a proxy, such as the existence of galaxies, that in principle could be determined by the details specified in defining the ensemble. Finally, the most contentious element is the assignment of probabilities to elements of this ensemble— either via a principle of indifference, or some other means. Considering a space of "possible models" is not unusual in physics. But it is essential to this type of argument that the ensemble is taken as actually existing, rather than merely possible. Weinberg's prediction for the value of Λ described above has this form.

The amount of ink devoted to discussions of the multiverse has increased substantially recently, because (some) string theorists and inflationary cosmologists regard the creation of a multiverse as an inevitable outcome of these theories.[44] By a multiverse, I mean (roughly) a single connected spacetime consisting of several quasi-isolated "pocket universes" whose properties vary in some specified manner. Within inflationary cosmology the same mechanism that produces a uniform, homogeneous universe on scales on the order of the Hubble radius, leads to a dramatically different global structure of the universe. Inflation is said to be "generically eternal" in the sense that inflationary expansion continues in different regions of the universe, constantly creating bubbles such as our own universe, in which inflation is followed by reheating and a much slower expansion. The individual bubbles are effectively causally isolated from other bubbles. The second line of thought relates to the proliferation of vacuums in string theory. Many string theorists now expect that there will be a vast landscape of vacuums, with no way to fulfill the original hope of finding a unique compactification of extra dimensions to yield low-energy physics.

Both of these developments suggest treating the low-energy physics of the observed universe as partially fixed by parochial contingencies related to the history of a particular pocket universe. Other regions of the multiverse may have drastically different low-energy physics because, for example, the inflaton field tunneled into a local minima with different properties. Here my main focus will be on a philosophical issue that is relatively independent of the details of implementation: in what sense does the multiverse offer satisfying explanations?

But, first, what do we mean by a "multiverse" in this setting?[45] These lines of thought lead to a multiverse with two important features. First, it consists of quasi-isolated pocket universes, and second, there is significant variation from one pocket universe to another. There are other ideas of a multiverse, such as that employed by Collins and Hawking (1973): an ensemble of distinct possible worlds, each in its own right a topologically connected, maximal spacetime, completely isolated from other elements of the ensemble. But in contemporary cosmology, the pocket universes are all taken to be effectively causally isolated parts of a single, topologically connected spacetime—the multiverse. Such regions also occur in some cosmological spacetimes in classical general relativity. In de Sitter spacetime, for example, there are inextendible timelike geodesics γ_1, γ_2 such that $\mathcal{J}^-(\gamma_1)$ does not intersect $\mathcal{J}^-(\gamma_2)$. In cases like this the definition of "effectively causally isolated" can be cashed out in terms of relativistic causal structure.

The example of pocket universes within de Sitter spacetime lacks the second feature, variation from one pocket universe to another. Multiverse proponents have discussed various types of variation: in the constants appearing in the standard models of cosmology and particle physics, to the laws themselves. Within the context of eternal inflation or the string theory landscape, what were previously regarded as "constants" may instead be fixed by the dynamics. For example, Λ is often treated as the consequence of the vacuum energy of a scalar field displaced from the minimum of its effective potential. The variation of Λ throughout the multiverse may then result from the scalar field settling into different minima. Greater diversity is suggested by the string theory landscape, according to which the details of how extra dimensions are compactified and stabilized are reflected in different low-energy physics.

In the multiverse some laws will be demoted from universal to parochial regularities. But presumably there are still universal laws that govern the mechanism that generates pocket universes. This mechanism for generating a multiverse with varying features may be a direct consequence of an aspect of a theory that is independently well-tested. Rather than treating the nature of the ensemble as speculative or conjectural, one might then have a sufficiently clear view of the multiverse to calculate probability distributions of different observables, for example. In this case, there is a direct reply to multiverse critics who object that the idea is "unscientific" because it is "untestable:" other regions of the multiverse would then have much the same status as other unobservable entities proposed by empirically successful theories.[46] Unfortunately for fans of the multiverse, the current state of affairs does not seem so straightforward. Although multiverse proposals are motivated by trends in fundamental physics, the detailed accounts of how the multiverse arises are typically beyond theoretical control. As long as this is the case, there is a risk that the claimed multiverse explanations are just-so stories where the mechanism of generating the multiverse is contrived to do the job. This strikes me as a legitimate worry regarding current multiverse proposals, but I will set this aside for the sake of discussion.

Suppose, then, that we are given a multiverse theory with an independently motivated dynamical account of the mechanism churning out pocket universes. What explanatory questions might this theory answer, and what is the relevance of the existence of the multiverse itself to its answers?[47] Here we can distinguish between two different kinds of questions. First, should *we* be surprised to measure a value of a particular parameter X (such as Λ) to fall within a particular range? Our surprise ought to be mitigated by a discussion of anthropic bounds on X, revealing various unsuspected connections between our presence and the range of allowed values for the parameter in question. But, as with Dicke's approach discussed above, this explanation can be taken to demystify the value of X without also providing evidence for a multiverse. The value of this discussion lies in tracing the connections between, for example, the timescale needed to produce carbon in the universe or the constraints on expansion rate imposed by the need to form galaxies. The existence of a multiverse is irrelevant to this line of reasoning.

A second question pertains to X, without reference to our observation of it: why does the value of X fall within some range in a particular pocket universe? The answer to this question offered by a multiverse theory will apparently depend on contingent details regarding the mechanism that produced the pocket universe. This explanation will be *historical* in the sense that the observed values of the parameter will ultimately be traced back to the mechanism that produced the pocket universe.[48] It may be surprising that various features of the universe are given this type of explanation rather than following as necessary consequences of fundamental laws. However, the success of historical explanations does not support the claim that other pocket universes must exist. Analogously, the success of historical explanations in evolutionary biology does not imply the existence of other worlds where pandas have more elegant thumbs.

Acknowledgments

It is a pleasure to thank Malcolm Longair for encouragement, and to thank Malcolm and Helge Kragh for comments on an earlier draft. I have benefitted from discussions with many people on the topics discussed here, but am particularly grateful to George Ellis and the members of the Western-UCI research group (JB Manchak, Jim Weatherall, Kevin Kadowaki, Mike Schneider; Yann Benétreau-Dupin, Craig Fox, Marie Gueguen, Marc Holman, Melissa Jacquart, and Adam Koberinski). Work on this paper was supported by a grant from the John Templeton Foundation. The views expressed here are those of the author and are not necessarily endorsed by the John Templeton Foundation.

....

NOTES

1. See also Ellis and Silk (2014), and the recent edited collection (Chamcham et al., 2017). My treatment of several topics below draws on my own earlier survey (Smeenk, 2013), as well as an encyclopedia entry co-authored with Ellis (Ellis and Smeenk, 2017).

2. I call the construction of cosmological models based on local physics the "standard approach" below, because it has been the dominant view for the last century, especially from the mid-1960s onward. But it is rarely spelled out explicitly, in part because it is regarded as simply applying an uncontroversial methodology from physics to cosmology. For discussions of this line of argument, in addition to the papers cited below, see McCrea (1962), Bergmann (1970), Pauri (1991), Ellis (2003), and Earman (2009).

3. Relativistic cosmological models represent the universe as a four-dimensional manifold equipped with a spacetime metric, which specifies the geometry. A manifold is connected if it cannot be broken into two (or more) non-overlapping, non-empty open sets, and a spacetime is maximal if it cannot be isometrically embedded as a proper subset into another spacetime. Hence the "whole universe" is a spacetime manifold, including a representation of the observable universe, which is as large as possible.

4. See, in particular, Scheibe (1991). Debates regarding Mach's principle are the most important case of this line of thought. Einstein at one point took it to be a foundational principle of general relativity, named to acknowledge the influence of Mach's proposal that the inertial properties of matter could be attributed to interactions with distant matter. The exact formulation and status of the principle has been a subject of ongoing dispute (see, for example, Sciama, 1953; Ellis and Sciama, 1972; Barbour and Pfister, 1995). For those who accept Mach's principle, the idea that the effects of distant stars can be treated as negligible in the treatment of a local system leads to a profound misunderstanding of basic dynamical concepts.

5. The steady-state theory was partly inspired by the earlier "deductive" approach to cosmology defended by E. A. Milne, which spurred the first round of methodological debates in cosmology (see Gale, 2017).

6. Hoyle (1948), by contrast, presented essentially the same theory as a solution to Einstein's field equations with the addition of a "creation" field (similar to a cosmological constant).

7. Several philosophers contributed to these debates, which also focused on the legitimacy of postulating the creation of matter, as the steady-state theory had to do to reconcile the constancy of the large-scale properties of the universe (such as average matter density) with expansion. See Kragh (1996), and his chapter in this volume, for further discussion and references. Balashov (1994, 2002) gives a detailed philosophical assessment of the steady-state theory.

8. Early estimates of Hubble's constant implied, at least in the simplest evolving models, that the universe is younger than some of its constituents. Steady state advocates criticized the account of structure formation in the evolving models as simply assuming a spectrum of initial perturbations that had all the right properties to seed later structures (Burbidge et al., 1963).

9. $\mathcal{J}^-(p)$ is the set of points q such that there is a future-directed curve from q to p with tangent vectors that are timelike or null. Note that an observer at p can make *some* claims regarding physics outside of $\mathcal{J}^-(p)$; as Ellis and Sciama (1972) emphasize, constraint equations in theories such as electromagnetism

restrict field values outside this region. Furthermore, although the entire interior of $\mathcal{J}^-(p)$ is in principle accessible, most of the data that we in fact use reaches us along the light cone (electromagnetic radiation from distant sources), or from regions close to the Earth's past worldline (in the form of geological "records") (Ellis, 1980).

10. The standard definitions of horizons rely on further structure present in cosmological models, such as cosmic time and a class of fundamental observers (moving along geodesics), whereas the causal past of a point is well-defined in any spacetime. These definitions trace back to Rindler's seminal paper, see also Ellis and Rothman (1993) for an accessible overview. Following Rindler, the particle horizon is a surface in a three-dimensional hypersurface of constant cosmic time t_0 dividing the fundamental particles which could have been observed by t_0, at a particular point, from those which could not, given some time of emission t_e. Subsequently others have introduced distinctions between different kinds of horizon reflecting choices of t_e.

11. The future domain of dependence of a region S in a relativistic spacetime, $D^+(S)$, is the region of spacetime to the future of S from which data on S, in conjunction with the field equations, determine a unique solution. The essential point is that in general relativity, it is typically the case that $D^+(\mathcal{J}^-(p)) = \mathcal{J}^-(p)$. There are some exotic spacetimes in which $D^+(\mathcal{J}^-(p))$ is larger, even encompassing the whole spacetime (Geroch, 1977).

12. The light cone is the boundary of the causal past; the field equations of general relativity can be used to determine the spacetime geometry in the causal past $\mathcal{J}^-(p)$ from this data set. More precisely, if the ideal data set extends back to redshift z^*, they can be used to determine the geometry of the lightcone up to that distance along with the past Cauchy development of the relevant part of $C^-(p)$. The result is also limited to the part of the lightcone which is free of caustics.

13. This observational cosmology program was initiated by Kristian and Sachs (1966). Subsequent work by Ellis, Stoeger, Nel, and various collaborators has considered what is feasible with more realistic data sets, and with the addition of weak assumptions regarding background spacetime geometry.

14. A *local* spacetime property is a property such that for any pair of locally isometric spacetimes, they either both have the property or neither does. The property of being a solution to the field equations of general relativity is a local property in this sense. *Global* properties, by contrast, can vary between locally isometric spacetimes. There are a hierarchy of conditions that characterize the global causal structure of spacetimes. See Manchak (2013) for further discussion and references.

15. See Ellis (1971b) for early work on these models, and Lachieze-Rey and Luminet (1995) for a more recent review.

16. A "copy" is an isometric embedding of $I^-(p)$ into the second manifold. This relation is not symmetric: there is no requirement that there is a "copy" for every point in the second manifold in the first. This is the weakest of several definitions of "observational indistinguishability" introduced by Malament (1977), but it is arguably most appropriate as a way of characterizing the cosmologists' situation.

The definition is formulated in terms of $I^-(p)$, the chronological past (rather than the causal past): the set of points q such that there is a future-directed, timelike curve from q to p. These are always topologically open sets, which makes the proofs and constructions more straightforward than if $\mathcal{J}^-(p)$ were used.

17. Malament vividly describes this as a "clothesline construction": the counterpart spacetime includes a collection of "pieces" $\{I^-(p)\}$ strung together, like clothes hung out to dry. Manchak (2009) establishes the generality of this construction.

18. These models are also sometimes attributed to Robertson and Walker (or some combination of the four), due to their contributions in clarifying their geometrical properties. See Realdi's Chapter 3 in this volume for a detailed assessment of their contributions.

19. This was a common refrain in discussions of the expanding universe models. McVittie (1965), for example, argues for a similar position three decades later.

20. Tolman studied the approach to a "singular state" in a closed FL solution in some detail, and he concluded that the idealizations of the model fail to hold as the singular state is approached (Tolman, 1934b, pp. 438–9, 484–6). On the history of the singularity theorems and Einstein's views, see Earman (1999) and Earman and Eisenstaedt (1999).

21. Misner recognized a clear obstacle to dynamical explanations of isotropy, the horizon problem (Misner, 1969). Points on the surface of last scattering from which we receive CMB photons at very close to the same temperature have non-overlapping past light cones. This apparently precludes a dynamical explanation of why the distant regions have the same temperature (and other physical properties).

22. This question is still the subject of active debates; see, for example, Buchert and Räsänen (2012) and Green and Wald (2014).

23. See Beringer et al. (2012), for example, for a review of constraints on these parameters. Typically 5–10 fundamental parameters are used to determine the best fit to a given data set, although there is some variation in how these are defined. (Specific models often require a variety of further "nuisance parameters.")

24. The CMB indicates that baryonic matter was very smooth at the time of decoupling because it was strongly coupled to radiation. Dark matter decouples from radiation earlier than baryonic matter, and can be much lumpier at the time the CMB is emitted; these lumps then generate perturbations in baryonic matter. The total amount of baryonic matter is also constrained by big-bang nucleosynthesis, since the light element abundances are sensitive to the value of Ω_b.

25. At the time of writing, there are no generally accepted candidates for successful detection of dark matter particles; instead, ongoing experimental searches have ruled out parts of the parameter space of candidate particles.

26. See Ellis and Madsen (1991) for the general procedure, and Lidsey et al. (1997) for its use as regarding reconstructing the inflaton potential. This is a version of what relativists call "Synge's G-method": given some spacetime geometry, it is always possible to define a stress-energy tensor, namely whatever tensor is required for this spacetime geometry to be a solution of the field equations.

27. Refocusing leads to the "onion" shape of the past light cone: it reaches a maximum radius at some finite time, and decreases at earlier times (Ellis, 1971a). See Ellis and Rothman (1993) for further discussion.

28. Penrose has emphasized this point; see Chapter 3 of Penrose (2016) for a recent discussion.

29. Particle horizons are discussed in Section 13.3.1. For a radiation-dominated FL model, the expression for horizon distance d_h is finite; the horizon distance at decoupling corresponds to an angular separation of $\approx 1°$ on the surface of last scattering, so observations of the CMB comprise many distinct, non-interacting regions if the FL models correctly describe causal structure.

30. It follows from the FL dynamics that $\frac{|\Omega-1|}{\Omega} \propto R^{3\gamma-2}(t)$. $\gamma > 2/3$ if the strong energy condition holds, and in that case an initial value of Ω not equal to 1 is driven rapidly away from 1. Observational constraints on $\Omega(t_0)$ can be extrapolated back to a constraint on the total energy density of the Planck time, namely $|\Omega(t_p) - 1| \leq 10^{-59}$.

31. The Hubble radius $d(H_0)$ is defined in terms of the instantaneous expansion rate $\dot{R}(t)$, by contrast with the particle horizon distance d_h, which depends upon the expansion history since the start of the universe. For radiation or matter-dominated solutions, the two quantities have the same order of magnitude.

32. See, for example, Carr (2007) for a recent entry point into these discussions, or Barrow and Tipler (1986) for an earlier comprehensive discussion.

33. See McGrew et al. (2001) and Colyvan et al. (2005) for challenges to justifying probabilities in this case, and Manson (2009) for a response and general discussion of fine-tuning.

34. See Callender (2004) and Price (2004) for a recent formulation of opposing views in this debate.

35. For recent discussions of fine-tuning problems in cosmology, see Carroll (2014) and Holman (2018).

36. Cyclic cosmologies have been pursued since the early days of relativistic cosmology. Recently, Steinhardt, Turok, and several co-authors have proposed a string-theory motivated cyclic cosmology (see Lehners, 2008, for a review), and Penrose has advocated a cyclic cosmology as well (Penrose, 2016). See Kragh (2011) for a more detailed history of the various proposals that have been pursued.

37. See, in particular, Chapter 9 of Kragh (2011) for a thorough historical discussion of these debates, as well as Ćirković's Chapter 12 in this volume.

38. There have been efforts to clarify what is at stake by formulating several distinct "anthropic principles" (see, in particular, Barrow and Tipler, 1986), as refinements of terminology originally introduced by Carter (1974). I will not use the terminology of the "weak" vs. "strong" anthropic principles (and etc.) below, for ease of exposition and because the standard definitions do not draw the correct contrast between evidential (related to selection effects) and explanatory considerations.

39. Although I will not pursue the topic here, Weinberg's argument is a special case that avoids some of the questions that arise in giving a general account of "anthropic prediction." For example, the argument concerns variation of a single parameter, whereas the general case requires considering the variation of several parameters

concurrently. See Aguirre (2007) for an account of the challenges and complications involved in carrying out anthropic predictions for a variety of parameters, and Starkman and Trotta (2006) for further problems with these methods.

40. There are more subtle questions, regarding whether, for example, planets might also have formed much earlier in dwarf galaxies (as emphasized by Abraham Loeb), and whether it is appropriate to consider varying only one parameter (as emphasized by Anthony Aguirre). See Kragh (2011), pp. 238–1 for further discussion.

41. More precisely, the assignment of probabilities depends on algebraic structure— the event algebra—defined on the sample space. Many different event algebras, corresponding to different ways of grouping elements of the sample space, can be assigned over the same sample space.

42. This is closely related to Vilenkin's (1995) "Principle of Mediocrity," and Bostrom's (2002) "Self-Sampling Assumption" (although he eventually argues for a principle applied to "observer-moments" rather than observers).

43. As Aguirre et al. (2007a) note, it is possible to choose some other object to conditionalize on in a Weinberg-style argument; but this leads to similar problems regarding the choice of reference class and appeal to indifference.

44. See Kragh (2011) for a careful discussion of the historical roots of this conception, and a contrast with other multiverse proposals (such as "many-world" interpretations of quantum theory, and Tegmark's proposals) with distinctive motivations.

45. See also Tegmark (2009) for an influential classification of four different types or levels of the multiverse.

46. This line of argument has appeared numerous times in the literature; see, for example, Livio and Rees (2005) for a clear formulation.

47. Here I am indebted to discussions with John Earman; see also Earman (2009).

48. The explanation may also be path-dependent in the sense of depending not just on an initial state, but on various stochastic processes leading to the formation of the pocket universe.

References

Aad, G., Abajyan, T., Abbott, B., et al. (2012). Observation of a new particle in the search for the standard model Higgs boson with the ATLAS detector at the LHC, *Physics Letters B*, **716**, 1–29.

Abazajian, K. N., Calabrese, E., Cooray, A., et al. (2011). Cosmological and astrophysical neutrino mass measurements, *Astroparticle Physics*, **35**, 177–84.

Abbe, C. (1867). On the distribution of nebulae in space, *Monthly Notices of the Royal Astronomical Society*, **27**, 257–64.

Abbott, B. P., Abbott, R., Abbott, T. D., et al. (2016). Observation of gravitational waves from a binary black hole merger, *Physical Review Letters*, **116**, 061102.

Abbott, B. P., Abbott, R., Abbott, T. D., et al. (2017). GW170817: observation of gravitational waves from a binary neutron star inspiral, *Physical Review Letters*, **119**, 161101.

Abell, G. O. (1958). The distribution of rich clusters of galaxies, *Astrophysical Journal Supplement*, **3**, 221–88.

Abell, G. O. (1962). Membership of clusters of galaxies, in *Problems of Extragalactic Research*, ed. McVittie, G. C., pp. 213–38. New York: Macmillan.

Abell, G. O., Corwin, H. G., Jr., and Olowin, R. P. (1989). A catalogue of rich clusters of galaxies, *Astrophysical Journal Supplement*, **70**, 1–138.

Abraham, R. G., Tanvir, N. R., Santiago, B., et al. (1996). Galaxy morphology to $I = 25$ mag in the Hubble Deep Field, *Monthly Notices of the Royal Astronomical Society*, **279**, L47–L52.

Adams, F. C. (2008). Stars in other universes: stellar structure with different fundamental constants, *Journal of Cosmology and Astroparticle Physics*, **8**, 010.

Adams, F. C. and Laughlin, G. (1997). A dying universe: the long-term fate and evolution of astrophysical objects, *Reviews of Modern Physics*, **69**, 337–72.

Adams, F. C. and Laughlin, G. (2016). *The Five Ages of the Universe: Inside the Physics of Eternity*. New York: Simon and Schuster.

Adams, F. C., Coppessa, K. R., and Bloch, A. M. (2015). Planets in other universes: habitability constraints on density fluctuations and galactic structure, *Journal of Cosmology and Astroparticle Physics*, **9**, 030.

Adams, W. S. (1941). Some results with the Coudé spectrograph of the Mount Wilson Observatory, *Astrophysical Journal*, **93**, 11.

Ade, P. A. R., Aghanim, N., Ahmed, Z., et al. (2015). Joint analysis of BICEP2/Keck Array and Planck data, *Physical Review Letters*, **114**, 101301.

Afonso, C., Albert, J. N., Andersen, J., et al. (2003). Limits on galactic dark matter with 5 years of EROS SMC data, *Astronomy and Astrophysics*, **400**, 951–6.

Afshordi, N. and Magueijo, J. (2016). Critical geometry and thermal big bang, *Physical Review D*, **94**, 101301.

Aguirre, A. (2007). Making predictions in a multiverse: conundrums, dangers, coincidences, in *Universe or Multiverse?*, ed. Carr, B., pp. 367–81. Cambridge: Cambridge University Press.

Aguirre, A. and Gratton, S. (2002). Steady-state eternal inflation, *Physical Review D*, **65**, 083507.

Aguirre, A., Gratton, S., and Johnson, M. C. (2007a). Hurdles for recent measures in eternal inflation, *Physical Review D*, **75**, 123501.

Aguirre, A., Johnson, M. C., and Shomer, A. (2007b). Towards observable signatures of other bubble universes, *Physical Review D*, 74, 063509.

Aharonian, F. A. and Akerlof, C. W. (1997). Gamma-ray astronomy with imaging atmospheric Cherenkov telescopes, *Annual Reviews of Nuclear Science*, 47, 273–314.

Albrecht, A. and Magueijo, J. (1999). A time varying speed of light as a solution to cosmological puzzles, *Physical Review D*, 59, 043516.

Albrecht, A. and Steinhardt, P. J. (1982a). Cosmology for grand unified theories with radiatively induced symmetry breaking, *Physical Review Letters*, 48, 1220–3.

Albrecht, A. and Steinhardt, P. J. (1982b). Reheating an inflationary universe, *Physical Review Letters*, 48, 1437–40.

Albrecht, A., Coulson, D., Ferreira, P., et al. (1996). Causality, randomness, and the microwave background, *Physical Review Letters*, 76, 1413–16.

Alcock, C., Akerlof, C. W., Allsman, R. A., et al. (1993a). Possible gravitational microlensing of a star in the Large Magellanic Cloud, *Nature*, 365, 621–3.

Alcock, C., Allsman, R. A., Alves, D. R., et al. (2000). The MACHO Project: microlensing results from 5.7 years of Large Magellanic Cloud observations, *Astrophysical Journal*, 542, 281–307.

Alcock, C., Allsman, R. A., Axelrod, T. S., et al. (1993b). The MACHO Project – a search for the dark matter in the Milky Way, in *Sky Surveys: Protostars to Protogalaxies*, ed. Soifer, T., pp. 291–6. San Francisco: Astronomical Society of the Pacific Conference Series.

Alfvén, H. (1966). *Worlds–Antiworlds: Antimatter in Cosmology*. San Francisco: W. H. Freeman.

Alfvén, H. (1983). On hierarchical cosmology, *Astrophysics and Space Science*, 89, 313–24.

Alfvén, H. (1984). Cosmology: myth or science?, *Journal of Astrophysics and Astronomy*, 5, 79–98.

Alfvén, H. and Herlofson, N. (1950). Cosmic radiation and radio stars, *Physical Review*, 78, 616.

Alfvén, H. and Klein, O. (1962). Matter–antimatter annihilation and cosmology, *Arkiv för Fysik*, 23, 187–94.

Aliu, E., Andringa, S., Aoki, S., et al. (2005). Evidence for muon neutrino oscillation in an accelerator-based experiment, *Physical Review Letters*, 94, 081802.

Allen, S. W., Rapetti, D. A., Schmidt, R. W., et al. (2008). Improved constraints on dark energy from Chandra X-ray observations of the largest relaxed galaxy clusters, *Monthly Notices of the Royal Astronomical Society*, 383, 879–96.

Almassi, B. (2009). Trust in expert testimony: Eddington's 1919 eclipse expedition and the British response to general relativity, *Studies in History and Philosophy of Modern Physics*, 49, 57–67.

Alpher, R. A. and Herman, R. C. (1948). Evolution of the universe, *Nature*, 162, 774–5.

Alpher, R. A. and Herman, R. C. (1949). Remarks on the evolution of the expanding universe, *Physical Review*, 75, 1089–95.

Alpher, R. A. and Herman, R. C. (1950). Theory of the origin and relative distribution of the elements, *Reviews of Modern Physics*, 22, 153–212.

Alpher, R. A. and Herman, R. C. (1958). On nucleon–antinucleon symmetry in cosmology, *Science*, 128, 904.

Alpher, R. A. and Herman, R. C. (1988). Reflections on early work on 'big bang' cosmology, *Physics Today*, 41, 24–34.

Alpher, R. A. and Herman, R. C. (1990). Early work on 'big-bang' cosmology and the cosmic blackbody radiation, in *Modern Cosmology in Retrospect*, ed. Bertotti, B., Balbinot, R., and Bergia, S., pp. 129–58. Cambridge: Cambridge University Press.

Alpher, R. A. and Herman, R. C. (2001). *Genesis of the Big Bang*. Oxford: Oxford University Press.

Alpher, R. A., Bethe, H., and Gamow, G. (1948). The origin of the chemical elements, *Physical Review*, 73, 803–4.

Alpher, R. A., Follin, J. W., and Herman, R. C. (1953). Physical conditions in the initial stages of the expanding universe, *Physical Review*, **92**, 1347–61.

Alpher, V. S. (2012). Ralph A. Alpher, Robert C. Herman, and the cosmic microwave background radiation, *Physics in Perspective*, **14**, 300–34.

Amaldi, E. (1989). The search for gravitational waves, in *Cosmic Gamma Rays, Neutrinos, and Related Astrophysics*, ed. Shapiro, M. M. and Wefel, J. P., pp. 563–607. Dordrecht: Springer-Verlag.

Anders, E. (1963). Meteorite ages, in *The Moon, Meteorites and Comets – The Solar System IV*, ed. Middelhurst, B. M. and Kuiper, G. P., pp. 402–95. Chicago: University of Chicago Press.

Anderson, J. D. and Williams, J. G. (2001). Long-range tests of the equivalence principle, *Classical and Quantum Gravity*, **18**, 2447.

Anderson, J. D., Esposito, P. B., Martin, W., et al. (1975). Experimental test of general relativity using time-delay data from Mariner 6 and Mariner 7, *Astrophysical Journal*, **200**, 221–33.

Anderson, J. D., Keesey, M. S., Lau, E. L., et al. (1978). Tests of general relativity using astrometric and radio metric observations of the planets, *Acta Astronautica*, **5**, 43–61.

Anderson, L., Aubourg, É., Bailey, S., et al. (2014). The clustering of galaxies in the SDSS-III Baryon Oscillation Spectroscopic Survey: baryon acoustic oscillations in the Data Releases 10 and 11 Galaxy samples, *Monthly Notices of the Royal Astronomical Society*, **441**, 24–62.

Anderson, W. (1929). Gewöhnliche Materie und Strahlende Energie als Verschiedene "Phasen" eines und Desselben Grundstoffes (Ordinary matter and radiation energy as different phases of the same underlying matter), *Zeitschrift für Physik*, **54**, 433–44.

Ansoldi, S. and Guendelman, E. I. (2007). Solitons as key parts to produce a universe in the laboratory, *Foundations of Physics*, **37**, 712–22.

Antoniadis, I. and Cotsakis, S. (2017). Infinity in string cosmology: a review through open problems, *International Journal of Modern Physics D*, **26**, 1730009.

Antonucci, R. R. and Miller, J. S. (1985). Spectropolarimetry and the nature of NGC 1068, *Astrophysical Journal*, **297**, 621–32.

Arp, H. C. (1989). *Quasars, Redshifts and Controversies*. Cambridge: Cambridge University Press.

Arrhenius, S. (1903). *Lehrbuch der kosmischen Physik*. Leipzig: Hirzel.

Arrhenius, S. (1908). *Worlds in the Making: The Evolution of the Universe*. New York: Harper & Brothers.

Arrhenius, S. (1909). Die Unendlichkeit der Welt, *Scientia*, **5**, 217–29.

Astashenok, A. V., Nojiri, S., Odintsov, S. D., et al. (2012). Phantom cosmology without Big Rip singularity, *Physics Letters B*, **709**, 396–403.

Atkinson, R. D. (1931a). Atomic synthesis and stellar energy I, *Astrophysical Journal*, **73**, 250–95.

Atkinson, R. D. (1931b). Atomic synthesis and stellar energy II, *Astrophysical Journal*, **73**, 308–47.

Aubourg, É., Bailey, S., Bautista, J. E., et al. (2015). Cosmological implications of baryon acoustic oscillation measurements, *Physical Review D*, **92**, 123516.

Auger, P., Ehrenfest Jr., P., Maze, R., et al. (1939). Extensive air showers, *Reviews of Modern Physics*, **11**, 288–91.

Avelino, P. P., de Carvalho, J. P. M., and Martins, C. J. A. P. (2001). Can we predict the fate of the Universe?, *Physics Letters B*, **501**, 257–63.

Aver, E., Olive, K. A., Porter, R. L., et al. (2013). The primordial helium abundance from updated emissivities, *Journal of Cosmology and Astroparticle Physics*, **11**, 017.

Axford, W. I., Leer, E., and Skadron, G. (1977). The acceleration of cosmic rays by shock waves, *Proceedings of the 15th International Cosmic Ray Conference*, 11, 132–5.

Azhar, F. (2014). Prediction and typicality in multiverse cosmology, *Classical and Quantum Gravity*, 31, 035005.

Baade, W. (1926). Über eine Möglichkeit, die Pulsationstheorie der δ-Cephei-Veränderlichen zu Prüfen (On a possible method of testing the pulsation theory of the variations of δ-Cephei), *Astronomische Nachrichten*, 228, 359–62.

Baade, W. (1952). A revision of the extra-galactic distance scale, *Transactions of the International Astronomical Union*, 8, 397–8.

Baade, W. and Gaposchkin, C. H. P. (1963). *Evolution of Stars and Galaxies*. Cambridge, MA: Harvard University Press.

Baade, W. and Minkowski, R. (1954). Identification of the radio sources in Cassiopeia, Cygnus A, and Puppis A, *Astrophysical Journal*, 119, 206–14.

Baade, W. and Zwicky, F. (1934a). Cosmic rays from super-novae, *Proceedings of the National Academy of Sciences*, 20, 259–63.

Baade, W. and Zwicky, F. (1934b). On super-novae, *Proceedings of the National Academy of Sciences*, 20, 254–9.

Baade, W. and Zwicky, F. (1938). Photographic light-curves of the two supernovae in IC 4182 and NGC 1003, *Astrophysical Journal*, 88, 411–21.

Babbedge, T. S. R., Rowan-Robinson, M., Vaccari, M., et al. (2006). Luminosity functions for galaxies and quasars in the Spitzer Wide-area Infrared Extragalactic Legacy Survey, *Monthly Notices of the Royal Astronomical Society*, 370, 1159–80.

Bahcall, N. A. (2000). Clusters and cosmology, *Physics Reports*, 333, 233–44.

Baker, T., Psaltis, D., and Skordis, C. (2015). Linking tests of gravity on all scales: from the strong-field regime to cosmology, *Astrophysical Journal*, 802, 63.

Balashov, Y. (1994). Uniformitarianism in cosmology: background and philosophical implications of the steady-state theory, *Studies in History and Philosophy of Science Part A*, 25, 933–58.

Balashov, Y. (2002). *Laws of Physics and the Universe*. Boston: Birkhäuser, pp. 107–48.

Balbus, S. A. and Hawley, J. F. (1991). A powerful local shear instability in weakly magnetized disks. I – Linear analysis. II – Nonlinear evolution, *Astrophysical Journal*, 376, 214–33.

Barbosa, D., Bartlett, J. G., Blanchard, A., et al. (1996). The Sunyaev–Zeldovich effect and the value of Ω_0, *Astronomy and Astrophysics*, 314, 13–17.

Barbour, J. B. (1990). The part played by Mach's principle in the genesis of relativistic cosmology, in *Modern Cosmology in Retrospect*, ed. Bertotti, B. et al., pp. 47–66. Cambridge: Cambridge University Press.

Barbour, J. B. and Pfister, H. (Eds.) (1995). *Mach's Principle: From Newton's bucket to Quantum Gravity, Einstein Studies Vol. 6*. Boston: Birkhäuser.

Bardeen, J. M. (1970). Kerr metric black holes, *Nature*, 226, 64–5.

Bardeen, J. M. (1980). Gauge-invariant cosmological perturbations, *Physical Review D*, 22, 1882–905.

Bardeen, J. M., Bond, J. R., Kaiser, N., et al. (1986). The statistics of peaks of Gaussian random fields, *Astrophysical Journal*, 304, 15–61.

Bardeen, J. M., Steinhardt, P. J. and Turner, M. J. (1983). Spontaneous creation of almost scale-free density perturbations in an inflationary universe, *Physical Review D*, 28, 679–93.

Barnes, L. A. (2012). The fine-tuning of the universe for intelligent life, *Publications of the Astronomical Society of Australia*, 29, 529–64.

Barnes, L. A. and Lewis, G. F. (2017). Producing the deuteron in stars: anthropic limits on fundamental constants, *Journal of Cosmology and Astroparticle Physics*, 7, 036.

Barrow, J. D. (2002). *From Alpha to Omega: The Constants of Nature*. London: Jonathan Cape.

Barrow, J. D. and Tipler, F. J. (1986). *The Anthropic Cosmological Principle*. New York: Oxford University Press.

Barrow, J. D., Juszkiewicz, R., and Sonoda, D. H. (1985). Universal rotation – how large can it be?, *Monthly Notices of the Royal Astronomical Society*, 213, 917–43.

Barthel, P. D. (1989). Is every quasar beamed?, *Astrophysical Journal*, 336, 606–11.

Barthel, P. D. (1994). Unified schemes of FR2 radio galaxies and quasars, in *First Stromlo Symposium: Physics of Active Galactic Nuclei*, ed. Bicknell, G. V., Dopita, M. A., and Quinn, P. J., pp. 175–86. San Francisco: ASP Conference Series, Vol. 54.

Bashinsky, S. and Seljak, U. (2004). Signatures of relativistic neutrinos in CMB anisotropy and matter clustering, *Physical Review D*, 69, 083002.

Basko, M. M. and Polnarev, A. G. (1980). Polarization and anisotropy of the RELICT radiation in an anisotropic universe, *Monthly Notices of the Royal Astronomical Society*, 191, 207–15.

Batchelor, R., Jauncey, D. L., Johnston, K. J., et al. (1976). First global radio telescope, *Soviet Astronomy Letters*, 2, 181–3.

Baum, W. A., Johnson, F. S., Oberly, J. J., et al. (1946). Solar ultraviolet spectrum to 88 kilometers, *Phyiscal Review*, 70, 781–2.

Baumann, D. (2007). On the quantum origin of structure in the inflationary universe. arXiv preprint arXiv:0710.3187.

Becker, B. (2011). *Unravelling Starlight: William and Margaret Huggins and the Rise of the New Astronomy*. Cambridge: Cambridge University Press.

Becker, K., Becker, M., and Schwarz, J. H. (2006). *String Theory and M-theory: A Modern Introduction*. Cambridge: Cambridge University Press.

Beckwith, S. V. W., Stiavelli, M., Koekemoer, A. M., et al. (2006). The Hubble Ultra Deep Field, *Astronomical Journal*, 132, 1729–55.

Beisbart, C. (2009). Can we justifiably assume the cosmological principle in order to break model underdetermination in cosmology?, *Journal for General Philosophy of Science*, 40, 175–205.

Belenkiy, A. (2013). 'The waters I am entering no one yet has crossed:' Alexander Friedman and the origins of modern cosmology, in *Origins of the Expanding Universe: 1912–1932*, ed. Way, M. and Hunter, D., pp. 71–96. Astronomical Society of the Pacific Conference Series.

Bell, A. R. (1978). The acceleration of cosmic rays in shock fronts. I, *Monthly Notices of the Royal Astronomical Society*, 182, 147–56.

Bell-Burnell, J. (1983). The discovery of pulsars, in *Serendipitous Discoveries in Radio Astronomy*, eds. Kellermann, K. I. and Sheets, B., pp. 160–70. Green Bank, WV: National Radio Astornomy Publications.

Belopolski, A. (1929). Die Fixsterne und extra-galaktischen Nebel, *Astronomische Nachrichten*, 236, 357.

Bennett, A. S. (1962). The revised 3C catalogue of radio sources, *Memoirs of the Royal Astronomical Society*, 67, 163–72.

Bennett, C. L., Banday, A. J., Gorski, K. M., et al. (1996). Four-year COBE DMR cosmic microwave background observations: maps and basic results, *Astrophysical Journal Letters*, 464, L1.

Bennett, C. L., Bay, M., Halpern, M., et al. (2003). The Microwave Anisotropy Probe Mission, *Astrophysical Journal*, 583, 1–23.

Berendzen, R. and Hart, R. (1973). Adriaan van Maanen's influence on the island universe theory, *Journal for the History of Astronomy*, **4**, 46–56 and 73–98.

Berendzen, R., Hart, R., and Seeley, D. (1984). *Man Discovers the Galaxies*. New York: Science History Publications.

Berendzen, R. and Hoskin, M. A. (1971). *Astronomical Society of the Pacific Leaflet, No. 504: Hubble's Announcement of Cepheids in Spiral Nebulae*. Astronomical Society of the Pacific Publications.

Bergia, S. and Mazzoni, L. (1999). Genesis and evolution of Weyl's reflections on de Sitter's universe, in *The Expanding Worlds of General Relativity. Einstein Studies Vol. 7*, ed. Goenner, H. et al., pp. 325–42. Boston: Birkhäuser.

Bergmann, P. G. (1970). Cosmology as a science, *Foundations of Physics*, **1**, 17–22.

Beringer, J., Arguin, J., Barnett, R., et al. (2012). Review of particle physics, *Physical Review D*, **86**, 280–8.

Bernal, J. L., Verde, L., and Riess, A. G. (2016). The trouble with H_0, *Journal of Cosmology and Astroparticle Physics*, **10**, 019.

Bersanelli, M., Witebsky, C., Bensadoun, M., et al. (1989). Measurements of the cosmic microwave background radiation temperature at 90 GHz, *Astrophysical Journal*, **339**, 632–7.

Bertschinger, E. (1996). Cosmological dynamics, in *Cosmology and Large Scale Structure*, ed. Schaeffer, R., Silk, J., Spiro, M., et al., pp. 273–346. Amsterdam: Elsevier.

BICEP2 & Keck Array and *Planck* Collaborations, Ade, P. A. R., Ahmed, Z., et al. (2016). Improved constraints on cosmology and foregrounds from BICEP2 and Keck Array cosmic microwave background data with inclusion of 95 GHz band, *Physical Review Letters*, **116**, 031302.

BICEP2 and Keck Array Collaborations, Ade, P. A. R., Ahmed, Z., et al. (2015). BICEP2/Keck Array V: measurements of B-mode polarization at degree angular scales and 150 GHz by the Keck Array, *Astrophysical Journal*, **811**, 126.

BICEP2 Collaboration, Ade, P. A. R., Aikin, R. W., et al. (2014). Detection of B-Mode polarization at degree angular scales by BICEP2, *Physical Review Letters*, **112**, 241101.

Birkinshaw, M., Gull, S. F., and Hardebeck, H. (1984). The Sunyaev–Zeldovich effect towards three clusters of galaxies, *Nature*, **309**, 34–5.

Blackett, P. M. S. (1948). A possible contribution to the light of the night sky from the Cherenkov radiation emitted by cosmic rays, in *The Emission Spectra of the Night Sky and Aurorae, Gassiot Committee Report*, pp. 34–5. London: Physical Society of London.

Blain, A. W. and Longair, M. S. (1993). Sub-millimetre cosmology, *Monthly Notices of the Royal Astronomical Society*, **264**, 509–21.

Blain, A. W. and Longair, M. S. (1996). Observing strategies for blank-field surveys in the sub-millimetre waveband, *Monthly Notices of the Royal Astronomical Society*, **279**, 847–58.

Blanchard, A. and Schneider, J. (1987). Gravitational lensing effect on the fluctuations of the cosmic background radiation, *Astronomy and Astrophysics*, **184**, 1–6.

Blandford, R. D. and Ostriker, J. P. (1978). Particle acceleration by astrophysical shocks, *Astrophysical Journal*, **221**, L29–L32.

Bludman, S. A. and Ruderman, M. A. (1977). Induced cosmological constant expected above the phase transition restoring the broken symmetry, *Physical Review Letters*, **38**, 255–7.

Blum, A. S., Lalli, R., and Renn, J. (2016). The renaissance of general relativity: how and why it happened, *Annalen der Physik*, **528**, 344–9.

Blumenthal, G. R., Faber, S. M., Primack, J. R., et al. (1984). Formation of galaxies and large-scale structure with cold dark matter, *Nature*, **311**, 517–25.

Boddy, K. K. and Carroll, S. M. (2013). Can the Higgs boson save us from the menace of the Boltzmann brains? arXiv preprint arXiv:1308.4686.

Boggess, N. W., Mather, J. C., Weiss, R., et al. (1992). The COBE mission – its design and performance two years after launch, *Astrophysical Journal*, **397**, 420–9.

Bohlin, K. (1907). Versuch einer Bestimmung der Parallaxe des Andromedanebels, *Astronomische Nachrichten*, **176**, cols 205–6.

Böhringer, H. (1994). Clusters of galaxies, in *Frontiers of Space and Ground-based Astronomy*, ed. Wamsteker, W., Longair, M. S., and Kondo, Y., pp. 359–68. Dordrecht: Kluwer Academic Publishers.

Bolte, M. (1997). Globular clusters: old, in *Critical Dialogues in Cosmology*, ed. Turok, N., pp. 156–68. Singapore: World Scientific.

Bolton, C. T. (1972). Identifications of CYG X-1 with HDE 226868, *Nature*, **235**, 271–3.

Bolton, J. G., Stanley, G. J., and Slee, O. B. (1949). Positions of three discrete sources of galactic radio-frequency radiation, *Nature*, **164**, 101–2.

Boltzmann, L. (1895). On certain questions in the theory of gases, *Nature*, **51**, 483–5.

Boltzmann, L. (1898). *Vorlesungen über Gastheorie, part 2*. Leipzig: Barth.

Bond, J. R. and Efstathiou, G. (1987). The statistics of cosmic background radiation fluctuations, *Monthly Notices of the Royal Astronomical Society*, **226**, 655–87.

Bond, J. R. and Efstathiou, G. (1991). The formation of cosmic structure with a 17 keV neutrino, *Physics Letters B*, **265**, 245–50.

Bond, J. R., Efstathiou, G., and Tegmark, M. (1997). Forecasting cosmic parameter errors from microwave background anisotropy experiments, *Monthly Notices of the Royal Astronomical Society*, **291**, L33–L41.

Bondi, H. (1948). Review of cosmology, *Monthly Notices of the Royal Astronomical Society*, **108**, 104–20.

Bondi, H. (1952). *Cosmology*. Cambridge: Cambridge University Press.

Bondi, H. (1960). *Cosmology*, second edition. Cambridge: Cambridge University Press.

Bondi, H. and Gold, T. (1948). The steady-state theory of the expanding universe, *Monthly Notices of the Royal Astronomical Society*, **108**, 252–70.

Bondi, H., Bonnor, W. B., Lyttleton, R. A., et al. (1960). *Rival Theories of Cosmology*. London: Oxford University Press.

Bonnor, W. (1954). The stability of cosmological models, *Zeitschrift für Astrophysik*, **35**, 10–20.

Bonnor, W. (1957). La formation des nébuleuses en cosmologie relativiste, *Annales de l'Institut Henri Poincaré*, **15**, 158–72.

Bonolis, L. (2017). Stellar structure and compact objects before 1940: towards relativistic astrophysics, *European Physical Journal H*, **42**, 311–93.

Bosma, A. (1981). 21-cm line studies of spiral galaxies. II. The distribution and kinematics of neutral hydrogen in spiral galaxies of various morphological types, *Astronomical Journal*, **86**, 1825–46.

Bostrom, N. (2002). *Anthropic Bias: Observation Selection Effects in Science and Philosophy*. New York: Routledge.

Bostrom, N. (2003). Are you living in a computer simulation?, *Philosophical Quarterly*, **53**, 243–55.

Bothe, W. and Kolhörster, W. (1929). The nature of the high-altitude radiation, *Zeitschrift für Physik*, **56**, 751–77.

Boughn, S. P. and Crittenden, R. (2004). A correlation between the cosmic microwave background and large-scale structure in the Universe, *Nature*, **427**, 45–7.

Boughn, S. P., Fram, D. M., and Partridge, R. B. (1971). Isotropy of the microwave background at 8-mm wavelength, *Astrophysical Journal*, **165**, 439–44.

Bouhmadi-López, M. and Madrid, J. A. J. (2005). Escaping the big rip?, *Journal of Cosmology and Astroparticle Physics*, **5**, 005.

Bousso, R. and Freivogel, B. (2007). A paradox in the global description of the multiverse, *Journal of High Energy Physics*, **6**, 018.

Bousso, R. and Polchinski, J. (2004). The string theory landscape, *Scientific American*, **291**, 60–9.

Bousso, R. and Susskind, L. (2012). Multiverse interpretation of quantum mechanics, *Physical Review D*, **85**, 045007.

Bousso, R., Hall, L. J., and Nomura, Y. (2009). Multiverse understanding of cosmological coincidences, *Physical Review D*, **80**, 063510.

Bouwens, R. J., Illingworth, G. D., Blakeslee, J. P., et al. (2006). Galaxies at $z \sim 6$: the UV luminosity function and luminosity density from 506 HUDF, HUDF parallel ACS field, and GOODS i-dropouts, *Astrophysical Journal*, **653**, 53–85.

Bowick, M. J., Smolin, L., and Wijewardhana, L. C. R. (1987). Does string theory solve the puzzles of black hole evaporation?, *General Relativity and Gravitation*, **19**, 113–19.

Bowyer, S., Byram, E. T., Chubb, T. A., et al. (1964). Lunar occulation of X-ray emission from the Crab Nebula, *Science*, **146**, 912–17.

Boyle, B. J., Shanks, T., Croom, S. M., et al. (2000). The 2dF QSO Redshift Survey – I. The optical luminosity function of quasi-stellar objects, *Monthly Notices of the Royal Astronomical Society*, **317**, 1014–22.

Boynton, P. E., Stokes, R. A., and Wilkinson, D. T. (1968). Primeval fireball intensity at $\lambda = 3.3$ mm, *Physical Review Letters*, **21**, 462–5.

Bracessi, A., Formiggini, L., and Gandolfi, E. (1970). Magnitudes, colours and coordinates of 175 ultraviolet excess objects in the field 13^h, $+36°$, *Astronomy and Astrophysics*, **5**, 264–79. Erratum: *Astronomy and Astrophysics*, **23**, 159.

Bracewell, R. N. (Ed.) (1959). *Paris Symposium on Radio Astronomy*. Stanford: Stanford University Press.

Bracewell, R. N. and Conklin, E. K. (1968). An observer moving in the 3deg K radiation field, *Nature*, **219**, 1343–4.

Braes, L. L. E. and Miley, G. K. (1971). Radio emission from Scorpius X-1 at 21.2 cm, *Astronomy and Astrophysics*, **14**, 160–3.

Braginskii, V. B. and Gertsenshtein, M. E. (1967). Concerning the effective generation and observation of gravitational waves, *ZhETF Pis'ma*, **5**, 348–50.

Branch, D. and Patchett, B. (1973). Type I supernovae, *Monthly Notices of the Royal Astronomical Society*, **161**, 71–83.

Branch, D. and Tammann, G. A. (1992). Type I supernovae as standard candles, *Annual Rreview of Astronomy and Astrophysics*, **30**, 359–89.

Brandt, W. N. and Hasinger, G. (2005). Deep extragalactic X-ray surveys, *Annual Review of Astronomy and Astrophysics*, **43**, 827–59.

Brans, C. H. (2014). Jordan–Brans–Dicke theory. http://www.scholarpedia.org/article/Jordan-Brans-DickeTheory.

Brans, C. H. and Dicke, R. H. (1961). Mach's principle and a relativistic theory of gravitation, *Physical Review*, **124**, 925–35.

Braude, S., Dubinskii, B. A., Kaidanovskii, N. L., et al. (Eds.) (2012). *A Brief History of Radio Astronomy in the USSR: A Collection of Scientific Essays (Vol. 382)*. Dordrecht: Springer.

Breitenberger, E. (1984). Gauss' geodesy and the axiom of parallels, *Archive for History of Exact Sciences*, **31**, 273–89.

Bridgman, P. W. (1955). *Reflections of a Physicist*. New York: Philosophical Library.

Bruggencate, P. T. (1930). The radial velocities of globular clusters, *Proceedings of the National Academy of Sciences*, **16**, 111–18.

Brush, S. G. (1978). A geologist among astronomers: the rise and fall of the Chamberlin–Moulton cosmogony, *Journal for the History of Astronomy*, **9**, 1–41, 77–104.

Brush, S. G. (1987). The nebular hypothesis and the evolutionary world view, *History of Science*, **25**, 245–78.

Brush, S. G. (1996). *Nebulous Earth: The Origin of the Solar System and the Core of the Earth*. New York: Cambridge University Press.

Brush, S. G. (2001). Is the Earth too old? The impact of geochronology on cosmology, 1929–1952, in The Age of the Earth: From 4004 BC to AD 2002, ed. Lewis, C. L. and Knell, S. J., pp. 157–75. London: Geological Society.

Brustein, R., Gasperini, M., Giovannini, M., et al. (1995). Relic gravitational waves from string cosmology, *Physics Letters B*, **361**, 45–51.

Bruzual, G. and Charlot, S. (2003). Stellar population synthesis at the resolution of 2003, *Monthly Notices of the Royal Astronomical Society*, **344**, 1000–28.

Buchert, T. and Räsänen, S. (2012). Backreaction in late-time cosmology, *Annual Review of Nuclear and Particle Science*, **62**, 57–79.

Buckley, J. H., Boyle, P., Burdett, A. et al. (1997). Multiwavelength observations of Markarian 421, *AIP Conference Proceedings* **410**, *Fourth Compton Symposium*, C.D. Dermer, M.S. Strickman and J.D. Kurfess (Eds.), 1381–5. New York: AIP Publications.

Burbidge, E. M., Burbidge, G. R., Fowler, W. A., et al. (1957). Synthesis of the elements in stars, *Reviews of Modern Physics*, **29**, 547–650.

Burbidge, E. M., Burbidge, G. R. and Hoyle, F. (1963). Condensations in the intergalactic medium, *Astrophysical Journal*, **138**, 873–88.

Burbidge, E. M., Burbidge, G. R., and Sandage, A. R. (1963). Evidence for the occurence of violent events in the nuclei of galaxies, *Reviews of Modern Physics*, **35**, 947–72.

Burbidge, G. R. (1959). Estimates of the total energy in particles and magnetic field in the non-thermal radio sources, *Astrophysical Journal*, **129**, 849–51.

Burbidge, G. R. (1967). On the wavelengths of the absorption lines in quasi-stellar objects, *Astrophysical Journal*, **147**, 851–5.

Burbidge, G. R. and Burbidge, E. M. (1967). *Quasi-Stellar Objects*. New York: Freeman and Company.

Burbidge, G. R. and Hoyle, F. (1957). Matter and antimatter, *Astronomical Journal*, **62**, 9.

Burgess, C. P., Cicoli, M., and Quevedo, F. (2013). String inflation after Planck 2013, *Journal of Cosmology and Astroparticle Physics*, **11**, 003.

Burigana, C., Danese, L., and de Zotti, G. (1991). Constraints on the thermal history of the universe from the cosmic microwave background spectrum, *Astrophysical Journal*, **379**, 1–5.

Butcher, H. and Oemler, Jr., A. (1978). The evolution of galaxies in clusters. I – ISIT photometry of C1 0024+1654 and 3C 295, *Astrophysical Journal*, **219**, 18–30.

Butcher, H. and Oemler, A., Jr. (1984). The evolution of galaxies in clusters. V – A study of populations since Z ~ 0.5, *Astrophysical Journal*, **285**, 426–38.

Butterfield, J. (2014). On under-determination in cosmology, *Studies in History and Philosophy of Science Part B: Studies in History and Philosophy of Modern Physics*, **46**, 57–69.

Calcagni, G. (2017). String cosmology, in *Classical and Quantum Cosmology*, pp. 701–821. New York: Springer.

Caldeira, K. and Kasting, J. F. (1992). The life span of the biosphere revisited, *Nature*, **360**, 721–3.

Caldwell, R. R. (2002). A phantom menace? Cosmological consequences of a dark energy component with super-negative equation of state, *Physics Letters B*, **545**, 23–9.

Caldwell, R. R., Kamionkowski, M., and Weinberg, N. N. (2003). Phantom energy: dark energy with $w < -1$ causes a cosmic doomsday, *Physical Review Letters*, **91**, 071301.

Callender, C. (2004). There is no puzzle about the low-entropy past, in *Contemporary Debates in Philosophy of Science*, ed. Hitchcock, C., pp. 240–56. Oxford: Blackwell Publishing.

Campbell, W. (1896). Review of Mr. Lowell's book on Mars, *Publications of the Astronomical Society of the Pacific*, **8**, 207–20.

Campbell, W. (1917). The nebulae: address of the retiring president of the American Association for the Advancement of Science, *Science*, **45**, 513–48.

Carlip, A. (2007). Transient observers and variable constants, or repelling the invasion of the Boltzmann's brains, *Journal of Cosmology and Astroparticle Physics*, **6**, 001.

Carlstrom, J. E., Holder, G. P., and Reese, E. D. (2002). Cosmology with the Sunyaev–Zeldovich effect, *Annual Review of Astronomuy and Astrophysics*, **40**, 643–80.

Carr, B. (Ed.) (2007). *Universe or Multiverse?* Cambridge: Cambridge University Press.

Carr, B. J. and Rees, M. J. (1979). The anthropic principle and the structure of the physical world, *Nature*, **278**, 605–12.

Carroll, S. M. (2006). Is our universe natural?, *Nature*, **440**, 1132–6.

Carroll, S. M. (2014). In what sense is the early universe fine-tuned? arXiv preprint arXiv:1406.3057.

Carroll, S. M., Press, W. H., and Turner, E. L. (1992). The cosmological constant, *Annual Review of Astronomy and Astrophysics*, **30**, 499–542.

Carter, B. (1971). Axisymmetric black hole has only two degrees of freedom, *Physical Review Letters*, **26**, 331–3.

Carter, B. (1974). Large number coincidences and the anthropic principle in cosmology, in *Confrontation of Cosmological Theories with Observational Data: Proceedings of the Symposium, Krakow, Poland, September 10–12, 1973*, ed. Longair, M. S., pp. 291–8. Dordrecht: D. Reidel Publishing Co.

Carter, B. (1983). The anthropic principle and its implications for biological evolution, *Philosophical Transactions of the Royal Society of London A*, **310**, 347–63.

Carter, B. (1993). The anthropic selection principle and the ultra-Darwinian synthesis, in *The Anthropic Principle: Proceedings of the Second Venice Conference on Cosmology and Philosophy*, ed. Bertola, F. and Curi, U., pp. 33–66. Cambridge: Cambridge University Press.

Case, S. (2015). 'Land-marks of the universe': John Herschel against the background of positional astronomy, *Annals of Science*, **72**, 417–34.

Cercignani, C. (1998). *Ludwig Boltzmann: The Man Who Trusted Atoms*. Oxford: Oxford University Press.

Chaboyer, B. (1998). The age of the universe, *Physics Reports*, **307**, 23–30.

Chae, K.-H., Biggs, A. D., Blandford, R. D., et al. (2002). Constraints on cosmological parameters from the analysis of the Cosmic Lens All Sky Survey radio-selected gravitational lens statistics, *Physical Review Letters*, **89**, 151301.

Chaisson, E. J. (1997). The rise of information in an evolutionary universe, *World Futures*, **50**, 447–55.

Chaisson, E. J. (2001). *Cosmic Evolution: The Rise of Complexity in Nature*. Cambridge, MA: Harvard University Press.

Chamberlin, T. C. (1899). Lord Kelvin's address on the age of the Earth as an abode fitted for life, *Science*, **10**, 11–18.

Chambers, R. (1844). *Vestiges of the Natural History of Creation*. London: John Churchill.

Chamcham, K., Silk, J., Barrow, J. D., et al. (Eds.) (2017). *The Philosophy of Cosmology*. Cambridge: Cambridge University Press.

Chandrasekhar, S. (1931). The maximum mass of ideal white dwarfs, *Astrophysical Journal*, **74**, 81–2.

Chandrasekhar, S. and Henrich, L. R. (1942). An attempt to interpret the relative abundances of the elements and their isotopes, *Astrophysical Journal*, **95**, 288–98.

Charlier, C. V. L. (1896). Ist die Welt endlich oder unendlich in Raum und Zeit?, *Archiv für systematische Philosophie*, **2**, 477–94.

Charlier, C. V. L. (1908). Wie eine unendliche Welt aufgebaut kann, *Arkiv för Matematik, Astronomi och Fysik*, **4**, 1–15.

Charlier, C. V. L. (1922). How an infinite world may be built up, *Arkiv för Matematik, Astronomi och Fysik*, **16**, 1–34.

Charlier, C. V. L. (1925a). An infinite universe, *Publications of the Astronomical Society of the Pacific*, **37**, 177–91.

Charlier, C. V. L. (1925b). On the structure of the universe, *Publications of the Astronomical Society of the Pacific*, **37**, 53–76.

Charlier, C. V. L. (1925c). On the structure of the universe, *Publications of the Astronomical Society of the Pacific*, **37**, 115–35.

Chavanis, P.-H. (2007). White dwarf stars in D dimensions, *Physical Review D*, **76**, 023004.

Chown, M. (1993). *Afterglow of Creation: From the Fireball to the Discovery of Cosmic Ripples*. London: Arrow Books.

Christian, D. (2004). *Maps of Time: An Introduction to Big History*. Berkeley: University of California Press.

Christianson, G. E. (1995). *Edwin Hubble: Mariner of the Nebulae*. New York: Farrar, Straus and Giroux.

Chwolson, O. D. (1910). Dürfen wie die physikalische Gesetze auf das Universum anwenden?, *Scientia*, **8**, 41–53.

Chwolson, O. D. (1924). Über eine mögliche Form fiktiver Doppelsterne, *Astronomische Nachrichten*, **221**, 329.

Cimatti, A., Daddi, E., Renzini, A., et al. (2004). Old galaxies in the young Universe, *Nature*, **430**, 184–7.

Ćirković, M. M. (2002). Laudatores temporis acti, or why cosmology is alive and well – a reply to Disney. *General Relativity and Gravitation*, **34**, 119–29.

Ćirković, M. M. (2003a). Resource letter PEs-1: physical eschatology, *American Journal of Physics*, **71**, 122–33.

Ćirković, M. M. (2003b). The thermodynamical arrow of time: reinterpreting the Boltzmann–Schuetz argument, *Foundations of Physics*, **33**, 467–90.

Ćirković, M. M. (2012). *The Astrobiological Landscape: Philosophical Foundations of the Study of Cosmic Life*. Cambridge: Cambridge University Press.

Ćirković, M. M. (2016). Anthropic arguments outside of cosmology and string theory, *Belgrade Philosophical Annual*, **30**, 91–114.

Clark, G. W., Garmire, G. P., and Kraushaar, W. L. (1968). Observation of High-Energy Cosmic Gamma Rays, *Astrophysical Journal Letters*, **153**, L203–L207.

Clarkson, C. (2012). Establishing homogeneity of the universe in the shadow of dark energy, *Comptes Rendus Physique*, **13**, 682–718.

Clausius, R. (1868). On the second fundamental theorem of the mechanical theory of heat, *Philosophical Magazine*, **35**, 405–19.

Clerke, A. M. (1890). *The System of the Stars*. London: Longmans, Green & Co.

Clerke, A. M. (1903). *Problems of Astrophysics*. London: Adam and Charles Black.

CMS and LHCb Collaborations, Khachatryan, V., Sirunyan, A. M., et al. (2015). Observation of the rare $B_s^0 \to \mu^+\mu^-$ decay from the combined analysis of CMS and LHCb data, *Nature*, **522**, 68–72.

Cockcroft, J. D. and Walton, E. T. S. (1932). Disintegration of lithium by swift protons, *Nature*, **129**, 649.

Cocke, W. J., Disney, M. J., and Taylor, D. J. (1969). Discovery of optical signals from pulsar NP 0532, *Nature*, **221**, 525–7.

Cole, S., Percival, W. J., Peacock, J. A., et al. (2005). The 2dF Galaxy Redshift Survey: power-spectrum analysis of the final data set and cosmological implications, *Monthly Notices of the Royal Astronomical Society*, **362**, 505–34.

Coles, P. and Lucchin, F. (1995). *Cosmology: the Origin and Evolution of Cosmic Structure*. Chichester: John Wiley & Sons.

Coles, P. and Lucchin, F. (2002). *Cosmology: the Origin and Evolution of Cosmic Structures*, second edition. Chichester: John Wiley & Sons.

Coles, P., Melott, A. L., and Shandarin, S. F. (1993). Testing approximations for non-linear gravitational clustering, *Monthly Notices of the Royal Astronomical Society*, **260**, 765–76.

Collins, C. B. and Hawking, S. W. (1973). Why is the universe isotropic?, *Astrophysical Journal*, **180**, 317–34.

Colyvan, M., Garfield, J. L., and Priest, G. (2005). Problems with the argument from fine tuning, *Synthese*, **145**, 325–38.

Conklin, E. K. (1969). Velocity of the Earth with respect to the cosmic background radiation, *Nature*, **222**, 971–2.

Conklin, E. K. (1972). *Anisotropy and inhomogeneity in the cosmic background radiation*. PhD thesis, Stanford University.

Conklin, E. K. and Bracewell, R. N. (1967). Limits on small scale variations in the cosmic background radiation, *Nature*, **216**, 777–9.

Connolly, A. J., Scranton, R., Johnston, D., et al. (2002). The angular correlation function of galaxies from early Sloan Digital Sky Survey data, *Astrophysical Journal*, **579**, 42–7.

Cooke, R. J., Pettini, M., Jorgenson, R. A., et al. (2014). Precision measures of the primordial abundance of deuterium, *Astrophysical Journal*, **781**, 31.

Corey, B. E. (1978). *The dipole anisotropy of the cosmic microwave background at a wavelength of 1.6 cm*. PhD thesis, Princeton University.

Corey, B. E. and Wilkinson, D. T. (1976). A measurement of the cosmic microwave background anisotropy at 19 GHz, *Bulletin of the American Astronomical Society*, **8**, 351.

Costa, E., Frontera, F., Heise, J., et al. (1997). Discovery of an X-ray afterglow associated with the gamma-ray burst of 28 February 1997, *Nature*, **387**, 783–5.

Couchot, F., Henrot-Versillé, S., Perdereau, O., et al. (2017). Cosmological constraints on the neutrino mass including systematic uncertainties, *Astronomy and Astrophysics*, **606**, A104.

Couderc, P. (1952). *The Expansion of the Universe*. London: Faber and Faber.

Cowan, J. J., Thielemann, F.-K., and Truran, J. W. (1991). Radioactive dating of the elements, *Annual Reviews of Astronomy and Astrophysics*, **29**, 447–97.

Cowie, L. L., Barger, A. J., and Kneib, J.-P. (2002). Faint submillimeter counts from deep 850 micron observations of the lensing clusters A370, A851, and A2390, *Astronomical Journal*, **123**, 2197–205.

Cowie, L. L., Lilly, S. J., Gardner, J., et al. (1988). A cosmologically significant population of galaxies dominated by very young star formation, *Astrophysical Journal*, **332**, L29–L32.

Cowie, L. L., Songaila, A., Hu, E. M., et al. (1996). New insight on galaxy formation and evolution from Keck spectroscopy of the Hawaii Deep Fields, *Astronomical Journal*, **112**, 839–64.

Crane, L. (2010). Possible implications of the quantum theory of gravity: an introduction to the meduso-anthropic principle, *Foundations of Science*, **15**, 369–73.

Crane, P., Hegyi, D. J., Kutner, M. L., et al. (1989). Cosmic background radiation temperature at 2.64 millimeters, *Astrophysical Journal*, **346**, 136–42.

Crane, P., Hegyi, D. J., Mandolesi, N., et al. (1986). Cosmic background radiation temperature from CN absorption, *Astrophysical Journal*, **309**, 822–7.

Crawford, L. (2013). Freak observers and the simulation argument, *Ratio*, **XXVI**, 250–64.

Crelinsten, J. (2006). *Einstein's Jury: The Race to Test Relativity*. Princeton: Princeton University Press.

Crittenden, R. G., Coulson, D., and Turok, N. G. (1995). Temperature–polarization correlations from tensor fluctuations, *Physical Review D*, **52**, R5402–6.

Crookes, W. (1886). On the nature and origin of the so-called elements, in *Report, British Association for the Advancement of Science*, pp. 558–76.

Crowe, M. J. (1986). *The Extraterrestrial Life Debate 1750–1900: The Idea of Plurality of Worlds from Kant to Lowell*. Cambridge: Cambridge University Press.

Curtis, H, D. (1915). Proper motions of the nebulae, *Publications of the Astronomical Society of the Pacific*, **27**, 214–18.

Curtis, H. D. (1917). New stars in spiral nebulae, *Publications of the Astronomical Society of the Pacific*, **29**, 180–2.

Curtis, H. D. (1918). Descriptions of 762 nebulae and clusters photographed with the Crossley reflector, *Publications of the Lick Observatory*, **13**, 11–42.

Curtis, H. D. (1933). The nebulae, in *Handbuch der Astrophysik, Vol. 5.2*, ed. Curtis, H. D. Berlin: Springer Verlag.

Dagkesamanskii, R. D. (2009). The Pushchino Radio Astronomy Observatory of the PN Lebedev Physical Institute Astro Space Center: yesterday, today, and tomorrow, *Physics-Uspekhi*, **52**, 1159.

Dahlen, A. (2010). Odds of observing the multiverse, *Physical Review D*, **81**, 063501.

Dalgarno, A. (1983). Rydberg atoms in astrophysics, in *Rydberg States of Atoms and Molecules*, ed. Stebbings, R. F. and Dunning, F. B., pp. 1–30. Cambridge: Cambridge University Press.

Danese, L. and de Zotti, G. (1982). Double Compton process and the spectrum of the microwave background, *Astronomy and Astrophysics*, **107**, 39–42.

Daniels, N. (1975). Lobatchewsky: some anticipations of later views on the relation between geometry and physics, *Isis*, **66**, 75–85.

Darwin, G. H. (1905). Presidential address, in *Report, British Association for the Advancement of Science*, pp. 33–2.

Dashevsky, V. M. and Zeldovich, Y. B. (1964). Propagation of light in a nonhomogeneous non-flat universe II, *Astronomicheskii Zhurnal*, **41**, 1071–4. Translation: (1965), *Soviet Astronomy*, **8**, 854–6.

Davidson, W. (1962). The cosmological implications of the recent counts of radio sources, I. Analysis of the results and their immediate interpretation, *Monthly Notices of the Royal Astronomical Society*, **123**, 425–35.

Davidson, W. and Davies, M. (1964). Interpretation of the counts of radio sources in terms of a 4-parameter family of evolutionary universes, *Monthly Notices of the Royal Astronomical Society*, **127**, 241–55.

Davies, P. C. W. (1994). *The Last Three Minutes*. New York: Basic Books.

Davis, M. and Peebles, P. J. E. (1983). A survey of galaxy redshifts. V – The two-point position and velocity correlations, *Astrophysical Journal*, 267, 465–82.

Davis, M., Efstathiou, G., Frenk, C., et al. (1992). The end of cold dark matter?, *Nature*, 356, 489–94.

Davis, M., Geller, M. J., and Huchra, J. (1978). The local mean mass density of the universe – new methods for studying galaxy clustering, *Astrophysical Journal*, 221, 1–18.

Davis, T. M. and Lineweaver, C. H. (2004). Expanding confusion: common misconceptions of cosmological horizons and the superluminal expansion of the universe, *Publications of the Astronomical Society of Australia*, 21, 97–109.

De la Rue, W. (1861). Proceedings of the Chemical Society, *Chemical News*, 4, 130–3.

De Simone, A., Guth, A. H., Linde, A., et al. (2010). Boltzmann brains and the scale-factor cutoff measure of the multiverse, *Physical Review D*, 82, 063520.

De Sitter, W. (1916). On Einstein's theory of gravitation, and its astronomical consequences. Second paper, *Monthly Notices of the Royal Astronomical Society*, 77, 155–84.

De Sitter, W. (1917a). On Einstein's theory of gravitation, and its astronomical consequences. Third paper, *Monthly Notices of the Royal Astronomical Society*, 78, 3–28.

De Sitter, W. (1917b). On the curvature of space, *Koninklijke Akademie van Wetenschappen, Proceedings*, 20, 229–43.

De Sitter, W. (1920). On the possibility of statistical equilibrium of the universe, *Koninklijke Akademie van Wetenschappen*, 23, 866–8.

De Sitter, W. (1930). On the magnitudes, diameters and distances of the extragalactic nebulae, and their apparent radial velocities, *Bulletin of the Astronomical Institute of the Netherlands*, 5, 157–71.

De Sitter, W. (1933). The astronomical aspect of the theory of relativity, *University of California Publications in Mathematics*, 2, 142–96.

De Vaucouleurs, G. (1970). The case for a hierarchical cosmology, *Science*, 167, 1203–12.

De Vaucouleurs, G. (1971). The large-scale distribution of galaxies and clusters of galaxies, *Publications of the Astronomical Society of the Pacific*, 83, 113–43.

Dehnen, H. and Hönl, H. (1968). Informationen über das Universum aus antipodisch beobachteten Radioquellen, *Naturwissenschaften*, 55, 413–15.

Dekel, A. (1986). Biased galaxy formation, *Comments on Astrophysics*, 11, 235–56.

Dekel, A. and Rees, M. J. (1987). Physical mechanisms for biased galaxy formation, *Nature*, 326, 455–62.

Dekel, A., Burstein, D., and White, S. D. M. (1997). Measuring Ω, in *Critical Dialogues in Cosmology*, ed. Turok, N., pp. 175–92. Singapore: World Scientific.

Dennett, D. (1995). *Darwin's Dangerous Idea: Evolution and the Meanings of Life*. New York: Simon and Schuster.

DeVorkin, D. (1998). The American astronomical community, *Journal for the History of Astronomy*, 29, 389–92.

Dewhirst, D. and Hoskin, M. A. (1991). The Rosse spirals, *Journal for the History of Astronomy*, 22, 257–66.

Dicke, R. H. (1961). Dirac's cosmology and Mach's principle, *Nature*, 192, 440–1.

Dicke, R. H. and Peebles, P. J. E. (1979). Big bang cosmology – enigmas and nostrums, in *General Relativity: An Einstein Centenary Survey*, ed. Hawking, S. W. and Israel, W., pp. 504–17. Cambridge: Cambridge University Press.

Dicke, R. H., Beringer, R., Kyhl, R. L., et al. (1946). Atmospheric absorption measurements with a microwave radiometer, *Physical Review*, 70, 340–8.

Dicke, R. H., Peebles, P. J. E., Roll, P. G., et al. (1965). Cosmic black-body radiation, *Astrophysical Journal*, 142, 414–19.

Dickey, J. O., Bender, P. L., Faller, J. E., et al. (1994). Lunar laser ranging: a continuing legacy of the Apollo program, *Science*, **265**, 482–90.

Dingle, H. (1937). Modern Aristotelianism, *Nature*, **139**, 784–6.

Dirac, P. A. M. (1937). The cosmological constants, *Nature*, **139**, 323.

Dirac, P. A. M. (1938). A new basis for cosmology, *Proceedings of the Royal Society A*, **165**, 199–208.

Dirac, P. A. M. (1974). Cosmological models and the large numbers hypothesis, *Proceedings of the Royal Society A*, **338**, 439–46.

Disney, M. J. (2000). The case against cosmology, *General Relativity and Gravitation*, **32**, 1125–34.

Disney, M. J. (2007). Modern cosmology: science or folktale?, *American Scientist*, **95**, 383–5.

Dodelson, S. (2003). *Modern Cosmology*. Amsterdam: Academic Press.

Dodelson, S., Gates, E. I., and Turner, M. S. (1996). Cold dark matter, *Science*, **274**, 69–75.

Doroshkevich, A. G. and Novikov, I. D. (1964). Mean density of radiation in the metagalaxy and certain problems in relativistic cosmology, *Dokladi Akademiya Nauk SSSR*, **154**, 809–11. Translation: (1964), *Soviet Physics Doklady*, **9**, 111–13.

Doroshkevich, A. G., Sunyaev, R. A., and Zeldovich, Y. B. (1974). The formation of galaxies in Friedmanian universes, in *Confrontation of Cosmological Theories with Observational Data, IAU Symposium No. 63*, ed. Longair, M. S., pp. 213–25. Dordrecht: D. Reidel Publishing Company.

Doroshkevich, A. G., Zeldovich, Y. B., Sunyaev, R. A., et al. (1980a). Astrophysical implications of the neutrino rest mass. Part II. The density-perturbation spectrum and small-scale fluctuations in the microwave background, *Pis'ma v Astronomicheskii Zhurnal*, **6**, 457–64.

Doroshkevich, A. G., Zeldovich, Y. B., Sunyaev, R. A., et al. (1980b). Astrophysical implications of the neutrino rest mass. Part III. The non-linear growth of perturbations and hidden mass, *Pis'ma v Astronomicheskii Zhurnal*, **6**, 465–9.

Douglas, A. V. (1956). *The Life of Arthur Stanley Eddington*. London: Thomas Nelson and Sons Ltd.

Draper, K., Draper, P., and Pust, J. (2007). Probabilistic arguments for multiple universes, *Pacific Philosophical Quarterly*, **88**, 288–307.

Dressler, A. (1980). Galaxy morphology in rich clusters – implications for the formation and evolution of galaxies, *Astrophysical Journal*, **236**, 351–65.

Dressler, A. (1984). The evolution of galaxies in clusters, *Annual Review of Astronomy and Astrophysics*, **22**, 185–222.

Dreyer, J. L. E. (1895). Index catalogue of nebulae in the years 1888 to 1894, with notes and corrections to the new general catalogue, *Memoirs of the Royal Astronomical Society*, **51**, 185–228.

Drieschner, M. (Ed.) (2014). *Carl Friedrich von Weizsäcker: Major Texts in Physics*. Heidelberg: Springer.

Duerbeck, H. W. and Seitter, W. (2005). The nebular research of Carl Wirtz, in *The Multinational History of Strasbourg Astronomical Observatory*, ed. Heck, A., pp. 167–87. Dordrecht: Springer.

Duerbeck, H. W. (2002). Extragalactic research in Europe and the United States in the early 20th century, *Astronomische Nachrichten*, **323**, cols 534–7.

Duhem, P. (1974). *The Aim and Structure of Physical Theory*. New York: Atheneum.

Dunlop, J. S. (2011). The cosmic history of star formation, *Science*, **333**, 178.

Dunlop, J. S. (2013). Observing the first galaxies, in *The First Galaxies*, ed. Wiklind, T., Mobasher, B., and Bromm, V., pp. 223–92. Berlin: Springer.

Dunlop, J. S. and Peacock, J. A. (1990). The redshift cut-off in the luminosity function of radio galaxies and quasars, *Monthly Notices of the Royal Astronomical Society*, **247**, 19–42.

Durrer, R. (2008). *The Cosmic Microwave Background*. Cambridge University Press, Cambridge.

Dyer, C. C. and Roeder, R. C. (1972). The distance-redshift relation for universes with no intergalactic medium, *Astrophysical Journal*, **174**, L115–17.

Dyson, F. J. (1979). Time without end: physics and biology in an open universe, *Reviews of Modern Physics*, **51**, 447–60.

Dyson, F. W., Eddington, A. S., and Davidson, C. (1920). A determination of the deflection of light by the Sun's gravitational field, from observations made at the total eclipse of May 29, 1919, *Philosophical Transactions of the Royal Society*, **220**, 291–333.

Dyson, L., Kleban, M., and Susskind, L. (2002). Disturbing implications of a cosmological constant, *Journal of High Energy Physics*, **10**, 011.

Earman, J. (1987). The SAP also rises: a critical examination of the anthropic principle, *American Philosophical Quarterly*, **24**, 307–17.

Earman, J. (1999). The Penrose–Hawking singularity theorems: history and implications, in *The Expanding Worlds of General Relativity. Einstein Studies Vol. 7*, ed. Goenner, H. et al., pp. 235–70. Boston: Birkhäuser.

Earman, J. (2001). Lambda: the constant that refuses to die, *Archive for History of Exact Sciences*, **55**, 189–220.

Earman, J. (2009). Cosmology: a special case? Unpublished manuscript.

Earman, J. and Eisenstaedt, J. (1999). Einstein and singularities, *Studies in History and Philosophy of Science Part B: Studies in History and Philosophy of Modern Physics*, **30**, 185–235.

Eastman, R. G. and Kirshner, R. P. (1989). Model atmospheres for SN 1987A and the distance to the Large Magellanic Cloud, *Astrophysical Journal*, **347**, 771–93.

Eddington, A. S. (1914). *Stellar Movements and the Structure of the Universe*. Cambridge: Cambridge University Press.

Eddington, A. S. (1916). The kinetic energy of a star cluster, *Monthly Notices of the Royal Astronomical Society*, **76**, 525–8.

Eddington, A. S. (1918). *Report on the Relativity Theory of Gravitation*. London: Fleetway Press.

Eddington, A. S. (1920a). *Space, Time and Gravitation. An Outline of the General Relativity Theory*. Cambridge: Cambridge University Press.

Eddington, A. S. (1920b). The internal constitution of the stars, *Observatory*, **43**, 341–58.

Eddington, A. S. (1923). *The Mathematical Theory of Relativity*. Cambridge: Cambridge University Press.

Eddington, A. S. (1926a). *The Internal Constitution of the Stars*. Cambridge: Cambridge University Press. Reprinted 1988.

Eddington, A. S. (1926b). The source of stellar energy, *Nature*, **117**, 25–32.

Eddington, A. S. (1930). On the instability of Einstein's spherical world, *Monthly Notices of the Royal Astronomical Society*, **90**, 668–78.

Eddington, A. S. (1931). The end of the world from the standpoint of mathematical physics, *Nature*, **127**, 447–53.

Eddington, A. S. (1933). *The Expanding Universe*. Cambridge: Cambridge University Press.

Edge, D. O., Shakeshaft, J. R., McAdam, W. B., et al. (1959). A survey of radio sources at a frequency of 159 Mc/s, *Memoirs of the Royal Astronomical Society*, **68**, 37–60.

Efstathiou, G. (1990). Cosmological perturbations, in *Physics of the Early Universe*, ed. Peacock, J. A., Heavens, A. F., and Davies, A. T., pp. 361–463. Edinburgh: SUSSP Publications.

Efstathiou, G. (1995). High-redshift galaxies: problems and prospects, in *Galaxies in the Young Universe*, ed. Hippelein, H., Meissenheimer, K., and Röser, H. J., pp. 299–314. Berlin: Springer-Verlag.

Efstathiou, G. and Rees, M. J. (1988). High-redshift quasars in the cold dark matter cosmogony, *Monthly Notices of the Royal Astronomical Society*, **230**, 5P–11P.

Efstathiou, G., Frenk, C. S., White, S. D. M., et al. (1988). Gravitational clustering from scale-free initial conditions, *Monthly Notices of the Royal Astronomical Society*, **235**, 715–48.

Eguchi, K., Enomoto, S., and 97 authors (2003). First results from Kamland: evidence for reactor anti-neutrino disappearance, *Physical Review Letters*, **90**, 021802.

Ehlers, J. (1988). Hermann Weyl's contributions to the general theory of relativity, in *Exact Sciences and their Philosophical Foundations*, ed. Deppert, W. et al., pp. 83–105. Frankfurt am Main: Verlag Peter Lang.

Ehlers, J., Geren, P., and Sachs, R. K. (1968). Isotropic solutions of the Einstein–Liouville equations, *Journal of Mathematical Physics*, **9**, 1344–9.

Einstein, A. (1915). Erklärung der Perihelbewegung des Merkur aus der allgemeinen Relativitäts theorie, Sitzungsberichte der Königlich Preussischen Akademie der Wissenschaften 831–39. Berlin.

Einstein, A. (1916a). *Die Grundlage der Allgemeinen Relativitätstheorie (The Foundation of the General Theory of Relativity)*. Leipzig: J. A. Barth. English translation: In *The Collected Papers of Albert Einstein, Vol. 6. The Berlin Years: Writings, 1914–1917 (English translation supplement)*, ed. Kox, A.J et al., pp. 146–200. Princeton: Princeton University Press.

Einstein, A. (1916b). Ernst Mach, in *The Collected Papers of Albert Einstein, Vol. 6. The Berlin Years: Writings, 1914–1917 (English translation supplement)*, ed. Kox, A. J. et al., pp. 141–5. Princeton: Princeton University Press.

Einstein, A. (1916c). Näherungsweise Integration der Feldgleichungen der Gravitation, *Sitzungs berichte der Königlich Preußischen Akademie der Wissenschaften (Berlin)*, 688–96.

Einstein, A. (1917a). Kosmologische Betrachtungen zur Allgemeinen Relativitätstheorie (Cosmo logical considerations in the general theory of relativity), *Sitzungsberichte, Königlich Preussische Akademie der Wissenschaften (Berlin)*, 1, 142–52. English translation: In *The Collected Papers of Albert Einstein, Vol. 6. The Berlin Years: Writings, 1914–1917 (English translation supplement)*, ed. Kox, A.J et al., pp. 421–32. Princeton: Princeton University Press.

Einstein, A. (1917b). *Über die spezielle und die allgemeine Relativitätstheorie, gemeinverständlich (On the Special and General Theory of Relativity: A Popular Account)*. Braunschweig: Vieweg.

Einstein, A. (1918a). Critical comment on a solution of the gravitational field equations given by Mr. de Sitter, in *The Collected Papers of Albert Einstein, Vol. 7. The Berlin Years: Writings, 1918–1921 (English translation supplement)*, ed. Janssen, M. et al., pp. 36–8. Princeton: Princeton University Press.

Einstein, A. (1918b). On the foundations of the general theory of relativity, in *The Collected Papers of Albert Einstein, Vol. 7. The Berlin Years: Writings, 1918–1921 (English translation supplement)*, ed. Janssen, M. et al., pp. 33–5. Princeton: Princeton University Press.

Einstein, A. (1919). Do gravitational fields play an essential role in the structure of elementary particles of matter?, in *The Collected Papers of Albert Einstein, Vol. 7. The Berlin Years: Writings, 1918–1921 (English translation supplement)*, ed. Janssen, M. et al., pp. 80–8. Princeton: Princeton University Press.

Einstein, A. (1922). Bemerkung zu der Arbeit von A. Friedman "Ueber die Kruemmung des Raumes" (Remark on the work of A. Friedman "On the curvature of space"), *Zeitschrift für Physik*, **11**, 326. English translation: (1986) in *Cosmological Constants: Papers in Mod ern Cosmology*, ed. Bernstein, J. and Feinberg, G., p. 66. New York: Columbia University Press.

Einstein, A. (1936). Lens-like action of a star by the deviation of light in the gravitational field, *Science*, **84**, 506–7.

Einstein, A. (1945). *The Meaning of Relativity*. Princeton: Princeton University Press.

Einstein, A. (1956). *The Meaning of Relativity*, revised edition. Princeton: Princeton University Press.

Einstein, A. and Straus, E. G. (1945). The influence of the expansion of space on the gravitation fields surrounding the individual stars, *Reviews of Modern Physics*, **17**, 120.

Eisenstaedt, J. (1989). Cosmology: a space for thought on general relativity, in *Foundations of Big Bang Cosmology*, ed. Meyerstein, W., pp. 271–95. Singapore: World Scientific.

Eisenstaedt, J. (1994). Lemaître and the Schwarzschild solution, in *The Attraction of Gravitation: New Studies in the History of General Relativity*, ed. Earman, J., Janssen, M., and Norton, J., pp. 353–89. Boston: Birkhäuser.

Eisenstein, D. J. and Hu, W. (1998). Baryonic features in the matter transfer function, *Astrophysical Journal*, **496**, 605–14.

Eisenstein, D. J., Zehavi, I., Hogg, D. W., et al. (2005). Detection of the baryon acoustic peak in the large-scale correlation function of SDSS luminous red galaxies, *Astrophysical Journal*, **633**, 560–74.

Elder, F. R., Gurewitsch, A. M., Langmuir, R. V., et al. (1947). Radiation from electrons in a synchrotron, *Physical Review*, **71**, 829–30.

Ellis, G. F. R. (1971a). Relativistic cosmology, in *General Relativity and Cosmology: Proceedings of the International School of Physics "Enrico Fermi"*, ed. Sachs, R. K., pp. 104–82. New York: Academic Press.

Ellis, G. F. R. (1971b). Topology and cosmology, *General Relativity and Gravitation*, **2**, 7–21.

Ellis, G. F. R. (1980). Limits to verification in cosmology, *Annals of the New York Academy of Sciences*, **336**, 130–60.

Ellis, G. F. R. (1984a). Alternatives to the big bang, *Annual Review of Astronomy and Astrophysics*, **22**, 157–84.

Ellis, G. F. R. (1984b). Relativistic cosmology: its nature, aims and problems, in *General Relativity and Gravitation*, ed. Bertotti, B., de Felice, F., and Pascolini, A., pp. 215–88. Dordrecht: D. Reidel Publishing Co.

Ellis, G. F. R. (1989). The expanding universe: a history of cosmology from 1917 to 1960, in *Einstein and the History of General Relativity*, ed. Howard, D. and Stachel, J., pp. 367–401. Boston: Birkäuser.

Ellis, G. F. R. (1990). Innovation, resistance and change. The transition to the expanding universe, in *Modern Cosmology in Retrospect*, ed. Bertotti, B. et al., pp. 97–113. Cambridge: Cambridge University Press.

Ellis, G. F. R. (2003). The unique nature of cosmology, in *Revisiting the Foundations of Relativistic Physics: Festschrift in Honor of John Stachel*, ed. Renn, J., Divarci, L., and Schröter, P. et al., pp. 193–220. Berlin: Springer Verlag.

Ellis, G. F. R. (2007). Issues in the philosophy of cosmology, in *Handbook for the Philosophy of Physics: Part B*, ed. Earman, J. and Butterfield, J., pp. 1183–286. Dordrecht: Elsevier.

Ellis, G. F. R. (2011a). Does the multiverse really exist?, *Scientific American*, **305**(2), 38–43.

Ellis, G. F. R. (2011b). Editorial note to: Brandon Carter, large number coincidences and the anthropic principle in cosmology, *General Relativity and Gravitation*, **43**, 3213–23.

Ellis, G. F. R. (2014). On the philosophy of cosmology, Studies in History and Philosophy of Science Part B: *Studies in History and Philosophy of Modern Physics*, **46**, 5–23.

Ellis, G. F. R. and Brundrit, G. B. (1979). Life in the infinite universe, *Quarterly Journal of the Royal Astronomical Society*, **20**, 37–41.

Ellis, G. F. R and Madsen, M. S. (1991). Exact scalar field cosmologies, *Classical and Quantum Gravity*, **8**, 667–76.

Ellis, G. F. R. and Penrose, R. (2010). Dennis William Sciama, *Biographical Memoirs of Fellows of the Royal Society*, **56**, 401–22.

Ellis, G. F. R. and Rothman, T. (1993). Lost horizons, *American Journal of Physics*, **61**, 883–93.

Ellis, G. F. R. and Sciama, D. W. (1972). Global and non-global problems in cosmology, in *General Relativity: Papers in Honour of J. L. Synge*, ed. O'Raifeartaigh, L., pp. 35–59. Oxford: Clarendon Press.

Ellis, G. F. R. and Silk, J. (2014). Scientific method: defend the integrity of physics, *Nature*, **516**, 321–3.

Ellis, G. F. R. and Smeenk, C. (2017). Philosophy of cosmology, in *The Stanford Encyclopedia of Philosophy*, ed. Zalta, E. N. Stanford: Metaphysics Research Lab, Stanford University.

Ellis, G. F. R. and Stoeger, W. (1987). The 'fitting problem' in cosmology, *Classical and Quantum Gravity*, **4**, 1697.

Ellis, G. F. R., Kirchner, U., and Stoeger, W. R. (2004). Multiverses and physical cosmology, *Monthly Notices of the Royal Astronomical Society*, **347**, 921–36.

Ellis, G. F. R., Nel, S. D., Maartens, R., et al. (1985). Ideal observational cosmology, *Physics Reports*, **124**, 315–417.

Epstein, E. E. (1967). On the small-scale distribution at 3.4-mm wavelength of the reported 3° K background radiation, *Astrophysical Journal Letters*, **148**, L157.

Fabbri, R., Guidi, I., Melchiorri, F., et al. (1980). Measurement of the cosmic-background large-scale anisotropy in the millimetric region, *Physical Review Letters*, **44**, 1563–6.

Faber, S. M. and Jackson, R. E. (1976). Velocity dispersions and mass-to-light ratios for elliptical galaxies, *Astrophysical Journal*, **204**, 668–83.

Fabian, A. C., Vaughan, S., Nandra, K., et al. (2002). A long hard look at MCG-6-30-15 with XMM-Newton, *Monthly Notices of the Royal Astronomical Society*, **335**, L1–L5.

Fabricant, D. G., Lecar, M., and Gorenstein, P. (1980). X-ray measurements of the mass of M87, *Astrophysical Journal*, **241**, 552–60.

Falb, R. (1875). Die Welten Bildung und Untergang, *Sirius*, **8**, 193–202.

Fan, X., Hennawi, J. F., Richards, G. T., et al. (2004). A survey of $z \geq 5.7$ quasars in the Sloan Digital Sky Survey. III. Discovery of five additional quasars, *Astronomical Journal*, **128**, 515–22.

Fan, X., Narayanan, V. K., Lupton, R. H., et al. (2001). A survey of $z \geq 5.8$ quasars in the Sloan Digital Sky Survey. I. Discovery of three new quasars and the spatial density of luminous quasars at $z \sim 6$, *Astronomical Journal*, **122**, 2833–49.

Fara, P. (2004). *Pandora's Breeches: Women, Science and Power in the Enlightenment*. London: Random House.

Faraoni, V. (2004). *Cosmology in Scalar-Tensor Gravity*. Dordrecht: Kluwer Academic.

Farhi, E., Guth, A. H., and Guven, J. (1990). Is it possible to create a universe in the laboratory by quantum tunneling?, *Nuclear Physics B*, **339**, 417–90.

Fath, E. A. (1909). The spectra of some spiral nebulae and globular star clusters, *Lick Observatory Bulletin*, **149**, 71–7.

Feast, M. W. and Catchpole, R. M. (1997). The Cepheid period–luminosity zero-point from HIPPARCOS trigonometrical parallaxes, *Monthly Notices of the Royal Astronomical Society*, **286**, L1–L5.

Fermi, E. (1949). On the origin of the cosmic radiation, *Physical Review*, **75**, 1169–74.

Fichtel, C. E., Simpson, G. A., and Thompson, D. J. (1978). Diffuse gamma radiation, *Astrophysical Journal*, **222**, 833–49.

Field, G. B. and Hitchcock, J. L. (1966). The radiation temperature of space at $\lambda = 2.6$ mm and the excitation of interstellar CN, *Astrophysical Journal*, **146**, 1–6.

Field, G. B., Arp, H., and Bahcall, J. N. (1974). *The Redshift Controversy. Papers from a Symposium, Washington, DC, December 1972.* Reading, MA: W. A. Benjamin (Addison-Wesley).

Finlay-Freundlich, E. (1954). Red shifts in the spectra of celestial bodies, *Philosophical Magazine,* 45, 303–19.

Fitch, W. S., Pacholczyk, A. G., and Weymann, R. J. (1967). Light variations of the Seyfert galaxy NGC 4151, *Astrophysical Journal,* 150, L67–L70.

Fixsen, D. J. (2009). The temperature of the cosmic microwave background, *Astrophysical Journal,* 707, 916–20.

Fixsen, D. J., Cheng, E. S., Gales, J. M., et al. (1996). The cosmic microwave background spectrum from the full COBE FIRAS data set, *Astrophysical Journal,* 473, 576–87.

Fixsen, D. J., Kogut, A., Levin, S., et al. (2011). ARCADE 2 measurement of the absolute sky brightness at 3–90 GHz, *Astrophysical Journal,* 734, 5.

Flauger, R., Hill, J. C., and Spergel, D. N. (2014). Toward an understanding of foreground emission in the BICEP2 region, *Journal of Cosmology and Astroparticle Physics,* 8, 039.

Flin, P. and Duerbeck, H. (2006). Silberstein, general relativity and cosmology, in *Albert Einstein Century International Conference,* ed. Alimi, J.-M. and Füzfa, A., pp. 1087–94. American Institute of Physics Conference Proceedings, Vol. 861.

Fomalont, E. B., Kellermann, K. I., Anderson, M. C., et al. (1988). New limits to fluctuations in the cosmic background radiation at 4.86 GHz between 12 and 60 arcsecond resolution, *Astronomical Journal,* 96, 1187–91.

Ford, H. C., Harms, R. J., Tsvetanov, Z. I., et al. (1994). Narrowband HST images of M87: evidence for a disk of ionized gas around a massive black hole, *Astrophysical Journal Letters,* 435, L27–L30.

Forgan, D., Dayal, P., Cockell, C., et al. (2017). Evaluating galactic habitability using high-resolution cosmological simulations of galaxy formation, *International Journal of Astrobiology,* 16, 60–73.

Forman, W., Jones, C., Cominsky, L., et al. (1978). The fourth UHURU catalog of X-ray sources, *Astrophysical Journal Supplement Series,* 38, 357–412.

Fowler, R. H. (1926). On dense matter, *Monthly Notices of the Royal Astronomical Society,* 87, 114–22.

Fowler, W. A. (1972). New observations and old nucleocosmochronologies, in *Cosmology, Fusion and Other Matters,* ed. Reines, F., p. 67. London: Hilger.

Francis, P. J., Hewett, P. C., Foltz, C. B., et al. (1991). A high signal-to-noise ratio composite quasar spectrum, *Astrophysical Journal,* 373, 465–70.

Frank, J., King A. and Raine, D. (2002). *Accretion Power in Astrophysics,* third edition. Cambridge: Cambridge University Press.

Frautschi, S. (1982). Entropy in an expanding universe, *Science,* 217, 593–9.

Freedman, W. L. (2017). Correction: cosmology at a crossroads, *Nature Astronomy,* 1, 0169.

Freedman, W. L., Madore, B. F., Gibson, B. K., et al. (2001). Final results from the Hubble Space Telescope key project to measure the Hubble constant, *Astrophysical Journal,* 533, 47–72.

Frenk, C. (1986). Galaxy clustering and the dark-matter problem, *Philosophical Transactions of the Royal Society,* A330, 517–41.

Friedman, A. (1922). Über die Krümmung des Raumes (On the curvature of space), *Zeitschrift für Physik,* 10, 377–86. English translation: (1986), in *Cosmological Constants: Papers in Modern Cosmology,* ed. Bernstein, J. and Feinberg, G., pp. 49–58. New York: Columbia University Press.

Friedman, A. (1924). Über die Möglichkeit einer Welt mit konstanter negativer Krümmung des Raumes (On the possibility of a world with constant negative curvature), *Zeitschrift für Physik,* 12, 326–32. English translation: (1986), in *Cosmological Constants: Papers in Modern Cosmology,* ed. Bernstein, J. and Feinberg, G., pp. 59–65. New York: Columbia University Press.

Friedman, H. (1986). *Sun and Earth*. New York: Scientific American Library.

Friedman, H., Lichtman, S. W., and Byram, E. T. (1951). Photon counter measurements of solar X-rays and extreme ultraviolet light, *Physical Review*, **83**, 1025–30.

Frieman, J., Turner, M., and Huterer, D. (2008). Dark energy and the accelerating universe. arXiv preprint arXiv:0803.0982.

Fritz, G., Henry, R. C., Meekins, J. F., et al. (1969). X-ray pulsar in the Crab Nebula, *Science*, **164**, 709–12.

Fukugita, M., Futamase, T., Kasai, M., et al. (1992). Statistical properties of gravitational lenses with a nonzero cosmological constant, *Astrophysical Journal*, **393**, 3–21.

Galbraith, W. and Jelley, J. V. (1953). Light pulses from the night sky associated with cosmic rays, *Nature*, **171**, 349–50.

Galbraith, W. and Jelley, J. V. (1955). Light-pulses from the night sky and Cherenkov radiation, Part 1, *Journal of Atmospheric and Terrestrial Physics*, **6**, 250–62.

Gale, G. (2017). Cosmology: methodological debates in the 1930s and 1940s, in *The Stanford Encyclopedia of Philosophy*, ed. Zalta, E. N. Stanford: Metaphysics Research Lab, Stanford University.

Gamow, G. (1937). *Atomic Nuclei and Nuclear Transformations*. Oxford: Oxford University Press.

Gamow, G. (1939). Physical possibilities of stellar evolution, *Physical Review*, **55**, 718–25.

Gamow, G. (1940). *The Birth and Death of the Sun*. New York: Viking Press.

Gamow, G. (1946). Expanding universe and the origin of the elements, *Physical Review*, **70**, 572–3.

Gamow, G. (1949). On relativistic cosmogony, *Reviews of Modern Physics*, **21**, 367–73.

Gamow, G. (1954). Modern cosmology, *Scientific American*, **190**, 55–63.

Gamow, G. (1956). The evolutionary universe, *Scientific American*, **195**, 136–56.

Gamow, G. (1967). Does gravity change with time?, *Proceedings of the National Academy of Sciences*, **57**, 187–93.

Gamow, G. (1970). *My World Line*. New York: Viking Press.

Gamow, G. and Fleming, J. A. (1942). Report on the eighth annual Washington conference of theoretical physics, *Science*, **95**, 579–81.

Garbell, M. A. (1963). Theses of the First Soviet Gravitation Conference held in Moscow in the summer of 1961. San Francisco: Garbell Research Foundation.

Gardiner, J. (2014). *The Law of Chaos: The Multiverse of Michael Moorcock*. New York: Headpress.

Gardner, A. and Conlon, J. P. (2013). Cosmological natural selection and the purpose of the universe, *Complexity*, **18**, 48–56.

Garnavich, P. M., Kirshner, R. P., Challis, P., et al. (1998). Constraints on cosmological models from Hubble Space Telescope observations of high-z supernovae, *Astrophysical Journal Letters*, **493**, L53–8.

Garriga, J., Mukhanov, V. F., Olum, K. D., et al. (2000). Eternal inflation, black holes, and the future of civilizations, *International Journal of Theoretical Physics*, **39**, 1887–900.

Gasperini, M. and Maharana, J. (Eds.) (2007). *String Theory and Fundamental Interactions: Gabriele Veneziano and Theoretical Physics: Historical and Contemporary Perspectives*. Berlin: Springer-Verlag.

Gasperini, M. and Veneziano, G. (1993). Pre-big-bang in string cosmology, *Astroparticle Physics*, **1**, 317–39.

Gasperini, M. and Veneziano, G. (2003). The pre-big bang scenario in string cosmology, *Physics Reports*, **373**, 1–212.

Gasperini, M. and Veneziano, G. (2015). String theory and pre-big bang cosmology, *Il Nuovo Cimento*, **38**, 160.

Gasperini, M., Giovannini, M., and Veneziano, G. (1995). Primordial magnetic fields from string cosmology, *Physical Review Letters*, 75, 3796–9.

Gerig, A. (2014). Are there many worlds? arXiv preprint arXiv:1406.7215.

Gerig, A., Olum, K. D., and Vilenkin, A. (2013). Universal doomsday: analyzing our prospects for survival, *Journal of Cosmology and Astroparticle Physics*, 5, 013.

Geroch, R. (1977). Prediction in general relativity, in *Foundation of Space-Time Theories*, ed. Earman, J. S., Glymour, C. N., and Stachel, J. J., pp. 81–93. Minneapolis: University of Minnesota Press.

Gershtein, S. S. and Zeldovich, Y. B. (1966). Rest mass of a muonic neutrino and cosmology, *Pisma v Zhurnal Eksperimentalnoi i Teoreticheskoi Fiziki*, 4, 174–7. Translation: Soviet Journal of Experimental and Theoretical Physics Letters, 1966, 4, pp. 120–2.

Gertsenshtein, M. E. (1962). Wave resonance of light and gravitional waves, *Soviet Physics JETP*, 14, 84–5.

Gertsenshtein, M. E. and Pustovoit, V. I. (1963). On the detection of low frequency gravitational waves, *Journal of Experimental and Theoretical Physics*, 16, 433–5.

Ghez, A. M., Morris, M., Becklin, E. E., et al. (2000). The accelerations of stars orbiting the Milky Way's central black hole, *Nature*, 407, 349–51.

Giacconi, R., Bechtold, J., Branduardi, G., et al. (1979). A high-sensitivity X-ray survey using the Einstein Observatory and the discrete source contribution to the extragalactic X-ray background, *Astrophysical Journal Letters*, 234, L1–L7.

Giacconi, R., Gursky, H., Kellogg, E., et al. (1971a). Discovery of periodic X-ray pulsations in Centaurus X-3 from UHURU, *Astrophysical Journal*, 167, L67–L73.

Giacconi, R., Gursky, H., Paolini, F. R., et al. (1962). Evidence for X-rays from sources outside the Solar System, *Physical Review Letters*, 9, 439–43.

Giacconi, R., Gursky, H., and van Speybroeck, L. P. (1968). Observational techniques in X-ray astronomy, *Annual Review of Astronomy and Astrophysics*, 6, 373–416.

Giacconi, R., Kellogg, E., Gorenstein, P., et al. (1971b). An X-ray scan of the galactic plane from UHURU, *Astrophysical Journal*, 165, L27–L35.

Gibbons, G. W. and Hawking, S. W. (1977). Cosmological event horizons, thermodynamics, and particle creation, *Physical Review*, D15, 2738–51.

Gibbons, G. W., Shellard, E. P. S., and Rankin, S. J. (Eds.) (2003). *The Future of Theoretical Physics and Cosmology*. Cambridge: Cambridge University Press.

Gillespie, R. (2011). *The Great Melbourne Telescope*. Melbourne: Museum Victoria.

Gillessen, S., Eisenhauer, F., Fritz, T. K., et al. (2013). The distance to the galactic center, in *Advancing the Physics of Cosmic Distances*, ed. de Grijs, R., volume 289 of *IAU Symposium Series*, pp. 29–35.

Gillessen, S., Eisenhauer, F., Trippe, S., et al. (2009). Monitoring stellar orbits around the massive black hole in the galactic center, *Astrophysical Journal*, 692, 1075–1109.

Gilli, R., Comastri, A., and Hasinger, G. (2007). The synthesis of the cosmic X-ray background in the Chandra and XMM-Newton era, *Astronomy and Astrophysics*, 463, 79–96.

Gillies, G. T. (1997). The Newtonian gravitational constant: recent measurements and related studies, *Reports on Progress in Physics*, 60, 151–223.

Gindilis, L. M. (2012). The development of radio astronomy at the Sternberg Astronomical Institute of Lomonosov Moscow State University and the Space Research Institute of the USSR Academy of Sciences, in *A Brief History of Radio Astronomy in the USSR: A Collection of Scientific Essays (Vol. 382)*, ed. Braude, S., Dubinskii, B. A., Kaidanovskii, N. L., et al., pp. 89–116. Dordrecht: Springer.

Ginzburg, V. L. (1951). Cosmic rays as a source of galactic radio-radiation, *Doklady Akademiya Nauk SSSR*, **76**, 377–80.

Gleiser, M. (2010). Drake equation for the multiverse: from the string landscape to complex life, *International Journal of Modern Physics D*, **19**, 1299–308.

Goenner, H. (2001). Weyl's contributions to cosmology, in *Hermann Weyl's 'Raum-Zeit-Materie' and a General Introduction to his Scientific Work*, ed. Scholz, E., pp. 105–37. Basel: Birkhäuser.

Goenner, H. (2012). Some remarks on the genesis of scalar-tensor theories, *General Relativity and Gravitation*, **44**, 2077–97.

Goenner, H. (2017). A golden age of general relativity? Some remarks on the history of general relativity, *General Relativity and Gravitation*, **49**, 42.

Gold, T. (1965). After-dinner speech, in *Quasi-stellar sources and gravitational collapse*, ed. Robinson, I., Schild, A., and Schucking, E. L., p. 470. Chicago: University of Chicago Press.

Gold, T. (1968). Rotating neutron stars as the origin of pulsating radio sources, *Nature*, **218**, 731–2.

Gold, T. and Pacini, F. (1968). Can the observed microwave background be due to a superposition of sources?, *Astrophysical Journal Letters*, **152**, L115.

Goldhaber, G., Boyle, B., Bunclark, P., et al. (1996). Cosmological time dilation using type Ia supernovae as clocks, *Nuclear Physics B Proceedings Supplements*, **51**, 123–7.

Goldhaber, M. (1956). Speculations on cosmogony, *Science*, **124**, 218–19.

González-Díaz, P. F. (2003). You need not be afraid of phantom energy, *Physical Review D*, **68**, 021303.

Goobar, A. and Perlmutter, S. (1995). Feasibility of measuring the cosmological constant Lambda and mass density Omega using type IA supernovae, *Astrophysical Journal*, **450**, 14–18.

Goodman, J. (1995). Geocentrism reexamined, *Physical Review D*, **52**, 1821–7.

Gorenstein, M. V., Muller, R. A., Smoot, G. F., et al. (1978). Radiometer system to map the cosmic background radiation, *Review of Scientific Instruments*, **49**, 440–8.

Gott, J. (1993). Implications of the Copernican principle for our future prospects, *Nature*, **363**, 315–19.

Gower, J. F. R. (1966). The source counts from the 4C Survey, *Memoirs of the Royal Astronomical Society*, **133**, 151–61.

Gower, J. F. R., Scott, P. F., and Wills, D. (1967). A survey of radio sources in the declination ranges -07 to 20 and 40 to 80, *Monthly Notices of the Royal Astronomical Society*, **71**, 49–144.

Graham, L. R. (1972). *Science and Philosophy in the Soviet Union*. New York: Alfred A. Knopf.

Graham Smith, F. (1951). An accurate determination of the positions of four radio stars, *Nature*, **168**, 555.

Green, S. R. and Wald, R. M. (2014). How well is our universe described by an FLRW model?, *Classical and Quantum Gravity*, **31**, 234003.

Greenstein, J. L. and Schmidt, M. (1964). Red-shifts of the radio sources 3C 48 and 3C 273, *Astrophysical Journal*, **140**, 1–43.

Gregory, A. (2007). *Ancient Greek Cosmogony*. London: Duckworth.

Gregory, J. (2005). *Fred Hoyle's Universe*. Oxford: Oxford Universtiy Press.

Greisen, K. (1966). End to the cosmic-ray spectrum?, *Physical Review Letters*, **16**, 748–50.

Gribbin, J. R. (1976). *Galaxy Formation. A Personal View*, New York: John Wiley and Sons.

Gribbin, J. R. and Rees, M. J. (1989). *Dark Matter, Mankind and Anthropic Cosmology*. New York: Bantam Books.

Grünbaum, A. (1952). Some highlights of modern cosmology and cosmogony, *Review of Metaphysics*, **5**, 481–98.

Gull, S. F. (1975). The X-ray, optical and radio properties of young supernova remnants, *Monthly Notices of the Royal Astronomical Society*, **171**, 263–78.

Gunn, J. E. (1978). The Friedman models and optical observations in cosmology, in *Observational Cosmology: 8th Advanced Course, Swiss Society of Astronomy and Astrophysics, Saas-Fee 1978*, ed. Maeder, A., Martinet, L., and Tammann, G., pp. 1–121. Geneva: Geneva Observatory Publications.

Gunn, J. E. and Peterson, B. A. (1965). On the density of neutral hydrogen in intergalactic space, *Astrophysical Journal*, **142**, 1633–6.

Gurevich, L. E. (1975). On the origin of the metagalaxy, *Astrophysics and Space Science*, **38**, 67–78.

Gursky, H., Giacconi, R., Paolini, F. R., et al. (1963). Further evidence for the existence of galactic X-rays, *Physical Review Letters*, **11**, 530–5.

Gush, H. P. (1981). Rocket measurement of the cosmic background submillimeter spectrum, *Physical Review Letters*, **47**, 745–8.

Gush, H. P., Halpern, M., and Wishnow, E. H. (1990). Rocket measurement of the cosmic-background-radiation mm-wave spectrum, *Physical Review Letters*, **65**, 537–40.

Guth, A. H. (1981). Inflationary universe: a possible solution to the horizon and flatness problems, *Physical Review D*, **23**, 347–56.

Guth, A. H. (1997). *The Inflationary Universe: The Quest for a New Theory of Cosmic Origins*. Reading, MA: Addison-Wesley.

Guth, A. H. (2007). Eternal inflation and its implications, *Journal of Physics A: Mathematical and Theoretical*, **40**, 6811–26.

Guth, A. H. and Pi, S.-Y. (1982). Fluctuations in the new inflationary universe, *Physical Review Letters*, **49**, 1110–13.

Guth, A. H. and Tye, S.-H. H. (1980). Phase transitions and magnetic monopole production in the very early universe, *Physical Review Letters*, **44**, 631–5.

Hale, G. E. (1928). The possibilities of large telescopes, *Harper's Magazine*, **156**, 639–46.

Hamilton, A. J. S., Kumar, P., Lu, E., et al. (1991). Reconstructing the primordial spectrum of fluctuations of the universe from the observed nonlinear clustering of galaxies, *Astrophysical Journal*, **374**, L1–L4.

Hannestad, S. and Mersini-Houghton, L. (2005). First glimpse of string theory in the sky?, *Physical Review D*, **71**, 123504.

Hanson, D., Hoover, S., Crites, A., et al. (2013). Detection of B-mode polarization in the cosmic microwave background with data from the South Pole Telescope, *Physical Review Letters*, **111**, 141301.

Hanson, N. (1963). Some philosophical aspects of contemporary cosmologies, in *Philosophy of Science: The Delaware Seminar, Vol. 2*, ed. Baumrin, B., pp. 465–82. New York: Interscience.

Häring, N. and Rix, H. (2004). On the black hole mass–bulge mass relation, *Astrophysical Journal Letters*, **604**, L89–L92.

Harms, R. J., Ford, H. C., Tsvetanov, Z. I., et al. (1994). HST FOS spectroscopy of M87: evidence for a disk of ionized gas around a massive black hole, *Astrophysical Journal Letters*, **435**, L35–8.

Harper, E. (2001). George Gamow: Scientific amateur and polymath, *Physics in Perspective*, **3**, 335–72.

Harrison, E. R. (1970). Fluctuations at the threshold of classical cosmology, *Physical Review*, **D1**, 2726–30.

Harrison, E. R. (1986). Newton and the infinite universe, *Physics Today*, **39**, 24–32.

Harrison, E. R. (1987). *Darkness at Night: A Riddle of the Universe*. Cambridge: Cambridge University Press.

Harrison, E. R. (1993). The redshift–distance and velocity–distance law, *Astrophysical Journal*, **403**, 28–31.

Harrison, E. R. (1995). The natural selection of universes containing intelligent life, *Quarterly Journal of the Royal Astronomical Society*, **36**, 193–203.

Harrison, E. R. (2000). *Cosmology. The Science of the Universe*. Cambridge: Cambridge University Press.

Hart, R. (1973). *Adriaan van Maanen's influence on the island universe theory*. PhD thesis, Boston University.

Hartle, J. B. (2003). *Gravity. An Introduction to Einstein's General Relativity*. San Francisco: Addison Wesley.

Harvey, B. and Zakutnyaya, O. (2011). *Russian Space Probes: Scientific Discoveries and Future Missions*. Dordrecht: Springer.

Harzer, P. (1908). Die Sterne und der Raum, *Jahresbericht der Deutschen Mathematiker-Vereinigung*, **17**, 237–67.

Hasinger, G., Burg, R., Giacconi, R., et al. (1993). A deep X-ray survey in the Lockman Hole and the soft X-ray log N–log S, *Astronomy and Astrophysics*, **275**, 1–15.

Hauser, M. G. and Dwek, E. (2001). The cosmic infrared background: measurements and implications, *Annual Review of Astronomy and Astrophysics*, **39**, 249–307.

Hausman, M. A. and Ostriker, J. P. (1977). Cannibalism among galaxies – dynamically produced evolution of cluster luminosity functions, *Astrophysical Journal*, **217**, L125–9.

Hawking, S. W. (1969). On the rotation of the Universe, *Monthly Notices of the Royal Astronomoical Society*, **142**, 129–41.

Hawking, S. W. (1972). Black holes in general relativity, *Communications in Mathematical Physics*, **25**, 152–66.

Hawking, S. W. (1975). Particle creation by black holes, in *Quantum Gravity: Proceedings of the Oxford Symposium*, ed. Isham, C. J., Penrose, R., and Sciama, D. W., pp. 219–67. Oxford: Clarendon Press.

Hawking, S. W. and Ellis, G. R. (1973). *The Large Scale Structure of Space-Time*. Cambridge: Cambridge University Press.

Hawking, S. W. and Mlodinow, L. (2010). *The Grand Design*. London: Bantam Press.

Hawking, S. W. and Penrose, R. (1969). The singularities of gravitational collapse and cosmology, *Proceedings of the Royal Society*, **A314**, 529–48.

Hayakawa, S. and Matsuoka, M. (1964). Part V. Origin of cosmic X-rays, *Supplement of Progress of Theoretical Physics (Japan)*, **30**, 204–28.

Hayashi, C. (1950). Proton–neutron concentration ratio in the expanding universe at the stages preceding the formation of the elements, *Progress of Theoretical Physics (Japan)*, **5**, 224–35.

Hazard, C. and Salpeter, E. E. (1969). Discrete sources and the microwave background in steady state cosmologies, *Astrophysical Journal Letters*, **157**, L87.

Hazard, C., Mackey, M. B., and Shimmins, A. J. (1963). Investigation of the radio source 3C 273 by the method of lunar occultations, *Nature*, **197**, 1037–9.

Hearnshaw, J. (2014). *The Analysis of Starlight: Two Hundred Years of Astronomical Spectroscopy*. Cambridge: Cambridge University Press.

Heckmann, O. and Schücking, E. (1959). Andere kosmologische Theorien, in *Handbuch der Physik*, *vol. 53*, ed. Flügge, S., pp. 320–57. Berlin: Springer.

Heller, M. and Szydlowski, M. (1983). Tolman's cosmological models, *Astrophysics and Space Science*, **90**, 327–35.

Henderson, A. (1925). Is the universe finite?, *The American Mathematical Monthly*, **32**, 213–23.

Henning, J. W., Sayre, J. T., Reichardt, C. L., et al. (2018). Measurements of the temperature and E-mode polarization of the CMB from 500 square degrees of SPTpol data, *Astrophysical Journal*, **852**, 97.

Henry, P. S. (1971). Isotropy of the 3 K background, *Nature*, **231**, 516–18.

Hentschel, K. (1997). *The Einstein Tower*. Stanford: Stanford University Press.

Hentschel, K. (2002). *Mapping the Spectrum: Techniques of Visual Representation in Research and Teaching*. Oxford: Oxford University Press.

Henyey, L. G. and Keenan, P. C. (1940). Interstellar radiation from free electrons and hydrogen atoms, *Astrophysical Journal*, **91**, 625–30.

Herschel, J. (1822). Address of the Society explanatory of their views and objects, *Memoirs of the Astronomical Society of London*, **1**, 1–7.

Herschel, J. (1826). Account of some observations made with a 20-feet Reflecting telescope ..., *Memoirs of the Royal Astronomical Society, Vol. II*.

Herschel, J. (1847). *Results of Astronomical Observations Made During the Years 1834, 5, 6, 7, 8, at the Cape of Good Hope ...*. London: Smith, Elder, and Co.

Herschel, W. (1802). Catalogue of 500 new nebulae, nebulous stars, planetary nebulae, and clusters of stars; with remarks on the construction of the heavens, *Philosophical Transactions of the Royal Society of London Series I*, **92**, 477–528.

Herschel, W. (1811). Astronomical observations relating to the construction of the heavens, arranged for the purpose of a critical examination ..., *Philosophical Transactions of the Royal Society of London*, **101**, 269–336.

Hertzsprung, E. (1913). Über die räumliche Verteilung der Veränderlichen vom δ-Cephei-Typus, *Astronomische Nachrichten*, **196**, cols. 201–10.

Herzberg, G. (1950). *Molecular Spectra and Molecular Structure. Vol. 1: Spectra of Diatomic Molecules*. New York: Van Nostrand Reinhold.

Hess, V. F. (1912). Über Beobachtungen der durchdringenden Strahlung bei sieben Freiballonfahrten, (Concerning observations of penetrating radiation on seven free balloon flights), *Physikalische Zeitschrift*, **13**, 1084–91.

Hesser, J. E., Harris, W. E., VandenBerg, D. A., et al. (1989). A CCD color-magnitude study of 47 Tucanae, *Publications of the Astronomical Society of the Pacific*, **99**, 739–808.

Hetherington, N. S. (1982). Philosophical values and observations in Edwin Hubble's choice of a model of the universe, *Historical Studies in the Physical Sciences*, **13**, 41–67.

Hetherington, N. S. (1990). *The Edwin Hubble Papers. Previously Unpublished Manuscripts on the Extragalactic Nature of Spiral Nebulae*. Tucson: Pachart.

Hetherington, N. S. (1996). *Hubble's Cosmology: A Guided Study of Selected Texts*. Tucson: Pachart Publishing House.

Hewish, A. (1986). The pulsar era, *Quarterly Journal of the Royal Astronomical Society*, **27**, 548–58.

Hewish, A., Bell, S. J., Pilkington, J. D. H., et al. (1968). Observations of a rapidly pulsating radio source, *Nature*, **217**, 709–13.

Hewitt, J. N., Turner, E. L., Burke, B. F., et al. (1987). A VLA gravitational lens survey, in *Observational Cosmology: IAU Symposium No. 124*, ed. Hewitt, A., Burbidge, G., and Fang, L. Z., pp. 747–50. Dordrecht: D. Reidel Publishing Company.

Hey, J. S. (1946). Solar radiations in the 4–6 metre radio wave-length band, *Nature*, **157**, 47–8.

Hey, J. S., Parsons, S. J., and Phillips, J. W. (1946). Fluctuations in cosmic radiation at radio-frequencies, *Nature*, **158**, 234.

Higgs, P. W. (1964). Broken symmetries, massless particles and gauge fields, *Physics Letters*, **12**, 132–3.

Hinshaw, G., Spergel, D. N., Verde, L., et al. (2003). First-year Wilkinson Microwave Anisotropy Probe (WMAP) observations: the angular power spectrum, *Astrophysical Journal Supplement*, **148**, 135–59.

Hirata, C. M., Ho, S., Padmanabhan, N., et al. (2008). Correlation of CMB with large-scale structure. II. Weak lensing, *Physical Review D*, **78**, 043520.

Hirsh, R. F. (1979). The riddle of the gaseous nebulae, *Isis*, **70**, 197–212.

Hjellming, R. and Wade, C. (1971). Further radio observations of Scorpius X-1, *Astrophysical Journal*, **170**, 523–8.

Hjorth, J., Sollerman, J., Møller, P., et al. (2003). A very energetic supernova associated with the γ-ray burst of 29 March 2003, *Nature*, **423**, 847–50.

Hobson, M. P., Jaffe, A. H., Liddle, A. R., et al. (2013). *Bayesian Methods in Cosmology*. Cambridge: Cambridge University Press.

Hogan, C. J. (2000). Why the universe is just so, *Reviews of Modern Physics*, **72**, 1149–61.

Hogg, D. W., Eisenstein, D. J., Blanton, M. R., et al. (2005). Cosmic homogeneity demonstrated with luminous red galaxies, *Astrophysical Journal*, **624**, 54–8.

Hohl, F. (1971). Numerical experiments with a disk of stars, *Astrophysical Journal*, **168**, 343–59.

Holberg, J. B. (2010). Sirius B and the measurement of the gravitational redshift, *Journal for the History of Astronomy*, **41**, 41–64.

Holder, R. and Mitton, S. (Eds.) (2012). *Georges Lemaître: Life, Science and Legacy*. Berlin: Springer-Verlag.

Holman, M. (2018). How problematic is the near-euclidean spatial geometry of the large-scale universe? *Foundations of Physics*, **48**, 1617–47.

Holmberg, G. (1999). *Reaching for the Stars: Studies in the History of Swedish Stellar and Nebular Astronomy 1860–1940*. Lund: Ugglan.

Holmes, R. (2008). *The Age of Wonder: How the Romantic Generation Discovered the Beauty and Terror of Science*. New York: Harper Collins Publishers.

Holton, G. (1988). *Thematic Origins of Scientific Thought: Kepler to Einstein*. Cambridge, MA: Harvard University Press.

Holton, G. (1993). *Science and Anti-science*. Cambridge, MA: Harvard University Press.

Hoskin, M. (1967). Apparatus and ideas in mid-nineteenth century cosmology, *Vistas in Astronomy*, **9**, 79–85.

Hoskin, M. (1976a). The 'Great Debate': what really happened, *Journal for the History of Astronomy*, **7**, 169–82.

Hoskin, M. (1976b). Ritchey, Curtis and the discovery of novae in spiral nebulae, *Journal for the History of Astronomy*, **7**, 47–53.

Hoskin, M. (1982). William Herschel's investigations of nebulae, in *Stellar Astronomy: Historical Studies*, Chalfont St. Giles, UK: Science History Publication, pp. 125–36.

Hoskin, M. (1987). John Herschel's cosmology, *Journal for the History of Astronomy*, **18**, 1–34.

Hoskin, M. (1990). Rosse, Robinson, and the resolution of the nebulae, *Journal for the History of Astronomy*, **21**, 331–44.

Hoskin, M. (2011). *Discoverers of the Universe: William and Caroline Herschel*. Princeton: Princeton University Press.

Hoskin, M. (2012). *The Construction of the Heavens: William Herschel's Cosmology*. Cambridge: Cambridge University Press.

Hoskin, M. (2013). *Caroline Herschel: Priestess of the New Heavens*. Sagamore Beach, MA: Watson Publishing International.

Howell, T. F. and Shakeshaft, J. R. (1966). Measurement of the minimum cosmic background radiation at 20.7-cm wave-length, *Nature*, **210**, 1318–19.

Howell, T. F. and Shakeshaft, J. R. (1967). Spectrum of the 3° K cosmic microwave radiation, *Nature*, **216**, 753–4.

Hoyle, F. (1948). A new model for the expanding universe, *Monthly Notices of the Royal Astronomical Society*, **108**, 372–82.

Hoyle, F. (1950). *The Nature of the Universe*. Oxford: Blackwell.

Hoyle, F. (1954). On nuclear reactions occurring in very hot stars. I. The synthesis of elements from carbon to nickel, *Astrophysical Journal Supplement*, **1**, 121–46.

Hoyle, F. (1955). *Frontiers of Astronomy*. London: William Heinemann.

Hoyle, F. (1975). *Astronomy and Cosmology: A Modern Course*. San Francisco: WH Freeman and Co.

Hoyle, F. (1982). Steady state cosmology revisited, in *Cosmology and Astrophysics*, ed. Terzian, Y. and Bilson, E., pp. 17–57.

Hoyle, F. (1990). An assessment of the evidence against the steady-state theory, in *Modern Cosmology in Retrospect*, ed. Bertotti, B. et al., pp. 221–31. Cambridge: Cambridge University Press.

Hoyle, F. and Fowler, W. A. (1962). On the nature of strong radio sources, *Monthly Notices of the Royal Astronomical Society*, **125**, 169–76.

Hoyle, F. and Fowler, W. A. (1963). Nature of strong radio sources, *Nature*, **197**, 533–5.

Hoyle, F. and Narlikar, J. V. (1966). A radical departure from the 'steady-state' concept in cosmology, *Proceedings of the Royal Society of London Series A*, **290**, 162–76.

Hoyle, F. and Narlikar, J. V. (1972). Cosmological models in a conformally invariant gravitational theory, *Monthly Notices of the Royal Astronomical Society*, **155**, 305–21, 323–35.

Hoyle, F. and Tayler, R. J. (1964). The mystery of the cosmic helium abundance, *Nature*, **203**, 1108–10.

Hoyle, F. and Wickramasinghe, N. C. (1967). Impurities in interstellar grains, *Nature*, **214**, 969–71.

Hoyle, F., Burbidge, G. R., and Narlikar, J. (2000). *A Different Approach to Cosmology*. Cambridge: Cambridge University Press.

Hoyle, F., Burbidge, G. R., and Sargent, W. L. W. (1966). On the nature of the quasi-stellar sources, *Nature*, **209**, 751–3.

Hoyt, W. G. (1976). *Lowell and Mars*. Tucson: University of Arizona Press.

Hu, W. and Dodelson, S. (2002). Cosmic microwave background anisotropies, *Annual Review of Astronomy and Astrophysics*, **40**, 171–216.

Hu, W. and White, M. (1997). A CMB polarization primer, *New Astronomy*, **2**, 323–44.

Hu, W. and White, M. (2004). The cosmic symphony, *Scientific American*, **290**, 44–53.

Hu, W., Sugiyama, N., and Silk, J. (1997). The physics of microwave background anisotropies, *Nature*, **386**, 37–43.

Hubble, E. (1920). Photographic investigations of giant nebulae, *Publications of the Yerkes Observatory*, **4**, 69–85.

Hubble, E. (1925). NGC6822, a remote stellar system, *Astrophysical Journal*, **62**, 409–33.

Hubble, E. (1926a). Extra-galactic nebulae, *Astrophysical Journal*, **64**, 321–69.

Hubble, E. (1926b). A spiral nebula as a stellar system, Messier 33, *Astrophysical Journal*, **63**, 236–74.

Hubble, E. (1929a). A relation between distance and radial velocity among extra-galactic nebulae, *Proceedings of the National Academy of Sciences*, **15**, 168–73.

Hubble, E. (1929b). A spiral nebula as a stellar system, Messier 31, *Astrophysical Journal*, **69**, 103–58.

Hubble, E. (1934a). The distribution of extra-galactic nebulae, *Astrophysical Journal*, **79**, 8–76.

Hubble, E. (1934b). *Red-Shifts in the Spectra of Nebulae*. Oxford: Clarendon Press.

Hubble, E. (1935). Angular rotations of spiral nebulae, *Astrophysical Journal*, 81, 334–5.

Hubble, E. (1936a). Effects of red shifts on the distribution of nebulae, *Astrophysical Journal*, 84, 517–54.

Hubble, E. (1936b). *The Realm of the Nebulae*. New Haven: Yale University Press.

Hubble, E. (1937). Red-shifts and the distribution of nebulae, *Monthly Notices of the Royal Astronomical Society*, 97, 506–13.

Hubble, E. (1938). Adventures in cosmology, *Astronomical Society of the Pacific leaflets*, 3, 120–3.

Hubble, E. and Humason, M. (1931). The velocity–distance relation among extra-galactic nebulae, *Astrophysical Journal*, 74, 43–80.

Hubble, E. and Tolman, R. C. (1935). Two methods of investigating the nature of the nebular red-shift, *Astrophysical Journal*, 82, 302–37.

Hudson, M. J., Dekel, A., Courteau, S., et al. (1995). Ω and biasing from optical galaxies versus POTENT mass, *Monthly Notices of the Royal Astronomical Society*, 274, 305–16.

Hufbauer, K. (1991). *Exploring the Sun: Solar Science since Galileo*. Baltimore: Johns Hopkins University Press.

Huggins, W. (1864). On the spectra of some of the nebulae, *Philosophical Transactions of the Royal Society of London*, 154, 437–44.

Huggins, W. (1865). On the spectrum of the Great Nebula in the sword-handle of Orion, *Proceedings of the Royal Society of London*, 14, 39–42.

Huggins, W. (1891). Presidential address, British Association for the Advancement of Science, Cardiff 1891, in *The Scientific Papers of Sir William Huggins (with M. L. Huggins)*, ed. Huggins, W. and Huggins, M. L., pp. 504–39. London: William Wesley and Son.

Huggins, W. (1897). The new astronomy: a personal retrospect, *The Nineteenth Century: A Monthly Review*, 41, 907–28.

Huggins, W. and Huggins, M. (1909). *The Scientific Papers of Sir William Huggins*. London: William Wesley and Son.

Hulse, R. A. and Taylor, J. H. (1975). Discovery of a pulsar in a binary system, *Astrophysical Journal Letters*, 195, L51–3.

Humason, M. (1934). The apparent velocity of a nebula in the Böotis cluster no. 1, *Publications of the Astronomical Society of the Pacific*, 46, 290–2.

Humason, M. L., Mayall, N. U., and Sandage, A. R. (1956). Redshifts and magnitudes of extra-galactic Nebulae, *Astronomical Journal*, 61, 97–162.

Hutten, E. H. (1962). Methodological remarks concerning cosmology, *The Monist*, 47, 104–15.

Ijjas, A., Steinhardt, P. J., and Loeb, A. (2014). Inflationary schism, *Physics Letters B*, 736, 142–6.

Isham, C. and Butterfield, J. (2000). On the emergence of time in quantum gravity, in *The Arguments of Time*, ed. Butterfield, J., pp. 111–68. Oxford: Oxford University Press.

Israel, W. (1987). Dark stars: the evolution of an idea, in *300 Years of Gravitation*, ed. Hawking, S. W. and Israel, W., pp. 199–276. Cambridge: Cambridge University Press.

Israelit, M. and Rosen, N. (1989). A singularity-free cosmological model in general relativity, *Astrophysical Journal*, 342, 627–34.

Jaki, S. L. (1969). *The Paradox of Olbers' Paradox*. New York: Herder and Herder.

Jaki, S. L. (1974). *Science and Creation. From Eternal Cycles to an Oscillating Universe*. Edinburgh: Scottish Academic Press.

Jaki, S. L. (1977). *Planets and Planetarians: A History of Theories of the Origin of Planetary Systems*. New York: John Wiley and Sons.

Jaki, S. L. (1979). Das Gravitations-Paradoxon des unendlichen Universums, *Sudhoffs Archiv*, **63**, 105–22.

James, W. (1895). Is life worth living?, *International Journal of Ethics*, **6**, 1–24.

Jansky, K. G. (1933). Electrical disturbances apparently of extraterrestrial origin, *Proceedings of the Institution of Radio Engineers*, **21**, 1387–98.

Janssen, M. (2014). No success like failure …, in *The Cambridge Companion to Einstein*, ed. Janssen, M. and Lehner, C., pp. 167–227. Cambridge: Cambridge University Press.

Jeans, J. H. (1902). The stability of a spherical nebula, *Philosophical Transactions of the Royal Society of London*, **199**, 1–53.

Jeans, J. H. (1917). Internal motions in spiral nebulae, *The Observatory*, **40**, 60–1.

Jeans, J. H. (1919). *Problems of Cosmogony and Stellar Dynamics*. Cambridge: Cambridge University Press.

Jenkins, A. and Perez, G. (2010). Looking for life in the multiverse, *Scientific American*, **302**(1), 42–9.

Jennison, R. C. and Das Gupta, M. K. (1953). Fine structure of the extra-terrestrial radio source Cygnus 1, *Nature*, **172**, 996–7.

Jiménez, J. B., Lazkoz, R., Sáez-Gómez, D., et al. (2016). Observational constraints on cosmological future singularities, *European Physical Journal C*, **76**, 631.

Jöeveer, M. and Einasto, J. (1978). Has the universe the cell structure?, in *The Large Scale Structure of the Universe*, ed. Longair, M. S. and Einasto, J., pp. 241–51. Dordrecht: D. Reidel Publishing Company.

Johnson, D. G. and Wilkinson, D. T. (1987). A 1 percent measurement of the temperature of the cosmic microwave radiation at $\lambda = 1.2$ centimeters, *Astrophysical Journal Letters*, **313**, L1–L4.

Jordan, P. (1952). *Schwerkraft und Weltall: Grundlagen der theoretischen Kosmologie*. Braunschweig: Vieweg & Sohn.

Jordan, P. (1971). *The Expanding Earth: Some Consequences of Dirac's Gravitation Hypothesis*. Oxford: Pergamon Press.

Jung, T. (2005). Franz Selety (1893–1933?): Seine kosmologische Arbeiten und der Briefwechsel mit Einstein, in *Einsteins Kosmos*, ed. Duerbeck, H. W. and Dick, W. R., pp. 125–42. Frankfurt am Main: Harri Deutsch.

Kahane, G. (2014). Our cosmic insignificance, *Nous*, **48**, 745–72.

Kaiser, C. R. and Alexander, P. (1997). A self-similar model for extragalactic radio sources, *Monthly Notices of the Royal Astronomical Society*, **286**, 215–22.

Kaiser, D. (2005). Making tools travel: pedagogy and the transfer of skills in postwar theoretical physics., in *Pedagogy and the Practice of Science: Historical and Contemporary Perspectives*, ed. Kaiser, D., pp. 41–74. Cambridge MA: MIT Press.

Kaiser, D. (2006). Whose mass is it anyway? Particle cosmology and the objects of theory, *Social Studies of Science*, **36**, 533–64.

Kaiser, N. (1984). On the spatial correlations of Abell clusters, *Astrophysical Journal*, **284**, L9–L12.

Kaiser, N. (1987). Clustering in real space and in redshift space, *Monthly Notices of the Royal Astronomical Society*, **227**, 1–21.

Kamionkowski, M., Kosowsky, A., and Stebbins, A. (1997). Statistics of cosmic microwave background polarization, *Physical Review D*, **55**, 7368–88.

Kant, I. (1981). *Universal Natural History and Theory of the Heavens*. Edinburgh: Scottish Academic Press.

Kantowski, R. and Sachs, R. K. (1966). Some spatially homogeneous anisotropic relativistic cosmological models, *Journal of Mathematical Physics*, **7**, 443–6.

Kassiola, A., Kovner, I., and Blandford, R. D. (1991). Bounds on intergalactic compact objects from observations of compact radio sources, *Astrophysical Journal*, **381**, 6–13.

Kauffmann, G., J. M. Colberg, J. M., Diaferio, A., et al. (1999). Clustering of galaxies in a hierarchical universe: I. Methods and results at $z = 0$, *Monthly Notices of the Royal Astronomical Society*, **303**, 188–206.

Kaufman, M. (1965). Limits on the density of intergalactic ionized hydrogen, *Nature*, **207**, 736–7.

Kazanas, D. (1980). Dynamics of the universe and spontaneous symmetry breaking, *Astrophysical Journal Letters*, **241**, L59–L63.

Keeler, J. E. (1900). On the predominance of spiral forms among the nebulae, *Astronomische Nachrichten*, **151**, cols. 165–81.

Kennedy, R. J. and Barkas, W. (1936). The nebular redshift, *Physical Review*, **49**, 449–52.

Kennedy, R. J. and Thorndike, E. M. (1932). Experimental establishment of the relativity of time, *Physical Review*, **42**, 400–18.

Kennefick, D. (2012). Not only because of theory: Dyson, Eddington, and the competing myths of the 1919 eclipse expedition, in *Einstein and the Changing Worldviews of Physics, Einstein Studies. Vol. 12*, ed. Lehner, C. et al., pp. 201–32. Boston: Birkäuser.

Kennicutt, R. C. (1989). The star formation law in galactic discs, *Astrophysical Journal*, **344**, 685–703.

Kenyon, I. R. (1990). *General Relativity*. Oxford: Oxford University Press.

Kerr, R. P. (1963). Gravitational field of a spinning mass as an example of algebraically special metrics, *Physical Review Letters*, **11**, 237–8.

Kerszberg, P. (1989). *The Invented Universe. The Einstein–de Sitter Controversy (1916–1917) and the Rise of Relativistic Cosmology*. Oxford: Clarendon Press.

Kiang, T. and Saslaw, W. C. (1969). The distribution in space of clusters of galaxies, *Monthly Notices of the Royal Astronomical Society*, **143**, 129–38.

Kibble, T. W. B. (1976). Topology of cosmic domains and strings, *Journal of Physics A: Mathematical and General*, **9**, 1387–98.

Kiepenheuer, K. O. (1950). Cosmic rays as the source of general galactic radio emission, *Physical Review*, **79**, 738–9.

Kilmister, C. (1994). *Eddington's Search for a Fundamental Theory*. Cambridge: Cambridge University Press.

King, H. C. (1979). *The History of the Telescope*. New York: Dover Publications.

Kirchhoff, G. R. (1861). On the chemical analysis of the solar atmosphere, *Philosophical Magazine*, **21**, 185–8.

Kirshner, R. P. and Kwan, J. (1974). Distances to extragalactic supernovae, *Astrophysical Journal*, **193**, 27–36.

Kirshner, R. P. and Oke, B. (1975). Supernova 1972e in NGC 5253, *Astrophysical Journal*, **200**, 574–81.

Kirzhnits, D. A. (1972). Weinberg model and the 'hot' universe, *Soviet Journal of Experimental and Theoretical Physics Letters*, **15**, 529–30.

Klebesadel, R. W., Strong, I. B., and Olson, R. A. (1973). Observations of gamma-ray bursts of cosmic origin, *Astrophysical Journal Letters*, **182**, L85–8.

Klein, O. (1954). Some considerations in connection with the problem of the origin of the elements, in *Les Processus Nucléaires dans les Astres*, ed. Ledoux, P., pp. 42–51. Louvain: Société Royale des Sciences de Liège.

Klein, O. (1971). Arguments concerning relativity and cosmology, *Science*, **171**, 339–45.

Kniffen, D. A., Chipman, E., and Gehrels, N. (1994). The gamma-ray sky according to Compton: a new window to the universe, in *Frontiers of Space and Ground-Based Astronomy*, ed. Wamsteker, W., Longair, M. S., and Kondo, Y., pp. 5–16. Dordrecht: Kluwer Academic Publishers.

Knobe, J., Olum, K. D., and Vilenkin, A. (2006). Philosophical implications of inflationary cosmology, *The British Journal for the Philosophy of Science*, 57, 47–67.

Knop, R. A., Aldering, G., Amanullah, R., et al. (2003). New constraints on Ω_M, Ω_Λ, and w from an independent set of 11 high-redshift supernovae observed with the Hubble Space Telescope, *Astrophysical Journal*, 598, 102–37.

Kochanek, C. S. (1996). Is there a cosmological constant?, *Astrophysical Journal*, 466, 638–59.

Koertge, N. (Ed.) (1998). *A House Built on Sand: Exposing Postmodernist Myths about Science*. Oxford: Oxford University Press.

Kogut, A., Bensadoun, M., de Amici, G., et al. (1990a). A measurement of the temperature of the cosmic microwave background at a frequency of 7.5 GHz, *Astrophysical Journal*, 355, 102–13.

Kogut, A., Smoot, G. F., Petuchowski, S. J., et al. (1990b). In situ measurement of the cosmic microwave background temperature at a distance of 7.5 kiloparsecs, *Astrophysical Journal Letters*, 348, L45–8.

Kogut, A., Spergel, D. N., Barnes, C., et al. (2003). First-year Wilkinson Microwave Anisotropy Probe (WMAP) observations: temperature–polarization correlation, *Astrophysical Journal Supplement*, 148, 161–73.

Kohler, M. (1933). Beitrage zum Kosmologischen Problem und zur Lichtausbreitung in Schwerefeldern, *Annalen der Physik*, 408, 129–61.

Kolb, E. W. and Turner, M. S. (1990). *The Early Universe*. Redwood City, CA: Addison-Wesley Publishing Co.

Kompaneets, A. (1956). The Establishment of thermal equilibrium between quanta and electrons, *Zhurnal Eksperimentalnoi i Teoreticheskoi Fiziki*, 31, 876–85. Translation: (1957), *Soviet Physics JETP*, 4, 730–7.

Koo, D. C. and Kron, R. (1982). QSO counts – a complete survey of stellar objects to $B = 23$, *Astronomy and Astrophysics*, 105, 107–19.

Koonin, E. V. (2007). The cosmological model of eternal inflation and the transition from chance to biological evolution in the history of life, *Biology Direct*, 2, 15.

Kormendy, J. and Richstone, D. O. (1995). Inward bound—the search for supermassive black holes in galactic nuclei, *Annual Review of Astronomy and Astrophysics*, 33, 581–624.

Kovac, J. M., Leitch, E. M., Pryke, C., et al. (2002). Detection of polarization in the cosmic microwave background using DASI, *Nature*, 420, 772–87.

Kragh, H. (1990). *Dirac: A Scientific Biography*. Cambridge: Cambridge University Press.

Kragh, H. (1991). Cosmonumerology and empiricism: the Dirac–Gamow dialogue, *Astronomical Quarterly*, 8, 109–26.

Kragh, H. (1995). Cosmology between the wars: the Nernst–MacMillan alternative, *Journal for the History of Astronomy*, 26, 93–115.

Kragh, H. (1996). *Cosmology and Controversy: The Historical Development of Two Theories of the Universe*. Princeton: Princeton University Press.

Kragh, H. (1997). The electrical universe: grand cosmological theory versus mundane experiments, *Perspectives on Science*, 5, 199–231.

Kragh, H. (1999). Steady-state cosmology and general relativity: reconciliation or conflict?, in *The Expanding World of General Relativity*, ed. Goenner, H. et al., pp. 377–402. Boston: Birkhäuser.

Kragh, H. (2001). From geochemistry to cosmochemistry: the origin of a scientific discipline, in *Chemical Sciences in the 20th Century*, ed. Reinhardt, C., pp. 160–90. Weinheim: Wiley-VCH.

Kragh, H. (2004). *Matter and Spirit in the Universe: Scientific and Religious Preludes to Modern Cosmology*. London: Imperial College Press.

Kragh, H. (2006). Cosmologies with varying speed of light: a historical perspective, *Studies in History and Philosophy of Modern Physics*, **37**, 726–37.

Kragh, H. (2007). *Conceptions of Cosmos. From Myths to the Accelerating Universe: A History of Cosmology*. Oxford: Oxford University Press.

Kragh, H. (2008a). *Entropic Creation. Religious Contexts of Thermodynamics and Cosmology*. London: Ashgate.

Kragh, H. (2008b). The origin and earliest reception of big bang cosmology, *Publications of the Astronomical Observatory of Belgrade*, **85**, 7–16.

Kragh, H. (2008c). Pierre Duhem, entropy, and Christian faith, *Physics in Perspective*, **10**, 379–95.

Kragh, H. (2009a). Contemporary history of cosmology and the controversy over the multiverse, *Annals of Science*, **66**, 529–51.

Kragh, H. (2009b). The solar element: a reconsideration of helium's early history, *Annals of Science*, **66**, 157–82.

Kragh, H. (2009c). Continual fascination: the oscillating universe in modern cosmology, *Science in Context*, **22**, 587–612.

Kragh, H. (2010). An anthropic myth: Fred Hoyle's carbon-12 resonance level, *Archive for History of Exact Sciences*, **64**, 721–51.

Kragh, H. (2011). *Higher Speculations: Grand Theories and Failed Revolutions in Physics and Cosmology*. Oxford: Oxford University Press.

Kragh, H. (2012a). Is space flat? Nineteenth-century astronomy and non-Euclidean geometry, *Journal of Astronomical History and Heritage*, **15**, 149–58.

Kragh, H. (2012b). The wildest speculation of all: Lemaître and the primeval-atom universe, in *Georges Lemaître: Life, Science and Legacy*, ed. Holder, R. D. and Mitton, S., pp. 23–38. Berlin: Springer.

Kragh, H. (2012c). Zöllner's universe, *Physics in Perspective*, **14**, 392–420.

Kragh, H. (2013a). Nordic cosmogonies: Birkeland, Arrhenius and fin-de-siècle cosmical physics, *European Physical Journal H*, **38**, 549–72.

Kragh, H. (2013b). Science and ideology: the case of cosmology in the Soviet Union, 1947–1963, *Acta Baltica Historiae et Philosophiae Scientiarum*, **1**, 35–66.

Kragh, H. (2013c). 'The most philosophically important of all the sciences': Karl Popper and physical cosmology, *Perspectives on Science*, **21**, 325–57.

Kragh, H. (2014a). Naming the big bang, *Historical Studies in the Natural Sciences*, **44**, 3–36.

Kragh, H. (2014b). Testability and epistemic shifts in modern cosmology, *Studies in History and Philosophy of Science Part B: Studies in History and Philosophy of Modern Physics*, **46**, 48–56.

Kragh, H. (2014c). The true (?) story of Hilbert's infinite hotel. arXiv preprint arXiv:1403.0059.

Kragh, H. (2015a). *Masters of the Universe: Conversations with Cosmologists of the Past*. Oxford: Oxford University Press.

Kragh, H. (2015b). Pascual Jordan, varying gravity, and the expanding Earth, *Physics in Perspective*, **17**, 107–34.

Kragh, H. (2016a). The source of solar energy, ca. 1840–1910: from meteoric hypothesis to radioactive speculations, *European Physical Journal H*, **41**, 365–94.

Kragh, H. (2016b). *Varying Gravity: Dirac's Legacy in Cosmology and Geophysics*. Basel: Birkhäuser.

Kragh, H. (2017a). Eddington's dream: a failed theory of everything, in *Information and Interaction: Eddington, Wheeler, and the Limits of Knowledge*, ed. Rickles, D. and Durham, I., pp. 45–58. Basel: Springer.

Kragh, H. (2017b). Fundamental theories and epistemic shifts: can history of science serve as a Guide? arXiv preprint arXiv:1702.5648.

Kragh, H. (2017c). The Nobel Prize system and the astronomical sciences, *Journal for the History of Astronomy*, **48**, 257–80.

Kragh, H. (2017d). Is the universe expanding? Fritz Zwicky and early tired-light hypotheses, *Journal of Astronomical History and Heritage*, **20**, 2–12.

Kragh, H. and Lambert, D. (2007). The context of discovery: Lemaître and the origin of the primeval-atom hypothesis, *Annals of Science*, **64**, 445–70.

Kragh, H. and Smith, R. (2003). Who discovered the expanding universe?, *History of Science*, **41**, 141–62.

Kraushaar, W. L., Clark, G. W., Garmire, G. P., et al. (1965). Explorer XI experiment on cosmic gamma rays, *Astrophysical Journal*, **141**, 845–63.

Krauss, L. M. and Scherrer, R. J. (2007). The return of a static universe and the end of cosmology, *General Relativity and Gravitation*, **39**, 1545–50.

Krauss, L. M. and Starkman, G. D. (2000). Life, the universe, and nothing: life and death in an ever-expanding universe, *The Astrophysical Journal*, **531**, 22–30.

Krauss, L. M. and Turner, M. S. (1999). Geometry and destiny, *General Relativity and Gravitation*, **31**, 1453–9.

Krige, J., Long, A., Maharaj, A., et al. (2013). *NASA in the World: Fifty Years of International Collaboration in Space*. New York: Palgrave.

Kristian, J. and Sachs, R. (1966). Observations in cosmology, *Astrophysical Journal*, **143**, 379–99.

Křížek, M. and Somer, L. (2016). Excessive extrapolations in cosmology, *Gravitation and Cosmology*, **22**, 270–80.

Kruskal, M. D. (1960). Maximal Extension of Schwarzschild Metric, *Physical Review*, **119**, 1743–5.

Krymsky, G. F. (1977). A regular mechanism for the acceleration of charged particles on the front of a shock wave, *Doklady Akademiya Nauk SSSR*, **234**, 1306–8.

Krzeminski, W. (1973). The identification and UBV photometry of the visible component of the Centaurus X-3 binary system, *International Astronomical Union Circular No. 2612*.

Krzeminski, W. (1974). The identification and UBV photometry of the visible component of the Centaurus X-3 binary system, *Astrophysical Journal Letters*, **192**, L135–L138.

Kuhn, T. S. (1962). *The Structure of Scientific Revolutions*. Chicago: University of Chicago Press.

Kuhn, T. S. (1970). *The Structure of Scientific Revolutions*, second edition. Chicago: University of Chicago Press.

Lachieze-Rey, M. and Luminet, J. (1995). Cosmic topology, *Physics Reports*, **254**, 135–214.

Lagache, G., Dole, H., and Puget, J.-L. (2003). Modelling infrared galaxy evolution using a phenomenological approach, *Monthly Notices of the Royal Astronomical Society*, **338**, 555–71.

Lagache, G., Dole, H., Puget, J.-L., et al. (2004). Polycyclic aromatic hydrocarbon contribution to the infrared output energy of the universe at $z \simeq 2$, *Astrophysical Journal Supplement*, **154**, 112–17.

Lambert, D. (2007). *L'Itinéraire Spirituel de Georges Lemaître*. Brussels: Lessius.

Lambert, D. (2015). *The Atom of the Universe: The Life and Work of Georges Lemaître*. Kraków: Copernicus Center Press.

Lambert, J. H. (1761). *Cosmologische Briefe über die Einrichtung des Weltbaues*. Augsburg: Eberhard Kletts Wittib.

Lanczos, K. (1922). Bemerkung zur de Sitterschen Welt (Remarks on de Sitter's world model), *Physikalische Zeitschrift*, **23**, 539–43.

Lanczos, K. (1923). Über die Rotverschiebung in der de Sitterschen Welt, *Physikalishe Zeitschrift*, 17, 168–89.

Lanczos, K. (1924). Über eine stationäre Kosmologie in Sinne der Einsteinschen Gravitationstheorie, (On a stationary cosmology in the sense of Einstein's theory of gravitation), *Physikalishe Zeitschrift*, 21, 73–110. English translation: (1997), *General Relativity and Gravitation*, 29, 363–99.

Lanczos, K. (1925). Über eine zeitlich periodische Welt und eine neue Behandlung des Problems der Ätherstrahlung, *Zeitschrift fur Physik*, 32, 56–80.

Landau, L. D. (1938). Origin of stellar energy, *Nature*, 141, 333–4.

Lane, K. (2011). *Geographies of Mars: Seeing and Knowing the Red Planet.* Chicago: University of Chicago Press.

Lang, K. and Gingerich, O. (Eds.) (1979). *A Source Book in Astronomy and Astrophysics, 1900–1975.* Cambridge, MA: Harvard University Press.

Lankford, J. (1997). *The American Astronomy: Community, Careers and Power, 1859–1940.* Chicago: University of Chicago Press.

Large, M. I., Vaughan, A. E., and Mills, B. Y. (1968). A pulsar supernova association?, *Nature*, 220, 340–1.

Last, C. (2017). Big historical foundations for deep future speculations: cosmic evolution, atechnogenesis, and technocultural civilization, *Foundations of Science*, 22, 39–124.

Latour, B. (2004). Why has critique run out of steam? From matters of fact to matters of concern, *Critical Inquiry*, 30, 225–48.

Laughlin, G., Bodenheimer, P., and Adams, F. C. (1997). The end of the Main Sequence, *The Astrophysical Journal*, 482, 420–32.

LaViolette, P. A. (1986). Is the universe really expanding?, *Astrophysical Journal*, 301, 544–53.

Layzer, D. (1968). Black-body radiation in a cold universe, *Astrophyslcal Letters*, 1, 99.

Layzer, D. and Hively, R. (1973). Origin of the microwave background..., *Astrophysical Journal*, 179, 361–70.

Le Roux, E. F. M. (1956). *Mesures absolues en radio astronomie. Analyse des résultats obtenus sur la longueur d'onde de 33 cm. Etude de certains spectres.* PhD thesis, Faculte des Sciences, Paris.

Leavitt, H. S. (1912). Periods of 25 variable stars in the Small Magellanic Cloud, *Harvard College Observatory Circular*, 173, 1–2.

Lebach, D. E., Corey, B. E., Shapiro, I. I., et al. (1995). Measurement of the solar gravitational deflection of radio waves using very-long-baseline interferometry, *Physical Review Letters*, 75, 1439.

Lehners, J.-L. (2008). Ekpyrotic and cyclic cosmology, *Physics Reports*, 465, 223–63.

Lemaître, G. (1925a). Note on de Sitter's universe, *Journal of Mathematics and Physics*, 4, 188–92.

Lemaître, G. (1925b). Note on de Sitter's universe, *Physical Review*, 6, 903.

Lemaître, G. (1927). Un univers homogène de masse constante et de rayon croissant, rendant compte de la vitesse radiale des nébuleuses extra-galactiques (A homogeneous universe of constant mass and increasing radius accounting for the radial velocity of extra-galactic nebulae), *Annales de la Société Scientifique de Bruxelles*, **A**47, 49–59. English translation: (1931), *Monthly Notices of the Royal Astronomical Society*, 91, 483–90.

Lemaître, G. (1931a). The beginning of the world from the point of view of quantum theory, *Nature*, 127, 706.

Lemaître, G. (1931b). The expanding universe, *Monthly Notices of the Royal Astronomical Society*, 91, 490–501.

Lemaître, G. (1931c). L'expansion de l'espace, *Revue des Questions Scientifiques*, 17, 391–410.

Lemaître, G. (1933). Spherical condensations in the expanding universe, *Comptes Rendus de L'Academie des Sciences de Paris*, **196**, 903–4.

Lemaître, G. (1934). Evolution of the expanding universe, *Proceedings of the National Academy of Sciences*, **20**, 12–17.

Lemaître, G. (1949). The cosmological constant, in *Albert Einstein: Philosopher–Scientist*, ed. Schilpp, P. A., pp. 437–56. Evanston IL: Library of Living Philosophers.

Lemaître, G. (1958). The primaeval atom hypothesis and the problem of the clusters of galaxies, in *La Structure et l'Évolution de l'Univers: Rapports et Discussions. Institute International de Physique Solvay, Onzième Conseil de Physique*, ed. Stoops, R., pp. 1–31. Bruxelles: Institute International de Physique Solvay.

Lemaître, G. (1997). The expanding universe, *General Relativity and Gravitation*, **29**, 641–80.

Lepeltier, T. (2007). Quand l'univers était plus jeune que la terre, *Archives Internationales d'Histoire des Sciences*, **57**, 137–56.

Lerner, E. J. (1991). *The Big Bang Never Happened*. London: Simon & Schuster.

Lerner, E. J. (1995). Intergalactic radio absorption and the COBE data, *Astrophysics and Space Science*, **227**, 61–81.

Leslie, J. (1992). Doomsday revisited, *The Philosophical Quarterly*, **42**, 85–9.

Leslie, J. (1996). *The End of the World: The Ethics and Science of Human Extinction*. London: Routledge.

Lévy-Leblond, J.-M. (1990). Did the Big Bang begin?, *American Journal of Physics*, **58**, 156–9.

Lewis, D. (1986). *On the Plurality of Worlds*. Oxford: Blackwell.

Liddle, A. R. and Lyth, D. (2000). *Cosmological Inflation and Large-Scale Structure*. Cambridge: Cambridge University Press.

Lidsey, J. E., Liddle, A. R., Kolb, E. W., et al. (1997). Reconstructing the inflaton potential – an overview, *Reviews of Modern Physics*, **69**, 373.

Lifshitz, E. (1946). On the gravitational stability of the expanding universe, *Journal of Physics, Academy of Sciences of the USSR*, **10**, 116–29.

Lilly, S. J. and Cowie, L. L. (1987). Deep infrared surveys, in *Infrared Astronomy with Arrays*, ed. Wynn-Williams, C. G. and Becklin, E. E., pp. 473–82. Honolulu: Institute for Astronomy, University of Hawaii Publications.

Lin, C. C., Mestel, L., and Shu, F. (1965). The gravitational collapse of a uniform spheroid, *Astrophysical Journal*, **142**, 1431–46.

Linde, A. D. (1974a). Is the cosmological constant really constant?, *Pis'ma Zhurnal Eksperimentalnii i Teoreticheskhii Fizika*, **19**, 320–2.

Linde, A. D. (1974b). Is the Lee constant a cosmological constant?, *Zhurnal Experimentalnoi i Teoretichseskikh Fizica (JETP) Letters*, **19**, 183–4.

Linde, A. D. (1982). A new inflationary universe scenario: a possible solution of the horizon, flatness, homogeneity, isotropy and primordial monopole problems, *Physics Letters*, **108B**, 389–93.

Linde, A. D. (1983). Chaotic inflation, *Physics Letters B*, **129**, 177–81.

Linde, A. D. (1986). Eternally existing self-reproducing chaotic inflationary universe, *Physical Letters, B*, **175**, 395–400.

Linde, A. D. (2007). Sinks in the landscape, Boltzmann brains and the cosmological constant problem, *Journal of Cosmology and Astroparticle Physics*, **1**, 022.

Linde, A. D. (2017). A brief history of the multiverse, *Reports of Progress in Physics*, **80**, 022001.

Linde, A. D. and Vanchurin, V. (2010). How many universes are in the multiverse?, *Physical Review D*, **81**, 083525.

Linde, A. D., Linde, D., and Mezhlumian, A. (1994). From the big bang theory to the theory of a stationary universe, *Physical Review D*, **49**, 1783–826.

Lindquist, R. W. and Wheeler, J. A. (1957). Dynamics of a lattice universe by the Schwarzschild-cell method, *Reviews of Modern Physics*, **29**, 432–43.

Lineweaver, C. H. (2005). Inflation and the cosmic microwave background, in *The New Cosmology*, ed. Colless, M., pp. 31–65. Singapore: World Scientific.

Lineweaver, C. H. and Egan, C. A. (2007). The cosmic coincidence as a temporal selection effect produced by the age distribution of terrestrial planets in the universe, *The Astrophysical Journal*, **671**, 853–60.

Lineweaver, C. H., Davies, P. C., and Ruse, M. (2013). *Complexity and the Arrow of Time*. Cambridge: Cambridge University Press.

Linsky, J. L., Diplas, A., Savage, B., et al. (1994). Deuterium in the local interstellar medium: its cosmological significance, in *Frontiers of Space and Ground-based Astronomy, 27th ESLAB Symposium*, ed. W. Wamsteker, W., Longair, M. S., and Kondo, Y., pp. 301–04.

Livio, M. (2011). Lost in translation: mystery of the missing text solved, *Nature*, **479**, 171–3.

Livio, M. and Rees, M. (2005). Anthropic reasoning, *Science*, **309**, 1022–3.

Lloyd, S. (2000). Ultimate physical limits to computation, *Nature*, **406**, 1047–54.

Lloyd, S. (2002). Computational capacity of the universe, *Physical Review Letters*, **88**, 237901.

Longair, M. S. (1966a). Evidence on the evolutionary character of the universe derived from recent red-shift measurements, *Nature*, **211**, 949–50.

Longair, M. S. (1966b). On the interpretation of radio source counts, *Monthly Notices of the Royal Astronomical Society*, **133**, 421–36.

Longair, M. S. (1971). Observational cosmology, *Reports of Progress in Physics*, **34**, 1125–248.

Longair, M. S. (Ed.) (1974). *IAU Symposium 63. Confrontation of Cosmological Theories with Observational Data*. Dordrecht: D. Reidel Publishing Company.

Longair, M. S. (1988). The new astrophysics, in *The New Physics*, ed. Davies, P. C., pp. 94–208. Cambridge: Cambridge University Press.

Longair, M. S. (1995). The physics of background radiation, in *The Deep Universe: Saas-Fee Advanced Course 23. Lecture Notes 1993. Swiss Society for Astrophysics and Astronomy*, ed. Binggeli, B. and Buser, R., p. 204. Berlin: Springer-Verlag.

Longair, M. S. (1997). The Friedman Robertson–Walker models: on bias, errors and acts of faith, in *Critical Dialogues in Cosmology*, ed. Turok, N., pp. 285–308. Singapore: World Scientific.

Longair, M. S. (2003). *Theoretical Concepts in Physics: An Alternative View of Theoretical Reasoning in Physics*. Cambridge: Cambridge University Press.

Longair, M. S. (2006). *The Cosmic Century: A History of Astrophysics and Cosmology*. Cambridge: Cambridge University Press.

Longair, M. S. (2008). *Galaxy Formation*, second edition. Berlin: Springer-Verlag.

Longair, M. S. (2011). *High Energy Astrophysics*, third edition. Cambridge: Cambridge University Press.

Longair, M. S. (2013). *Quantum Concepts in Physics*. Cambridge: Cambridge University Press.

Longair, M. S. and Scheuer, P. A. G. (1970). The luminosity–volume test for quasi-stellar objects, *Monthly Notices of the Royal Astronomical Society*, **151**, 45.

Longair, M. S., Ryle, M., and Scheuer, P. A. G. (1973). Models of extended radio sources, *Monthly Notices of the Royal Astronomical Society*, **164**, 253–70.

López-Corredoira, M. (2014). Non-standard models and the sociology of cosmology, *Studies in History and Philosophy of Modern Physics*, **46**, 86–96.

López-Corredoira, M. (2017). Tests and problems of the standard model in cosmology, *Foundations of Physics*, **47**, 711–68.

Louis, T., Grace, E., Hasselfield, M., et al. (2017). The Atacama Cosmology Telescope: two-season ACTPol spectra and parameters, *Journal of Cosmology and Astroparticle Physics*, **6**, 031.

Lovejoy, A. (1936). *The Great Chain of Being*. Cambridge, MA: Harvard University Press.

Lovelace, R. V. E. and Romanova, M. M. (2003). Relativistic Poynting jets from accretion disks, *Astrophysical Journal*, **596**, L159–62.

Lubin, P., Villela, T., Epstein, G., et al. (1985). A map of the cosmic background radiation at 3 millimeters, *Astrophysical Journal Letters*, **298**, L1–L5.

Lundmark, K. (1924). The determination of the curvature of space-time in de Sitter's world, *Monthly Notices of the Royal Astronomical Society*, **84**, 747–70.

Lundmark, K. (1925). The motions and the distances of spiral nebulae, *Monthly Notices of the Royal Astronomical Society*, **85**, 865–94.

Lundmark, K. (1927). Studies of anagalactic nebulae, first paper, *Meddelanden från Astronomiska Observatorium, Uppsala*, Series C, i, No. 8, 50–3.

Lynden-Bell, D. (1967). Statistical mechanics of violent relaxation in stellar systems, *Monthly Notices of the Royal Astronomical Society*, **136**, 101–21.

Lynden-Bell, D. (1969). Galactic nuclei as collapsed old quasars, *Nature*, **223**, 690–4.

Lynden-Bell, D., Faber, S. M., Burstein, D., et al. (1988). Spectroscopy and photometry of elliptical galaxies, *Astrophysical Journal*, **326**, 19–49.

Lyne, A. G. and Graham Smith, F. (1998). *Pulsar Astronomy*. Cambridge: Cambridge University Press.

Lyubimov, V. A., Novikov, E. G., Nozik, V. Z., et al. (1980). An estimate of the ν_e mass from the β-spectrum of tritium in the valine molecule, *Physics Letters*, **138**, 30–56.

MacMillan, W. D. (1925). Some mathematical aspects of astronomy, *Science*, **62**, 63–72, 96–9, 121–7.

Madau, P., Ferguson, H. C., Dickinson, M. E., et al. (1996). High-redshift galaxies in the *Hubble Deep Field*: colour selection and star formation history to $z \sim 4$, *Monthly Notices of the Royal Astronomical Society*, **283**, 1388–404.

Maddox, S. J., Efstathiou, G., Sutherland, W. G., et al. (1990). Galaxy correlations on large scales, *Monthly Notices of the Royal Astronomical Society*, **242**, 43P–7P.

Maeder, A. (1994). A selection of 10 most topical stellar problems, in *Frontiers of Space and Ground-Based Astronomy*, ed. Wamsteker, W., Longair, M. S., and Kondo, Y., pp. 177–86. Dordrecht: Kluwer Academic Publishers.

Magorrian, J., Tremaine, S., Richstone, D., et al. (1998). The demography of massive dark objects in galaxy centers, *Astronomical Journal*, **115**, 2285–305.

Malament, D. (1977). Observationally indistinguishable space-times., in *Foundation of Space-Time Theories*, ed. Earman, J. S., Glymour, C. N., and Stachel, J. J., pp. 61–80. Minneapolis: University of Minnesota Press.

Malmquist, K. G. (1920). A study of stars of spectral type A, *Meddelanden från Lunds Astronomiska Observatorium*, Series II, 22, 1–69.

Manchak, J. (2009). Can we know the global structure of spacetime?, *Studies In History and Philosophy of Science Part B: Studies In History and Philosophy of Modern Physics*, **40**, 53–6.

Manchak, J. B. (2013). Global spacetime structure, in *Oxford Handbook of Philosophy of Physics*, ed. Batterman, R. W., pp. 587–606. Oxford: Oxford University Press.

Manly, S. (2011). *Visions of the Multiverse*. Pompton Plains, NJ: New Page Books.

Mann, M. (1962). The march of science – Big Bang up, steady-state down, *Popular Science*, **180**, 29.

Manson, N. A. (2009). The fine-tuning argument, *Philosophy Compass*, **4**, 271–86.

Manson, N. A. and Thrush, M. J. (2003). Fine-tuning, multiple universes, and the "This Universe" objection, *Pacific Philosophical Quarterly*, **84**, 67–83.

Margon, B. and Ostriker, J. P. (1973). The luminosity function of galactic X-ray sources: a cut-off and a "Standard Candle"?, *Astrophysical Journal*, **186**, 91–6.

Markarian, B. E. (1967). Galaxies with an ultraviolet continuum, *Astrofizica*, **3**, 24–38.

Markarian, B. E., Lipovetsky, V. A., and Stepanian, D. A. (1981). Galaxies with ultraviolet continuum XV, *Astrofizica*, **17**, 619–27. Translation: (1982), *Astrophysics*, **17**, 321–32.

Martin, H. M., Partridge, R. B., and Rood, R. T. (1980). Interferometric limits on very small-scale fluctuations in the cosmic microwave background, *Astrophysical Journal Letters*, **240**, L79–L82.

Martin, J., Ringeval, C., and Vennin, V. (2013). Encyclopaedia inflationaris. arXiv preprint arXiv:1303.3787.

Marx, G. and Szalay, A. S. (1972). Cosmological limit on neutretto mass, in *Neutrino '72*, Volume 1, pp. 191–5. Budapest: Technoinform.

Marx, W. and Bornmann, L. (2010). How accurately does Thomas Kuhn's model of paradigm change describe the transition from the static view of the universe to the big bang theory in cosmology?, *Scientometrics*, **84**, 441–64.

Mascall, E. L. (1956). *Christian Theology and Natural Science*. London: Longmans, Green and Co.

Mash, R. (1993). Big numbers and induction in the case for extraterrestrial intelligence, *Philosophy of Science*, **60**, 204–22.

Mather, J. C. and Boslough, J. (1996). *The Very First Light*. New York: Basic Books.

Mather, J. C., Cheng, E. S., Eplee, J., et al. (1990). A preliminary measurement of the cosmic microwave background spectrum by the Cosmic Background Explorer (COBE) satellite, *Astrophysical Journal Letters*, **354**, L37–L40.

Matsumoto, T., Hayakawa, S., Matsuo, H., et al. (1988). The submillimeter spectrum of the cosmic background radiation, *Astrophysical Journal*, **329**, 567–71.

Matt, G., Fabian, A. C., and Reynolds, C. S. (1997). Geometrical and chemical dependence of K-shell X-ray features, *Monthly Notices of the Royal Astronomical Society*, **289**, 175–84.

Matthews, T. A. and Sandage, A. R. (1963). Optical identification of 3C 48, 3C 196 and 3C 286 with stellar objects, *Astrophysical Journal*, **138**, 30–56.

Maudlin, T. (2007). *The Metaphysics within Physics*. Oxford: Oxford University Press.

Maunder, E. W. (1908). *The Astronomy of the Bible*. London: Epworth Press.

Mayer-Hasselwander, H. A., Kanbach, G., Bennett, K., et al. (1982). Large-scale distribution of galactic gamma radiation observed by COS-B, *Astronomy and Astrophysics*, **105**, 164–75.

Maynard-Smith, J. and Szathmáry, E. (1996). On the likelihood of habitable worlds, *Nature*, **384**, 107.

McAllister, L. and Silverstein, E. (2008). String cosmology: a review, *General Relativity and Gravitation*, **40**, 565–605.

McCarthy, P. J. (2006). Galaxy formation and cosmology in the ELT era, in *Scientific Requirements for Extremely Large Telescopes: IAU Symposium No. 232*, ed. Whitelock, P., Dennefeld, M., and Leibundgut, B., pp. 119–29. Cambridge: Cambridge University Press.

McConnell, C. S. (2002). Twentieth-century cosmologies, in *Science and Religion: A Historical Introduction*, ed. Ferngren, G. B. pp. 314–21. Baltimore: Johns Hopkins University Press.

McCoy, C. D. (2015). Does inflation solve the hot big bang model's fine-tuning problems?, *Studies in History and Philosophy of Modern Physics*, **51**, 23–36.

McCrea, W. H. (1935). Observable relations in relativistic cosmology, *Zeitschrift für Astrophysik*, 9, 290–314.

McCrea, W. H. (1950). The steady-state theory of the expanding universe, *Endeavour*, 9, 3–10.

McCrea, W. H. (1960). The interpretation of cosmology, *Nature*, 186, 1035.

McCrea, W. H. (1962). Information and prediction in cosmology, *Monist*, 47, 94–103.

McCrea, W. H. (1970). A philosophy for big bang cosmology, *Nature*, 228, 21–4.

McGrew, T., McGrew, L., and Vestrup, E. (2001). Probabilities and the fine-tuning argument: a sceptical view, *Mind*, 110, 1027–38.

McKellar, A. (1941). Molecular lines from the lowest states of diatomic molecules composed of atoms probably present in interstellar spaces, *Publications of the Dominion Astrophysical Observatory (Victoria)*, 7, 251–72.

McLeod, J. M. and Andrew, B. H. (1968). The radio source VRO 42.22.01, *Astrophysical Letters*, 1, 243.

McMullin, E. (1993). Indifference principle and anthropic principle in cosmology, *Studies in the History and Philosophy of Science*, 24, 359–89.

McNally, S. J. and Peacock, J. A. (1995). The small-scale clustering power spectrum and relativistic decays, *Monthly Notices of the Royal Astronomical Society*, 277, 143–51.

McVittie, G. C. (1965). *General Relativity and Cosmology, second edition*. International Astrophysics Series. London: Chapman and Hall.

Merkowitz, S. M. (2010). Tests of gravity using lunar laser ranging, *Living Reviews in Relativity*, 13, 7.

Merleau-Ponty, J. (1965). *Cosmologie du XX Siècle*. Paris: Gallimard.

Merleau-Ponty, J. (1983). *La Science de l'Univers à l'âge du positivism*. Paris: Vrin.

Merrill, P. W. (1933). Cosmic chemistry, *Leaflet of the Astronomical Society of the Pacific*, 2, 25–8.

Mersini-Houghton, L. (2006). Do we have evidence for new physics in the sky?, *Modern Physics Letters A*, 21, 1–21.

Mészáros, P. (1975). Primeval black holes and galaxy formation, *Astronomy and Astrophysics*, 38, 5–13.

Mészáros, P. and Rees, M. J. (1993). Gamma-ray bursts: multiwaveband spectral predictions for blast wave models, *Astrophysical Journal*, 418, L59–L62.

Metcalfe, N., Shanks, T., Campos, A., et al. (1996). Galaxy formation at high redshifts, *Nature*, 383, 236–7.

Meyer, D. M., Roth, K. C., and Hawkins, I. (1989). A precise CN measurement of the cosmic microwave background temperature at 1.32 millimeters, *Astrophysical Journal Letters*, 343, L1–L4.

Michell, J. (1767). An inquiry into the probable parallax, and magnitude of the fixed stars, from the quantity of light which they afford us, and the particular circumstances of their situation, *Philosophical Transactions of the Royal Society*, 57, 234–64.

Mihos, J. C. and Hernquist, L. (1994). Triggering of starbursts in galaxies by minor mergers, *Astrophysical Journal*, 425, L13–L16.

Mihos, J. C. and Hernquist, L. (1996). Gasdynamics and starbursts in major mergers, *Astrophysical Journal*, 464, 641–63.

Millea, M. F., McColl, M., Pedersen, R. J., et al. (1971). Cosmic background radiation at $\lambda = 3.3$ mm, *Physical Review Letters*, 26, 919–22.

Miller, R. H., Prendergast, K. H., and Quirk, W. J. (1970). Numerical experiments on spiral structure, *Astrophysical Journal*, 161, 903–16.

Mills, B. Y. and Slee, O. B. (1957). A preliminary survey of radio sources in a limited region of the sky at a wavelength of 3.5 m, *Australian Journal of Physics*, 10, 162–94.

Milne, E. A. (1933). World-structure and the expansion of the universe, *Zeitschrift für Astrophysik*, 6, 1–96.

Milne, E. A. (1935). *Relativity, Gravitation and World-Structure*. Oxford: Clarendon Press.

Milne, E. A. (1937). On the origin of laws of nature, *Nature*, 139, 997–9.

Milne, E. A. (1938). On the equations of electromagnetism, *Proceedings of the Royal Society of London A*, 165, 313–57.

Milne, E. A. (1952). *Modern Cosmology and the Christian Idea of God*. New York: Oxford University Press.

Minkowski, R. (1941). Spectra of supernovae, *Publications of the Astronomical Society of the Pacific*, 53, 224–5.

Minkowski, R. (1960). A new distant cluster of galaxies, *Astrophysical Journal*, 132, 908.

Mirabel, I. F. and Rodrigues, L. F. (1994). A superluminal source in the galaxy, *Nature*, 371, 46–8.

Mirabel, I. F. and Rodrigues, L. F. (1998). Microquasars in our galaxy, *Nature*, 392, 673–6.

Misner, C. W. (1968). The isotropy of the universe, *Astrophysical Journal*, 151, 431–57.

Misner, C. W. (1969). Mixmaster universe, *Physical Review Letters*, 22, 1071–4.

Misner, C. W., Thorne, K. S., and Wheeler, J. A. (1973). *Gravitation*. San Francisco: W.H. Freeman and Co.

Mitchell, J. L., Keeton, C. R., Frieman, J. A., et al. (2005). Improved cosmological constraints from gravitational lens statistics, *Astrophysical Journal*, 622, 81–98.

Miyoshi, M., Moran, J., Herrnstein, J., et al. (1995). Evidence for a black-hole from high rotation velocities in a sub-parsec region of NGC4258, *Nature*, 373, 127–9.

Moffat, J. W. (2011). Taking the multiverse on faith, *Physics World*, 24, 46–7.

Moffat, J. W. (2014). Inflationary schism, *Physics Letters B*, 736, 142–6.

Monaco, P. (1998). The cosmological mass function, *Fundamentals of Cosmic Physics*, 19, 157–317.

Monaco, P. (1999). Dynamics in the cosmological mass function (or, why does the Press & Schechter work?), in *Observational Cosmology: The Development of Galaxy Systems*, ed. Giuricin, G., Mezzetti, M., and Salucci, P., pp. 186–97. San Francisco: Astronomical Society of the Pacific Conference Series No. 176.

Moorcock, M. (1995). *The Eternal Champion*. Clarkston, GA: White Wolf.

Moroz, V. I. (2001). Spectra and spacecraft, *Planetary and Space Science*, 49, 173–90.

Mortonson, M. J. and Seljak, U. (2014). A joint analysis of Planck and BICEP2 B modes including dust polarization uncertainty, *Journal of Cosmology and Astroparticle Physics*, 10, 035.

Moschella, U. (2006). The de Sitter and anti-de Sitter sightseeing tour, in *Einstein 1905–2005: Poincaré Seminar 2005*, ed. Damour, T. et al., pp. 120–33. Basel: Birkhäuser Verlag.

Moulton, F. R. (1905). On the evolution of the solar system, *Astrophysical Journal*, 22, 165–81.

Muehlner, D. and Weiss, R. (1970). Measurement of the isotropic background radiation in the far infrared, *Physical Review Letters*, 24, 742–6.

Mukhanov, V. F. (2005). *Physical Foundations of Cosmology*. Cambridge: Cambridge University Press.

Mukhanov, V. F. and Chibisov, G. V. (1981). Quantum fluctuations and a nonsingular universe, *JETP Letters*, 33, 532–5.

Munitz, M. K. (1952). Scientific method in cosmology, *Philosophy of Science*, 19, 108–30.

Munitz, M. K. (1962). The logic of cosmology, *British Journal for the Philosophy of Science*, 13, 34–50.

Murphy, M. T. *et al.* (2001). Possible evidence for a variable fine-structure constant from QSO absorption lines, *Monthly Notices of the Royal Astronomical Society*, 327, 1208–22.

Murphy, T., Adelberger, E. G., Battat, J. B. R., et al. (2011). Laser ranging to the lost Lunokhod, *Icarus*, **211**, 1103–8.

Myers, S. T., Baker, J. E., Readhead, A. C. S., et al. (1997). Measurements of the Sunyaev–Zeldovich effect in the nearby clusters A478, A2142, and A2256, *Astrophysical Journal*, **485**, 1–21.

Narlikar, J. V. and Wickramasinghe, N. C. (1967). Microwave background in a steady-state universe, *Nature*, **216**, 43–4.

Naselsky, P. D., Novikov, D. I., and Novikov, I. D. (2006). *The Physics of the Cosmic Microwave Background*. Cambridge: Cambridge University Press.

Nasim, O. (2014). *Observing by Hand: Sketching the Nebulae in the Nineteenth Century*. Chicago, University of Chicago Press.

Nath, B. B. (2013). *The Story of Helium and the Birth of Astrophysics*. New York: Springer.

Nazaretyan, A. P. (2005). Big (universal) history paradigm: versions and approaches, *Social Evolution and History*, **4**, 61–86.

Neal, R. M. (2006). Puzzles of anthropic reasoning resolved using full non-indexical conditioning. arXiv preprint math/0608592.

Negroponte, J. and Silk, J. (1980). Polarization of the primeval radiation in an anisotropic universe, *Physical Review Letters*, **44**, 1433–7.

Newcomb, S. (1898). The philosophy of hyper-space, *Science*, **7**, 1–7.

Newcomb, S. (1906). *Side-Lights on Astronomy: Essays and Addresses*. New York: Harper and Brothers.

Newman, E. T., Couch, K., Chinnapared, K., et al. (1965). Metric of a rotating charged mass, *Journal of Mathematical Physics*, **6**, 918–9.

Neyman, J., Scott, E. L., and Shane, C. D. (1954). The index of clumpiness of the distribution of images of galaxies, *Astrophysical Journal Supplement*, **1**, 269–93.

Nichol, J. P. (1848). *Thoughts on Some Important Points Relating to the System of the World*, second edition. Edinburgh: Johnstone.

Nicholson, J. W. (1913). The physical interpretation of the spectrum of the corona, *Observatory*, **36**, 103–12.

Nomura, Y. (2015). A note on Boltzmann brains, *Physics Letters B*, **749**, 514–18.

Norberg, P., Baugh, C. M., Hawkins, E., et al. (2001). The 2dF Galaxy Redshift Survey: luminosity dependence of galaxy clustering, *Monthly Notices of the Royal Astronomical Society*, **328**, 64–70.

Norberg, P., Baugh, C. M., Hawkins, E., et al. (2002). The 2dF Galaxy Redshift Survey: the dependence of galaxy clustering on luminosity and spectral type, *Monthly Notices of the Royal Astronomical Society*, **332**, 827–38.

Nordtvedt, K. (1991). Lunar laser ranging reexamined: the non-null relativistic contribution, *Physical Review D*, **43**, 3131.

Nordtvedt, K. (1998). Optimizing the observation schedule for tests of gravity in lunar laser ranging and similar experiments, *Classical and Quantum Gravity*, **15**, 3363.

Nordtvedt, K., Jr. (1968). Testing relativity with laser ranging to the moon, *Physical Review*, **170**, 1186.

North, J. D. (1965). *The Measure of the Universe: A History of Modern Cosmology*. Oxford: Oxford University Press.

North, J. D. (1990). *The Measure of the Universe: A History of Cosmology*, new edition. New York: Dover Publications.

North, J. D. (2008). *Cosmos: An Illustrated History of Astronomy and Cosmology*. Chicago: University of Chicago Press.

Norton, J. D. (1999). The cosmological woes of Newtonian gravitation theory, in *The Expanding Worlds of General Relativity*, ed. Goenner, H., et al., pp. 271–324. Boston: Birkhäuser.

Norton, J. D. (2000). How we know about electrons, in *After Popper, Kuhn and Feyerabend: Recent Issues in Theories of Scientific Method*, ed. Nola, R. and Sankey, H., pp. 67–97. Dordrecht: Kluwer Academic Publishers.

Norton, J. D. (2010). Cosmic confusions: not supporting versus supporting not, *Philosophy of Science*, 77, 501–23.

Norton, J. D. (2011). Observationally indistinguishable spacetimes: a challenge for any inductivist, in *Philosophy of Science Matters: The Philosophy of Peter Achinstein*, ed. Morgan, G. J., pp. 164–6. Oxford: Oxford University Press.

Noterdaeme, P., Petitjean, P., Srianand, R., et al. (2011). The evolution of the cosmic microwave background temperature. Measurements of T_{CMB} at high redshift from carbon monoxide excitation, *Astronomy and Astrophysics*, 526, L7.

Novikov, I. D. (1964). On the possibility of appearance of large scale inhomogeneities in the expanding universe, *Journal of Experimental and Theoretical Physics*, 46, 686–9.

Novikov, I. D. (1968). Expected anisotropy of cosmological radio emission in homogeneous anisotropic models, *Soviet Astronomy*, 12, 427–8.

Nowotny, H. and Rose, H. (1979). *Counter-Movements in the Sciences*. Dordrecht: Reidel.

Nussbaumer, H. and Bieri, L. (2009). *Discovering the Expanding Universe*. Cambridge: Cambridge University Press.

O'Connell, D. (Ed.) (1958). *Semaine d'Etude sur le Probleme des Populations Stellaires*. Vatican City: Pontificia Academia Scientiarum.

Oda, M., Gorenstein, P., Gursky, H., et al. (1971). X-Ray pulsations from Cygnus X-1 observed from UHURU, *Astrophysical Journal*, 166, L1–L7.

O'Dell, C. R., Peimbert, M., and Kinman, T. D. (1964). The planetary nebulae in the globular cluster M15, *Astrophysical Journal*, 140, 119–29.

Ogilvie, M. (1975). John Herschel's cosmology: Robert Chambers and the nebular hypothesis, *British Journal for the History of Science*, 30, 214–32.

Ohm, E. A. (1961). Project Echo receiving system, *Bell System Technology Journal*, 40, 1065–94.

Olbers, H. W. (1826). Ueber die Durchsichtigkeit des Weltraums, *Berliner Astronomisches Jahrbuch*, 51, 110–21.

Olive, K. A. (1990). Inflation, *Physics Reports*, 190, 307–403.

Oliver, S. J., Rowan-Robinson, M., and Saunders, W. (1992). Infrared background constraints on the evolution of IRAS galaxies, *Monthly Notices of the Royal Astronomical Society*, 256, 15P–22P.

Olson, S. J. (2015). Homogeneous cosmology with aggressively expanding civilizations, *Classical and Quantum Gravity*, 32, 215025.

Olum, K. (2004). Conflict between anthropic reasoning and observation, *Analysis*, 64, 1–8.

Omnès, R. (1969). Possibility of matter–antimatter separation at high temperatures, *Physical Review Letters*, 23, 38–40.

Oort, J. H. (1932). The force exerted by the stellar system in the direction perpendicular to the galactic plane and some related problems, *Bulletin of the Astronomical Institutes of the Netherlands*, 6, 249–87.

Oort, J. H. (1958). Distribution of galaxies and density in the universe, in *Solvay Conference on The Structure and Evolution of the Universe*, pp. 163–81. Brussels: Institut International de Physique Solvay.

Opal Collaboration (1990). A combined analysis of the hadronic and leptonic decays of the Z^0, *Physics Letters*, **B240**, 497–512.

Oppenheimer, J. R. and Volkoff, G. M. (1939). On massive neutron cores, *Physical Review*, 55, 374–81.

Oppy, G. (2001). Physical eschatology, *Philo*, 4, 148–68.

O'Raifeartaigh, C. and McCann, B. (2014). Einstein's cosmic model of 1931 revisited: an analysis and translation of a forgotten model of the universe, *European Physical Journal H*, 39, 63–85.

O'Raifeartaigh, C., McCann, B., Nahm, W., et al. (2014). Einstein's steady-state theory: an abandoned model of the cosmos, *European Physical Journal* H, 39, 353–67.

O'Raifeartaigh, C., O'Keeffe, M., Nahm, W., et al. (2017). Einstein's 1917 static model of the universe: a centennial review, *European Physical Journal H*, 42, 431–74.

Orosz, J. A. (2007). Home-pages of Jerome A. Orosz. http://mintaka.sdsu.edu/faculty/orosz/web/.

Orosz, J. A., McClintock, J. E., Narayan, R., et al. (2007). A 15.65-solar-mass black hole in an eclipsing binary in the nearby spiral galaxy M 33, *Nature*, 449, 872–5.

Ørsted, H. C. (1998). *Selected Scientific Works of Hans Christian Ørsted*, ed. K. Jelved, A. Jackson, and O. Knudsen. Princeton: Princeton University Press.

Osmer, P. S. (1982). Evidence for a decrease in the space density of quasars at $z \geq 3.5$, *Astrophysical Journal*, 253, 28–37.

Osterbrock, D. E. (1984). *James E. Keeler: Pioneer American Astrophysicist and the Early Development of American Astrophysics*. Cambridge: Cambridge University Press.

Osterbrock, D. E. (1991). The observational approach to cosmology: U.S. observatories pre-World War II, in *Modern Cosmology in Retrospect*, ed. Bertotti, B. et al., pp. 247–90. Cambridge: Cambridge University Press.

Osterbrock, D. E. (2001). *Walter Baade: A Life in Astrophysics*. Princeton: Princeton University Press.

Osterbrock, D. E. and Rogerson, J. B. (1961). The helium and heavy-element content of gaseous nebulae and the Sun, *Publications of the Astronomical Society of the Pacific*, 73, 129–34.

Ostriker, J. P. and Cowie, L. L. (1981). Galaxy formation in an intergalactic medium dominated by explosions, *Astrophysical Journal*, 243, L127–31.

Ostriker, J. P. and Peebles, P. J. E. (1973). A numerical study of the stability of flattened galaxies: or, can cold galaxies survive?, *Astrophyiscal Journal*, 186, 467–80.

Ostriker, J. P. and Steinhardt, P. J. (1995). The observational case for a low-density Universe with a non-zero cosmological constant, *Nature*, 377, 600–2.

Oswalt, T. D., Smith, J. A., Wood, M. A., et al. (1996). A lower limit of 9.5 Gyr on the age of the galactic disk from the oldest white dwarf stars, *Nature*, 382, 692–4.

Pachner, J. (1965). An oscillating isotropic universe without singularity, *Monthly Notices of the Royal Astronomical Society*, 131, 173–6.

Pacini, F. (1967). Energy emission from a neutron star, *Nature*, 216, 567–8.

Page, D. N. (2006). The lifetime of the universe, *Journal of the Korean Physical Society*, 49, 711–14.

Page, D. N. (2008a). Return of the Boltzmann brains, *Physical Review D*, 78, 063536.

Page, D. N. (2008b). Typicality derived, *Physical Review D*, 78, 023514.

Pagels, H. R. (1998). A cozy cosmology, in *Modern Cosmology and Philosophy*, ed. Leslie, J., pp. 180–6. Amherst: Prometheus Books.

Panagia, N., Gilmozzi, R., Macchetto, F., et al. (1991). Properties of the SN 1987A circumstellar ring and the distance to the Large Magellanic Cloud, *Astrophysical Journal*, 380, L23–6.

Pariiskii, Y. N. (1968). On the origin of the blackbody radiation of the universe, *Soviet Astronomy*, **12**, 219–24.

Pariiskii, Y. N. (1973). Detection of hot gas in the Coma cluster of galaxies., *Soviet Astronomy*, **16**, 1048.

Partridge, R. B. (1969). The primeval fireball today, *American Scientist*, **57**(1), 37–74.

Partridge, R. B. (1980a). Flucutations in the cosmic microwave background radiation at small angular scales, *Physica Scripta*, **21**, 624–9.

Partridge, R. B. (1980b). New limits on small-scale angular fluctuations in the cosmic microwave background, *Astrophysical Journal*, **235**, 681–7.

Partridge, R. B. (1995). *3K: The Cosmic Microwave Background Radiation*. Cambridge: Cambridge University Press.

Partridge, R. B. and Wilkinson, D. T. (1967). Isotropy and homogeneity of the universe from measurements of the cosmic microwave background, *Physical Review Letters*, **18**, 557–9.

Partridge, R. B., Cannon, J., Foster, R., et al. (1984). Automated measurement of the temperature of the atmosphere at 3.2 cm, *Physical Review D*, **29**, 2683–5.

Paul, E. R. (1993). *The Milky Way and Statistical Cosmology 1890–1924*. New York: Cambridge University Press.

Pauli, W. (1996). Letter to A. Jaffé, 3 December 1951, in *Wissenschaftlicher Briefwechsel, vol. 4, part 1*, ed. von Meyenn, K. Berlin: Springer.

Pauri, M. (1991). The universe as a scientific object, in *Philosophy and the Origin and Evolution of the Universe*, ed. Agazzi, E. and Cordero, A., pp. 291–339. Dordrecht: Kluwer Academic Publishers.

Peacock, J. A. (2000). *Cosmological Physics*. Cambridge: Cambridge University Press.

Peacock, J. A. and Dodds, S. J. (1994). Reconstructing the linear power spectrum of cosmological mass fluctuations, *Monthly Notices of the Royal Astronomical Society*, **267**, 1020–34.

Peacock, J. A. and Heavens, A. F. (1985). The statistics of maxima in primordial density perturbations, *Monthly Notices of the Royal Astronomical Society*, **217**, 805–20.

Peacock, J. A., Cole, S., Norberg, P., et al. (2001). A measurement of the cosmological mass density from clustering in the 2dF Galaxy Redshift Survey, *Nature*, **410**, 169–73.

Pearce, J. (2017). The unfolding of the historical style in modern cosmology: emergence, evolution, entrenchment, *Studies in History and Philosophy of Modern Physics*, **57**, 17–34.

Pearson, T. J., Unwin, S. C., Cohen, M. H., et al. (1981). Superluminal expansion of quasar 3C273, *Nature*, **290**, 365–8.

Pearson, T. J., Unwin, S. C., Cohen, M. H., et al. (1982). Superluminal expansion of 3C273, in *Extragalactic Radio Sources*, ed. Heeschen, D. S. and Wade, C. M., pp. 355–6. Dordrecht: D. Reidel Publishing Company.

Peebles, P. J. E. (1966). Primeval helium abundance and the primeval fireball, *Physical Review Letters*, **16**, 410–13.

Peebles, P. J. E. (1968). Recombination of the primeval plasma, *Astrophysical Journal*, **153**, 1–11.

Peebles, P. J. E. (1971a). *Physical Cosmology*. Princeton, NJ: Princeton University Press.

Peebles, P. J. E. (1971b). Two old cosmological tests, *Comments on Astrophysics and Space Physics*, **3**, 173–7.

Peebles, P. J. E. (1976). A cosmic virial theorem, *Astrophysics and Space Science*, **45**, 3–19.

Peebles, P. J. E. (1980). *The Large-Scale Structure of the Universe*. Princeton: Princeton University Press.

Peebles, P. J. E. (1982). Large-scale background temperature and mass fluctuations due to scale-invariant primeval perturbations, *Astrophysical Journal*, **263**, L1–L5.

Peebles, P. J. E. (1993). *Principles of Physical Cosmology*. Princeton: Princeton University Press.

Peebles, P. J. E. (2005). Probing general relativity on the scales of cosmology, in *General Relativity and Gravitation: Proceedings of the 17th International Conference. Held 18–23 July 2004 in Dublin*, ed. Florides, P., Nolan, B., and Ottewill, A., pp. 106–17. Singapore: World Scientific.

Peebles, P. J. E. (2014). Discovery of the hot big bang: what happened in 1948, *European Physical Journal H*, **39**, 205–24.

Peebles, P. J. E. (2017). Robert Dicke and the naissance of experimental gravity physics, 1957–1967, *European Physical Journal H*, **42**, 177–260.

Peebles, P. J. E. and Wilkinson, D. T. (1968). Comment on the anisotropy of the primeval fireball, *Physical Review*, **174**, 2168.

Peebles, P. J. E. and Yu, J. T. (1970). Primeval adiabatic perturbation in an expanding universe, *Astrophysical Journal*, **162**, 815.

Peebles, P. J. E., Page, L. A., and Partridge, R. B. (2009). *Finding the Big Bang*. Cambridge: Cambridge University Press.

Penrose, R. (1965). Gravitational collapse and space-time singularities, *Physical Review Letters*, **14**, 57–9.

Penrose, R. (1969). Gravitational collapse: the role of general relativity, *Rivista Nuovo Cimento*, **1**, 252–76.

Penrose, R. (1979). Singularities and time-asymmetry, in *General Relativity: An Einstein Centenary Survey*, ed. Hawking, S. W. and Israel, W., pp. 581–638. Cambridge: Cambridge University Press.

Penrose, R. (2016). *Fashion, Faith, and Fantasy in the New Physics of the Universe*. Princeton: Princeton University Press.

Penzias, A. A. (1979). The origin of the elements, *Reviews of Modern Physics*, **51**, 425–32.

Penzias, A. A. and Wilson, R. W. (1965). A measurement of excess antenna temperature at 4080 MHz, *Astrophysical Journal*, **142**, 419–21.

Penzias, A. A. and Wilson, R. W. (1967). A measurement of the background temperature at 1415 MHz, *Astronomical Journal*, **72**, 315.

Peratt, A. L. (1995). Introduction to plasma astrophysics and cosmology, *Astrophysics and Space Science*, **227**, 3–11.

Perley, R. A., Dreher, J. W., and Cowan, J. J. (1984). The jet and filaments in Cygnus A, *Astrophysical Journal*, **285**, L35–8.

Perlmutter, S., Aldering, G., della Valle, M., et al. (1998). Discovery of a supernova explosion at half the age of the universe, *Nature*, **391**, 51–4.

Perlmutter, S., Boyle, B., Bunclark, P., et al. (1996). High-redshift supernova discoveries on demand: first results from a new tool for cosmology and bounds on q_0, *Nuclear Physics B*, **51**, 20–9.

Perlmutter, S., Gabi, S., Goldhaber, G., et al. (1997). Measurements of the cosmological parameters Omega and Lambda from the first seven supernovae at $z > 0.35$, *Astrophysical Journal*, **483**, 565–81.

Perlmutter, S., Aldering, G., Goldhaber, G. et al. (1999). Measurements of Ω and Λ from 42 high-redshift supernovae, *Astrophysical Journal*, **517**, 565–86.

Perrin, J. (1923). *Atoms*. New York: Van Nostrand. Translated by D. L. Hammick.

Peruzzi, G. and Realdi, M. (2011). The quest for the size of the universe in early relativistic cosmology, *Archive for History of Exact Sciences*, **65**, 659–89.

Peterson, J. B., Richards, P. L., and Timusk, T. (1985). Spectrum of the cosmic background radiation at millimeter wavelengths, *Physical Review Letters*, **55**, 332–5.

Pickering, E. (1912). *Harvard College Observatory Circular, No. 173: Periods of 25 Variable Stars in the Small Magellanic Clouds.* Cambridge, MA: Harvard College Observatory.

Piel, G. et al. (Eds.) (1956). *Cosmology and Science.* New York: Simon and Schuster.

Pilkington, J. D. H. and Scott, P. F. (1965). A survey of radio sources between declinations 20 and 40, *Monthly Notices of the Royal Astronomical Society,* **69**, 183–224.

Pinch, T. (1985). Towards an analysis of scientific observation: The externality and evidential significance of observational reports in physics, *Social Studies of Science,* **15**, 3–36.

Planck Collaboration (2014). Planck 2013 results. XVI. Cosmological parameters, *Astronomy and Astrophysics,* **571**, A16.

Planck Collaboration (2016a). Planck 2015 results. I. Overview of products and scientific results, *Astronomy and Astrophysics,* **594**, A1.

Planck Collaboration (2016b). Planck 2015 results. XIII. Cosmological parameters, *Astronomy and Astrophysics,* **594**, A13.

Planck Collaboration (2016c). Planck 2015 results. XV. Gravitational lensing, *Astronomy and Astrophysics,* **594**, A15.

Planck Collaboration (2016d). Planck 2015 results. XVII. Constraints on primordial non-Gaussianity, *Astronomy and Astrophysics,* **594**, A17.

Planck Collaboration (2016e). Planck 2015 results. XX. Constraints on inflation, *Astronomy and Astrophysics,* **594**, A20.

Planck Collaboration (2016f). Planck intermediate results. XLVI. Reduction of large-scale systematic effects in HFI polarization maps and estimation of the reionization optical depth, *Astronomy and Astrophysics,* **596**, A107.

Pochoda, P. and Schwarzschild, M. (1964). Variation of the cosmological constant and the evolution of the sun, *Astrophysical Journal,* **139**, 587–93.

Poincaré, H. (1911). *Leçons sur les Hypothèses Cosmogoniques.* Paris: Hermann et fils.

Polanyi, M. (1967). The growth of science in society, *Minerva,* **5**, 533–45.

POLARBEAR Collaboration, Ade, P. A. R., Aguilar, M., et al. (2017). A measurement of the cosmic microwave background B-mode polarization power spectrum at subdegree scales from two years of Polarbear data, *Astrophysical Journal,* **848**, 121.

Polites, M. E. (1998). An assessment of the technology of automated rendezvous and capture in space. NASA Technical Report NASA/TP-1998-208528, M-877, NAS 1.60:208528.

Popper, K. R. (1959). *The Logic of Scientific Discovery.* London: Hutchinson and Co.

Porpora, D. V. (2013). How many thoughts are there? Or why we likely have no Tegmark duplicates $10^{10^{115}}$ m away, *Philosophical Studies,* **163**, 133–49.

Pozdnyakov, L. A., Sobol, I. M., and Sunyaev, R. A. (1983). Comptonization and the shaping of X-ray source spectra: Monte Carlo calculations, *Astrophysics and Space Science Reviews,* **2**, 189–331.

Prendergast, K. H. and Burbidge, G. R. (1968). On the nature of some galactic X-ray sources, *Astrophysical Journal,* **151**, L83–8.

Press, W. H. and Schechter, P. (1974). Formation of galaxies and clusters of galaxies by self-similar gravitational condensation, *Astrophysical Journal,* **187**, 425–38.

Price, H. (1996). *Time's Arrow and Archimedes' Point.* Oxford: Oxford University Press.

Price, H. (2004). On the origins of the arrow of time: Why there is still a puzzle about the low entropy past, in *Contemporary Debates in the Philosophy of Science,* ed. Hitchcock, C., pp. 219–39. Oxford: Blackwell Publishing.

Price, M. E. (2017). Entropy and selection: life as an adaptation for universe replication. *Complexity* **2017**, 4745379.

Proctor, R. (1869a). Distribution of the nebulae, *Monthly Notices of the Royal Astronomical Society*, **39**, 337–44.

Proctor, R. A. (1869b). A new theory of the universe, *The Student and Intellectual Observer*, 3, 1–10, 110–19, 177–89.

Proctor, R. A. (1896). *Other Worlds than Ours*. New York: Appleton.

Rappaport, S., Doxsey, R., and Zaumen, W. (1971). A search for X-ray pulsations from Cygnus X-1, *Astrophysical Journal*, **168**, L43–7.

Rappaport, S. A. and Joss, P. C. (1983). X-ray pulsars in massive binary systems, in *Accretion Driven Stellar X-ray Sources*, ed. Lewin, W. H. G. and van den Heuvel, E. P. J., pp. 1–39. Cambridge: Cambridge University Press.

Readhead, A. C. S., Lawrence, C. R., Myers, S. T., et al. (1989). A limit on the anisotropy of the microwave background radiation on arc minute scales, *Astrophysical Journal*, **346**, 566–87.

Realdi, M. and Peruzzi, G. (2009). Einstein, de Sitter and the beginning of relativistic cosmology in 1917, *General Relativity and Gravitation*, **41**, 225–47.

Reasenberg, R. D., Shapiro, I. I., MacNeil, P. E., et al. (1979). Viking relativity experiment – verification of signal retardation by solar gravity, *Astrophysical Journal*, **234**, L219–21.

Reber, G. (1940). Cosmic static, *Astrophysical Journal*, **91**, 621–4.

Reber, G. (1944). Cosmic static, *Astrophysical Journal*, **100**, 279–87.

Rees, M. J. (1966). Appearance of relativistically expanding radio sources, *Nature*, **211**, 468–70.

Rees, M. J. (1967). Studies in radio source structure – I. A relativistically expanding model for variable quasi-stellar radio sources, *Monthly Notices of the Royal Astronomical Society*, **135**, 345–60.

Rees, M. J. (1968). Polarization and spectrum of the primeval radiation in an anisotropic universe, *Astrophysical Journal Letters*, **153**, L1–L5.

Rees, M. J. (1969). The collapse of the universe: an eschatological study, *Observatory*, **89**, 193–8.

Rees, M. J. (1971). New interpretation of extragalactic radio sources, *Nature*, **229**, 312–17.

Rees, M. J. (1976). Beam models for double sources and the nature of the primary energy source, in *The Physics of Non-thermal Radio Sources*, ed. Setti, G., pp. 107–20. Dordrecht: D. Reidel Publishing Company.

Rees, M. J. (1984). Black hole models for active galactic nuclei, *Annual Review of Astronomy and Astrophysics*, **22**, 471–506.

Rees, M. J. (1997). *Before the Beginning: Our Universe and Others*. Cambridge, MA: Helix.

Rees, M. J. (2001). *Our Cosmic Habitat*. (Princeton: Princeton University Press).

Rees, M. J. (2013). Snowflakes and the multiverse, in *This Explains Everything*, ed. Brockman, J., pp. 59–60. New York: HarperCollins.

Rees, M. J. and Ostriker, J. P. (1977). Cooling, dynamics and fragmentation of massive gas clouds – clues to the masses and radii of galaxies and clusters, *Monthly Notices of the Royal Astronomical Society*, **179**, 541–59.

Reinganum, M. R. (1986). Is time travel impossible? A financial proof, *The Journal of Portfolio Management*, **13**, 10–12.

Reiser, O. L. (1954). The field theory of matter in a pantheistic cosmology, *Scientia*, **89**, 211–8, 251–9.

Renn, J. (2007). The third way to general relativity: Einstein and Mach in context, in *The Genesis of General Relativity*, ed. Renn, J. et al., pp. 945–1000. Dordrecht: Springer.

Renn, J. and Schemmel, M. (Eds.) (2007). *The Genesis of General Relativity Vol. 3: Gravitation in the Twilight of Classical Physics. Between Mechanics, Field Theory, and Astronomy*. Dordrecht: Springer.

Richards, G. T., Strauss, M. A., Fan, X., et al. (2006). The Sloan Digital Sky Survey quasar survey: quasar luminosity function from Data Release 3, *Astronomical Journal*, **131**, 2766–87.

Richmond, A. (2006). Recent work: the doomsday argument, *Philosophical Books*, **47**, 129–42.

Rickles, D. (2014). *A Brief History of String Theory: From Dual Models to M-Theory*. Berlin: Springer-Verlag.

Riemann, B. (1873). On the hypotheses which lie at the bases of geometry, *Nature*, **8**, 15–17, 36–7.

Riess, A. G., Macri, L., Casertano, S., et al. (2011). A 3% solution: determination of the Hubble constant with the Hubble Space Telescope and Wide Field Camera 3, *Astrophysical Journal*, **730**, 119.

Riess, A. G., Macri, L. M., Hoffmann, S. L., et al. (2016). A 2.4% determination of the local value of the Hubble constant, *Astrophysical Journal*, **826**, 56.

Rindler, W. (1956). Visual horizons in world models, *Monthly Notices of the Royal Astronomical Society*, **116**, 662–77.

Robertson, D. S., Carter, W., and Dillinger, W. (1991). New measurement of solar gravitational deflection of radio signals using VLBI, *Nature*, **349**, 768–70.

Robertson, H. P. (1928). On relativistic cosmology, *Philosophical Magazine*, **5**, 835–48.

Robertson, H. P. (1929). On the foundations of relativistic cosmology, *Proceedings of the National Academy of Sciences*, **15**, 822–9.

Robertson, H. P. (1932). The expansion of the universe, *Science*, **76**, 221–6.

Robertson, H. P. (1933). Relativistic cosmology, *Review of Modern Physics*, **5**, 62–90.

Robinson, I., Schild, A., and Schucking, E.L. (Eds.) (1965). *Quasi-stellar sources and gravitational collapse*. Chicago: University of Chicago Press.

Robinson, T. R. (1845). Lord Rosse's telescopes, *Proceedings of the Royal Irish Academy*, **iii**, 114–33.

Rogerson, J. B. and York, D. G. (1973). Interstellar deutrium abundance in the direction of Beta Centauri, *Astrophysical Journal*, **186**, L95–8.

Roll, P. G. and Wilkinson, D. T. (1966). Cosmic background radiation at 3.2 cm – support for cosmic black-body radiation, *Physical Review Letters*, **16**, 405–7.

Roll, P. G. and Wilkinson, D. T. (1967). Measurement of cosmic background radiation at 3.2-cm wavelength, *Annals of Physics*, **44**, 289–321.

Rosenband, T., Hume, D. B., Schmidt, P. O., et al. (2008). Frequency ratio of Al^+ and Hg^+ single-ion optical clocks: metrology at the 17th decimal place, *Science*, **319**, 1808–12.

Roseveare, N. T. (1982). *Mercury's Perihelion: From Le Verrier to Einstein*. Oxford: Clarendon Press.

Rosse, Earl of (1846). On the nebula 25 Herschel, or 51 of Messier's catalogue, in *Notices and Abstracts of Miscellaneous Communications to the Sections, Report of the Fifteenth Meeting of the British Association for the Advancement of Science*. John Murray.

Rosse, Earl of (1850). Observations on the nebulae, *Philosophical Transactions of the Royal Society of London*, **140**, 499–514.

Rossi, B. (1970). An X-ray pulsar in the Crab Nebula, in *Non-Solar X- and Gamma-Ray Astronomy, IAU Symposium No. 37*, ed. Gratton, L., pp. 183–4. Dordrecht: D. Reidel Publishing Company.

Rothman, T. and Ellis, G. F. R. (1987). Has cosmology become metaphysical?, *Astronomy*, **15**, 6–22.

Rothman, T. and Ellis, G. F. R. (1993). Smolin's natural selection hypothesis, *Quarterly Journal of the Royal Astronomical Society*, **34**, 201–12.

Roush, S. (2003). Copernicus, Kant, and the anthropic cosmological principles, *Studies In History and Philosophy of Modern Physics*, **34**, 5–35.

Rovelli, C. and Vidotto, F. (2014). *Covariant Loop Quantum Gravity: An Elementary Introduction to Quantum Gravity and Spinfoam Theory*. Cambridge: Cambridge University Press.

Rowan-Robinson, M. (1968). The determination of the evolutionary properties of quasars by means of the luminosity-volume test, *Monthly Notices of the Royal Astronomical Society*, **141**, 445–58.

Rowan-Robinson, M. (1985). *The Cosmological Distance Ladder: Distance and Time in the Universe.* New York: W. H. Freeman and Co.

Rowan-Robinson, M. (1988). The extragalactic distance scale, *Space Science Reviews*, **48**, 1–71.

Rowe, D. (2016a). A snapshot of debates on relativistic cosmology, 1917–1924 (Part I), *The Mathematical Intelligencer*, **38**, 46–58.

Rowe, D. (2016b). A snapshot of debates on relativistic cosmology, 1917–1924 (Part II), *The Mathematical Intelligencer*, **38**, 52–60.

Royal Astronomical Society (1888). Meeting of the Royal Astronomical Society. Friday, December 14, 1888, *The Observatory*, **12**, 51–9.

Royal Astronomical Society (1930). Meeting of the Royal Astronomical Society. January, 1930, *The Observatory*, **53**, 33–44.

Rubin, V. C., Thonnard, N., and Ford, W. K. (1980). Rotational properties of 21 Sc galaxies with a large range of luminosities and radii from NGC 4605 ($R = 4$ kpc) to UGC2885 ($R = 122$ kpc), *Astrophysical Journal*, **238**, 471–87.

Rugh, S. and Zinkernagel, H. (2000). The quantum vacuum and the cosmological constant problem, arXiv preprint. hep-th/0012253.

Rugh, S. and Zinkernagel, H. (2011). Weyl's principle, cosmic time and quantum fundamentalism, in *Explanation, Prediction and Confirmation*, ed. Dieks, D. et al., pp. 411–24. Dordrecht: Springer.

Russell, M. J., Nitschke, W., and Branscomb, E. (2013). The inevitable journey to being, *Philosophical Transactions of the Royal Society of London, B*, **368**, 20120254.

Ryle, M. (1955). Radio Stars and their cosmological significance, *The Observatory*, **75**, 137–47.

Ryle, M. and Graham Smith, F. (1948). A new intense source of radio-frequency radiation in the constellation of Cassiopeia, *Nature*, **162**, 462–3.

Sachs, R. K. and Wolfe, A. M. (1967). Perturbations of a cosmological model and angular variations of the microwave background, *Astrophysical Journal*, **147**, 73.

Sahu, K. C., Livio, M., Petro, L., et al. (1997). The optical counterpart to gamma-ray burst GRB 970228 observed using the Hubble Space Telescope, *Nature*, **387**, 476–8.

Sakharov, A. D. (1967). Violation of CP invariance, C asymmetry, and baryon asymmetry of the universe, *Zhurnal Experimentalnoi i Teoretichseskikh Fizica (JETP) Letters*, **5**, 32–5.

Salpeter, E. E. (1964). Accretion of interstellar matter by massive objects, *Astrophysical Journal*, **140**, 796–800.

Sànchez-Ron, J. M. (2005). George McVittie, the uncompromising empiricist, in *The Universe of General Relativity*, ed. Kox, A. J. and Eisenstaedt, J., pp. 189–222. Boston: Birkhäuser.

Sandage, A. R. (1958). Current problems in the extra-galactic distance scale, *Astrophysical Journal*, **127**, 513–26.

Sandage, A. R. (1961a). The ability of the 200-inch telescope to discriminate between selected world models, *Astrophysical Journal*, **133**, 355–80.

Sandage, A. R. (1961b). *The Hubble Atlas of Galaxies.* Publication 618. Washington, DC: Carnegie Institution of Washington.

Sandage, A. R. (1965). The existence of a major new constituent of the universe: the quasistellar galaxies, *Astrophysical Journal*, **141**, 1560–78.

Sandage, A. R. (1968). Observational cosmology, *The Observatory*, **88**, 91–106.

Sandage, A. R. (1970). Cosmology: a search for two numbers, *Physics Today*, 34–41.

Sandage, A. R. (1995). Practical cosmology: inventing the past, in *The Deep Universe, by* Sandage, A.R., Kron, R.G., and Longair, M.S., ed. Binggeli, B. and Buser, R., pp. 1–232. Berlin: Springer.

Sandage, A. R. (1998). Beginnings of observational cosmology in Hubble's time: historical overview, in *The Hubble Deep Field: Proceedings of the Space Telescope Science Institute Symposium*, ed. Livio, S. M., Fall, S. M., and Madau, P., pp. 1–26. Cambridge: Cambridge University Press.

Sandage, A. R. and Hardy, E. (1973). The redshift–distance relation. VII. Absolute magnitudes of the first three ranked cluster galaxies as function of cluster richness and Bautz–Morgan cluster type: the effect on q_0, *Astrophysical Journal*, **183**, 743–58.

Sandage, A. R. and Schwarzschild, M. (1952). Inhomogeneous stellar models II. Models with exhausted cores in gravitational contraction, *Astrophysical Journal*, **116**, 463–76.

Sargent, W. L. W., Young, P. J., Lynds, C. R., et al. (1978). Dynamical evidence for a central mass concentration in the galaxy M87, *Astrophysical Journal*, **221**, 731–44.

Sato, K. (1981). First-order phase transition of a vacuum and the expansion of the Universe, *Monthly Notices of the Royal Astronomical Society*, **195**, 467–79.

Schaffer, S. (1989). The nebular hypothesis and the science of progress, in *History, Humanity and Evolution: Essays for John C. Greene*, ed. Moore, J. R. New York: Cambridge University Press.

Schaffer, S. (1998). The leviathan of Parsonstown: literary technology and scientific representations, in *Inscribing Science: Scientific Texts and the Materiality of Communication*, ed. Lenoir, T., pp. 182–222. Stanford: Stanford University Press.

Schechter, P. L. (2002). Tales within tales and cutoffs within cutoffs: what sets the mass scale for galaxies?, in *Lighthouses of the Universe: The Most Luminous Objects and Their Use for Cosmology*, ed. Gilfanov, M., Sunyaev, R., and Churazov, E., pp. 3–12. Berlin: Springer-Verlag.

Scheibe, E. (1991). General laws of nature and the uniqueness of the universe, in *Philosophy and the Origin and Evolution of the Universe*, ed. Agazzi, E. and Cordero, A., pp. 341–60. Dordrecht: Kluwer Academic Publishers.

Scheiner, J. (1899). On the spectrum of the Great Nebula in Andromeda, *Astrophysical Journal*, **9**, 149–50.

Schemmel, M. (2005). An astronomical road to general relativity: the continuity between classical and relativistic cosmology in the work of Karl Schwarzschild, *Science in Context*, **18**, 451–78.

Scheuer, P. A. G. (1957). A statistical method for analysing observations of faint radio stars, *Proceedings of the Cambridge Philosophical Society*, **53**, 764–73.

Scheuer, P. A. G. (1965). A sensitive test for the presence of atomic hydrogen in intergalactic space, *Nature*, **207**, 963.

Scheuer, P. A. G. (1974). Models of extragalactic radio sources with a continuous energy supply from a central object, *Monthly Notices of the Royal Astronomical Society*, **166**, 513–28.

Scheuer, P. A. G. (1975). Radio astronomy and cosmology, in *Stars and Stellar Systems*, ed. Sandage, A. R., Sandage, M., and Kristian, J., volume 9 (Galaxies and the Universe), pp. 725–60. Chicago: University of Chicago Press.

Schiff, L. I. (1961). A report on the NASA conference on experimental tests of theories of relativity, *Physics Today*, **14**, 42.

Schmidhuber, J. (2002). Hierarchies of generalized Kolmogorov complexities and nonenumerable universal measures computable in the limit, *International Journal of Foundations of Computer Science*, **13**, 587–612.

Schmidt, B. P., Kirshner, R. P., and Eastman, R. G. (1992). Expanding photospheres of type II supernovae and the extragalactic distance scale, *Astrophysical Journal*, **395**, 366–86.

Schmidt, M. (1963). 3C 273: a star-like object with large red-shift, *Nature*, **197**, 1040.

Schmidt, M. (1965). Large redshifts of five quasi-stellar sources, *Astrophysical Journal*, **141**, 1295–300.

Schmidt, M. (1968). Space distribution and luminosity functions of quasi-stellar sources, *Astrophysical Journal*, **151**, 393–409.

Schmidt, M. and Green, R. F. (1983). Quasar evolution derived from the Palomar Bright Quasar Survey and other complete quasar surveys, *Astrophysical Journal*, **269**, 352–74.

Schmidt, M., Schneider, D. P., and Gunn, J. E. (1995). Spectrscopic CCD surveys for quasars at large redshift. IV. Evolution of the luminosity function from quasars detected by their Lyman-alpha emission, *Astronomical Journal*, **110**, 68–77.

Schramm, D. N. (1997). The age of the Universe, in *Critical Dialogues in Cosmology*, ed. Turok, N., pp. 81–91. Singapore: World Scientific.

Schramm, D. N. and Wasserburg, G. J. (1970). Nucleochronologies and the mean age of the elements, *Astrophysical Journal*, **162**, 57–69.

Schreier, E., Levinson, R., Gursky, H., et al. (1972). Evidence for the binary nature of Centaurus X-3 from UHURU X-ray observations, *Astrophysical Journal*, **172**, L79–L89.

Schröder, K.-P. and Connon Smith, R. (2008). Distant future of the Sun and Earth revisited, *Monthly Notices of the Royal Astronomical Society*, **386**, 155–63.

Schrödinger, E. (1957). *Expanding Universes*. Cambridge: Cambridge University Press.

Schulmann, R., Kox, A. J., Janssen, M., et al. (Eds.) (1998a). *The Collected Papers of Albert Einstein, Vol. 8. The Berlin Years: Correspondence, 1914–1918*. Princeton: Princeton University Press.

Schulmann, R., Kox, A. J., Janssen, M., et al. (Eds.) (1998b). *The Collected Papers of Albert Einstein, Vol. 8. The Berlin Years: Correspondence, 1914–1918 (English translation supplement)*. Princeton: Princeton University Press.

Schuster, A. (1898). Potential matter. A holiday dream, *Nature*, **58**, 367.

Schuster, P. M. (2005). *Moving the Stars: Christian Doppler, His Life, His Works and Principle, and the World After*. Pöllauberg, Austria: Living Edition.

Schwarzschild, K. (1900). Über das zulässige Krümmungsmass des Raumes (On an upper limit to the curvature of space), *Vierteljahrsschrift der Astronomischen Gesellschaft*, **35**, 337–47.

Schwarzschild, K. (1916). Über das Gravitationsfeld einis Massenpunktes nach der Einsteinschen Theorie (On the gravitational field of a point mass according to Einsteinian theory), *Sitzungs-berichte der Königlich Preussischen Akademie der Wissenschaften zu Berlin*, **1**, 189–96.

Schwarzschild, M. (1970). Stellar evolution in globular clusters, *Quarterly Journal of the Royal Astronomical Society*, **11**, 12–22.

Schweber, S. (1990). Auguste Comte and the nebular hypothesis, in *The Presence of the Past*, ed. Bienvenu, R. T. and Finegold, M. pp. 131–91. Dordrecht: Kluwer.

Schwinger, J. (1946). Electron radiation in high energy accelerators, *Physical Review*, **70**, 798–9.

Schwinger, J. (1949). On the classical radiation of accelerated electrons, *Physical Review*, **75**, 1912–25.

Sciama, D. W. (1953). On the origin of inertia, *Monthly Notices of the Royal Astronomical Society*, **113**, 34–42.

Sciama, D. W. (1959). *The Unity of the Universe*. Garden City, NY: Doubleday and Co.

Sciama, D. W. (1965). Radio astronomy and cosmology, *Science Progress*, **53**, 1–16.

Sciama, D. W. (1966). On the origin of the microwave background radiation, *Nature*, **211**, 277–9.

Sciama, D. W. (1971). *Modern cosmology*. Cambridge: Cambridge University Press.

Scranton, R., Johnston, D., Dodelson, S., et al. (2002). Analysis of systematic effects and statistical uncertainties in angular clustering of galaxies from early Sloan Digital Sky Survey data, *Astrophysical Journal*, **579**, 48–75.

Secord, J. (2000). *Victorian Sensation: The Extraordinary Publication, Reception and Secret Authorship of Vestiges of Natural History of Creation*. Chicago: University of Chicago Press.

Seeley, D. and Berendzen, R. (1972a). The development of research in interstellar absorption, c. 1900–1930: Part I, *Journal for the History of Astronomy*, **3**, 52–64.

Seeley, D. and Berendzen, R. (1972b). The development of research in interstellar absorption, c. 1900–1930: Part II, *Journal for the History of Astronomy*, **3**, 75–86.

Seeliger, H. v. (1895). Über das Newtonsche Gravitationsgesetz, *Astronomische Nachrichten*, **137**, 129–36.

Seeliger, H. v. (1898a). On Newton's law of gravitation, *Popular Astronomy*, **5**, 544–51.

Seeliger, H. v. (1898b). Betrachtungen über die räumliche Verteilung der Fixsterne, *Abhandlungen der Mathematisch–Physikalischer Klasse der K. Bayerischen Akademie der Wissenschaften zu München*, **19**, 565–629.

Seiffert, M., Fixsen, D. J., Kogut, A., et al. (2011). Interpretation of the ARCADE 2 absolute sky brightness measurement, *Astrophysical Journal*, **734**, 6.

Seldner, M., Siebars, B., Groth, E. J., et al. (1977). New reduction of the Lick catalog of galaxies, *Astronomical Journal*, **82**, 249–56.

Seljak, U. and Zaldarriaga, M. (1997). Signature of gravity waves in the polarization of the microwave background, *Physical Review Letters*, **78**, 2054–7.

Seyfert, C. K. (1943). Nuclear emission in spiral nebulae, *Astrophysical Journal*, **97**, 28–40.

Shakeshaft, J. R., Ryle, M., Baldwin, J. E., et al. (1955). A radio survey of radio sources between declinations -38 and $+83$, *Memoirs of the Royal Astronomical Society*, **67**, 106–54.

Shakura, N. and Sunyaev, R. A. (1973). Black holes in binary systems. Observational appearance, *Astronomy and Astrophysics*, **24**, 337–55.

Shane, C. D. and Wirtanen, C. A. (1957). The distribution of galaxies, *Publications of the Lick Observatory*, **22**, 1–60.

Shapiro, I. I. (1958). *The Prediction of Ballistic Missile Trajectories from Radar Observations (Vol. 129)*. New York: McGraw-Hill.

Shapiro, I. I. (1964). Fourth test of general relativity, *Physical Review Letters*, **13**(26), 789–91.

Shapiro, I. I., Ash, M. E., Ingalls, R. P., et al. (1971). Fourth test of general relativity: new radar result, *Physical Review Letter*, **26**, 1132.

Shapiro, I. I., Pettengill, G. H., Ash, M. E., et al. (1968). Fourth test of general Relativity: preliminary results, *Physical Review Letters*, **20**, 1265–9.

Shapiro, I. I., Reasenberg, R. D., MacNeil, P. E., et al. (1977). The Viking relativity experiment, *Journal of Geophysical Research*, **82**, 4329–34.

Shapiro, S. I. and Teukolsky, S. A. (1983), *Black Holes, White Dwarfs and Neutron Stars: The Physics of Compact Objects*. New York: Wiley Interscience.

Shapiro, S. S., Davis, J. L., Lebach, D. E., et al. (2004). Measurement of the solar gravitational deflection of radio waves using geodetic very-long-baseline interferometry data, 1979–1999, *Physical Review Letters*, **92**, 121101.

Shapley, H. and Ames, A. (1926). A study of a cluster of bright spiral nebulae, *Harvard College Observatory Circular no. 294*, pp. 1–8.

Shapley, H. and Ames, A. (1932). A survey of the external galaxies brighter than the thirteenth magnitude, *Annals of the Astronomical Observatory of Harvard College*, **2**, 42–75.

Shaviv, G. (2009). *The Life of Stars: The Controversial Inception and Emergence of the Theory of Stellar Structure*. Berlin: Springer-Verlag.

Shellard, P. (2003). The future of cosmology: observational and computational prospects, in *The Future of Theoretical Physics and Cosmology*, ed. Gibbons, G. W., Shellard, E. P. S., and Rankin, S. J., pp. 755–80. Cambridge: Cambridge University Press.

Shimasaku, K., Ouchi, M., Furusawa, H., et al. (2005). Number density of bright Lyman-Break galaxies at $z \sim 6$ in the Subaru Deep Field, *Publications of the Astronomical Society of Japan*, **57**, 447–58.

Shivanandan, K., Houck, J. R., and Harwit, M. O. (1968). Preliminary observations of the far-infrared night-sky background radiation, *Physical Review Letters*, **21**, 1460–2.

Shklovskii, I. S. (1966). Untitled letter. *Astronomical Circular*, No. 364.

Shklovskii, I. S. (1967). The nature of the X-ray source Sco X-1, *Astronomicheskii Zhurnal*, **44**, 930–8. Translation: (1967), *Soviet Astronomy*, **11**, 749–55.

Shmaonov, T. A. (1957). A method for measuring the absolute effective radiation temperature of radio emission at low equivalent temperatures, *Pribory i Tekhnika Experimenta (Instruments and Experimental Methods)*, **1**, 83–6.

Silberstein, L. (1924a). Determination of the curvature invariant of space-time, *Philosophical Magazine*, **47**, 907–18.

Silberstein, L. (1924b). Further determination of the curvature radius of space-time, *Nature*, **113**, 602–3.

Silberstein, L. (1924c). The radial velocities of globular clusters and de Sitter's cosmology, *Nature*, **113**, 350–1.

Silberstein, L. (1929). *The Size of the Universe: Attempts at a Determination of the Curvature Radius of Spacetime*. Oxford: Oxford University Press.

Silk, J. (1968). Cosmic black-body radiation and galaxy formation, *Astrophysical Journal*, **151**, 459–71.

Silk, J. and Wyse, R. F. G. (1993). Galaxy formation and Hubble sequence, *Physics Reports*, **231**, 293–365.

Singal, J., Fixsen, D. J., Kogut, A., et al. (2011). The ARCADE 2 instrument, *Astrophysical Journal*, **730**, 138.

Sironi, G., Limon, M., Marcellino, G., et al. (1990). The absolute temperature of the sky and the temperature of the cosmic background radiation at 600 MHz, *Astrophysical Journal*, **357**, 301–8.

Skobeltsyn, D (1929). Über eine neue art sehr schneller β-strahlen (On a new type of very fast β-rays), *Zeitschrift für Physik*, **54**, 686–702.

Slipher, V. M. (1913). The radial velocity of the Andromeda Nebula, *Lowell Observatory Bulletin*, **2**, 56–7.

Slipher, V. M. (1917). A spectrographic investigation of spiral nebulae, *Proceedings of the American Philosophical Society*, **56**, 403–9.

Smail, I., Ivison, R. J., and Blain, A. W. (1997). A deep sub-millimeter survey of lensing clusters: a new window on galaxy formation and evolution, *Astrophysical Journal Letters*, **490**, L5–L8.

Smeenk, C. (2005). False vacuum: early universe cosmology and the development of inflation, in *The Universe of General Relativity, Einstein Studies, Vol. 11*, ed. Kox, A. J. and Eisenstaedt, J., pp. 223–57. Boston: Birkhäuser.

Smeenk, C. (2013). Philosophy of cosmology, in *The Oxford Handbook of Philosophy of Physics*, ed. Batterman, R., pp. 607–52. Oxford: Oxford University Press.

Smeenk, C. (2014). Einstein's role in the creation of relativistic cosmology, in *The Cambridge Companion to Einstein*, ed. Janssen, M. and Lehner, C., pp. 228–69. Cambridge: Cambridge University Press.

Smeenk, C. (2018). Inflation and the origins of structure, in *Beyond Einstein, Einstein Studies, Vol. 14*, ed. Rowe, D., Walter, S., and Sauer, T. Berlin: Springer.

Smith, G. E. (2014). Closing the loop, in *Newton and Empiricism*, ed. Biener, Z. and Schliesser, E., pp. 262–351. Oxford: Oxford University Press.

Smith, H. J. and Hoffleit, D. (1963). Light variations in the superluminous radio galaxy 3C 273, *Nature*, **198**, 650–1.

Smith, K. M., Zahn, O., and Doré, O. (2007). Detection of gravitational lensing in the cosmic microwave background, *Physical Review D*, **76**, 043510.

Smith, M. G. and Partridge, R. B. (1970). Can discrete sources produce the cosmic microwave radiation?, *Astrophysical Journal*, **159**, 737–44.

Smith, R. W. (1982). *The Expanding Universe: Astronomy's Great Debate 1900–1931*. Cambridge: Cambridge University Press.

Smith, R. W. (2003). The remaking of astronomy, in *Cambridge History of Science. Vol. 5. Modern Physical and Mathematical Sciences*, ed. Nye, M. J., pp. 154–73. Cambridge: Cambridge University Press.

Smith, R. W. (2006). Beyond the big galaxy: the structure of the stellar System 1900–1952, *Journal for the History of Astronomy*, **37**, 307–42.

Smith, R. W. (2008). Beyond the galaxy: the development of extragalactic astronomy, 1885–1965, Part I, *Journal for the History of Astronomy*, **39**, 91–119.

Smith, R. W. (2009). Beyond the galaxy: the development of extragalactic astronomy, 1885–1965, Part II, *Journal for the History of Astronomy*, **40**, 71–107.

Smith, R. W. (2013). The road to radial velocities: V. M. Slipher and mastery of the spectrograph, in *Origins of the Expanding Universe: 1912–1932*, ed. Way, M. and Hunter, D., pp. 143–64. Astronomical Society of the Pacific Conference Series.

Smith, R. W. (2014). The 'Great Plan of the Visible Universe': William Huggins, evolutionary naturalism and the nature of the nebulae, in *The Age of Scientific Naturalism*, ed. Lightman, B. and Reidy, M., pp. 113–36. London: Pickering and Chatto.

Smith, R. W. (2015). Hubble's law, in *Discoveries in Modern Science: Exploration, Invention, Technology: Volume 2*, ed. Trefil, J., pp. 511–21. New York: Macmillan.

Smolin, L. (1992). Did the universe evolve?, *Classical and Quantum Gravity*, **9**, 173–91.

Smolin, L. (1997). *The Life of the Cosmos*. Oxford: Oxford University Press.

Smolin, L. (2004). Cosmological natural selection as the explanation for the complexity of the universe, *Physica A: Statistical Mechanics and its Applications*, **340**, 705–13.

Smolin, L. (2015). Temporal naturalism, *Studies in the History and Philosophy of Modern Physics*, **52**, 86–102.

Smoot, G. F., Bennett, C. L., Weber, R., et al. (1990). COBE differential microwave radiometers – instrument design and implementation, *Astrophysical Journal*, **360**, 685–95.

Smoot, G. F., Bennett, C. L., Kogut, A., et al. (1992). Structure in the COBE differential microwave radiometer first-year maps, *Astrophysical Journal Letters*, **396**, L1–L5.

Smoot, G. F., de Amici, G., Friedman, S. D., et al. (1985). Low-frequency measurements of the cosmic background radiation spectrum, *Astrophysical Journal Letters*, **291**, L23–7.

Smoot, G. F., Gorenstein, M. V., and Muller, R. A. (1977). Detection of anisotropy in the cosmic blackbody radiation, *Physical Review Letters*, **39**, 898–901.

Sneden, C., McWilliam, A., Preston, G. W., et al. (1992). The ultra-metal-poor, neutron-capture-rich giant star CS 22892-052, *Astrophysical Journal*, **467**, 819–40.

Soddy, F. (1904). *Radio-Activity: An Elementary Treatise from the Standpoint of the Disintegration Theory*. London: The Electrician.

Sokal, A. D. and Bricmont, J. (1998). *Intellectual Impostures*. London: Profile.

Solheim, J.-E. (1968a). Radio sources in opposite directions as a cosmological test, *Nature*, **217**, 41–3.

Solheim, J.-E. (1968b). Radio sources opposite quasars as a cosmological test, *Nature*, **219**, 45–7.

Sommerville, D. (1911). *Bibliography of Non-Euclidean Geometry Including the Theory of Parallels, the Foundations of Geometry, and the Space of n Dimensions*. London: Harrison and Sons.

Sonnenschmidt, H. (1880). *Kosmologie: Geschichte und Entwicklung des Weltbaues*. Cologne: E. H. Mayer.

Spergel, D. N., Bean, R., Doré, O., et al. (2007). Three-year Wilkinson Microwave Anisotropy Probe (WMAP) observations: implications for cosmology, *Astrophysical Journal Supplement*, 170, 377–408.

Spergel, D. N., Verde, L., Peiris, H. V., et al. (2003). First-year Wilkinson Microwave Anisotropy Probe (WMAP) observations: determination of cosmological parameters, *Astrophysical Journal Supplement Series*, 148, 175–94.

Spite, F. and Spite, M. (1982). Abundance of lithium in unevolved halo stars and old disk stars – interpretation and consequences, *Astronomy and Astrophysics*, 115, 357–66.

Springel, V., White, S. D. M., Jenkins, A., et al. (2005). Simulations of the formation, evolution and clustering of galaxies and quasars, *Nature*, 435, 629–36.

Stachel, J. (1995). History of relativity, in *Twentieth Century Physics*, ed. Brown, L. M., Pais, A., and Pippard, A. B., pp. 249–356.

Staelin, D. H. and Reifenstein, E. C. (1968). Pulsating radio sources near the Crab Nebula, *Science*, 162, 1481–3.

Staniszewski, Z., Ade, P. A. R., Aird, K. A., et al. (2009). Galaxy clusters discovered with a Sunyaev–Zeldovich effect survey, *Astrophysical Journal*, 701, 32–41.

Stanley, M. (2013). The varieties of universal expansion: Eddington and the complexities of early cosmology, in *Origins of the Expanding Universe: 1912–1932*, ed. Way, M. and Hunter, D., pp. 39–44. Astronomical Society of the Pacific Conference Series.

Starkman, G. D. and Trotta, R. (2006). Why anthropic reasoning cannot predict Λ, *Physical Review Letters*, 97, 201301.

Starobinsky, A. A. (1979). Spectrum of relict gravitational radiation and the early state of the universe, *Journal of Experimental and Theoretical Physics Letters*, 30, 682–5.

Starobinsky, A. A. (1980). A new type of isotropic cosmological models without singularity, *Physics Letters B*, 91, 99–102.

Stecker, F. W. (1968). Effect of photomeson production by the universal radiation field on high-energy cosmic rays, *Physical Review Letters*, 21, 1016–18.

Stecker, F. W. (1974). The role of antimatter in big-bang cosmology, *Bulletin of Atomic Scientists*, 1974, 12–16.

Steckline, V. S. (1983). Zermelo, Boltzmann, and the recurrence paradox, *American Journal of Physics*, 51, 894–7.

Štefancic, H. (2004). Generalized phantom energy, *Physics Letters B*, 586, 5–10.

Steidel, C. C. (1998). Galaxy evolution: has the "Epoch of Galaxy Formation" been found?, in *Eighteenth Texas Symposium on Relativistic Astrophysics and Cosmology*, ed. Olinto, A. V., Frieman, J. A., and Schramm, D. N., pp. 124–35. River Edge, NJ: World Scientific.

Steidel, C. C., Adelberger, K. L., Giavalisco, M., et al. (1999). Lyman-break galaxies at $z \geq 4$ and the evolution of the ultraviolet luminosity density at high redshift, *Astrophysical Journal*, 519, 1–17.

Steigman, G. (1978). A crucial test of the Dirac cosmologies, *Astrophysical Journal*, 221, 407–11.

Steigman, G. (2004). Big bang nucleosynthesis: probing the first 20 minutes, in *Measuring and Modeling the Universe*, ed. Freedman, W. L., pp. 169–95. Cambridge: Cambridge University Press.

Steigman, G. (2006). Primordial nucleosynthesis: successes and challenges, *International Journal of Modern Physics E*, 15, 1–35.

Steigman, G., Schramm, D. N., and Gunn, J. E. (1977). Cosmological limits to the number of massive leptons, *Physics Letters B*, 66, 202–4.

Steinhardt, P. J. (2004). The endless universe: a brief introduction, *Proceedings of the American Philosophical Society*, 148, 464–70.

Steinhardt, P. J. and Turner, M. S. (1984). Prescription for successful new inflation, *Physical Review D*, **29**, 2162–71.

Steinhardt, P. J. and Turok, N. (2007). *Endless Universe*. New York: Doubleday.

Stokes, R. A., Partridge, R. B., and Wilkinson, D. T. (1967). New measurements of the cosmic microwave background at λ = 3.2 cm and λ = 1.58 cm – evidence in support of a black-body spectrum, *Physical Review Letters*, **19**, 1199–202.

Stoner, E. C. (1929). The limiting density in white dwarf stars, *Philosophical Magazine*, **7**, 63–70.

Stoops, R. (Ed.) (1958). *La Structure et l'Évolution de l'Univers*. Brussels: Coudenberg.

Story, K. T., Reichardt, C. L., Hou, Z., et al. (2013). A measurement of the cosmic microwave background damping tail from the 2500-square-degree SPT-SZ survey, *Astrophysical Journal*, **779**, 86.

Strömberg, G. (1925). Analysis of radial velocities of globular clusters and non-galactic nebulae, *Astrophysical Journal*, **61**, 353–62.

Strominger, A. and Vafa, C. (1996). Microscopic origin of the Bekenstein–Hawking entropy, *Physics Letters B*, **379**, 99–104.

Strukov, I. A. and Skulachev, D. P. (1984). Deep-space measurements of the microwave background anisotropy – first results of the Relikt experiment, *Soviet Astronomy Letters*, **10**, 1–4.

Strukov, I. A., Brukhanov, A. A., Skulachev, D. P., et al. (1993). Anisotropy of relic radiation in the RELICT-1 experiment and parameters of grand unification, *Physics Letters B*, **315**, 198–202.

Strukov, I. A., Kulachev, D. P., Klypin, A. A., et al. (1988). The anisotropy of the microwave background: space experiment RELICT, in *International Astronomical Union Symposium No. 130*, ed. Audouze, J., Pelletan, M.-C., and Szalay, A., pp. 27–34. Cambridge: Cambridge University Press.

Subrahmanyan, R., Ekers, R. D., Sinclair, M., et al. (1993). A search for arcminute scale anisotropy in the cosmic microwave background, *Monthly Notices of the Royal Astronomical Society*, **263**, 416.

Sullivan, W. T., III (1990). The entry of radio astronomy into cosmology: radio stars and Martin Ryle's 2C survey, in *Modern Cosmology in Retrospect*, ed. Bertotti, B. et al., pp. 309–30. Cambridge: Cambridge University Press.

Sunyaev, R. A. and Zeldovich, Y. B. (1970a). Small-scale fluctuations of relic radiation, *Astrophysics and Space Science*, **7**, 3–19.

Sunyaev, R. A. and Zeldovich, Y. B. (1970b). The interaction of matter and radiation in the hot model of the Universe, II, *Astrophysics and Space Science*, **7**, 20–30.

Sunyaev, R. A. and Zeldovich, Y. B. (1972). The observations of relic radiation as a test of the nature of X-ray radiation from the clusters of galaxies, *Comments on Astrophysics and Space Physics*, **4**, 173.

Sunyaev, R. A. and Zeldovich, Y. B. (1980). Microwave background radiation as a probe of the contemporary structure and history of the universe, *Annual Review of Astronomy and Astrophysics*, **18**, 537–60.

SuperCDMS Collaboration, Agnese, R., Aramaki, T., et al. (2017). Results from the Super Cryogenic Dark Matter Search (SuperCDMS) experiment at Soudan. arXiv preprint arXiv: 1708.8869.

Susskind, L. (2003). The anthropic landscape of string theory. preprint arXiv arXiv:hep-th/0302219.

Susskind, L. (2006). *The Cosmic Landscape: String Theory and the Illusion of Intelligent Design*. New York: Back Bay Books.

Suzuki, S. (1928). On the thermal equilibrium of dissociation of atom-nuclei, *Proceedings of the Physico-Mathematical Society of Japan*, **10**, 166–9.

Szalay, A. S. and Marx, G. (1976). Neutrino rest mass from cosmology, *Astronomy and Astrophysics*, **49**, 437–41.

Tananbaum, H., Gursky, H., Kellogg, E. M., et al. (1972). Discovery of a periodic binary X-ray source in Hercules from UHURU, *Astrophysical Journal*, **174**, L144–9.

Tassoul, J.-L. and Tassoul, M. (2004). *A Concise History of Solar and Stellar Physics*. Princeton: Princeton University Press.

Tavakol, R. and Gironi, F. (2017). The infinite turn and speculative explanations in cosmology, *Foundations of Science*, **22**, 785–98.

Tayler, R. J. (1995). Helium in the universe, *Contemporary Physics*, **36**, 37–48.

Taylor, J. H. (1992). Pulsar timing and relativistic gravity, *Philosophical Transactions of the Royal Society*, **341**, 117–34.

Tegmark, M. (1998). Is 'The Theory of Everything' merely the ultimate ensemble theory?, *Annals of Physics*, **270**, 1–51.

Tegmark, M. (2003). Parallel universes, *Scientific American*, **288**, 41–51.

Tegmark, M. (2005). What does inflation really predict?, *Journal of Cosmology and Astroparticle Physics*, **4**, 001.

Tegmark, M. (2008). The mathematical universe, *Foundations of Physics*, **38**, 101–50.

Tegmark, M. (2009). The multiverse hierarchy. arXiv preprint arXiv:0905.1283.

Tegmark, M., Aguirre, A., Rees, M. J., et al. (2006). Dimensionless constants, cosmology, and other dark matters, *Physical Review D*, **73**, 023505.

Teller, E. (1948). On the change of physical constants, *Physical Review*, **73**, 801–2.

Thaddeus, P. and Clauser, J. F. (1966). Cosmic microwave radiation at 2.63 mm from observations of interstellar CN, *Physical Review Letters*, **16**, 819–22.

Thomson, W. (1882–1911). *Mathematical and Physical Papers*. 6 vols. Cambridge: Cambridge University Press.

Thomson, W. (1891). *Popular Lectures and Addresses, Vol. 1*. London: Macmillan.

Thomson, W. (1901). On ether and gravitational matter through infinite space, *Philosophical Magazine*, **2**, 160–77.

Thorne, K. S. (1967). Primordial element formation, primordial magnetic fields, and the isotropy of the universe, *Astrophysical Journal*, **148**, 51.

Thorne, K. S. (1994). *Black Holes and Time Warps: Einstein's Outrageous Legacy*. New York: W.W. Norton and Company.

Thorne, K. S., Price, R. H., and Macdonald, D. A. (1986). *Black Holes: The Membrane Paradigm*. New Haven: Yale University Press.

Timbie, P. T. and Wilkinson, D. T. (1990). A search for anisotropy in the cosmic microwave radiation at medium angular scales, *Astrophysical Journal*, **353**, 140–4.

Tinsley, B. M. (1976). The evolution of chemical abundances and stellar populations, in *Saas-Fee Advanced Course 6: Galaxies*, ed. Martinet, L. and Mayor, M., p. 155.

Tinsley, B. M. (1980). Evolution of the stars and gas in galaxies, *Fundamentals of Cosmic Physics*, **5**, 287–388.

Tinsley, B. M. and Gunn, J. E. (1976). Luminosity functions and the evolution of low-mass population I giants, *Astrophysical Journal*, **206**, 525–35.

Tipler, F. J. (1988). Johann Mädler's resolution of Olbers' paradox, *Quarterly Journal of the Royal Astronomical Society*, **29**, 313–25.

Tolman, R. (1929a). On the astronomical implications of the de Sitter line element for the universe, *Astrophysical Journal*, **69**, 245–74.

Tolman, R. (1929b). On the possible line elements for the universe, *Proceedings of the National Academy of Sciences*, **15**, 297–304.

Tolman, R. C. (1934a). Effect of inhomogeneity on cosmological models, *Proceedings of the National Academy of Sciences*, **20**, 169–76.

Tolman, R. C. (1934b). *Relativity, Thermodynamics and Cosmology*. Oxford: Oxford University Press. Reprinted by Dover, New York (1987).

Tonry, J. L., Schmidt, B. P., Barris, B., et al. (2003). Cosmological results from high-z supernovae, *Astrophysical Journal*, **594**, 1–24.

Totsuji, H. and Kihara, T. (1969). The correlation function for the distribution of galaxies, *Publications of the Astronomical Society of Japan*, **21**, 221–9.

Trabert, W. (1911). *Lehrbuch der kosmischen Physik*. Leipzig: Teubner.

Trautman, A. (1965). Foundations and current problems of general relativity, in *Lectures on General Relativity: Vol. 1 of* Brandeis Summer Institute in Theoretical Physics 1964, ed. Trautman, A., Pirani, F. A. E., and Bondi, H., pp. 1–248. Englewood Cliffs, NJ: Prentice Hall.

Tremaine, S. and Gunn, J. E. (1979). Dynamical role of light neutral leptons in cosmology, *Physical Review Letters*, **42**, 407–10.

Treumann, R. A. (1993). Evolution of the information in the universe, *Astrophysics and Space Science*, **201**, 135–47.

Trimble, V. (1996). H_0: the incredible shrinking constant 1925–1975, *Publications of the Astronomical Society of the Pacific*, **108**, 1073–82.

Trotta, R. (2007). Dark matter: probing the Arche-Fossil (interview), *Collapse*, **II**, 83–170.

Trotta, R. (2008). Bayes in the sky: Bayesian inference and model selection in cosmology, *Contemporary Physics*, **49**(2), 71–104.

Trotta, R. and Starkman, G. D. (2006). What's the trouble with anthropic reasoning?, in *2nd International Conference on The Dark Side of the Universe, AIP Conference Proceedings*, ed. Munoz, C. and Yepes, G., volume 878, pp. 323–9. Amherst, NY: Prometheus Books.

Trumpler, R. J. (1930). Preliminary results on the distances, dimensions, and space distribution of open star clusters, *Lick Observatory Bulletin*, **14**, 154–88.

Tucker, W. and Giacconi, R. (1985). *The X-ray Universe*. Cambridge, MA: Harvard University Press.

Tully, R. B. and Fisher, J. R. (1977). A new method of determining distances to galaxies, *Astronomy and Astrophysics*, **54**, 661–73.

Turner, M. S. and Wilczek, F. (1982). Is our vacuum metastable?, *Nature*, **298**, 633–4.

Turok, N. (Ed.) (1997). *Critical Dialogues in Cosmology*. Singapore: World Scientific.

Udaltsov, V. A. (1963). Polarization of 21-cm radiation of the Crab Nebula, *Soviet Astronomy*, **6**, 665–9.

Unger, R. M. and Smolin, L. (2015). *The Singular Universe and the Reality of Time*. Cambridge; Cambridge University Press.

Urani, J. and Gale, G. (1994). E. A. Milne and the origins of modern cosmology: an essential presence, in *The Attraction of Gravitation: New Studies in the History of General Relativity*, ed. Earman, J., Janssen, M., and Norton, J., pp. 390–419. Boston: Birkhäuser.

Urry, C. M. and Padovani, P. (1994). Unification of BL Lac objects and FR1 radio galaxies, in *First Stromlo Symposium: Physics of Active Galactic Nuclei*, ed. Bicknell, G. V., Dopita, M. A., and Quinn, P. J., pp. 215–26. San Francisco: Astronomical Society of the Pacific Conference Series, Volume 34.

Uson, J. M. and Wilkinson, D. T. (1984). Small-scale isotropy of the cosmic microwave background at 19.5 GHz, *Astrophysical Journal*, **283**, 471–8.

Uzan, J.-P. (2010). Dark energy, gravitation and the Copernican principle, in *Dark Energy: Observational and Theoretical Approaches*, ed. Ruiz-Lapuente, P., pp. 3–47. Cambridge: Cambridge University Press.

Vaas, R. (2012). Cosmological artificial selection: creation out of something?, *Foundations of Science*, 17, 25–8.

Vacanti, G., Cawley, M. F., Colombo, E., et al. (1991). Gamma-ray observations of the Crab Nebula at TeV energies, *Astrophysical Journal*, 377, 469–79.

Van der Kruit, P. and van Berkel, K. (Eds.) (2000). *The Legacy of J. C. Kapteyn*. Dordrecht: Kluwer Academic Publisher.

Van Der Marel, R. P., Besla, G., Cox, T. J., et al. (2012). The M31 velocity vector. III. Future Milky Way M31-M33 orbital evolution, merging, and fate of the Sun, *Astrophysical Journal*, 753, 9.

Van Flandern, T. C. (1975). A determination of the rate of change of G, *Monthly Notices of the Royal Astronomical Society*, 170, 333–42.

Van Helden, A. (1984). Building large Telescopes, 1900–1950, in *Astrophysics and Twentieth Century Astronomy to 1950. Part A. Volume 4. The General History of Astronomy*, ed. Gingerich, O., pp. 134–52. Cambridge: Cambridge University Press.

Van Maanen, A. (1916a). Preliminary evidence of internal motions in the spiral nebula Messier 101, *Proceedings of the National Academy of Sciences*, 2, 386–90.

Van Maanen, A. (1916b). Preliminary evidence of internal motions in the spiral nebula Messier 101, *Astrophysical Journal*, 44, 210–28.

Van Maanen, A. (1935). Internal motion in spiral nebulae, *Astrophysical Journal*, 81, 336–7.

Veneziano, G. (1998). A simple/short introduction to pre-big-bang physics/cosmology. arXiv preprint hep-th/9802057.

Verde, L., Heavens, A. F., Percival, W. J., et al. (2002). The 2dF Galaxy Redshift Survey: the bias of galaxies and the density of the Universe, *Monthly Notices of the Royal Astronomical Society*, 335, 432–40.

Vidal-Madjar, A., Laurent, C., Bonnet, R. M., et al. (1977). The ratio of deuterium to hydrogen in interstellar space. III – The lines of sight to Zeta Puppis and Gamma Cassiopaeia, *Astrophysical Journal*, 211, 91–107.

Vilenkin, A. (1983). Birth of inflationary universes, *Physical Review D*, 27, 2848–55.

Vilenkin, A. (1995). Predictions from quantum cosmology, *Physical Review Letters*, 74, 846–9.

Vilenkin, A. (2006a). *Many Worlds in One*. London: Hill and Wang.

Vilenkin, A. (2006b). On cosmic natural selection. arXiv preprint hep-th/0610051.

Vilenkin, A. (2007). Freak observers and the measure of the multiverse, *Journal of High Energy Physics*, 01, 092.

Vilenkin, A. and Shellard, E. P. S. (2000). *Cosmic Strings and Other Topological Defects*. Cambridge: Cambridge University Press.

Vincenti, W., Boyd, J., and Bugos, G. E. (2007). H. Julian Allen: an appreciation, *Annual Review Fluid Mechanics*, 39, 1–17.

Vukotić, B., Steinhauser, D., Martinez-Aviles, G., et al. (2016). 'Grandeur in this view of life': N-body simulation models of the galactic habitable zone, *Monthly Notices of the Royal Astronomical Society*, 459, 3512–24.

Wagoner, R. V. (1973). Big-bang nucleosynthesis revisited, *Astrophysical Journal*, 179, 343–60.

Wagoner, R. V. (1990). Deciphering the nuclear ashes of the early universe: a personal perspective, in *Modern Cosmology in Retrospect*, ed. Bertotti, B., Balbinot, R., and Bergia, S., p. 159–85. Cambridge: Cambridge University Press.

Wagoner, R. V., Fowler, W. A., and Hoyle, F. (1967). On the synthesis of elements at very high temperatures, *Astrophysical Journal*, 148, 3–49.

Wall, J. V. (1996). Space distribution of radio source populations, in *Extragalactic Radio Sources, IAU Symposium No. 175*, ed. Ekers, R., Fanti, C., and Padrielli, L., pp. 547–52. Dordrecht: Kluwer Academic Publishers.

Walsh, D., Carswell, R. F., and Weymann, R. J. (1979). 0957+561A, B – twin quasistellar objects or gravitational lens, *Nature*, **279**, 381–4.

Wandel, A. and Mushotzky, R. F. (1986). Observational determination of the masses of active galactic nuclei, *Astrophysical Journal*, **306**, L61–6.

Way, M. J. and Hunter, D. (Eds.) (2013). *Origins of the Expanding Universe: 1912–1932*. Astronomical Society of the Pacific Conference Series No. 471.

Way, M. J. (2013). Dismantling Hubble's legacy, in *Origins of the Expanding Universe: 1912–1932*, ed. Way, M. and Hunter, D., pp. 97–132. Astronomical Society of the Pacific Conference Series. No. 471.

Weber, J. (1961). *General Relativity and Gravitational Waves*. New York: Interscience, Interscience Tracts on Physics and Astronomy.

Weber, J. (1966). Observation of the thermal fluctuations of a gravitational-wave detector, *Physical Review Letters*, **17**, 1228–30.

Weber, J. (1969). Evidence for discovery of gravitational radiation, *Physical Review Letters*, **22**, 1320–4.

Weber, J. (1970). Anisotropy and polarization in the gravitational-radiation experiments, *Physical Review Letters*, **25**, 180–4.

Webster, B. L. and Murdin, P. (1972). Cygnus X-1: a spectroscopic binary with a heavy companion?, *Nature*, **235**, 37–8.

Weekes, T. C., Cawley, M. F., Fegan, D. J., et al. (1989). Observation of TeV gamma rays from the Crab Nebula using the atmospheric Cerenkov imaging technique, *Astrophysical Journal*, **342**, 379–95.

Weinberg, S. (1972). *Gravitation and Cosmology: Principles and Applications of the General Theory of Relativity*. New York: John Wiley & Sons.

Weinberg, S. (1977). *The First Three Minutes*. New York: Basic Books. Second edition published in 1993.

Weinberg, S. (1987). Anthropic bound on the cosmological constant, *Physical Review Letters*, **59**, 2607–10.

Weinberg, S. (1989). The cosmological constant problem, *Reviews of Modern Physics*, **61**, 1–23.

Weinberg, S. (1997). Theories of the cosmological constant, in *Critical Dialogues in Cosmology*, ed. Turok, N., pp. 195–203. Singapore: World Scientific.

Weinberg, S. (2007). Living in the multiverse, in *Universe or Multiverse?*, ed. Carr, B., pp. 29–42. Cambridge: Cambridge University Press.

Weinheimer, C. (2001). Neutrino mass from tritium β-Decay, in *Dark Matter in Astro- and Particle Physics, Proceedings of the International Conference DARK 2000*, ed. Klapdor-Kleingrothaus, H. V., pp. 513–19. Berlin: Springer-Verlag.

Weiss, R. (1980). Measurements of the cosmic background radiation, *Annual Review of Astronomy and Astrophysics*, **18**, 489–535.

Weizsäcker, C. F. v. (1938). Element transformation inside stars. II, *Physikalische Zeitschrift*, **39**, 633–46.

Wesselink, A. J. (1947). The observations of brightness, colour and radial velocity of δ-Cephei and the pulsation hypothesis, *Bulletin of the Astronomical Institutes of the Netherlands*, **10**, 91–9. Errata, **10**, 258 and 310.

Wesson, P. S. (1978). *Cosmology and Geophysics*. Bristol: Adam Hilger.

Wetterich, C. (2013). Universe without expansion, *Physics of the Dark Universe*, **2**, 184–7.

Weyl, H. (1923a). *Raum, Zeit, Materie: Vorlesungen Über Allgemeine Relativitätstheorie*. Berlin: Springer.

Weyl, H. (1923b). Zur Allgemeinen Relativitätstheorie (On the general relativity theory), *Physikalishe Zeitschrift*, **24**, 230–2. English translation: (2009), *General Relativity and Gravitation*, **41**, 1661-6.

Weyl, H. (1924). Observations on the note of Dr. L. Silberstein: determination of the curvature invariant of space-time, *Philosophical Magazine*, **48**, 348–9.

Weymann, R. (1966). The energy spectrum of radiation in the expanding universe, *Astrophysical Journal*, **145**, 560.

Wheeler, J. A. (1968). Our universe: the known and the unknown, *American Scientist*, **56**, 1–20.

Wheeler, J. A. (1977). Genesis and observership, in *Foundational Problems in the Special Science*, ed. Butts, R. E. and Hintikka, J., pp. 3–33. Dordrecht: D. Reidel Publishing Company.

White, S. D. M. and Rees, M. J. (1978). Core condensation in heavy halos – a two-stage theory for galaxy formation and clustering, *Monthly Notices of the Royal Astronomical Society*, **183**, 341–58.

Whitehead, A. N. (1933). *Adventures of Ideas*. Cambridge: Cambridge University Press.

Whitrow, G. J. (1953). A query concerning the steady-state theory of the homogeneous expanding universe, *The Observatory*, **73**, 205–6.

Whitrow, G. J. and Bondi, H. (1954). Is physical cosmology a science?, *The British Journal for the Philosophy of Science*, **4**, 271–83.

Wilczek, F. (2013). Multiversality, *Classical and Quantum Gravity*, **30**, 193001.

Wilkinson, D. T. (1967). Measurement of the cosmic microwave background at 8.56-mm wavelength, *Physical Review Letters*, **19**, 1195–8.

Wilkinson, D. T. and Peebles, P. J. E. (1983). Discovery of the 3 K radiation, in *Serendipitous Discoveries in Radio Astronomy*, ed. Kellermann, K. and Sheets, B., pp. 175–84. Greenbank WV: National Radio Astronomy Observatory Publishers.

Wilkinson, P. N., Henstock, D. R., Browne, I. W., et al. (2001). Limits on the cosmological abundance of supermassive compact objects from a search for multiple imaging in compact radio sources, *Physical Review Letters*, **86**, 584–7.

Will, C. M. (2006). The confrontation between general relativity and experiment, *Living Reviews in Relativity*, **9**, 3.

Williams, J. G., Anderson, J. D., Boggs, D. H., et al. (2001). Solar system tests for changing gravity, *Bulletin of the American Astronomical Society*, **33**, 836.

Williams, J. G., Newhall, X., and Dickey, J. O. (1996). Relativity parameters determined from lunar laser ranging, *Physical Review D*, **53**, 6730.

Williams, J. G., Turyshev, S. G., and Boggs, D. H. (2004). Progress in lunar laser ranging tests of relativistic gravity, *Physical Review Letters*, **93**, 261101.

Wilson, B. and Kaiser, D. (2014). Calculating times: radar, ballistic missiles, and Einstein's relativity, in *Science and Technology in the Global Cold War*, ed. Oreskes, N. and Krige, J., pp. 273–316. Cambridge, MA: MIT Press.

Wilson, C. T. R. (1901). On the ionisation of atmospheric air, *Proceedings of the Royal Society of London*, **68**, 151–61.

Wolf, M. (1912). Die Entfernung der Spiralnebel, *Astronomische Nachrichten*, **190**, cols. 229–32.

Wolfe, A. M. and Burbidge, G. R. (1969). Discrete source models to explain the microwave background radiation, *Astrophysical Journal*, **156**, 345–71.

Woody, D. P. and Richards, P. L. (1979). Spectrum of the cosmic background radiation, *Physical Review Letters*, **42**, 925–9.

Yang, J., Turner, M. S., Schramm, D. N., et al. (1984). Primordial nucleosynthesis – a critical comparison of theory and observation, *Astrophysical Journal*, **281**, 493–511.

Young, C. A. (1893). *A Text-Book of General Astronomy*. Boston: Ginn & Co.

Young, P. J., Westphal, J. A., Kristian, J., et al. (1978). Evidence for a supermassive object in the nucleus of the galaxy M87 from SIT and CCD area photometry, *Astrophysical Journal*, 221, 721–30.

Zaldarriaga, M., Spergel, D. N., and Seljak, U. (1997). Microwave background constraints on cosmological parameters, *Astrophysical Journal*, 488, 1–13.

Zanstra, H. (1957). On the pulsating or expanding universe and its thermodynamical aspect, *Proceedings of the Royal Dutch Academy of Sciences, Series B*, 60, 285–307.

Zehnder, L. (1897). *Die Mechanik des Weltalls in ihren Grundzügen dargestellt*. Freiburg: Mohr.

Zeldovich, Y. B. (1964a). The fate of a star and the evolution of gravitational energy upon accretion, *Soviet Physics Doklady*, 9, 195–7.

Zeldovich, Y. B. (1964b). Observations in a universe homogeneous in the mean, *Astronomicheskii Zhurnal*, 41, 19–24. Translation: (1964), *Soviet Astronomy*, 8, 13–16.

Zeldovich, Y. B. (1965). Survey of modern cosmology, *Advances of Astronomy and Astrophysics*, 3, 241–379.

Zeldovich, Y. B. (1968). The cosmological constant and the theory of elementary particles, *Uspekhi Fizisheskikh Nauk*, 95, 209–30.

Zeldovich, Y. B. (1970a). Gravitational instability: an approximate theory for large density perturbations, *Astronomy and Astrophysics*, 5, 84–9.

Zeldovich, Y. B. (1970b). The universe as a hot laboratory for the nuclear and particle physicist, *Comments on Astrophysics and Space Physics*, 2, 12.

Zeldovich, Y. B. (1972). A hypothesis, unifying the structure and the entropy of the universe, *Monthly Notices of the Royal Astronomical Society*, 160, 1P–3P.

Zeldovich, Y. B. and Khlopov, M. Y. (1978). On the concentration of relic magnetic monopoles in the universe, *Physics Letters B*, 79, 239–41.

Zeldovich, Y. B. and Novikov, I. D. (1970). A hypothesis for the initial spectrum of perturbations in the metric of the Friedman model universe, *Soviet Astronomy*, 13, 754.

Zeldovich, Y. B. and Novikov, I. D. (1983). *Relativistic Astrophysics. Volume 2. The Structure and Evolution of the Universe, revised edition*. Chicago: University of Chicago Press. The original Russian edition was published in 1967 by Nauka, as the second of a two volume work entitled *Relavtivistic Astrophysics*.

Zeldovich, Y. B. and Sunyaev, R. A. (1969). The interaction of matter and radiation in a hot-model universe, *Astrophysics and Space Science*, 4, 301–16.

Zeldovich, Y. B. and Sunyaev, R. A. (1980). Astrophysical implications of the neutrino rest mass. I. The universe, *Pis'ma v Astronomicheskii Zhurnal*, 6, 451–6.

Zeldovich, Y. B., Illarionov, A. F., and Sunyaev, R. A. (1972). The effect of energy release on the emission spectrum in a hot universe, *Soviet Journal of Experimental and Theoretical Physics*, 35, 643.

Zeldovich, Y. B., Kobzarev, I. I., and Okun, L. B. (1975). Cosmological consequences of a spontaneous breakdown of a discrete symmetry, *Zhurnal Eksperimentalnoi i Teoreticheskoi Fiziki*, 67, 3–11.

Zeldovich, Y. B., Kurt, D., and Sunyaev, R. A. (1968). Recombination of hydrogen in the hot model of the universe, *Zhurnal Eksperimentalnoi i Teoreticheskoi Fiziki*, 55, 278–86. Originally submitted December 27, 1967. Translation: (1969), *Soviet Physics – JETP*, 28, 146–50.

Zinkernagel, H. (2002). Cosmology, particles, and the unity of science, *Studies in the History and Philosophy of Modern Physics*, 33, 493–516.

Zöllner, J. K. F. (1872). *Über die Natur der Cometen: Beiträge zur Geschichte und Theorie der Erkenntnis*. Leipzig: Engelmann.

Zwicky, F. (1929). On the red shift of spectral lines through interstellar space, *Proceedings of the National Academy of Sciences*, **15**, 773–9.

Zwicky, F. (1933). Rotverschiebung von Extragalaktischen Nebeln, (The redshift of extragalactic nebulae), *Helvetica Physica Acta*, **6**, 110–18.

Zwicky, F. (1935). Remarks on the redshift from nebulae, *Physical Review*, **48**, 802–6.

Zwicky, F. (1937). On the masses of nebulae and of clusters of nebulae, *Astrophysical Journal*, **86**, 217–46.

Zwicky, F. and Zwicky, M. A. (1971). *Catalogue of Selected Compact Galaxies and of Post-eruptive Galaxies*. Guemligen, Switzerland: Zwicky.

Subject index

The items highlighted in bold are the titles of the sections of the chapters of this book. Subsections and subsubsections are also included with their page ranges, but these are not highlighted in bold. The other items are subjects recommended for inclusion by the authors of the chapters.

Author index